NANOSCALE DEVICE PHYSICS

This text is the fourth volume of the series entitled Electroscience.

1. Sandip Tiwari. *Quantum, Statistical and Information Mechanics: A Unified Introduction*, Electroscience Series, Volume 1. Oxford University Press
    ISBN 978-0-19-875985-0

2. Sandip Tiwari. *Device Physics: Fundamentals of Electronics and Optoelectronics*, Electroscience Series, Volume 2. Oxford University Press
    ISBN 978-0-19-875984-3

3. Sandip Tiwari. *Semiconductor Physics: Principles, Theory and Nanoscale*, Electroscience Series, Volume 3. Oxford University Press
    ISBN 978-0-19-875986-7 and

4. Sandip Tiwari. *Nanoscale Device Physics: Science and Engineering Fundamentals*, Electroscience Series, Volume 4. Oxford University Press
    (2017) ISBN 978-0-19-875987-4

These volumes comprise a sequence of undergraduate to graduate textbooks on the underlying physical foundations leading up to advanced devices of the nanometer scale. Teaching slides are available on the book's companion website at http://www.oup.co.uk/companion/nanoscaledevices2017, and the solutions manual may be requested at https://global.oup.com/academic/category/science-and-mathematics/physics/solutions.

Nanoscale Device Physics

Science and Engineering Fundamentals

ELECTROSCIENCE SERIES, VOLUME 4

Sandip Tiwari

OXFORD

UNIVERSITY PRESS

OXFORD
UNIVERSITY PRESS

Great Clarendon Street, Oxford, OX2 6DP,
United Kingdom

Oxford University Press is a department of the University of Oxford.
It furthers the University's objective of excellence in research, scholarship,
and education by publishing worldwide. Oxford is a registered trade mark of
Oxford University Press in the UK and in certain other countries.

First Edition published in 2017

Impression: 1

Published in the United States of America by Oxford University Press
198 Madison Avenue, New York, NY 10016, United States of America

British Library Cataloguing in Publication Data
Data available

Library of Congress Control Number: 2017935051

ISBN 978–0–19–875987–4

Printed and bound by
CPI Group (UK) Ltd, Croydon, CR0 4YY

To

Mari, Nachiketa, Kunal,

Sadhana, Rajeev, Lakshmi, Anandilal, Joy and Edward,

past and future, old and young, departed and living,

for all the love, care, teaching, learning and patience.

You gave this life meaning and purpose.

हिरण्मयेन पात्रेण सत्यस्यापिहितं मुखम् ।

तत् त्वं पूषन्नपावृणु सत्यधर्माय दृष्टये ॥

ईशोपनिषद्

THE FACE OF TRUTH IS COVERED WITH GOLD.

REMOVE IT SO THAT YOU MAY SEE WHAT IS REAL.

Ish Upanishad

Acknowledgments

One's approach to life and, more specifically, to the choices one makes in the course of it, the questions one asks and the irritants one learns to ignore have a cumulative outcome. In this, one's family and friends, early and advanced education, the company one keeps, and the approaches one learns from the specially gifted people that one has the good fortune of befriending all make a difference. So do people who mentor and support through the vicissitudes of life. I was lucky to have had a good share of exemplary teachers in my early years, and during my undergraduate education at the Indian Institute of Technology, Kanpur, as well as particularly enlightened colleagues during my years at the research laboratory of International Business Machines Corporation (*IBM*). This latter was a community with a preponderance of exceptional people. Every encounter had something to learn from. Science ruled. Nonsense was unacceptable. All these benefactors have influenced choices I have made, especially the nature of questions I became interested in and asked and the multifaceted approaches of different perspectives I used in answering them. This has given me great satisfaction in my pursuits in science and engineering. The loss of the research institutions, particularly the *IBM* and Bell laboratories, and the reduction in the diversity of thoughts and approaches that has followed, has changed the American story. These institutions were exemplary for most of their lives in their approach to discovery, research and development, understood their interconnectedness and had an intellectual commitment that overrode palace intrigues. The test of good management is in how an institution responds and negotiates through hard times. *IBM* maneuvered remarkably well for a fair period.

Foremost among the people I wish to acknowledge are the colleagues I had at Yorktown during Research's golden period. Among them, Paul Solomon, David Frank, Steve Laux, Massimo Fischetti, Wen Wang, Frank Stern, Rolf Landauer, Peter Price, Bob Dennard, Subu Iyer, Tom Jackson, Supratik Guha, Pat Mooney, Arvind Kumar, Chuck Black, Jeff Welser, Doug Buchanan, Dan DiMaria, Jim Stathis, Emilio Mendez, Leo Easaki, Jim Misewich, Ravi Nair, Charles Bennett, Jerry Woodall, Peter Kirchner, Jeff Kash, Jimmy Tsang, Dennis Rogers, Marshall Nathan, Frank Fang, Alan Fowler, Reuben Collins, Dick Rutz and many others shared technical and life-enriching wisdom.

The early thoughts on how to organize this writing occurred during a sabbatical leave at Harvard in 2006–2007. I even wrote some preliminary notes. But, work could not begin in earnest until I managed to relinquish several responsibilities that work-life brings. The first drafts started in Ithaca around 2010, but serious work had to wait for my next sabbatical leave in 2012–2013, which also gave a chance to try the material with different student audience. At the Indian Institute of Science, my hosts included Professors Navakant Bhat, Rudra Pratap and S. Shivashankar; at Stanford University, Professor Roger Howe; and, at Technische Universität München, Prof. Paolo Lugli. The environment at these institutions was ideal for what I had in mind. And in addition, it provided an opportunity to be in the company of several other faculty with a joyful outlook to science and life: Prof. Ambarish Ghosh, Srinivasan Raghavan, Philip Wong, Yoshio Nishi, Walter Harrison, Christian Jirauschek, Wolfgang Porod, and Peter and Johannes Russer. My immense gratitude to them for a stimulating year. The students who participated in these courses around the world have provided invaluable feedback that is reflected in the writing and rewriting.

Many colleagues have read and commented on parts or the whole and this has helped with the exposition. To Rob Ilic, Christian Jirauschek, Yoshio Nishi, Arvind Kumar, Sam Bader, Damien Querlioz, Ed Yu, Nader Engheta, Mike Stopa, Wei Lu, Seth Putterman, Kunal Tiwari and Evelyn Hu, my thanks for sharing their precious time and suggestions. Jack and Mary East—my dear friends—have given crucial support through their constant interest in the progress of this work. My wife Mari has kept a careful eye on the goings-on, helped keep my centrifugal propensities in check and has given invaluable advice on the presentation. The very constructive exchange with Sonke Adlung, Harriet Konishi and Elizabeth Farrell at Oxford University Press has been immensely valuable in the creation of the final form. The LATEX class for Edward Tufte's style suggestions has been largely followed in these texts. The authors of such open source resources, which here also includes Python for the calculations in the exercises and figures, perform an immense service to the society.

Over the years, I have been fortunate to have had generous and understanding mentors and supporters who have made the research and academic pursuits fulfilling. The research environment of those early years was a reflection of the focus on research with a perspective. The management ably fostered this. In my academic years, I

One notable aphorism from those days: "Beware the *PF9*s and *PF10*s." Before the internet, or the bitnet before that, there existed within *IBM* an internal network (*VNET*) for electronic mail, for accessing repositories of useful codes and technical documents that individuals had written, and for other network-wide computing tasks. *PF9* and *PF10* were programmable function keys for the "Receive" and the "Send" tasks in the mailing program. "*PF9*s and *PF10*s" was the euphemism for the gatekeepers, speed breakers and messengers—folks whose technical careers were short but who had the wherewithal to generate pointless work—from whom one needed to protect oneself to be a good scientist and engineer.

have similarly benefited from the long view of many at the National Science Foundation and others at Defense Advanced Projects Agency. For this shepherding, to Jim McGroddy Venky Narayanamurti, Larry Goldberg, Rajinder Khosla and Usha Varshney, I express gratitude for their commitment over the fruitful years.

In days past, on the board outside Prof. Les Eastman's office, there used to be a press clipping circa the late 1970s, from the Universal Press Syndicate, with the following quote: "Our futures almost certainly depend less on what Ronald Reagan and Walter Mondale say and do than on what is going on inside the head of some young Cornell graduate waiting for a plane in Pittsburgh." The routes are now through Philadelphia, but the thought is still right. To students who have interacted with me through the classes goes the ultimate tip of the hat.

Science touches us all through its beauty. My colleagues, teachers, family, students, but also many others who have, through the occasional talks, conversations, writings, the way an argument was framed, a clever twist of reasoning, or even plowing through when a situation demands, have enlightened this recognition.

Sandip Tiwari

Ithaca, Capstick, Southport, Bengaluru, Stanford and München

Contents

Introduction to the series   xv

Introduction   xix

1   Information mechanics   1

    1.1   Information is physical   4

    1.2   The Church-Turing thesis, and state machines   5

    1.3   The mechanics of information   9

    1.4   Probabilities and the principle of maximum entropy   17

    1.5   Algorithmic entropy   22

    1.6   Conservation and non-conservation   25

    1.7   Circuits in light of infodynamic considerations   33

    1.8   Fluctuations and transitions   40

    1.9   Errors, stability and the energy cost of determinism   43

    1.10   Networks   48

    1.11   Information and quantum processes   52

    1.12   Summary   53

    1.13   Concluding remarks and bibliographic notes   54

    1.14   Exercises   57

2   Nanoscale transistors   63

    2.1   Transistors as dimensions shrink   64

2.2 *Geometries and scaling* 70

2.3 *The off state of a nanoscale transistor* 72

2.4 *Conduction at the nanoscale* 91

2.5 *The on state of a nanoscale transistor* 111

2.6 *Zero bandgap and monoatomic layer limits* 132

2.7 *Parasitic resistances* 133

2.8 *Summary* 143

2.9 *Concluding remarks and bibliographic notes* 146

2.10 *Exercises* 147

3 *Phenomena and devices at the quantum scale and the mesoscale* 155

3.1 *Quantum computation and communication* 157

3.2 *At the mesoscale* 180

3.3 *Single and many electrons in a nanoscale dot* 207

3.4 *Summary* 230

3.5 *Concluding remarks and bibliographic notes* 233

3.6 *Exercises* 234

4 *Phase transitions and their devices* 237

4.1 *Phase transitions* 243

4.2 *Ferroelectricity and ferroelectric memories* 260

4.3 *Electron correlations and devices* 281

4.4 *Spin correlations and devices* 306

4.5 *Memories and storage from broken translational symmetry* 379

4.6 *Summary* 400

4.7 *Concluding remarks and bibliographic notes* 404

4.8 *Exercises* 407

5 *Electromechanics and its devices* 411

5.1 *Mechanical response* 412

5.2 *Coupled analysis* 426

5.3 *Acoustic waves*   455

5.4 *Consequences of nonlinearity*   456

5.5 *Caveats: Continuum to nanoscale*   468

5.6 *Summary*   473

5.7 *Concluding remarks and bibliographic notes*   474

5.8 *Exercises*   475

6   *Electromagnetic-matter interactions and devices*   481

6.1 *The Casimir-Polder effect*   482

6.2 *Optomechanics*   492

6.3 *Interactions in particle beams*   502

6.4 *Plasmonics*   512

6.5 *Optoelectronic energy exchange in inorganic and organic semiconductors*   526

6.6 *Lasing by quantum cascade*   543

6.7 *Summary*   554

6.8 *Concluding remarks and bibliographic notes*   558

6.9 *Exercises*   560

Appendices   563

A   *Information from the Shannon viewpoint*   565

B   *Probabilities and the Bayesian approach*   571

C   *Algorithmic entropy and complexity*   577

D   *Classical equipartition of energy*   583

E   *Probability distribution functions*   585

E.1 *The Poisson distribution*   586

E.2 *The Gaussian normal distribution*   587

F   Fluctuations and noise   589
    F.1   Thermal noise   595
    F.2   1/f noise   598
    F.3   Shot noise   598

G   Dimensionality and state distribution   603

H   Schwarz-Christoffel mapping   609

I   Bell's inequality   615

J   The Berry phase and its topological implications   619

K   Symmetry   623

L   Continuum elasticity   629

M   Lagrangian dynamics   633

N   Phase space portraiture   641

O   Oscillators   649
    O.1   Relaxation oscillators   651
    O.2   Parametric oscillators   653

P   Quantum oscillators   655

Glossary   659

Index   669

Introduction to the series

These books are a labor of love and love of labor. They reflect a personal philosophy of education and affection for this small but vital subject area of electroscience, one that has given me satisfaction. This subdiscipline is also a domain where knowledge has evolved rapidly, leaving in its wake an unsatisfactory state in the coherence of content tying mathematical and physical descriptions to its practice.

Engineering-oriented science education, even though not really that different from science education itself, is difficult for two reasons. It aims to provide strong scientific foundations and also to make the student capable of practicing it for society's benefit. Adequate knowledge of design and technology to invent and optimize within constraints demands a fundamental understanding of the natural and physical world we live in. Only then can we create usefully with these evolving tools and technology. Three hundred years ago, calculus and Kepler's and Newton's laws may have been adequate. Two hundred years ago, this basic foundation was expanded to include a broader understanding of thermodynamics; the Lagrangian approach to classical mechanics; probability; and the early curiosity about the compositional origins of lightening. A hundred years ago, the foundational knowledge had expanded again to include a fair understanding of the periodic table, Hamiltonians, electricity, magnetism, and statistical mechanics, and, yet again, it was incomplete, as the development of new, non-classical approaches, such as Planck's introduction of quantum of action, and relativity, indicated. Our understanding still remains very incomplete today even as evolution is gathering pace via science and engineering with non-carbon forms of intelligent machines quite imaginable. Reductive and constructive approaches, as before, pervade the pursuit of science and engineering. Understanding singularities, whether in black holes, in phase transitions with their information mechanics implications, or for solving near-infinite differential equations in near-infinite variables and constraints and the networks they form, is central to science and engineering problems—connecting back the two ends of the string.

All this evolving knowledge and its usage would be deficient were we incapable of adequately using the tools available, which in their modern forms include software for the implementation of mathematics and their computational, observational and experiment-stimulating machinery and their operating software for designing and optimizing suitable answers to the questions posed. A physical understanding of the connections between different interactions, as well as the reasoning that leads one to identify the most primary of these interactions, is essential to utilizing them gainfully.

Engineering education, with these continuing changes in fundamental understanding and its practice, raises difficult questions of content and delivery too under the constraint of a fixed time period for education. It has also raised serious humanist questions of affordability, even as engineering education claims to aim at frugality through less expensive scaled delivery mechanisms. Engineering, more than science, is beholden to societal needs. In growing fields, particularly the ones that that have the most immediate societal relevance, this rapidly brings content and finances into conflict. As the amount of engineering and science knowledge required rapidly increases, with the rapidly evolving technology, training becomes obsolete just as rapidly. In technical areas, whose educational needs expand suddenly because of their societal use and consequent professional needs, specialized course offerings proliferate rapidly. This puts pressure on the teaching of the foundational knowledge of the disciplines, and the time available for it. The inclusion of broad skill sets into the core curriculum is threatened by the need to teach an expanding number of specialized topics in an ever-shrinking amount of time. The pace of and need for change through new offerings or modifications to courses risk introducing disjointedness and decreasing rigor, because modifying and harmonizing a curriculum is a difficult and time-consuming task.

This series of books is an experiment in attempting to answer today's needs in my areas of interest while preserving thoroughness and rigor. It is an attempt at coherent systematic education with discipline, while maintaining reverence and a healthy disrespect for received wisdom.

The books aim to be conceptual not mechanical. This series is aimed ultimately at the electroscience of the nanoscale—the current interest of the semiconductors and devices stream—but which is also far more interdisciplinary than the norms suggest. Its objective is to have students understand electronic devices, in the modern sense of that term, which includes magnetic, optical, mechanical, and informational devices, as well as the implications of the use of such devices. It aims, in four semester-scale courses, to introduce the under-

A hallmark of the present times is introduction of new words when older ones lose their apparent luster or ``branding.″ ``Multidisciplinary″ evolved to ``interdisciplinary″ with an expansion of indiscipline. ``Transdisciplinary″ must be trying to birth itself. Richard Feynman's statement, ``In these days of specialization there are too few people who have such a deep understanding of two departments of our knowledge that they do not make a fools of themselves in one or the other″ (from R. P. Feynman, ``The meaning of it all: Thoughts of a citizen-scientist,″ Perseus ISBN 0-7382-0166-9 (1989), p. 9), is not inappropriate here.″

lying science, starting with the fundamentals of quantum, statistical and informational mechanics and connecting these to an exposition of classical device physics, then dive deeper into the condensed matter physics of semiconductors, and finally address advanced themes regarding devices of nanometer scale: so, starting with the basics and ending with the integration of electronics, optics, magnetics and mechanics at the nanoscale.

The first book[1] of the series explores the quantum, statistical and information mechanics foundations for understanding semiconductors and the solid state. The second[2] discusses microscale electronic, optical and optoelectronic devices, for which mostly classical interpretation and understanding suffice. The third[3] builds advanced foundations utilizing quantum and causal approaches to explore electrons, phonons and photons and their interaction in the solid state, particularly in semiconductors, as relevant to devices and to the properties of matter used in devices. The fourth book[4] is a treatment of the nanoscale-specific physics of electronic, optical, magnetic and mechanical devices of engineering interest. The second and the third volumes are for subjects that can be taught in parallel but are necessary for the fourth. The value of this approach is that this sequence can be completed by the first year of graduate school or even the senior year of undergraduate studies, for a good student, while leaving room for much else that the student must learn. For those interested in electrosciences, this still includes electromagnetics, deeper understanding of lasers, analog, digital and high frequency circuits, and other directions. The fourth book is the first to come out because of the urgency I have felt.

I have always admired simplicity of exposition with a thorough discussion that even if simplified, is devoid of propaganda or the much too common modern practice of using templates where depth and nuances are lost and doubts and questions are not addressed. Also consistency is easily lost when modern tools, instead of a pencil and paper, are employed. The style of these books follows these beliefs. Notations, figures, the occasional use of color and other stylistic choices are consistent across the book series.

From early years, I have been a devotee of marginalia—much of the learning and independent thought have come from doodling in the margins and the back pages of notebooks. These books are organized so that the reader will feel encouraged to do so.

A list of very readable, in-depth sources, with my perspectives serving as a trigger for different contents within the book, is to be found at the end of each chapter, in the section titled "Concluding remarks and bibliographic notes." No attempt has been made to credit original discoverers or authors. These remarks and notes ascribe

[1] S. Tiwari, "Quantum, statistical and information mechanics: A unified introduction," Electroscience 1, Oxford University Press, ISBN 978-0-19-875985-0 (forthcoming).

[2] S. Tiwari, "Device physics: Fundamentals of electronics and optoelectronics," Electroscience 2, Oxford University Press, ISBN 978-0-19-875984-3 (forthcoming).

[3] S. Tiwari, "Semiconductor physics: Principles, theory and nanoscale," Electroscience 3, Oxford University Press, ISBN 978-0-19-875986-7 (forthcoming).

[4] S. Tiwari, "Nanoscale device physics: Science and engineering fundamentals," Electroscience 4, Oxford University Press, ISBN 978-0-19-875987-4 (2017).

them, or they are to be found by following the references in these notes to their origins.

The exercises are formulated for use in self-study and in the class-room. A subject cannot really be learned by simply reading. Problems requiring application of the information learned and encouraging further thinking and learning are necessary. When we discover for ourselves, we learn best. The exercises here are meant to inform and to be instructive. They are also ranked for difficulty— those that need only a short time but test fundamental understanding are marked as (S), for simple; those requiring considerable effort, bordering on being research problems, are rated (A), for advanced; and those that are intermediate are rated (M), for moderate.

Teaching slides are available on the companion website recorded in the front. The solutions manual may also be requested by providing information at the second link furnished in the front. Slides, when in the modern template-based style, can seriously hinder teaching when they become a tool for filtering key information and explanation while emphasizing summary points. The available presentation material is a tool to avoiding mistakes in writing out equations and to carefully and graphically explain the relationships that science and engineering unfold. They do not substitute for the book and the instructor needs to be diligent in making sure that important themes of teaching—probing, questioning, reasoning, explaining, exploring evidence—come out credibly. I am also happy to hear and discuss the subtleties and the different viewpoints of principles, approaches and the deeper meanings of a derived result.

The books could have been shorter and crisper had there been more time. But, what time there was has given enormous pleasure— a time out for integrity in the presence of the incessant pressure of existence, particularly of life in modern academe. For this escape, my gratitude to this world. For making possible the following of my wishes to produce these songs as the shadow of a life in research, teaching and writing, I thank the Hitkarini Foundation.

सर्वजन हिताय । सर्वजन सुखाय ॥

Introduction

Microscale devices, whether transistors in their various forms, memories or others, have been the foundation upon which information technology's success has been built. Nanoscale devices are distinguishable from these microscale devices in their specific dependence on physical phenomena and effects that are central to their operation. The size change manifests itself through changes in importance of the phenomena and effects that become dominant, and the changes in scale of underlying energetics and response. Examples of these include classical effects, such as single electron effects; quantum effects, such as the confined states accessible as well as their properties; ensemble effects ranging from the consequences of the laws of numbers to changes in properties arising from the different magnitudes of the interactions, and others. These interactions, with the limits placed on size, make not just electronic, but also magnetic, optical and mechanical behavior interesting, important and useful.

Connecting these properties to the behavior of devices is the focus of this textbook.

These nanometer-scale devices are the components of the data- and information-centric edifice of information technology that detects signals, moves these signals over distances, amplifies and manipulates them in digital or analog form, stores them and performs mathematical transformations individually or collectively in an integrated form. In short, these devices compute the data that the signals represent, interface with the world via these signals and otherwise manipulate them towards a useful purpose. Nanoscale devices also convert energy into suitable forms and so themselves have to be energy efficient. Since digital computation is an important part of this mix, storage or memory is also an important task that nanoscale devices are useful in. Since analog signal transformations are also part of this mix, noise and nonlinearities are also important.

The nanoscale phenomena that these electronic, magnetic, photonic and mechanic devices utilize arise from the behavior of electrons, photons and phonons in solids, principally semiconductors, or in small structures such as molecules where molecular vibrations may be involved. Electron charge, electron and nuclear spin and electromagnetic and mechanical interactions at the atomic scale and the nanoscale underlie information processing and communication. Electronic devices including different transistors and memories that draw on few electrons, quantum confinement, and phase transitions such as ferroelectric, metal–insulator, and structural phase transitions; magnetic devices employing field switching, spin torque, and spin Hall effects; photonic devices using photon-matter interactions, photonic bandgaps and nonlinearities; and mechanical devices employing electromechanical deflection, torsion and resonance at nanometer and quantum scale, are discussed. The physical phenomena that connect these devices to are electron-phonon effects in high permittivity dielectrics, single electron phenomena, phase transition theory, tunneling, magnetic switching, spin torque effect, quantum entanglement, mesoscopic interactions, the Casimir-Polder effect, plasmonics and their coupled interactions.

The text intends to provide the reader with an in-depth discussion of nanoscale physics and engineering as relevant to information manipulation, emphasizing both science and engineering, as well as their interplay.

Readers who are familiar with the equations at the end of the Glossary would benefit from this text, which assumes prior learning and familiarity with quantum and statistical mechanics, classical electronics and solid-state physics. For students who understand the contents of the earlier books of the series, this text can be employed in one semester, as I have done at Cornell. But, for others, it is best used in a two-semester course; alternatively, selective parts from it can be used in a one-semester course. The large number of appendices are meant to provide the necessary knowledge from other areas of science or to refresh the reader's memory.

1

Information mechanics

THE PRINCIPAL USE OF NANOSCALE DEVICES is in the manipulation, communication, storage, and acquisition of information. Energy is intimately connected to information: energy interaction is needed to obtain information, and information, in turn, is needed to transform energy. Hundreds of years ago, the development of the mathematical description of the conversion of thermal to mechanical energy was the genesis of thermostatics and thermodynamics, where pressure, volume, temperature, et cetera, are the variables of interests and heat exchange and temperature related through entropy, a macroscopic state variable, which unlike pressure, volume and temperature, cannot be directly measured.

Understanding the classical mechanics of particles in ensembles led to the development of statistical-mechanical description of particle behavior over a distribution of states. In an attempt to describe the effects of deviations from equilibrium, entropy again was used to quantify deviations from equilibrium, where it is at a maximum. This approach to entropy led to the probabilistic ensemble description of particle behavior—a statistical view. Both the macroscopic mechanical and the microscopic statistical-mechanical approaches represent attempts to describe a system and the flow of energy through it when it deviates from thermal equilibrium. To define the entropy of a single microscopic state, one has to consider an equilibrium ensemble of identical systems and then identify the microscopic configurations resulting in macroscopic properties identical to those of the single microscopic state whose entropy is desired. In a system evolving dynamically as described by a Hamiltonian, the number of microscopic states is a constant, and each microstate evolves to another microstate in time. The phase space is incompressible. The Copenhagen quantum view, in which the collapse of a system, when an observation is made of an observable, to a definite state with an eigenvalue and an eigenfunction, leads to its own set of questions. According to

Information is dynamic. When an external event in time and space causes a transition from a large number of *a priori* possibilities to a small number of allowable events, it increases information—a term we have not quite defined precisely—so that precision is increased and entropy is reduced. However, information is both objective, via its use of symbolic references, and subjective, via the meaning attached to the patterns of connections one observes. Knowledge—an equally nebulous term—is embodied in connections that are established from past experiences in time and space and which exist between entities belonging to different sets of domains. An ensemble of connections itself cannot provide any information or entropy reduction, although it does allow the transfer of entropy reduction from one set to another. However, while one of these sets will be observable, the other may not necessarily be so. A reservoir is not an observable but a system using which an attempt is made to quantify entropy change. Searches, graphs, networks, et cetera, are useful/informative, and energy does some useful work for us when it facilitates this kind of connection making.

Nanoscale device physics: Science and engineering fundamentals. Sandip Tiwari.
© Sandip Tiwari 2017. Published 2017 by Oxford University Press.

this view a multitude of possibilities reducing to one results in increased precision, while observation, if conducted reversibly, need not consume energy. However, this view is in contradiction with the second law of thermodynamics. One may also ask "Whose observation?", and probabilities and statistical expectations at any moment are based on the information the observer has at that moment. These are but some of the fundamental questions that both the classical and quantum interpretations raise. Yet, the success of the thermodynamic and the statistical-mechanical approaches speak to their near-completeness in describing the essentials. The areas lacking here are those related to information. We have already filled in many of the missing pieces of this puzzle via the following: first, a probabilistic description which encapsulates some of the essentials of likelihoods in both deterministic classical mechanics and non-deterministic quantum mechanics; second, through developments in communications theory, also drawing on probabilistic expectations; and, more recently algorithmic complexity, and developments in quantum computing. These are the physical foundations of the informational origins in nanoscale devices, and of their use at dimensions where the energy tied to this information mechanics is a very key parameter. This chapter is an exploration of these themes and relates the classical, microscopic, quantum-mechanical and informational mathematical description to the implementation in circuits, energy flows, errors, and both the reversible and the irreversible transformations that implementations undergo in devices and circuits. Our goal in this chapter is to sketch an understanding and some general interpretations that will underlie the following chapters, where we explore a variety of devices.

When we employ the word information, it usually refers to content, message or thought related to a signal. Mathematically, we reduce this concept to a measure. The Boltzmann-Gibbs entropy helps us describe the physical disorder in a system. The Shannon entropy provides a similar statistical measure in a bit stream of a communication channel. How much information should be given to a computer to calculate a particular binary string or to describe the state of a system is also a measure of information in defining the system. The human interpretation of information is a result of both the patterns observed in space and time, and the past accumulation of experiences that allows one to interpret. With these complexities, it is not surprising that mathematically quantifying the meaning of information still remains beyond our reach. Knowledge and thought are even more ill-understood and ill-defined than information is. Consequently, providing mathematical descriptions for these concepts is even more challenging than doing so for information. Here lies the

The quote "You cannot get something for nothing, not even an observation" is relevant here. It implies that the measurement process inextricably interlinks irreversibility, information and entropy. Strictly speaking, the process of observation can be reversible and so no energy need be irreversibly lost. As we shall see, it is the process of destruction of the information that leads to the dissipation of energy.

Apropos, the motto of the Royal Society is "Words alone are nothing."

Our cosmos is a kind of pattern—a fusion of colors, motion, feelings, thoughts, actions, et cetera. Mathematics is the science used to study it. Number, space, logic, infinity and information allow us to classify these patterns. A hand as a symbol for 5 is a number. It is also an object in space—convex, concave, made up of multiple other shapes and connected to the body. It is a logic machine operating with logical patterns: for example, tendons cause bones to move, and touching a hot surface causes the hand to jerk back. It is an infinity of endless complex fractals and a continuing structure at diminishing scales. It is also information, for example, in the instructions coded in the *DNA* needed to create it. These are all complementary classifications intimately related to the human activity of perception, emotion, thought, intuition and communication. Saying that a theory of information is a theory of everything would not be too far from the truth.

conflict between the objectiveness that mathematics desires and the subjectiveness that human interpretation brings. However, early developments of thermodynamics from viewpoints of disturbance from equilibrium, statistical mechanics, quantum mechanics and computational procedure approaches have given us major mathematical tools to analyze and explore information. Quantification of both the disorder and missing information is necessary to describe the mechanics of the information system.

A signal is any quantity, obtained through a measurement, that provides data regarding a physical system. It can vary in the physical coordinates of time and space. Through its processing, it allows us to communicate and extract behavioral characteristics of the physical system from which it appeared. Voice, a vibrational signal coupling the mouth of one person and the ears of another, allows the two to communicate. A thermocouple provides an electrical signal representing the temperature. A manometer measures pressure. Data are quantities, characters, or symbols that may represent a signal or signals. Data, in digital or analog form, comprise the mathematical representation of the signal and can thus be manipulated and stored electrically, optically, mechanically, magnetically or by other means in which we can employ force for action. Information is the result of processing, manipulating and organizing data in a way that adds to the knowledge of the observer, that is, the receiver. It gives us means to distinguish. The context of the data is also important since it also adds knowledge. Information is therefore not data alone. The context in which the data was taken is important. Data is also not signal alone. In the process of codifying it in digital or analog form, one may drop off parts of it, for example, the high frequency components in an analog recording, or the analog or variations within a band in a digital one. Information represents a degree of choice that is exercised in the choice of a particular set of data out of a number of possible options.

Signals, data and information represent the physical characteristics of a system, and inherent in this representation is the physical process of representation. Physical processes conform to the principles of conservation of energy, so energy is a central element in any discussion. Information thus has a meaning akin to that of entropy, which in statistical physics, because of the statistical distribution of microstates, is related to the logarithm of the number of states at the energy of the system. This is the statistical view, initiated by Ludwig Boltzmann and Josiah Willard Gibbs, of thermodynamics.

The connections between physics, computation and the physical devices in which we implement them are deep and broad. They bring together thermodynamics, emergent phenomena, phase transitions,

In 1929 Leo Szilard—Szilárd Leó in his native Hungarian—in exploring the problem of Maxwell's demon, noted this correspondence of information and entropy. Thermodynamics has its antecedents in the understanding of heat. The state quantity entropy, introduced by Rudolf Clausius in 1850, is the amount of heat *reversibly* exchanged, at a temperature T, between a reservoir and a system. In an isolated system in thermodynamic equilibrium, the entropy is constant, and since all non-equilibrium processes lead to an increase in entropy, the isolated system has the maximum entropy. The laws of thermodynamics, in one of many forms, are as follows:

0th law: An isolated system, left to itself, eventually comes to a final state—a thermal equilibrium state—that does not change.

1st law: When a system changes state, the sum of the work, the heat and any action quantity it receives from its surroundings is determined only by the initial and final state and not by intermediate processes.

2nd law: A process which involves no change other than the transfer of heat from a hotter to a cooler body is irreversible.

3rd law: The entropy of a chemically uniform body of finite density approaches a limiting value as the temperature goes to absolute zero, regardless of pressure, density or phase.

The *0th law* posits the existence of thermal equilibrium, and temperature T. The *1st law* is a statement of the conservation of energy. The *2nd law* leads to the conclusion that the entropy of a closed system can only increase with time and that thus natural phenomena are irreversible. Implicitly, this law establishes the arrow of time through evolution toward equilibrium without an external influence. The *3rd law* is a statement of absolute zero temperature as a limiting condition.

the energetics and time scales of processes, their stochastic variations
in small ensembles, quantum effects and classical manifestations
and implementation in devices. Thus, they tie together mathematics,
physics, computation science and engineering both in the theoretical description and in the practical consequences. The mathematical
foundations of computational processes and optimization, that is,
logic in large ensembles, such as in solving large number of equations with variables and constraints, have universal classes of behavior, self-organized criticality and phase transitions. The approaches
here gain from the successes of statistical mechanics. Entropy, as an
idea originating in classical mechanics study of heat engines, turns
out to be only one manifestation—a physical entropy—and one can
view it as a subset of logic. What these directions indicate is that laws
of physics are just as much laws of logic. In quantum mechanics, a
wavefunction, which allows one to determine the observables, has
to be deeply connected to information, given that observation is information gathering. This chapter is a discussion bringing some of
these themes together in the context of signals, as engineers think of
them; information, as engineers and physicists perceive it; and the
classical and the quantum perspectives that connect approaches to
computation in the respective regimes where these effects dominate.
Since representation, energy transfer and information transformation
are what we do with devices, we start our discussion of nanoscale
devices with an examination of data and information; the importance of energy in information processing; the connection of this idea
to the thermodynamic concepts of reversible processes, irreversible
processes , adiabatic processes and isothermal processes; the role
of probabilities and the ways these ideas relate to power, speed and
errors.

1.1   *Information is physical*

WHEN WE MANIPULATE INFORMATION, we connect the physical
world to mathematical representations. Physical resources are required to represent the information and for the computation and
communication of the mathematical representations. With physical
states representing the information, manipulating information content becomes a symbolic mathematical transformation which has its
correspondence in the evolution of the physical system. In this sense,
information is at the origin of physical reality.

This connection between the physical and the mathematical is
manifest in the reversibility of information manipulation and in superposition;; the former has ties to thermodynamics and statistical

``Information is physical´´ is a phrase
attributed to Rolf Landauer who, at
the dawn of semiconductors-based
electronic computing, as early as 1962,
stressed the importance of physical
principles, such as the 2nd law of thermodynamics, to theories of information
processing. Marshall McLuhan in *Understanding Media: The Extensions of
Man,* uses the phrase ``the medium
is the message´´ implying that the
physical form has a direct hold on our
behavior. The concepts of reversibility
and irreversibility; used for centuries
in the study of the thermodynamics
of mechanical engines, have a corresponding connection to the information
engine. John Wheeler used the pithy
expression ``It from bit´´ to stress the
role of information in the very origin
of physical reality. Information here is
more than just a quantum-mechanical
wavefunction that contains the various eigenvalues representing different
realities associated with the Hermitian operator representing a physical
quantity.

mechanics, while the latter is tied to quantum information and computational theory. Information in this sense is not an abstract entity but is tied to physical representation and embodiment. A different state of voltage or spin, a printed page, or the code in a *DNA* sequence, all provide information through a representation that arises from the physical embodiment. Measuring, communicating and computing are all exchanges of information. These exchanges therefore involve energy.

1.2   The Church-Turing thesis, and state machines

REDUCING THE SOLUTION PROCEDURE of any problem to a mathematical form that can then be transformed into a logical procedure manipulating symbols and then into a machine-implementable form allows us to formulate a general technique for solving problems. This approach has an attractive appeal. It provides a way to convert information processing and decision making to a construction that is general for all problems that are computable, that is, solvable, and to implement it in a machine form—a physical embodiment of devices and circuits woven in an architecture.

The Church-Turing thesis concerns the notion of the method of logic, that is, an effective procedure that sets out a finite number of precise instructions using a finite number of symbols and reaches the result in a finite number of steps. Alonzo Church demonstrated this approach in recursive function theory by using recursiveness as a way of reducing functions to a lower-order and an end point. If we know the solution procedure to the end point, we can then find the solution of the higher order complex problem. Alan Turing provided an alternative and equivalent form for writing instructions that step through the procedure in a Turing machine. In principle, this procedure—an algorithm—then can be carried out by a human or a machine without any additional resource—physical, informational or of any other form of intelligence. The thesis's physical interpretation is that any function that can be computed by any physical system, for example, by a human, can be computed by a machine via the recursive function or a Turing machine, which we will presently discuss. There do exist functions, of course, that are incomputable, for example, a variety of nonlinear differential equations, or the halting problem, that is, determining whether a given program will eventually halt or instead enter an infinite loop. These approaches give us the means to explore the limits on resources for the computable functions: for example, the number of operations, and therefore time needed to compute a function, measurable as the order of polynomial

If one knows that one problem is unsolvable, such as the "halting" problem, one can show that a second problem is also unsolvable by showing that if one can solve the second problem, one has solved the first one. This is the method of reduction. The halting problem is the problem of distinguishing the initial conditions that lead a program to compute forever from those that cause it eventually to halt. It can be proved that there exists no such procedure.

Complexity theory defines complexity classes to capture the order of growth in resources and execution. For example, $TIME(n^2)$ is the set of problems that can be solved with complexity $\mathcal{O}(n^2)$ in input size n. The class P is the union of all $TIME(n^k)$ for all k, that is, problems that are computable in polynomial time as the input size n is varied. One could write an equivalent class for the spatial need $SPACE(n^k)$. If one knew the solution, what order of growth in resources and execution would be needed to verify the solution? The set of problems for which this answer is verifiable efficiently, that is, in polynomial time, is in the class NP, for non-deterministic polynomial time. NP-complete problems are problems that are provable to be as hard as any problem in NP, the implication being that if one could solve one of these quickly, one could do the same with the other problems in the class. The traveling salesman problem, that is, given a list of cities and the distances between each pair of cities, what is the shortest possible route that can be used to visit each city exactly once and then return to the the origin city, or determining the satisfiability (SAT) of Boolean expressions to which all logic problems can be reduced, are NP-complete problems. Another group of problems consists of those which are not known to be in P but are also not known to be NP-complete. So, an unsolved problem is, is $P = NP$, or, is $P \subset NP$?

time growth $\mathcal{O}(n^k)$, where n is the input size, and k the power; or the amount of computational data that one needs to keep track of, that is, memory, which represents a spatial need.

The machine construction that allows one to project the logical procedure into a physical execution is enabled through representation of the different states that the machine can be in and the transitions between those states as the procedure is executed—the state representation of the machine. It corresponds to the phase space or configuration space of classical mechanics. Figure 1.1(a) shows an example of a state transition representation for a finite state automaton, in this case a 4-state machine; a specific example—a delay machine —is shown in Figure 1.1(b). If a 4-state machine is in state q_1 and receives a stimulus s_1, it transitions to state q_2 and generates a response r_1 after a unit time step. If it had received a stimulus s_2, it would have transitioned to q_3 and generated response r_2. For stimulus s_3, it goes to state q_4 and generates response r_3. This example has 4 different states, 3 different stimulus and 3 different responses. It has a finite number of states, and the diagram in Figure 1.1(a) describes all the possible, responses, and states that the machine can have. This diagram is useful, as it graphically gives a concise description that a table could not easily provide.

This state machine diagram can also be seen to have a correspondence to the phase space description of classical mechanics. The state is the equivalent of a position coordinate, and the response and stimulus prescribe the change in the state—a conjugate coordinate that corresponds to the flow.

The simple binary state machine in Figure 1.1(b), for example, can be understood by looking at it as a delay machine. It provides a response that is identical to the stimulus after a delay. The state of the machine is uniquely defined by the stimulus one cycle before; thus, the machine has a memory of one elapsed cycle. Since there are two different stimulus possibilities, the machine needs two states. If the system is in state q_1, and receives a stimulus 0, it stays in q_1 and responds with a 0. In the next time step, this 0 serves as a stimulus for the new state q_1, which happens to be the same as in the previous step; so, the system remains in q_1 and responds with a 0. On the other hand, if the state q_1 had received a stimulus 1, it would have transitioned to q_2 and responded with a 0. In the next step, this 0 stimulus would have caused a response of 1 and a return to the state q_1. So, when this state q_1 received a 0, it would return a 0 after a delay and remain in q_1; when it received a 1, after a delay, it would return a 1 and then return to q_1 after the transition to q_2. The reader should convince herself that if the machine had been in state q_2 at the beginning, it would have done the same in responding with the

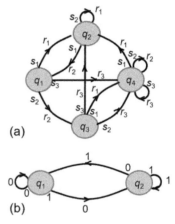

(a)

(b)

Figure 1.1: (a) shows an example of a small four state machine—and its responses to stimuli—an exemplar of a small finite state machine. (b) shows a finite state machine that delays its memory response by one cycle.

memory of the original stimulus signal and a return to the state q_2; this result follows from mirror symmetry. We have, in this representation, the state of the machine reflected in q_is, where i represents the different number of states possible. This description of the finite state machine, an automaton—a self-operating machine—is complete. Since all the states of the state machine have been prescribed, they are robust and self-contained. Unlike the large number of machine states of a general-purpose or a special-purpose computer, potentially leading the machine to unforeseen conditions, the finite state machine is capable of maintaining robustness. It enunciates precisely the time stepping of the state and the response, given a stimulus.

Mathematically, what we have described is a machine behavior in discrete time given by

$$q(t+1) = u[s(t), q(t)], \text{ and}$$
$$r(t+1) = v[s(t), q(t)]. \tag{1.1}$$

A multi-dimensional space describes the deterministic discrete time evolution of the machine. It is analogous in discrete time to the continuous evolution of state in the classical view. In classical mechanics, that is, the deterministic view, the state of a system is completely prescribed. All its particles are identified and describable by employing as canonical variables the position and velocity of the particles in the Lagrangian description, and position and momentum in the Hamiltonian description. The phase space or configuration space that the canonical variables define for the n particles is a multi-dimensional representation space where the evolution of the state of an ensemble can be described. In a finite state machine, the state of the machine is the canonical parameter, with the stimulus as the force under which it evolves, and the state diagram provides a complete description of the machine state space that results. The state-to-state change can be represented in this multi-dimensional space. The logical procedure can be projected onto this form, and since all problems are reducible to logical representation, the finite state machine gives a form with which to explore computational problems mathematically.

A Turing machine, sketched in Figure 1.2, is another example of a finite state machine, in this case a pushdown automaton that is a serial machine. A Turing machine is capable of tackling all the solvable computational problems by using unlimited and unrestricted memory. For our interests here, it exemplifies the sequential approach by which information can be manipulated, and hence the complexities and, through them, the entropic and energy considerations underlying it. The Turing model employs an infinite tape for its memory. A tape is just a spatial representation of memory. We could just as well imagine an infinite silicon memory column, row, or array. A tape

It will come as no surprise that finite state machines are used in nuclear machinery and defense machinery, neither of which that should be fooled with. Finite state machines provide the control and containment of states that safety and prudence require.

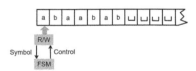

Figure 1.2: A simplified view of a Turing machine using a bitstream and a finite state machine to compute.

The term "pushdown" in pushdown automaton refers to the employment of a stack where only one position of the stack can be worked with.

head reads and writes symbols and moves around on the tape, which has the input string on it as its initial condition. The rest of the infinite tape is blank. The head moves 1 step at a time, going backward to read, or forward to write. The machine ends computation, that is, halts, by entering an accepting or a rejecting state—in logical terms, for example, true or false. If this doesn't happen, the machine never halts.

Again, as with the state-machine diagrams in Figure 1.1, one can see the correspondence with the classical mechanics description of phase space. Between the tape and the finite state machine, one has a collective description of position and its canonical conjugate—momentum.

An example of non-pushdown automaton, which in this case is also non-serial, is the cellular automaton, the simple conceptual view of which, for a one-dimensional form, is drawn in Figure 1.3, where multiple nearest neighbor interactions are shown occurring in a cellular structure simultaneously. A cellular automaton is a discrete dynamical system whose behavior is completely specified in terms of a local relation. The computational element g interacts only with its neighbors, and the state of the system evolves in time. Each of the interactions occurs via simple rules. The system evolves in time to other states by following these rules, but it rapidly evolves from a simple pattern to very complex forms exhibiting stability, oscillations, randomness and other characteristics that we associate with complexity.

This discussion has constructed for us a mathematical description of computational problems. The Church-Turing thesis establishes that solvable problems can be reduced to a symbolic manipulation form, that is, a procedure can be written to describe how to solve the problem. Machines may take many forms. We here restrict ourselves to the digital form. Machines can then be described by the state, which must include all the information that is necessary to reproduce the complete description of the logical state of the machine. In a finite state automaton, where we employed time stepping, receiving a stimulus causes the machine to enter a new state and produce a response in the next time step. So, the mathematical description has now been translated into a symbolic description. For a binary digital computation machine, 0s and 1s are used. A cellular automaton is a machine form where the computation proceeds spatially in time, so involving neighbor interactions. In a Turing machine, a sequential machine that is a push down automaton, computation occurs under the influence of the finite state machine description that provides the rules, receives the stimulus and responds so that the tape head may move, read or write on the tape that is the memory. Any solvable

The machine state of a classical computer is also an amalgam of position and flow. The machine state is a precise description of what state the computer is in now and a prescription of what state to step to next.

Martin Gardner's *Scientific American* column (see M. Gardner, "Mathematical games: The fantastic combinations of John Conway's new solitaire game 'life'," Scientific American, **223**, 120–123 [1970]) introducing John Conway's cellular automaton game called "life" is one of the most read scientific articles of all times. "Life" is played on a two-dimensional lattice, for example, a screen, ruled paper, or a go board, and shows complexity arriving from simple rules for interaction—patterns changing rapidly, blank fields appearing, sometimes patterns showing stability, and sometimes two or more patterns oscillating—just as in life. See S. Wolfram, "A new kind of science," Wolfram Media, ISBN 978-1579550080 (2002) for a thorough discussion of this theme. Conway's offer of a prize for a starting pattern that continues to expand was won by an MIT student, Bill Gosper, who showed a "glider gun" ejecting a continuous stream of gliders that sailed across the lattice. The glider gun is an example of "life" becoming a Turing machine. The cellular automaton is a good example of how complexity can appear from simple rules of interaction.

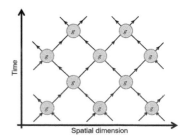

Figure 1.3: A space-time representation of a simple one-dimensional cellular automaton; g is a logic gate of the combinatorial network representing the passage of each cell through one clock cycle.

What a human can compute, a machine can compute, so long as we restrict the definition for computation to symbolic manipulation.

problem can be reduced to suit a universal Turing machine.

Both the information, in the sense discussed so far, and its manipulation in the physical world, that is, the machine, have to be tackled mathematically to understand the general relationships that constrain their behavior and hence to understand the physical constraints, such as energy, speed or others, in putting them to use. This observation leads us into a discussion of entropy which is so intimately connected to energy.

Analog approaches, amplifiers with resistors and capacitors, can be put together to solve individual differential equations quite efficiently, but not as a general approach.

1.3 The mechanics of information

A SOLVABLE PROBLEM can be reduced to a symbolic form. This form is then a machine-computable representation, and it brings together the mathematical and the physical mechanics of manipulation of information. To gain insight into the mechanics of information, we will have to understand the mathematical and the physical both separately and together.

First, we explore the properties of the mathematical symbols representing information, and the manipulation of these symbols to solve problems. This discussion will help us draw conclusions regarding the dynamics of this part of the mechanics of information, and through it, the general properties of the physical systems. These conclusions then can be connected to the physical functioning of the nanoscale devices described in the subsequent chapters.

1.3.1 Shannon, Boltzmann and von Neumann entropy

WHEN NEW INFORMATION IS PROVIDED, it reduces uncertainty. We know more. Information increases with a decrease in probability from maximum randomness—an uncertainty of $1/2$. In two independent events, the information content of the two together is the sum of information of the individual events. Information is additive. Claude Shannon posited the measure of information as I in units known as bits (b), for *binary digits*, as

If it were to rain in the Sahara, such an event is informative. If it were to rain in Ithaca, such an event would be par for the course, and so not that informative. The occurrence of events that are hard to predict has more information than that of events that are easy to predict.

$$I = -k \ln \mathfrak{p}, \qquad (1.2)$$

where k is some constant, and $\mathfrak{p}$ is the probability. This mathematical articulation is essential for building a scientific edifice around this idea. But, to do this, these probabilities must be possibilities in the data, not connections between them, as certainly are what a human being would also consider important in interpreting. Information, these digits, can be physically in any form. Information is physical.

The mathematics is built using the symbolic representation of this information.

Consider a written page. If there are m possible writable positions—the states of this system—and there are N distinguishable informational entities that can occupy these positions—the possible arrangements—then the total number of configurations that this system of the page can have is $\Omega = N^m$. This system need not be a page. Consider a 2-input $NAND$ gate. At its input, it has 2 positions, each of which can be a logical "0" or "1"; so there are 4 possible configurations. It has only one output, which can be either a "0" or "1", so 2 possible configurations. Thus, the information content between the input and the output changes. The gate, as an engine in thermodynamic sense, performs work by consuming energy and reducing the entropy of the system. Returning to our page example with its $\Omega = N^m$ possible arrangements, in independent random processes, any possible pattern of arrangement of this page has a probability of $\mathfrak{p} = 1/\Omega = 1/N^m$. With all these probabilities being equal, the information capacity of the page by including all the possible m positions is

$$I_m = km \ln N. \tag{1.3}$$

Typically, a page has about 3000 print positions, that is, $m = 3000$. If we set $N = 27$, for the 26 alphabet characters plus 1 for the space between words, a page has $27^{3000} \approx 10^{4294}$ possible configurations. For language, with its preferred arrangements of letters in words (e.g., u usually follows q) and then words in sentences, the number of possible configurations is significantly less than 10^{4294}. The arrangement of letters on a page, and our use of language, use a small subset of the possible arrangements in order to communicate. But, it is this ordered arrangement that gives the written symbols of language the capacity to communicate. If the arrangements were all random, that is, maximally disordered, they would not be understood.

How do we build a measure of the possible choices we could make and which thus represents the uncertainty of the outcome? Let $\mathfrak{p}_1, \mathfrak{p}_2, \ldots, \mathfrak{p}_n$ be the probabilities of occurrences of possible events. Let $H(\mathfrak{p}_1, \mathfrak{p}_2, \ldots, \mathfrak{p}_n)$ be a function that describes the choice in the selection of events, or how uncertain we are of the outcome. This function is a continuous function of these probabilities of the n different choices that exist. If all these probabilities are equal, then $\mathfrak{p}_i = 1/n$. This function must be a monotonically increasing function of n since, for equally likely events, an increase in choices, that is, of the n events, also means an increase in uncertainty. In addition, in breaking down successive choices through intermediate steps, this measure must remain the weighted sum of the individual measures of the choices broken down. This fact implies that, in Figure 1.4, the

The Oxford English Dictionary, Second Edition, contains about 172,000 words spread over about 22,000 pages. Computational analysis implies that the frequency of occurrence of words in spoken language drops precipitously beyond about 2,000 common words. William Shakespeare's writing has about 20,000 different words. One wonders what the mass adoption of Twitter and other tools at the confluence of social networks and social media will do to human development, human capabilities and human information transmission.

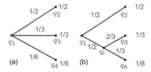

Figure 1.4: (a) Starting from state q_1, let there be a probability of $1/2$ of ending up in state q_2, of $1/3$ of ending up in state q_3, and of $1/6$ of ending up in state q_4. Now consider the $q_1 \mapsto q_3$ and the $q_1 \mapsto q_4$ state transitions, where q_1 is the starting state, and q_3 and q_4 are the final states. The final states really arise through an intermediate state q_i, and q_i alone; q_i occurs with a probability of $1/2$, and $q_i \mapsto q_3$ then occurs with a probability of $2/3$, and $q_i \mapsto q_4$ occurs with a probability of $1/3$.

information content

$$H(\frac{1}{2},\frac{1}{3},\frac{1}{6}) = H(\frac{1}{2},\frac{1}{2}) + \frac{1}{2}H(\frac{2}{3},\frac{1}{3}), \qquad (1.4)$$

and the only mathematical form of the measure that satisfies these requirements is

$$H = -k \sum_{i=1}^{N} \mathfrak{p}_i \ln \mathfrak{p}_i. \qquad (1.5)$$

This H is the averaged content of information contained in a collection of N random events, where the ith event has a probability of occurrence $\mathfrak{p}_i$. Strictly speaking, this equation measures the lack of information. If all the i events are known, the summation is zero. Our uncertainty of information has gone to zero. There is no lack of information. For a derivation of this relationship, see Appendix A. Self-information is the information content of an event that only depends on the probability of that event. $k \ln \mathfrak{p}_i$ is the self-information of the event i. With N such events possible, the averaged information is H, which is weighted over the self-information of all the events. In binary code, for any state, there are two possible configurations—0 and 1—with a probability of $1/2$ for each, and the averaged information in binary code per state is 1 b. If we know whether the state is 0 or 1, there is no uncertainty left. So, by knowing 1 b of information, we can completely describe this 1 state system. With the bit as our unit, we can now determine the normalization constant k:

$$1 = -k\left(\frac{1}{2}\ln\frac{1}{2} + \frac{1}{2}\ln\frac{1}{2}\right) \quad \therefore k = \frac{1}{\ln 2}, \qquad (1.6)$$

and hence

$$H = -k \sum_{i=1}^{N} \mathfrak{p}_i \ln \mathfrak{p}_i = - \sum_{i=1}^{N} \mathfrak{p}_i \log_2 \mathfrak{p}_i. \qquad (1.7)$$

This averaged information is a maximum when all the probabilities are equal, that is.

$$H_{max} \equiv I = -k\ln \mathfrak{p}_i = -\log_2 \mathfrak{p}_i. \qquad (1.8)$$

Now consider binary coding and decoding of information. We have, say, m positions in a string. If only one of the positions of the set of m positions is occupied and if it could only be in one of the two binary (0 or 1) states, then we have $\Omega = m$ as the number of configurations possible. There are m possibilities for this m-positions string. Generally, if a fixed number n could be in these binary states, then the number of configurations is ^mC_n, and if these n position choices are allowed to vary between 0 and m, that is, all the possibilities, then

$$\Omega = \sum_{n=0}^{m} {}^mC_n = 2^m. \qquad (1.9)$$

This is the statistical view of information in bits in the absence of any relationships between the bits. More accurately, it speaks to "missing information." Negentropy, meaning negative entropy, is a term that is sometimes employed to signify the capacity to do the useful work that is available when the system is not at thermodynamic equilibrium—the state where the entropy is maximum. Having information gives the ability to do useful work. If there are several possibilities, and one doesn't know which one is true, information is missing. Ignorance and entropy are high, information is low, and the possibility of doing useful work is reduced. The Shannon H can be viewed as the expectation over the discrete distribution of of $-k\ln \mathfrak{p}_i$s.

Classical information theory, strictly speaking, is a theory of data, the mathematical handling of randomness and the efficiency of handling data. In this sense, it achieves objectivity. The numbers π and e, are transcendental numbers whose digits are asymptotically random—take a longer and longer sequence of digits in these two transcendental numbers and you will see the frequencies of the digits converging to a single value. This measure is a maximum. However, one of the most beautiful equations that we come across early in education, $\exp(i\pi) + 1 = 0$, attributed to Euler is a very simple and compact representation of a relational property between the two. It provides context, or connections. Similarly, a picture contains a lot of information from the context it provides and the connections found therein; for example, a picture of a tree in nature shows the species of the tree, its environment and even its state of health. However, a binary representation of the picture does not provide the same level of information. Subjectivity matters.

This set of possibilities is identical to that of our starting example of a written page, where N is the number of distinguishable information units (in this case, 2, consisting of states 0 and 1), and m is the number of possible positions.

Now let us see what these possibilities reflect in terms of the communication and manipulation of information. Shannon's message theorem states that for a coded message, that is, a message consisting of original data of length M, and coding bits resulting in a message of length M_c, with ε being the probability of error in any single bit, the residual error rate $f(\varepsilon)$ is subject to

$$\frac{M}{M_c} \le f(\varepsilon) = 1 - \left[\varepsilon \log_2 \left(\frac{1}{\varepsilon} \right) + (1 - \varepsilon) \log_2 \left(\frac{1}{1 - \varepsilon} \right) \right]. \quad (1.10)$$

Thus, this theorem provides error bounds. Appendix A sketches a derivation of this significant theoretical prediction; this mathematical relationship places limits on error, given that coding bits are added to data, by drawing on the data to improve the correctness of the decoded message. Shannon's theorem also provides insight into constructing of switching circuits, in the presence of stochastic variability, to place limits on hardware-based error correction. Equation 1.10 implies that if there is no limit on the length of the code and the number of coding bits added for error correction, the residual error rate can be made arbitrarily close to zero. Errors arise for a variety of reasons. Limited bandwidth and the presence of noise can give rise to errors. The Shannon-Hartley theorem, Equation 1.11, describes the maximum rate of information transmission over a communications channel with limited bandwidth and with noise:

$$C = \int_0^B \log_2 \left(1 + \frac{S}{N} \right) dv = B \log_2 \left(1 + \frac{S}{N} \right), \quad (1.11)$$

where C is the channel capacity, B is the bandwidth of the channel, v is the frequency and S/N is the signal-to-noise ratio; the right hand side of this expression is a simplification of the integral, for constant, frequency-independent signal and noise. This theorem tells us the limits on what can be accomplished in transmission in the presence of noise. If the signal-to-noise ratio is large ($S/N \gg 1$), then the channel capacity is $C \approx 0.332 \times B \times SNR|_{dB}$, where the capacity is in b/s and SNR is the signal-to-noise ratio expressed in the logarithmic unit of dB. If $S/N \ll 1$ and constant, $C \approx 1.44 \times B \times S/N$.

We can now relate these thoughts to those from statistical mechanics. In the Boltzmann picture of a large classical ensemble of particles, such as an ideal gas consisting of n molecules and confined in a volume V in which the gas molecules can occupy m different positions—regions that can be occupied by one molecule—the entropy of the system is

$$S = k_B \ln \Omega, \qquad (1.12)$$

where Ω is the number of possible configurations, k_B is Boltzmann's constant $(1.38 \times 10^{-23} \; J/K)$ and S is the Boltzmann entropy. Although an interpretation of entropy as a measure of disorder or as a measure of freedom is only partially correct, Boltzmann entropy is the only function that satisfies the requirements for a function to measure the uncertainty of classical particles subject to certain constraints. This approach applies also to the string of bits that are the message. In the case of a binary information string, we set the constant k to precisely 1, without losing generality. So, Equation 1.7 and 1.12 are equivalent—except for Shannon's sign choice—as they describe the possible configurations and the ability of these configurations to describe the possibilities of making choices. In information, the entropy is a measure of the uncertainty over the true content of a message, but the task is complicated by the fact that successive bits in a string are not random and therefore not mutually independent in a real message.

The implication of this connection is direct. In an ideal gas, if the volume is reduced from V to $V/2$ under isothermal conditions, the entropy is also reduced by half because the number of possible configurations for the gas atoms has been reduced by half. Reduction of this entropy happens via the dissipation of the thermal energy $\Delta Q = T\Delta S$. Now consider a single binary bit information system. It can be in one of two states. If they are both equally probable, the Boltzmann entropy is $S = k_B \ln \Omega = k_B \ln 2$. When the state of this single bit is ascertained, that is, the observer reduces the number of possible states to one, a dissipation of $\Delta Q = T\Delta S = T(k_B \ln 2 - k_B \ln 1) = k_B T \ln 2$ happens. The amount of dissipation is is about $10^{-21} \; J/b$ at 300 K, in energy, or about $10^{-23} \; J/K$, in entropy change, per erasure of a bit. Why is this fact significant? In processing information by performing irreversible logic, a loss of information occurs, and, in isothermal conditions, $k_B T \ln 2$ is the minimum energy dissipation per bit of information loss. Thus, it is the fundamental lowerbound and sets fundamental limits to the traditional technology of information processing.

In the $NAND$ (Figure 1.5[a]) operation $(a, b) \mapsto \overline{a \wedge b}$, suppose the 4 initial states—$(0,0), (0,1), (1,0)$ and $(1,1)$—have the same probability $(1/4)$, the starting information entropy is

$$H_{initial} = -4 \times \left(\frac{1}{4} \log_2 \frac{1}{4} \right) = 2 \, b, \qquad (1.13)$$

The inscription $S = k \ln W$ is carved on Boltzmann's grave at Central Cemetery in Vienna. Boltzmann struggled hard to convince leading compatriots—Ernst Mach, Wilhelm Ostwald, Georg Helm and others—with his notions of tackling disorder. Ostwald was a strong believer that energy fluxes and transformations suffice to explain physics and chemistry and didn't believe in atoms and molecules. So strong were these beliefs that he is quoted to having said about atoms and molecules, "We have little right to expect from them, as from symbols of algebra, more than we put into them, and certainly not more enlightenment than from experience itself." For a collection of independent classical particles, Boltzmann's H-factor is

$$H(t) = \int_0^\infty f(E,t) \left[\ln \left(\frac{f(E,t)}{E^{1/2}} \right) - 1 \right] dE,$$

where $f(E, t)$ is the particle energy distribution function in time. In an isolated collection—gas molecules, for example, as an approximation—this H-factor is at a minimum when the particles obey a Maxwell-Boltzmann distribution and that of any other distribution with the same total kinetic energy will be higher. If collisions are allowed, any starting particle distribution will asymptotically approach the minimum H and a Maxwell-Boltzmann distribution. Shannon's H—the averaged information content with its negative sign—has its antecedents in the uncertainty that is represented in the discrete counterpart of Boltzmann's H-factor.

Entropy is maximum in thermal equilibrium. This fact establishes the direct correspondence between temperature and entropy. Since the temperature scale is defined separately—in practical matters, it is referenced through the triple point of water—the normalization constant appears as the Boltzmann constant in units of J/K. The thermodynamic entropy has units of energy over temperature. The choice of bit (b) as a unit defined the normalization constant in Shannon information entropy.

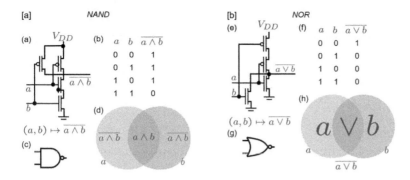

Figure 1.5: *NAND* and *NOR* logic and its implementation in static *CMOS* circuits. In the top set of figures, (a) and (e) show the *CMOS* implementations. For *NAND*, in (a), only when logical variables a and b are 1 physically through a high voltage do the $nMOS$ transistors pull the output node voltage down to a low level, a logical 0. During this set of conditions the $pMOS$ transistors, whose function is to pull the output node to the high voltage, V_{DD}, are off. The logical transformation that this gate represents is for the sets a and b, each of which has 0 and 1 as its members; the output is $\overline{a \wedge b}$. The truth table for this mathematical operation is shown in (b). (c) shows the circuit schematic representation, and (d) the Venn diagram. Venn diagrams show all the possible logical relations between a finite collection of sets. In this diagram, $a \wedge b$ is the intersection, and therefore all the region outside is $\overline{a \wedge b}$. In the Venn diagram, $a \wedge b$ is the intersection, and therefore all the region outside is $\overline{a \wedge b}$. The second set of figures show the same information for a *NOR* gate.

and the ending information entropy is

$$H_{final} = -\left(\frac{3}{4}\log_2\frac{3}{4} + \frac{1}{4}\log_2\frac{1}{4}\right) = 2 - \frac{3}{4}\log_2 3 \; b. \quad (1.14)$$

The information loss in this 2-input *NAND* gate is $(3/4)\log_2 3 \; b$. The state space has been compressed and the input may not be recovered from the output. The same holds true for the *NOR* gate shown in Figure 1.5[b]. Here, there is now a region in state space which arises from a multiplicity of states—an overlap representing compression. The Venn diagram captures and pictorially represents this loss.

Now consider another gate, $(a, b) \mapsto (a \vee b, a \wedge b)$, a logical operation where the 0th binary position is the logical *AND*, and the 1st binary position is the logical *OR*. The ending information entropy for this gate is

$$H_{final} = -\left(2\frac{1}{4}\log_2\frac{1}{4} + \frac{1}{2}\log_2\frac{1}{2}\right) = 1.5 \; b. \quad (1.15)$$

Since only $0.5 \; b$ of information loss occurred in this operation, compared to $\sim 1.19 \; b$ for the *NAND* gate, the *NAND* gate is more irreversible that this one is.

The entropy of information, a Shannon notion, and the entropy of thermodynamics, a Boltzmann-Gibbs notion, are related by constant factors. The sign of Shannon entropy is opposite of that employed in the Boltzmann-Gibbs definition. Shannon entropy, strictly speaking is, negative entropy, termed negentropy, which we discussed earlier. Both Shannon entropy and Boltzmann-Gibbs entropy are both connected to the degrees of randomness or freedom that the system has. The logical operations on information—on the distinct logic states— are operations on distinct physical states. A state of $n \; b$ has n degrees of freedom. When $n \; b$ are erased, say set to 0, then $2n$ logical states have been compressed to 1 state with no degree of freedom. Erasure of information is dissipative, and the irreversible loss of information occurs through heat dissipation.

This correspondence between an information engine and the thermodynamic understanding of a mechanical engine is broad. Logical operations compress the phase space spanned by the computer's information-bearing degrees of freedom. In order for this compression to occur, a corresponding expansion of entropy into other degrees of freedom must occur.

We emphasize this point through a discussion of the correspondence of this approach with that of classical mechanics. Phase space in classical mechanics is the space of all possible states of a system, that is, a state space. For a collection of particles, this space would encompass all the positions, that is, the physical space, and momenta, that is, the momentum space, of the particles. Since position and momentum are canonical conjugates, one requires both of these to describe the future state of the system, that is, to completely define the dynamical system. Recall that position q and momentum p are related to equations of motion by

$$\frac{d}{dt}q(t) = \frac{\partial \mathcal{H}(q,p)}{\partial p}, \text{ and}$$

$$\frac{d}{dt}p(t) = -\frac{\partial \mathcal{H}(q,p)}{\partial q}, \tag{1.16}$$

where $\mathcal{H}(q,p)$ is the Hamiltonian of the system. Information systems are also dynamical systems. The state space includes all the possible machine states that evolve in time under evolution rules that are programmed into the machine. In a computing engine, as a closed system with its own power source, the state/phase space consists of states that contain both information-bearing degrees of freedom, and irrelevant ones, such as thermal vibration. This phase space in the conservative system cannot be compressed. It is a closed system. So, when information is discarded during a logic-processing step, the irrelevant degrees of freedom must expand. Entropy must increase.

An illustration of this equivalence is Maxwell's demon. Figure 1.6 shows a gedanken experiment originally formulated by James Clark Maxwell in 1867 and restated in a modern form by Szilard. An isolated system consisting of a distribution of hot gas molecules and cold gas molecules has a partition whose opening and closing a demon can control through a frictionless process, that is, no energy is lost in the mechanical mechanism. The system consisting of the chamber walls and the demon is in thermal equilibrium and is isolated. The 2nd law of thermodynamics states that entropy always increases in a thermally isolated system, that is, $\Delta S \geq 0$. For reversible changes, the entropy change vanishes, and for irreversible changes, entropy increases. $dS = dQ/T$, where T is the absolute

Ambiguity underlies much of the physical and informational world. Heat is energy in degrees of freedom which we can't completely describe. Thus, this entity is statistical in nature, and capturing this characteristic by using the concept of $k_B T/2$ per degree of freedom in a classical distribution quite often suffices. Statistics also underlies our mental processes. We are just not consciously aware of the ambiguities. Logical processes and statistical processes are related to one another. Thus, the conscious logical mental processes are connected with the unconscious ones. This relationship, not yet understood, is in the background of the success achieved in artificial intelligence, pattern recognition and vision processing when these fields adopted statistical approaches at the end of the 20th century. It also shows up here in the connection between statistical notions of information, and physical ones. Probability has both objective and subjective connotations. In a random physical system—a fair dice, for example—physical probability is a stable asymptotic limit. Probability exists even without randomness when it represents the degree to which a piece of evidence supports a belief. This interpretation of probability is subjective. The Shannon measure is an objective quantification of information. However, this measure does not capture the subjective meanings and associations of information.

temperature. The demon lets only the swift, that is, hot gas molecules pass from Chamber A to B, and only the laggards, that is, the cool molecules egress from B to A. Chamber A therefore decreases in temperature, and Chamber B increases in temperature. We started in thermal equilibrium. However, now T_A is dropping, and T_B is rising. dQ_A is negative, as hot molecules leave and cold molecules enter, whereas dQ_B is positive, with the opposite set of events. So, as the change takes place, apparently

$$dS = dS_A + dS_B = \frac{dQ_A}{T_A} + \frac{dQ_B}{T_B} < 0. \qquad (1.17)$$

In this system, $dQ_A = -dQ_B$, since the only exchange allowed takes place through the partition. But if the demon functions as described, then T_A is being lowered, and T_B is being raised, which would mean that, in the sum in Equation 1.17, the first term, which is negative because heat is being transferred from Chamber A, and is larger in magnitude than the second, the sum would be less than 0. The entropy apparently decreases.

A modern $k_B T \ln 2$ version of this thought experiment is to imagine an isolated system with just one molecule. The demon inserts a massless, infinitesimally thin and adiabatic partition trapping the molecule on one side or the other. She performs a measurement to determine the side that the molecule is on and then allows the molecule to expand isothermally to fill back the entire volume of the isolated system. In expanding to fill the volume, the molecule performs $k_B T \ln 2$ of work in moving the partition to one end from the middle, doubling the volume and, in the process, doing work on a weight that was attached to this partition. Entropy is now reduced by $k_B T \ln 2$ ($= \Delta Q/T$). This conundrum was resolved a hundred plus years later. The demon needs to store the information of the measurement. She could put it in the memory. The demon now has increased informational entropy. If she erases it, because she has only a $1\ b$ memory and needs to make another measurement, then this energy must be dissipated. In either of these situations, there is a change of $k_B T \ln 2$ in work done. In the former case, the work was done with an increase in entropy in the form of the demon's information entropy. In the latter, it is the energy needed to erase the bit of information. While the measurement itself can be performed reversibly, an increase in entropy occurs because of information held by the demon.

From a quantum-mechanical viewpoint, the Schrödinger or Heisenberg representations describe a reversible and deterministic solution through linear theory. It is in the process of measurement, or equivalently in the use of Hermitian operators, that one obtains specific eigenvalues with probabilities arising in the coefficients of the eigenfunctions. The non-deterministic results are a consequence of prob-

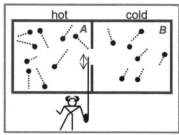

Figure 1.6: Szilard's version of the paradox of Maxwell's demon gedanken experiment. The demon controls a loss less partition through which she allows only the hot molecules to move from A to B, and only the cold ones from B to A. The 2nd law is not violated since the demon's entropy is part of this closed system.

The demon was eventually exorcised in 1982 following nearly a hundred years of conceptual developments, with the final one of reversible computing, building on progressive thinking from Leo Szilard, Leon Brillouin, Denis Gabor, Rolf Landauer and Charles Bennett: it is Maxwell's demon that gets hot from dissipation. In a class once, I was asked, "If `God´ keeps track of all these states ...," to which one response may very well be to imagine "God" as the ultimate Maxwell's demon.

abilities. The irreversible process occurs because of wavefunction collapse. The process of measurement unfolds in lack of causality and state reduction. The equivalent of the Boltzmann relationship in quantum mechanics is due to von Neumann:

$$H = -\text{Tr}(\rho \ln \rho), \qquad (1.18)$$

where ρ is the density matrix of the quantum system, and Tr (for trace) is the summation over the diagonals, that is, the summation of the expectation probabilities of the eigenstates.

A few important theorems relate to the nature of the quantum wavefunction of a single unknown system: it is not possible to measure it (the no-measurement theorem), copy it via a unitary process (the no-cloning theorem) or eliminate it via a unitary process (the no-deletion theorem). Some of these details will appear in our discussion in Chapter 3. It is through these properties that one sees that the quantum wavefunction has only a statistical meaning and no objective reality, or determinism, that exists at the fundamental level of the world. However, the unitary evolution of the observed $|\psi\rangle$, which is a superposition quantum state, and the observer $|\phi\rangle$ can still be described through the wavefunction $|\psi\rangle|\phi\rangle \mapsto \sum_n c_n |\psi_n\rangle|\phi_n\rangle$, based on the orthonormal states. The source of dissipation is the interaction with the environment; this interaction, which introduces decoherence (i.e., unknown degrees of freedom) where the dissipation occurs. Without environment interactions, the unitary evolution of the system of observer and observed maintains information. Measurement or copying does not cause dissipation. The dissipation only happens when information is erased. Thus, $k_B T \ln 2$ is the thermodynamic cost of decoherence and its associated entropy change in unknown degrees of freedom.

1.4   Probabilities and the principle of maximum entropy

THE EQUILIBRIUM AND NON-EQUILIBRIUM BEHAVIOR that we are interested in understanding in information manipulation in machines have their foundations in statistical mechanics as we have seen through the Shannon- and Boltzmann-centric discussion in Section 1.3. Entropy appeared in both as a measure of uncertainty, thus introducing the idea of probability.

The principle of maximum entropy provides us with a predictive tool. The principle states that, given a precisely stated *prior*, the probability distribution that best represents current knowledge is the one with the largest entropy.

A coin can be called fair or not fair only after an infinite number of tosses. The earth was flat in the European world to a probability somewhere close to 1 until Columbus's observations, when the probability flipped to close to 0. New information arising from observations (Columbus being credited here, though I wonder who figured this out in Europe first) changes this expectation that is expressed by probability; objective probability change arising in rational logical thought's progress. Probability as a personal belief has a subjective connotation. This objective-subjective conflict is inherent to probability, information, entropy and randomness through the ambiguity that incompleteness brings. Johann Carl Friedrich Gauss, the great German mathematician protested a friend's use of infinity—while appreciating the misuse—by saying, "I object to the use of infinity as something completed. This is impermissible. Infinity is a way of speaking. Its true meaning is of a limit approached by certain ratios indefinitely, or of others that are allowed to increase without constraints. Infinity is be used as `a manner of speaking.' It is not a mathematical value."

Recall our discussion of Boltzmann's H-factor. There is a correspondence between likelihood and the function $f(E,t)$'s form, given the constraint on the energy of the closed system. The Maxwell-Boltzmann distribution arises as the one with maximum entropy and can be derived from this principle of maximum entropy. This approach is a particularly useful tool for the field of artificial intelligence, where one must make inferences based on an incomplete collection of information.

Entropy, as discussed, is tied to statistics and thus is connected to probability. Probabilities are hypothesized—are subjective—unless an experiment is carried out *ad infinitum*. Bayes's theorem allows information based on any new observation to be incorporated into the description of a system. Bayes's theorem states that if $\mathfrak{p}(A|C)$ denotes the probability of a proposition A given a hypothesis C, which itself is a proposition, then additional information B leads to a reassessment of the probability to the *posterior* probability of

$$\mathfrak{p}(A|BC) = \mathfrak{p}(A|C)\frac{\mathfrak{p}(B|AC)}{\mathfrak{p}(B|C)}. \tag{1.19}$$

If proposition C is the prior information, that is, a *prior* in the vernacular, $\mathfrak{p}(A|C)$ is the prior probability of A based on only the information C. With the additional information B, the posterior probability $\mathfrak{p}(A|BC)$ is modified from the prior probability by the ratio in Equation 1.19. See Appendix B for a proof of Bayes's theorem.

With the introduction of this Bayesian notion and notation, we can restate the fundamental conditions that lead to the Shannon relationship. The key point here is that one wants certain conditions to be met by an entropy function so that its form as proposed by Boltzmann and Gibbs is the only choice and is not arbitrary. These fundamental conditions are as follows:

- For a given n, and $1 = \sum_{i=1}^{n} \mathfrak{p}_i$, the function $H(\mathfrak{p}_1, \ldots, \mathfrak{p}_n)$ is a maximum when $\forall i \ \mathfrak{p}_i = 1/n$.

- The function $H(\mathfrak{p}_1, \ldots, \mathfrak{p}_n) = H(\mathfrak{p}_1, \ldots, \mathfrak{p}_n, 0)$. Including an impossible event does not change H.

- If A and B are two finite set of events, not necessarily independent, then the entropy $H(A, B)$ for joint events A and B is given by the entropy for the set A expanded by the weighted average of the conditional entropy $H(B|A)$ for set B, given the occurrence of the ith event A_i of A, that is,

$$H(A, B) = H(A) + \sum_i \mathfrak{p}_i H(B|A_i), \tag{1.20}$$

where $\mathfrak{p}_i$ is the probability of the occurrence of event A_i. The

The Monty Hall problem is a historic example of what this statement says. Probabilities reflect on what is not known. They depend on the observer and her information which is dynamic. *Let's Make a Deal*, a 1970s television game hosted by Monty Hall (Maurice Halprin), engendered much public debate since many mathematicians, including the great Paul Erdös, got it wrong. The prize to be won is behind a door in a set of many. The contestant picks one of the doors, and then the showman opens all the doors except the one chosen and an additional one, to show that the prize is not behind the opened doors. Should the contestant stay with or change her choice? She should change because additional information has now been provided—information that the showman had and then used in opening the doors. Her probability of being incorrect now drops to 1/2 if she changes. Convince yourself of this fact by thinking through a large number of doors, or by just writing the truth table with 3 doors as the initial condition of the problem. An earlier version of this informational puzzle appeared in Martin Gardner's 1959 Mathematical Games column titled "The Three Prisoner Problem." M. Gardner, "Mathematical Games: Problems involving questions of probability and ambiguity," Scientific American, **201**, 174–182. How a story is told also matters.

relationship is a general statistical relationship independent of whether particles or bits are involved.

In a system with Avogadro's number of molecules in a volume, or oodles of transistors in an area or volume, we must resort to probability theory because neither specifying the initial conditions nor following the trajectories of states of particles or a machine is realistic. Statistical mechanics teaches us how to work with this complexity but only based on *priors*. There are missing informational elements in this description. We do not update estimates. Initial data consisting of few pieces of macroscopic information—temperature or pressure, for example—is all we have. There will be more on this thought in Section 1.5, which discusses algorithmic entropy, which also attempts to capture informational content.

If a macroscopic variable is in the form of a function $f(x)$ at discrete values of variable x, with the expectation of $f(x)$ over n mutually exclusive and alternatives of the variable $\{x_i\}$, then

But not entropy, which, as we will discuss, lacks the correspondence between a measurable of the macroscopic state and a measurable of the microscopic state, unless we invoke information. Temperature follows from the distribution of kinetic energy, and pressure from the flux.

$$\langle f(x) \rangle \equiv \sum_{i=1}^{n} \mathfrak{p}_i f(x_i) \tag{1.21}$$

constrained by

$$\sum_{i=1}^{n} \mathfrak{p}_i = 1, \text{ and } \mathfrak{p}_i \equiv \mathfrak{p}(x_i) \geq 0. \tag{1.22}$$

Since this set of equations describes a very unconstrained problem, we resort to assigning probabilities. Shannon's relation ascribes an optimum measure of uncertainty for this weakly constrained problem in the form of entropy given by the logarithmic relationship in Equation 1.7. We restate it, using the conventional symbol S to denote entropy but understanding that Shannon is really dealing with the negative of the Boltzmann-Gibbs view, as

$$S(\mathfrak{p}_1, \ldots, \mathfrak{p}_n) \equiv -k \sum_{i=1}^{n} \mathfrak{p}_i \ln \mathfrak{p}_i = - \sum_{i=1}^{n} \mathfrak{p}_i \log_2 \mathfrak{p}_i. \tag{1.23}$$

The principle of maximum entropy gives the means to determine the set $\{\mathfrak{p}_i\}$ that satisfies these constraints, by using the method of undetermined Lagrange multipliers:

$$\mathfrak{p}_i = \frac{1}{Z(\lambda)} \exp[-\lambda f(x_i)], \text{ where}$$
$$Z(\lambda) \equiv \sum_{i=1}^{n} \exp[-\lambda f(x_i)]. \tag{1.24}$$

This partition function is obtained by defining through the constraint equation, Equation 1.22, that maximizes entropy through Equation 1.23, with the Lagrange multiplier in the expectation equation,

Equation 1.21, as

$$F \equiv \langle f(x) \rangle = -\frac{\partial}{\partial \lambda} \ln Z(\lambda). \tag{1.25}$$

The known expectation value of the function $F \equiv \langle f(x) \rangle$ is obtained by adjusting λ. This adjustment is all that we may do. The expectation of any other function $g(x)$ then follows as $\langle g(x) \rangle \equiv \sum_1^n \mathfrak{p}_i g(x_i)$. Results due to any additional constraints now follow from additional Lagrangian constraints. So, for example, when data is specified for $m < n$, for functions $f_r(x)$,

$$F_r = \langle f_r(x) \rangle = \sum_i^n \mathfrak{p}_i f_r(x_i), \tag{1.26}$$

where $r = 1, \ldots, m$, we define

$$\lambda \cdot f(x_i) \equiv \lambda_1 f_1(x_1) + \cdots + \lambda_m f_m(x_i), \tag{1.27}$$

for all constants leading to the probability distribution that maximizes entropy, subject to the expectation that

$$\begin{aligned} \mathfrak{p}_i &= \frac{1}{Z(\lambda)} \exp[-\lambda \cdot f(x_i)], \text{ and} \\ Z(\lambda) &\equiv \sum_i \exp[-\lambda f(x_i)]. \end{aligned} \tag{1.28}$$

The Lagrange multipliers follow from the coupled differential equations:

$$F_r = -\frac{\partial}{\partial \lambda_r} \ln Z(\lambda_1, \ldots, \lambda_m), \quad r = 1, \ldots, m. \tag{1.29}$$

The maximum entropy, using this relationship in Equation 1.23, is

$$K_I = k \ln Z + k\lambda \cdot F, \tag{1.30}$$

with

$$\frac{\partial K_I}{\partial \lambda_r} = 0, \text{ and } \lambda_r = \frac{1}{k} \frac{\partial K_I}{\partial F_r}, \tag{1.31}$$

the Legendre transformations involving the variables $\{F_r\}$ and $\{\lambda_r\}$. If the functions f_r depend on external parameters, say α, so that $f_r = f_r(x; \alpha)$ then,

$$Z = Z(\lambda_1, \ldots, \lambda_m; \alpha), \text{ and } K_I = K_I(F_1, \ldots, F_m; \alpha). \tag{1.32}$$

Defining $\langle df_r \rangle = \langle \partial f_r / \partial \alpha \rangle d\alpha$ then leads to

$$dK_I = K\lambda \cdot dM, \text{ where } dM_r = d \langle f_r \rangle - \langle df_r \rangle, \tag{1.33}$$

which are inexact differentials.

These are examples of making predictions from incomplete and sparse data. We are attempting to estimate the cause of a phenomenon. This is an inverse problem in which, from $F = Kf$, with F not known completely, one wishes to find f. Drawing an inference based on incomplete information raises the Bayesian statistical quandary discussed earlier. If there are m possible outcomes of a trial, and n trials are carried out, there are m^n possible outcomes. If the ith result occurs n_i times, it has a frequency $f_i \equiv n_i/n$ for $1 \leq i \leq n$. If we are provided data F_j in terms of M numbers, that is,

$$F_j = \sum_{i=1}^{m} K_{ij} f_i, \ \ 1 \leq j \leq M < m, \tag{1.34}$$

where K_{ij} is known, and then asked for the true frequencies producing this data, we will proceed based on the *priors*. The number of ways a particular set of occupation numbers $\{n_i\}$ can be obtained is the multinomial coefficient

$$\Omega \equiv \frac{n!}{(nf_1)! \cdots (nf_m)!}, \tag{1.35}$$

the multiplicity factor. Maximizing Ω subject to the given data lets us determine the set $\{n_i\}$ realizable in the maximum number of ways. Sterling relations leads us to $(1/n) \ln \Omega = -\sum_i f_i \ln f_i$, subject to Equation 1.34.

The maximum entropy then leads to

$$f_i = \frac{1}{Z} \exp \left(\sum_j \lambda_j K_{ji} \right). \tag{1.36}$$

The implications of this approach in thermal equilibrium now follow. Measurements lead us to an energy $\langle E \rangle$. If system energy levels are E_i and there is only one external parameter, say V, the system volume, then the constraints are

$$\langle E \rangle = \sum_i p_i E_i, \ \ \text{with} \ \ \sum_i p_i = 1. \tag{1.37}$$

Consequently,

$$p_i = \frac{1}{Z} \exp(-\beta E_i), \ \ \text{and} \ \ Z(\beta) = \sum_i \exp(-\beta E_i), \tag{1.38}$$

with the Lagrangian multiplier β following from

$$\langle E \rangle = -\frac{\partial}{\partial \beta} \ln Z(\beta). \tag{1.39}$$

The maximum entropy is then

$$K_I = \kappa \ln Z + \kappa \beta \langle E \rangle, \tag{1.40}$$

where κ is a constant, that we will soon see to be the Boltzmann constant, k_B. From Equation 1.33,

$$dK_I = \kappa \beta dQ, \tag{1.41}$$

with

$$dQ = d\langle E\rangle - \langle dE\rangle. \tag{1.42}$$

The second term is the work performed, that is,

$$dW \equiv \langle dE\rangle = \sum_i \mathfrak{p}_i \left(\frac{\partial E_i}{\partial V}\right) dV = -PdV, \tag{1.43}$$

this result following from the definition of pressure. This equation directly leads to Equation 1.42 in the form

$$-dE = dQ + dW, \tag{1.44}$$

the 1st law of thermodynamics; dQ must correspond to the heat introduced, an inexact differential, in classical thermodynamics. The Lagrange multiplier β follows as the integrating factor of dQ. As κ can now be seen as the Boltzmann constant k_B, one can rewrite Equation 1.40 in the form:

$$E - TS = -k_B T \ln Z \equiv \mathcal{F}(T, V), \tag{1.45}$$

where $\mathcal{F}$ is the Helmholtz free energy. Pressure is now determinable as $P = (1/\beta)(\partial/\partial V)\ln Z$. From the Legendre transformation parameters,

$$\frac{1}{T} = \left.\frac{\partial S}{\partial E}\right|_V. \tag{1.46}$$

The principle of maximum entropy has directly led us to the classical thermodynamic relations. The Boltzmann relationship states $S = k_B \ln \Omega$, where Ω is the number of microstates available to the system. Since $S = (E - F)/T$, it follows that

$$\exp(-\beta F) = \Omega \exp(-\beta E). \tag{1.47}$$

As Ω changes rapidly with energy, and both β and E are positive, the probability is sharply peaked about the equilibrium energy.

The implication of Bayes's theorem here is as follows. No probabilistic theory can assure prediction. If that happens, the theory tells us, there are constraints operating of which we do not know. The theorem thus points to the possible existence of additional information that needs to be incorporated.

1.5   Algorithmic entropy

THE CLASSICAL AND THE QUANTUM-MECHANICAL definitions of entropy, as discussed up to this point, connected through the probabilities of what is missing, are still incomplete. A string of length n b, produced by tossing a fair coin, is just as random as another string produced by the fair coin. This fact reflects the statistical definition of entropy applicable to ensembles but not individual events.

In statistical mechanics, in equilibrium, it is not possible to describe the macroscopic entropy as an ensemble average of microscopic variables where m is the mass and v are the velocities. The probabilistic definition $S(\mathfrak{p}) = -\sum_i \mathfrak{p}_i \log_2 \mathfrak{p}_i$ also implies that there is no function $f(x)$ that satisfies $S(\mathfrak{p}) = -\sum_i \mathfrak{p}_i f(q_i, v_i)$ over the phase space states. Individual configurations have no direct measure of entropy. Entropy needs indirect methods. Another way of looking at this problem is that a stream of data will capture the time path but will not extract any information content in the form of spatially extended structures—connections represented, for example, in words, programs or pictures. So, patterns and their representation remain poorly incorporated. In a physical system, the measurement of states leads to the identification of states in informational form. A program of this physical system, in its computational form, is complete if the description it provides of the system contains sufficient information for the reconstruction of the system. In classical mechanics, this information represents the numbers associated with the canonical variables (position and velocity in the Lagrangian approach, or position and momentum in the Hamiltonian approach) of all the particles provided as bits of information.

Algorithmic information content/algorithmic randomness attempts to capture the part of entropy that is not within the Shannon formulation. Algorithmic entropy is a measure defined for a bit string s as the size in bits for the shortest computer program with the output s. The procedure for describing random states requires more bits of information than that for ordered states. It is the number of bits in an absolute and precise sense to describe an output string of a universal machine. Where there is more order, the algorithm to reconstruct the regular state is more concise than that to reconstruct a more random state. The definition of algorithmic entropy makes intuitive sense in this light. In our Turing machine picture, the algorithmic randomness $K(s)$ of the binary string s is the shortest program s^* that produces the output s and then halts, that is,

$$K(s) \equiv |s^*|. \tag{1.48}$$

Figure 1.7 shows a program for writing a string of decimal integers starting with 0. It is the informational representation that gives the context of the bit string output.

This Turing machine–usable definition of algorithmic entropy is also appropriate for a universal computer—a universal Turing machine. A program p operating on any other computer, say C, requires

Contrast this with temperature for classical particles, which is expressible through averaging the kinetic energy $mv^2/2$. There exists no microscopic quantity corresponding to entropy.

That Shannon or Boltzmann entropy are incomplete picture of the entire entropy can be seen to partially lie in the lack of connections or context within their entirely statistical individual bit or state representation. A short algorithm can describe the writing of π to unlimited precision. Yet, the digits in its string occur with equal probability. There is a connection in this arrangement of bits that the Shannon or Boltzmann viewpoint misses but is relevant to entropy.

```
def intstring(n):
    i = n - 1
    list = []
    while i > 0:
        list.append(i)
        i = i - 1
    print list
intstring(10)
```

[9, 8, 7, 6, 5, 4, 3, 2, 1]

Figure 1.7: A program loop for writing a string of decimal integers.

an additional component c, in so that cp may execute on a universal Turing machine. When s is a string which is a binary representation of an integer, $K(s) = \log_2 s$; using a referencing to universal Turing machine for this measure makes it unambiguous. Microstates, such as those of the collection of particles in classical mechanics, are specifiable w.r.t. encoding schemes that satisfy criteria for accuracy, such as in the Shannon form. One could also specify the microstates as cells—regions—which only one particle can dwell in, to align it with our interpretation in classical mechanics. The algorithmic randomness of the microstates then is determined as the length of the shortest program to reproduce the state to the resolution. We can now observe the correspondences between information in the universal Turing machine, and particles in classical mechanics. In a statistical ensemble($\mathscr{S}$), consisting of the collection of microstates $\{s_k\}$ of probabilities $\mathfrak{p}_k$, the statistical entropy is

$$H(\mathscr{S}) = -\sum_{s_k \in \mathscr{S}} \mathfrak{p}_k \log_2 \mathfrak{p}_k. \tag{1.49}$$

To include information in this representation, the ensemble must be defined precisely and concisely. Let a concise algorithm ε exists that describes all the microstates of $\mathscr{S}$. It provides as its output the probabilities to the accuracy desired. Now, it may be that ε may not halt for reasons of accuracy or because of the number of microstates. When we demand a certain accuracy δ, the program uses finite steps after which all the states of $\mathscr{S}$ that have probability larger than δ are known. The length of this smallest program ε then defines the algorithmic information content of the ensemble; $K(\mathscr{S}) \equiv |\varepsilon^*| \ll H(\mathscr{S})$ is the thermodynamic representation of this ensemble. Appendix C provides proof that

$$\langle K \rangle_{\mathscr{S}} = \sum_{s_k \in \mathscr{S}} \mathfrak{p}_k K(s_k), \tag{1.50}$$

which closely follows the Shannon entropy

$$\langle K \rangle_{\mathscr{S}} \approx -\sum_{s_k \in \mathscr{S}} \mathfrak{p}_k \log_2 p_k. \tag{1.51}$$

Algorithmic entropy resolves the problems which, because of the probability viewpoint, are encountered in macroscopic to microscopic connections. It is the microscopic analog of Shannon or Boltzmann-Gibbs statistical entropy. If a macrostate is precisely and concisely describable such as through the equations of motions and boundary conditions definable through a small number of bits, then the statistical entropy of the macrostate is equal to the ensemble average of the microstates' algorithmic entropy:

$$-\sum_i \mathfrak{p}_i \log_2 \mathfrak{p}_i < \sum_i \mathfrak{p}_i H_i \leq -\sum_i \mathfrak{p}_i \log_2 \mathfrak{p}_i H(\mathfrak{p}) + \mathscr{O}(1). \tag{1.52}$$

Any fixed length of the Euler's constant e or of π can be computed from a small procedural description, our program. These transcendental numbers are only pseudorandom. In contrast, there is no concise procedure to describe a sequence of fair coin tosses. The latter sequence is incapable of doing work. The former is.

In this equation, i identifies a specific microstate of the macrostate whose distribution is $\mathfrak{p}$; H_i is the algorithmic entropy of the microstate i, $H(\mathfrak{p})$ the entropy of the macrostate described by $\mathfrak{p}$, and $\mathscr{O}(1)$ is asymptotic order of the additional bits required, that is, a very small number.

Cryptography, the science of making messages difficult to spy on but readable to those with keys, and of quantifying the complexity inherent in this process, lends itself easily to this algorithmic viewpoint. If a message is encoded by changing the characters of the alphabet via a simple rule, say $a \mapsto b, b \mapsto c, \cdots$, the program sequence is short, and the code is easy to break. If the code is highly randomizing, it is harder to break than the simple code, but the sequence is longer than it would be for the simple code.

This discussion establishes the direct relationship between algorithmic information grounded in the microstates, and the statistical description grounded in the macrostate. It thus gives us a way to examine the entropy discussion within a machine context.

1.6 Conservation and non-conservation

WHEN WE SAY THAT A PROPERTY IS CONSERVED we mean that it is a constant—an invariant of the system. The term energy conservation means that the total energy of the system, that is, a collection of particles, or even one particle, is maintained through all the interactions. Likewise, the term momentum conservation means that the total momentum is conserved. Energy comes in many forms. It can be stored electrically, magnetically, as heat, or in motion, et cetera. We invoke non-conservation since some of the forms in which energy exists have such a large degree of freedom that we cannot keep track of it. Heat is one such form.

Conservative forces are those that recover the energy if one returns to the starting state and are independent of path. Such forces arise from forms of energy that are potentially available all along the path and the force can be expressed as a gradient of a potential. Electrostatic forces and gravitational forces are conservative forces. Friction is not. The energy is lost in the degrees of freedom in the randomization events that we cannot and do not keep precise track of. It is the existence of these, and not keeping track of their information, that lead to irreversibility, even though quantum mechanics, itself being linear theory, describes reversible processes.

The physical manifestation of conservative forces, non-conservative forces, and dissipation should now be clearer than when we started. When states couple tightly, reversing the process is an easier and

The equation $E = \left(p^2c^2 + m^2c^4\right)^{1/2} \approx mc^2 + p^2/2m + \cdots$ gives the kinetic energy form in this relation. Buried in the first term are all forms of other energies, some of which are easier to exchange than others. In any problem, we are interested in only the energies that change, not the ones that remain constant. And most of the former are much smaller in magnitude than the totality of the energy reflected in this form.

more probabilistically likely path than when a state couples to a broadband of states, which themselves couple to a cascade of states. So, the degrees of freedom of these states, as well as of all the other states they can couple to, matter. Friction and heat are examples of a broadband of states whose energy is not recoverable in its entirety. We will see that electrical circuits, also subject to this reasoning, can be operated in a form where one may reduce the energy dissipation through a reduction of coupling to the broadband and thus keep more of the energy recoverable.

A speculation: irreversibility, arrow of time, and the Golden rule have a very intimate link. In drawing out the Golden rule from time-dependent perturbation theory, when one replaces the sinc function by the Dirac δ function, one transforms the reversibility of quantum mechanics to irreversibility of thermodynamics.

1.6.1 Adiabatic, isothermal, reversible and irreversible processes

CERTAIN CONDITIONS OF INTERACTIONS within systems and between systems and their surroundings have special import in thermodynamics since they establish limits. A process in which no heat is exchanged is an *adiabatic* process. A thermally isolated system, that is, one across whose boundaries no heat exchange takes place, undergoes adiabatic process. Adiabatic transformation is a succession of states of equilibrium, that is, quasistatic, but with the additional condition of an absence of heat exchange. A process in which temperature is maintained at a constant level is an *isothermal* process. Such processes may be performed so slowly that the temperature is maintained in the midst of heat flow.

Processes that cannot reverse themselves are *irreversible* processes. Irreversible processes proceed over non-equilibrium states. Isolated systems reach equilibrium through irreversible processes. For example, a free-swinging pendulum loses its mechanical energy to heat and the surrounding air through friction and eventually stops. A stopped pendulum doesn't swing even if the surroundings are cooled. Processes that can reverse themselves are *reversible* processes. The net entropy change is zero. The system and the environment exchange entropy. A reversible process is quasistatic, that is, infinitesimally near thermodynamic equilibrium, and is not accompanied by dissipative effects. All reversible processes are quasistatic, but not all quasistatic processes are reversible. The process here proceeds over equilibrium states. All this is an idealization since the system is in the equilibrium state and hence all variables of the state are time independent. What we mean here is that the reversible changes can be simulated by infinitesimally small changes of the variables of state so that equilibrium is only infinitesimally disturbed. This quasi-reversible process allows one to proceed incrementally over equilibrium states so that the variables of state can be obtained by integrating over the infinitesimal reversible steps. In irreversible pr-

Energies represent the thermodynamic potential of a system. Work performed and heat generated are forms of energy, so the energies of interest, free energies, occur in a variety of forms which can be converted into each other. The internal energy of the system, U, can be thought of as the energy required to create a system in the absence of temperature or volume changes. So, it includes the various forms that may undergo change, for example, electrostatic or electromagnetic energies, except those due to temperature, that is, thermal energy and the occupation of a volume, that is, mechanical movement. If the environment has a temperature T, then some energy is spontaneously transferred from the environment to the system, with a concomitant change in the entropy of the system. The system, which is now in a state of higher disorder than before, now has less energy to lose. This is the Helmholtz free energy $\mathcal{F} = U - TS$. The system occupies a volume V in the environment at temperature T by requiring additional work of PV. This work is the Gibbs free energy $\mathcal{G} = U - TS + PV$. Enthalpy, the fourth thermodynamic potential, is an energy measure that is useful when systems release energy, such as in an exothermic reaction. Enthalpy, $\mathcal{H} = U + PV$, is the energy change associated with internal energy and the work done by the system. In the information systems discussion, we can safely assume that pressure P and volume V are constant and that only changes in Helmholtz energy through changes in internal energy, temperature and entropy are of concern.

ocesses, this cannot happen.

Energy is a central quantity in this discussion, as it is elsewhere. It takes the form of kinetic and potential energy, with the latter showing up, for example, as electromagnetic/electrostatic energy, the quantum-mechanical/chemical energy of bonds, or in another form. In thermodynamics, it is a system's total energy, which is a macroscopic quantity, that is, a quantity representative of the average over an ensemble, that is important. In a cyclic thermodynamic process—the basis of thermodynamic engines, including the information engine—if the system regains its initial state after a series of changes of state, then for the system's total internal energy U, which is a state function, during a completed cycle,

$$\oint dU = 0. \tag{1.53}$$

The state function differential is a total differential and is thus independent of the path followed. Figure 1.8 shows the Carnot cycle, an idealized limiting case of a real cycle, which will make clear to us the thermodynamic relationships and their connections to the properties that are of interest to us at the nanoscale. And, by understanding these concepts, we will be able to interpret the information cycle.

The equation $\partial(T,S)/\partial(p,V) = 1$ contains the various thermodynamic interrelationships—the Maxwell relations—and it directly states that, in any cyclic process, the work done, which is the area enclosed under the curve in the (p,V) or the (T,S) plane, is equal to the heat exchanged.

No engine operating between two heat reservoirs can be more efficient than a Carnot engine operating between the same reservoirs.

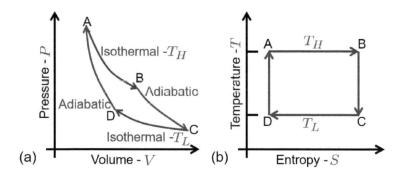

Figure 1.8: (a) Pressure-volume diagram of the Carnot cycle of a mechanical engine. (b) Temperature-entropy diagram of the Carnot cycle of a mechanical engine. Pressure, volume, temperature and entropy are all macroscopic quantities. The transition from state A to state B—an expansion—and the transition from state C to state D—a compression—are under isothermal conditions at a high temperature (T_H) and a low temperature (T_L), respectively. The B-to-C transition is an adiabatic expansion, and the D-to-A transition is an adiabatic compression. The same transitions are shown in (b). Note that adiabatic changes happen with no heat exchange and that the entropy remains constant.

During isothermal changes ($A \mapsto B$, and $C \mapsto D$), the internal energy is a constant in the ideal gas, since temperature is a constant: $\Delta U = \Delta W + \Delta Q = 0$, where ΔU is the change in internal energy, ΔW is the work done by the system, and ΔQ is the heat extracted from the thermal bath surrounding the system. During adiabatic changes, no heat exchange happens and $\Delta U = \Delta W$, with the change in internal energy reflected in the change in temperatures between T_H for the high temperature, and T_L for the low. Over the cycle,

$$\begin{aligned} \Delta U &= \Delta Q_{A \mapsto B} + \Delta W_{A \mapsto B} + \Delta W_{B \mapsto C} \\ &\quad + \Delta Q_{C \mapsto D} + \Delta W_{C \mapsto D} + \Delta W_{D \mapsto A} = 0. \end{aligned} \tag{1.54}$$

In isothermal transitions, $\Delta Q_{A \mapsto B} + \Delta W_{A \mapsto B} = 0$, and $\Delta Q_{C \mapsto D} + \Delta W_{C \mapsto D} = 0$; for the adiabatic transitions, $\Delta W_{B \mapsto C} = -\Delta W_{D \mapsto A}$, with the work proportional to the difference in temperature and a proportionality constant. For the ideal gas example, this constant is the heat capacity at constant volume, $C_V = (3/2)Nk_B$ where N is the the number of molecules and where

$$\Delta W_{B \mapsto C} = -\frac{3}{2}Nk_B(T_H - T_L), \quad \text{and}$$

$$\Delta W_{D \mapsto A} = \frac{3}{2}Nk_B(T_L - T_H). \tag{1.55}$$

During transitions $A \mapsto B$, and $C \mapsto D$, both of which occur at constant temperatures, the work is related to changing pressure and volume ($\delta W = -pdV - Vdp$), and the heat exchanged is also known:

$$\Delta Q_{A \mapsto B} = -\Delta W_{A \mapsto B} = Nk_B T_H \ln\left(\frac{V_B}{V_A}\right), \quad \text{and}$$

$$\Delta Q_{C \mapsto D} = -\Delta W_{C \mapsto D} = Nk_B T_L \ln\left(\frac{V_D}{V_C}\right). \tag{1.56}$$

The volumes can be related—$V_C/V_B = V_D/V_A$—since adiabatic changes occur with the change of temperature; so, for example, the left hand sides of the expressions in Equation 1.56 is proportional to $(T_H/T_L)^{3/2}$. The equations in 1.56 then reduce to

$$\frac{\Delta Q_{A \mapsto B}}{T_H} + \frac{\Delta Q_{C \mapsto D}}{T_L} = 0. \tag{1.57}$$

This equation, which holds true for any reversible cyclic process and not just the Carnot cycle, says that, in reversible processes, $\Delta Q_{rev}/T$ is independent of the path and is an exact differential. The applicability for arbitrary cycle follows from dividing any cycle into infinitesimally small Carnot cycles. We associate entropy S with this extensive state function. It is proportional to the amount of the material. Pressure or temperature are an intensive state function and are independent of the amount of material in the system. It is an extensive state function. In a reversible step,

$$dS = \frac{\delta Q_{rev}}{T}. \tag{1.58}$$

An entropy change, but not the absolute entropy, can be determined by measuring the amount of heat reversibly exchanged by the system at a temperature T. The other important implication of this statement follows from $\delta Q_{irrev} < \delta Q_{rev} = TdS$. In an isolated system, $\delta Q_{rev} = 0$. Therefore, entropy is a constant in thermodynamic equilibrium. This is so because of the reversibility in the system. It is also a maximum because $dS = 0$. All irreversible processes in an isolated system leading toward equilibrium are tied to an increase in entropy until it reaches its maximum when the system achieves equilibrium. This is another form in which the 2nd law of thermodynamics may be stated.

How do these thermodynamic facts reflect in the processing of information through digital logic and other alternatives that the physical embodiments of information processing take? By defining the symbolic and physical basis for information, any human element has been eliminated. We have also established the correspondence between information and entropy, and the thermodynamic and statistical considerations that should be applied. However, some aspects of this topic remain to be elucidated. Thermodynamics and statistical mechanics have their origins in understanding classical mechanics, that is, deterministic mechanics. But, these considerations are just as vital, even if only a part and an addendum to quantum, non-deterministic behavior such as in many aspects of nanoscale.

From our abstract mathematical view, data, whether from gossip, rumor or a real measurement, are all equivalent. We have ignored the human value.

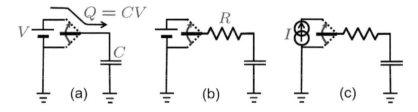

Figure 1.9 diagrams three conditions under which a capacitor can be charged. These are circuit simplifications which idealize conditions that may exist in electronic circuits. Electronics employs devices in close proximity, and the interactions within the device are almost always, by design, strongly coupled. The input of a MOS transistor is the gate which capacitively controls the channel, so its gate capacitance dominates at the input. Interconnect lines that are not long are also capacitive. So, this figure reflects the environment of transistor gates. The capacitor is the load in such environments. Here, an idealized voltage source is charging a transistor. If the charging occurs because of the instantaneous turning on of a switch, as in Figure 1.9(a), charge $Q = CV$ builds on the capacitor from the voltage source at potential V before equilibrium is reached. In this process, the ideal source at potential V supplies the charge Q, that is, expends the energy $U_{suppl} = QV = CV^2$. The energy stored in the capacitor is $U_{stor} = \int_0^Q v \, dq = \int_0^Q qC \, dq = CV^2/2$. By an instantaneous connecting of the capacitor to the voltage source, $CV^2/2$ of energy is lost.

Now consider Figure 1.9(b). The capacitor is charged through a resistance R; therefore, the voltage V does not instantaneously appear across the capacitor. The current flowing through the circuit

Figure 1.9: Three ways of charging a capacitor. In (a), a switch is turned on instantaneously connecting an ideal battery to a capacitor. In (b) the capacitor is charged through a resistor. In (c), a current source charges a capacitor through a resistor. These panels represent idealizations of the situations that happen in digital circuits. When a pass transistor is used, the transistor provides resistance in the transmission path. Transistors can be configured in circuits as voltage sources or current sources.

charging the capacitor is $i(t) = (V/R)\exp[-(t/RC)]$, which is derivable from Kirchoff's equations. The dissipation in the resistor is $R\int_0^\infty i^2(t)dt = CV^2/2$, and again, as in the previous case, an energy of $CV^2/2$ is stored in the capacitor. Half of the energy is lost whether the process of charging is with or without the resistor. This loss is due to the lost degrees of freedom—the randomization when energy exchange occurs rapidly. In the former case, which is resistanceless, the dissipation happens in the capacitor. In the latter case, the dissipation happens via randomization in the resistor.

To elaborate this point, consider Figure 1.9(c), where the charging is from a current source and through the resistor. The capacitor keeps charging up in potential, and the energy dissipation over time T is $U_{dissip} = \int_0^T i^2 R\, dt = I^2 RT = (Q/T)^2 RT = CV^2(RC/T)$. The dissipation, instead of the $CV^2/2$ for the instantaneously turned on voltage source, has the factor of $1/2$ replaced by RC/T. Because the charging is from a current source, the instantaneous voltage on the capacitor builds slowly, and the dissipation is normalized by the ratio of the electrical time constant of the charging source with the time constant of charging the capacitor. If the voltage V builds over $T \to \infty$, $U_{dissip} \to 0$. A slow building up of the charge over a long time, and a very small disturbance through the infinitesimally small current through the resistor creates conditions where the system remains close to equilibrium through the change, is adiabatic—that is, there is no heat exchange, in our classical thermodynamic interpretation— and no energy is lost. Infinitesimally small current needs to flow for this charging of the capacitor so that the dissipation in the resistor or the capacitor remains negligible.

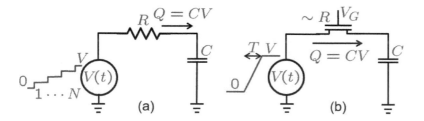

(a)   (b)

Figure 1.10: The charging of a capacitor by a voltage source that steps to the maximum voltage V in N steps is diagrammed in (a). This is multi-step charging to the same voltage as in Figure 1.9(a). (b) shows a circuit idealization of low-dissipation charging that is charging via a ramp of voltage through a pass transistor that resists the flow of charge through a resistance R.

The importance of this small deviation from equilibrium during change is clarified and emphasized by breaking up the charging process into successive small steps and, in the limit, as a gradient through a pass transistor. The voltage source charges in N steps, each of them V/N, to the same final voltage V, as in Figure 1.9(a) and (b).

The total heat dissipation in the resistor is $(1/2)C(V/N)^2 N = (1/2)(CV^2/N)$, and the source supplies $(1/2)CV^2(1+1/N)$ of energy. The increase in entropy of this closed system, assuming steps are slow and small enough to not raise the temperature, is $\Delta S = \Delta Q/T = (CV^2/2T)(1/N)$. If resistance is absent in this circuit, the supplied energy is the same as in the previous case, and the energy dissipated remains $(1/2)(CV^2/N)$, now dissipating as randomization in the capacitor. The entropy change decreases as the number of steps increases and, in the limiting case of $N \rightarrow \infty$, vanishes. As a result of our slowing down the process and making small changes to it, the process disturbs the equilibrium less than before and is closer to being reversible and quasistatic. In addition, less entropy is generated through heat loss or other losses involving unaccounted degrees of freedom, such as radiation loss. The process is thus made adiabatic. The voltage ramp in the system is the idealization of stepping in its asymptotic limit.

Transistors, when operating in saturation, have low output conductance, that is, are nonideal current sources. When operating at low drain-to-source voltages, they operate with low series resistances. For a constant gate voltage V_G, this pass transistor allows current to flow to the capacitor. For the part of transition where the drain-to-source voltage (drain at the power supply potential, and source at the capacitor potential) is in the saturation current region of the transistor, a constant current flows. For parts when the transistor is in the linear region, it is through a low resistance. For a voltage source rising from 0 to V in time T, charging a capacitor C through a resistance R, the dissipation is

$$U_{dissip} = \eta \left\{ 1 + \mathscr{D} \left[\exp\left(-\frac{1}{\eta} \right) - 1 \right] \right\} CV^2. \qquad (1.59)$$

Here, $\eta = RC/t$, and $\mathscr{D}$ is a function of the ratios of time constants. In the limit $T \ll RC$, that is, with a rapidly turning on driving potential, the dissipation $U_{dissip} \rightarrow (1/2)CV^2$; in the opposite limit, that is, with a very slow turn-on process where $T \gg RC$, $U_{dissip} = \eta CV^2$, with the dissipation asymptotically decreasing. A very slow turn-on process with an infinitely small current eliminates dissipation in this ideal circuit, even when resistance in present. This situation represents the adiabatic limit. There is no dissipation, and the entropy remains constant. All the energy remains in the system while some is now stored in the capacitor, no energy is lost in the unaccounted degrees of freedom.

The current that flows in the capacitor because of a change in voltage over time is displacement current. It is a real current and is on the same footing as current due to moving charge. This current flow in capacitor is associated with the changing displacement field in the capacitor across which a voltage drop exists. The dissipation

In the thermodynamic view with its mechanical antecedents, as discussed earlier, an adiabatic process is an unfolding of events where heat transfer is absent at the system boundary. In the equation relating change in internal energy to useful work performed (a high-quality energy) and heat dissipated (a waste energy), $dU = dW + dQ$, we must also take a broad view of dW as representing useful work and dQ as representing all the wasteful energy loss in the other degrees of freedom. The dQ category includes, for electrical systems, radiation loss resulting from any rapid change in fields which is not being reharnessed. Radiation from antenna arises in dipole oscillation, some of this energy is reharnessed and is useful work, but quite a bit of it is lost. With the antenna's objective as radiation, all the radiation is work, even if much of it is wasted and not useful work.

within the capacitor, arising from the flow of displacement current, is $\int v i \, dt = \int v C (dv/dt) dt$. If the current is infinitesimally small, then there is no dissipation; if the change is instantaneous, then the integral reduces to $\int_0^V v C \, dv = CV^2/2$. With the particle flow current and the displacement current, the dissipation in the resistor and that in the capacitor are equivalent. For an inductor, where there is no dissipation, that is, under adiabatic conditions, this argument reduces to requiring that the current change in time be negligible so that the voltage drop that occurs across the inductor is negligible. In each of these instances, the dissipation occurs via energy transfer to various degrees of freedom: heat in the form of vibration, but also radiation, as well as other forms of energy.

Irreversibility, in the mechanistic view, arises through the randomness and the loss of details that would have allowed one to recover the original state. Both the $NAND$ and the NOR gates, as discussed in Section 1.3, lose some of the information and, as a result, the ability to reverse the information process and recover its energy content. Conservation and reversibility are interlinked in this sense. Reversibility makes it possible to imagine computing machines that do not dissipate or that dissipate very little. A discussion of this idea will be instructive, since it ties with our entropy and randomness discussions. Conservative forces have the property that the work they perform is independent of the path. In classical mechanics, when a particle returns to the same point, as in a closed loop, the net work performed is zero. The potential energy change of the particle arises from the force but is independent of the path taken. Forces such as gravitation, electric forces in a time-independent magnetic field, et cetera, are conservative forces. Ohmic loss or friction, because of the randomness and entropy change they engender, or radiation, because of its time-dependent potentials, are non-conservative—that is, path-dependent. Dissipation arises from non-conservation. In these processes, energy is converted into a non-recoverable form—the randomness that we discussed. When one has one-to-one mapping between inputs and outputs, the input state can be reconstructed from the output state. This is the case with a reversible gate. A conservative system preserves all the states of the system and hence is reversible. In such an information system, in principle, one is able to avoid dissipation since no information is eliminated. This case is interesting as a limit since, in it, $k_B T \ln 2$ per bit energy losses are being avoided.

Figure 1.11 shows conceptual examples of logic operations using frictionless billiard balls; in these examples, information is not erased and all possible states are accessible. Let the presence or absence of a trajectory describe the state of the system. Figure 1.11(a) shows

Power lines on a chip are generally inductive. A constant current flow, even if resistively dissipative, will be inductively nondissipative. However, spikes in current, arising from switching in functional blocks, and even from clocking, can have a fair-sized dissipative inductive component.

the four logic operations possible for two logical inputs, a and b, which can each be 0 or 1. With the set of logical functions available, the introduction of mirrors suffices to simulate any conservative logic function. The implication of this fact is that an infinite system consisting of two-dimensional hard-sphere gas in an appropriate periodic potential, that is, mirrors, is computationally universal. It can be programmed through its initial condition to simulate any digital computation. Figure 1.11(b) shows the observation step of such a reversible computer. The dark ball detects the light ball and then lets it go back to its original trajectory. Dissipation in this system occurs only when the dark ball is stopped so that it may be reused.

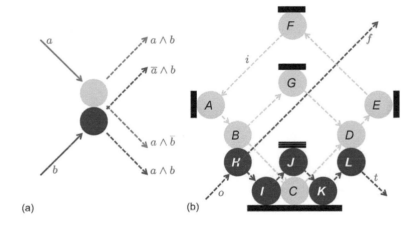

(a)   (b)

Figure 1.11: Two examples of reversible logic operations using frictionless billiard balls; here, presence or absence of a trajectory determines the state. In (a), four output logic functions are computed based on inputs a and b. Logical a and b imply the presence of light or dark balls, respectively. (b) A reversible measurement operation. If a light ball is directed along i, it follows the path $ABCDEFA$ when the dark ball is absent. The light ball can be detected by directing the dark ball along o; if the dark ball encounters the light ball, the light ball will follow the path $ABGDEFA$ and then return to the original track. The dark ball responds by following the path $HIJKL$. The system dissipates when the dark ball is stopped so that it can be reused. The detection of the light ball by the dark ball is dissipationless.

If the information content of the output is kept the same as information content of the input, information loss such as that which occurs in the $NAND$ and the NOR gates can be avoided. In transistor-based approaches, the statistical entropy changes manifested in Boltzmann entropy prevail by a large order of magnitude. But, in principle between reversibility and preservation of information content, we have here created a means to lower energy dissipation.

There is more on this topic in the discussion related to quantum computation, to which this billiard ball model has special relevance.

1.7   Circuits in light of infodynamic considerations

FIELD-EFFECT TRANSISTORS ARE SPECIALLY WELL-SUITED for circuits that are adiabatic or reversible. They are bilateral devices. Reversing the drain and the source terminals in these transistors reverses the direction of the current flow, and they can have negligible source-drain voltages for the flow of the signal, that is, have low resistances and, therefore, low RC/T ratios, where T is the time scale of change. The factor $1/T$ gives a limiting $1/f$ scaling in power dependence on frequency.

In the discussion of charging a capacitor through small steps and slow ramps, we have the first example of attempting adiabatic low-dissipation charging. There are other ways that energy can be dissipated, and these too need to be avoided even if the circuit is operated slowly.

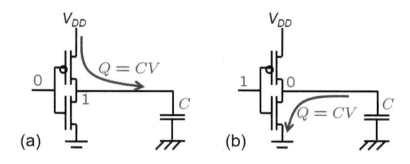

Figure 1.12: The charging and discharging of a capacitive load by a static $CMOS$ gate. (a) shows the transition of logic $1 \mapsto 0$ at the input; this transition causes the output to be a logic 1 by turning on the $pMOS$ gate (and turning off the $nMOS$ gate) and raising the output node to close to the power supply's high voltage. (b) shows the transition of logic $0 \mapsto 1$ at the input; this transition turns on the $nMOS$ gate and turns off the $pMOS$ gate, lowering the output node to close to the ground potential.

Figure 1.12 is a schematic view of a static $CMOS$ inverter charging and discharging a capacitive output. When the input is a logic low (0) with a low voltage, the output is a logic high (1) with a high voltage. This high voltage arises because the low voltage shuts off the $nMOS$ transistor—the lower transistor of this stack—and turns on the $pMOS$ transistor—the transistor higher in the stack—which connects the power supply to the output node. For the dominantly capacitive load, a charge $Q = CV$ flows and charges up the output node. If the switching is fast, this process is rapid and, following our discussion of the energy transport for Figure 1.9(a) and (b), there is dissipation, both in the presence and absence of resistance. An adequate first order description for the $pMOS$ in this system is that it acts as a resistance while the power supply charges the capacitor. When the input is switched from a logic low (0) to a logic high (1), the $nMOS$ transistor is turned on, and the charge $Q = CV$ that had charged the capacitive output is discharged to the ground. So, in the process of switching to a logic low (0) from a logic high (1), the charge CV is transferred from the supply to the circuit; when it switches back to the logic high (1), this charge is sent to the ground. Thus, a total energy of $CV \times V = CV_{DD}^2$ is dissipated because this charge is removed from the power supply and grounded. The up-and-down transition dissipates CV^2 of energy in static logic because the process transfers the CV charge to the ground independently of the speed in the information cycle.

This is energy lost to the environment—the classical reservoir for this circuit—that the ground represents.

Having *pMOS* charging the output node (input $1 \mapsto 0$ with output $0 \mapsto 1$) is quite akin to the ramped charging process, if one raises the power supply voltage slowly after having changed the input slowly. This method would not create dissipative losses. The *pMOS* transistor here acts as a transmission gate. The reverse of this logic process, however, discharges the output to the ground and that is where the principal problem of dissipation—even in a slow process here—lies.

Each combination of logic transition, in all the static logic forms— inverters, *NAND*, *NOR* and others—is an operation of dissipation of $CV \times V = CV^2$ of energy. We take this energy from the supply and put it into the thermalizing ground reservoir. If we do this step often, that is, when the frequency ν is high, we dissipate the energy rapidly. The amount of power dissipated is νCV^2. It directly depends on the rate at which we take energy from the supply and dump it into the ground.

Ideally, one would like to not waste this energy.

Adiabatic circuits, in various forms, depend on the slow changes in clocked power supplies and signal transitions, together with the recovery of energy rather than its loss to the ground reservoir for reducing power. Dissipation in the discussion up to this point happens because of

- rapid transitions leading to dissipative currents,

- significant current in paths of large resistances, for example, in transistors in the presence of voltage drops across which current flows and

- dissipation of the charge from the power supply to the ground.

Each of these items can be addressed in the design and operation of circuits. Gate voltages should be changed only when the source-drain voltage drops are absent, source-drain voltages should be changed only when the device is shut off and changes should be gradual. Energy should be recirculated instead of sending it off to the ground. Slow ramping transitions reduce dissipation by a factor of RC/T. Currents can be kept low when resistances are large, for example, when transistors are turning on or off. They can be allowed to rise when the resistances in the path are low. Energy recovery, that is, the movement of the charge back to the power supply or another storage element prevents the dissipation of the charge to the ground. From the thermodynamic view, energy flow, energy storage, and the loss of energy to other degrees of freedom is minimized or avoided by these practices. Having no heat flow or loss to radiation or other

The power-supply—the battery— may be viewed as an internal energy source that is part of the circuit. It is the ground—the circuit's environment— that acts as the reservoir. In principle, it is possible to return the energy to the battery without involving the ground, thus keeping the circuit-battery system closed to the environment/reservoir for particle exchange.

Much of biology's electrochemoin-fothermomechanics happens with each critical step that is sub-$100 k_B T$ in energy. Replication, neural signaling and even the stepping of biological motors are low energy processes. Errors become correctable in a sequential process when it is possible to reverse the process. This fact is what makes stochastic processes very powerful— including possibly those in our physical computing approaches.

degrees of freedom implies that there has been no entropy change. As an isolated system, the circuit then has maximum entropy, the entropy of the equilibrium state. In electrical circuits, the canonical conjugates for the configuration space will be a normalized voltage and current. Voltage is the equivalent of position, and current is the equivalent of momentum. So, if from the configuration space view, an adiabatic process follows a trajectory where no heat flows in or out of the parts of the system, whose degrees of freedom carry out the motion along the trajectory. A resistor or a capacitor needs to have a small current flowing through it to limit the heat dissipation to the degrees of freedom that create the trajectory.

Current is the flux of charge and is naturally related to momentum. One can make a fair correspondence between voltage and mechanical force, and similarly between current and momentum.

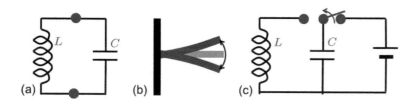

This generalized coordinate discussion is emphasized in the idealized electrical and mechanical examples in Figure 1.13. In a steady state, in an ideal LC tank, energy is exchanged between the magnetic fields in the inductor and the electrical fields in the capacitor. This circuit is the equivalent of an ideal, lossless vibrating cantilever (Figure 1.13(b)), where energy is exchanged between the kinetic form in the no-displacement position, and the potential form at the extreme end of the cantilever's displacement.

In our energy picture, the canonical conjugates are different but equivalent. The potential energy in the cantilever arises from the tensile strain that in the elastic regime is due to stretching of the bonds, which themselves are quantum mechanical in origin and are derived from the electromagnetic forces under quantum constraints. The energy can oscillate between the inductor and the capacitor at very high frequencies, many tens of gigahertz and more, and at reasonable power levels, that is, with high voltages and currents, if the LC network has a high quality factor $Q = (\Delta U/U_{stored})^{-1}$ per cycle, that is, if the value of the inverse of the energy dissipated within the system per cycle as a fraction of energy stored is high. This is because circuits with a high quality factor allow no sudden changes in energy—only the changes permitted within the fundamental mode of oscillation where it stays. This is an example of strong coupling and the consequent absence of transfer of energy to a broadband of states from which it cannot be brought back. However, if the capacitor is

Figure 1.13: (a) shows an ideal LC circuit where energy is exchanged between magnetic and electrical fields, (b) shows an ideal vibrating cantilever, where the energy exchange is between the kinetic and the potential forms, and (c) shows the LC circuit where instantaneous changes are introduced. This last diagram shows an LC circuit where the capacitor is charged first by a battery and then is connected in the classical LC tank form. Rapid changes occur in the beginning and are followed by oscillation of energy at a resonant frequency that can be very high. Once the initial voltage and current transient ends, the charge is transferred back and forth between the capacitor and the inductor. The circuit is naturally adiabatic when it operates in its resonant eigenmode.

Transformers and generators are analogous examples in which this strong coupling takes place at low frequencies. Their efficiencies come close to being nearly 100 % because of tight coupling. Good thing too. If they didn't, we would be in big trouble, both at the point of generation and near the point of usage. Photon—as the elementary particle for quantum excitation of the electromagnetic field—is the ultimate example of strong electromagnetic coupling.

first charged and then the *LC* combination circuit established, as in Figure 1.13(c), the circuit will dissipate some of the stored energy because it does allow for sudden change in current as it gets going. A large number of frequencies/energy modes are allowed, and these will couple out to the physical medium in the form of heat, radiation and other broadband vibrational forms. Impedance mismatches also cause energy flow reflections. A sudden change in the inductor current of the *LC* circuit causes the inductor to rapidly build a voltage causing dissipative losses. A sudden change in voltage causes the capacitor of the *LC* circuit to be dissipative because of the current, as it did for our earlier capacitor discussion. These two inductor and capacitor situations are equivalent. Its equivalent for a cantilever circuit is the impulse initiation of vibrations. It too dissipates because of the multiple modes that will arise and the loss of energy within the beam and its surroundings as it settles into its resonant mode.

In digital circuits, one does not want the transitions to be so slow that they are of little practical use. Therefore, circuits that are adiabatic like, or quasiadiabatic, that is, that dissipate some energy and thus increase the speed of the transition, are also of interest. We now look at a few examples of circuits in their practical quasiadiabatic form to explore their energy recovery and how dissipative loss is prevented. Figure 1.14 shows a split-level charge recovery logic (*SCRL*) circuit, which is an adiabatic circuit.

The *SCRL* inverter transforms our static inverter into a quasiadiabatic inverter. The pass transistor gate isolates the internal output node from external output to prevent current flow in the presence of voltage drops. The pass transistors are driven to the maximum potentials through clocks ($P, \overline{P}$) at ground and V_{DD}. The inverter part is driven at the input end by trapezoidal power supply clocks $\phi, \overline{\phi}$ that change the rail voltages between the ground and V_{DD}, but from a quiescent bias of $\phi, \overline{\phi} = V_{DD}/2$. Initially, the values of the input, the rail clocks, the output and all the nodes are $V_{DD}/2$. The pass transistor clocks ($P, \overline{P}$) are set at ground and V_{DD} respectively, that is., the pass transistors are shut off. Then, the input is set to the valid state logical 0 or 1 corresponding to the ground or the V_{DD} potential, respectively. Next, the pass transistors are ramped on. Next, the clock rails are ramped to V_{DD} and ground, respectively. Depending on the input voltage, the internal node now swings through a transistor to one of the clock rail potentials. If the input is high, it goes to the ground potential; if the input is low, it goes to the V_{DD} potential. The inversion operation appears at the output, is and passed on by the on pass transistors and is available for use by the next stage. The circuit is returned to the quiescent state by ramping off the pass transistors and then return ramping the rail clocks to $V_{DD}/2$. The inputs can

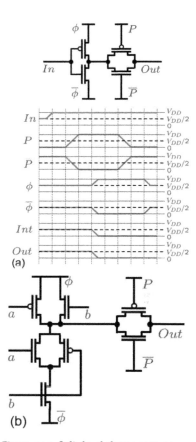

Figure 1.14: Split-level charge recovery logic (*SCRL*) circuit, whose simplest form for inversion is shown together with a timing diagram in (a). (b) An *SCRL NAND* gate. The initial powering condition is all potentials at $V_{DD}/2$. The input is turned on, following which the power supply and ground potentials are ramped. This is followed by ramping on the pass transistor to pass on the internally computed inversion to the output. An extension of this approach is shown in (b), where *NAND* is computed.

now be ramped to a new condition, and the computation cycle repeated. By splitting levels, the circuit recovers the charge through the replacement of the ground of a static inverter by a clocked supply. The circuit also incorporates the loss minimization themes discussed previously in this section.

Figure 1.14(b) shows a $NAND$ gate. The difference between it and the static inverter analog is the additional $pMOS$ transistor that is tied to the logic input b. If this transistor is missing, when the supply clocks are split, for logical $a = 1$ and for logical $b = 0$, the upper clock supply voltage is passed through the $pMOS$ load attached to the logical b input and through the $nMOS$ attached to the logical a input, appearing as a large potential drop across the $nMOS$ transistor of the logical b input. The presence of the $pMOS$ in the lower arm of the circuit prevents the internal node from dissipating energy. This result follows from the rule of low current through large resistances. If a potential difference exists across a device, it will not be be turned on. Energy transfer through devices takes place while they are on and in a low resistance state, so that the factor RC/T can be kept small. The transition is stretched over the clock cycle, and the energy dissipation reduced. There is a spatial penalty for reducing this energy. These circuits employ about twice the number of transistors needed for static $CMOS$ logic.

Figure 1.15 shows another example of a quasiadiabatic approach, the 2N-2P inverter buffer—a buffer circuit—together with the timing diagram. It again has reduced energy by following some of the rules for reducing dissipation. Like the $SCRL$ circuit, this circuit too can be expanded to multi-input circuits of increasing complexity. It too replaces the single fixed power rail to achieve reduction in energy consumption. Like the prior example, the charge transferred from the supply rail to the circuit is recovered by the rail when it discharges a charged node and is not allowed to go to the ground. This circuit uses differential logic. Logic function and its complement are computed. Computation starts with a *Reset* phase with the inputs set to low. Since the switch is differential, the $pFET$ corresponding to a high output, is driven by a low (complementary) input. It therefore follows the ramp voltage so that, at the end of this phase, both outputs are low. In the *Wait* phase, supply rails are low and maintain the output at low while input is evaluated. This scenario sets the conditions for the following gate, which is a quarter cycle behind, to perform its *Reset* operation. Since the gate is powered down, the evaluation of the input does not affect the state of the gate. In the third phase—*Evaluate*—the supply rails are ramped up. The output now establishes the complementary state. The gate section with high input will have the output go to low, and the complementa-

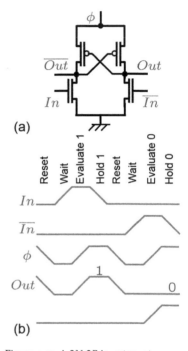

Figure 1.15: A 2N-2P inverter gate, an example of a quasistatic gate. (a) shows the circuit diagram of the buffer gate, and (b) shows a timing diagram for the changes in input, clock supply and output. The various timing phases for resetting, waiting, evaluating and holding are shown. Even though there is a ground to which there is an energy loss, there is less dissipation with a slower ramp and its timing, compared to a fast switching.

ry part ride to high. The *Evaluate* phase ends with complementary output through its use of a cross-coupled pair. In the fourth phase—*Hold*—the rail stays high while the inputs ramp down to low, even as outputs remain valid during the phase. Four quadrature clocks are required because of the four phases of the computation cycle. Each logic phase holds its output valid while its following stage evaluates and the prior stage is resetting.

All these examples of quasiadiabatic circuits have the attribute that the output data is valid only in a part of the computation cycle, unlike the case in a static *CMOS*, which allows the data to be valid for the entire *Hold* time period. Circuits of additional complexity can now follow from this discussion.

We have discussed how the movement of charge from the power supply to the ground is the single most dissipative aspect of static logic. The place where this has the severest consequences is in synchronous logic. Synchronous logic employs clocks to time operation and to realign the timing mismatches arising in the different sequential logic circuits requiring different times to complete operation, the variability in these times resulting from the variability of devices and the presence of noise. Clocks are approaches to achieving precise timing. Clocks, therefore contain no information—in theory they are precisely known. Thermodynamically, they comprise one part of the design, which while it serves no informational purpose, consumes a significant fraction of the power. Energy of the clocks can be recovered if it is not sent to the ground.

Clocks are essential for complex designs where timing in the presence of variability and an emphasis on fast operation serve as gate keepers assuring that the worst of numerous computational transformations take place to completion, so that the next step needing these may ensue without error.

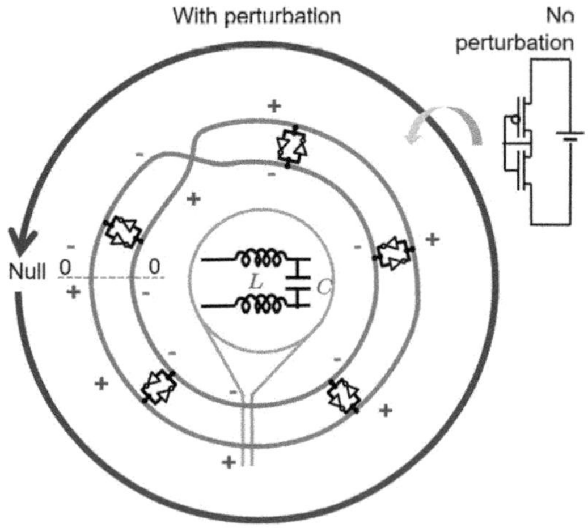

Figure 1.16: A rotary clock that loops as in a Möbius strip along a transmission line whose two folded lines are connected by inverters in shunt. Neither of the lines being grounded moves the charge and energy along the folded lines. In noise less conditions, the *CMOS* inverters will stay at the $V_{in} = V_{out}$ intersection at the internal node. The presence of noise causes the back-to-back inverters, because of the gain, to rapidly move away—with reinforcement—from this condition. In the steady state, a square wave rotates around the Möbius strip, where any losses are compensated by energy from the supply.

Figure 1.16 shows the use of transmission line/*LC*-type movement of energy to reduce dissipation in clocks. The transmission line as a lumped model consisting of inductor-capacitor (*LC*) network where the energy is exchanged continuously as the wave travels. With the single fold shown in this figure, the oscillation frequency is $f_{osc} = 1/2(L_T C_T)^{1/2}$, where the index T stands for total and where the factor of 2 arises from the two laps required. The inverters are shunt connected in the transmission loop without a ground and add energy coherently while clamping the voltages.

1.8 Fluctuations and transitions

ERRORS IN PHYSICAL COMPUTATION can arise from a multitude of sources. Noise arising in energetic interaction with the environment can be thermal in origin. The multitudes of degrees of freedom, and their randomness, which is not accounted for, interact with a system and lead to fluctuations whose energy is uncorrelated with the signal of interest and therefore interferes with the information of interest. In a resistor of resistance R, the noise arising from thermal fluctuations is white. The noise voltage has the characteristic $\langle V_\theta^2 \rangle = 4k_B TR\Delta\nu$, where $\Delta\nu$ is the frequency band, that is, the bandwidth of the system. This function measures the fluctuations in the thermal motion of the ensemble of electrons. Statistical fluctuation in electrons as they pass boundaries of systems is also unpredictable and causes noise. This is shot noise, which too is white, and it is characterized by $\langle I_{shot}^2 \rangle = 2q \langle I \rangle \Delta\nu$. This noise also is characteristic of random behavior.

Our statistical treatment in classical conditions reflects the distributions in the Boltzmann functions. In forming deterministic gates, we form regions of stability in energy. The inverter is a good starting point to see energy relationships with other behaviors of the devices in light of our thermodynamic discussion. The inverter is our bounded system, which is tied to the environment. The electrical supply may or may not be part of the environment, depending on the source, but the ground is for electrical particle exchange, and the physical environment is for the heat exchange. The latter two are certainly the equivalents of classical reservoirs.

Figure 1.17(a) shows the circuit schematic and input/output and output/input characteristics—the butterfly curve—of the inverter—a transistor circuit—that make the inversion possible. Inversion happens because, when the input voltage is low, the n-channel device is off because of the low gate to the source voltage—which is lower than threshold voltage. The p-channel device is on. Note that the p-channel device can be viewed with the source connected to the power supply and therefore gate-to-source voltage biased for the device to be turned on. When such gates are tied in sequence, stable voltages for high and low are achieved as shown by the output

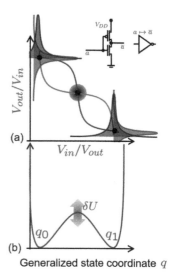

Figure 1.17: Diagram of the operation of a *CMOS* inverter gate emphasizing the consequences of fluctuations arising from noise as well as from statistical effects in devices. (a) shows a schematic of the inverter, together with an eye diagram showing the variations therein. (b) shows the corresponding generalized energy-coordinate diagram identifying the system's two states (low in, resulting in high out, or high in, resulting in low out). Switching from low-in to high-in state is caused by changing the input signal, which transitions the gate through the voltage gain region by drawing enough energy to overpower the barrier and, as a result, causes the system to settle in the second state. Barrier modulation happens because static energy is supplied by the power signal, which is under the influence of the gain of the transistors.

voltage driving the input voltage of the following stage, that is, the V_{in} and V_{out} are flipped in successive stages. The stable low voltage and high voltage, as well as the cross over point in the gain region, have statistical variations arising from the fluctuations represented by the noise as well as in the ensemble, that is, within-device variations and device-to-device variations, the latter occurring when a large collection of these devices are put together. Figure 1.17(b) is the energy-generalized coordinate relationship that the inverter's behavior corresponds to. Two stable states exist so long as devices have gain and the voltage is large enough to delineate them. But, the statistical effects also bring out a distribution of possibilities around an equilibrium. The input-output voltage relationship, which is a continuous relationship, is subject to the Shannon-Nyquist-Hartley constraints, and the statistical effects are part of these constraints. The change in state is accomplished by modulating the barrier through the use of the input voltage. The statistical variations arising in the 2 stable states and the gain regions affect the behavior of the switch and its limits, even if this switch is designed to be deterministically correct, that is, with a probability of error limiting to 0. In theory, this gate is reversible, although by itself it cannot accomplish the minimum set of Boolean functions required. Our thermodynamic notions can be applied to the energy-general coordinate diagram to extract engineering constraints for the problem.

Shannon, since informational entropy is involved; Nyquist, since temperature is a source of energetic fluctuations leading to noise in signals; and Hartley, because an inverter, at its simplest, can be viewed as an inversion transmission channel in which noise and signal interact.

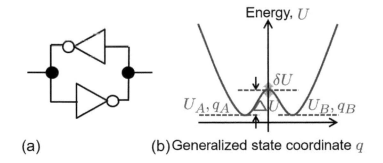

(a)

(b) Generalized state coordinate q

Figure 1.18: (a) Inverters connected back-to-back form a static memory, a property arising from its bistability, that is, two stable states exist with the two energy minimums shown in (b) so long as energy flows through the system. Again, the change in the generalized coordinate of the state is made possible by providing energy from the static supply under the influence of the input and the gain of the gate.

Cross-coupling inverters leads to the creation of a memory device: a system with two stable states. This is a four transistor-based cross-coupled pair which form the innards of a static random access memory ($sRAM$). This is shown in Figure 1.18. The energy-general coordinate diagram here has two stable states that we identify as $q_A = (0,1)$, and $q_B = (1,0)$, respectively, and which are associated with a small range of voltages at the outside accessible terminals. Its state is changed by changing the potential barrier, raising one of the potential wells and lowering another to move the system into the other state. The stability occurs in this cell through the two stable minimums in energy, but they exist only in the presence of the applied potential and the energy flow. Remove the bias, and the cell loses its information. Trapping of the state in the potential minimum depends on there existing a barrier high enough so that flipping is of extremely low probability and thus depends on the size of the barrier ΔU that separates the two states. The system shown in this energy state coordinate picture describes a number of different embodiments of bistable memory devices. Where the energy barrier arrives from, or even if energy flow is needed, depends on the type of memory. For example, in magnetic storage, the confinement well arises from magnetic polarization—the well is a consequence of the thermodynamic phase transition arising in magnetic polarization because of spin. Ferroelectric memory employs electric polarization—in this case, the well is a consequence of the phase transition arising from electric polarization because of charge. In each of these kinds of memory, the statistical fluctuations of the barrier and of the participating entities give us an idea of the probabilities of events that may give rise to an error. In the case of back-to-back inverters, the barrier is formed because of the stable high or low state of the inverter and the reinforcement of the state via back-to-back coupling with these same inverters. Each of the transistors has a significant ensemble of carriers (electrons for the n-channel device, or holes for the p-channel device), and the transistors are arranged in a mirrored asymmetry. When one deliberately causes a change in the state of the system, say by increasing the potential of one of the ``low'' node, the change in signal, reflected in turn in the mean and fluctuation of the properties of the associated ensemble, causes both the barrier to be lowered and energy to be coupled from the static power supply. The energy barrier itself is lowered, and the lowest energy of the states is changed, allowing the transition process to take place. These are is the basic features of any state transition.

It is easy to see this process conceptually by just looking at how one logic state transitions to another logic state. Figure 1.19 shows four snapshots in time of the energy configuration dynamics in this transition. Initially, at $t = t_1$, the system is in state q_0. Let us say this state corresponds to a logical output of 1, and a logical input of 0, that is, a high voltage at the output, and a low voltage at the input. As one raises the input voltage, the minimum in the energy is raised, and the barrier energy is lowered, slowly at first. But, as the input voltage is raised enough to reach the voltage gain region of the inverter, that is, the threshold of the transition, it rapidly snaps from the configuration at $t = t_2$ to that at $t = t_3$. The voltage gain in this structure also occurs with power gain because of the impedance

The back-to-back inverter is quite an interesting configuration. It is stable, the output signal can directly drive another input without any translation or sensing or amplification, and this same structural form can also be used to sense a state since the cross-coupled arrangement is also an amplifier with voltage compatibility. No signal conversion is required. It is almost as if it is self-aware.

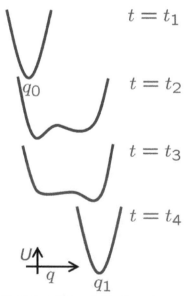

Figure 1.19: The evolution of an energy state coordinate during a logic state transition in static random access memory ($sRAM$). The same configuration picture can be drawn for the transition of an inverter.

transformation in the inverter's and $sRAM$'s circuit operation, that is, their characteristic impedance, and employs the power from the supply in moving the system toward the final state at $t = t_4$: a logical 0 at the output, with a low output voltage, and a logical 1 at the input, with a high voltage.

Consider the forward transition process shown in Figure 1.20. The barrier for going from a state with the energy minimum U_1 to the other with the energy minimum U_2 is $U' - U_1$, where U' is the peak barrier energy related to this transition. It is a function of the applied potentials and the ensemble represented by the inverter gate. The forward rate is proportional to the Boltzmann factor $\exp[-(U' - U_1)/k_B T]$, and the backward rate to $\exp[-(U' - U_2)/k_B T]$. These rates, proportional to the Boltzmann exponential factor, simply connect the energy U' (the minimum energy needed for the transition) and the energy of the two states. Each of the activation energies may be manipulated by external force. The prefactor term of the transition rate is related to the fluctuations, which follow from entropy-based argument, and the specifics of the energy-related properties of the state, such as the scattering rate of the particles. We can draw a number of inferences from this dependence. The ratio of the forward rate to the backward rate is $\exp[-(U_1 - U_2)/k_B T]$; $U_1 \gg U_2$ implies that the forward process will be faster than the backward process. If the rate is r, given that the prefactors are same, the energy expended is

$$k_B T \ln r = U_1 - U_2. \qquad (1.60)$$

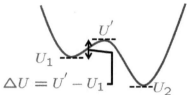

Figure 1.20: The rate of transition between two energy minimums is controlled by the random processes at play in the wells and by the energy barriers preventing these processes from causing a transition of the individual contributing components to the other well. The rates therefore depend both on activities within the wells and on the energy barriers to overcoming the confinement of the wells.

1.9 Errors, stability and the energy cost of determinism

THE STATISTICAL FLUCTUATIONS in the presence of the limited confinement energies of states limit the stability of the state that one wishes the system to stay in. What is the probability that the this state, the state of the memory, or a logic gate confined to a state, remains the same?

This issue, conceptually, is the situation shown in Figure 1.21. The expectation values of the transition rates in the two directions are the same. But, statistical fluctuations also exist, for example, the on state of a n-channel transistor, or the on state of a p-channel transistor, is a resistive states subject to noise such as thermal noise. The rate of the transition between the two states depends on the Boltzmann exponential, which is tied to the difference between the barrier energy and the energy minimum of the states (an activation energy), and a coefficient that represents the fluctuation attempts.

Electrons are the moving parts of the electronic machine. They are also a cause of breakdown in the machine—just as with the mechanical moving parts we see in daily life.

The time that a memory device stays in a particular state is

$$T_s = \tau_s \exp\left(\frac{\Delta U}{k_B T}\right), \qquad (1.61)$$

where τ_s is a time arising in fluctuations as well as a source flux. It represents the inverse of the rate of attempts to overcome the barrier that is keeping the state confined and stable. For an electron ensemble, the expression has an exponential dependence related to the expectation on electrons with sufficient energy to surmount the barrier. So, there exists a temperature dependence through the Boltzmann factor and a prefactor related to thermal fluctuations, and the details of the method of achieving the state, for example, the electron's scattering characteristics, which guide the carrier energy distribution as well as the fluctuations and the number of attempts at surmounting the barrier, et cetera. In a back-to-back inverter, the peak energy of the barrier is determined by the bias voltages and the properties of the transistors, so it too itself has variations. In magnetic memory, which is based on the polarization of domains, the barrier is now related to anisotropy terms and magnetic polarization and is degraded in the paramagnetic limit, something we discuss in Chapter 4. The τ_s here is related to fluctuations in precession or random walk arising from in strain and the magnetic coupling within the system. τ_s in this respect is the dwell time within the system. The probability that the state may flip, that is, have an error ε while sitting around, is simply proportional to $1/T_s$, that is, $\varepsilon \propto \exp(-\Delta U/k_B T)$. So, the error rates of such a memory device is proportional to an exponential of the degree of confinement of the energy in the device.

Figure 1.21 represents the energy picture of a bistable memory structure. In either of the two stable states, any small disturbance from the energy minimum average results in a restoring force back toward the minimum. This is because conservative forces of magnitude $-\nabla U$ dominate; they are restorative because the potential well has a even-power dependence on the disturbance in abscissa. Ferroelectrics too lead to such bistable memories.

But, there are also memories that do not have natural energetic stability. Such memory structures rely on the restorative process being slow. Figure 1.22 is a pictorial energy and structural representation of such memory structure. Dynamic random access memory ($dRAM$), flash memory ($FLASH$), or their charge-trapping incarnations, and electrochemical memory cells, which use reduction-oxidation reactions to change the lengths of conducting paths, are examples of random walk memory. In such types of memory, the process that leads toward the natural equilibrium state of the structure is suppressed; $dRAM$ does this by having a threshold voltage for the access tran-

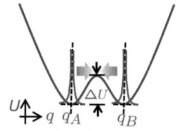

Figure 1.21: The stability of a bistable memory with the energy barrier ΔU between the two states. As drawn, with the symmetry, the rates of transition from one state to the other are, on average, equal, but errors, that is, transitions caused by statistical fluctuations in the system, are also possible.

sistor that is so high so that the charge on the capacitor leaks very slowly. The current flowing through the transistor in the off state is diffusive, that is, the charge carriers undergo random walks. In *FLASH* structures, which have polysilicon floating gates for charge storage, the presence of charge raises the energy of the well over that of the gate or the channel, both of which are also made of silicon. This charge wants to leak out, but the leakage is suppressed because of the high barrier height of the insulators and their low defect quality. Again, the leakage consists of charge carriers undergoing random walks through percolation paths or undergoing tunneling. Such types of memory devices, because of their lack of inherent stability, will be subject to phenomena that are not present in bistable memory devices with restorative energetic stability. So, defects and other competing mechanisms for leakage become important. In *dRAM*, a single electron trapped in a defect may cause one *dRAM* cell to be very different from another and show associated effects/random telegraph signal noise due to single-defect-induced random events. In *FLASH* memory, individual cells may show unusual defect-mediated leakage paths through the insulators.

For a charge of $\sim 30\ fC$ stored on the capacitor, an average storage time of $1\ s$ implies a leakage current of $3 \times 10^{-14}\ A$, a current that is achievable, although it becomes incredibly hard to do so when the length scale is $\sim 10\ nm$. The devices have to be longer than that, and the threshold voltage has to be close to the bandgap of the material for the device, even when back biasing is used. Because of the statistical distribution of the threshold voltage, this leakage demands refresh times that are considerably faster than this average storage time.

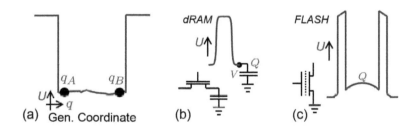

Figure 1.22: (a) shows the energy-generalized coordinate representation of random walk memory. In a dynamic random access memory (*dRAM*) device, schematically shown in (b), the leakage current is suppressed because the high threshold voltage of the access transistor creates a large barrier for the leakage of charge in the hold state. In a flash memory (*FLASH*) device, schematically shown in (c), the charge trapped in the well raises its energy but has difficulty leaking out because of the high barriers that suppress tunneling and because of low defect-mediated currents.

Fluctuations place limits on how much signal should be in the system for a certain error rate. Even if a system was ideal and resistance-less, temperature and, through temperature, randomization would still exist in capacitors and inductors. One can see the constraints of this behavior in a switch. Take an ideal switch, which is resistance less, turning on or off a connection to a load represented as a simple capacitance C. The input loading of a transistor—a quintessential load—is capacitive. The energy per operation is $U = P\tau_{sw}$, relating the power consumed during switching, and the switching time. The mean thermal energy associated with these fluctuations, second order in the primary coordinate—the voltage v—is $(1/2)C\langle v^2 \rangle = k_B T/2$.

The probability of state 0, ideally at $0\ V$, and state 1, ideally at a potential of $V\ V$, with these random fluctuations, is a normal distri-

The principle of $k_B T/2$ of energy per degree of freedom is a general conclusion from the equipartition of energy and the virial of Clausius—valid for systems whose stability arises from harmonic behavior and which exhibit classical distribution. See S. Tiwari, "Quantum, statistical and information mechanics: A unified introduction," Electroscience 1, Oxford University Press, ISBN 978-0-19-875985-0 (forthcoming), and the fluctuations discussion in Chapter 5 and Appendix D.

bution:

$$p(0) = \frac{1}{\sqrt{2\pi}} \exp\left(-\frac{v^2 C}{2k_B T}\right),$$

$$\text{and } p(1) = \frac{1}{\sqrt{2\pi}} \exp\left(-\frac{(v-V)^2 C}{2k_B T}\right). \tag{1.62}$$

This result allows us to determine the error rates (see Figure 1.23) by defining a normalized switching threshold of $S = V/(2\sqrt{2k_B T/C})$:

$$\varepsilon = \frac{1}{2}(\varepsilon_{01} + \varepsilon_{10}) = \frac{2}{\sqrt{2\pi}} \int_S^\infty \exp\left(-\frac{v^2}{2k_B T}\right) d\frac{v}{2\sqrt{2k_B T/C}}. \tag{1.63}$$

Here, ε_{01} and ε_{10} are the errors: ε_{01} when 0 is transmitted but a 1 is detected because the measurement is above the switching threshold S, and ε_{10} for the reverse. This integral can be inverted to relate error rates in this ideal but noisy system. This relationship is log-linear over a very large range of parameters. For a single device switching at a GHz, one error in ten years ($\varepsilon \approx 3 \times 10^{-18}$) corresponds to an energy of $152k_B T$ that must be put in the switch. This energy is what is needed to exceed the loss to the degrees of freedom sufficiently to achieve the desired correctness. If we integrate 10^{10} of such ideal devices in an ensemble, the energy needed per switch increases to ensure that only one of them has the likelihood of an error in ten years. For this error probability of $\varepsilon \approx 3 \times 10^{-28}$, an energy consumption of $250k_B T$ is required, as can be seen in Figure 1.23. Noise present in the signals associated with states, even stable states, causes a normal distribution imposed by the fluctuations whose effect must be overcome to assure correctness.

Unlike the case for the reverse-coupled inverter system, the storage of states in magnetic media, based on magnetic polarization in two orientations, is relatively simple, devoid as it is of any amplification effects from the static supply. The potential well picture still holds, and the transitions are affected by changing transition rates through external fields. The Néel-Brown relationship of thermal fluctuations of magnetic moment in such structures begets similar arguments, Néel's to vibrational precession over the barrier, and Brown's to magnetic fluctuations. So, ignoring any asymmetry effects, the collective ensemble description of magnetic systems leads to a similar order-of-magnitude energies for magnetic systems. Similar considerations would prevail also in other spin and mechanical switching arrangements.

Overcoming noise requires energy to achieve determinism with quantifiable correctness. This situation is true even in the adiabatic conditions that we have discussed. Transistors make non ideal switches, so, to overcome noise, the idealized description should be

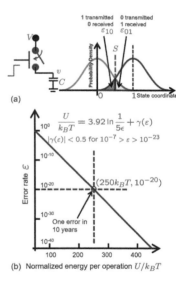

Figure 1.23: (a) shows an ideal switch driving a capacitor in the presence of noise. An error occurs when the state detected is opposite to the one attempted, because of the normal distribution from the noise's energy of $k_B T/2$ in the unaccounted degrees of freedom. (b) shows the inversion of this error relationship, to extract error versus energy for an ensemble of 10^{10} such ideal devices.

modified. For an idealized inverter, the errors become

$$\varepsilon = \frac{1}{2} - \frac{1}{4}\text{erf}\left(\frac{V_m}{\sqrt{2}\sigma}\right) - \frac{1}{4}\text{erf}\left(\frac{V_{DD} - V_m}{\sqrt{2}\sigma}\right), \qquad (1.64)$$

where V_m is the voltage at which input and output are the same, that is, the crossover point of the butterfly curve, and σ is the standard deviation—the root mean square value—of the noise, and

$$U = 4C\sigma^2[\text{Inverf}(1 - 2\varepsilon)]^2, \qquad (1.65)$$

which includes only noise. Here, the inverter is idealized and devoid of parasitic resistances. As should be expected, the appearance of additional independent sources of fluctuations and variability will make the requirements more onerous than they had been before.

Fluctuations arise from many sources in this ensemble. There are fluctuations in the physical device structure. A transistor is a collection of atoms forming a conducting region using dopant atoms and capacitively coupled by voltages of regions whose thicknesses fluctuate. One could think of many other sources of fluctuation, so noise due to thermal effects, in the signal and appearing in time, is only one of these sources. An input noise is amplified by the transistor. The spatial fluctuations are represented in how the gate switches. For an assembly of transistors, if one measured a distribution of threshold voltages, one would measure a threshold voltage that too has a normal distribution, ignoring for now systematic effects that arise from controlled or uncontrolled process effects. In the *CMOS* switch, the effect that dominates switching variability arises in the threshold voltage of the device, because it in turn affects the *CMOS* gate's switching threshold. The distribution of this threshold voltage follows

$$\wp(V_T) = \frac{1}{\sqrt{2\pi}\sigma_{V_T}} \exp\left[-\frac{(V - V_T)^2}{2\sigma_{V_T}^2}\right], \qquad (1.66)$$

where σ_{V_T} is the standard deviation of the threshold voltage. In a *CMOS* inverter, this variability leads to an error, given a supply voltage V_{DD}, of

$$\varepsilon \propto \exp\left[-\frac{2(V_{DD} - V_T)}{\sigma_{V_T}}\right]. \qquad (1.67)$$

Normal random variations arising from independent variations are additive. The variabilities in the squares add, that is, $\sigma^2 = \sigma_1^2 + \cdots + \sigma_n^2$, for n independent causes of similar propensities.

The conclusion also has implications for adiabatic logic or when information is not erased—two ways of lowering energy. In both of these approaches when employing classical systems, the noise and stochastic distributions must still be overcome to assure correctness. When a path between the high potential and the ground is avoided

Since the inverter has gain in the transition region, it has consequences for error due to input amplification. It increases error propensity.

If there is a flux variation in the ion-implantation process that incorporates dopants by using a beam of ions, there will be systematic variations because the characteristics of the beam at the ends of the traversal region may be slightly different from those in the middle of it.

This result also implies that circuits with a large number of transistors, that is, high fan-in or fan-out circuits, will have higher variability than those with less number of transistors. Gain matters too, because a gate amplifies in its transition region. Effects then get coupled. Delays and time variations are amplified. Circuits have to be slowed to overcome the variabilities introduced in the time domain as the result of variations in the spatial domain.

and the changes are slow enough to reduce displacement and resistive losses, and/or when information is not erased, the penalty is extracted in error correction through additional switches and in delay dictated by the error relationships and the swapping of energy. Noise is the penalty of coupling to the environment. One can introduce nonidealities and other probabilistic effects to this discussion to make this calculation more realistic than the idealized version. Random independent variations are multiplicative in probabilities for correctness and additive in the squares of variance. Error reduction requires larger and larger energy consumption for all the errors arising from random effects. The consequence of additional probabilistic variations is to increase the energy necessary to assure correctness. If Arrhenius behavior prevails, and most electronic device systems can be viewed in a classical energy-generalized coordinate relationship, with a barrier, the general feature of these error-energy relationships is that the error probability has a relationship of $\varepsilon \propto \exp(-\Delta U / k_B T)$. This relationship, as we see, also holds for bistable memories, which similarly achieve their stability through the harmonic behavior in classical structures. The general feature of most electronic devices is the use of a reasonable ensemble of electrons or its quasiantiparticle hole or spin in magnetization to achieve robustness. The Boltzmann factor represents the distribution probability of the ensemble of physical objects in the presence of thermal energy and its random noise, which leads to the error.

1.10 Networks

THE BAYESIAN DISCUSSION in Section 1.4 addressed the changes in nature of informational content given the availability of new information. Information also arrives from numerous interactions: a particle interacting with other particles through numerous perturbation processes such as, for example, long-range charge interactions, or in the different pathways by which information is delivered and passed along in circuits. A network, such as a network of internet nodes or interlinked computational blocks, or the network of cells in a cellular automaton, describes the relationships between the nodes through which the information flows. For example, the dissemination of information—such as a memorandum being distributed—is a diffusive flow of information with short-range interactions that may be dominated by node-to-node hopping as the information spreads from a starting node to all others. A search may be a complex form of information flow, with long-range jumps and short-range interactions with local information to reach a target. Such networks may

The circuit that we draw is a graph that describes signal (and information) flow where the nodes and the connectors have prescribed relationships through laws such as those of Kirchoff.

have high-degree nodes, that is, nodes that are connected to a high number of nodes, low-degree nodes, long-range connections and short-range connections. These networks are also therefore an important part of information mechanics. A cellular automaton generally has short range/nearest neighbor interactions. But, one can imagine long-range interactions too, for example, between energy-coupling mechanisms. Mechanical excitations traveling along on a substrate can mediate long-range interactions, while electromechanical interactions can cause local changes in a bias voltage range.

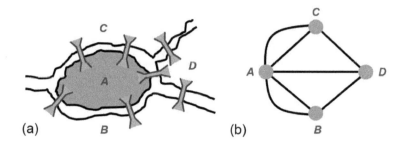

(a)   (b)

Figure 1.24: (a) shows a physical layout of the Königsburg bridge problem. Two islands, A and D, are connected through seven bridges. (b) shows the graphical representation of this problem.

Statistical mechanics, graph theory and other developments from past centuries have significant implications for the behavior of networks. Graph theory describes many consequences from the nature of the network. The classical example, solved by Euler, is of the Königsburg bridge problem, as seen in Figure 1.24. There are two islands—A and D—linked to each other over the Pregel River by seven bridges. The question is, if starting at any land area—A, B, C or D—can one walk across each bridge exactly once and return to the starting point in Figure 1.24(a)? The graph theory equivalent is shown in Figure 1.24(b) with four nodes, also called vertices or points, and seven connectors, which can also be called arcs, edges or lines. This question is a combinatorial problem of computation. Euler's general theorem states that, for a graph to be traversable without repeating, it must be connected and every point must be incident with an even number of arcs. The answer to the Königsburg problem, therefore, is ``no'' since this even rule is not satisfied.

Another classic example of graph theory and combinatorial mathematics, for the complexity it represents and the opportunity it presents for finding higher order ``truths,'' is Ramsey's theorem on the properties of a collection of elements, as represented by the party problem. The theorem can be paraphrased as, for any sufficiently large set of elements, there always exists a large subset with a special property—an island of special order. In its dinner-party form, the Ramsey question for an island of order 3 is the following. What is

Königsburg, a city located in what used to be Prussia, is today's Kaliningrad. The river is Pregel, now known as reka Pregolya. Even though the main island in the middle of the city still exists, modern construction and thoroughfares have made the description in the problem obsolete.

Figure 1.25: A graph showing why a group of 5 is insufficient for the formation of a minimum ordered group of 3. Solid lines and dashed lines are meant to represent acquaintance (``yes'') and absence of acquaintance (``no''), respectively, and this figure shows a counter-example.

We will see this order in phase transitions. Spontaneous polarization leading to ferroelectricity, ferromagnetism or solid-liquid-gas phases are examples of changes in order. In these cases, it is for the entire assembly.

the minimum number of people present in a party so that either at least 3 of them are strangers to each other or 3 of them know each other? The answer here is 6. You, as the host, are one of these people. First, consider yourself. You either know or don't know each of the other 5 people. This means that you either know or don't know at least 3 of them. Take the first case. The argument is precisely the same for the second case. Consider the relationships between these 3 of your acquaintances. If 2 of them know each other, then you and these 2 form a group of 3, our minimum size requirement. If none of them know each other, than they too constitute a group of 3 people of interest—ones that don't know each other. This too is the minimum size. So, you and 5 guests in this party of 6 form the minimalist set that assures that a group of 3 either knows or doesn't know each other. The graph representation of this problem shows the answer clearly. Let 6 vertices represent the 6 people with edges showing the relationships—of knowing each other or not. There are 15 edges to this graph, and no matter how one connects—"yes" or "no"— one cannot avoid forming at least one triangle that contains either all "yes" edges or all "no" edges. This result follows from the fact that there are five edges to each vertex. At least 3 of these must be either "yes" or "no." Now the vertices at the end of this group of 3 too are connected to each other through this choice of relationship. If one of them is the same, then we have our condition of 3. If they are not, then they too form a condition of 3. That having 5 people at the party is not sufficient can be seen in Figure 1.25, where solid and dashed lines represent the "yes" and "no" relationships. For a Ramsey problem for an island of order 4, the minimum number of guests must be 18. Here, graph theory—its representation of a network of relationships—provides a powerful computational tool for combinatorial problems.

Another example of a networking problem emphasizing the Bayesian connection of information is shown in Figure 1.26. This is a classic adapted from Judea Pearl. From a human point of view, probability is a state of mind with respect to the assertion of an event that may happen or of other matters on which absolute knowledge doesn't exist. If one receives a message that the house alarm is ringing, one would think that a burglary is in progress or has happened. We may call this rule $A \mapsto B$. But, if one receives news that there has been an earthquake, one would also recognize that too to be a cause of the alarm. If one considers an earthquake to be a more credible cause of the alarm than the burglary would be, it makes the possibility that the alarm is ringing because a burglary has happened less credible than the possibility that an alarm is ringing because an earthquake has happened. In Figure 1.26, the implication of a non-

The philosophical implication of this Ramsey theory combinatorics is that complete mathematical disorder is impossible. Order's appearance is a matter of scale. Carl Sagan, in the television series *Cosmos*, invokes Ramsey in probing why the fact that eight stars are lined up in the sky does not imply that they were placed there. Take a large-enough group of stars, and one should see almost anything one wishes.

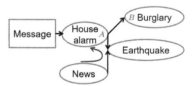

Figure 1.26: An inferencing problem with both local and global informational relationships in time and space with inferencing implications.

This is a paraphrase from Augustus De Morgan who is known for formulating the logical rules for conjunction $(\overline{A \wedge B} = \overline{A} \vee \overline{B})$ and disjunction $(\overline{A \vee B} = \overline{A} \wedge \overline{B})$.

local cause is to reduce the inference of burglary versus earthquake. This is a nonlocal belief that has informational consequence; so here, this nonlocal belief and Bayes's arguments are germane.

An example of the statistical mechanics connection is illustrated in Figure 1.27. Consider a structured grid network, for example, a two-dimensional network where each node is connected through a short-range link to each adjacent node. One would like to find what additional long-range links one might want to establish so that the time to go from any arbitrary node to any other arbitrary node would be minimized. A long range-percolation model—a model to find the paths of easy flow—applies to this situation. For each node v on the $n \times n$ grid, add a directed link to a random node w with a probability $p \propto d_{vw}^{\eta}$, where d_{vw} is the lattice distance from v to w, and η is some power that we will allow to vary. With varying η, Figure 1.27(b) shows the lowerbound time for delivery time. There exists an order parameter η that describes the probability distribution of random connections between v and w for the minimum time. By placing the links conforming to this random probability, a network with an orderly local structure has been reduced to one with a small diameter, where information can be transported with optimal timing between any nodes of the network—a two-dimensional planar mesh network now becomes a network on a spherical surface, with time shortened.

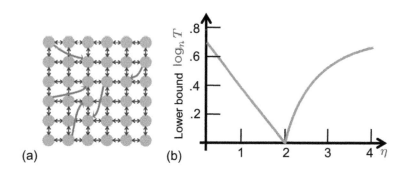

(a)   (b)

Figure 1.27: (a) shows a structured grid network with nearest neighbor connections and long-range links added. (b) shows the lower bound on delivery time as the order parameter η—the exponent for probability $p \propto d_{vw}^{\eta}$, where d_{vw} is the lattice distance between v and w—is varied.

This order parameter, which gives a specific property to the network is a feature of phase transitions that statistical mechanics describes. An order parameter in phase transitions, such as the appearance of magnetism, ferroelectricity or other property changes, is a result of the nearest- and farther-neighbor interactions, as in the information network example shown in Figure 1.27. With a breaking of symmetry, a property appears spontaneously. Magnetism is due to spin correlations—magnetic polarization, while ferroelectricity arises from charge correlations—electric polarization. Many of the conclusions derived in statistical mechanics, such as for spin glasses, have been useful in information networks as well as in the computing of complex interactions. This subject is an important area in computer sciences and artificial intelligence and of much interest, with several important questions still unanswered.

1.11 Information and quantum processes

IN EXPLAINING THE BEHAVIOR OF MICROSCOPIC BODIES, where we must invoke quantum principles and quantum properties of particles—individual photons or electrons or others—information cannot be read or copied without disturbing the system. In the classical view, information provides a means to distinguish observation or data. In this, it is an abstract property that in digital logic is reducible to binary bits which through processing reveals implicit truths. It can be copied undisturbed, it can be erased and it cannot move faster than speed of light. While information may never travel faster than the speed of light, unlike the classical view of communications of packets of information being sent between and being entirely separately in the possession of two different observers, it is possible in the quantum view to connect the two separate observers by a strong correlation—entanglement. This quantum state, even as a superposition state, is one that cannot be described as the product of two distinguishable quantum states. For example, in a two-photon quantum system, where the composite is from $|\uparrow\downarrow\rangle$ and $|\downarrow\uparrow\rangle$ states, up-down or down-up polarization can be generated through physical processes, such as a two-photon-generating process. But, it is not the product of two separate photons, each of which may be in the $|\uparrow\rangle$(up) or the $|\downarrow\rangle$(down) state. Quantum information here is embedded in a quantum bit—a qbit using a two-state quantum system.

This entanglement may still not send information faster than light. But, this entangled state has hidden within it information. So, it too has an entropy notion intimately tied to it. This is entanglement entropy. We will discuss this concept in Chapter 3, where we will explore the device thoughts in the mesoscale regime, with quantum and classical notions intersecting.

Even where entanglement, that is, strong correlation in the quantum state, is absent, so, with two or more photons, or other such quantum objects with quantum properties such as spin, which contain and encode information but are entirely independent of each other, entropy arises through the known and the unknown elements of the coefficient matrix. This is the von Neumann entropy

We will not follow the network theme in any other chapters, except in the statistical mechanics aspects that the Ising model brings to the understanding of magnetics and phase transitions. But, it is incredibly important nearly everywhere—in sciences, ecology and economics, as well as in humanity in general. For example, groups of altruists will prevail. Good thing, too, given how selfishness drives so much of an individual's behavior.

And determining the state of these objects gives us two bits of information.

$H = -\text{Tr}(\rho \ln \rho)$, where ρ is the density matrix of the quantum system, as we have mentioned earlier.

This von Neumann entropy is a description of entropy, similar to the Shannon form, and measures statistical randomness. The unentangled quantum system reduces only to possibilities along the diagonal. The off-diagonal terms represent the incompleteness of knowledge. We will dwell on these quantum-specific properties in Chapter 3.

The information context must incorporate all these classical and quantum-mechanic views of the nature of the world. Just as, in the classical view, Bayesian notions are important both for information and for inference, which is a form of work from the information, Bayesian notions also are relevant to the quantum-mechanical description of information.

1.12   *Summary*

THIS CHAPTER ESTABLISHED the salient characteristics of information and its mechanics, providing a perspective that we should keep in discussing nanoscale devices.

Information is physical. It only exists through an observation on a physical system and only exists in that physical form. Spin orientation, a high or low voltage, the presence or absence of charge, or a pencil mark on a piece of paper are all forms in which information can be supplied. We also established that a solvable problem can be reduced to a symbolic notation that then can be processed in a machine, which is our tool for executing mathematical operations. Solvable problems can therefore be placed in an information mechanics form for implementation and for a study of their characteristics. The Turing machine provides us one way to do this. The mathematical representation of this information, in turn, provides statistical representations such as that of Shannon entropy on the potential content extractable from collections of bits, as well as of the properties of their movement. We also established that both this description and the traditional thermodynamic description are incomplete, since they do not completely define the system and, in particular, because in neither of these is there a configurational property of a microstate that leads to the macrostate's description. Algorithmic entropy, described in terms of the minimum length program, provides this description and helps make the description less incomplete than it was before.

Information, because it is physical, when erased involves the consumption of energy. A degree of freedom has been lost. Information is not abstract; it stands together with other macroscopic variables in the information mechanics framework. This is pictured in Figure 1.28, which shows a ``model T'': a nearly friction-free vehicle driven by a Turing machine drives a nearly friction-free vehicle. Here, our example of a molecule in a volume—our vehicle's cylinder— serves as the physical form for the information bit. If the sequence is random, no useful work comes out of the Turing vehicle, and it ends up returning to the same mean position it started from. But, if the sequence is nonrandom, that is, a defined problem that is solvable, the vehicle moves precisely to achieve the precise objective, which is reflected here in the movement to a precise position. Even pseudorandom numbers, such as π or Euler's constant, will do precise work in this machine, since they are not truly random, as algorithmic entropy showed.

The mechanics ideas of our discussions lead us to relating entropy, randomness and energy, and these in turn to errors, noise, reversibility and non-reversibility. Information engines can be operated both reversibly and non-reversibly. Under adiabatic conditions, the energy dissipation can be reduced in non-reversible and removed entirely in ideal reversible systems. Billiard balls computing, for example., showed that the dissipation limit is zero if no information is erased. In *CMOS*-based circuits, it is possible to lower the dissipation by slowing the process in the circuits, and by removing conditions that lead to large dissipation, such as transitions that move charge from a high potential to the ground reservoir or where, significant current flows through voltage drops, and by channeling the energy flow in the circuit through its exchange between different storage elements, be they distributed or lumped inductors and capacitors, or power supplies.

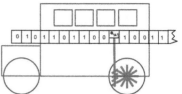

Figure 1.28: A nearly friction-free Turing vehicle—model T—showing how the information of a Turing tape has energy content that can be physically observed. For the bits in the example, the presence in one half or the other half of an atom in a chamber is used for the information storage. As in the discussion of Maxwell's demon, a partition is moved in, and the vehicle moves one way or the other, leaving a random condition in the bit cell as the partition is moved fully one way or the other.

1.13 Concluding remarks and bibliographic notes

MORE THAN IN COMMUNICATIONS ALONE, recognition of information as an essential concept, with physical foundations, has become a pervasive element of scientific and engineering thinking since the last decade of the 20th century. Information is useful for the improvement of the ability to predict when using it. Information is just as relevant to sequences of 0s and 1s as it is to stories. Great books, that is, sequences of 0s and 1s in some text symbol–to–binary symbol translation code, bring out a new meaning in a person every time she reads it. Today, the reach of information and of its interpretation extends to biology, with the enormous amounts of signal and data arising in genomics, to business and social and economic decision-

There is a Leo Rosten story about a rabbi who had a perfect parable for any subject. When asked how this happened to be, the rabbi tells a parable about a recruiter of the Tsar's army. Riding through a village, this recruiter sees that there are many chalked targets on a barn, each with a bullseye bullet hole in the center. When the recruiter goes looking to rope in this ``sharp-shooter,'' a villager tells him, ``Oh, that's Shepsel. He is the shoemaker's son. He is a little peculiar.'' The recruiter is not discouraged until he is told, ``You see, Shepsel shoots first, and then draws the circle.'' The rabbi then said with a smile, ``I only introduce subjects for which I have parables.''

making with the interconnectedness of information embedded in the traditional measures of different disciplines. It is at least an approach to finding higher order connections in a collection, making it pervasive. Understanding the number of possibilities, probabilities, Bayes inferencing and the smallest ways of capturing this informational essence have much to say about finding this higher order.

Even in quantum mechanics, the wavefunction as a mathematical object is a creation with information embedded within it. It has taken some time to understand the Kolmogorov algorithmic complexity and its information measure. Entanglement hides within it information and therefore entropy that is still not entirely clear. So, the thought process started by Maxwell through his gedanken demon experiment, Shannon and his mathematical measure for streams of data, analogous to the Boltzmann measure, and Szilard introducing negentropy and information in thinking through resolution of the problem of Maxwell's demon, flourished as computing drove the understanding of reversibility and irreversibility, and energy dissipation became an important consideration in the information edifice that the Western society increasingly became at the turn of the century. These are ideas across the spectrum of engineering and science, and particularly grounded in applied mathematics as well as physical reasoning. This is the rationale behind making this swath through the subject be the introductory chapter to what is to follow.

Landauer seeded the early interest in information, dissipation and its physical origin with the early emphasis and discussion of irreversibility and heat generation when manipulating information[1]. The definitive unraveling of Maxwell's demon is due to Bennett[2]. Both Landauer and Bennett are paragons of lucidity in their writings, which border on the ideal form that mathematicians aspire to in making an argument complete.

Shannon's ideas about information, and the Church-Turing construction of symbolic manipulation and recursiveness, together with their implications for information machines, have been staples of information theory as taught in electrical engineering for far more than half-a-century. Most classical texts elaborate the established insights of what these concepts mean. Emmanuel Desurvire's[3] is an excellent modern text that connects classical and quantum approaches and remains rigorous while emphasizing the essential concepts.

Shannon's[4] foundational publication is readable even today for its clarity of thought as well as its awareness of the constraints inherent in the treatment.

Probabilities are the natural mathematical means for the exploration of interactions in an ensemble. Physical systems can often be simplified to a level of simplicity, that is useful for gaining intuitive

[1] R. Landauer, "Irreversibility and heat generation in the computing process," IBM Journal of Research and Development, 5, 261–269 (1961)

[2] C. Bennett, "Logical reversibility in computation," IBM Journal of Research and Development, 17, 525–532 (1973)

[3] E. Desurvire, "Classical and quantum information theory," Cambridge, ISBN 0-521-88171-5 (2009)

[4] C. E. Shannon, "A mathematical theory of communication," Bell Systems Technical Journal, 27, 379–423 and 623–656 (1948)

understanding, and for making predictions. Grandy[5] provides an extensive discussion of these connections between probabilities and physical systems. The principle of maximum entropy is best seen through its leading light—Edwin Jaynes. Jaynes[6] started on this track with two publications. A book edited by Skilling[7], and other such edited volumes in the series of workshops on the subject, with Jaynes contributing, are a good source for seeing the application technique of rational inference from the observations that we draw our conclusions from.

Algorithmic complexity has its origins in the work of Solomonoff[8] and Kolmogorov[9]. Solomonoff's interest in this direction was from a machine-learning and artificial-intelligence perspective. Kolmogorov, one of the many remarkable Russian mathematicians with contributions in many directions, particularly complexity, also arrived at the notions independently and did much more to advance understanding and to clarify. The text by Li and Vitányi[10] dwells in depth on this subject.

A discussion of the quantum aspects of the algorithmic complexity theme is to be found in the text edited by Hey[11] and in the article by Zurek[12]. The volume edited by Wheeler and Zurek[13] is also very worthwhile reading. It contains a number of papers by numerous others whom we have mentioned or will mention in the ensuing chapters.

The early discussion of dissipation and its physical underpinnings, as well as what happens in practice, took place in the articles by Landauer and Bennett referenced above and many others that followed as the subject cycled in time. But, the last decade of the 20th century saw the first practice of many of these thoughts in silicon technology. An early publication[14] in the journal of record on solid-state circuits started this approach to practicing adiabatic logic. One of the logic circuits we discussed—the $2N$-$2N2P$—appears in Kramer et al.'s[15] publication.

Judea Pearl[16] explores the connections between graphs, Bayesian probabilistic reasoning and drawing inference. The reader will find this very readable book an excellent source on information computation approaches and the rationale underlying them for drawing inference by reasoning in presence of uncertainty.

The connection between network theory and statistical mechanics, together with its implications for information, is discussed in detail by Albert and Barábasi[17]. This writing is quite mathematical, even for a physical science perspective.

Social networks, the the myriad of two-way and multi-way couplings in economics and the markets, the behavior of crowds in wars and peace, and other such subjects, provide a fertile ground for both

[5] W. T. Grandy, "Entropy and the time evolution of macroscopic systems," Oxford, ISBN 0-19-954617-6 (2008)

[6] E. T. Jaynes, "Information theory and statistical mechanics, I," Physical Review, **106**, 620–630 (1957) and "Information theory and statistical mechanics, II," Physical Review, **108** 171–190 (1957)

[7] J. Skilling (ed.), "Maximum entropy and Bayesian methods," Kluwer, ISBN 0-7923-0224-9 (1988)

[8] R. Solomonoff, "A formal theory of inductive inference, Part I, "Information and Control, **7**, 1–22 (1964) and "A formal theory of inductive inference, Part II," Information and Control, **7**, 224-254 (1964)

[9] A. N. Kolmogorov, "Logical basis for information theory and probability theory," IEEE Transactions on Information Theory, **IT14**, 662–664 (1968)

[10] M. Li and P. Vitányi, "An introduction to Kolmogorov complexity and its applications," Springer, ISBN 0-38-733998-1 (2008)

[11] A. J. G. Hey (ed.), "Feynman and computation: Exploring the limits of computers," Perseus, ISBN 0-81-334039-5 (1998)

[12] W. H. Zurek, "Thermodynamic cost of computation, algorithmic complexity and the information metric," Nature, **341**, 119–124(1989)

[13] J. A. Wheeler and W. H. Zurek (eds), "Quantum theory and measurement," Princeton University Press, ISBN 978-0691613161 (1983)

[14] A. G. Dickinson and J. S. Denker, "Adiabatic dynamic logic," IEEE Journal of Solid-State Circuits, **30**, 311–315 (1995)

[15] A. Kramer, J. S. Denker, B. Flower and J. Moroney, "2nd order adiabatic computation with $2N$-$2P$ and $2N$-$2N2P$ logic circuits," International Symposium on Low Power Design, 1–6 (1995)

[16] J. Pearl, "Probabilistic reasoning in intelligent systems: Networks of plausible inference," Morgan Kaufmann, ISBN 1-55-860479-7 (1988)

[17] R. Albert and A.-L. Barábasi, "Statistical mechanics of complex networks," Reviews of Modern Physics, **74**, 47–97 (2002)

a mathematical analysis, where reasonable assumptions are required to make a problem tractable, as well as computational analysis via simulations. A computer science perspective for some of the important societal areas for this set of problems can be found in the book by Easley and Kleinberg[18]. For those interested in seeing the statistical mechanics connections to networks and other advanced information problems of interest in computer science, a highly recommended source is the book by Mézard and Montanari[19].

A good introduction to the quantum-specific tackling of information—a vibrant current subject—with connections to cryptography, secure communications and perhaps computation, as well as to the disciplines of engineering and sciences that it integrates can be found in the two textbooks: Barnett[20] integrates classical engineering information pedagogy with the modern quantum view, while Vedral's text[21] provides a quantum-mechanics-drawn perspective.

[18] D. Easley and J. Kleinberg, "Networks, crowds and markets: Reasoning about a highly connected world," Cambridge, ISBN 0-52-119533-0 (2010)

[19] M. Mézard and A. Montanari, "Information, physics and computation," Oxford, ISBN 0-19-857083-X (2009)

[20] S. M. Barnett, "Quantum information," Oxford, ISBN 978-0-19-852762-6 (2009)

[21] V. Vedral, "Introduction to quantum information science," Oxford, ISBN 0-19-921570-7 (2006)

1.14 Exercises

1. Explain how the cellular automaton example of the chapter computes and solves the problem. Argue succinctly. **[S]**

2. Design a finite state machine that delays the binary stimulus by precisely 2 time delays. **[M]**

3. A fair dice has 8 faces. What is its information entropy? **[S]**

4. A random variable X can take values $(0, 1)$ with probabilities $p(0) = \varepsilon$, and $p(1) = 1 - \varepsilon$. What is the entropy of this system? **[S]**

Note the implications of this value in error correction.

5. Assume that all the characters of English—consider only 26 letters plus the character of space—were equally likely. What is the entropy per character? If one needed to transmit a message of 10 symbols, what would be the minimum number of bits needed? **[S]**

6. If all the characters of the alphabet were equally likely on a printed page containing 3000 characters, what is the Shannon information entropy, and how much information can the page convey? Reason your answer for the last part in one sentence. **[S]**

7. Not all characters are equally likely in written English. Some occur more often. Table 1.1 gives an approximate distribution. What is the entropy per character? What assumption did you make in calculating the value? **[S]**

8. If entropy is a measure of "redundancy" in the message, show that if the information content of the message is equal to the size of the message, then there is no redundancy. **[M]**

Letter	Frequency	Letter	Frequency	Letter	Frequency
a	0.0182	j	0.0015	s	0.0633
b	0.0149	k	0.0077	t	0.0906
c	0.0278	l	0.0403	u	0.0276
d	0.0425	m	0.0241	v	0.0098
e	0.1270	n	0.0675	w	0.0236
f	0.0223	o	0.0751	x	0.0015
g	0.0202	p	0.0193	y	0.0197
h	0.0610	q	0.0010	z	0.0007
i	0.0697	r	0.0599		

Table 1.1: An approximate frequency of occurrence of the different letters of modern English.

9. Can a Turing machine determine if a string is truly random?   [M]

10. We have a simple radio frequency amplifier which has a bandwidth of 10 GHz and a signal-to-noise power ratio of 10. If we reduce the bandwidth to 5 GHz, how much should the signal power be increased, in dBs to maintain the same channel capacity?

   [S]

11. Light is used to transmit information and is physical: the photon is the particle of light. Does a light channel conveying information have entropy? If so, is the entropy a Shannon entropy, a Boltzmann entropy, neither or both? What does this result imply?   [M]

12. We determined exponential error relationships in logical computation through switches. What happens in adiabatic logic, where the energy theoretically is claimed to be made vanishingly small if we also avoid erasure of information? Does noise place a constraint on the minimum energy, given a certain probability of error desired?

   [M]

13. In an inverter, no information is erased, so there is no $k_B T \ln 2$ of dissipation. Why, then, do most inverters consume energy? Could we design one that would not, and how would it work?   [M]

14. Why does does an inductor appear to open when there is an instantaneous change in voltage across it? If power lines are inductive, what does it mean when a circuit is connected to power supply instantaneously?   [M]

An answer saying Ldi/dt will not do.

15. Are all adiabatic circuits also reversible?   [S]

16. If we slow a static logic gate down, that is, slowly ramp the voltages of input, then, do the same for the power supply in some appropriate sequence, can we reduce dissipation in it?   [S]

17. Where does dissipation take place when current is passing through the transistor when it is sitting in a static condition? **[S]**

18. How should a pass transistor be operated so that there is minimum dissipation in it? **[S]**

19. Compare how long it takes to get a signal from one end of the chip to another (say, a distance of 2 *cm*) without any logic processing, to how much time it takes to process a signal through 10 gate operations, each about 50 *ps* long, which is typical in a processor environment. What does this say for clock frequency on the chip? **[M]**

20. Consider a *CMOS* inverter formed from an *nMOS* silicon transistor and, a *pMOS* silicon transistor with threshold voltages of V_{Tn} and V_{Tp}, respectively, as shown in Figure 1.29.

   The drain-to-source current in the bias region above threshold is approximated by

$$I = \frac{W\mu C_{ox}}{L}\left[(V_{GS} - V_T)V_{DS} - \frac{1}{2}V_{DS}^2\right],$$

   where W is the width, μ is the mobility, C_{ox} is insulator capacitance per unit area, L is the channel length, V_{GS} is the gate-to-source voltage, V_T is the threshold voltage and V_{DS} is the drain-to-source voltage. This is the simplest and crudest of equations in the common simulation tools. This equation applies to both *n*-channel and *p*-channel transistors, and we may identify each through an additional subscript to each of the transistor parameters. For the *n*-channel transistor, this equation implies

$$I_n = \frac{W_n\mu_n C_{oxn}}{L_n}\left[(V_{GSn} - V_{Tn})V_{DSn} - \frac{1}{2}V_{DSn}^2\right],$$

   and there exists a similar equation with a *p* subscript for the *p*-channel transistor. The inverter has characteristics that look like those in Figure 1.30, which shows 2 output curves from consecutive inverters in a chain of inverters when a bias voltage of V_{DD} and a ground is applied with a stable high voltage of V_H, and a low voltage of V_L; V_{in} is input voltage, and V_{out} is output voltage; in a chain, they are consecutively applied. If two inverters are tied back-to-back, they form the innards of a static *RAM* device (also, a high-gain amplifier), whose stable point in absence of noise is V_x.

   We wish to determine some of the static parameters that result from this idealized inverter.

   • Determine V_H and V_L in terms of the known transistor parameters. At least write the procedure—algorithm—for determining these values if you cannot determine them explicitly.

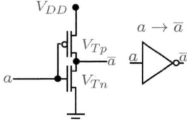

Figure 1.29: A *CMOS* inverter with some of its transistor parameters and applied bias voltages.

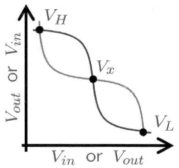

Figure 1.30: The output-to-input voltage behavior when *CMOS* inverters drive other *CMOS* inverters. Stable high-state (V_H) and low-state (V_L) voltages result. When inverters drive each other back-to-back, that is, in a loop rather than a chain, one gets the stable point V_x under an idealized no-noise condition.

- Determine V_x, which is the metastable point that the back-to-back inverter will sit at.

- Now, let us assume that threshold voltages are variable, with a variance of $\sigma_{V_{Tn}}$ and $\sigma_{V_{Tp}}$ for the two transistors. Determine the variance in the cross over V_x, that is, σ_{V_x} in terms of the threshold variances. At least set it up if you cannot solve it explicitly.

- What does σ_{V_H} or σ_{V_L} look like owing to the variances in threshold voltage?   [A]

21. Consider one inverter as a channel with the probability of ε of error in its logic processing. It operates on an input bit string M. For $\varepsilon = 10^{-20}, 10^{-10}$ and 10^{-5}, what is the limit on the residual error rate, and how long, in terms of percentages, must the string be in each case in order to detect this rate? If you had a string of 10 inverters, how would the error rate bound change?   [M]

Note that $1 - f(\varepsilon)$ is the residual error bound; $f(\varepsilon)$, in our notation, denotes the probability of correctness.

22. Why does static $CMOS$ dissipate $\sim \nu C V^2$ power, where C is the total capacitive load, V is the supply voltage and ν the frequency? How much dissipation occurs in a pass transistor gate, which is a component of pass transistor logic, operating under these same conditions?   [S]

23. Is there a magnetic field locally around an integrating circuit that is operating? If yes, where is it mostly, and does it affect the transistor's operation?   [S]

24. Consider the combinatorial gate, a sorter, whose truth table is provided in Table 1.2. There are the same number of 0s and 1s in the input as in the output.

In		Out	
0	0	0	0
0	1	1	0
1	0	1	0
1	1	1	1

Table 1.2: Truth table of a sorter.

- What is the total logical entropy change? What is the value for logical entropy in the beginning, and in the end?

- How much heat is dissipated during this operation at room temperature?

- Draw a gate and a $CMOS$ circuit implementation for this sorter.

- Where exactly in the circuit is entropic dissipation, if any, happening?   [A]

25. A mobility of 100 $cm^2/V \cdot s$ corresponds to approximately 100 fs of relaxation time—the time between energy-losing scattering events. This value is not too far from what one gets in silicon at extreme dimensions—that is, in devices of the type employed in the back-to-back inverters used for static random access memory ($sRAM$). Estimate the minimum energy barrier necessary to achieve a storage time of 10 $years$.   [M]

Note the implications of this for $sRAM$ stability and reliability. And for arrays of $sRAM$ devices!

26. Figure 1.31 shows the 2N-2N2P 2 input *NAND* circuit, which is quasi-adiabatic. Explore how it may work in light of the discussion of principles that reduce dissipation. Explain its behavior in a simple paragraph, with arguments on how dissipation is minimized by the time evolution of different changes. **[M]**

27. We discussed the use of energy recovery to generate clock signal in a rotary clock. We have a 2×2 cm^2 processor chip on which we want to employ such a clock, and we wish to have it run at 2 *GHz*. Design a clock system to achieve this. You will have to make assumptions. Make judicious ones and defend them. But, in the end also determine the total wire length of the transmission line, the loading capacitance and inductance on the line, and the line's impedance, and explain how you arrived at these values. **[A]**

28. We have said that a rapid change in charging causes the energy losses that one sees in charging of the capacitor. But, in a resonator, such in an *LC* circuit, even those at *GHz*, the energy loss is very small (the Q can be very large). Is it possible to have a small lost at any frequency? At optical frequencies? Explain. **[M]**

29. A typical small transistor in an integrated chip environment (2×2 cm^2 area, about 10^9 transistors, about 10% of which are switching, that is, are active, at any time) at best switches in ~ 10 *ps*. Estimate the amount of charge and current being manipulated by each transistor, starting with justifiable assumptions. **[M]**

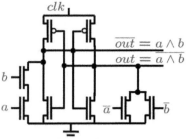

Figure 1.31: Circuit schematic of a 2N-2N2P 2-input *NAND* circuit.

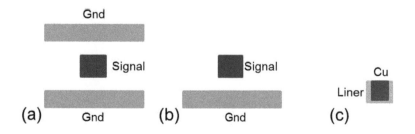

Figure 1.32: A variety of lines that are used for transmitting signals on a chip. (a) A ground-signal-ground (*G-S-G*), or stripline, configuration. (b) A signal-ground (*S-G*), or microstripline. These configurations localize signal and have better signal transmission characteristics than (c) does. (c) is a classical interconnection line. Copper, the interconnection metal, is clad with a liner that prevents reliability-degrading interactions with the surrounding medium; *Cu*, copper; *Gnd*, ground.

30. Figure 1.32 shows three ways that a signal may be transported electrically in an integrated circuit. The point of this exercise is to estimate some of the chip's device and interconnect parameters. Make assumptions suitable for 2×2 cm^2 chip that utilizes 8 levels of metal interconnects in a 65 *nm* physical gate length technology with about 100 solder bumps for the power supply and the ground, and about 1000 for input/output signals. Give short justifications in your answers for the choices you have made.

- Estimate the number of gates and transistors on the chip, assuming that they have, on average, a fan-in of 2 and a fan-out of 2.

- Estimate the inductance per unit length in the power delivery lines.

- Design a 50 Ω transmission line for use on the chip. Just give the dimensions of the wires, and the line's layout.

- On average, how much resistance exists from the wiring from the supply to the gate? Assume the microbump to be about 25 μm in diameter and with a 10 μm opening. Where does all the resistance come from?

- On average, how much resistance and capacitance exists in connecting two gates 1 mm apart, that is gates that are in two different functional blocks. What is the time constant for this connection?

- What is the approximate resistance that connects the output of one gate to the input of an adjacent gate if it is connected through the first level of metal?   **[A]**

31. Can an observation be made without energy dissipation? Give a gedanken example if your answer is in the affirmative.   **[S]**

Look up the design of transmission lines in any standard text or on the Web.

2
Nanoscale transistors

TRANSISTORS ARE USEFUL IN TRADITIONAL ELECTRONICS because
they work well as digital switches with isolation and as analog elements with amplification, as well as providing other functions—all
with integration in vast numbers. The high integration capability
makes the digital technology possible. When ``on,″ they pass useful
currents that can charge and discharge capacitances of connected
switches or long interconnection lines; when ``off,″ they are a useful approximation of the absence of conduction, they isolate output
from input through the voltage gain, they restore signals through the
power gain and they are sufficiently reproducible to allow large-scale
integration. These properties make digital and analog technology
succeed.

What is uniquely different and significant when transistors are
reduced to nanoscale dimensons? When we make devices extremely
small, we reduce the ensemble size. Multitudes of effects result. The
electrostatic characteristics of the transistors—the device behavior
under static conditions, and the electrodynamics characteristics—
change over time, undergoing significant transformations. A transistor, whether in the off state or the on state, is biased at potentials
appropriate to the circuit configuration. These are significantly different for static versus dynamic circuits. In each of these states, the
ensemble size affects the importance of different interactions. The
dominating mechanisms change and, together as an ensemble, their
behavioral manifestations such as those of ``off state″ conduction—
the leakage current—or those of ``on state″ conduction—-which
drives the next stage by charging or discharging the capacitance—
change, as do the phenomena that occur during these transitions.
Applied potentials in the confined nanoscale geometries couple, so
electrostatics is multi-dimensional. At the least, the out-of-plane gate-
dominated and the in-plane source-to-drain-dominated variations
couple. The statistical effects of dopants employed intentionally or

Charge control in field-effect transistors
is through gate fields. Input voltages
exercise this control, which changes
the output current. Charge control
in bipolar transistors is by charge
injection. Input current exercises this
control, which changes output current.

For many decades, almost from the
time time when transistors were invented, predicting the smallest possible
dimensions, that is, the end of scaling,
for transistors, has been a challenge that
few of the best and the brightest have
been able to resist. It has been a mas-
sacre of all. In the 1960s, the dimension
predictions was a couple of μms, gradually shrinking, till in the 1990s, getting
to be several 10s of nm. At some point
in the 1990s, the dimension prediction
in the time-evolution curve was over-
taken by research demonstrations, and
later on by manufacturing practice. The
limits, if any—or perhaps ``difficulties″
would be a more suitable term—have
largely arisen from practical issues, that
is, problems arising from dissipation,
the large variances in the controllability
of the device, or other problems with
performance characteristics or even cost
of production. The development of integrated technology is a good example of
how needs can lead to reinventions for
solving problems.

Nanoscale device physics: Science and engineering fundamentals. Sandip Tiwari.
© Sandip Tiwari 2017. Published 2017 by Oxford University Press.

occurring unintentionally influence the off state behavior. Confinement geometries also affect the states of the carriers. The turning on or turning off of the devices is rapid and occurs in conditions where few or no scattering events occur because the length scales of the devices are close to the length scales over which scattering changes the momentum or energies of the carriers. Since the off state and the on state state are the two steady states of the device, these will be the principal focus of the discussions of this chapter. We will examine these through explorations of two-dimensional scaling of the off state, ballistic and near-ballistic transport in the on state, the effects of confined geometries, and contact resistances in structure with minimum scattering. We assume that reader understands the classical device physics discussion at the level of Volume 2 of this series.

2.1 Transistors as dimensions shrink

CHARGE CARRIERS MOVING FROM THE SOURCE ELECTRODE TO THE DRAIN ELECTRODE biased for attracting potential will move faster as the electrodes are brought closer together. In a semiconductor, with these distances long and electric fields small, the carriers continuously lose the energy that they gain as they move under the force exerted by the field. A constant mobility characterizes this linear velocity-field relationship. As the field increases further, the carriers, the electrons, pick up more energy between the scattering processes that cause the loss of energy. The effective velocity of this charge cloud is larger, and for moderate fields, linear with field. In the phase space, the carrier-velocity distribution is shifted and centered at this effective velocity. As the field keeps getting increased in this device, in which the electrodes are still quite far apart, electrons begin to lose nearly as much energy as they are gaining between scattering events because the scattering rate increases nonlinearly. The scattering events, that is, the interactions with the environment of the charge, happen at rates dependent on the characteristics of the electron's and the environment's state. Optical phonon emission, for example, begins above a threshold: the lowest energy of that lattice vibration mode. Electrons occupying higher energy states can also scatter into more states since the density of states generally increases with energy. The velocity now saturates. In low mobility materials, this may require large fields. In high mobility materials, this may happen at quite low fields.

What we have just described are some of the possibilities affecting charge flow when the length between electrodes, L, is much larger than a scale length that characterizes scattering—$\langle \lambda_{scatt} \rangle$, a distance

S. Tiwari, "Device physics: Fundamentals of electronics and optoelectronics," Electroscience 2, Oxford University Press, ISBN 978-0-19-875984-3 (forthcoming) explores the character of and the internal statics and dynamics of devices emphasizing the engineering implications.

In detail, the scattering rates bounce around, as the interactions and the details of states are quite bandstructure dependent. See the scattering rate discussion in S. Tiwari, "Semiconductor physics: Principles, theory and nanoscale," Electroscience 3, Oxford University Press, ISBN 978-0-19-875986-7 (forthcoming).

There will be a few different length scales arising from scattering processes. The phase, the momentum and the energy are all parameters—specific to a particle or averaged over an ensemble, that is, denoted by $\langle \rangle$—that will have a scale length associated with their march towards a steady state. For now, we leave this as a general scale length whose details must be found from the scattering-related length scale of the parameter in question.

that the electron travels on an average between the various scattering events that cause it to change momentum and arising from different causes at different rates. What happens at the other limit, when $\langle \lambda_{scatt} \rangle \gg L$? If the carriers do not lose their energy and momentum at all due to scattering, the electron maintains its forward motion dictated by the electron's state evolution under the influence of the field.

We will find that the conductance of such an electron channel is a constant—$2e^2/h$, where e is electron charge, and h is Planck's constant—when both a spin-up and a spin-down electron occupy a specific $(E, \mathbf{k})$ state for the electron, where E is the energy, and $\mathbf{k}$ is the wavevector of the state. The ratio of current change per unit energy when an additional channel is occupied by both the spin-up and the spin-down electrons is given by a constant. $2e^2/h$ is a conductance of 77.27 μS and a resistance of 12.95 $k\Omega$. When the spin degeneracy is broken, for example, when magnetic interactions such as from a magnetic impurity prevail, the conductance quantum of e^2/h can be observed. Later on, we see that for single electron tunneling events to be observable, conductances of barriers have to be less than these limits. In superconductive Cooper-pair tunneling, the corresponding conductance is $4e^2/h = 0.1545$ mS or 6.47 $k\Omega$. We will see this ratio e^2/h appear often since it is a fundamental conductance parameter for the wave conductance of electrons, so in any related quantum phenomena—from the flux quantum consequences to the quantized Hall effect.

Nanoscale transistors, in their limits of small dimensions, approach the limit of no scattering. We will call this the ballistic limit. But, nanoscale transistors also are three-dimensional structures where all different dimensions are being shrunk. These transistor dimensions and the applied potentials influence the conditions under which the electron moves. So, it becomes a two-dimensional and three-dimensional problem.

First, let us take a brief look at the behavior one should expect as one changes the device dimension from $L \gg \langle \lambda_{scatt} \rangle$ to $L \ll \langle \lambda_{scatt} \rangle$. Figure 2.1 shows transistors of different lengths. (a) is what we call a long-channel transistor, and (b) is a short-channel transistor where the high field region exists at the drain end; with its reduced number of scattering events, this high field region becomes considerably important to the short device's behavior. (c) is a near-ballistic transistor where scattering events are sufficiently small in number along the whole device, and (d) is a ballistic transistor where no scattering occurs during the transit from source to drain.

In the classical long-channel transistor—a bulk nMOS transistor as shown in Figure 2.1(a)—the dimensions are large. Our theory of

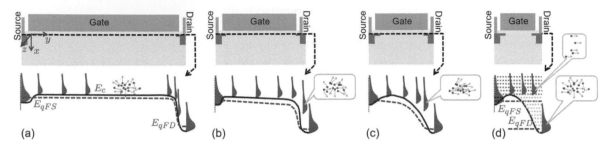

Figure 2.1: Schematic illustration of charge and charge control behavior in a transistor in its on state as dimensions decrease. (a) shows the long-channel transistor operating in a low, y-directed electric field and charge transport dominated by the mobility of a drifted carrier distribution. (b) shows changes as device length is shortened, lateral fields increase as one approaches the higher fields at the drain end. The carrier distribution function is shifted and distorted, with overshoot in velocity because fewer randomizing scattering events occur during the short transit time across this region and the electron energy gain and loss cannot equilibrate. (c) further shortening of the gate length, leading to a significant reduction in the number of scattering processes across the whole device and thus placing more importance on what happens at the source end and the velocity overshoot. (d) shows the limiting case of no scattering. Carriers now travel at velocities appropriate to their state, with the quantum conductance of each of these channels of conduction being constant.

the description of this large device is through a linear velocity-field relationship enshrined in the mobility parameter. The electron charge cloud here travels the distances hugging the bandedge in energy with a distribution that is defined through the states available and their occupation probabilities. We employ gradual channel approximation to connect the charge to the electrostatic potential ψ or the electric field $\mathcal{E} = -\nabla\psi$, that is, that the charge density is defined by the gate field alone, but the mobile portion of this charge moves with minimum perturbation arising in the lateral field that exists because of the drain-source bias that has been applied.

The electron charge cloud undergoes inelastic scattering that is also randomizing. The charge cloud is a shifted distribution in momentum in the reciprocal space whose averaged velocity is related by the mobility—a constant—to the electric field. The distribution in the figure is shown as a function of energy, with the Boltzmann tail for a three-dimensional density of states distribution. If the mobility were very high, such as in compound semiconductors, then this velocity becomes a constant, the saturated velocity (v_{sat}). Current is just flux of charge. What we need to know is the charge and the velocity. This long-channel device largely follows this description when a bias is applied between source and drain, and the carriers are induced through the gate field. Now, we employ the gradual channel approximation, which decouples the dependences of the y-directed, slowly varying component along the channel from the x-directed—transverse to the conducting plane—component. The

In the Maxwell displacement relationship, which is also Gauss's law, or its Poisson relationship embodiment, that is, $\nabla \cdot \mathbf{D} = \rho$ or $\nabla \cdot (-\epsilon\nabla\psi) = \rho$, where $\mathbf{D}$ is the displacement, ρ is the charge density, ϵ is the permittivity and ψ is the electrostatic potential, the gradual channel approximation is a statement of the importance of only one of the directions—the gate-to-channel direction—for the purposes of estimating the charge. A multi-dimensional equation is reduced to one dimension. The movement of the mobile part of this charge between the drain and the source leads to the current of interest.

charge at any position is then determined by the field directed from the gate—the stronger field direction. It is then easy for the current to be determined self-consistently—a condition of validity in this approximation—since it is the product of charge density, which is dependent on the field in one direction, and the charge's velocity, which is dependent on the field in the orthogonal direction. Through any cross-section in the xz plane bisecting the source-to-drain axis, the same current flows through the cross-section plane if only drain-to-source current exists. This is current continuity. The lateral y-directed field as a function of position along the channel is the result of this constraint under the applied bias voltage. In silicon, in long-channel devices, high fields appear only at the drain end. Current continuity means that there are fewer carriers at the this end, so both drift and diffusion currents exists, but because of current continuity and because this length is much smaller than that of the rest of the device, one can find the current at which this high field region begins to appear. The current beyond this bias is saturated due to pinch-off of the channel, that is, we approximate that an unphysically large velocity and an unphysically small carrier concentration exist in the pinched-off region to maintain the current continuity. A saturated current ensues. Our first order perturbation correction to the current at bias voltages beyond pinch-off is through the modulation of the channel length. The excess voltage beyond pinch-off is approximated to drop entirely across the high field region. The effective channel length is shortened, and current continuity is maintained, with a larger current flowing at the higher bias voltage across this shortened effective channel length. The effective long-channel region of low field then decreases.

If the material has high mobility, so that saturation in velocity at v_{sat} occurs even when low field exists in the long-channel portion of the structure, then the saturation in current is due to velocity saturation. Current continuity and saturated velocity give a nearly constant charge density across the length of the long-channel region. The difference between this behavior, where high velocities become possible close to the source, versus the previous case, where the velocities were low, is that, at the drain end, in the latter case, the channel is pinched off before the carriers can reach their saturated velocity, so the current is smaller.

As one shrinks the dimension, as shown in Figure 2.1(b), a short-channel behavior ensues. We are assuming silicon-like small mobility, where the saturated velocity behavior does not happen in long-channel conditions. Now, although the drain field region is short, it is a significant enough fraction of the entire device. The off-equilibrium consequences from the finite, small number of randomizing scatter-

Strictly speaking, gradual channel approximation states that the rate of change of field in one direction is much larger than the other direction. Since the gate-directed field, by Gauss's law, is related to charge, and Poisson's equation relates the rate of change of displacement to the charge, the gradual channel approximation is also a statement of slow variation of the charge density in the channel from the source to the drain.

ing events in this high field region cannot be ignored. The carriers overshoot in velocity, as well as the distribution function of carriers, is shifted higher in energy, that is, w.r.t. the bottom of the conduction bandedge. The high field region dimension is short, comparable to the scattering length scales, and the number of scattering events is small and finite, with the consequence that instead of carrier velocities settling at the saturated velocities expected of energy equilibrium conditions, the velocity overshoots. Current continuity forces a reconfiguration of fields and transport within the device. There is significant deviation from the long-channel behavior even in the lower field region towards the source end. The device, within the constraints of current continuity, now carries higher current than it did at long-channel-dominated dimensions, even if scaled by the length scale L. The limited scattering in the drain high field region causes the velocity to exceed the velocity-field behavior that dominates the carrier velocity with field relationships at long length scales and that one might measure in a long semiconductor sample across which that uniform field may be applied. This is the short-channel behavior where the velocity overshoots the velocity-field curves in the high field drain region. It is a region in which the drain bias has expanded to have an effect over a significant fraction of the channel length so that the current continuity and the field existent in the device cause the amount of current flowing to be higher than the $1/L$ dependence that one would get from the velocity-field relationship that one gets in constant mobility long-channel behavior.

Further shrinking of dimensions to those represented in Figure 2.1(c) leads to a near-ballistic behavior of the device, a behavior arising out of finite scattering events across the whole device, and an even weaker connection to the velocity-field relationship of long-channel length. This is the condition in which the length scale of the source-to-drain transport (L) is of an order of magnitude that is similar to that of the length scale of scattering ($\langle \lambda_{scatt} \rangle$). Electrons lose some of the kinetic energy and forward momentum that they would have gained as a result of the potential energy acquired through the drain-to-source bias, but not as much as they would have if the same field had existed across a long region, with continuous gain and loss of energy. This region is short, and the time taken to traverse it is short, comparable to the scattering times. Velocity overshoot now spreads across the whole device transport length between the ohmic-conducting regions of the source and the drain. The device now has a velocity that is higher than that due to velocity saturation limited transport across the whole device. The drain field has penetrated nearly all the way to the source, so the gate fields and the drain fields have deviated significantly from the gradual channel approximation.

Injection at the source starts to become a limiting factor in current, unlike the case in a long-channel device, where one may assume that the amount of current extracted from the source is smaller than what the source is capable of.

The limit of this scaling of dimensions is $L \ll \langle \lambda_{scatt} \rangle$ (Figure 2.1(d)), where nearly no scattering events occur during the transit of the electron from the source to the drain. In the absence of scattering that changes energy or momentum or both, an electron picks up all the potential as kinetic energy as it traverses from the source to the drain. This is the case for the ballistic transistor, where, absent scattering, the electron can be viewed as a wave traveling the short source-to-drain nanoscale distance. The dissipation here occurs during the inelastic scattering induced thermalization in the contacting drain and source regions. Now, the source injection conditions, what states are occupied, whether the electrons in them are stationary and confined or mobile, and the direction those electrons are traveling become important in determining the current. We will view this collective description through the quantized conductance, referred to earlier, and one will have to be careful in considering the statistics of occupation and the channels of conduction available. Figure 2.1(d), because it shows discretized energy states, is drawn for structures with quantum confinement in the xz plane, that is, in both the x and the z directions, with only the y direction available for movement.

This description brings out two essential features in the transit of the electron: the path that electrons take as they move from the source to the drain and which has a spatial context, and the velocity, which is related to scattering times and which has an additional, material-dependent time context. The former feature is strongly dependent on the dimensionality of this three-dimensional structure— at the least, two of the dimensions; the x-directed and the y-directed dimensions are important, even if one could assume that the z direction has invariant characteristics and so could be ignored. The former feature strongly influences the electrostatics of the structure; the latter strongly affects not only the dynamics but also the statics, through continuity.

So far, this description has focused specifically on the on state of the transistor. However, the off state of the device is just as important in the switch. A device that is only weakly off has large dissipation without performing any useful work. In the transistor, whether long-channel, short-channel, near-ballistic or ballistic, we also have to worry about the off state of the device. The physical picture is now not as dependent on the charge flow effects since, presumably, only a small current exists. The electrostatics, determined by the structure and its dopings, workfunctions and other material characteristics

When we draw the quasi-Fermi level at the source and its gradient absent and identical to the Fermi level in the source at thermal equilibrium in the band diagram, we are effectively saying that the current density from the source, $\mathbf{J} = n\mu \nabla E_{qFs}$ is much smaller in magnitude than the potential maximum current density $J = qnv_\theta$ where v_θ is the thermal velocity.

By this, we mean that the edges, at least, have no perturbation effect, and that quantum-confinement effects are absent in this z direction because it is wide enough.

are important. Figure 2.2 shows schematically the influence of application of bias and its consequences on the current. The device
achieves its off state by building a large barrier for injection of carriers from the source. When the voltage applied between the drain and
the source is increased, the drain field penetrates further toward the
source, lowering the barrier that limits the injection of electrons from
the source. Consequently, the current increases. Since this injection
is dominated by the tail of the distribution, it is exponentially related
to the gate-to-source voltage, which is tightly tied to the conduction
bandedge at the source end of the channel via the use of high gate
fields. This exponential change is usually characterized by the subthreshold swing (SS), which is the gate-to-source bias voltage needed
to change drain current by a decade in this exponential response region. However, penetration of the drain field reduces this barrier by
$\sim \Delta \varphi_B$, and hence the current increases in the exponential turning-
off region. This is drain-induced barrier lowering ($DIBL$). The other
field effect important in the off state is gate-induced drain lowering
($GIDL$). At small gate voltages, and as the drain voltage is raised,
for example, when the n-transistor of a $CMOS$ inverter is off and the
p-transistor is on, a high field exists at the drain end and tunneling
current can flow from the drain. This picture of the transistor points
to the importance of the off state, where the two dimensionality, the
electrostatics and the high fields of the state can cause significant off
state current to flow.

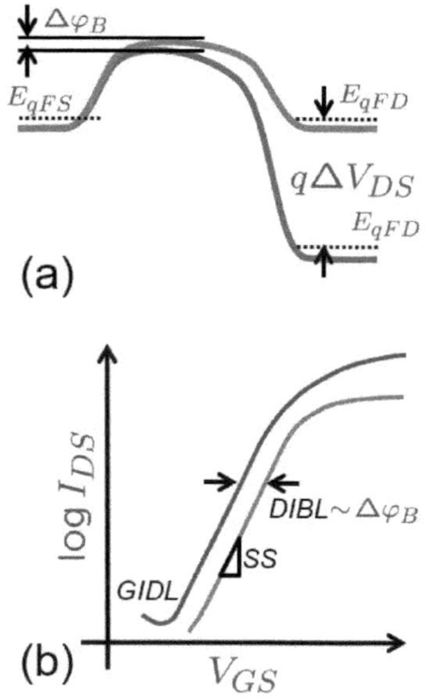

Figure 2.2: Schematic band diagram
(a) and the drain-to-source current as a
function of controlling gate voltage (b)
in a transistor.

2.2   Geometries and scaling

SHRINKING THE TRANSISTOR DIMENSIONS, we have now seen,
changes both the static and the dynamic characteristics of the device. The classical transistor's success owes much to scalability of the
transistor. The scaling, based on fields, allowed the dimensions and
the voltages to be shrunk, power per device decreased, and speed
increased, while keeping the mathematical description invariant. In
Chapter 1, we found numerous factors that limit this procedure as
one gets to the nanoscale. Variabilities, error rates, et cetera, are all
dependent on the bias voltages, and the limit to the voltages is tied
to the error rate desirable in a noisy and variable environment where
circuits must remain correct. So, we need to think of forms that maintain voltages and yet produce useful nanoscale transistors, that is,
ones where off state currents are manageable, and on state currents
are attractive. This is a constant potential scaling, where the device
appears invariant as length scales are reduced. Poisson's equation,
$\nabla^2 \psi = -\rho/\epsilon$, where ψ is the electrostatic potential in voltage units,

Again, we refer to the constant field
(Dennard) scaling section of S. Tiwari,
"Device physics: Fundamentals of
electronics and optoelectronics," Elec-
troscience 2, Oxford University Press,
ISBN 978-0-19-875984-3 (forthcoming).

and ϵ the permittivity, determines this scaling's relationships too, just as in constant field scaling. We can look at the homogeneous equation to draw the implications of fields and potentials in this scaling.

Figure 2.3(a) shows the desired potential variation that one wishes to be invariant as dimensions are scaled. In rectangular geometry, as shown in Figure 2.3(a), in

$$\nabla^2\psi = \frac{\partial^2\psi}{\partial x^2} + \frac{\partial^2\psi}{\partial y^2} + \frac{\partial^2\psi}{\partial z^2} = \frac{\partial^2\psi}{\partial x'^2} + \frac{\partial^2\psi}{\partial y'^2} + \frac{\partial^2\psi}{\partial z'^2} = 0 \quad (2.1)$$

under the transformations $x \mapsto x'$, $y \mapsto y'$, and $z \mapsto z'$, we will assume that z-direction variations, that is , the variations in device width, are invariant, no new effects appear and no field exists along the width of the device.

We desire for the scaling to leave the device potential shape invariant in all directions. For the device to behave similarly to those of larger dimensions, each of the directions where the terms are significant must scale similarly. With z as an invariant direction, the x and the y dependences must be similar; that is, under scaling λ, or with $x' = \lambda x$, and $y' = \lambda y$, the Poisson equation remains invariant. This requires, at constant potentials, the gate field dependence $\mathcal{E}_x = \partial\psi/\partial x' = (1/\lambda)\partial\psi\partial x$ to continue to scale at the same rate as the lateral field $\mathcal{E}_y = -\partial\psi/\partial y' = -(1/\lambda)\partial\psi/\partial y$ does. With the effective insulator thicknesses limited, and the fields in SiO_2 near tunneling limits, the fields cannot be scaled while keeping the potentials constant. Rectangular geometries, with an orthogonal Cartesian form as shown in Figure 2.3(b), do not lend themselves well to constant potential scaling unless the fields remain scalable. On the other hand, in cylindrical geometries, as in Figure 2.3(c), the homogeneous Poisson equation,

$$\nabla^2\psi = \frac{1}{r}\frac{\partial}{\partial r}\left(r\frac{\partial\psi}{\partial r}\right) + \frac{\partial^2\psi}{\partial y^2} = \frac{1}{r'}\frac{\partial}{\partial r'}\left(r'\frac{\partial\psi}{\partial r'}\right) + \frac{\partial^2\psi}{\partial y'^2} = 0 \quad (2.2)$$

under the transformation $r' = \lambda r$, and $y' = \lambda y$ remains invariant, even if $\partial\psi/\partial r'$ is allowed to increase in the conducting medium while the potentials are constant.

In cylindrical geometries, if the aspect ratio is retained, that is, if there is scaling of physical dimensions, the physics described by the Poisson equation remains invariant even with a constant potential

Figure 2.3: (a) shows a potential profile of a transistor, in the direction from source to drain, (b) the Cartesian geometry in an orthogonally patterned device, and (c) the polar geometry of the cylinder—rod or a wire geometry of the conducting regions of a transistor.

as the scaling constraint. When the same potential profile is applied, absent conducting carriers, the device looks the same. This is the off state condition—absent carriers. And as the carriers appear with bias, the effects of lateral fields, and others, result in various improvements in the characteristics one desires: increases in the device's speed as well as an increase in the density arising from the decrease in the device's dimensions. This sets the importance of the off state in scaling considerations of devices at constant potential.

We now explore the off state electrostatics in rectangular geometries that are relatively easy to achieve and are the common form of usage. They may not be the most suitable, given the above discussion, but with a y-directed lateral field for transport, z-directed invariance and x-directed confinement, they provide a suitable geometry to explore our long-channel, short-channel, near-ballistic, and ballistic transistor behaviors.

Of course, we have now ignored what happens within the material. If dimensions of the radial direction are in quantum-confinement limits, there will be consequences through reduction and discretization of the states that the carriers can occupy. In turn, this has implications for useful potentials' magnitudes in the conduction limits.

2.3   The off state of a nanoscale transistor

THE DOUBLE-GATE TRANSISTOR IS A SUITABLE ARCHETYPE for the exploration of scaling lengths. It is also a realistic model for transistor geometries. We will assume that the z direction—the width direction—is wide enough to be considered an invariant, that is, that no z-dependences exist.

Figure 2.4 contains a drawing of the cross-section that we wish to analyze and a potential profile of a cut through the cross-section showing the x-directed variations. We assume that the semiconductor doping is a constant, that no charge exists in the insulator or at insulator-semiconductor interfaces, that the gates, source and drain are perfectly conducting planes in contact and with no electrochemical energy change and that the x-directed fields are larger than the y-directed fields and their rates of change are higher enough to be similar to the gradual channel approximation of bulk devices. While this assumption is a good first start, we will not have to make it later; however, by shrinking the thickness of the semiconductor region where the conduction is allowed to take place, and bringing the gates together on both sides in this symmetric structure, we have created conditions similar to that of gradual channel approximation structurally. With this setting up of the problem, we should be able to explore scaling lengths and the threshold voltage that we have argued to be essential parameters for characterizing the off state.

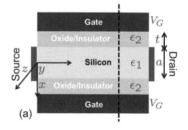

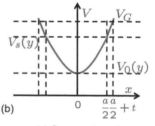

Figure 2.4: (a) Cross-section of a symmetric double-gate transistor, together with the coordinate system; ($x = 0$, $y = 0$) is the symmetry point at the source end, the semiconductor is a thick, and the insulator is t thick. (b) shows the potential profile along the cut, with the lowest potential at $x = 0$ in the symmetry plane of $V_0(y)$.

2.3.1   Scaling lengths and threshold

THE POISSON EQUATION lets us sketch the electrostatics of the device shown in Figure 2.4. For the Poisson equation

$$\nabla^2 V(x,y,z) = \frac{\partial^2 V}{\partial x^2} + \frac{\partial^2 V}{\partial y^2} + \frac{\partial^2 V}{\partial z^2} = -\frac{\rho(x,y,z)}{\epsilon}, \qquad (2.3)$$

the boundary conditions are listed in Table 2.1.

Position	Boundary condition
$x = 0$	$V = V_0(y)$
$x = \pm a/2$	$V = V_s(y)$
$x = \pm(a/2 + t)$	$V = V_G$
$y = 0, -a/2 \leq x \leq a/2$	$V = V_S = 0$
$y = L, -a/2 \leq x \leq a/2$	$V = V_D$

Table 2.1: Boundary conditions for the symmetric double-gate transistor's electrostatic potential determination.

The potential at the interface of the insulator, of permittivity ϵ_2, with the semiconductor, of permittivity ϵ_1, is $V_s(y)$, not to be confused with the source potential V_S, which is a terminal that is our reference at $y = 0$. The perturbation in the potential within the body is assumed to be of the form $k(y)x^2$, that is, the y dependence is separated from the parabolic x dependence. This follows from the assumed graduality of the channel and the assumption that the charge density is constant in the semiconductor, so that the field varies linearly in x, and the potential varies as the square of x. The function $k(y)x^2$ then represents the separation of variables form befitting a solution for the Poisson equation. Figure 2.4 shows this change at an arbitrary y, which is parabolic in the semiconductor and linear in the insulator, where there is no charge. It also shows the different potentials at important reference points. We have

$$V(x,y) = V_0(y) + k(y)x^2 \text{ for } |x| \leq \frac{a}{2}, \qquad (2.4)$$

which implies

$$\frac{\partial V}{\partial x} = 2k(y)x, \text{ and } \frac{\partial^2 V}{\partial x^2} = 2k(y). \qquad (2.5)$$

The potential at the insulator-semiconductor interface is

$$V(\pm\frac{a}{2}, y) = V_s(y) = V_0(y) + k(y)\frac{a^2}{4}. \qquad (2.6)$$

The field at this interface in the semiconductor is

$$\mathcal{E}_s = -\left.\frac{\partial V}{\partial x}\right|_{x=\mp a/2} = \pm k(y)a. \qquad (2.7)$$

The continuity of the normal component of the displacement, with no interface charge at the insulator-semiconductor interface ($\epsilon_1 \mathcal{E}_s = \epsilon_2 \mathcal{E}_{ins}$), then leads to

$$\epsilon_1 k(y) a = \epsilon_2 \frac{V_G - V_s(y)}{t}, \quad \therefore \quad k(y) = \frac{\epsilon_2}{\epsilon_1} \frac{1}{at} [V_G - V_s(y)]. \qquad (2.8)$$

Using this parabolic-in-semiconductor and linear-in-insulator relationship, we can now write the insulator/semiconductor interface potential as

$$V_s(y) = V_0(y) + \frac{\epsilon_2}{\epsilon_1} \frac{a}{4t} [V_G - V_s(y)], \qquad (2.9)$$

which leads to, in compact notations,

$$V_s(y) = \frac{V_0(y) + \beta V_G}{1 + \beta}, \qquad (2.10)$$

where $\beta = \epsilon_2 a / 4 \epsilon_1 t$, and

$$k(y) = \frac{\epsilon_2}{\epsilon_1} \frac{1}{at} [V_G - V_s(y)] = \frac{4\beta}{a^2(1 + \beta)} [V_G - V_0(y)]. \qquad (2.11)$$

We can now look at the scale length relationships of potentials in the semiconductor by normalizing parameters and using them in the homogeneous part of the Poisson relationship:

$$
\begin{aligned}
V(x, y) &= V_0(y) + k(y) x^2 \\
&= V_0(y) + \frac{4\beta}{a^2(1 + \beta)} \\
&\quad \times [V_G - V_0(y)] x^2 \quad \text{for } |x| \leq \frac{a}{2}, \qquad (2.12)
\end{aligned}
$$

which, upon substitution in the homogeneous part, gives

$$\frac{\partial^2 V}{\partial y^2} + \frac{\partial^2 V}{\partial x^2} = \frac{\partial^2 V_0}{\partial y^2} + \frac{8\beta}{a^2(1 + \beta)} [V_G - V_0(y)] = 0. \qquad (2.13)$$

This equation can be recast in the form

$$\frac{\partial^2 U_0}{\partial y^2} - \frac{8\beta}{a^2(1 + \beta)} U_0 = 0, \quad \text{where } U_0(y) = V_0(y) - V_G, \qquad (2.14)$$

that is, we have found that the potential of the bottom of the parabolic well in the semiconductor referenced to the constant gate potential, $U_0(y) = V_0(y) - V_G$ varies along the source-to-drain axis as

$$\frac{\partial^2 U_0}{\partial y^2} = \frac{U_0}{\lambda^2}, \qquad (2.15)$$

where

$$\lambda = \sqrt{\frac{a^2(1 + \beta)}{8\beta}} = \frac{a}{2\sqrt{2}} \sqrt{1 + \frac{1}{\beta}} = \frac{a}{2\sqrt{2}} \sqrt{1 + \frac{4\epsilon_1 t}{\epsilon_2 a}}; \qquad (2.16)$$

λ is our scaling length of transistor in this off state approximation. It characterizes how the bottom of the potential shifts or how the potential applied at the drain penetrates into the semiconductor region. As Equation 2.15 is second order differential equation, its solution is composed of exponentials, in this case, an exponential that causes the potential to shift down by V_D, the drain potential applied. It has to be a decaying exponential. If the length scale is small, the drain penetration is small, and since the field has λ in the denominator, the field at the drain end is large, and the potential change is sharp. For the scaling length to be small, both the thickness of the insulator and the thickness of the semiconductor must be small. In the hypothetical limit of $t \to 0$, but with no leakage conduction, $\lambda \to a/2\sqrt{2}$. It is also decreased by a higher permittivity ratio for the insulator to the semiconductor. A higher insulator permittivity couples the gate more strongly. The other interesting attribute of this second term within the square root of Equation 2.16 is the following: when one writes the full form of the perturbation term increasing the scaling length from $a/2\sqrt{2}$ scaling, it takes on a physical meaning. Increasing the thickness of the semiconductor increases the two-dimensional effects, emphasizes drain field penetration and also causes reduced gate coupling. Thus, semiconductor thickness and gate coupling are interlinked. So, this analysis shows both the importance of decreasing the thicknesses and the importance of gate coupling. For devices of the order of 2λ or more in dimension, the short-channel effect will be small, and it will be in the subthreshold swing (SS) that one may observe the clear signs of degradation. At 2λ in from the drain end, the drain potential effect has been reduced by about e^{-2} or about 13 %, so a $DIBL$ and an SS degradation in accord with this magnitude of drain potential effect are also present.

In this analysis, we employed parabolic dependence. This is an approximation embedded in $k(y)x^2$ and represents the gradual change in the potential from the source to the drain. Our argument for its legitimacy is that a small semiconductor thickness is employed and it confines due to the proximity of the constant potential gate in a way similar to what gradual channel approximation does in the long-channel classical devices. We now work towards a more accurate description. For this, first we will seek a general approach to solving a boundary value problem for a second order differential equation—the Laplace equation. A bias voltage V_G at $x = \pm a/2$, V_S at $y = 0$, and V_D at $y = b$ form the boundary conditions on the region of permittivity ϵ_1 in Figure 2.5. For now, we will ignore the insulator, as including it makes the solution mathematically cumbersome and distracts from achieving physical insight. Since the Laplace equation is a second order differential equation, its solution is composed of

Pierre-Simon Laplace was the first to study the properties of the second order homogeneous differential equation. Poisson's equation $\nabla^2\psi = f$ is the more general form of the Laplace equation; in it, ψ and f are real and complex-valued functions on any manifold, not just Euclidean ones.

exponentials, that is, one may write it from the trigonometric and hyperbolic functions that are also expressible in exponentials. There are $^4C_2$ possibilities, but we can limit these to $\cos kx \sinh ky$ and $\cos kx \cosh ky$ since, in these, the potential in the y direction can be formed as exponentially tailing, and one can maintain symmetry w.r.t. the $x = 0$ plane. So, the series solution is of the form

$$V(x,y) = V_G + \sum_{i=1}^{\infty} \cos k_i x (a_i \sinh k_i y + b_i \cosh k_i y). \qquad (2.17)$$

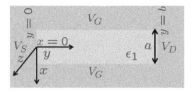

Figure 2.5: Schematic diagram, with boundary conditions, for which we desire to find a series expansion solution for the Laplace equation in the material with permittivity ϵ_1. The boundary conditions for the cross-section are the applied voltages at the gate and the drain. For now, we have ignored the existence of gate insulators.

The boundary conditions necessitate, from symmetry and from the potentials of the source and the drain,

$$k_i = \frac{2(n-1)\pi}{a},$$

$$V_s = V(x,0) = V_G + \sum_{i=1}^{\infty} b_i \cos k_i x, \text{ and}$$

$$V_D = V(x,b) = V_G + \sum_{i=1}^{\infty} \cos k_i x$$
$$\times (a_i \sinh k_i b + b_i \cosh k_i b). \qquad (2.18)$$

In this series solution, the exponential hyperbolic terms dominate, so the potential contribution in the equation decays as $\exp(-k_i y) = \exp(-y/\lambda)$ for each term. If the first term is dominant, as is common, this equation yields a single scaling length. Designed rationally, the first terms in the series should be dominant and important; and if they prevail, the length scale is $\lambda = 1/k_1 = a/2\pi$. So, absent the insulator, a drain penetration length of this magnitude dominates. We had found the length scale in Equation 2.16 to be

$$\lambda = \frac{a}{2\sqrt{2}}\sqrt{1 + \frac{1}{\beta}} = \frac{a}{2\sqrt{2}}\sqrt{1 + \frac{4\epsilon_1 t}{\epsilon_2 a}} \qquad (2.19)$$

for the parabolic approximation with an insulator of thickness t.

Instead of a factor of 2π, we found a factor of $2\sqrt{2}$, which is the smaller of the two denominators and so produces a larger length; although these factors are within a fraction of each other, we thus find that our parabolic approximation produced a larger estimate that was better. If one includes a higher order term, these two approaches—of gradual channel approximation and of two dimensionality via series expansion—will be closer. The higher terms do make a difference. But, just considering the first term, if one wants to attenuate the drain potential penetration by a factor α at any position close to the drain, one needs a length $b \approx (a/2\pi)\ln \alpha$. A factor of 10 suppression requires a device length of $\sim 0.74a$, so about the width of the semiconductor layer in this zero-insulator-thickness approximation.

We now reintroduce the insulator, as shown in Figure 2.6. Here, the potentials in the semiconductor and the insulator can be written as

$$
\begin{aligned}
V_1(x,y) &= V_G + \sum_{i=1}^{\infty} c_{1i} \cos k_i x \\
&\quad \times (\sinh k_i y + b_{1i} \cosh k_i y), \quad \text{and} \\
V_2(x,y) &= V_G + \sum_{i=1}^{\infty} c_{2i} \sin k_i (x - h - t) \\
&\quad \times (\sinh k_i y + b_{2i} \cosh k_i y). \quad (2.20)
\end{aligned}
$$

The reference points can be understood from the figure.

The normal component of the displacement at the interface is continuous, while the potential at the interface must have the same lateral dependence so that y-directed field remains continuous. This forces $b_{1i} = b_{2i}$. The constraints are

$$
V_1(\pm\tfrac{a}{2}, y) = V_2(\pm\tfrac{a}{2}, y), \quad \text{and}
$$

$$
\epsilon_1 \left. \frac{V_1(x,y)}{\partial x} \right|_{x-\pm a/2} = \epsilon_2 \left. \frac{V_2(x,y)}{\partial x} \right|_{x-\pm a/2}. \quad (2.21)
$$

So, the coefficients are now related as

$$
c_{1i} \cos k_i \tfrac{a}{2} = -c_{2i} \sin k_i t, \quad \text{and}
$$

$$
\epsilon_1 c_{1i} k_i \sin k_i \tfrac{a}{2} = -\epsilon_2 c_{2i} k_i \cos k_i t, \quad (2.22)
$$

which may be written as finding the parameter $k_i = \pi/\lambda_i$ through the solution of the transcendental equation

$$
\frac{\epsilon_1}{\epsilon_2} \tan k_i t \tan k_i \frac{a}{2} = 1. \quad (2.23)
$$

By plotting the tangent and the cotangent dependences with the permittivities, we can find the intersection and hence the parameter $\lambda_i = \pi/k_i$. Figure 2.7 shows these intersections for a set of parameters, and one can see from this figure the change in the first term's scaling length parameter.

The length scales for each term are not regularly placed, dependent as they are on the thickness of the insulator and the semiconductor. The coefficients that define the relative amplitude of the exponentially scaled effects will need to be found using the relationships in Equation 2.22, but the first term, because it is the longest in length has special importance. However, as opposed to the zero-insulator-thickness case, the intersection point now shifts to make the length scale slightly larger and move towards the parabolic solution. The effect of the insulator thickness can be viewed in Figure 2.8 as the

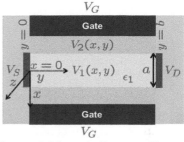

Figure 2.6: Schematic diagram, with boundary conditions and the insulator, approximating the double-gate transistor geometry for which we want an accurate solution of the potential in both the semiconductor and the insulator.

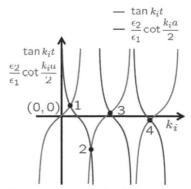

Figure 2.7: An example of a transcendental equation solution for the different terms of the series expansion for a double-gate transistor under specific parametric conditions. The plot shows the intersection points of $\tan k_i t = (\epsilon_2/\epsilon_1) \cot k_i a/2$.

ratio of the insulator thickness to the semiconductor thickness is varied. The abscissa here is normalized to the physical thickness of the region in question, and the ordinate is the ratio of the thicknesses of these regions. Note that the length scale of the first term dominates; by the 10th term, the length scale decreases to below a half, at its worst. The figure also shows that, at large insulator thickness, the scaling length goes close to twice the thickness, and, at the smallest thicknesses, it approaches the thickness of the semiconductor region. Both of these results are physically intuitive. If the semiconductor is thin, the potential changes occur in the insulator and are naturally two dimensional, with the insulator thickness as the dominating parameter. If the insulator is very thin, the potential change is dominantly dependent on the semiconductor thickness.

The precise scaling lengths from this Laplace solution for the double-gate geometry are shown in Figure 2.9. The contours in this figure follow the earlier two-dimensionality comments on thickness dependences. The change in the slope in the dependence at any length scale shows the increasing importance of the larger parameter: either the thickness of silicon along the abscissa or that of the insulator along the ordinate. A scale length of 50 nm, producing a device that may be good at about 100 nm lengths, can be reasonably designed in the double-gate form at 5 to 6 nm oxide thickness and 25 to 30 nm silicon thickness. At the other end, to make an effective 10 nm gate length device, so an ~ 5 nm scale length, a very thin silicon thickness, of the order of a few nms, is necessary. Such a length scale is relatively more forgiving w.r.t. the insulator thickness than the larger one is. This is important. The double-gate geometry, by restricting the silicon semiconductor region, allows the effective thickness of the insulator to be relaxed if drain-induced barrier lowering and related short-channel effects were the primary consideration. But, if the insulator is thick, of course, it will affect the on state conduction.

This approach of series expansion is general. So, it is also appropriate to a bulk transistor geometry.

Figure 2.10 shows a geometry for this analysis; this geometry is equivalent to double-gate geometry. The constant potential substrate here is subsumed and represented by the body contact. As in the analysis of the double-gate geometry, the series expansion in this situation entails finding the intersection condition in the transcendental equation of

$$\epsilon_1 \tan (k_i t) + \epsilon_2 \tan [k_i(t + a)] = 0; \qquad (2.24)$$

k_i and therefore λ_i can now be found, following the procedure used for Figure 2.7. Two parameterized results from this are quite instructive, reinforcing the physical conclusions drawn earlier. In Fig-

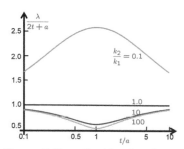

Figure 2.8: Normalized length scales as a function of the ratio of the insulator thickness to the semiconductor thickness, in the series solution of the Laplace equation for various ratios of k_i.

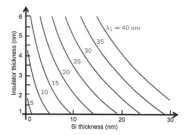

Figure 2.9: For an SiO_2 insulator, that is, where $\epsilon_2/\epsilon_1 \approx 3$, the contours of the constant length scales as a function of SiO_2 thickness and the silicon semiconductor thickness. The figure is adapted from the article by D. J. Frank et al. (1998); the full reference is cited at the end of this chapter.

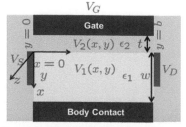

Figure 2.10: Schematic diagram, with boundary conditions approximating a bulk transistor geometry.

ure 2.3.1(a) is plotted for SiO_2 the insulator thickness relationship with depletion depth for constant length scales. This figure bears comparison with the earlier, similar figure for double-gate geometry (Figure 2.9).

Figure 2.3.1(b) draws the scale length implications for dielectric constant at constant equivalent oxide thickness and depletion width. Figure 2.3.1(a) shows a similar relationship between insulator and silicon thicknesses.

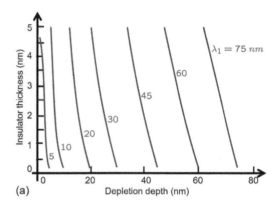

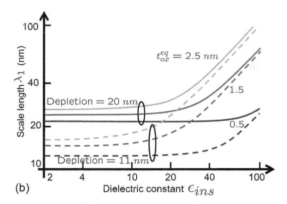

(a)

(b)

Figure 2.11: (a) shows, for an SiO_2/Si bulk transistor, the relationship between insulator thickness and depletion depth. (b) shows the scale length's change with dielectric constant for three different equivalent insulator thicknesses and two different depletion depths. These figures are adapted from the article by D. J. Frank et al. (1998); the full reference is cited at the end of this chapter.

The slopes and therefore the interplay between the insulator thickness and the depletion width change as one goes from long device lengths to short device lengths. At long scaling lengths— $\lambda_1 \approx w + (\epsilon_{Si}/\epsilon_{ins})t$—and with thicker silicon, there exists a more linear dependence on the depletion width in silicon. The effective insulator can be changed by several factors with small changes in silicon thickness while maintaining the scale length. On the other hand, at small lengths, the width has to have a relatively larger change to maintain scale length. The nature of the relationship now changes to $\lambda_1 \approx t + (\epsilon_{ins}/\epsilon_{Si})w$. When the silicon thickness is extremely small, the flexibility in varying the thickness is still maintained, at least for the purposes of determining how strongly and how far in the drain the potential penetration can be suppressed. The slope changes in the curves plotted in Figure 2.3.1(a) indicate the penalty associated with increasing oxide thickness. But, they also emphasize that if the conducting layer is extremely small, say a monolayer, the insulator can be relaxed while still allowing one to observe small transistor behavior. This point is emphasized in Figure 2.3.1(b). For any given length scale and an equivalent oxide thickness, that is, one that is electrically

Of course, we are ignoring all the transport and confinement consequences for allowed states in the conduction layer here.

equivalent to that of SiO_2, $t_{ox}^{equiv} = (\epsilon_{ox}/\epsilon_2)t$, consider the change in dielectric constant over which the scale length remains constant for the different equivalent thicknesses of oxides. If one makes the equivalent insulator thickness large, the scale lengths are higher. This is, of course, not surprising. Nor is it surprising that, to get smaller scale lengths, one needs to have the silicon thickness be shorter. But, what this figure also shows is that a flat region in scale length occurs over a range of dielectric constants. In this flat region, one can achieve suitable devices of certain lengths, but if one increases the dielectric constant beyond this, for the same equivalent insulator thickness, the scale length degrades. Physically thicker insulators with higher dielectric constants can degrade the scale length beyond a limit in the dielectric constant. What this is saying, shown as the bend in Figure 2.3.1(a), is that when the physical thickness of the insulator becomes large, at any given finite semiconductor thickness, two-dimensional effects will still come in and should be avoided. So, both the physical and electrical insulator thickness have to be below limits, even if the smaller thickness of the semiconductor does allow one to relax the insulator requirements for any length of the device.

The next consideration we turn to in the off state is that of effects arising from the small statistics in the ensemble at nanoscale. The dopants employed to establish the threshold voltage or the number of carriers in the device during its on state are now limited. The consequences for the off state of the device are acute. The variation in the small number of dopant atoms, whose device parametric effect is reflected in the threshold voltage (V_T), shows the consequences of the character of small numbers and also illuminates the complex physical considerations that the classical descriptions fail to describe. The limited number of carriers in the transport dynamics of the device will mean a change in the behavior of current noise. The electron's wave nature, with a scale length similar to the size of the device, would also result in numerous other consequences that are not consequential at the classical scale.

2.3.2 Dopant effects

CONSIDER A HYPOTHETICAL EXAMPLE: a bulk transistor of length 30 nm, width 30 nm, and a scale length 15 nm, and, to make this transistor work reasonably as a device, we choose a depletion of 12 nm consistent with the flat part of Figure 2.3.1(b). To get about a 0.8 eV potential barrier on the back in the on state of the device and with conduction close to the insulator-silicon interface, classically, a constant doping of $\sim 10^{18}$ cm^{-3} is required. In a volume of 30 nm ×

An insulator made with higher permittivity then allows a physically thicker insulator while maintaining an electrical thickness equivalent to that of an oxide. This has its advantages. Tunneling depends exponentially on a power of the physical thickness—and therefore charge control in the channel can be maintained without a corresponding increase in gate tunneling currents. We discuss the implications of such materials in S. Tiwari, ``Semiconductor physics: Principles, theory and nanoscale,'' Electroscience 3, Oxford University Press, ISBN 978-0-19-875986-7 (forthcoming) and S. Tiwari, ``Device physics: Fundamentals of electronics and optoelectronics,'' Electroscience 2, Oxford University Press, ISBN 978-0-19-875984-3 (forthcoming).

Inconsiderate increase in permittivity is unproductive.

The drift-diffusion equation or the hydrodynamic equations are arrived at as moments of the Boltzmann transport equation, which is itself a semi-classical approximation of the quantum Liouville equation. When device dimensions are of similar magnitude as the electron wavelength, phase breaking scattering, scattering as an independent event and other, similar transport-central issues all need to be treated with care. Drift diffusion is an incredible success. That electrons as particles, devoid of any many-body effects, follow classical mechanics laws independent of what the semiconductor and its characteristics are, with only the mobility and diffusivity capturing all the transport physics, is a success truly to be marveled at. But, it is a success of appearance, a success that is cosmetic.

$30\ nm \times 12\ nm$, with an average of $10^{18}\ cm^{-3}$ acceptors, there is a mean of only 11 acceptors. In normal distributions, an ensemble of N has a half width of $N^{1/2}$, so the normalized relative width varies as $N^{-1/2}$. The larger the ensemble, the sharper the peak, and the more likely any sampling would be close to this mean. Eleven acceptors correspond to a half width of nearly 50 %. With the device length, the inter-dopant spacing, the Coulombic screening lengths, and the phase coherence lengths of the electrons, effects will manifest in a variety of ways due to this statistical discreteness and related interaction. One important consequence is the effect on threshold voltage. It is a suitable starting point.

The first questions that the finiteness of the number of dopants raises are, how does the dopant impurity behave in the semiconductor and how do our models incorporate this behavior. From this, one may assess the impact of discreteness.

A shallow hydrogenic dopant has an influence within a length scale of the effective Bohr radius in the semiconductor. A representative schematic of this description of the semiconductor with the dopant impurities and the electrons is shown in Figure 2.12. The Coulomb potential, whether due to the shallow dopant atom, as in this figure or due to an electron, has short-range and long-range components that are quite different from each other.

The long-range part, with its spread, is consequential in many-body effects; it is also the reason for drawing the bands and band diagrams with shallow donor energy, since the Bohr radius is large. It allows us to draw the macroscopic picture where there is a continuum of atomic dopants and electronic charge, and the macroscopic current continuity equations are relevant. This macroscopic picture, with its long-range basis, and use of averages as if they were continuous everywhere, allows us to use Poisson's equation, the band picture, and the current equations to describe the classical macroscopic transistor. Figure 2.13(a) for the transistor shows the different doping regions and the constant potential surfaces under this approximation. This is the jellium picture.

But, there are only a discrete number of ionized acceptors in this device in the depletion region at the nanoscale. If it has only five acceptor ions, as in Figure 2.13(b), which is drawn in a cross-section as though these acceptors are across the length and width of the device within the depletion in the substrate underneath the gate and contributing by providing the confinement for carriers towards the insulator-semiconductor interface, how accurate is this description? First, the dopants really cannot be represented as being arranged continuously. Their positioning affects both their short- and long-range effects, so Figure 2.13(a), which would suffice if the device

In depletion approximation, which is only marginally acceptable here because the depth is so small and doping so high that presence of mobile carriers cannot be entirely ignored, an electrostatic potential change of $\phi = qN_Aw^2/2\epsilon_{Si}$, for a $\phi = 0.8\ V$ gives $N_A \approx 10^{18}\ cm^{-3}$. The Debye screening length at this doping, $\lambda_D^2 = \epsilon_{Si}k_BT/q^2N_A$, is $\sim 4\ nm$. This is 1/3rd of the "depletion depth," which is clearly not depleted, and is the same order of magnitude as the inter-dopant spacing ($N_A^{-1/3} = 10\ nm$ for $N_A = 10^{18}\ cm^{-3}$).

Using the free space, or per the Bohr model of hydrogen, the Bohr radius $a_B = 4\pi\epsilon_0\hbar^2/m_0q^2 = 0.0529\ nm$, one imagines that the electron in the crystal environment samples the interactions that are captured in the permittivity for the polarization and effective mass due to the forces exerted by the atoms of the periodic crystal. This leads to a Bohr radius of $a = (\epsilon/\epsilon_0)(m_0/m^*)a_B$ around the shallow dopant due to electrostatic attraction. For silicon, $m^* \approx 0.26m_0$, and $\epsilon = 11.9\epsilon_0$, so $a_B^* = 2.42\ nm$—about 5 unit cells.

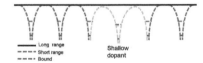

Figure 2.12: A schematic view of the electric potential in a semiconductor in the presence of ionized shallow dopants. While the potential itself changes rapidly at the atomic scale length, we employ spatially smoothened bandedges as well as spatially smoothened shallow states.

were 100s of *nm*s long and wide—a macroscopic description—is not as good anymore.

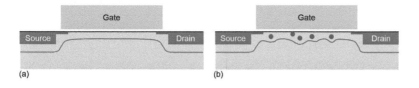

(a)   (b)

Figure 2.13: (a) show a diagram of a bulk nanotransistor, with the traditional way of drawing doping and constant potential edges for depletion, that is, as if the dopants have a uniform distribution and a spatial effect that removes short-range fluctuations so that the band structure, ionized dopants and carriers and their energies behave as if only long-range effects remain. (b) shows a modification of this diagram, using atomistic dopants. The length scale of the distances between the dopants is similar to that of the device as well as that of the Bohr radius. The constant potential surface representing the depletion region edge is drawn only for pictorial purposes. The dots represent the net number of dopants underneath the entire gate region within the depletion region in the substrate.

The microscopic potential variations are important. But, if one utilizes the full Coulombic potential, as represented in Figure 2.12, one is also stating that majority carriers will be localized by the attractive potential of each dopant, and therefore this dopant potential will be screened independent of bias voltages. Such carriers are largely absent in this region of interest. Dopants are barely screened by carriers in depletion conditions. So, this majority carrier screening argument too is incorrect in depletion conditions. Resolving the implications of these long-range and short-range interactions and of the macroscopic and microscopic views and determining how to rationally employ them physically represent a difficult challenge. Our treatment of these problems is through approximations suitable for the question being addressed.

For evaluating the effect of the discreteness of dopants establishing the threshold voltage of a device, the problem is that of introducing microscopic nonuniformity inside a microscopically small device. Although, at the microscopic level, the number of discreet dopants within the device is small, at a macroscopic level, there is an average number of ions—the mean doping density—and if the ions are randomly distributed, their distribution follows a Poisson distribution

We determine the threshold voltage distribution in this Poisson limit, neglecting mobile charges, and we also assume that only the potential dependence transverse to the conducting channels is important. The lateral dependences will be ignored. The threshold voltage then, using the bulk potential definition, is

$$V_T = V_{FB} + 2\psi_B + \frac{Q_{dep}}{C_{ox}}. \qquad (2.26)$$

Here, V_{FB} is the flatband voltage, ψ_B the bulk potential, and Q_{dep} the areal space charge density in the depletion region. The areal charge density arises from the discrete dopants, whose doping concentration is

$$N_A(\mathbf{r}) = \sum_{i=1}^{N} \delta(\mathbf{r} - \mathbf{R}_i), \qquad (2.27)$$

The Poisson distribution describes the probability of k discrete events, where $k = 0, 1, 2, \ldots$, occurring in a fixed interval, with a known average rate $\lambda > 0$ and independent of prior, as

$$f(k, \lambda) = \frac{\lambda^k}{k!} \exp(-\lambda). \qquad (2.25)$$

This is the law of small numbers. The Gaussian distribution is the law of large numbers. Both follow from the binomial distribution. Different distribution functions—binomial, Poisson, Gaussian and others—are discussed in Appendix E.

If a dopant is near the source or the drain, it will have a different potential perturbation than when it is in the middle. If a device is nanoscale, the lateral and transverse lengths are similar, as we have discussed. So, we are assuming that the fields in the gate-substrate direction comprise the dominant component. Any effect related to channel shortness will be ignored.

where N_A is the dopant density (acceptors), and $\mathbf{R}_i$ is the position of the ith dopant.

Figure 2.14 is a representation of this discrete dopant dominated threshold description of the device which has a depletion with W_d that is positionally varying. We assume that the width direction is wide, that short-channel effects are absent and that threshold voltage is a parameter meant to measure onset of conduction but is one that will have an ill-defined condition. Theoretically, Equation 2.26 is the long-channel classical definition of the threshold voltage. However, this definition has its own issues. One of the exercises at the end of the chapter pursues this incompleteness.

The x-positional dependence of the acceptors is

$$N_A(x) = \int \frac{1}{L} \, dy \int \frac{1}{W} N_A(\mathbf{r}) \, dz = \frac{1}{LW} \sum_{i=1}^{N} \delta(x - X_i), \qquad (2.28)$$

where X_i is the x position of the ith dopant, which is averaged out, with other direction effects, such as percolation of the mobile charge, ignored. The space charge density follows as

$$Q_{dep} = \int_0^{W_d} N_A(x) \, dx = q \int_0^{\overline{W}_d + \delta W_d} \left[\overline{N}(x) + \delta N(x) \right] dx, \qquad (2.29)$$

where $\overline{N}(x)$ is the macroscopic mean doping, $\delta N(x)$ is the microscopic fluctuation, and the depletion depth $W_d = \overline{W}_d + \delta W_d$ results from the average $\overline{N}(x)$ and its fluctuation $\delta N(x)$:

$$
\begin{aligned}
2\psi_B &= \frac{q}{c} \int_0^{\overline{W}_d + \delta W_d} x \left[\overline{N}(x) + \delta N(x) \right] dx \\
&= \frac{q\overline{W}_d^2}{\epsilon} \left[\frac{1}{2} N_{eff} + \frac{\overline{N}(\overline{W}_d) + \delta N(\overline{W}_d)}{\overline{W}_d} \delta W \right. \\
&\quad \left. + \int_0^1 t \delta N(\overline{W}_d) \, dt \right],
\end{aligned}
\qquad (2.30)
$$

where the effective doping $N_{eff} = (2/\overline{W}_d^2) \int_0^{\overline{W}_d} x \overline{N}(x) \, dx$ is the first moment average. The implication of Equation 2.30 is that the closer the impurity is to the insulator-semiconductor interface, the stronger its effect is through the field coupling to the gate and therefore the potential. Equation 2.30 has validity up to an order represented in the perturbations included. This is $(\delta W_d/\overline{W}_d) \times \delta N(\overline{W}_d)/\overline{N}(\overline{W}_d)$. The depletion depth parameters are

$$\overline{W}_d = \sqrt{\frac{2\epsilon(2\psi_B)}{q\overline{N}}} \qquad (2.31)$$

and

$$\delta W = \frac{\overline{W}_d}{\overline{N}(\overline{W}_d) + \delta N(\overline{W}_d)} \int_0^t \delta N(\overline{W}_d) \, dt. \qquad (2.32)$$

Experimentally, threshold is defined *ad hoc* in a manner suitable for the application of interest. It is quite commonly defined as being the gate-to-source voltage necessary for a specific current or current density. There also exist text book definitions based on extrapolations of conduction characteristics. None of these definitions is really satisfying.

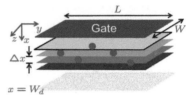

Figure 2.14: Schematic cross-section showing the geometry and approach for determining threshold voltage at a doping of N_A, with the Poisson distribution for acceptors, and a potential determined entirely by a one-dimensional x-directed treatment.

The depletion region charge and its fluctuations are

$$Q_{dep} = \overline{Q} + q \int_0^{\overline{W}} \left(1 - \frac{x}{W_d}\right) \delta N(x), \tag{2.33}$$

which has this perturbation term emphasizing the positioning of the fluctuating charge. Since the dopants are locally random, the correlation over the ensemble

$$\langle \delta N(x) \delta N(x') \rangle = \frac{\overline{N}(x)}{LW} \delta(x - x') \tag{2.34}$$

is representative of the consequence of the fluctuations embodied in the Poisson distribution. The variance of the threshold voltage can now be determined as

$$\Delta V_T^2 = \frac{q^2}{C_{ox}^2} \frac{1}{LW} \int_0^{\overline{W}_d} \left(1 - \frac{x}{W_d}\right)^2 \overline{N}(x)\,dx. \tag{2.35}$$

This relationship is within the following approximations:

- the dependences are one dimensional in the out-of-plane direction and

- the corrections in fluctuating depletion depths and space charge are incorporated as a prefactor relationship in the depletion depth of Equation 2.32.

Following Figure 2.14, we now incorporate the effect of dopants across the Poisson distribution through the slices. Let m_i be the number of dopants in the ith slice at $x = x_i$ of the thickness dx; $\overline{n} = \int LW\overline{N}(x)\,dx$ is the mean of the number of dopants in this slice, and m_i is the Poisson distribution to the mean $\overline{n}$ in our discretized states of the distribution. The threshold voltage variation in V_T due to the ith slice at $x = x_i$ is also a random variable resulting from this probability distribution. One can write it as

$$\Delta V_T = \sum_i V_i = \sum_i \frac{q}{C_{ox}} \left(1 - \frac{x}{W_d}\right) \frac{m_i - \overline{n}}{LW}, \tag{2.36}$$

and the distribution in the potential V_i, which represents the contribution in the fluctuation from each of the slices as

$$F_i(V_i) = \sum_{m_i=0}^{\infty} \delta \left[V_i - \frac{q}{C_{ox}} \left(1 - \frac{x}{W_d}\right) \frac{m_i - \overline{n}}{LW}\right] f(m_i, \overline{n}), \tag{2.37}$$

where $f(m_i, \overline{n}) = (\overline{n}^{m_i}/m_i!) \exp(-\overline{n})$ is our Poisson expression for discrete states m_i with the mean $\overline{n}$. To proceed further, we need to approximate. The characteristic function $\zeta_i(k)$ for this distribution function of threshold fluctuation is

$$\zeta_i(k) = \exp[\overline{n}(\exp(ik\gamma_i) - 1 - ik\gamma_i)], \tag{2.38}$$

See Appendix F for a discussion of fluctuations and noise. Characteristic functions are useful tools in tackling distributions and in determining moments that are essential to understanding fluctuations.

where

$$\gamma_i = \frac{q}{C_{ox}}\left(1 - \frac{x}{W_d}\right)\frac{1}{LW}. \tag{2.39}$$

The spatial dependence is embodied in these relations. The average and the variances of the threshold fluctuations from each slice V_i are, respectively,

$$\langle V_i \rangle = \left.\frac{\partial \zeta_i(k)}{\partial(ik)}\right|_{k=0} = 0, \text{ and}$$

$$\langle \Delta V_i^2 \rangle = \left.\frac{\partial^2 \zeta_i(k)}{\partial(ik)^2}\right|_{k=0} = \bar{n}\gamma_i^2$$

$$= \left(\frac{q}{C_{ox}}\right)^2 \frac{1}{LW}\left(1 - \frac{x}{W_d}\right)^2 \bar{N}(x_i)\,dx. \tag{2.40}$$

The first part of this equation is identical to the statement of Equation 2.29. No threshold shift occurs under the single dimensionality approximation adopted here. The dopant interaction if employed rationally, that is, without the artifacts of short-range and long-range Coulomb interactions, leaves the mean threshold voltage undisturbed in the Poisson distribution limit so long as the one dimensionality of the charge fields coupling to the gate and the substrate holds and lateral effects are absent.

The threshold voltage variation can be recast in a more useful form. The probability distribution for the ith slice can be determined as

$$F_i(V_i) = \frac{1}{2\pi}\int_\infty^\infty \exp[-ik(V_i + \gamma_i\bar{n})]\exp[\bar{n}(\exp(ik\gamma_i) - 1)]$$

$$\approx \exp\left(\frac{V_i}{\gamma_i}\right)\left(\bar{n} + \frac{V_i}{\gamma_i}\right)^{-1/2}\left(\frac{\bar{n}}{\bar{n} + V_i/\gamma_i}\right)^{n+V_i/\gamma_i}. \tag{2.41}$$

This local probability distribution is shifted towards the lower threshold voltage and skewed towards higher threshold voltage. The fluctuation reduces with thicker depletion depth. This is shown in Figure 2.15.

Equation 2.41 has as its first term in the expansion, using logarithms, the form

$$F_i(V_i) = F_0 \exp\left[-\frac{(V_i + V_0)^2}{2\sigma^2}\right], \tag{2.42}$$

with F_0 as a normalization constant $V_0 = \bar{n}\gamma_1/(2\bar{n} - 1)$, and $\sigma^2 = 2\bar{n}^2\gamma_1^2/(2\bar{n} - 1) \approx \bar{n}\gamma_1^2$ when $\bar{n} \gg 1$.

Equation 2.42 is a useful form, but it also has inaccuracies. It doesn't capture the first order skewing effect. It also has a shift V_0 that is unphysical so long as the one-dimensional effect prevails within this Poisson distribution of impurities in the depletion region.

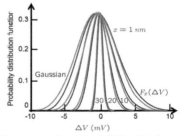

Figure 2.15: Local probability distribution of threshold fluctuation in a device with length and width of 100 nm at a few different depth positions below the interface. The slices are 1 nm thick, with an average of $\bar{n} = 10$ dopants for a substrate doping of 10^{18} cm^{-3}, and a mean depletion depth of 35 nm. From N. Sano et al., Japan Journal of Applied Physics, **41**, L552 (2002).

Note that in the limit $\bar{n} \gg 1$, $V_0 \approx \gamma_i/2$, but finite σ also implies vanishing γ_1. So, V_0 disappears, and this Gaussian probability distribution approximation does not reflect it. The Gaussian distribution will therefore give an incorrect threshold shift and not have the skew, which is seen in Equation 2.41 towards positive threshold variation.

Since the relative fluctuation in the number of dopants is dependent on the size of the ensemble, threshold voltage variation should be expected to have an inverse relationship with the gate area, that is, $\sqrt{LW}$, and a direct dependence on some parameter that relates inter-dopant distance and the depletion region over which these have an integrated effect—a direct dependence on a sub-integer power of substrate doping and depletion depth. Experimentally, these are observed.

These consequences of Poisson distribution appear in numerous places in nanoscale structures because of their reduced dimensions. The effects of doping can be seen not only from the substrate but also from the gate as well as from the drain and source regions. Since polysilicon depletion is an important phenomenon and occurs with very small dimensions—small fractions of a nm—it can have a noticeable effect even with very high doping. The doping employed in the source and the drain will give rise to lateral variations that will also unfold in device behavior. The effective insulator thickness too contributes a source of variation that appears in the threshold voltage.

Equally importantly, even the techniques employed for fabricating devices, for example, those using optical, electron or ion beams, employ limited number of particles to define dimensions, and these show up as fluctuations. The finite photon, electron or ion counts used to define dimensions show the effect of shot noise in patterns. Effects also appear from the similarity of lengths of the polymers and other molecules employed in forming the particle energy–sensitive resists. Table 2.2 shows the variance arising in defining a 10 nm dimension through photon-, electron- and ion-based techniques.

Examples of fitting relationships for the variance of the threshold include

$$\sigma = \frac{q}{C_{inv}}\sqrt{\frac{N_A W_d}{3LW}},$$

$$\sigma = \left[\frac{k_B T}{q N_D} + \frac{q t_{ox}}{\epsilon_{ox}}\right]\frac{1}{\sqrt{LW}} \times$$
$$\sqrt{\int_0^{W_d} N_A(x)\left(1 - \frac{x}{W_d}\right)dx},$$

$$\sigma = \frac{(q^3 \epsilon_{Si}\psi_B)^{1/4}}{\sqrt{2}\epsilon_{ox}}\frac{t_{ox}N_A^{1/4}}{\sqrt{LW}},$$

$$\sigma = \frac{q}{C_{ox}}\sqrt{\frac{N_A \overline{W_d}}{3LW}}\left(1 - \frac{x_s}{\overline{W_d}}\right)^{3/2},$$

and a fit to simulations of

$$\sigma = 3.2 \times 10^{-8}\frac{t_{ox}N_A^{0.4}}{\sqrt{LW}}.$$

The $1/\sqrt{L}$ dependence is reflective of the one dimensionality, so it works only if there are no lateral effects, including no lateral doping changes. Since devices are made with complex profiles meant to provide good short channel behavior, they do have lateral doping changes, such as shallow doping extensions and counter doping halos, and these provide variations that differ from the analytic treatment given here.

Lithographic technique	Wavelength	Particle energy	Resist dose sensitivity	Particle count (N) in $10 \times 10\ nm^2$	Variance ($1/\sqrt{N}$)
	(nm)	(eV)	(mJ/cm^2)		(%)
Deep ultraviolet	193	6.4	20	20000	0.71
Extreme ultraviolet	13.5	92	2	136	8.58
X-ray	1.3	920	40	272	6.06
Electron beam		50000	150	19	22.94
Ion beam		100000	50	3	57.73

Table 2.2: Dose ensembles for defining dimensions at the nanoscale for some of the common patterning techniques, and the variances in particle count when these techniques are employed for a $10 \times 10\ nm^2$ area.

When these patterns are transferred through etching and other processes, they leave behind a variance in lateral dimensions such as the length L of the transistor gate. Since for any function $f(x,y,z,\ldots)$ of independent variables $x,y,z,\ldots$, the net change Δf for small perturbations in all the variables is

$$df = \left.\frac{\partial f}{\partial x}\right|_{y,z,\ldots} dx + \left.\frac{\partial f}{\partial y}\right|_{x,z,\ldots} dy + \left.\frac{\partial f}{\partial z}\right|_{x,y,\ldots} dz + \cdots; \qquad (2.43)$$

when fluctuations arise from random independent sources, these cause the variance to be related through

$$\sigma_f^2 = \left(\frac{\partial f}{\partial x}\right)^2 \sigma_x^2 + \left(\frac{\partial f}{\partial y}\right)^2 \sigma_y^2 + \cdots. \qquad (2.44)$$

The net variance in threshold voltage thus arises from random doping, line edge roughness, gate roughness and others—all acting independently. Since the transistor off state currents, measured usually at zero gate bias voltage, are exponentially related to the threshold voltage, the leakage current depends exponentially on this gate-to-threshold voltage spread. Variance in threshold voltage arising from this multitude of causes increases this leakage current through the exponential dependence accentuating the higher current. The average off state current from threshold and thermal causes in the presence of the subthreshold swing parameter η, which characterizes the effectiveness of gate control, can then be written as

$$
\begin{aligned}
\bar{I}_{off} &= \int_{-\infty}^{\infty} \mathfrak{p}(V_T) I_{off}(V_T)\, dV_T \\
&= \int_{-\infty}^{\infty} \frac{1}{\sqrt{2\pi}\sigma} \exp\left(-\frac{(V_T - \overline{V}_T)^2}{2\sigma^2}\right) I_0 \exp\left(\frac{(V_T - \overline{V}_T)}{\eta k_B T}\right) dV_T \\
&= I_0 \exp\left[\frac{\sigma^2}{2(\eta k_B T)^2}\right].
\end{aligned}
\qquad (2.45)
$$

The variance in threshold, if it exceeds $\eta k_B T$, causes a rapidly increasing leakage. The subthreshold swing and the thermal voltage are important for limiting the excessive leakage currents arising in ensembles of devices.

A fluctuation effect, reflective of the small numbers and quite important for small transistors, is the appearance of random telegraph noise, an illustration of which is shown in Figure 2.16. A single isolated defect in this figure, near the insulator-semiconductor interface, is capable of capturing and emitting carriers. A capture process eliminates a carrier available for conduction or, equivalently, screens the region in its neighborhood, so that the total charge available for conduction reduces. The drain current is smaller. An emission process, in contrast, raises the current. Defects within tunneling lengths of mobile charge at the insulator's interface with the semiconductor cause the capture and emission of mobile charge. Random telegraph signal, such as in the drain current of a transistor, shows the discreteness because the time in captured and emitted state is significant—larger than scattering or transit times in devices. In a small device, this effect can be significant and quite noticeable.

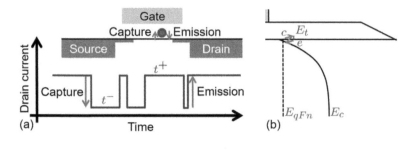

Figure 2.16: Random telegraph signal in current fluctuation in a device as a result of capturing and emission of carriers at defects at low time constants. (a) A hypothetical signal in the drain current of a transistor, arising from the capture and emission of an electron from the inversion layer, in a trap defect state in the insulator at the insulator-semiconductor interface. (b) The trapping is a thermally activated process that may also couple with tunneling and so is a process that occurs within a very short distance of the interface. The current of the transistor is a response to the local screening changes that accompany trapping and detrapping.

So, even a single defect can provide plenty of complexity. The defect can be in a multitude of charged states. It may couple to other physically proximate states. The behavior of this multitude of charge states may be influenced by the state of the physically proximate defect state. We will keep the problem simple to extract basic physical features. We will consider an isolated singly charged state, random and independent, to understand the physical character important to the nanoscale. Bias voltage and time-dependent measurements allow one to characterize locale, as well as capture cross-section and other parameters in individual devices. We referred to the importance of defects, their Poisson distribution and their consequence in random-walk memory, such as $dRAM$ or $FLASH$ memory, whose lowest energy state is not the memory state. But, single defect-mediated consequences can be observed in transistors operating in low leakage conditions, where these single events become individually significant. These single events with their current fluctuation manifestation represents a shot noise in this respect, but, being triggered by a slow capture and emission, the shot noise shows up as low frequency $1/f$ noise in small structures.

This emission or capture in time is a discrete event of low probability while the mean is finite. In a symmetric system, the system is on average in each of these states half of the time. Poisson statistics describes the event distribution of this process. The event probability is

$$\mathfrak{p}(k, \nu T) = \frac{(\nu T)^k}{k!} \exp(-\nu T), \tag{2.46}$$

where ν is the mean rate of transitions per second, T is the time interval, and νT is the net average time in the state. If τ_+ and τ_- are the average times spent in the emitted and captured states, the probability distribution of the dwell time in the emitted state (t^{+}) or the captured state (t^{-}) is

$$p(t^{\pm}) = \frac{1}{\tau^{\pm}} \exp\left(-\frac{t^{\pm}}{\tau^{\pm}}\right). \tag{2.47}$$

We will call the high current state—the "emitted" state—e, and the low current state—the "captured" state—c. The probability per unit time of transition from state e to state c through the process of capture is given by $1/\overline{\tau}_e$, and for the reverse process of emission, the probability is given by $1/\overline{\tau}_c$. The transitions are assumed to be instantaneous; $p_e(t)\,dt$ is the probability that the state does not transition from e to c during the time interval $(0,t)$ but does so in $(t, t+dt)$. This implies that

$$p_e(t) = \frac{A(t)}{\overline{\tau}_e}, \tag{2.48}$$

where $A(t)$ is the probability that the state has not made a transition during $(0,t)$ and that $1/\overline{\tau}_e$ is the probability of the transition per unit time at time t;

$$A(t+dt) = A(t)\left(1 - \frac{dt}{\overline{\tau}_e}\right), \tag{2.49}$$

which states that the product of not making a transition during $(0,t)$ and $(t, t+dt)$ is the probability of not making a transition during time interval $(0, t+dt)$. This equation can be rewritten in the simple form

$$\frac{d}{dt}A(t) = -\frac{A(t)}{\overline{\tau}_e}, \tag{2.50}$$

whose solution is

$$A(t) = \exp\left(-\frac{1}{\overline{\tau}_e}\right), \tag{2.51}$$

with $A(0) = 1$. Then,

$$p_e(t) = \frac{1}{\overline{\tau}_e} \exp\left(-\frac{t}{\overline{\tau}_e}\right), \tag{2.52}$$

with

$$\int_0^{\infty} p_e(t)\,dt = 1. \tag{2.53}$$

The equivalent expression for the captured state c is

$$p_c(t) = \frac{1}{\overline{\tau}_c} \exp\left(-\frac{t}{\overline{\tau}_c}\right). \tag{2.54}$$

When the emission and capture transitions happen at a single characteristic attempt rate, the times should be exponentially distributed. As a result, the mean time in state e is

$$\int_0^{\infty} t p_e(t)\,dt = \overline{\tau}_e, \tag{2.55}$$

and the standard deviation is

$$\left[\int_0^\infty t^2 \mathfrak{p}_e(t)\, dt - \overline{\tau}_e^2\right]^{1/2} = \overline{\tau}_e. \qquad (2.56)$$

A similar expression applies for the captured state.

The standard deviation of the mean time spent in any state is the same as the mean time spent in the state for this simple trapping model. But, trapping and energetics can be quite complicated, and, experimentally, a variety of random telegraph signals are observed, particularly since amorphous materials, such as the gate insulators, allow a variety of metastable states and opening and closing of pathways. If trapping at a defect causes local deformation, the energy is in elastic and Coulombic form. This configuration allows multiple metastable states which may be reflected in different time constants and signal intensities during different periods. Coupled charge states of defects may be active, resulting in turning on and off of different behaviors.

The power spectral density, that is, the autocorrelation of measured macroscopic parameters, allows one to look at the spectral correlations: the randomness and the net energy reflected in the phenomena behind the measured parameter. An increase in this energy, such as in the noise arising from the fluctuations, causes a higher probability of errors in digital devices, and a higher noise floor in analog and high frequency–use devices. The power spectral density is

$$S(\omega) = \int_{-\infty}^\infty \exp(-\omega\tau)\langle x(t+\tau)x(t)\rangle \, d\tau. \qquad (2.57)$$

A random process with the characteristic time τ, such as the one discussed, has a Lorentz spectrum:

$$S(\omega) = \frac{\tau}{1 + \omega^2\tau^2}. \qquad (2.58)$$

This approach can be extended, for example, when there is a distribution of characteristic times, or when the time arises from thermally activated defects:

$$S(\omega) = \int \frac{\tau}{1 + \omega^2\tau^2}\mathscr{C}(\tau)\, d\tau, \qquad (2.59)$$

for example, extends this behavior to a distribution of traps. If this distribution function $\mathscr{C}(\tau) \propto 1/\tau$ for $\tau_1 \leq \tau \leq \tau_2$, then $S(\omega) \propto \omega^{-1}$ for $\tau_2^{-1} \leq \tau \leq \tau_1^{-1}$.

This power spectral dependence reduces to a $1/f$ dependence only for specific conditions of the distribution being constant for $k_BT\ln(\tau_1/\tau_0) \leq ET\ln(\tau_2/\tau_0)$, where E is the activation energy. If the characteristic time is thermally activated, that is, $\tau = \tau_0\exp(E/k_BT)$, then the frequency dependence of the spectrum is more complex

Spectral density will be discussed in detail in the discussion of mechanics where fluctuations due to Brownian motion is coupled to the ensemble response of mechanical resonators and determine sensing limits. This is similar to the fluctuations represented in low frequency noise critical to mixing of electrical signals in nanoscale. Appendix F is a short summary of the mathematical description of noise. The spectral relationship is also discussed at length for sensitivity of force measurements in Chapter 5.

than that for a simple single defect with a constant characteristic time. What this implies is that a variety of frequency dependences may be observed in the spectrum, all increasing as the frequency is lowered, depending on the nature of the trapping processes that are active.

2.4 Conduction at the nanoscale

In preparation for a discussion of the on state of the nanotransistor in the ballistic and semi-ballistic transport limit, we discuss first, summarily, the nature of the movement of charge carriers in semiconductors: the scattering events' effect, the nature of drift and diffusion, the effect of velocities, and the effect of dimensionality to build a coherent picture between the long classical scale to the small—nano, meso and ballistic—scale. The simplest description of this conduction process is the treatment of the carrier as a classical particle that moves independently, with average properties. This is the Drude model.

The Boltzmann transport equation is a refinement of the classical Drude description, somewhat *ad hoc*, where the carriers are still particles whose ensemble distribution over energy and momentum is considered and where additional considerations—even those drawn from quantum implications such as characteristics of different scattering types or confinement and other bandstructure effects—can be folded in. One can now include each specific scattering process in detail in this particle transport. This is to include the quantum-mechanically derived details of scattering. The carriers, though, are still independent particles. The Boltzmann transport equation can describe the change of different moments of the distribution function, so one can write equations from it for the carrier density; current, that is, the product of charge and carrier flux; energy, that is, of the carrier energy flux; and higher moments. The common form of the drift-diffusion equation follows from the Boltzmann transport equation under several approximations. Finally, we will discuss what happens when the carriers do not undergo any scattering at all and when the dimensions of the device are similar in order of magnitude to the wavelength of the carrier.

In the Drude model, the simplest of descriptions, the carriers—independent particles—move under the influence of fields, undergoing scattering. A momentum scattering time τ_k characterizes the events that affect the momentum and velocity of the particles;

$$\frac{d\langle p \rangle}{dt} = q\mathcal{E} - \frac{\langle p \rangle}{\langle \tau_k \rangle} \qquad (2.60)$$

See S. Tiwari, "Semiconductor physics: Principles, theory and nanoscale," Electroscience 3, Oxford University Press, ISBN 978-0-19-875986-7 (forthcoming) for a detailed discussion of this subject.

The collective behavior of the ensemble, including screening by ensemble, which is an approach that we discussed in the statistical discussion of threshold, and tunneling, which is purely derived from quantum-mechanical and nonlocal effects, becomes increasingly harder to incorporate in our model because of the *ad hoc* aspects that arise from the classical origins of the model. The quantum Boltzmann equation, also called the Fokker-Planck equation, is more rigorous than the Boltzmann transport equation but is beyond our scope.

Here, $\langle \tau_k \rangle$ and τ_k will be used nearly synonymously to avoid symbol clutter. The specific τ_k and its dependences, even lack of randomization characteristics, et cetera—if a specific process or effect is being discussed—will be clear from the context. Otherwise, we mean τ_k to be an average over the ensemble and its characteristics, that is, $\langle \tau_k \rangle$.

describes the change in averaged momentum of the particle ensemble ($\langle p \rangle$), which has a force $q\mathcal{E}$ acting on it from an electric field $\mathcal{E}$ representing the mean of the effect of the entire environment. The subscript $\mathbf{k}$ refers to the scattering time being for momentum; $\mathbf{p} = \hbar\mathbf{k}$ in reduced momentum form for the momentum in the solid. In the steady state,

$$0 = q\mathcal{E} - \frac{\langle p \rangle}{\langle \tau_\mathbf{k} \rangle}$$
$$\therefore q\langle \tau_\mathbf{k} \rangle \mathcal{E} = \langle p \rangle = m^* \langle v \rangle, \tag{2.61}$$

where m^* is an effective mass, and we define the parameter mobility as relating the velocity to the electric field as

$$\mu \equiv \frac{\langle v \rangle}{\mathcal{E}} = \frac{q\langle \tau_\mathbf{k} \rangle}{m^*}. \tag{2.62}$$

The distance that carriers travel on average before scattering is a mean free path $\lambda_{mfp} = \langle \lambda_\mathbf{k} \rangle = \langle v \rangle / \langle \tau_\mathbf{k} \rangle$. This model, despite its enormous simplicity, is quite successful at the very classical dimensions in both semiconducting and metallic materials. Much of the model's success stems from the errors that compensate each other within its simple description .

If we assume a particle distribution function, for example, the Maxwell-Boltzmann distribution, for the classical statistical mechanics of the particles, we can derive numerous useful properties. In thermal equilibrium, the expectation value of the velocity, the thermal velocity v_θ, is

$$v_\theta = \langle v \rangle = \frac{\int vn(E)\, dE}{\int n(E)\, dE} = \sqrt{\frac{8k_B T}{\pi m^*}}, \tag{2.63}$$

which allows one to write the mobility as

$$\mu = q\frac{\langle \tau_\mathbf{k} \rangle}{m^*} = \frac{q\langle \lambda_\mathbf{k} \rangle}{v_\theta m^*} = q\langle \lambda_\mathbf{k} \rangle \sqrt{\frac{\pi}{8k_B T m^*}} \tag{2.64}$$

for three-dimensional transport in non-degenerate conditions in terms of the mean free path.

For changing dimensionality of systems, with the change in the distribution, the expectation values of velocity, such as v_θ, etc., will change.

If the material is degenerate, one must use degenerate statistics, that is, the Fermi-Dirac distribution. Energy states that are a few $k_B T$ below the Fermi energy level E_F are largely occupied, with the occupation probability being 1/2 at the Fermi energy; at a few $k_B T$ above, the states are mostly empty. So, only electrons near the Fermi energy contribute to the momentum and thus influence transport-related processes. In these situations, the Fermi velocity v_F, which is

The Drude model goes back to the year 1900 during the period when the electron had just been discovered, Boltzmann's statistical picture was being incorporated and the various short-comings of the classical description were coming to the fore. In addition, 1900 is the year when Max Planck hypothesized the $E = h\nu$ relationship for radiation—a starting point for quantum understanding.

In 1900, effective mass as a reflection of the forces within the crystal didn't exist in our understanding. At thermal energy—$k_B T$—momentum redirections are small angle and so are small in magnitude. Large momentum changes also exist in practice, for example, at Fermi energy, with optical phonon interactions into the high density of states available farther away in the Brillouin zone. A balancing of the errors of momentum redirection of small and large magnitude shows up as not an unreasonable estimation of the momentum relaxation time τ_k in the mobility relationship.

the velocity that carriers have at Fermi energy, is one of the relevant parameters.

The importance of dimensionality needs emphasis. Appendix G discusses this issue from an electron state perspective. Changes in the dimensionality of the system affect the available state distributions and the processes that influence the changing of the states. States, through the E-$\mathbf{k}$ relationship, also determine the velocity associated with the carrier in that state. So, the dimensionality, the scattering processes, and the properties of the state determine the velocity of an individual carrier as well as of an ensemble of carriers. Briefly, some important relationships of occupation in a parabolic central valley, that is, with $E = \hbar^2 k^2 / 2m^*$—a single isotropic and parabolic well minimum with an effective mass m^*—are

$$3D: \quad \mathscr{G}_{3D}(E) = 2\frac{1}{(2\pi)^3}\frac{4\pi k^2 dk}{dE} = \frac{1}{\pi^2\hbar^3}\sqrt{2m^{*3}E},$$

$$2D: \quad \mathscr{G}_{2D}(E) = 2\frac{1}{(2\pi)^2}\frac{2\pi k_\| dk_\|}{dE} = \frac{m^*}{\pi\hbar^2}, \quad \text{and}$$

$$1D: \quad \mathscr{G}_{1D}(E) = 2\frac{1}{2\pi}\frac{2dk_y}{dE} = \frac{1}{\pi\hbar}\sqrt{\frac{2m^*}{E}}. \tag{2.65}$$

These are the density of states for $3D$, $2D$ and $1D$ systems. If degenerate conditions exist, at absolute zero temperature, one can relate the carrier density by integrating the occupied state density to the Fermi energy E_F. With ν representing dimensionality in a general notation, since $n_{\nu D} = \int_0^{E_F} \mathscr{G}_{\nu D}\, dE$, we have

$$3D: \quad n_{3D} = \frac{1}{3\pi^2}\left(\frac{2m^* E_\Gamma}{\hbar^2}\right)^{3/2} \quad \therefore E_F = \frac{\hbar^2}{2m^*}(3\pi^2 n_{3D})^{2/3},$$

$$2D: \quad n_{2D} = \frac{m^* E_F}{\pi\hbar^2} \quad \therefore E_F = \frac{\pi\hbar^2}{m^*}n_{2D}, \quad \text{and}$$

$$1D: \quad n_{1D} = \left(\frac{8m^* E_F}{\pi^2\hbar^2}\right)^{1/2} \quad \therefore E_F = \frac{\pi^2\hbar^2}{8m^*}n_{1D}^2, \tag{2.66}$$

where n_{3D}, n_{2D} and n_{1D} are, respectively, the $3D$, $2D$ and $1D$ densities of the carriers, that is, the electrons that occupy the states. These are in normalized volume, area and length densities. These equations also connect the Fermi wavevector and the Fermi velocity associated with the Fermi energy. Under the degenerate absolute zero temperature conditions,

$$3D: \quad k_F = (3\pi^2 n_{3D})^{1/3} \quad \therefore v_F = \frac{\hbar}{m^*}(3\pi^2 n_{3D})^{1/3},$$

$$2D: \quad k_F = \sqrt{2\pi n_{2D}} \quad \therefore v_F = \frac{\hbar}{m^*}\sqrt{2\pi n_{2D}}, \quad \text{and}$$

$$1D: \quad k_F = \frac{1}{2}\pi n_{1D} \quad \therefore v_F = \frac{\hbar}{m^*}\frac{1}{2}\pi n_{1D}. \tag{2.67}$$

Of the degenerate distribution attempting to inject across a barrier, only the occupied states with energy and momentum suitable for injection to occur may contribute. In this case, if one were looking for a measure of the averaged velocity of the injecting stream, similar to the thermal velocity (v_θ), one would have to find it by averaging it over the occupied states. So, one form of this equivalent average for degenerate injection may be

$$\langle v \rangle|_\perp = \frac{\int_0^\infty \sqrt{\frac{2E}{m^*}}\mathscr{G}_{\nu D}(E)\mathcal{F}\left(\frac{E-E_F}{k_B T}\right)dE}{\int_0^\infty \mathscr{G}_{\nu D}(E)\mathcal{F}\left(\frac{E-E_F}{k_B T}\right)dE},$$

where ν is the dimensionality of freedom of movement. Different injection conditions will have different relevant velocities, different boundary conditions, et cetera. All, however, will be constrained by the maximum group velocity of the band structure and any asymmetry of orientation which this equation has avoided in its writing. For example, the lowerbound will be affected, and masses are different in different direction of movement.

What effective mass to use is not so straightforward. The relationships here are for the simplest of textbook examples—quite good for useful compound semiconductors with low effective masses—with their $3D$, $2D$ and $1D$ extensions. But when multiple valleys exist, bands are anisotropic, and additional symmetries are broken, then both density of states and conduction are subject to different constraints, and the masses differ. So, the problem gets more complex when dimensionality is reduced. The states of the six silicon ellipsoids and the electrons occupying them will behave quite differently both for occupation and for transport to forces and constraints of different directions. More on this shortly.

Note that there are many books and internet resources that get the $1D$ density of states wrong—underestimating it by a factor of 2. It is the two opposite k values that build a standing wave. There is a tendency in science and engineering for errors to propagate through cutting and pasting, and the Matthew effect. This shows up particularly acutely in the referencing of historically important papers.

This relationship is also a direct statement of the uniform spacing of wavevectors in the reciprocal space that the electrons may occupy in this constant mass isotropic approximation. The velocity of electrons at Fermi energy then is directly related to these energies or wavevectors.

Characteristic	Relationship	Dimensionality	Magnitude with conditions
Group velocity, v_g	$(1/\hbar)\nabla_k E$	3D, 2D, 1D	Si maximum: $\sim 8 \times 10^7 \; cm/s$
			GaAs maximum: $\sim 8 \times 10^7 \; cm/s$
			$Ga_{0.47}In_{0.53}As$ maximum: $\sim 8 \times 10^7 \; cm/s$
			InAs maximum: $\sim 10^8 \; cm/s$
			Graphene: $\sim 10^8 \; cm/s$
Kinetic energy, v_{KE}	$(2E/m^*)^{1/2}$	3D	Si: $8 \times 10^7 \; cm/s$, for $E = 0.5 \; eV$, and $m^* = 0.26 m_0$
Optical phonon, $\langle v \rangle_{op}$	$(2\hbar\omega_{op}/m^*)^{1/2}$	3D, 2D, 1D	Si: $2.8 \times 10^7 \; cm/s$, with $\hbar\omega_{op} \approx 60 \; meV$, and $m^* = 0.26 m_0$
			GaAs: $4.3 \times 10^7 \; cm/s$, with $\hbar\omega_{op} = 35.4 \; meV$, and $m^* = 0.067 m_0$
			$Ga_{0.47}In_{0.53}As$: $5.4 \times 10^7 \; cm/s$, with $\hbar\omega_{op} = 34 \; meV$, and $m^* = 0.041 m_0$
			InAs: $6.7 \times 10^7 \; cm/s$, with $\hbar\omega_{op} = 29.6 \; meV$, and $m^* = 0.023 m_0$
Thermal velocity, $\langle v \rangle_\theta$	$(8k_B T/\pi m^*)^{1/2}$	3D	Si: $1.96 \times 10^7 \; cm/s$, at $T = 300 \; K$, and $m^* = 0.36 m_0$
			GaAs: $4.16 \times 10^7 \; cm/s$, at $T = 300 \; K$, and $m^* = 0.067 m_0$
	$(\pi k_B T/2m^*)^{1/2}$	2D	Si: $1.54 \times 10^7 \; cm/s$, at $T = 300 \; K$, and $m^* = 0.26 m_0$
			GaAs: $3.27 \times 10^7 \; cm/s$, at $T = 300 \; K$, and $m^* = 0.067 m_0$
	$(2k_B T/\pi m^*)^{1/2}$	1D	Si: $0.98 \times 10^7 \; cm/s$, at $T = 300 \; K$, and $m^* = 0.26 m_0$
			GaAs: $2.08 \times 10^7 \; cm/s$, at $T = 300 \; K$, and $m^* = 0.067 m_0$

Table 2.3: A table of some approximate velocity magnitudes of relevance to carrier transport. Group velocity is the band structure-defined constraint for any state of the particle. Kinetic energy–defined velocity is the velocity at any given kinetic energy of a particle. Optical phonon–limited velocity is the ensemble average velocity that the particles would have if they were continuously losing all the excess energy by optical phonon emission. Thermal velocity is the ensemble root mean square average in non-degenerate conditions. For degenerate conditions, the velocity of carriers at Fermi energy, that is the Fermi velocity v_F, will matter. Dimensionality restricts the degrees of freedom of movement, and hence some of these velocities change through the constraints from states available allowing movement. Note, for example, the decrease in thermal velocity with dimensionality. More on this in a subsequent table.

It is instructive to understand how this dimensionality, the carrier population, the velocities and the energies relate for semiconductors. Table 2.3 attempts to provide a feel for this through examples and relationships. The group velocity represents the constraint that the bandstructure—the behavior of the electron in the crystal—speaks to. In semiconductors, group velocities are all of the order of $10^8 \; cm/s$ or less. The kinetic energy–limited velocity is meant to represent a velocity that the carrier would have should it reach that kinetic energy. This table indicates that, assuming that effective mass for conductivity in silicon doesn't change significantly—a major assumption—it will be of the order of magnitude of maximum group velocity. Carriers have to be this far up in the band for this maximum velocity to be possible. Well before that, with transport undergoing scattering, carriers will lose energy through phonon emission—a significant process—among many others. This places a constraint on how much kinetic energy the electron can realistically pick up. For the motion

However, one may be able to design structures, where the carriers are injected such as by tunneling into the semiconductor, at this energy. The carriers then don't have to pick up this energy through transport in a field.

of the electrons arising in their thermal energy under non-degenerate conditions with carriers occupying mostly states at the bottom of the band, since the carriers only sample the space to which they are restricted, this root mean square velocity changes with dimensionality. Fewer degrees of freedom result in less ability to move around. When going from $3D$ to $2D$ dimensionality, there are fewer higher energy states to be occupied under the Maxwell-Boltzmann distribution constraints. This argument of reduced access of space, due to dimensionality constraints, is also largely true for the other examples in this table.

Silicon, of course, does not quite fit well with this isotropic approximation, though for a number of parameters, specially in $3D$ conditions, Table 2.3 is quite appropriate. We have used $m^* = 0.26m_0$ as an approximation for conduction in silicon here. This follows from the following consideration. The conduction band of silicon consists of six [100]-oriented ellipsoids of revolution for the constant energy surface. This is $g_v = 6$ of valley degeneracy. The conduction minimum is about 15 % inside the X point in the first Brillouin zone, so all are counted in full. Germanium, by contrast, has a minimum precisely at the L point. So, the contribution is from four [111] ellipsoids of revolution of the constant energy surface and accounts for the sharing of 8 minima at zone boundaries. In silicon, $m_l^* = 0.916m_0$, and $m_t^* = 0.19m_0$. The conduction effective density of mass then, accounting for the occupation in six valleys where the motion is determined by m_l^*, m_t^* and m_t^* for the three different orientations, is

$$m_c^* = \frac{g_n - 6}{(2/m_l^*) + (2/m_t^*) + (2/m_t^*)} \approx 0.26m_0. \qquad (2.68)$$

For completeness, the density of state effective mass (m_d^*) for silicon— the isotropic single valley density of state approximation, representing the reality of six valleys—is

$$m_d^* = g_v^{2/3}(m_l^* m_t^* m_t^*)^{1/3} \approx 1.06m_0. \qquad (2.69)$$

On a silicon surface, the confinement quantization makes this picture considerably more complicated even in this simple approximation, where anisotropy changes with energy in all the bands. Table 2.4 summarizes these consequences for the masses for a silicon inversion layer—a $2D$ environment—on different surfaces. Because longitudinal mass is large, confinement along its direction forms a subband ladder that is lower, while a transverse mass confinement leads to a higher ladder. Except for the (111) orientation, two different ladders of subbands exist with different degeneracy. The $0.26m_0$ conductivity mass therefore is an approximation. In inversion layers, the position of the Fermi energy and the occupation of these subbands will matter. The effective conductivity and density of states

In describing various effective masses, under conditions of confinement, it is more appropriate to leave degeneracy of valleys as a separate parameter throughout, since these degeneracies change and thus any effective mass force-fitted to a single valley equivalent changes. With this convention, one may write $\mathscr{G}_{2D} = g_v m/\pi \hbar^2$, where m is a density of state effective mass that is not normalized for the number of equivalent valleys. So, a note of caution: when looking at effective masses, keep track of the convention that was employed.

mass will change since states that may be occupied and their freedom
of movement has changed.

Surface	(001)		(110)		(111)
Valley	Lower	Higher	Lower	Higher	Same
Degeneracy (g_v)	2	4	4	2	6
Normal mass					
m_z^*	$0.916m_0$	$0.190m_0$	$0.315m_0$	$0.190m_0$	$0.258m_0$
In-plane masses					
m_x^*	$0.190m_0$	$0.190m_0$	$0.190m_0$	$0.190m_0$	$0.190m_0$
m_y^*	$0.190m_0$	$0.916m_0$	$0.553m_0$	$0.916m_0$	$0.674m_0$
Conductivity mass[†]					
m_c^*	$0.190m_0$	$0.315m_0$	$0.283m_0$	$0.315m_0$	$0.296m_0$
Density of state mass[‡]					
m_d^*	$0.190m_0$	$0.417m_0$	$0.324m_0$	$0.417m_0$	$0.358m_0$

[†]$m_c^* = m_x m_y/(m_x + m_y)$; [‡]$m_d^* = (m_x m_y)^{1/2}$; these masses have not been normalized for valley degeneracy.

Table 2.4: Effective masses for silicon inversion layers. The surface is oriented in the z direction, with the confined direction's mass denoted as m_z^* to represent the collective effect of contributing valleys. In-plane masses likewise are then m_x^* and m_y^*. Note that these masses have not been normalized for valley degeneracies g_v.

We now look at what happens under degenerate conditions. These
are related to the E-**k** constraints arising in dimensionality, and we
have derived some of the major relationships in the isotropic approximation. Table 2.5 explores the implications of different masses and
therefore the different densities of states and the velocities that the
electron can have under the degenerate conditions that the variety
of semiconductors at our disposal potentially span. These in turn
will determine the relationships between bias voltage, which connects to accessible energies, and the transport properties, which are
determined through the material parameters. This table assumes
degenerate conditions and complete filling of states up to the Fermi
surface, that is, a temperature $T = 0\ K$.

The three-dimensional characteristics seem reasonable. The higher
mass material—which is silicon-like—becomes degenerate at a higher
carrier density than most compound semiconductors with their lower
effective mass would. Silicon inversion—the two-dimensional form—
supports a larger charge density than most compound semiconductors would. The larger effective mass approximates that of silicon
and the lower effective mass is representative of high mobility compound semiconductors. And these characteristics also translate to
the one-dimensional case. If one had a one-dimensional wire of a
silicon-like material with the masses assumed here, it will have about
$4 \times 10^6\ cm^{-1}$ carriers. Make it 10 nm long, such as the length of a

E_F	60 meV		120 meV		300 meV	
m^*	$0.26m_0$	$0.04m_0$	$0.26m_0$	$0.04m_0$	$0.26m_0$	$0.04m_0$
n_{3D} (cm^{-3})	8.8×10^{18}	5.3×10^{17}	2.5×10^{19}	1.5×10^{18}	9.8×10^{19}	5.9×10^{18}
n_{2D} (cm^{-2})	6.5×10^{12}	1.0×10^{12}	1.3×10^{13}	2.0×10^{12}	3.2×10^{13}	5.0×10^{12}
n_{1D} (cm^{-1})	4.1×10^{6}	1.6×10^{6}	5.8×10^{6}	2.3×10^{6}	9.1×10^{6}	3.6×10^{6}
k_F (cm^{-1}) for 3D, 2D and 1D	6.4×10^{6}	2.5×10^{6}	9.0×10^{6}	3.5×10^{6}	1.4×10^{7}	5.6×10^{6}
v_F (cm/s) for 3D, 2D and 1D	2.8×10^{7}	7.3×10^{7}	4.0×10^{7}	1.0×10^{8}	6.4×10^{7}	1.6×10^{8}

Table 2.5: A hypothetical look at the density of carriers and the Fermi velocity for low to high Fermi energies representing low, moderate and heavy degeneracy in two different materials, in an isotropic, single effective mass approximation. One has a heavier mass and therefore a high density of states, while the other has a low mass and therefore a low density of states but potentially higher velocities. The $0.26m_0$ effective mass is chosen to be representative of high effective mass. Silicon's density of states effective mass is $1.08m_0$ in 3D, and its effective conduction mass is $0.26m_0$. Table 2.4, however, shows that, under conditions of confinement, the different effective mass magnitudes lead to quite a different spread in the carrier density. The lower of the two masses in each column is representative of $Ga_{0.47}In_{0.53}As$.

transistor, it will have ~ 4 electrons in it on average at any instant of time if Fermi energy was 60 meV up in the band. It increases to ~ 10 300 meV up in the subband. Thus, in this wire, there are essentially two channels open for conduction with two electrons of opposite spin in each channel at the lower Fermi energy, which increases about 2-fold with a 240 meV rise in Fermi energy. If the effective mass is lower than that for silicon, the density of states is also lower, and the carrier population reduces to ~ 2. Now, there is only one channel open. Lower mass semiconductors need a significant raising of the electrochemical potential to achieve a carrier population. This degeneracy effect—a quantum capacitance consequence of charge in the layer requiring higher potential—will have consequences for nanoscale device functioning. This we will see presently. This table, of course, is an attempt at getting a feel for order of magnitude, and we should exercise reasonable skepticism in looking at it. For example, the lower mass at Fermi energy of 300 meV shows a Fermi velocity of 1.6×10^8 cm/s. Bandstructure prescribes what group velocity any state may have. Anisotropy will prohibit such a velocity in a long-lived state of the system.

The diffusion coefficient is related to the mobility, arising from the random walk of carriers undergoing the scattering events. The Einstein relationship relates the two:

$$\frac{\mathcal{D}}{\mu} = \frac{k_B T}{q} \text{ or}$$

$$= \frac{k_B T}{q} \frac{\mathscr{F}_{1/2}((E - E_F)/k_B T)}{\mathscr{F}_{-1/2}((E - E_F)/k_B T)} \tag{2.70}$$

for non-degenerate and degenerate conditions for classical isotropic random scattering, where $\mathscr{F}_\nu$ is the Fermi integral—here, of order $1/2$.

Diffusion, being a random walk problem, is invariant with the geometry dimensionality. Take, for example, the $1D$ case. The particle, absent bias, has an equal probability of traversing $\langle \lambda_{\mathbf{k}} \rangle$ either side of any position at each time step $\langle \tau_{\mathbf{k}} \rangle$. So, the first moment of the position, that is, $\langle x \rangle$, is invariant. It is the origin, say $x = 0$. After time $N \langle \tau_{\mathbf{k}} \rangle$, that is, N scattering events,

$$x^2(N) = \left(\sum_{i=1}^{N} s_i \right)^2 = \sum_{i=1}^{N} s_i \sum_{j=1}^{N} s_j = \sum_{i=1}^{N} s_i^2 + \sum_{i=1}^{N} \sum_{j=1, j \neq i}^{N} s_i s_j, \qquad (2.71)$$

where s is the step at each scattering event with $s = \pm \langle \lambda_{\mathbf{k}} \rangle$ at a probability of $1/2$ each. The second term is zero, because $\pm \langle \lambda_{\mathbf{k}}^2 \rangle$ each occur with $1/2$ probability, and therefore average to zero. And only the first term remains. So, the second moment, $\langle x^2 \rangle$, increases in proportion to the number of scattering events. It spreads out. This is true independent of the dimensionality, since this description is also true for each one of the degrees of freedom. If the dimensionality is 0, then the particle is not allowed to move, that is, $\langle \lambda_{\mathbf{k}} \rangle = 0$, which reduces the second moment to 0. The mean square distance traveled is proportional to the time traveled, characterized by a constant: the diffusion constant. So,

$$\mathcal{D} = \frac{\langle \lambda_{\mathbf{k}}^2 \rangle}{\tau_{\mathbf{k}}}. \qquad (2.72)$$

The Boltzmann transport equation is a more sophisticated description of this kinetic picture that employs the distribution function $f(\mathbf{r}, \mathbf{k}; t)$—in units normalized to real space, and reciprocal space volume—with $(\mathbf{r}, \mathbf{k}; t)$ coordinates, and following its evolution,

$$\frac{df(\mathbf{r}, \mathbf{k}; t)}{dt} = -\hbar \dot{\mathbf{k}} \cdot \nabla_{\mathbf{k}} f - \dot{\mathbf{r}} \cdot \nabla_{\mathbf{r}} f + \left. \frac{\partial f}{\partial t} \right|_{scatt}. \qquad (2.73)$$

Since the phase space is incompressible, if a relaxation time approximation is valid, that is, if scattering is randomizing characterized by a time constant τ, then

$$\frac{\partial f}{\partial t} = -\hbar \dot{\mathbf{k}} \cdot \nabla_{\mathbf{k}} f - \dot{\mathbf{r}} \cdot \nabla_{\mathbf{r}} f + \frac{f - \overline{f}}{\tau}, \qquad (2.74)$$

where $\overline{f}$ is the unperturbed distribution in thermal equilibrium. The moments of this equation provide the different conservation equations because the distribution function includes the ensemble statistics and the right hand side includes all the interactions.

The various moment equations from the distribution function, whose evolution is described by the Boltzmann transport equation, describe the evolution of parameters of interest to us in devices.

If the walk is non-random, that is, biased, then the 0th moment is non-zero, and the invariant is velocity. This is our ballistic transport case, with its absence of scattering and randomizing events.

These are all the properties of moments arising in fluctuations. See Appendix F for a discussion of moments in distributions in randomizing circumstances.

In the deterministic view, the coordinates $(\mathbf{r}, \mathbf{k}; t)$ for each particle in the collection of particles follows a deterministic trajectory in this phase space. Our system is incompressible in phase space. That flow in phase space is incompressible is the general statement of Liouville's theorem. We have reformed it as the Boltzmann transport equation using the distribution function. The quantum view, as the nondeterministic corollary, is represented in the quantum Liouville equation, which incorporates the mixed states of the ensemble. Our information mechanics discussion touched on this quantum view. The trajectory in that approach too is incompressible if algorithmic information is included.

Integrating the product of a parameter of interest with distribution function over **k**-space gives the time evolution of the parameter in real space. These are the various moments of the transport equation. For the 0th moment,

$$\varphi(\mathbf{k}) = 1 \quad \therefore \langle \varphi f \rangle = \int f(\mathbf{r}, \mathbf{k}; t)\, d^3\mathbf{k} = n(\mathbf{r}, t), \qquad (2.75)$$

the carrier concentration.

For the 1st moment,

$$\varphi(\mathbf{k}) = -q\frac{\hbar \mathbf{k}}{m^*} \quad \therefore \langle \varphi f \rangle = \int -q\frac{\hbar \mathbf{k}}{m^*} f(\mathbf{r}, \mathbf{k}; t)\, d^3\mathbf{k} = J(\mathbf{r}; t), \qquad (2.76)$$

the current density.

The 2nd moment, using $\varphi(\mathbf{k}) = \hbar^2 k^2 / 2m^*$, gives

$$\varphi(\mathbf{k}) = \frac{\hbar^2 k^2}{2m^*} \quad \therefore \langle \varphi f \rangle = \int \frac{\hbar^2 k^2}{2m^*} f(\mathbf{r}, \mathbf{k}; t)\, d^3\mathbf{k} = W(\mathbf{r}; t), \qquad (2.77)$$

the kinetic energy density.

The expressions that this leads to are as follows:

$$
\begin{aligned}
\text{0th moment:} \quad & \frac{\partial n}{\partial t} = \frac{1}{q}\nabla \cdot \mathbf{J} + \left.\frac{\partial n}{\partial t}\right|_{scatt} \quad (= \mathcal{G} - \mathcal{R}); \\[2mm]
\text{1st moment:} \quad & \frac{\partial \mathbf{J}}{\partial t} = \frac{q}{m^*}n\mathbf{F} + \frac{2q}{m^*}\nabla W + \left.\frac{\partial \mathbf{J}}{\partial t}\right|_{scatt}; \quad \text{and} \\[2mm]
\text{2nd moment:} \quad & \frac{\partial W}{\partial t} = \frac{1}{q}\mathbf{F}\cdot\mathbf{J} - \nabla_{\mathbf{r}}\cdot(\dot{\mathbf{r}}W) - \nabla_{\mathbf{r}}\cdot\mathcal{Q} \\[2mm]
& \qquad + \left.\frac{\partial W}{\partial t}\right|_{scatt},
\end{aligned}
\qquad (2.78)
$$

where $\mathcal{Q}$ is the heat energy flux density—one of the forms in which the kinetic energy is lost. The 0th moment is the continuity equation—it is the particle balance equation. The 1st moment is the drift-diffusion equation—the momentum balance equation. This is the Navier-Stokes equation for this fluid. Recall that Boltzmann transport equation is a description of classical particles with classical distribution function. The 2nd moment is the energy transport equation, so it has heat energy flow in it too.

The scattering term here refers to the interaction processes that change the distribution function of particles due to passage of time, where processes not included are the momentum response due to applied forces, such as electric and magnetic fields, and the positional response due to movement of particles. These are the other two terms of Equation 2.74. If the scattering response can be accurately described, such as when averaged into an accurate time constant, these equations can be simplified. For random processes, or when specific dependences average out so that a constant results, one can write the scattering—creation or generation of particles, changes in momentum reflected in the current, changes in energy—as an appropriate expectation value through a time constant: so, energy scattering with a time constant $\langle \tau_w \rangle$ in the 3rd moment equation, current scattering with a time constant $\langle \tau_{\mathbf{k}} \rangle$ in the 2nd moment equation, and generation and recombination through lifetimes τ_n and τ_p in the 1st moment particle equation—the carrier lifetimes. The energy relaxation time is usually larger than the momentum relaxation time since elastic scattering processes can cause momentum change through redirection without affecting the energy.

The reformed equations, under these briefly discussed constraints, reduce, for the 0th moment, to

$$\frac{\partial n}{\partial t} = \frac{1}{q} \boldsymbol{\nabla} \cdot \mathbf{J} + \frac{n - \bar{n}}{\tau_n}, \qquad (2.79)$$

which is the carrier continuity equation, and for the 1st moment equation, assuming constant temperature, non-degenerate carrier distribution and a steady state,

$$0 = \frac{1}{m^*} n\mathbf{F} - \frac{\mathbf{J}}{\langle \tau_{\mathbf{k}} \rangle}, \quad \therefore \quad \mathbf{J} = \frac{q\langle \tau_{\mathbf{k}} \rangle}{m^*} n\boldsymbol{\mathcal{E}} + \frac{q\langle \tau_{\mathbf{k}} \rangle}{m^*} k_B T \boldsymbol{\nabla}_{\mathbf{r}} n. \qquad (2.80)$$

This last equation is the momentum balance equation, which can be restated as

$$\mathbf{J} = q\mu n\boldsymbol{\mathcal{E}} + q\mathcal{D}\boldsymbol{\nabla}_{\mathbf{r}} n, \qquad (2.81)$$

the drift-diffusion equation, which we employ in classical analysis. It is the 2nd moment equation, where energies of the carriers change and are a function of position through the energy gaining and losing processes between and in scattering processes that we ignore in introductory texts assuming that carrier distribution is essentially in equilibrium with the lattice throughout even if current flows. This is to say that carriers don't carry energy across real or reciprocal space. So, we forget this third equation, as well as other higher orders of moment, when we do analysis of classical size scale. We assume there that the carrier distribution has no off-equilibrium effects.

This discussion highlights the approximations, even with the particle distribution picture, for drift diffusion. We will return to the nature and alleviation of these approximations. This equation's accuracy is tied to the many different approximations in the change in distribution function in the phase space due to externally applied forces and the interactions within, and it will become important as soon as we shrink dimensions below the long-channel limit of the transistor. The linear velocity-field relationship represented in the mobility embodies in it large randomizing scattering that assures that the temperature of the particle cloud is identical to that of the semiconductor. If the mobility is $1000 \; cm^2/V \cdot s$, an effective mass of $0.1m_0$

Generation and recombination, when described by a constant parameterized time constant, can be quite inaccurate. Generation in avalanche processes has field dependence that time constants fail at; recombination can be accurate with a lifetime time constant but may fail for direction-dependent processes that occur at high fields, Auger recombination, for example. But, if it is accurate, for example, recombination dominated by a single carrier lifetime, as for Hall-Shockley-Read recombination under certain constraints such as in quasineutral regions, then it simplifies this equation tremendously. This is of tremendous advantage in the carrier equation, leading to a simple diffusion-length-based analysis in devices with bipolarity.

In drift diffusion, the carriers, even under the applied forces—electric and magnetic—remain in equilibrium with the crystal, that is, the material in which they exist. Both the carrier and the material statistical distributions, of the charged particles and of phonons, for example, are described by the same temperature—T.

gives an estimate of momentum relaxation time of $\langle \tau_{\mathbf{k}} \rangle \approx 60 \; ps$. For carriers moving at thermal velocity, $\sim 10^7 \; cm/s$, this effective mass gives a path length of $\langle \lambda_{\mathbf{k}} \rangle \approx 5.7 \; nm$, a nanoscale device dimension, between scattering. If the mobility is $10^6 \; cm/s$, so a very long $\tau_{\mathbf{k}}$, such as in high mobility heterostructures at low temperatures, this path length is $\sim 5.7 \; \mu m$. Since scattering is limited, we should also compare the device dimensions to length scales related to the wave nature of the electron. The de Broglie wavelength for an electron moving thermally, λ_{deB}, is equal to $h/p \approx 7 \; nm$. The phase of the electron wave can also change due to interactions. For this, the phase coherence length $\lambda_\phi \approx 10 \; nm$ at low mobilities.

An electron as a wave exhibits all wave characteristics, similar to a photon, with the added twist of electron charge. So, in addition to interference, et cetera, the electron wave is manipulatable through fields. In high mobility materials, these characteristics are clearly observable. Along with creating modifications to electron states, and interactions resulting from quantum constraints by causing confinement dimensions that are of the order of magnitude of wave dimensions, one can build structures where electron beams can be focused, can be steered, and show a quantized Hall effect in response to a magnetic field.

In transistors, the implications of this manipulation of the electron wave are in a transistor-like control through the gate of conductance arising from wave motion: of channels opened and closed by gate voltage, that is, an effective mode filtering. To analyze this for the on state, we discuss the consequences of the wave nature.

Consider an electron waveguide, with the free direction for motion in y of length L, as in Figure 2.17. Positive y-directed states dN_n in the nth subband within the energy interval dE, moving with a velocity v and a distance of L of the waveguide, carry a current of

$$dI_n = ev \frac{dN_n}{L}, \qquad (2.82)$$

where dN_n/L is the effective carrier density. This velocity is defined by the E-$\mathbf{k}$ relationship of the material as

$$\mathbf{v}(n,k) = \frac{1}{\hbar} \mathbf{\nabla_k} E(n,k)$$

$$\therefore \quad v = \frac{1}{\hbar} \frac{\partial E(n,k)}{\partial k_y}. \qquad (2.83)$$

In a one-dimensional geometry, in the free direction of y, the k_y states are separated by $L/2\pi$. The degeneracy due to spin, g_s, is 2, hence the number of conducting electrons with positive k_y in the subband n with energy less than an energy E_0 is

$$N_n = g_s \frac{L}{2\pi} k(E_0) = 2 \frac{L}{2\pi} k(E_0). \qquad (2.84)$$

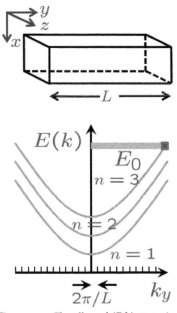

Figure 2.17: The allowed $(E,\mathbf{k})$ states in an electron waveguide, with electron confinement in the x and the z directions, and freedom of movement in the y direction.

See Appendix G for a detailed discussion of 3D, 2D and 1D E-$\mathbf{k}$ behavior, the density of states, and other attributes.

This gives the density of the total number of states as

$$\frac{dN_n}{dE} = 2\frac{L}{2\pi}\left(\frac{\partial E(n,k)}{\partial k}\right)^{-1} = 2\frac{L}{2\pi}\frac{1}{\hbar v}. \tag{2.85}$$

So, the current in Equation 2.82, from all the electrons with positive $\mathbf{k}$ in the subband n in the energy span dE at energy at E_0, is

$$dI_n = \frac{2e}{h}dE = \frac{2e^2}{h}dV. \tag{2.86}$$

The conductance per unit change in energy units of A/eV is $2e/h$, and conductance per unit change in voltage units of A/V is $2e^2/h$. Each open channel causes a conductance of $2e^2/h$ S, or $\sim 80\ \mu S$, where the factor of 2 arose from the presence of a spin-up and a spin-down electron in each of these channels. We introduce here a quantized conductance as $g_q = 2e^2/h$ per open channel, and a total conductance of $G_n = ng_q$, where n is the number of open channels.

Figure 2.18 shows this wave nature in a constricted geometry formed in a high mobility two-dimensional electron gas. Just as in a microwave waveguide, as more channels are opened, which is the equivalent of fitting additional z-confined waves, more electron modes are permitted and states occupied. The lateral constriction is modulated electrostatically by the bias voltage applied. One sees a step increase in conductance in units of g_q as the constriction w is made larger; in the limit of $w \gg \lambda_{deB}$, the de Broglie wavelength, a linear conductance relationship returns, and the normal ohmic behavior is observed. One may view this step behavior as ohmic conductance in quantized conductance limits.

What would happen if one had several constrictions in series? Instead of the resistances adding, the constriction with the smallest number of conductance channels would determine the total conductance, assuming that other wave effects were small.

The behavior of the smallest conductance prevailing is a statement that the structure acts as a mode filter. Only the smallest band of waves that passes through all the constrictions is allowed to proceed, just as with electromagnetic waves.

Let us now consider what happens if there is scattering. The wave nature allows us to consider this scattering problem by employing the well-developed techniques of electromagnetics. Figure 2.19 shows two examples. Reservoirs, at electrochemical potentials $E_{qFi}, i = 1,2,\ldots,$ are connected to the quantized channels. The scattering region is a target where carriers are transmitted or reflected. The scattering region is connected to the reservoirs for the incoming and outgoing states through perfect electron waveguides with adiabatic transitions. The electron reservoirs are connected to

The density of the total number of states is not normalized to the physical dimension of the system. Usually we will employ the density of states $\mathscr{G}_{3D}$, $\mathscr{G}_{2D}$ and $\mathscr{G}_{1D}$ as a function of energy E normalized to the physical dimension, that is, volume, area, or length. These are also expressible as a function of the wavevector k also.

An equivalent way of arriving at this conclusion is to look at the conduction that is due to the electron in a one-dimensional channel as being a function of position. Its velocity increases in proportion to $\sqrt{E}$ as the electron moves under the applied electrochemical energy change in the structure. But, the density of states is also decreasing in the one-dimensional system as $1/\sqrt{E}$. The two balance out, so conductance is constant.

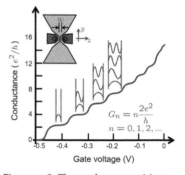

Figure 2.18: The conductance arising from the filling of the allowed $(E, \mathbf{k})$ states in an electron waveguide, with electron confinement in the x and the z directions, and freedom of movement in the y direction.

Examples of these restrictions are many. All reservoirs to channel constriction boundaries should be slowly varying, that is, adiabatic, slower than the reciprocal of wavevector which corresponds to the wavelength of the electron ($\lambda \approx 2\pi/k$). Else, quantum-mechanical reflections will occur. This is only one of many possibilities of quantum wave interferences. Any roughness could cause phase changes, or inelastic effects, and these will appear in conductance.

each of these electron waveguides. The electrochemical potential, the Fermi energy, characterizes the states filling in these reservoirs through the Fermi-Dirac function for the more general degenerate conditions. The reservoirs fill the channels of the waveguides according to the Fermi-Dirac function.

If the temperature is absolute zero ($T = 0\ K$), and $E_{qF1} \approx E_{qF2}$, that is, there is only a small change in the energy, then, in Figure 2.19(a), each of the perfect waveguides has N_1 and N_2 channels occupied with positive momentum and positive velocities (the incoming states), and similarly N_1 and N_2 channels occupied with negative momentum and negative velocities (the outgoing states). If $E_{qF1} > E_{qF2}$, then only the states between E_{qF1} and E_{qF2} contribute to current. States below E_{qF2} are all filled and contribute a net zero current. The current from each channel j is

$$I_j = \frac{2e}{h}(E_{qF1} - E_{qF2}) = \frac{2e}{h}\Delta E_{qF}. \tag{2.87}$$

This is the case at absolute zero.

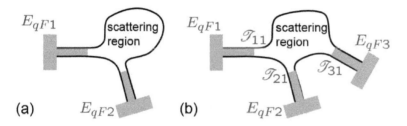

(a)

(b)

We will prefer electrical engineering notation where sensible. The Fermi energy—electrochemical potential energy (E_F)—is often the same as the chemical potential μ in may statistical mechanics texts, except that we must include the charge-related energies, which statistical mechanics was unaware of or ignored by stressing only the chemical potential at its inception. We introduce q for quasi in this notation when equilibrium is disturbed.

Figure 2.19: Two examples of treating scattering in a quantum channel systems through couplings. In (a), which shows a two terminal system, scattering causes a wave to be reflected back or be transmitted to one of the two terminals. In (b), a third terminal exists, and one needs to now consider all the different possibilities, as one does in network approaches to wave propagation. Terminals here are representative of the quantized states between which electron transitions occur through scattering.

At finite temperature, we will have broadening as a result both of the occupation of higher energy states and of the states below the quasi-Fermi energies being only partially occupied. The conductance steps will not be abrupt and will be characterized by a broadening in energy of $k_B T$, reflected from the occupation statistics. In the presence of scattering, we can again employ the transmittance and reflectance approach of electromagnetics. Current incident in Channel n of Waveguide 1, because of scattering, causes a current in Channel m of Waveguide 2, the latter current being given by

$$I_{mn} = \frac{2e}{h}\mathscr{T}_{mn}(E_{qF1} - E_{qF2}). \tag{2.88}$$

Therefore, the total current is

$$I = \frac{2e}{h}\mathscr{T}\Delta E_{qF}, \tag{2.89}$$

where $\mathscr{T}$ represents the transmission coefficient between states.

$$\mathscr{T} = \sum_{m=1,n=1}^{m=N_2,n=N_1} \mathscr{T}_{mn} \tag{2.90}$$

and the conductance is

$$G = \frac{eI}{\Delta E_{qF}} = \frac{2e^2}{h}\mathcal{T}. \qquad (2.91)$$

So, the conductance is simply the product of $2e^2/h$ and transmittance over the open channels. For the case of more than two waveguides, as in Figure 2.19(b), we can now employ scattering matrices, as in electromagnetics, with this conductance replacing the characteristic admittances. This approach ensures conservation and allows one to employ the approaches of electromagnetics.

We have now established the unit of quantum conductance g_q reflecting the charge of the electron wave, as well as seen mode filtering—an energy aperture in the transmission conductance of these electron wave systems. We will employ both of these in understanding transistors, where the gate bias voltage is a means of increasing and decreasing the quantum conductance channels and thus a means to mode filtering that leads to current modulation in the transistor. The channel region is where this modulation takes place, and the reservoirs for the source and the drain are connected to this quantum conductance waveguide region.

It is instructive to look at thermionic emission—emission from the thermal tail due to energetic species that can emit across a barrier, which is a metal-semiconductor interface barrier, for our case. This transport is where ballistic transport—in this case, in a small region across the interface—first appears. We will transplant the lessons from this case to a semiconductor-only system and then in turn to the semiconductor transistor, where quantum conductance, mode filtering and ballistic and non-ballistic behavior will all be incorporated.

In Figure 2.20, we assume a simple isotropic bandstructure for two materials that have a conduction band discontinuity of φ_B at their heterostructure interface. Conservation of energy and momentum must hold if no scattering is assumed in the transport across the interface. Carriers exist at both sides of the interface, and as they move from one side to the other, they do so with conservation. Although an abrupt barrier is shown, we assume that no scattering—quantum-mechanical reflection—takes place because of it. An electron from a "from" state, an occupied state, crosses under the rules of conservation to "to" states, which must be unoccupied. The Fermi energy is considerably below the energies of interest where the injection takes place across the barrier, the "to" states are essentially all the states, and they are all unoccupied. One may therefore only consider the source function arising in "from" states in the transmission process. The "from" states must have a y-directed momentum and sufficient energy to cross the barrier.

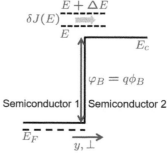

Figure 2.20: Injection across a semiconductor-semiconductor interface.

While the system of interest to us is that of the semiconductors alone, the metal-vacuum and the metal-semiconductor junctions are the systems for which thermionic theory was originally developed. The earliest success in modeling these systems as thermionic emission systems, achieved by Hans Bethe, dates back to 1935.

Often in science and engineering, one finds that units and symbols need to be understood from context, because of the need for so many and because of the historic unit evolution. For us, in bandedge changes, φ_B—the change, such as at a sharp junction—is in units of energy (e.g., eV). But, electrical engineers like to use $\phi_B = \varphi_B/q$, which is in V.

Again, we assume that $2\pi/k$, the de Broglie wavelength of the electrons, is smaller than the graded adiabatic barrier. This certainly will be true in transistors.

Figure 2.20 shows a slice in energy between E and $E + \Delta E$ of carriers where current flows from the first semiconductor into the second. Only if an electron has $k_y = k_\perp > 0$ is it available for possible transmittal across the barrier. The total energy of the carrier is $E = (\hbar^2/2m^*)(k_\perp^2 + k_\parallel^2)$, and this energy must exceed φ_B for the carrier to transmit over. The large availability of "to" states implicitly implies that states exist to match the momentum conservation requirement. This is certainly a good assumption where states are close together in reciprocal space, with energy separation much lower than $k_B T$. All we now have to do is match the energy requirements.

This constraint is shown in Figure 2.21. Only when $\hbar^2 k_\perp^2/2m^* \geq \varphi_B$ is there enough energy and forward momentum for the carrier to cross from the first to the second semiconductor. Only half of the carriers of an isotropic distribution, that is, one where carriers with $\pm k_\perp$ exist with equal probability, satisfy this condition. At the limit equality condition, $k_\parallel = 0$, all the carrier energy is associated with the perpendicular injection momentum towards the barrier and just sufficient enough so that upon crossing over into the second semiconductor, its kinetic energy and momentum vanish. At energies where $\hbar^2 k_\perp^2/2m^* < \varphi_B$, the carrier is reflected back. So, beyond this limit, any additional energy associated with motion parallel to the interface leaves the carrier unable to traverse into the second semiconductor. This is the filled block region of Figure 2.21. One must integrate the possibilities over the entire energy spectrum to calculate the total injected current from the first to the second semiconductor.

Following the descriptions in Figures 2.20 and 2.21, the current density that can arise at energy E is

$$
\begin{aligned}
\delta J &= q\frac{\hbar k_\perp}{m^*}2\frac{d^3k}{(2\pi)^3}\mathcal{F}\left(\frac{E_\parallel + E_\perp}{k_B T}\right) \\
&= \frac{2q}{h}dE_\perp\frac{d^2k_\parallel}{(2\pi)^2}\mathcal{F}\left(\frac{E_\parallel + E_\perp}{k_B T}\right),
\end{aligned}
\tag{2.92}
$$

where $\mathcal{F}$ is the Fermi-Dirac function defining the occupation of the states. Here, in the first formulation, the first two terms are the charge times the velocity of the carriers, and the remaining are the electrons available for conduction as a product of the two spin orientations, the density of states in k-space and their occupation probability described by the Fermi-Dirac function. The last relation, unconnected to our earlier quantized channel discussion, restates that for all the states within the energy range $dE_\perp$, that is, those with perpendicular momentum $k_\perp$, obtained by integrating over the parallel momenta, the conductance per unit energy is $2q/h$.

All occupied channels available for conduction contribute a conductance of $2q/h$ per unit of energy.

We will return to issues related to this assumption of unlimited "to" states—where precisely the difference between the continuity of the classical view versus the discreteness and the uncertainty of the quantum view becomes distinguishable—in Chapter 5. This argument just as well applies to the equipartition of energy and its breakdown, to Brownian motion and to other places where nanoscale and mesoscale dimensions appear at the boundaries of the applicability of these approaches.

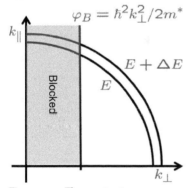

Figure 2.21: The constant energy surface, in $(k_\perp, k_\parallel)$, showing the blocked region for injection from Semiconductor 1 to Semiconductor 2.

Thermionic emission theory had this fundamental result embedded in it: that the quantum of conductance, in per energy units, associated an with electron moving ballistically is $2q/h$. This is nearly a century ago, very soon after the early rounds of quantum understanding.

We can integrate over all occupied states contributing to conduction, that is, $E_\perp$ varying from φ_B to ∞. Since $d^2k_\parallel = 2\pi k_\parallel dk_\parallel = 2\pi m^*/\hbar^2 dE_\parallel$, we obtain

$$
\begin{aligned}
J &= \frac{2q}{h} \frac{m^*}{(2\pi)\hbar^2} \int_{\varphi_B}^{\infty} dE_\perp \int_{E_\parallel} dE_\parallel \mathcal{F}\left(\frac{E_\parallel + E_\perp - E_F}{k_B T}\right) \\
&= \frac{2qm^* k_B T}{(2\pi)^2 \hbar^3} \int_{\varphi_B}^{\infty} dE_\perp \ln\left[1 + \exp\left(-\frac{E_\perp - E_F}{k_B T}\right)\right]. \quad (2.93)
\end{aligned}
$$

In the Maxwell-Boltzmann limit, this reduces to the thermionic current (J_{te}) relationship of

$$
J_{te} = \frac{2qm^* k_B^2}{(2\pi)^2 \hbar^3} T^2 \exp\left(-\frac{\varphi_B - E_F}{k_B T}\right) = \mathscr{A}^* T^2 \exp\left(-\frac{\varphi_B - E_F}{k_B T}\right), \quad (2.94)
$$

where $\mathscr{A}^* = 2qm^* k_B^2/(2\pi)^2\hbar^3$ is the Richardson constant. We have arrived at the classical thermionic emission current-voltage relationship.

This relationship considered transport only from the first semiconductor to the second at the heterojunction. We need to consider the other direction too, as only then can detailed balance can be assured. Transport in the other direction arises from the second material, which has a different effective mass and E-$\mathbf{k}$ relationship, so another prefactor—the Richardson constant —must be used. So, how is the detailed balance maintained, and what did we miss in this analysis? For transistor-like junctions, the situation we are interested in—in simplified form—is shown in Figure 2.22, where we have taken some liberties with the details of barriers, charges and therefore band bending, but it captures the transport essentials that need to be modeled in the thermionic framework. A barrier region, the channel region in which ballistic or near-ballistic conduction occurs, separates an injecting source and a collecting drain electrode. Here, the materials employed may have different E-$\mathbf{k}$ constitutive relationships. In the transistor, all the three regions are usually of the same material.

We need to take into account the carrier flux from both sides, and then we need to integrate over all the carriers that have the necessary momentum and energy, that is, the direction for momentum and the minimum magnitude of the perpendicularly directed energy, to overcome the barriers. The earlier energy barrier discussion can be translated to a maximum angle in the momentum space as a function of the energy in the constant energy section. Only when the carrier has sufficient energy and a directed angle of $\theta < \theta_{max}$ does the carrier satisfy both the momentum and the energy conditions to inject across. This argument is pictorially described in Figure 2.23.

For a $d^3k = 2\pi k^2 \sin\theta \, d\theta \, dk$ elemental volume in momentum space, where $\theta = \arcsin(k_\parallel/k) = \arccos(k_\perp/k)$, there exists a maximum

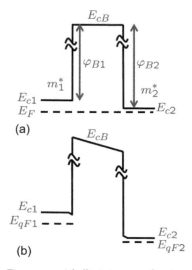

Figure 2.22: A ballistic transport barrier region formed from semiconductor heterojunctions of two different barrier heights, thus introducing different allowed modes in different regions. (a) shows a hypothetical structure in thermal equilibrium, and (b) is the same structure with a bias voltage of $(E_{qF1} - E_{qF2})/q$.

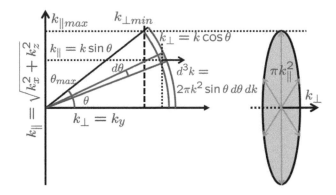

Figure 2.23: A reciprocal space pictorial representation of the slice of states in a band of energy dE at E that has both the properly directed momentum and the momentum and energy magnitude to inject across a barrier.

angle θ_{max} corresponding to $k_{\perp min} = [(2m^*/\hbar^2)(E_B - E_c)]^{1/2}$, and $k_{\parallel max} = (k^2 - k_{\perp min}^2)^{1/2}$, beyond which the wave will be reflected. The carriers will bounce back. This is the condition under which, for the energy E, the carrier has just enough forward momentum to overcome the barrier. In any other elemental volume with $\theta < \theta_{max}$ of this constant energy slice of width dE, the carrier has enough energy and momentum to appear in the second region ballistically. This maximum angle can be viewed as a window in the reciprocal space, a maximum in the parallel momentum $k_{\parallel}$ aligned to the plane parallel to the spatial plane of the barrier—an opening in the angle of the funnel—as shown in Figure 2.23. We can rewrite our equation of current from this elemental volume corresponding to the energy section of Figure 2.20 as

$$\delta J = \frac{2q}{(2\pi)^3} \frac{\hbar k_{\perp}}{m^*} 2\pi k^2 \, dk \, \mathcal{F}\left(\frac{E + E_c}{k_B T}\right) \int_0^{\theta_{max}} \cos\theta \sin\theta \, d\theta$$

$$= \frac{2q}{h} \frac{\pi k_{\parallel max}^2}{2\pi} \mathcal{F}\left(\frac{E + E_c}{k_B T}\right) dE. \tag{2.95}$$

The current flowing in the energy range dE at energy E is the quantum conductance times the number of channels that are filled, that is, the density of channels multiplied by the occupation probability described by the Fermi-Dirac function. In this slice, only states with $\theta < \theta_{max}$ provide conduction. This term can now be integrated over energy.

We can now resolve the issue of detailed balance. When entering from Semiconductor 1 into Semiconductor 2, carrier waves incident at the interface with $\theta < \theta_{max}$ enter Semiconductor 2. These carriers have the y-directed momentum and sufficient energy to overcome the barrier and have states that allow them to be in the 2nd semiconductor. Beyond this angle, they do not have sufficient perpendicular

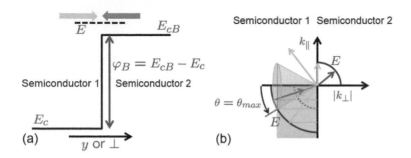

Figure 2.24: Injection across the interface accounting for the momentum and energy requirements in the angular picture. (a) reiterates the earlier description of the problem of flux from carriers at energy E in the two materials. (b) shows the condition of maximum angle from the conservation laws, with constant total energy E on the two sides of different semiconductors at the interface, showing that, at angles $\theta > \theta_{max}$, the carriers will reflect back in the first semiconductors. All carriers from Semiconductor 2 with a perpendicular momentum directed towards Semiconductor 1 will enter it. This assumes no quantum reflections.

momentum to enter. They reflect back. The constant energy surface in Semiconductor 2, which is drawn here as having an effective mass that is higher than that for Semiconductor 1 is such that for all the carriers occupying states in it, that is, for $E > E_{cB}$, if they have a $-y$-directed momentum, they will enter Semiconductor 1. The detailed balance is satisfied, that is, at each and every energy, there is a balance when in thermal equilibrium.

This behavior of incident particles within a solid angle being able to cross a boundary is a reciprocal space funneling effect. Only carriers in a narrow spread around the orthogonal injection path are allowed. Here this effect arose from the conservation conditions within a ballistic motion picture, so there is no scattering. Scattering will become considerably more important for injection, such as for ohmic contacts to monolayer materials. There, again, only a narrow $\mathbf{k}_\perp$ is allowed, and scattering that permits some momentum change through exchange makes for a narrow spread in $\mathbf{k}_\parallel$ to satisfy the injection conditions, even as most of the motion is aligned closely with $\mathbf{k}_\perp$. A corollary of this funneling effect takes place in real space under conditions of large scattering, such as in the shallow source contact to inversion layer of a silicon transistor, where carriers must cross the boundary between three dimensions and two dimensions. The real space funneling, a quite classical effect, occurs in the lower-confinement medium—the shallow junction region. We will explore the resistances arising in this situation later.

This description in Figure 2.24 is the equivalent of the wave picture from optics and is further clarified in Figure 2.25. Total internal reflection occurs at θ_{max}. Only propagating modes within the cone of this funnel, whose angle is a function of the energy component associated with perpendicular momentum are allowed to pass through. The funneling cone's angle disappears when this perpendicular energy is the barrier energy. The first semiconductor is the equivalent of the higher index of refraction optical material. An electron wave that enters the second semiconductor undergoes a redirection similar

Reciprocity holds. Note we have used an entirely linear wave propagation formalism in the crystal here.

to that described by Snell's law in optics. The reciprocal wavevector describes the positional dependence and the direction of travel. A magnitude of $k = [2m_1^*(E - E_c)/\hbar^2]^{1/2}$ in the first semiconductor changes to $k = [2m_2^*(E - E_{cB})/\hbar^2]^{1/2}$ in the second. The equivalent of the index of refraction ratio is the ratio of the product of the square root of excess energy and effective mass. This square-root term is proportional to the momentum perpendicular to the interface. This wave travels with a velocity $v = \hbar k/m^*$. If it slows down more, it is also bending more.

Mass and energy together determine the wave transmission characteristics in this ballistic condition. This is mode filtering and a statement equivalent to the one earlier emphasizing that the conductance in ballistic quantized channels is determined by the smallest conductance. Figure 2.26 describes this situation for a bit more complicated assembly of materials of long mean free paths with various masses and conduction bandedge energies. When regions with different properties adjoin, the region with the smaller aperture of allowed lateral ($k_\parallel$) modes with forward momentum ($k_\perp$) determines the wave transmission. When a barrier is encountered, a reduction in kinetic energy is reflected in a changed perpendicular momentum. In the parallel plane, the energy associated with the parallel momentum, transverse to this direction of motion, is conserved. Beyond a critical angle, total reflection happens. The electron wave, akin to an optical wave, undergoes a change in angle of refraction as it traverses the barrier. The circle defined by $k = [2m^*(E - E_{cB})/\hbar^2]^{1/2}$ is an aperture that filters the modes allowed to transmit through up to the maximum allowed. If there are multiple barriers and regions, with differing conduction band minimums and masses, it is the smallest aperture that will determine the ballistic current through the quantized channels. Note the consequence of this condition in the specific situation of the 2nd and the 3rd material region of Figure 2.26. In non-degenerate conditions, current flows due to carriers within a few $k_B T$s of the top of the barrier. So, that is the mass that is of relevance to the Richardson constant.

Having clarified thermionic emission, it is pertinent to ask what the criterion is likely to be that determines the rate-limiting step for carrier flow in the structure. Is it the ballistic motion at the barrier interface or does the random walk away from the barrier, embodied in the diffusion coefficient, determine the current flowing through a structure that is quasineutral and has no field? In other words, we are asking what determines the rate-limiting step in the conduction at junctions. Is it the flux of carriers to the junction, that is, diffusion or drift diffusion in the presence of fields, or is it the flux of carriers

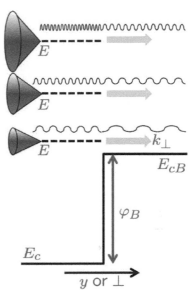

Figure 2.25: Carriers incident at the barrier pass through only if the energy associated with the perpendicular energy momentum is sufficient to enter the barrier. These are electrons that are in the cone where the incident wave has sufficient $k_\perp$. At higher energies of the incident electron wave, the cone has a larger aperture, that is, a larger angle for modes that are allowed to pass through. Reciprocal space funneling of carriers determines the transport.

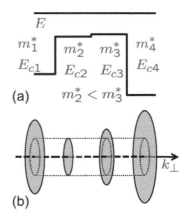

Figure 2.26: Mode filtering in a ballistic conduction structure with four materials whose properties are described in (a). Circles in (b) show the maximum $k_\parallel$ allowed for the energy E of the conducting electron wave in the different materials.

at the junction, that is, thermionic emission at the interface? From our discussion of drift diffusion, we have the tools to determine this. We write diffusion current in low level injection conditions as

$$J_{diff} = \frac{qD\mathcal{N}_c}{\mathcal{L}} \exp\left(-q\frac{\phi_B - V}{k_B T}\right),$$ (2.96)

where $\mathcal{N}_c$ is the effective density of states in the conduction band, that is, for electrons in this case, and $\mathcal{L}$ is the diffusion length. The relationship is still exponential in bias voltage because the carrier concentration is related through the Maxwell-Boltzmann relationship here, and the random walk characterized by diffusion occurs from a higher concentration to a lower one. The diffusion constant is

$$
\begin{aligned}
\mathcal{D} &= \frac{k_B T}{q}\mu = \frac{k_B T}{q}\frac{q \langle \tau_k \rangle}{m^*} = \frac{k_B T}{q}\frac{q \langle \lambda_k \rangle}{v_\theta m^*} \\
&= \frac{k_B T}{q} q \langle \lambda_k \rangle \sqrt{\frac{\pi}{8 k_B T m^*}}.
\end{aligned}
$$ (2.97)

We can guess the answer. If the diffusion constant is small, the process of this injection and extraction of carriers will limit current. If it is large, then the transport limitation at the junction will limit current. The diffusion current is

$$
\begin{aligned}
J_{diff} &= \frac{qD\mathcal{N}_c}{\mathcal{L}} \exp\left(-q\frac{\phi_B - V}{k_B T}\right) \\
&= \frac{k_B T}{\mathcal{L}} \mu \mathcal{N}_c \exp\left(-q\frac{\phi_B - V}{k_B T}\right) \\
&= \frac{k_B T}{\mathcal{L}} q \lambda_k \sqrt{\frac{\pi}{8 k_B T m^*}} 2 \left(\frac{m^* k_B T}{2\pi\hbar^2}\right)^{3/2} \exp\left(-q\frac{\phi_B - V}{k_B T}\right) \\
&= \frac{q m^* (k_B T)^2}{\hbar^3} \frac{1}{4\pi} \frac{\lambda_k}{\mathcal{L}} \exp\left(-q\frac{\phi_B - V}{k_B T}\right).
\end{aligned}
$$ (2.98)

The thermionic emission current is

$$J_{te} = \frac{2q m^* k_B^2}{(2\pi)^2 \hbar^3} T^2 \exp\left(-q\frac{\phi_B - V}{k_B T}\right),$$ (2.99)

which gives the simple criterion

$$\frac{J_{diff}}{J_{te}} = \frac{\pi}{2}\frac{\lambda_k}{\mathcal{L}}.$$ (2.100)

When $\langle \lambda_k \rangle \gg \mathcal{L}$, thermionic emission is the rate-limiting step. Carriers are swept away fast because of short diffusion lengths, and hence the transport at the interface is rate limiting. When $\langle \lambda_k \rangle \ll \mathcal{L}$, diffusion is the rate-limiting step. Carriers are swept away swiftly at the interface, but the approach to or away from the interface is rate limiting.

We now have the tools together for describing ballistic and near-ballistic transport in transistors. From density of states and specific conditions, that is, 3D, 2D or 1D conditions, one can determine the number of channels available filled with electrons. These are the channels that conduct with a conductance of $2q^2/h$ per channel in units of per V and $2q/h$ in units of per eV—the quantized conductance. The current is simply

$$J = \frac{2q}{h} \int_{occupied} \frac{1}{(2\pi)^2} d^2k_{\parallel},$$

(2.101)

where the integral is over all the momentum states in the plane transverse to the motion direction, that is, the parallel plane.

2.5 The on state of a nanoscale transistor

BALLISTIC TRANSPORT IN A QUANTUM WIRE, that is, with two dimensions in confinement and at the temperature of absolute zero, is perhaps the simplest system to begin understanding ballistic and near-ballistic transistors and the roles of the quantized conductance g_q and the number of channels n_q that conduct. We will introduce scattering into this understanding by employing the scale and potential behavior explored earlier.

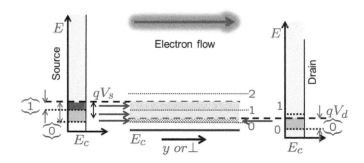

Figure 2.27: A schematic for the discussion of ballistic transport in confined conditions across an adiabatic geometry with a small bias voltage applied between the source end and the drain end. Subband edges are identified by 0 and 1 for the first two subbands. Filled states in the subbands are identified by braces.

In Figure 2.27, a small potential is applied between the source and the drain in a ballistic geometry; for now, consider this system to be at absolute zero temperature, so with degenerate conditions and with a step function for occupation at the quasi-Fermi level. The quasi-Fermi level in the drain is slightly below that in the source. The first two subbands, formed due to confinement and labeled 0 and 1, respectively, at their bandedge, are shown. Only the 0th and the 1st subbands are occupied at the source end of the transport region, for the quasi-Fermi level shown. Only the 0th subband is occupied at the drain end, as the subband edge of the 1st band is higher than the quasi-Fermi level at the drain end. The occupied subbands at the source and the drain ends are identified through the symbols $\overbrace{0}$

and $\overbrace{1}$, respectively.

We assume that the semiconductor is isotropic with a constant effective mass. Because of the confinement of the channel, the minimum energy is higher than the conduction bandedge of the unconfined material. Because of the choice of the drain quasi-Fermi level, across the entire channel, both forward and reverse moving states are occupied up to the drain quasi-Fermi level. These are in the 0th subband. In the 0th subband, there are additional states, of forward and reverse momentum, filled up in energy up to the source quasi-Fermi level at the source end. However, at the drain end across this spread in energy, that is, above the drain quasi-Fermi level, the states are unoccupied. What this says for the 0th subband is that, up to the drain quasi-Fermi energy, in the channel there are equal forward and reverse currents arising from the equal occupancy of the forward and the reverse momenta. However, above the drain quasi-Fermi level, occupied states with forward momentum conduct current that is not balanced by any reverse flux, because those states are asymptotically empty at the drain end. So, in the 0th subband, at any position along the channel, one finds that, up to the drain quasi-Fermi level, states occupying the forward and the reverse momenta balance out, and so does the associated current. Above this drain quasi-Fermi level and up to the source quasi-Fermi level, the occupied forward momentum states at the source end cause a current flux that is not balanced by the flux from the drain. A net current due to electrons transporting from the source to the drain arises as a result of this. The 1st subband states are occupied at the source end from the bottom of the subband to the source quasi-Fermi level, and this subband is unoccupied at the drain end. The current arising from all the forward momentum states occupied at the source end of the channel are not balanced by any reverse current from the drain. So, these electrons also contribute to the current in the structure.

The electron wave that extends from the source to the drain arises in the forward momentum states. Its momentum $\mathbf{k}_\perp$ is smaller at the source than at the drain end, where it has higher energy. At the source end, other states near it in momentum or energy are also filled, but at the drain end they are largely unfilled. For determining steady-state current under static bias conditions, current continuity implies one may determine this current at any cross-section of the device. We choose the source end, since we can determine the occupation there conveniently. Our discussion is still assuming absolute zero as the operating temperature; therefore, there exist no occupied tails. The channels that contribute to the current are (a) those between the source and the drain quasi-Fermi energies with forward momentum in the 0th subband, and (b) those between the bandedge of the 1st subband, and the source quasi-Fermi energy with forward momentum. Each of the channels available for this conduction contributes a conductance of $2q^2/h$ S or $2q/h$ A/eV. The current is

$$
\begin{aligned}
I_S \;=\; & I_D = I_{ch} = g_q n_q = g_q(n_{q,0th} + n_{q,1st}) \\
\;=\; & g_q \left[\sum_{0th,E=qV_D}^{E=qV_S} \mathscr{G}_0(E)\, dE + \sum_{1st,E=E_1}^{E-qV_S} \mathscr{G}_1(E)\, dE \right], \quad (2.102)
\end{aligned}
$$

where n_q is the number of occupied channels that contribute to conduction, and the sum is over the two different subbands that have different energy ranges over which these conducting channels exist. We can relax the absolute zero requirement by introducing the Fermi distribution functions. So,

$$
I_S = I_D = I_{ch} = \frac{2q}{h} \sum_i \int \{\mathscr{G}_i\,[\mathcal{F}(E - qV_S) - \mathcal{F}(E - qV_D)]\}\, dE \quad (2.103)
$$

is the current, including the band degeneracy and the occupation. This same approach can also be effective under Maxwell-Boltzmann distribution conditions.

We now translate this picture into conditions when the bands are not quasi-flat, that is, in low fields and at low drain-to-source bias voltage. In Figure 2.28, we look at a ballistic-transistor-like condition with a large positional dependence of the subbands and the conditions that the electron wave encounters. Barriers exist for the different subbands for electron traversal in both the source-to-drain and the drain-to-source directions; thus, the energy associated with the perpendicular momentum must be sufficient to surmount these barriers. This figure is for a quantum wire, so the $k_\parallel$ is quantized, that is, the x and the z components are strongly discretized. The density of states distribution plotted in this figure is the density of states for any of these subbands, where $k_\parallel$ has a specific value, and associated with it is the subband edge energy. The decay in the density of the states in the subband arises from dependence on $k_\perp$, which is the momentum for the source-to-drain direction. But, they still must have enough energy to surmount this barrier. Consider the states of the 0th subband. For states that are occupied below the barrier energy for the 0th subband, the forward momentum electron waves are reflected back. If the energy is higher than the barrier, they propagate. The same is true for each of the subbands, including the 1st as shown.

Now we can look at this in terms of the occupation of states and from that write the resulting current. Figure 2.29 shows the band

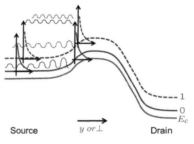

Figure 2.28: Band diagram together with electron behavior under conditions of transmission and reflection in a two-dimensionally confined, that is, quantum wire geometry. The figure shows two subbands for the confined region of transport.

and representative E-$k_\perp$ diagram, with occupied states, in the source reservoir, at the barrier at the source-channel junction region and in the drain reservoir for the 0th subband and 1st subband. This transistor, made of a hypothetical isotropic material, is operating in saturation under a gate bias voltage (V_G) that modulates the barrier, and source (V_S) and drain (V_D) bias voltages that establish the quasi-Fermi energies E_{qFS} and E_{qFD} for the source and the drain regions, respectively.

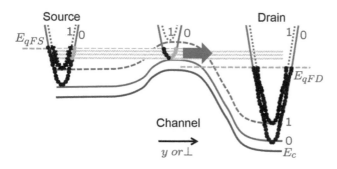

Figure 2.29: A hypothetical ballistic transistor made in an isotropic semiconductor. Transistor bandedges and the occupied and the unoccupied states are shown during operation in saturation conditions at absolute zero temperature. A composite of the E-$k_\perp$ diagrams for the two lowest subbands and the conduction bandedge for three-dimensional conditions are shown.

In the source reservoir, of the states occupied in the 0th subband, only the states that have a momentum $\hbar k_\perp$, that is, positive and directed towards the barrier, can potentially reach the drain reservoir. But, some of these have energy lower than the barrier energy for the 0th subband. These occupied states are the ones that reflect back, as was the case for the lowest wave shown in Figure 2.28 at the source. The $k_\perp$ states that have an energy higher than the barrier and have this positive momentum are the propagating modes. These are the conducting channels. Only half of the states between the barrier energy and E_{qFS} contribute, since only this half has the barrier-directed momentum. In the 1st subband, in the source reservoir, there exist no states that have sufficient energy to overcome the barrier for the 1st subband as drawn. So, only half of the occupied states in the 0th subband—those with forward propagating momentum and whose energy is between E_{qFS} and the 0th subband's maximum energy in the barrier—contribute to the current from the source. In the drain, any electron waves associated with and in the 0th and 1st subbands below the drain quasi-Fermi level (E_{qFD}) are reflected back from the barrier, since there exist barriers to both in the channel as drawn. So, these negative $k_\perp$ states contribute no current. The positive $k_\perp$ states do not connect the drain to the source. They too do not contribute any current. So, there is a sliver of energy from the filled positive $k_\perp$ states that propagate between the source and the drain. These are the y-directed filled positive $k_\perp$ states in the band indicated by the block arrow's band in Figure 2.29. They are also equivalently the positive $k_\perp$ half of the E-$k_\perp$ diagram drawn for the 0th subband in this figure. The E-$k_\perp$ relationship at any position shows the reciprocal k-space modes of the waves. So, the states shown as occupied, and propagating, whether in the source region or at the barrier, represent the same waves, just at different positions.

If this is a quantized wire, in the source, they have a higher momentum, but less density. At the barrier, they have a smaller momentum, but higher density. If this is a one-dimensionally confined structure, then the density of states is a constant in each of the subbands. But, now there are modes with a wavevector component orthogonal to the source-to-drain direction, and some of these that belong in the cone up to the θ_{max} will contribute. In either of these situations, in the steady state, current continuity is still maintained in a static steady state. This is the idealized picture at absolute zero temperature. Since the number of channels that are available for conduction is a constant, and no more open up as the drain bias is increased, the channel current is simply the product of quantum conductance g_q and the number of channels available for conduction with the positive source-to-drain electron passage, and it is a constant. The current is saturated and not dependent on drain bias voltage.

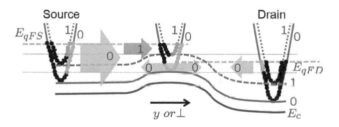

Figure 2.30: The hypothetical ballistic transistor operating in the linear conditions at absolute zero temperature; $E - k_\perp$ diagrams for the two lowest subbands, together with the conduction bandedge for the barrier-modulated low bias conditions, are shown.

How does one reach this saturation current at increased drain bias voltage through a linear conducting state? Figure 2.30 shows this condition, the one of low drain bias voltage.

Because of the small change in quasi-Fermi levels between the source and the drain, both of the subbands from our previous example have occupancy in the source and the drain regions. However, for Subband 1, as drawn here, there exists a barrier for propagation from the drain to the source for the $-y$-direction momentum states. So, for the 1st subband, the positive $k_\perp$ states shown in the source or at the barrier of the 1st subband contribute to conduction. In this energy range, there are no negative $k_\perp$ states in the drain contributing. As in our previous discussion, there exist negative $k_\perp$ states in the source, or at the barrier, that do not participate in the drain-directed flux or current because they are not tied to the drain. The corresponding drain states are empty. Occupied states in the source and the drain at energies below the corresponding subband barrier energy are blocked from propagating. There also exist occupied propagating states of the 0th subband that contribute flux from the source to the drain and from the drain to the source. Below the drain quasi-Fermi level, this flux exists in both directions, so the currents in the conducting channels precisely cancel out. The net current from the 0th subband arises from the positive $k_\perp$ states between the source quasi-Fermi level $E_{q\Gamma S}$ and the drain quasi-Fermi level E_{qFD}. For the 1st subband, the current contribution arises from the occupied states between E_{qFS} and the maximum energy that forms the barrier for this band in the channel. So, two subbands contribute current. The current is an integration of all the states that contribute through source-drain coupling, including in this calculation the sign of the momentum flux. Thus, Equations 2.102 and 2.103 should be modified to account for conductance in the opposing direction, with such conductance arising from the states that have negative momentum.

When the drain bias voltage is changed, and let us assume for now no drain barrier lowering, that is, that the barrier is very tightly coupled to the gate, then the flux from the drain to the source changes. A larger drain bias lowers the E_{qFD}. There are fewer states between this drain-end energy and the maximum energy of the 0th subband in the channel. The source flux remains unchanged. The flux from the drain has decreased. The net current increases. And it stops increasing when E_{qFD} reaches the maximum in the 0th subband's peak energy. The current now saturates.

So, this ballistic transistor behaves much like a transistor at large dimensions. If channels are open for conduction, via a gate bias voltage, raising the drain-to-source bias from a zero bias condition gradually decreases the back injection flux of the electrons from the drain to the source until all the back injection channels are closed because the quasi-Fermi energy of drain is below the peak in the barrier in the channel. The current keeps increasing as the back injection keeps decreasing, and then it saturates. This is a condition in which the conductance contribution arises from of all the open channels between the source and the drain. It is a condition in which we must take into account the position of the barrier in the channel at the source-channel interface region. In the saturation region, channels exist for conductance only from the source to the drain. In the linear region, channels exist for conductance in both directions, and any change in the drain bias voltage modulates the back-conducting channels. Our total current is found by determining the number of channels $n_b = \sum_j \int g_{bj} \mathscr{G}_j(E) dE$ available for forward and backward conduction. Here, g_{bj} is the band degeneracy of the jth subband, whose density of states is given by $\mathscr{G}_j(E)$, and the summation is over all the subbands j. The product of this number of channels and the quantum conductance g_q gives the current.

Is the resistance arising in the quantum conductance, that is, the conductance's inverse integrated over the filled conducting channels, dissipative? No. It arises in the wave nature of the electron and this current will flow back and forth in any open channel so long as the electron doesn't undergo an inelastic scattering event. In a ballistic transistor, the dissipation happens in the contact and contacting regions. It is only in the process of the interaction between the reservoir and its thermalized electrons and other scatterers coupling to this electron wave that the dissipation happens.

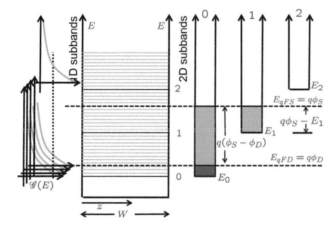

Figure 2.31: A subband view looking from the source to the drain of the states in a 2D ballistic transistor at absolute zero degenerate conditions. The device has a width W in the z direction. Each one of the 2D subbands at the right can be viewed as an accumulation of the closely packed 1D subbands at the left. The figure also shows the source and the drain potentials.

For devices of width W, that is, when the z-directed confinement constraint is relaxed, there exists freedom of movement in both the y and the z directions. The y-directed states are very close to each other. The z-directed states too are close if the width is much more than the quantum confinement condition, that is, $W \gg \lambda_{deB}$. As discussed in Appendix G, the 2D subband, that is, the one-dimensionally confined subband, is an ensemble built out of the 1D subbands (see Figure 2.31 for this 2D transistor example), that is, an accumulation of all the quantum wire subbands, each with a density of states $\mathscr{G}_{1D} = (1/\pi\hbar)\sqrt{2m^*/E}$, gives us the 2D subband density. The quantum wire description is generalizable. The aperture of the modes allowed is determined by the barrier energy. If one knows the open channel count, one can determine the current. There are states that are blocked. There are states that conduct. Depending on the bias voltage applied and occupation in the reservoirs, states from more than one subband may provide conduction.

Usually though, the 0th subband is the one most involved in the

conductance, and the discussion is easily generalizable to additional
subbands. We need to connect the applied bias voltages to the different potential energies that exist inside the device. Figure 2.32 shows
a relevant pictorial view in equivalent voltages at a position (y, z) in
the x direction, that is, directed into the substrate. This figure shows
an inversion layer formed at an insulator-semiconductor interface.
We are employing here the same approach as used in the classical
analysis. Solve the problem first in terms of internal potentials, then
apply the boundary conditions that relate the external conditions to
the internal conditions.

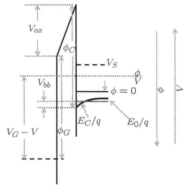

Figure 2.32: Band bending and potential
in voltage units along x at a position
(y, z) of the transistor.

We can write the potential in voltage units relationships:

$$
\begin{aligned}
\phi &= -V + V_G - V_{ox} - \phi_G + \phi_C - V_{bb} - E_0/q, \\
V_S &= -\phi_S, \quad V_D = -\phi_D, \quad \text{and} \\
V_{GS} &= \phi_S + V_{ox} + \phi_G - \phi_C + V_{bb} + E_0/q.
\end{aligned}
\tag{2.104}
$$

In this equation, $-\phi_G + \phi_C - V_{bb} - E_0/q$ represents a perturbation
term, approximately equivalent to V_{FB}, the flatband voltage, or the
metal-semiconductor (ϕ_{MS}) voltage in classical transistor analysis.
The total current, arising from 2 two-dimensional subbands, with
$E_{qFS} = q\phi_S$ and $E_{qFD} = q\phi_D$, can be written for the $0\ K$ degenerate
example as

$$
\begin{aligned}
J &= J_D \\
&= 8q\frac{\sqrt{2m^*q}}{h}\left[\int_0^{\phi_D} \frac{\phi_S - \phi_D}{\phi^{1/2}}\, d\phi + \int_{\phi_D}^{\phi_S} \frac{\phi_S - \phi}{\phi^{1/2}}\, d\phi\right] \\
&= \mathcal{K}_B\left(\phi_S^{3/2} - \phi_D^{3/2}\right), \quad \text{where } \mathcal{K}_B = 8q\frac{\sqrt{2m^*q}}{h},
\end{aligned}
\tag{2.105}
$$

a relationship applicable if all the states providing conduction are
counted. In Figure 2.33, these are the bands for the 0th and 1st subbands, and the arrows represent the momentum direction of the
states that contribute. In this equation, the first term represents the
contribution from the 0th subband, and the second term is from the
1st subband. The square root of the denominator in the integral is
from the inverse square-root dependence on energy of the density of
states, and the prefactor in the relationship arises from the rest of the
constants.

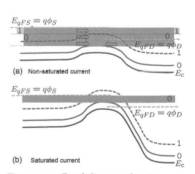

Figure 2.33: Band diagram showing
the integration limits in energy for
the different states that contribute to
conduction.

We can determine the charge density contributing to the current.
In the $q\phi < q\phi_D$ energy range, all the occupied channels contribute
to the current. For $\phi_D \le \phi \le \phi_S$, only the positive momentum
states contribute to the current, the others balance out, so only half
occupancy must be considered. This gives the charge density of
conducting electrons as

$$
Q = \frac{m^* q^2}{2\pi\hbar^2}\left(\phi_S + \phi_D\right) = \frac{1}{2}C_Q\left(\phi_S + \phi_D\right),
$$

where $C_Q = \dfrac{m^* q^2}{\pi \hbar^2}$ (2.106)

is a capacitance, the quantum capacitance, normalized to area.

	m^*/m_0	$\Delta E\ (meV)$ at $10^{12}\ cm^{-2}$	C_Q (F/cm^2)
$GaAs$	0.067	35.8	4.5×10^{-6}
$Ga_{0.43}In_{0.57}As$	0.041	58.6	2.7×10^{-6}
$InAs$	0.023	104.4	1.5×10^{-6}

Table 2.6: Energy spread above the bottom of thesubband needed in occupied states for achieving $1 \times 10^{12}\ cm^{-2}$ electron density, together with quantum capacitance (C_Q) for a few examples of low effective mass semiconductors. These calculations are for single-dimensionally confined two-dimensional electron gas. For comparison of magnitudes, the capacitance across an ideal SiO_2 capacitor that is 1.5 nm thick—a not unreasonable gate oxide thickness—is 2.4×10^{-6} F/cm^2. The quantum capacitance will dominate any insulated gate control that employs charge partitioning.

The quantum capacitance relates the density of states and their energy dependence in the semiconductor. To obtain increasing charge density within the semiconductor, higher energy states must be occupied because of the limits placed by the density of states in energy. Confinement spreads these states out in energy. If the density of states is large, the channel charge can be large without high occupancy in energy within the conducting medium. For the same bias voltage, more charge can exist and be available for conduction. Table 2.6 shows examples of a few materials with isotropic bandedge masses in Γ valleys, together with the energy up to which the conduction band must be filled to achieve $1 \times 10^{12}\ cm^{-2}$ of electrons. The quantum capacitance follows from this. When an effective mass is small, such as in $InAs$, the energy spread needed is high, and the quantum capacitance is low.

Ladder	Degeneracy g_v	Confining mass m^*/m_0	Conduction mass m^*/m_0	DOS^* mass m^*/m_0	Subband DOS $\mathscr{G}$/valley $(cm^{-2} \cdot eV^{-1})$	$\mathscr{G}$/ladder $(cm^{-2} \cdot eV^{-1})$	Subband C_Q (F/cm^2)
Lower	2	0.916	0.190	0.190	7.9×10^{13}	1.58×10^{14}	2.53×10^{-5}
Higher	4	0.19	0.315	0.417	1.7×10^{14}	6.95×10^{14}	1.11×10^{-4}

* DOS is density of states. The DOS mass here does not include ladder's minima/valley degeneracy. Its use will result in a density of states variable $\mathscr{G}$ that needs to be multiplied by degeneracy to obtain the total subband density of states for that ladder.

Figure 2.7: Silicon, following Table 2.4, forms multiple subbands with two ladders for a (100) surface. The density of states per ladder is $\mathscr{G} = g_v m_d^* / \pi \hbar^2$. To achieve a $1 \times 10^{12}\ cm^{-2}$ electron density, only 6.3 meV of the lower ladder needs to be occupied. The higher ladder, which is significantly displaced—by mass ratios—is quite unpopulated.: C_Q from this lower ladder subband is 2.53×10^{-5} F/cm^2—much larger than the controlling 2.4×10^{-6} F/cm^2 1.5 nm oxide capacitance. Subband occupation has minimal effect. At higher densities, more subbands from both ladders contribute, and their C_Q contributions appear in parallel.

Table 2.7 shows a similar calculation for (100) silicon. This calculation is now a little more complicated since silicon conduction minima are near the X point in the reciprocal space, and there exist six such valleys. When confinement is in the $[100]$ direction, two valleys—the out-of-plane ones with the larger longitudinal mass—form a lower ladder of states available for in-plane conduction. There also exists a second ladder, from the other four valleys, where subband confinement energy is determined by the transverse mass. Silicon has a larger effective mass. For non-confined geometries, it is $\sim 0.36 m_0$, but even with confinement, the larger density of states results in a lower energy spread for the carrier density, and the quantum capacitance is larger. We will see that the interaction between effective mass, density of states, and velocities make this quantum capacitance quite important, and integral to the current magnitudes since a limited energy spread means more field connecting the gate to the mobile charge.

In this two-dimensional ballistic transistor operating at $0\ K$, the current saturates when $\phi_D = 0$. This is when all the back channels are shut off. The current and charge at this saturation are, respectively,

$$J = J_{DSat} = \mathcal{K}_B \phi_S^{3/2}, \text{ and}$$
$$Q = Q_{Sat} = \frac{1}{2} C_Q \phi_S = \frac{m^* q^2}{2\pi\hbar^2} \phi_S. \tag{2.107}$$

Since one knows the current density and charge density, the effective velocity for the transistor follows as

$$v_S = \frac{J_{DSat}}{Q_{Sat}} = \frac{4}{3\pi} \sqrt{\frac{3 q \phi_S}{m^*}} = \frac{4}{3\pi} v_F. \tag{2.108}$$

The effective velocity, in the absence of scattering and due to quantum conductance, is the velocity at any cross-section in the channel region. Since the barrier appears at the source end because the drain field penetration pulls the other end down, the velocity is an effective velocity at the source.

The Fermi velocity—v_F—is the velocity of carriers of momentum k_F at the Fermi energy. This is where states occupied and unoccupied coexist for conduction to take place in a degenerate condition. In our case, we have created conditions so that carriers below the Fermi energy are also available for conduction arising in carriers from the half of the Fermi circle with positive momentum. This leads to an averaging that reduces the mean carrier velocity in a conducting degenerate gas by a factor of $4/3\pi$.

Half of the Fermi circle means and assumes a large W—width of the transistor—and $\theta_{max} = \pi/2$ for a full half of the Fermi circle.

We may now use boundary conditions to relate the internal dynamics to the external parameter. We will ignore the perturbation term, as it makes the relationship a little unwieldy, but as in classical analysis, we can put it back in later on. The boundary conditions are as follows:

$$\phi = -V + V_G - V_{ox} - \phi_G + \phi_c - V_{bb} - E_0/q,$$
$$V_S = -\phi_S,$$
$$V_D = -\phi_D, \text{ and}$$
$$V_{GS} = \phi_S + V_{ox} + \phi_G - \phi_c + V_{bb} + E_0/q. \tag{2.109}$$

Here, we have crossed out the perturbation—which is a constant determined by the material and the geometry—as it can be reintroduced as a linear change later. The voltage drop across the insulator, the oxide, is $V_{ox} = Q/C_{ox}$, which lets us relate the external potential voltages to internal potential voltages as

This process of perturbation term removal and reintroduction is just like the use of flatband voltage in classical transistor analysis. It is a correction term that is linear in the analysis but makes the equations look far more unwieldy than they really are.

$$V_{GS} = \phi_S + \frac{C_Q}{2C_{ox}}(\phi_S + \phi_D), \text{ and}$$
$$V_{DS} = \phi_S - \phi_D, \tag{2.110}$$

which lead to the following current and charge density relationships with voltage:

$$J_D = \frac{\mathcal{K}_B}{(1+2\gamma)^{3/2}}\left[(V_{GS}+\gamma V_{DS})^{3/2} - (V_{GS}-(1+\gamma)V_{DS})^{3/2}\right], \text{ and}$$
$$Q = \frac{1}{2}C_Q\frac{2V_{GS}-V_{DS}}{1+2\gamma} = \frac{C_Q C_{ox}}{C_Q+C_{ox}}\left(V_{GS}-\frac{1}{2}V_{DS}\right). \tag{2.111}$$

Here,

$$V_{GS} = \phi_S + \gamma(\phi_S + \phi_D), \text{ and}$$
$$V_{DS} = \phi_S - \phi_D, \tag{2.112}$$

with $\gamma = C_Q/2C_{ox}$ relating any potential change in the channel. This reflects a charge ratio effect that is caused by the gate-to-source potential and capacitance that induces the charge, represented in the capacitance C_{ox}, and the degradation of the charge control due to the loss of gate-to-source potential to the higher energy states that must be occupied in the build up of this charge represented in the capacitance C_Q. This ratio γ, which represents this degradation, is a factor in the transforms that describe the external and the internal potentials:

$$\begin{bmatrix} V_{GS} \\ V_{DS} \end{bmatrix} = \begin{bmatrix} 1+\gamma & \gamma \\ 1 & -1 \end{bmatrix}\begin{bmatrix} \phi_S \\ \phi_D \end{bmatrix}, \text{ and}$$
$$\begin{bmatrix} \phi_S \\ \phi_D \end{bmatrix} = \frac{1}{1+2\gamma}\begin{bmatrix} 1 & \gamma \\ 1 & -(1+\gamma) \end{bmatrix}\begin{bmatrix} V_{GS} \\ V_{DS} \end{bmatrix}. \tag{2.113}$$

This gives the current and channel charge density as

$$J_D = \frac{\mathcal{K}_B}{(1+2\gamma)^{3/2}}\left\{(V_{GS}+\gamma V_{DS})^{3/2} - [V_{GS}-(1+\gamma)V_{DS}]^{3/2}\right\}, \text{ and}$$
$$Q = \frac{1}{2}C_Q\frac{2V_{GS}-V_{DS}}{1+2\gamma} = \frac{C_Q C_{ox}}{C_Q+C_{ox}}\left(V_{GS}-\frac{1}{2}V_{DS}\right), \tag{2.114}$$

respectively. Saturation in current occurs at a drain bias of $V_{DS} = V_{GS}/(1+\gamma)$ (recall that $\gamma = C_Q/2C_{ox}$), resulting in a saturated current and a saturated charge density of

$$
J_{DSat} = \frac{\mathcal{K}_B}{(1+2\gamma)^{3/2}} V_{GS}^{3/2}, \text{ and}
$$

$$
Q_{Sat} = \frac{C_Q C_{ox}}{C_Q + C_{ox}}\left(V_{GS} - \frac{1}{2}\frac{V_{GS}}{1+\gamma}\right) = \frac{C_Q C_{ox}}{C_Q + C_{ox}}\frac{1+2\gamma}{2(1+\gamma)}V_{GS}
$$

$$
= \frac{C_Q C_{ox}}{C_Q + 2C_{ox}} V_{GS} = \frac{1}{2}\frac{1}{1+\gamma}C_Q V_{GS}, \tag{2.115}
$$

respectively.

There are a number of comments related to the questions that these equations elicit. First, the power law relationship is now a 3/2 power law, not the square law that exists in long-channel conditions, and not the linear law that exists when, even in long-channel conditions, the carriers move across most of the channel with approximately constant saturated velocity because the low field mobility is high. Our effective velocity of the carriers here under the degenerate conditions is close to the source-end Fermi velocity of the carriers in the channel. What the power law is stating is that, unlike the long-channel power law conditions, where the carriers move in energetic steady state with the crystal, hugging the conduction bandedge, the ballistic transistor's power law changes because of two reasons: first, the state density varies with an inverse half power dependence in energy, and second, there exists the linear energy dependence of the conducting states themselves. It is as if the long-channel saturated velocity situation has been modified with an additional power of half because of the higher energy states that the carriers occupy.

The second interesting feature in these equations is related to the interaction between C_{ox}—the gate's controlling capacitance—and C_Q—the semiconductor's ability to respond. If γ is large, that is, $C_Q \gg C_{ox}$, the charge is saturated charge density, which is just the product of the C_{ox} and the gate-to-source bias voltage. High quantum capacitance means that the charge appearing in the channel for conduction is simply like that of a parallel plate capacitor. All the fields of channel charge terminate in the gate. High C_Q means that there are a large number of states available for occupation in the channel, so the quasi-Fermi level does not rise significantly higher. The source-end Fermi energy is ratioed by the series combination of the oxide and the quantum capacitances represented in the factor $(1+\gamma)$. Since the Fermi energy rise in the semiconductor is related to this factor, and energy and velocity have a square law relationship, the Fermi velocity in the source is now reduced. For small γ, this changes by $1/(1+2\gamma)$ and, for large γ, by $1/(2\sqrt{2\gamma})$. A small γ means fewer carriers, because higher energy state occupation is necessary as more charge is induced in the channel.

This forcing of more electrons in the system, and these electrons then occupying higher energies as a consequence of quantization, is reflected in $\Delta E = qQ/C_Q = q\Delta V$ internally in the device. Here, ΔV is the amount of bias voltage that is lost to carriers needing to occupy higher energy. Figure 2.34 is a pictorial representation of this effect and its internal device consequences for the the states available for populating. This effect does not scale with layer thicknesses.

For subband occupancy in a $2D$ system, the density of states is a constant as a function of energy until a new subband opens up for occupation. A step increase will occur here. For a quantum wire system, the density of states decreases with energy in each individual doubly quantized wire subband. Silicon, with its higher valley degeneracy and larger effective mass, has a higher C_Q. $InAs$, with its low effective mass, has a low C_Q. Less charge can exist in an energy range of occupied states in $InAs$ than in silicon.

This description of $0\ K$ and degenerate statistics is quite a reasonable description for many materials: materials in which ballistic conduction exists because of the dimensions employed and the long mean free paths, and which have a small effective mass and density of states. $InAs$, graphene, carbon nanotubes, even $Ga_{1-x}In_xAs$ are reasonably described by it under idealized surface and interface conditions.

Silicon and germanium, however, have a high density of states. At finite temperature, depending on the conditions of bias voltage, either degenerate or non-degenerate conditions may prevail. In these materials, the channel region contains states with occupation probabilities considerably less than unity. Assuming that the ballistic description is still valid, we then just need to include the consequences of this distribution in the thermionic emission description. If the channel is two dimensional, we need to integrate over all the states that can allow transport between the source and the drain.

Figure 2.35 shows a pictorial representation of a two-dimensional, that is, one-dimensionally confined, ballistic transistor. Cones of injection exist at both the source end and the drain end because filled states exist, even if with low probability, at both ends, with momentum pointed in the right direction. The flux in this sliver of energy δE is

$$
\begin{aligned}
\delta J(E) &= g_q \Delta E \left(\int_0^{E-E_0} \frac{1}{\pi\hbar} \sqrt{\frac{2m^*}{E}}\, dE_\parallel \right) \\
&\quad \times \left[\mathcal{F}\left(-\frac{E - E_{qFS}}{k_B T} \right) - \mathcal{F}\left(-\frac{E - E_{qFD}}{k_B T} \right) \right] \\
&= 2 g_q \Delta E \frac{\sqrt{2m^*(E - E_0)}}{h}
\end{aligned}
$$

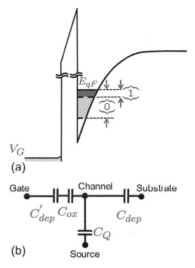

(a)

(b)

Figure 2.34: (a) A two-dimensional electron gas–based ballistic transistor with two subbands occupied. (b) A circuit diagram representing capacitances denoting the charge-potential effects; C'_{dep}, denotes gate depletion; C_{dep}, substrate depletion; C_{ox}, the insulator capacitance; and C_Q, the quantum capacitance arising from occupation of the quantized states.

$$\times \left[\mathcal{F}\left(-\frac{E - E_{qFS}}{k_B T}\right) - \mathcal{F}\left(-\frac{E - E_{qFD}}{k_B T}\right) \right], \quad (2.116)$$

where the total number of occupied channels has been determined by integrating the one-dimensional density of states. The prefactor term is thus the total number of channels multiplied by the quantum conductance. Since this is a ballistic conduction condition, this picture is also the equivalent of a distribution at the barrier's peak, where some of the states are forward propagating and some of the states are backward propagating. We now need to integrate this current flux over the energy range.

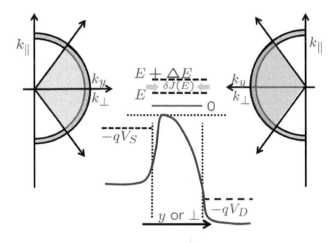

Figure 2.35: A hypothetical ballistic transistor operating at finite temperature. Cones of the allowed $k_\perp$ are shown consistent with the carriers having sufficient energy to cross the barrier and with properly oriented motion. These cones are another instance of reciprocal space funneling.

For Boltzmann conditions, the Fermi-Dirac function $\mathcal{F}$ can be simplified to a simple exponential, and the current is

$$
\begin{aligned}
J = J_D &= 2 g_Q (k_B T)^{3/2} \frac{\sqrt{2m^*}}{h} \exp\left(-\frac{E_0}{k_B T}\right) \\
&\quad \times \left[\exp\left(-\frac{qV_S}{k_B T}\right) - \exp\left(-\frac{qV_D}{k_B T}\right) \right] \int_0^\infty \sqrt{\zeta} \exp(-\zeta)\, d\zeta \\
&= g_q \frac{\sqrt{2\pi m^*}(k_B T)^{3/2}}{h} \exp\left(-\frac{E_0}{k_B T}\right) \\
&\quad \times \left[\exp\left(-\frac{qV_S}{k_B T}\right) - \exp\left(-\frac{qV_D}{k_B T}\right) \right].
\end{aligned}
\quad (2.117)
$$

Using the same procedure as before, the charge density is

$$ Q = \frac{1}{2} C_Q k_B T \left[\exp\left(-\frac{qV_S}{k_B T}\right) - \exp\left(-\frac{qV_D}{k_B T}\right) \right]. \quad (2.118) $$

The back propagation drops faster as the drain bias voltage is applied and current saturates. The current now has no drain bias voltage

The current becomes very weakly dependent. Within the various approximations that we have employed, in then 0th order, we can treat it as having saturated.

dependence. Under these conditions,

$$J_{DSat} = g_q \frac{\sqrt{2\pi m^*}(k_B T)^{3/2}}{h} \exp\left(-\frac{E_0 + qV_S}{k_B T}\right), \text{ and}$$

$$Q_{Sat} = \frac{1}{2} C_Q k_B T \exp\left(-\frac{qV_S}{k_B T}\right), \tag{2.119}$$

which gives the effective velocity as

$$v_S = \frac{J_{DSat}}{Q_{Sat}} = \sqrt{\frac{2 k_B T}{\pi m^*}} = v_\theta, \tag{2.120}$$

the thermal velocity for a two-dimensional electron gas. In degenerate conditions, we found the velocity to be related to the Fermi velocity as $(4/3\pi)v_F$; for non-degenerate conditions, it is v_θ, a result that we should find satisfying. It matches physically the description and the condition analyzed.

The expressions of current can be reformed as follows:

$$J_D = \mathscr{A}_\theta v_\theta \exp\left(-\frac{E_0}{k_B T}\right)$$
$$\times \left[\exp\left(-\frac{qV_S}{k_B T}\right) - \exp\left(-\frac{qV_D}{k_B T}\right)\right], \text{ and}$$

$$Q = \frac{1}{2} C_Q k_B T) \exp\left(-\frac{E_0}{k_B T}\right)$$
$$\times \left[\exp\left(-\frac{qV_S}{k_B T}\right) - \exp\left(-\frac{qV_D}{k_B T}\right)\right]. \tag{2.121}$$

To write it in terms of applied bias voltages, we employ the relationship $V_{GS} - V_{ox} + (1/q)E_0$, which leads to

$$J_D = \mathscr{A}_\theta v_\theta \exp\left(-q\frac{V_{GS} - Q/C_{ox}}{k_B T}\right)$$
$$\times \left[1 - \exp\left(-\frac{qV_{DS}}{k_B T}\right)\right], \text{ and}$$

$$Q = \frac{1}{2} C_Q k_B T \exp\left(-q\frac{V_{GS} - Q/C_{ox}}{k_B T}\right)$$
$$\times \left[1 - \exp\left(-\frac{qV_{DS}}{k_B T}\right)\right]. \tag{2.122}$$

These are not explicit relationships but are sufficient to describe the current-voltage behavior implicitly. Note the drain bias dependence, which describes the decrease in back injection as V_{DS} is increased. It is the equivalent of the drain barrier lowering that one observes in the classical transistor description. In these expressions, the charge equation may be rewritten in the form

$$Q = C_{ox}\left\{\frac{qV_{GS}}{k_B T} + \ln\left[1 + \exp\left(-q\frac{V_{DS}}{k_B T}\right)\right] - \ln\left(\frac{Q}{\mathscr{A}_\theta}\right)\right\}, \tag{2.123}$$

which provides the fastest way to find a converging solution.

This ballistic transistor, lacking dissipation in the active region, is equivalent to the billiard ball model discussed in Chapter 1. The dissipation happens when the information of each electron in each channel is erased in the thermalization process. In the ballistic picture, because of the quantization effect, the effective velocities of the electrons change as the bias voltage changes. This simply states that the electron states available for conduction are changing. But, irrespective of the energy or the momentum of the state, if the state is occupied and the channel associated with that state is open, it will contribute a conductance of g_q associated with the spin $\pm 1/2$ electrons that can have that energy and momentum. The opening and closing of the channels lead to a $3/2$ power dependence in the current as a joint effect arising as a direct consequence of the energy aperture being opened and the distribution of states inside it. In the long-channel model of constant mobility, the velocity is low at the source end and keeps increasing without any physical meaning towards the drain end except to maintain current continuity with a mobility model—constant or otherwise. The dependence on gate bias is square law. In saturated velocity operation, the gate voltage dependence is linear, and the device operates with a constant charge moving at the saturated velocity.

The effective mass and the density of states are important in this discussion because they establish what statistics should be employed and what the consequences of these statistics are on the fraction of the applied bias being sacrificed to the occupation of higher energy states in the limited density of states available. In a material such as $Ga_{1-x}In_xAs$ or $InAs$, with low effective mass, the quantum capacitance (C_Q) is low. If the insulator is thin, even as thin as 1 nm, $C_Q \ll C_{ox}$. For small dimensions, the Maxwell-Boltzmann model will be in quite some error since a low density of states means that the statistics of occupation will be wrong. It will place more carriers in conductance channel than allowed. The ballistic model in the Fermi-Dirac limit would be more correct. Note that thermal velocities and saturated velocities are not too off from each other near room temperature. But, mean thermal velocity is lower than mean Fermi velocity under degenerate conditions. Thermal velocity averages over all occupied states. Fermi velocity is the velocity of states associated with the Fermi energy, which is higher up in the band. Even then, the current in the Boltzmann approximation, which is wrong, will be higher, and resistances in the non-saturated region will be lower. If the insulator is thickened, and the mass is higher, the two will come closer. Charge is now under insulator control rather than being controlled by the density of states, and the mean thermal and Fermi velocities are similar.

We now introduce scattering in this ballistic description of degenerate and non-degenerate conditions, thus describing the conditions shown in Figure 2.1(c) and completing the entire picture. We will have to make several assumptions. Carriers undergo scattering through a variety of mechanisms; some of them cause a loss in energy and momentum, and some of them may be elastic and only cause a change in momentum. An acoustic phonon or an optical phonon scattering is an example of the former—the different forms of these will have their own characteristics, and some of these may not even fit with the randomizing scattering time-constant description that we have relied on up to this point. Electron-electron scattering causes two electrons to flip momentum, but the total energy remains constant. In such situations, a statistical simulation, for example, a Monte Carlo approach incorporating all these conditions and adequate sampling, would be a more comprehensive description. But, it too misses the importance of the wave nature—the nonlocality of that description—and treats the electron only as a particle. So, all the different descriptions have assumptions that limit their validity, or more appropriately, their exactness. One should keep this in mind and consider these models as useful guidelines for understanding what important phenomena are relevant and how they can best be incorporated so that one can make good judgments in predicting the general nature of the physical description. Quite a number of scattering events are randomizing, for example, due to Coulombic fluctuations at the insulator-semiconductor interface, even several of the phonon processes, and many of these are the major ones in limiting carrier transport. Therefore, a time-constant picture—even a single time constant—is useful in describing a near-ballistic device. The electron wave picture gives us a good tool for incorporating this time constant when including the reflection processes. This is a subject for which the electromagnetic scattering matrix description and its reflection coefficient viewpoint provide us with tools for analysis.

Figure 2.36 pictorially describes electron wave interactions with a single scatterer and with two scatterers. If the wave undergoes one scattering, it has a transmission probability/coefficient of $\mathcal{T}$ and a reflection probability/coefficient of $\mathcal{R}$. The sum of the amplitudes is a constant, so $\mathcal{T} + \mathcal{R} = 1$. If there are two sequential scattering events, one may view the scattering as a cascade of multiple reflections caused by two scatterers. Figure 2.36(b) shows this cascade, which forms the infinite series that the net scattering embodies. The wave that is reflected back does not appear at the output, that is, the drain. The wave that is transmitted with the probability amplitude $\mathcal{T}_1$ at the first scattering site transmits out past the sec-

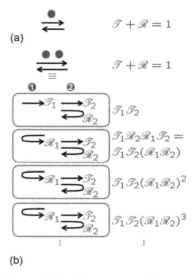

Figure 2.36: An electron wave undergoing a scattering event at a single scattering site (a) and at two scattering sites sequentially (b) in an idealized single channel system.

Recall that a quarter-wave transformer with a geometric mean for characteristic impedance matches impedances between a source and a load. This can be viewed through the wave picture and impedance transformation. It can also be viewed through scattering at the two ports. The matched impedance occurs with a standing wave in the quarter wave line. A standing wave is the result of balancing counter-propagating waves.

ond site with an accumulated probability of $\mathscr{T}_1\mathscr{T}_2$ and reflects back with an accumulated probability of $\mathscr{T}_1\mathscr{R}_2$. This reflected wave can again reflect back at the first scattering site and transmit at the second with a probability amplitude of $\mathscr{T}_1\mathscr{R}_2\mathscr{R}_1\mathscr{T}_2 = \mathscr{T}_1\mathscr{T}_2(\mathscr{R}_1\mathscr{R}_2)$. Or it can reflect back at the second site with an accumulated probability of $\mathscr{T}_1\mathscr{R}_2\mathscr{R}_1\mathscr{R}_2$ and then have a smaller accumulated probability, $\mathscr{T}_1\mathscr{T}_2(\mathscr{R}_1\mathscr{R}_2)^2$, of transmitting when it reaches the second scattering site again. This scattering cascade leads to a net transmission of $\mathscr{T}_{12}$ in the sequential two scattering events process of

$$
\begin{aligned}
\mathscr{T}_{12} &= \mathscr{T}_1\mathscr{T}_2 + \mathscr{T}_1\mathscr{T}_2(\mathscr{R}_1\mathscr{R}_2) \\
&+ \mathscr{T}_1\mathscr{T}_2(\mathscr{R}_1\mathscr{R}_2)^2 + \cdots = \frac{\mathscr{T}_1\mathscr{T}_2}{1 - \mathscr{R}_1\mathscr{R}_2}.
\end{aligned}
\tag{2.124}
$$

Since $\mathscr{T}_1 + \mathscr{R}_1 = 1$, and $\mathscr{T}_2 + \mathscr{R}_2 = 1$, it follows that

$$
\frac{1 - \mathscr{T}_{12}}{\mathscr{T}_{12}} = \frac{1 - \mathscr{T}_1}{\mathscr{T}_1} + \frac{1 - \mathscr{T}_2}{\mathscr{T}_2}.
\tag{2.125}
$$

This last relationship hints at the generalization. If there are N scattering events, then transmission through the N scattering events is

$$
\frac{1 - \mathscr{T}_{1N}}{\mathscr{T}_{1N}} = N\frac{1 - \mathscr{T}}{\mathscr{T}}
$$
$$
\therefore \quad \mathscr{T}_{1N} = \frac{\mathscr{T}}{\mathscr{T} + N(1 - \mathscr{T})},
\tag{2.126}
$$

assuming identical transmission coefficients for each scattering event. In multiple scattering events, the ratio of reflection to transmission is the product of the ratios for each of the events, and one can write the transmission coefficients in a simple relationship. This is the simple statement of Ohm's law in the presence of scattering in the electron wave picture. No scattering, perfect transmission, and the conductance is g_q for an open channel. Single scattering—transmission and reflection—and this relationship gives us the decrease in conductance and hence the current that follows, because we know the transmission coefficient. This is the scenario that was depicted in Figure 2.19. We have now made it more detailed for a two-port system.

This generalized wave relationship can be employed in the form of a transmission coefficient ($\mathscr{T}$) and reflection coefficient ($\mathscr{R}$) arising from multiple scattering events in transport through the near-ballistic transistor channel, as shown in Figure 2.37. For the source electron wave flux with positively directed forward momentum, represented by the current J_S, there is finite transmission ($\mathscr{T}_S$) and reflection ($\mathscr{R}_S$) at the source-channel interface. This is the input port, in network terminology. Electron waves that transmit beyond may undergo scattering. Some of this transmit through with a transmission coefficient

($\mathscr{T}_c$) and some reflect back ($\mathscr{R}_c$). Only some of the carriers that scatter in the channel actually return to the source. If the carrier loses energy beyond a certain amount, it doesn't have enough energy to surmount the barrier at the source-channel interface and so remains in the drain. Carriers that do not scatter at all also continue on and are collected. Our Port 2 is this boundary, crossing which ends the possibility of return. So, the reflection coefficient $\mathscr{R}_c$ is to be determined by parameters that describe the number of scattering events and a potential energy change. The former is reflected in length scale of the travel and the mean free path, and the latter in change that is sufficient to affect the back injection.

In this approach, we have introduced scattering, whether due to the barrier or due to scattering centers, through transmission and reflection coefficients. The accumulation of scattering events in the channel causes a forward scattering transmission of $\mathscr{T}_c$ and a backscattering reflection of $\mathscr{R}_c = 1 - \mathscr{T}_c$. The net current flux at the drain is

$$J_D = \mathscr{T}_S \mathscr{T}_c J_S. \tag{2.127}$$

The current density at the source end, with carriers moving at thermal velocity, so assuming non-degenerate conditions, is the sum of the incident and the reflected fluxes. This gives

$$
\begin{aligned}
Q &= qn_S = J_S \mathscr{T}_S (1 + \mathscr{R}_c) \frac{1}{v_\theta} \\
\therefore\ J_D &= qn_S v_\theta \frac{\mathscr{T}_c}{1 + \mathscr{R}_c} = qn_S v_\theta \frac{1 - \mathscr{R}_c}{1 + \mathscr{R}_c}.
\end{aligned} \tag{2.128}
$$

We could have written this equation under degenerate conditions too from the integration of Equation 2.116 and introduced these transmission and reflection coefficients.

The question we face is, how do we estimate the backscattering coefficient ($\mathscr{R}_c$) arising from the accumulation of the scattering events? The mean free path $\langle \lambda_{\mathbf{k}} \rangle$ is a length scale that estimates the distance traversed without scattering. We introduce an effective length ℓ_{eff} as the estimate of the distance that the electrons travel after which they cannot return back because they don't have enough energy. In his analysis of thermionic emission across a metal-semiconductor junction, the topic that seeded the start of our discussion, Bethe posited $k_B T$ as this energy. So, ℓ_{eff} is a length scale for $k_B T$ energy change. It is the energy scale of the distribution, and as the energy drops below this, the channels vanish for back injection. The magnitude is a measure of the means involved in the open channels, their occupation and the loss of energy. Because the density of occupation decreases at this energy scale length, its loss at this same scale length in the scattering process eliminates the back injection.

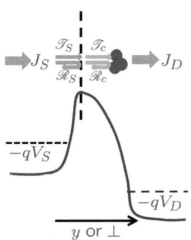

Figure 2.37: Transmission and reflection, including scattering, in a near-ballistic transistor, processes dominated by events at the source side injection interface of the channel. A carrier may reflect back into the channel because it still has sufficient energy for back transmission. If it loses more energy than a threshold, it cannot go back to the source. The electron waves of carriers that cannot return to the source, as well as of those that do not scatter at all, contribute to the drain current.

When is a carrier truly in the drain and not going back? The Bethe condition states that this is when the potential is dropped by $k_B T/q$ below the peak of the barrier. But, tunneling also occurs across thin barriers increasing the forward flux and there are lucky scattering events where energy can be gained— such as via absorbing an optical phonon together with a flipping of the direction, for example. So, it is not that simple. But, this is good enough for this analysis, which is meant to give us a good approximation of the magnitude and a picture of the description it embodies.

The relationship of transmission and reflection, as discussed for multiple scattering, can be employed easily using

$$\mathscr{R}_c = \frac{\ell_{eff}}{\ell_{eff} + \langle \lambda_{\mathbf{k}} \rangle},\tag{2.129}$$

which relates the reflection coefficient as a parameter and interpolates between the two limits. It is accurate in the limits and interpolated for convenience in between. A fraction of the electron wave stream is reflected. In the limit of $\langle \lambda_{\mathbf{k}} \rangle \gg \ell_{eff}$, that is, very few scattering events, the reflection coefficient is proportional to the probability of encountering a scattering center. In the limit $\langle \lambda_{\mathbf{k}} \rangle \ll \ell_{eff}$, scattering reflection from energy change dominates. Only a small fraction of incident source electron flux makes it through, that is, is extracted. This is the same as saying that the source is an infinite source of carriers, and the channel and drain are extracting a small number consistent with the movement of carriers and their density, that is, velocity, mobility, diffusivity and fields. This is what happens in long-channel structures. This relationship is certainly reflective of these two limits. But, it is only a forced fit to the dynamics between them. When scattering is limited, and the scattering events have different characteristics, including absence of randomization in some instances, then the use of this relationship is an *ad hoc* implementation where the physical meaning of some of these parameters is lost.

Using this relationship, we can now write the current. Assume that the quantum capacitance is large, that its effect in degrading charge induction is insignificant, and therefore that the design of the device is extremely efficient in coupling the gate to the channel; then, the drain current is

$$J_D = C_{ox}(V_G - V_T)v_\theta \frac{1 - \mathscr{R}_c}{1 + \mathscr{R}_c}.\tag{2.130}$$

We could have written this equation more rigorously because we have found the charge relationships and they certainly can be self-consistently employed.

We have not really said much yet about how to get the barrier right. It is modulated by the potentials through a number of device and material characteristics that are tied to the electrostatics and the material degeneracy. We have a good tool for this from the double-gate discussion: the Poisson solution employing hyperbolic and sinusoidal basis functions, because they satisfy the 2nd order differential equation. This is a good starting point for us. We found that scaling length was a good tool for analyzing the more complicated problem. In addition, we found that the decay towards the drain is hyperbolic and therefore exponentially changing. So, there exists a barrier along the y direction, estimable through these exponentials; and then, as we go further out, there is an ohmic field that develops, reflective of a quasineutral contact connecting to a controlled channel region.

In Figure 2.38, the origin is chosen as the position of the peak of the barrier. This is the channel point that determines all the open conductance channels. It is the source-channel interface, where our major shortcoming has been in ignoring the tunneling through the barrier. The potential near the source, based on the arguments of the Poisson relationship and its hyperbolic and sinusoidal basis functions, for $y \geq 0$ is

$$V(y) = V_0 \left[1 - \exp\left(-\frac{y}{\lambda_B} \right) \right] - \mathcal{E}_B y. \qquad (2.131)$$

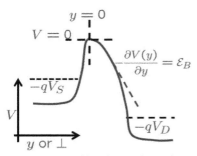

Figure 2.38: Band bending and coordinate origins for the potential voltage in the channel. The linear first term in the field relationship is also shown.

The electric field near the source is

$$-\frac{\partial V(y)}{\partial y} = -\frac{V_0}{\lambda_B} \exp\left(-\frac{y}{\lambda_B} \right), +\mathcal{E}_B \qquad (2.132)$$

where we have employed λ_B as a device barrier length scale parameter related to the effective field $\mathcal{E}_B$, which relates to introducing the Bethe condition. The barrier is a virtual cathode from which the emission takes place at a vanishing field at $y = 0$; $V_0 = \mathcal{E}_B \lambda_B$, and

$$V(y) = \mathcal{E}_B \lambda_B \left[1 - \exp\left(-\frac{y}{\lambda_B} \right) - \frac{y}{\lambda_B} \right]. \qquad (2.133)$$

We can expand this equation in Taylor series terms to the next higher order, so that

$$V(y) = \mathcal{E}_B \lambda_B \left[1 - 1 + \frac{y}{\lambda_B} - \frac{1}{2}\left(\frac{y}{\lambda_B} \right)^2 - \frac{y}{\lambda_B} \right] = -\mathcal{E}_B \frac{y^2}{2\lambda_B}, \qquad (2.134)$$

where a parabolic term appears again. The Bethe condition for the effective scale length, the condition when carriers should be considered not returnable from the drain, gives

$$\frac{k_B T}{q} = \mathcal{E}_B \frac{\ell_{eff}^2}{2\lambda_B} \quad \therefore \quad \ell_{eff} = \sqrt{\frac{2 k_B T \lambda_B}{q \mathcal{E}_B}}. \qquad (2.135)$$

We can now fill in the parameters. The field is $\mathcal{E}_B = J_D / \sigma_{ch}$, where the conductivity is

$$\sigma_{ch} = \mu C_{ox}(V_G - V_T) = \frac{q \langle \lambda_\mathbf{k} \rangle v_\theta}{2 k_B T} C_{ox}(V_G - V_T). \qquad (2.136)$$

By substituting the Drude expression in the far field, which extends into the drain region, where this field dominates with the injected carriers responding to it, we can write

$$\ell_{eff} = \sqrt{\frac{2 k_B T \lambda_B}{q} \frac{\mu C_{ox}(V_G - V_T)}{J_D}}$$

$$= \sqrt{\frac{\lambda_B \langle \lambda_\mathbf{k} \rangle v_\theta C_{ox}(V_G - V_T)}{J_D}}. \qquad (2.137)$$

This equation then leads to the following expression for the reflection ratio:

$$\frac{1 - \mathscr{R}_c}{1 + \mathscr{R}_c} = 1 + \xi - \sqrt{1 + 2\xi}, \tag{2.138}$$

where

$$\xi = \frac{2\mu k_B T}{q \lambda_B v_\theta} = \frac{\langle \lambda_\mathbf{k} \rangle}{\lambda_B} \tag{2.139}$$

is a parameter that relates the mean free path to the barrier length scale of travel. This equation leads to the following current-voltage relationship in the near-ballistic transistor in the Boltzmann limit:

$$\begin{aligned} J_D &= C_{ox}(V_G - V_T)v_\theta \frac{1 - \mathscr{R}_c}{1 + \mathscr{R}_c} \\ &= C_{ox}(V_G - V_T)v_\theta(1 + \xi - \sqrt{1 + 2\xi}). \end{aligned} \tag{2.140}$$

When $\xi \ll 1$, that is, $\lambda_\mathbf{k} \ll \lambda_B$, that is, there exists significant scattering, the current is small, with the effective velocity reducing to

$$v_S \approx \frac{1}{2}\xi^2 v_\theta = \frac{1}{2}\left(\frac{\langle \lambda_\mathbf{k} \rangle}{\lambda_B}\right)^2 v_\theta \text{ for } \xi \ll 1. \tag{2.141}$$

The source-end velocity is directly related to the mobility and the effective field at the source end in the high scattering limit for injection across the barrier.

In S. Tiwari, "Device physics: Fundamentals of electronics and optoelectronics," Electroscience 2, Oxford University Press, ISBN 978-0-19-875984-3 (forthcoming), we will find these correspondences in velocities when we analyze short-channel devices by approaching their analysis from our long-channel understanding.

2.6 Zero bandgap and monoatomic layer limits

Inversion layers are a few nm thick. Quantum well systems are in the limits of thickness $t \approx \lambda_{deB}$. Our discussion up to this point has been at these nanoscale dimensions. What happens when one gets to the atomic limit of the two-dimensional system? This is possible in carbon-based structures such as graphene and nanotubes and in compounds such as MoS_2, WSe_2, et cetera. Some of these, such as graphene, have no bandgap. Others, such as semiconducting nanotubes, and the two-dimensional compounds may have a bandgap.

Absent bandgap (see Figure 2.39) and when $\nabla_k^2 = 0$, one can work with a reformulation of our traditional semiconductor interpretations.

The material is always degenerate. If n_s is the sheet charge density, the Fermi reciprocal vector is $k_F = (2\pi n_s)^{1/2}$.

$$\mathbf{p} = m^* \mathbf{v}_F = \hbar \mathbf{k}_F \quad \therefore \quad m^* \equiv \frac{\hbar k_F}{v_F} = \frac{\pi^{1/2} \hbar}{v_F} n_s^{1/2}, \tag{2.142}$$

with the Dirac particle moving at a speed of v_F, the Fermi velocity, which is the equivalent of the speed of light c. With this, since $E =$

The semiconductor condensed matter physics view of crystalline carbon materials will be dwelt on in S. Tiwari, "Semiconductor physics: Principles, theory and nanoscale," Electroscience 3, Oxford University Press, ISBN 978-0-19-875986-7 (forthcoming).

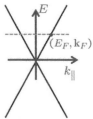

Figure 2.39: Energy-wavevectors near the Dirac point in atomic sheet planar materials such as graphene.

pv_F for the massless condition, the energy-wavevector relationship can be written as

$$E = \hbar v_F k_\parallel = \hbar v_F \sqrt{k_y^2 + k_z^2}. \tag{2.143}$$

Bandstructure calculations show that $v_F \approx 10^8 \ cm/s$ for graphene—about $1/300$ of the speed of light. For a sheet carrier density of $1 \times 10^{12} \ cm^{-2}$, $k_F \approx 1.8 \times 10^6 \ cm^{-1}$, and there is a "mass" of $0.02m_0$, and a Fermi energy of $116 \ meV$. The carrier density in these monoatomic layer materials is therefore small, and they sample a small part of the reciprocal space at energies of interest, but they have the compelling attributes of small scattering and high mobility under controlled conditions and as a result exhibit a variety of the phenomena—quantum conductance, ballistic conduction, quantum capacitance, et cetera—that we have discussed. Nanotubes also naturally have the cylindrical form that we have identified as ideal for constant potential scaling, while providing a bandgap in semiconducting varieties at small diameter.

2.7   Parasitic resistances

IF A TRANSISTOR PROVIDES superior conduction, then gaining the benefits of it requires superior access to the intrinsic characteristics of the device. This accessibility through the parasitic resistances of the source reservoir, the drain reservoir and the gate in atomically thin planar materials and in other semiconductors needs to be considered in some detail. When scattering is small and the conductance in the natural active controlled intrinsic behavior of the device is improved, then the conductance of the access to this intrinsic region must also improve proportionally. This is demonstrated in Figure 2.40, which shows a transistor circuit model that is useful for a relatively broad frequency range, as it includes channel signal phase delay effects in the form of the time constant τ_d, with the variety of regions over which fields exist being represented by capacitors and resistors under field control. Since transistors modulate the output current via the input voltage, it is the voltage across the part of the capacitance that controls the barrier in the channel (that is, v'_{gs} across C_{gs}) that is important, but this is only a fraction of the applied gate-to-source voltage v_{gs}. Because of the current flowing in the transistor as a gain device, there is a negative feedback that arises from the lost energy, particularly in the source resistance R_s. This is the resistance arising from the reservoir that allows the connection to the intrinsic device. The device also has a series resistance in the output that prevents coupling of all the power generated in the intrinsic device to the

Zero bandgap and a linear E-$\mathbf{k}$ relationship are photon-like, except for the nuances of charge, spin of $\pm 1/2$, and a velocity quite different in magnitude from that of a photon but which is still a constant. Such materials have been called Dirac materials, and the degeneracy point, $\mathbf{K}$ or $\mathbf{K}'$ point in the reciprocal space for graphene, for example, the Dirac point because of the equivalence of this to the relativistic description by the Dirac equation for the spin $\pm 1/2$ particles. They allow exploration of relativistic behavior at small velocities in condensed matter. Also recall the relativistic equation $E^2 = p^2 c^2 + m^2 c^4$ discussed earlier.

In silicon inversion layers, a carrier density of $1 \times 10^{12} \ cm^{-2}$ occurs with the Fermi energy about $30 \ meV$ below the conduction bandedge at $3 \times 10^{17} \ cm^{-3}$ acceptor doping on a (100) surface. Silicon remains non-degenerate. At oxide breakdown fields ($10 \ MV/cm$), silicon can sustain $2.2 \times 10^{13} \ cm^{-2}$ electrons, which would require a Fermi energy of $\sim 100 \ meV$—conditions that are degenerate.

output. Even if one assumes operation at static conditions, that is, capacitors as open devices, these resistances prevent coupling of all the power.

These losses in the access to the intrinsic device are traditionally reflected in two figures of merit. The first is the frequency at which the current gain between the output and the input, with the output shorted, goes to unity. This is the unity current gain frequency f_T. The second is the frequency at which the maximum power gain f_{max}, when a two-port device has been made unilateral using only lossless reciprocal elements, goes to unity. The current gain, in the model of Figure 2.40, is

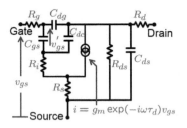

Figure 2.40: A generic circuit model for a transistor including delay effects due to signal transit time in the channel and the electrostatic effects that control the current flow.

$$\frac{i_d}{i_g} = \frac{1}{g_m}\left\{\omega(C_{gs}+C_{dg})[1+(R_d+R_s)/R_{ds}]\right.$$
$$\left. +\omega C_{dg}g_m(R_d+R_s)\right\}, \qquad (2.144)$$

which results in the short-circuit unity current gain frequency being

$$f_T = \frac{g_m}{2\pi}\left\{(C_{gs}+C_{dg})[1+(R_d+R_s)/(R_{ds})]\right.$$
$$\left. +C_{dg}g_m(R_d+R_s)\right\}^{-1}. \qquad (2.145)$$

The unity power gain frequency in the conditions described is

$$f_{max} = \frac{g_m}{2\pi(C_{gs}+C_{dg})}\left\{4\frac{R_g+R_s+R_i}{R_{ds}}\left[1\right.\right.$$
$$+4\pi f_T R_{ds}C_{dg}\left(1+\frac{R_s}{R_g+R_s+R_i}\right.$$
$$\left.\left.\left.+\frac{2\pi\tau_d}{(R_g+R_s+R_i)C_{gs}}\right)\right]\right\}^{-1/2}. \qquad (2.146)$$

The ratio $(R_d+R_s)/R_{ds}$ originates in the division of return current between the paths provided by the output resistance of the device and the parasitic resistances of the transistor. This feedback effect is increased by the Miller feedback gain factor arising from the inversion between the input and output. So, if the output resistance is large, that is, the conductance channels once the device reaches saturation are unaffected by any drain field penetration, then the consequences only arise from the interaction between the parasitic resistances and the parasitic resistances from the source and the drain. Current gain is a particularly important parameter for current-controlled current output devices such as the bipolar transistor. The field effect transistor's output current is controlled by the input voltage, and its ultimate utility at the highest frequencies is more closely aligned to the f_{max}. Here the parasitics show important direct effects arising from the dissipation in them, and these in turn limit the available power gain from the device in digital circuits. If the output resistance is poor and so are the source resistance and the drain resistance, the device will be of limited utility.

These consequences of negative feedback extend to lower frequencies too, the most important being the limiting of the current drive capability of the device and thus the time constants of charging the capacitive loads of interconnects and other transistors. For example, in Figure 2.41(a), which is stripped to bare essentials, the intrinsic transconductance of the device is g_{mi}, which causes this ideal device to have an intrinsic capability of $i_{di} = g_{mi}v_{gsi}$ in drive current for a v_{gsx} gate-to-source external voltage bias. In Figure 2.41(b), with the introduction of the source resistance, the current capability of the extrinsic device is now $i_{dx} = g_{mi}(v_{gsi} - i_{dx}R_s)$, that is, $i_{dx} = g_{mi}v_{gs}/(1 + g_{mi}R_s)$. The external transconductance of the device g_{mx} degrades by a factor of $1/(1 + g_{mi}R_s)$. The higher the current capability and, hence, conductance channel capability of the device, the lower its source resistance needs to be so that the capability is gainfully employed.

We should, therefore, explore resistances in nanoscale devices.

First, consider the situation where a thicker conducting region connects to a thinner conducting region. A two-dimensional electron gas plane connected to a shallow junction or a deep junction in a transistor is a good example of this. Most transistor geometries, double-gate, quantum wire, planar, etc., reflect variations on placing a highly conducting reservoir in particle and energy exchange contact with the intrinsic transistor region. The current flux has to enter the thin region and, even ignoring the effects in resistances arising from funneling of three-dimensional **k** states to two-dimensional ones in a one-dimensionally confined plane, or one-dimensional conduction in the case of a quantum wire—the reciprocal space funneling effect discussed earlier—there are crowding effects that should be considered as these dimensions shrink. This arises entirely from the dimensionality and the use of small dimensions of the semiconductor. In the zeroth order, this is entirely classical and is a real space funneling effect.

In addition to the basis set technique employed earlier in analysis of scaling in the off state, conformal mapping techniques allow one to transform problems into a more tractable form—analytic or computational. The Schwarz-Christoffel transformation technique is a particularly powerful form useful in electricity and magnetism and many problems with abrupt geometries. Appendix H discusses the power of these transformations. For the problem of transistor channel contact outlined in Figure 2.42, the solution to the Laplace equation, without resorting to the orthonormal basis expansion we employed

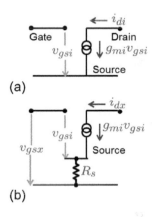

Figure 2.41: A very simplified model of a voltage-controlled current source as a model simplification of the transistor. (a) shows the intrinsic device with no parasitics. (b) shows the device with a source resistance introduced.

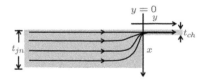

Figure 2.42: A thicker conducting region of depth t_{jn} contacting a thinner conducting transistor channel region of depth t_{ch}, illustrating current crowding and the consequent spreading-based resistive effects in nanoscale geometries.

earlier under abrupt transition conditions, follows relatively easily through this approach of conjugate functions. The surface potential $\psi_s(y)$ for the geometry of Figure 2.42 can be written as

$$\psi_s(y) = \frac{\mathcal{E}(y = -\infty)t_{jn}}{\pi}\left[\ln\left(\frac{\zeta^2 - 1}{\zeta_0^2 - 1}\right) + \ln\left(\frac{1 - \gamma^2\zeta_0^2}{1 - \gamma^2\zeta^2}\right)\right], \quad (2.147)$$

where ζ is the ratio of the surface to at-depth current flux, t_{jn} is the junction depth, and γ is the depth ratio, in the forms

$$\zeta = \frac{n(0,y)v(0,y)}{n(t_{jn},y)v(t_{jn},y)}, \quad \gamma = \frac{t_{ch}}{t_{jn}}, \quad \text{and} \quad \zeta_0 = \zeta(y=0), \quad (2.148)$$

which satisfy

$$\frac{\zeta_0 + 1}{\zeta_0 - 1}\left(\frac{1 + \gamma\zeta_0}{1 - \gamma\zeta_0}\right)^{\gamma} = 1. \quad (2.149)$$

For the source, $\zeta \leq \zeta_0$. For the channel, $\zeta \geq \zeta_0$, and in this transformation for the thin conducting sliver of the two-dimensional layer, the surface potential relationship of Equation 2.147 is modified so that the field $\mathcal{E}(y = -\infty)$ is replaced by the asymptotic low field of the source end of the channel, and the thickness is replaced by the channel thickness t_{ch}. The calculation of this surface potential, hence the carrier concentration, allows the source spreading resistance R_{ss} to be written as

$$\begin{aligned} R_{ss} &= \frac{\rho_{jns}t_{jn}}{\pi W}\ln\left[\frac{(1+\gamma)^{\gamma+1}}{4(1-\gamma)^{\gamma-1}}\frac{\zeta_0^2 - 1}{1 - \gamma^2\zeta_0^2}\right] \\ &\quad - \frac{\rho_{chs}}{\pi W}\ln\left[\frac{4\gamma^2(1-\gamma)^{1/(\gamma-1)}}{(1+\gamma)^{1/(\gamma+1)}}\frac{\zeta_0^2 - 1}{1 - \gamma^2\zeta_0^2}\right], \quad (2.150) \end{aligned}$$

where ρ_{jns} and ρ_{chs} are the sheet resistances of the source and the channel, respectively, and W is the device width. Transistors, with their thin channels and deeper junctions, have $\gamma \ll 1$, which causes $\gamma\zeta_0$, its product with the flux ratio at the channel interface, to approach a constant of ~ 0.83. This leads to a simple relationship for the resistance caused by the doped junction region and its interface with the channel:

$$R_{ss} = \frac{2\rho_{jns}t_{jn}}{\pi W}\ln\left(\frac{3t_{jn}}{4t_{ch}}\right) - \frac{0.21\rho_{chs}t_{ch}}{\pi W}. \quad (2.151)$$

The second term here represents the contribution of effective high conductivity in the doped source current spreading region to the shortening of the channel. This is the length scale for the potential change that takes place at the source-channel interface, for example, the rising barrier in Figure 2.38, where $y = 0$ was chosen as the cold cathode source. Since the channel is thin in an inversion layer, only a

The power of Schwarz-Christoffel mapping stems from the observation that if the derivative of the conformational transformation function can be expressed as the product of another set of functions, then these functions, if they are step functions, lead to a mapping in which coordinate axes transform to polygons. So, step changes and piecewise continuous changes become related through the transformation. This shape and form change allows a more tractable solution, where corners, lines and other such step features that create mathematical difficulties in solving differential equations are remapped to palatable conditions. In our problem here, we are transforming simple rectangular geometries.

few *nm*, this penetration is short, less than one tenth of the channel thickness, and a very, very small fraction of the gate length or the mean free paths of the geometries. It can effectively be ignored, and so we are left with the first term. In devices with inversion layers, where $t_{ch} \approx 2.0$ *nm*, such as those with planar or multi-gate geometries where the conduction is still via the surface inversion layer, the first term dominates. Even in double-gate with bulk inversion or quantum wire geometries, because the dimensions are of the order of λ_{deB}, this is a reasonably accurate description with other approximations more dominant.

For a silicon inversion layer that is ~ 2 *nm* thick, contacted by silicon doped at 2×10^{20} cm^{-3}, and a 20 *nm*-thick region with a sheet resistance of ~ 250 Ω per square, the spreading resistance is ~ 6.5 $\Omega \cdot \mu m$. The sheet resistance of the conducting channel region, at its highest, is ~ 2500 Ω per square, so the conductivities of the two regions are only a factor of 10 apart, about the limit of acceptability with fractional degradation even in scattering-dominated limits. In atomically thin planar geometries, such as with graphene or other monatomic layer semiconductors, this consideration becomes quite important. Assuming just classical spreading in identical transport conditions, a thinning of the t_{ch} by a factor of 10 to atomic dimensions results in an increase in this resistance contribution of a factor of $\ln 10$, that is, about 2.303. But, in structures of atomic thickness scale, one particularly needs to consider the contact metallurgy to the conducting contact region also. In silicon structures, such contacts employ tunneling that arises from a metal to highly doped semiconductor junction with a contact resistance that can be considerably small. In graphene and others, this may not be so because of the need to connect a three-dimensional **k**-state distribution to quantized conductance channels, where the degrees of freedom are significantly less.

We now look at the resistances in such metal-planar semiconducting regions. The simplest model of this is two conducting regions connected to each other using planar contacts, a classical geometry, as shown in Figure 2.43. This is an example of a distributed transmission line based on lumped parameters, a resistive line, where the contact resistance or contact conductance of the interface is ρ_c or g_c normalized to length units, and the sheet resistance or conductance of the semiconducting region is ρ_s or g_s.

In writing a sheet resistance, we have made a two-dimensional conducting plane approximation where the spreading effects discussed earlier have been ignored. At any position y along the distributed line, there is a potential $V(y)$ and a current flow of $I(y)$. The

See S. Tiwari, "Device physics: Fundamentals of electronics and optoelectronics," Electroscience 2, Oxford University Press, ISBN 978-0-19-875984-3 (forthcoming) for the classical contact theory.

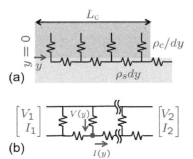

Figure 2.43: A highly conducting film contacting a resistive sheet through a resistive contact. This is a prototypical example of tunneling contacts, that is, bidirectional ohmic contacts, such as those encountered in silicon transistors using a metal silicide. (a) shows the origin of transmission line resistances, and (b) shows current and voltage definitions in a model in terms of impedance.

current is *normalized to width*, and so is in A/cm. We have

$$\frac{dV}{dy} = -\rho_s I = \frac{1}{g_s}I, \text{ and}$$

$$\frac{dI}{dy} = -\frac{1}{\rho_c}V = -g_c V. \tag{2.152}$$

These lead to the resistive transmission line equation, which can be written in voltage or current as

$$\frac{d^2 V}{dy^2} = \rho_s g_c V, \text{ or}$$

$$\frac{d^2 I}{dy^2} = \rho_s g_c I, \tag{2.153}$$

respectively. As these are second order differential equations, the solution is in terms of exponentials. Since the coefficient is real, it is hyperbolic. One form is

$$V(y) = V_+(0)\exp(-\sqrt{\rho_s g_c}y) + V_-(0)\exp(+\sqrt{\rho_s g_c}y), \text{ and}$$

$$I(y) = I_+(0)\exp(-\sqrt{\rho_s g_c}y) + I_-(0)\exp(+\sqrt{\rho_s g_c}y), \tag{2.154}$$

where

$$\frac{V_\pm(y)}{I_\pm(y)} = \rho_0 = \frac{1}{g_0} = \sqrt{\frac{\rho_s}{g_c}} = \sqrt{\rho_s \rho_c}. \tag{2.155}$$

Referring to Figure 2.43, for a contact of length L_c, we have, with reference at $y = 0$,

$$\begin{bmatrix} V_1 \\ I_1 \end{bmatrix} = \begin{bmatrix} 1 & 1 \\ 1/g_0 & -1/g_0 \end{bmatrix}\begin{bmatrix} V_+ \\ V_- \end{bmatrix}, \text{ and at } y = L_c,$$

$$\begin{bmatrix} V_2 \\ I_2 \end{bmatrix} = \begin{bmatrix} \exp(-\sqrt{\rho_s g_c}L_c) & \exp(\sqrt{\rho_s g_c}L_c) \\ -\frac{1}{g_0}\exp(-\sqrt{\rho_s g_c}L_c) & \frac{1}{g_0}\exp(\sqrt{\rho_s g_c}L_c) \end{bmatrix}\begin{bmatrix} V_+ \\ V_- \end{bmatrix}. \tag{2.156}$$

The two-port parameters can now be written as

$$\begin{bmatrix} V_1 \\ I_1 \end{bmatrix} = \begin{bmatrix} \cosh(\sqrt{\rho_s g_c}L_c) & -\frac{1}{g_0}\sinh(\sqrt{\rho_s g_c}L_c) \\ -g_0\sinh(\sqrt{\rho_s g_c}L_c) & -\cosh(\sqrt{\rho_s g_c}L_c) \end{bmatrix}\begin{bmatrix} V_2 \\ I_2 \end{bmatrix}, \tag{2.157}$$

which captures the behavior of the contact between the input and output. If a contact has voltage applied at one end, for example, V_1 here referenced to the constant potential of the conductive metal contact, with a current I_1 flowing into the semiconductor with an open circuit at the other end ($I_2 = 0$), one gets the resistance for the length L of the contact (R_c) as

$$R_c \equiv \frac{-V_1}{I_1}\bigg|_{I_2=0} = \rho_0 \coth(\sqrt{\rho_s g_c}L_c)$$

$$= \sqrt{\frac{\rho_s}{g_c}}\coth(\sqrt{\rho_s g_c}L_c) = \sqrt{\rho_s \rho_c}\coth\left(\sqrt{\frac{\rho_s}{\rho_c}}L_c\right). \tag{2.158}$$

This is the basis for the traditional transmission line method for measuring contact resistance. When two such contacts abut, or one extracts that abutment, one measures a resistance of $2R_c$. The scale length $L_T = 1/(g_c \rho_s)^{1/2} = (\rho_c/\rho_s)^{1/2}$ describes the exponential current transfer capability of the contact's resistive transmission line. A contact that is a few times this length scale allows one to achieve the maximum capability of the contact.

We now consider contacts to the semiconductor where transport is under ballistic and semi-ballistic conditions, as summarized in Figure 2.44, with the abscissa referenced to the abutment.

For purposes of convenience, we will consider the cylindrical geometry. This is where constant potential scaling is best achieved. We will define the quasi-Fermi levels E_{qFf} and E_{qFr} associated with the population of carriers with forward momentum and reverse momentum, respectively, in the semiconductor. Let

$$V_f = \frac{1}{q} E_{qFf},$$

$$V_r = \frac{1}{q} E_{qFr}, \text{ and}$$

$$V = \frac{1}{2}(V_f + V_r) \tag{2.159}$$

describe the quasi-Fermi voltages of these quasi-Fermi energies. If degeneracy is assumed, then the currents are

$$I_f = \frac{V_f}{R_Q}, \text{ and}$$

$$I_r = \frac{V_r}{R_Q}, \tag{2.160}$$

respectively, where R_Q is the net resistance of the quantized channels:

$$R_Q = \frac{h}{2n_{ch}e^2} = \frac{h}{2e^2} \left[\int_{E_{qFr}}^{E_{qFf}} \mathscr{G}_{1D}(E)\,dE \right]^{-1}. \tag{2.161}$$

From this, the current at any position in the channel is

$$I(y) = \frac{V_f - V_r}{R_Q}. \tag{2.162}$$

We assume that the contact and the semiconductor are at the same electrostatic potential. What this says is that the electrochemical potential and therefore the current do change with position as injection and extraction of charge occurs between the contact sheath and the conductor, but that the electrostatic potential radially remains constant. This is reflected in Figure 2.44. Now, we introduce, as in the previous constant resistance transmission line calculation, a contact

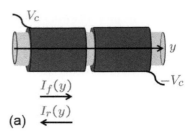

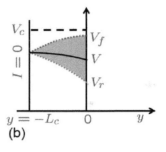

(b)

Figure 2.44: Contact, of length L_c, to a ballistic/quasi-ballistic semiconductor. (a) shows idealization of geometry where a conducting contact has a potential V_c applied in a symmetric arrangement of contacts. (b) shows the quasi-Fermi potentials associated with the forward and the reverse momenta with one of the contacts. Figure adapted from P. Solomon, IEEE Electron Device Letters, **32**, 246–248 (2011).

This use of quasi-Fermi energies to quantify the statistics with the system disturbed is analogous to the use of E_{qFn} and E_{qFp} for the quasi-Fermi levels for electron and hole populations, respectively, under conditions off from thermal equilibrium, that is, off equilibrium. In the case of E_{qFf} and E_{qFr}, it is simply the quasi-Fermi energy that suitably describes the filled forward momentum E-**k** states and the reverse momentum states that describe our statistical expectation.

coupling conductance g_c. We assume that this normalized conductance, or its inverse contact coupling resistance ρ_c, is a constant, whose magnitude will depend on the specifics of the physical processes at the sheath-quantum channels interface. The forward and reverse currents and the coupling between them due to scattering are reflected through the equations

$$\left.\frac{\partial I_f}{\partial y}\right|_c = \frac{1}{2}g_c\left(V_c - V_f\right),$$

$$\left.\frac{\partial I_r}{\partial y}\right|_c = -\frac{1}{2}g_c\left(V_c - V_r\right) \quad \text{at the contact, and}$$

$$\left.\frac{\partial I_f}{\partial y}\right|_{ch} = \left.\frac{\partial I_r}{\partial y}\right|_{ch} = -\alpha\left(I_f - I_r\right) \qquad (2.163)$$

in the channel. The latter equation describes the exchange of carriers between states of forward and reverse momenta because of scattering, which is characterized by the parameter α. This is similar to what we do in modeling impact ionization, the carrier generation process, or in modeling recombination process. In these, the electron and hole current equations are coupled through a length-dependent ionization or recombination parameter characterizing the coupling as a scattering process. The factor $1/2$ in the equations reflects that in the energy range of the E-$\mathbf{k}$ state distribution, half of the states have forward momentum, and the other half, reverse momentum. These considerations lead to the equations

$$\frac{dI}{dy} = -\frac{2\beta}{R_Q}\left(V - V_c\right), \quad \text{and}$$

$$\frac{dV}{dy} = -R_Q\left(\alpha + \frac{\beta}{2}\right)I, \quad \text{where } \beta = \frac{1}{2}g_c R_Q. \qquad (2.164)$$

This is another set of second order differential equations, similar to the resistive transmission line equation of before and with a similar form of solution of forward and backward propagating mode. The equations are

$$\frac{d^2 I}{dy} = (\beta^2 + 2\alpha\beta)I = \gamma^2 I, \quad \text{and}$$

$$\frac{d^2 V}{dy} = (\beta^2 + 2\alpha\beta)V = \gamma^2 V,$$

$$\text{where } \gamma = (2\alpha\beta + \beta^2)^{1/2}. \qquad (2.165)$$

The solutions for these equations are composed of $\exp(\pm\gamma y)$ functions with amplitudes subject to the boundary conditions of the problem. One of this set is that the current vanish, that is, $I = 0$, at the beginning of the contact ($y = -L_c$) and the other of this set is that $V = 0$ at the end of the contact, that is, at $y = 0$. Here, we are assuming a symmetric structure: two back-to-back contacts connected through the ballistic or near-ballistic conductor and separated by an infinitesimally small gap.

The current and voltage solutions are, respectively,

$$I(y) = \frac{2\beta V_c}{\gamma R_Q} \left\{ \frac{\exp(\gamma y) - \exp[-\gamma(y + 2L_c)]}{1 + \exp(-2\gamma L_c)} \right\}, \text{ and}$$

$$V(y) = V_c \left\{ 1 - \frac{\exp(\gamma y) - \exp[-\gamma(y + 2L_c)]}{1 + \exp(-2\gamma L_c)} \right\}. \quad (2.166)$$

The general contact resistance relationship is

$$R_c = \frac{V_c}{I(0)} = \frac{\gamma R_Q}{2\beta} \coth(\gamma L_c), \quad (2.167)$$

and it can be simplified in different limits of the existent scattering and of contacts. Diffusive transport is the high scattering ($\alpha \to \infty$) limit. Diffusive transport also implies that, instead of quantum conductance limited transport, we are now limited by transport characterized by a scattering-limited resistivity, such as of large dimension limit. Let a resistance R_s characterize this limit. It is in normalized units, so for this one-dimensional conductor, it is in units of resistance per length dimension. In general, we write this factor as

$$\alpha = \frac{R_s}{R_Q}, \quad (2.168)$$

where R_s is the resistance for the semiconductor per unit length and R_Q is the quantum resistance over all the conducting channels. α is thus in units of $1/cm$ which characterizes the exchange of forward momentum and reverse momentum states. These two comparable semiconductor properties, one in the scattering-dominated drift limit, the other in the scattering-absent ballistic limit, allow us to understand what happens in the contact when a conductivity g_c is sufficient in characterizing the contact interface. Therefore we can compare it to the physical description from Equation 2.158, which is for a contact to a semiconductor where scattering prevails.

The contact resistances over these different conditions, idealized and normalized, are shown in Figure 2.45, which shows calculated contact resistances over a few different conditions that are meant to reflect conditions of λ_k, and contact lengths in nanoscale geometries, assuming a few different possibilities for the contact-semiconductor interface properties. We look at this figure in light of what Equations 2.167 and 2.158 say.

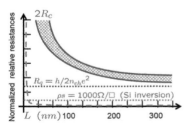

Figure 2.45: Contact resistance for different ballistic-to-quasi-ballistic conditions that are realistic in nanotubes and high mobility semiconductors. The figure is a normalized plot showing as reference the conditions of a classical silicon transistor.

Ballistic limit:

$$\lim_{\alpha \to 0} R_c = \frac{R_Q}{2} \coth\left(\frac{1}{2} g_c R_Q L_c\right). \quad (2.169)$$

This equation has the same hyperbolic cotangent limit that the classical transmission line contact equation 2.158 does, so increasing contact length decreases resistance in the ballistic or the classical case, of course. The prefactor however has no dependence on the contact interface's properties, that is, g_c or ρ_c. In the limit of a long contact or no limitations from the interface, a residual resistance of $R_Q/2$ still exists in this ballistic contact. For any current to flow, a certain energy window of open channels must be opened for that current to flow in the ballistic condition. This results in the asymptotic limit ballistic contact resistance of $R_Q/2$. Place two of these contacts back-to-back and pass a current through with no constraint from the contact interface, that is, unlimited g_c, and the total resistance will be R_Q caused by the open channels. In the case of drift transport, if the contact interface, that is, ρ_c or g_c, provides no limitation, then the contact resistance will asymptotically go to zero. Transport is assumed in the drift picture, with the ρ per square sheet resistance characterizing it. While both contacts have a contact length dependence through the hyperbolic function, the dependences on the contact interface and semiconductor resistance are different. This difference is also reflected in the prefactor.

Diffusive limit:

$$\lim_{\alpha \to \infty} R_c = \left(\frac{R_s}{g_c}\right)^{1/2} \coth\left(\sqrt{g_c R_s} L_c\right). \tag{2.170}$$

In the diffusive limit, we find the relationships to be identical. Normalized R_s is the sheet resistance. A classical and the quasi-ballistic large scattering limit is the diffusive limit.

Short contact limit:

$$\lim_{L_c \to 0} R_c = \frac{1}{g_c L_c}. \tag{2.171}$$

When contact is short, the resistance is entirely dominated by the interface conductance of the contact to the ballistic, or quasi-ballistic, channel.

Long contact limit:

$$\lim_{L_c \to \infty} R_c = \left(\frac{R_s}{g_c}\right)^{1/2}. \tag{2.172}$$

When the contact is long, in the quasi-ballistic limit, the resistance arises from the geometric mean arising from resistance in the interface and the resistance of the semiconductor at long lengths, that is, in scattering-controlled limit. The classical and the quasi-ballistic calculations lead to the same result.

These relationships can be viewed in light of what these conditions mean and the description we have incorporated in the derivation of

Equation 2.166. This is shown in Figure 2.46 for the applied, quasi-forward and quasi-backward potentials. This figure describes some key points. When there is no scattering, and the ballistic condition prevails ($\alpha = 0$), as one traverses further down the contact in the direction of current flow, the forward quasi-Fermi potential rises. Quantized channels contributing to forward flux are accumulating electrons up to the contact edge. For the reverse flux, they rapidly decrease, since these carriers are being swept into the contact. At the farthest edge of the contact, the forward and reverse quasi-Fermi voltages merge as constrained by the boundary condition. There is no current flowing here. The injection current at the contact begins to accumulate from this position along the contact, and this is why the number of open conducting channels is the maximum at the other end where it abuts the symmetric opposite contact. So, the figure is symmetric. Now, if there is scattering, a quasi-ballistic condition, such that $\alpha = 1\ cm^{-1}$, that is, a unity exchange per unit distance of travel down the contact, for the forward current, a quanta of $(\alpha = 1) \times I_f$ of current decrease due to removal of electrons from forward channels occurs, but a quanta of $(\alpha = 1) \times I_r$ is also gained. A symmetric process holds true for the reverse current. The total number of channels open and conducting at $y = 0$, the butting edge of the symmetric geometry, is smaller than when there is no scattering in transport.

If there is separation between contacts, then one can physically describe what this descriptive picture will look like. This is shown in Figure 2.47. Underneath the contacts, the events are similar to the ones just described. In the conducting region, if the transport is ballistic, the forward and reverse flux quasi-Fermi levels remain flat since these states are not interacting. In the quasi-ballistic case, there is interchange between the forward- and the reverse-directed states. The quasi-Ferm levels bend. But, for the quasi-Fermi levels band, at the contact edges, this is still a symmetric condition. The separation between the forward-directed and reverse-directed quasi-Fermi energies is increasingly smaller and vanishes at the farthest edge. At any position along the channel, the current is less, proportional to the difference between these two quasi-Fermi energies.

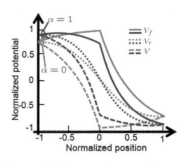

Figure 2.46: Contact voltage and the forward and reverse quasi-Fermi voltages of two back-to-back contacts to a quantized channels conductor. Figure adapted from P. Solomon, IEEE Electron Device Letters, **32**, 246–248 (2011).

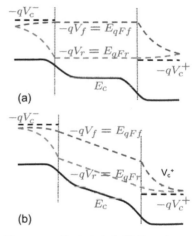

Figure 2.47: Contacts to ballistic and quasi-ballistic conducting channels under the application of a bias voltage of $2V_c$ between the contacts. (a) shows the quasi-Fermi energy—both the forward and the reverse quasi-Fermi energies—for the ballistic channel, and (b), for the quasi-ballistic channel. Adapted from P. Solomon, IEEE Electron Device Letters, **32**, 246–248 (2011)

2.8   Summary

THIS CHAPTER EXPLORED the behavior of transistors in their nanoscale limits under the non-conducting, that is, off, state, and conducting, that is, ballistic and quasi-ballistic, conditions.

The off state, which is important both by reason of the dissipative current flow, as well as for the random variations in the threshold voltage characterizing it and which in turn limit the smallest voltages that devices can be employed with, gave us a tool to understand the scaling laws for small geometries. We noted that, under constant potential scaling, which is necessary since constant field scaling reaches practical extremes of material characteristics, cylindrical geometries with a wrapped-around gate provide an invariant form. The off state analysis in double-gate geometries provided us with tools to understand length scaling. A good control of the threshold voltage and the drain bias independence of gate-control behavior in double-gate geometries led us to an understanding of the effective dimensional connections and hence good design rules for insulators to be employed for gate dielectrics—their physical and effective thicknesses—and the interrelationship of these thicknesses with drain-to-source spacing. We also employed this off state condition to explore how random dopants influence the statistical distribution of threshold voltage in device structures. This distribution too is important to the voltages that devices must employ in light of the error and energy constraints that were discussed in Chapter 1.

We established the description of how the state of rest and of motion of carriers, electrons specifically, but also holes, can be described within the wave description, and we related it to the classical particle picture. This let us employ the appropriate description tailored to the specific question to achieve a good understanding with an acceptable level of accuracy quantitatively. The ballistic electron requires us to explore it as a wave. In a confined geometry, there are only so many states that the electrons can occupy. This establishes their wavevector and energy that may be position dependent reflecting their kinetic energy. These can be stationary standing waves. An example is a low energy electron in the source region of a ballistic transistor, as it remains confined in the source. A traveling wave electron, with its charge, can be incorporated in the transistor description through conductance channels. These conductance channels are the allowed modes analogous to any other wave, such as, for example, an electromagnetic wave. The flux of this traveling wave, because of its charge, is associated with a current. A quantum conductance of $g_q = 2e/h \ A/eV$ or $2e^2/h \ S$ describes the conductance of each one of these channels where two electrons, of up ($\uparrow$) and down spin ($\downarrow$), respectively, exist. So, a channel of conductance gives rise to a quantum of conductance between two connected reservoirs. When two reservoirs are connected through, say, a quantum wire, the total conductance is the number of channels that exist between the extent filled states of the two reservoirs. The current arises from the balance of filled states from one reservoir conducting current to empty states that exist in the other reservoir. The net current is the difference of the flow from the two reservoirs. The current can flow, that is, the wave may travel in many directions defined by the E-$\mathbf{k}$ states. In a quantum wire, it travels along the wire. If one places a gate to control the electron population in these allowed states, it opens and closes the number of channels that exist between the source and drain reservoirs that the gate intervenes between. In the wave description, this opening and closing of channels is equivalent to mode filtering. The number of modes allowed to exist is changed by the gate potential, which changes the barrier for injection at the source. The gate effectively controls the mode filtering. When no channels with electrons are available for conduction, such as when the source and the drain filled states are at an energy below the barrier, then no current exists. When this barrier is lowered, and one channel opened, then a conductance of g_q exists with an $\uparrow$ and a $\downarrow$ electron wave propagating. If the electrochemical potential of the drain is changed w.r.t. the source, but only this channel remains extent, that is, the gate barrier effectively controls the modes allowed, then the current will not change. It will remain a constant. This was our simple description of the ballistic transistor where the linear current region arises because current flow in both directions.

This description can be modified in the near-ballistic limit to account for scattering by assuming randomizing scattering that causes reflections and transmissions. We could modify the behavior and see how the effective velocity of the injecting carrier at the source changes from a fraction of the Fermi velocity in degenerate or thermal velocity in non-degenerate conditions. Scattering length scale $\langle \lambda_{\mathbf{k}} \rangle$ and the Bethe condition could be folded in here through a transmission-reflection perspective borrowing from electromagnetics and thermionic emission theory.

Contacting nanoscale structures is an interesting problem since it concerns linking states whose dimensionality is changing. Many materials, such as carbon nanotubes, achieve their properties through reduced interactions between their charge cloud and the surroundings. Connecting unmatched states and dimensionality leads to resistances in contacting. Even conventional materials such as silicon, for example, at nanoscale, with their enhancement in current capability, place increased burden on the contact. After all, the contact should only cause a small fraction of loss in any active process. It is a parasite—a parasitic resistance. When two conducting regions are connected together so that current can flow between them, extreme dimension changes can cause resistance because of current crowding. An example of this is in the source-to-channel junction of a silicon transistor, even when a high doping is relatively easy to achieve in the source reservoir. So, extremes of dimensions matter and show up in these resistances. We found what the limits are likely to be in an idealized condition where the contact-semiconductor interface is characterized by a normalized conductance g_c. In practice, this may arise from tunneling, thermionic emission or other processes. This exploration let us understand the limits of resistance in connecting to quantum conductance channels as well as classical channels in their nanoscale limits.

2.9   Concluding remarks and bibliographic notes

TRANSISTORS, THEIR SCALING, their integration and their multifaceted use was one of the remarkable industrial achievement of the latter half of the 20th century. The improvement of device properties—in speed, area occupied, energy consumed—and, as a result, how many one could assemble in an integrated environment led to a fair social transformation. Computers, communication and information became ubiquitous, making many tasks easy, and commerce and the running of enterprises more streamlined. Like any invention and technology, it has had its beneficial effects and its deleterious effects. But, it is inarguable that scaling and the ability to practice the scaling through technology made this transformation possible. Our thought process here has been that when dimensions become so small, the traditional field scaling found in most microelectronics texts, including that from the author, must change. One needs to include the unique nanoscale and quantum effects in a way that is not simply *ad hoc*.

Scaling, however, may bring different conclusions at the nanoscale when one views the entire gamut of operation in the electrodynamic environment of the device and the collection of the device for a function. For the scaling length implications by threshold and gate control, Frank's publication describes the two-dimensional approach[1]. It also references several other earlier publications aimed at this subject.

Threshold variation is a limiting consideration because of the effect of noise margin. Sano's[2] treatment of this property is very complete.

These and other fluctuations, because of the smallness, and hence the consequence of the importance of single events, manifest in many ways. I view thermal noise and shot noise as complement of each other. One is time-centric and the other is space-centric. It is therefore not too unfair to say that there really are two and half—not three– types of noise of interest to us. The thermal noise and the shot noise constitute as one and a half.

Random telegraph signal and its manifestation in phase noise is

See the analysis of the various device structural forms, material interdependence and their physical underpinnings in S. Tiwari, "Device physics: Fundamentals of electronics and optoelectronics," Electroscience 2, Oxford University Press, ISBN 978-0-19-875984-3 (forthcoming).

[1] D. J. Frank, Y. Taur and H.-S. P. Wong, "Generalized scale length for two-dimensional effects in MOSFETs," IEEE Electron Device Letters, **10**, 385–387 (1998)

[2] N. Sano, K. Matsuzawa, A. Hiroki and N. Nakayama, "Probability distribution of threshold voltage fluctuations in metal-oxide-semiconductor field-effect-transistors," Japanese Journal of Applied Physics, **41**, L552–L554 (2002)

one example of what we have discussed. It is often misunderstood and it is particularly important at the nanoscale. A good review of low frequency noise in condensed matter is an old paper[3] that is still relevant today. A good discussion of random telegraph signal can be found in the publication of Cobden et al.[4]

For the device in its on state, operating under ballistic or quasi-ballistic conditions, we utilized the Bethe condition as an important part of the discussion of thermionic injection. A sharp discussion of the Bethe condition, including its history, may be found in Berz's[5] paper.

For ballistic and quasi-ballistic transport and its inclusion in description of the transistor's output characteristics, Natori's publication[6] is a fairly thorough early foray. Likharev[7] too idealizes these characteristics to extract the key underlying physical thoughts. Two other papers are also of note. Paul Solomon and Steve Laux[8] explore different materials with a rigorous Boltzmann transport equation–based simulation while including the presence of limited scattering and realistic bandstructure.

Contacts—particularly, trying to connect a diffusive region unconfined region to a confined ballistic region—have important implications, as we have emphasized. They must not dominate for the nanoscale improvement in device characteristics to be useful. Here, we have followed, in part, Paul Solomon's[9] treatment of the subject. In multi-dimensionally confined transistors, that employ ballistic properties of the material, the contact-length scale, rather than the gate-length scale, is the limiting feature because of funneling's rate-limiting consequences in the real space and the reciprocal space.

[3] P. Dutta and P. M. Horn, "Low-frequency fluctuations in solids: $1/f$ noise," Reviews of Modern Physics, **53**, 497–516 (1981)

[4] D. H. Cobden and M. J. Uren, "Random telegraph signals from liquid helium to room temperature," Micro-electronic Engineering, **22**, 163–170 (1993)

[5] J. Berz,, "The Bethe condition for thermionic emission near an absorbing boundary," Solid State Electronics, **28**, 1007–1013 (1985)

[6] K. Natori, "Ballistic metal-oxide-semiconductor field effect transistor," Journal of Applied Physics, **76**, 4879–4890 (1994)

[7] Y. Naveh and K. K. Likharev, "Modeling of 10-*nm*-scale ballistic MOSFETs," IEEE Electron Device Letters, **21**, 242–244 (2000)

[8] P. M. Solomon and S. E. Laux, "The ballistic FET: Design, capacitance and speed limit," Proceedings of the International Electron Devices Meeting, 5.1.1–5.1.4 (2001)

[9] P. Solomon, "Contact resistance to a one-dimensional quasi-ballistic nanotube/wire," IEEE Electron Device Letters, **32**, 246–248 (2011)

2.10   Exercises

1. A breakdown field of 10 MV/cm in SiO_2 corresponds to about 2×10^{13} cm^2 of elementary charge supporting the displacement. Yet, we have used the continuity of displacement $\epsilon\mathcal{E}$ at this same interface, that is, $\nabla \cdot \mathbf{D} = 0$, to derive the double-gate characteristics. How are these two consistent?   [S]

2. If no field is applied, $\mathbf{v} = \mu\mathcal{E}$ states that the velocity is zero. Are electrons in silicon, when no field exists in it, moving or standing still? Do they have any energy? Short explanation, please.   [S]

3. In a silicon transistor operating at a drain-to-source voltage of 1 V, a gate voltage of 1 V, and a current of 1 $mA/\mu m$, what is the average energy of each electron that is collected in the drain?   [S]

4. Let us apply a combination of classical and ballistic pictures in

a metal—copper. Calculate the time it takes for an electron to travel 10 *nm* in a nanowire of copper, assuming that the motion is ballistic. Justify why you chose the specific properties of that electron. How much would this time be in silicon going in the $\langle 100 \rangle$ direction on a (100) face for the same conditions. For this latter, has enough information been provided? If not, make a suitable assumption, and compare and remark on the two cases.   **[S]**

5. To a first approximation, in double-gate geometries at and below threshold conditions, the electrostatic potential is parabolic in the transverse direction, that is, between the two gates, and exponential longitudinally, that is, source to drain. Why? What would it look like in inversion when the channel thickness is very small?

     [S]

6. Which should be larger for electrons in semiconductors: the mean free path for energy relaxation, or the mean free path for momentum relaxation?   **[S]**

7. If it were possible to make high permittivity insulator transistors with both bulk and double-gate geometries, which one is likely to be scalable to the smaller dimension with good subthreshold and drive current characteristics?   **[S]**

8. Small transistors work with electrons moving about at the saturated velocity 1×10^7 *cm/s*. Estimate the frequency up to which the transistor will have a gain (current or voltage, make your choice) at 100 *nm*, 50 *nm* and 10 *nm* physical gate lengths.   **[S]**

This defines frequencies up to which linear circuits can be made to function.

9. This question is to get you to understand the practicalities of a small transistor. We did not discuss much related to this and assumed that we have the ability to work through the estimations. So, this question is to get this transistor understanding under your belt. Plenty of the undergraduate books will relate to this. Some of it can really be done through just geometric reasoning.

   • A 30 *nm* gate length transistor of 100 *nm* gate width uses modern via technology in its fabrication with a 1.8 *nm* SiO_2 insulator. Estimate its intrinsic gate capacitance and the parasitic gate capacitance due to the interconnection to it. Assume that a via contact is dropped across the entire source and drain contact region.

   • Take the example of the 30 *nm* gate length transistor of this last question. Assume that the threshold is defined by an acceptor doping of the order of 1×10^{18} *cm*$^{-3}$. Assume that two-dimensional effects do not dominate. Estimate the subthreshold swing of the transistor.   **[M]**

10. Employing the more rigorous scaling analysis of two-dimensional effects in a transistor, design approximately a decent 30 nm gate length transistor in bulk $CMOS$ and in double-gate geometry. Argue your choices. What are the insulator thickness, permittivity, silicon doping and thicknesses? Ignore source and drain design. Assume that they are good and we know how to achieve shallow and deep junctions. What do you estimate will be the subthreshold swing?   **[M]**

11. Optical phonons in silicon have an energy of about 60 meV. Electrons at high fields in silicon lose energy by optical phonon emission. Silicon has a low field mobility of about 400 $cm^2/V \cdot s$. Calculate the approximate saturated velocity and the distance traveled by an electron between optical phonon emissions at high fields. What is the distance traveled by electrons between scattering at low fields?   **[S]**

An important point to consider in this, and we have not, is that even if there is no field, these electrons are moving around. They have an average speed that we can calculate.

12. Let us explore what a threshold voltage means. This is a long question mathematically, so attempt it as much as you can. Let us assume that the material we work with is silicon, and the mobility of the electrons for the n-channel device is 300 $cm^2/V \cdot s$. It is a long-channel device. We have a p-type silicon substrate doped 10^{18} cm^{-3}. Consider only the component arising from band bending in the semiconductor. A few "definitions" of threshold voltage that you may have come across are as follows:

A good reference for understanding the intricacies of threshold voltage is S. Tiwari, *Compound semiconductor device physics*, Chapter 6, which is titled "Insulator and heterostructure field effect transistors." Do this question approximately, as you should the others too. Conceptual understanding is what these are attempting to emphasize. Here notice all the different threshold magnitudes. Note how all these different definitions are ignored and the threshold voltage distributions drawn! Should one expect different sensitivities for each one of these?

- A surface band bending of twice the bulk potential, that is, $\psi_s = 2\psi_B$. This condition just states that electron density at the surface reached the magnitude of the hole density in the bulk.

- A surface band bending where the inversion charge is equal to the depletion charge, that is, $\psi_s = \psi_H$. This is moderate inversion.

- A certain amount of sheet charge in the inversion region, as when one can start observing the onset of conduction. Let us say this is about mid-10^{10} cm^2 of electron density.

- A certain amount of current starts flowing, say, 1 $\mu A/\mu m$, which is a many-decades-smaller fraction of the current capability of a good transistor.

Determine the contribution arising from each of the above to threshold voltage. For the last, assume a reasonable transistor design and find the contribution arising from the substrate at that current density. We just want to compare how far apart the values obtained by using the different definitions are from each other. Now assume that the material is germanium, not silicon, but with the same mobility and doping. What do the numbers say and imply? [A]

13. Consider a hypothetical symmetric double-gate transistor with undoped silicon 20 nm in thickness, a heavily doped polysilicon gate and a high permittivity insulator. You would like to make a good 20 nm transistor out of this. What is the insulator permittivity, thickness, and scale length (or its range) that one should work with? What kind of $DIBL$ and subthreshold swing would you expect at 20 nm gate length? [M]

Ignore threshold voltage considerations for now. We should be able to tackle that with workfunctions.

14. Consider a single atomic sheet plane of doping σ_D^+ in per unit area at $x = 0$ in silicon $(-\infty < x < \infty)$ doped to N_D per unit volume and all ionized. Determine the carrier distribution as a function of position as well as the electrostatic potential ψ, assuming that $\psi \to 0$ as $x \to \pm\infty$. What is the characteristic length of the decay of the carrier density? This is the Debye length. [M]

15. We discussed cylindrical (nanowire based) geometries because of the scalability reflected in threshold voltage invariance with gate length. But how would one achieve a specific threshold voltage that we may desire? [S]

16. If fluctuations, such as in threshold voltage or in noise, arise from a large number of different sources, how is the net effect of fluctuations related to the individual sources? If another parameter is related to the parameter whose fluctuation you determined, for example, current is related to voltage, will fluctuations in that parameter, for example, current, have the same relationship to the sources of fluctuations? [S]

17. The Gaussian normal distribution is a limit case for finite probabilities in binomial distributions. If random dopants have a Poisson distribution, that is, a low probability of being in any compartmentalized cell in the semiconducting region where charge contribution in the threshold voltage arises from, then can the threshold voltage distribution that one gets with transistors really be a normal distribution? Argue. [M]

18. Would Fermi velocity be higher or lower in a one-dimensional semiconductor versus a two-dimensional semiconductor, assuming both are degenerate and at the same Fermi Energy? What if they have similar magnitudes of carrier density? [S]

19. The wavefunction of the ground state in a triangular potential can be approximated by a simple envelope function—the Fang-Howard function—as the amplitude of the Bloch function. The

Fang-Howard function is

$$\varsigma(z) = \sqrt{\frac{b^3}{2}} z \exp\left(-\frac{bz}{2}\right).$$

Here, z is the depth coordinate, with its origin at the interface where the triangular potential of the *MOSFET* inversion layer forms. What is the expectation value of the electron location? Discuss the implications of this for the silicon *MOSFET*. **[M]**

20. Why is conductance a constant in a channel in the ballistic limit?

    [S]

21. Why does the current saturate with drain voltage in ballistic transistor? **[S]**

22. For identical geometries, which should have higher quantum capacitance: silicon or *InAs*? Whose current dependence on applied gate voltage will be more nonlinear in the ballistic limit? **[S]**

23. If in graphene or a Dirac point nanotube, the electron has no mass, then how come it has a momentum and moves with a finite speed?

    [S]

24. In a perfect nanotube ballistic transistor with transistor operation purely through the quantum conductance channels, does the presence of quantum conductance imply a dissipation in the channel? Or is the dissipation elsewhere? Explain. **[S]**

25. Assume *GaAs* has a mobility of 4000 $cm^2/V \cdot s$, that of silicon is 500 $cm^2/V \cdot s$. and amorphous silicon's is 1 $cm^2/V \cdot s$. They are all doped at 1×10^{17} cm^{-3}. We form metal-semiconductor barriers on these. *GaAs*'s has a barrier height of 0.8 eV, silicon's is 0.65 eV, and amorphous silicon's is 0.6 eV. Draw on a single plot the current density(log)-voltage characteristics for bias voltage varying between 0^+ V and 0.5 V for these devices. What mechanism dominates the current in each case? Thermionic emission or diffusion? What is the depletion width in thermal equilibrium in these devices? What is it at 0.5 V? How does it compare at the forward bias with the mean free path? **[M]**

26. Estimate the quantum capacitance when states up to 0.3 eV are filled in a 5 $nm \times 5$ nm cross-section quantum wire for *InAs*, $Ga_{0.47}In_{0.53}As$, *GaAs*, *Ge* and *Si*? Assume that the gate wraps around the quantum wire. How many conductance channels are available for conduction? **[A]**

27. Calculate the number of channels per unit length (say, in nm) of the width direction for graphene ribbons. What is it for a semiconducting $(10, 0)$ nanotube? What bandgap would you expect for the $(10, 0)$ nanotube?   [A]

28. We would like to use the $(10, 0)$ nanotube for a ballistic transistor. So, maybe a 10 nm gate length device. To have it be suitable in digital technology, we need to pack it at high density, so we use a contact that is 10 nm in gate contact length. Calculate the contact resistance in the ballistic limit and in the diffusive limit. Does this contact behave more like a short contact or long contact? Imagine a contact that is 100 nm in length. What is the contact resistance for that contact? How do these resistances or their inverse (conductances) compare to the conductance expected from the intrinsic device operating in the quantum conductance limit? What are the implications?   [A]

29. Let us determine the relevant (non-degenerate or degenerate) formulations that we should employ in modeling ballistic transistors in a few different materials. Take the equivalent effective mass and effective density of states, and determine the thermal velocity (2D gas), Fermi velocity, if relevant, and the effective density of states, at $T = 300$ K, $T = 77$ K, and $T = 4$ K, for $InAs$, $Ga_{0.47}In_{0.53}As$, $GaAs$, Ge and Si for a carrier concentration of 1×10^{12} cm^{-2}, 3×10^{12} cm^{-2} and 5×10^{12} cm^{-2}, that is, in 2D layers. Give your answer in the form of a table listing the parameters for the multiple materials, multiple temperatures and multiple carrier densities. Comment on what the numbers say and which of the different ballistic models would be appropriate to use in each of the cases.

I suggest you set this problem up as a suitable software procedure in a simulation so that the spewing of numbers doesn't overwhelm you.

                                                      [M]

30. Let us consider a hypothetical silicon wire transistor. It has a 5×5 nm^2 square cross-section, a 10 nm channel length and an idealized non-conducting oxide insulator that is 1 nm thick. Assume that the transistor turns on at a reasonable gate bias voltage so that the metal workfunction for an undoped channel is appropriately chosen. Assume no scattering and that the source and the drain have 1×10^{20} cm^3 doped contact regions that are adiabatically flared out from the wire. Estimate the current in this device when a gate-source bias of 1 V and a drain-source bias of 1 V are applied. Make judicious approximations, with justification for any other data that you may wish to have.   [M]

31. The quantities in Table 2.8 are experimental measurements relevant to transistors and Si and $GaAs$ based on two-dimensional electron gas. These are in systems with high mobility. Calculate

At $T = 4.2\,K$	GaAs	Si
Electron density (cm^{-2})	4×10^{11}	0.7×10^{11}
Electron mobility $(cm^2/V \cdot s)$	10^6	4×10^4
Effective mass (m_0)	0.067	0.19
Dephasing time (s)	$30 \times 10^{-12}\,s$	$10 \times 10^{-12}\,s$

Table 2.8: Experimentally measured parameters at 4.2 K in $GaAs$ and Si two-dimensional systems pertinent to transistors with high mobility.

from these parameters the scattering time τ_k, the diffusion coefficient $\mathcal{D}$, the Fermi wavelength λ_F, the phase coherence length λ_ϕ, the thermal wavelength λ_θ and the interaction parameter r_s. The Fermi wavelength is the de Broglie wavelength of the electrons at the Fermi surface. The thermal length is the length scale over which thermal effects take place. The interaction parameter is a measure—a ratio—of the unscreened Coulomb energy between two electrons at their average distance and the kinetic energy of the Fermi surface.   [M]

3
Phenomena and devices at the quantum scale and the mesoscale

MESOSCALE IS THE SIZE SCALE, somewhat imprecise, where the classical description characterized by continuum approximations and the quantum description, for example, a behavior characterized by ``discretization´´, uncertainty and the probabilistic nature intersect. The classical description breaks down, and the quantum description remains too complicated because of the number of interactions that have to be accounted for. The nanotransistor of Chapter 2, in its ballistic and finite scattering limit, is an example of a mesoscale device. As discussed earlier, properties such as mobilities and diffusion coefficients, make sense only at a large scale with so much scattering that its effects can be captured in certain parameters. But if the number of scattering events, for example, for the electron, is small, and this may happen because the size of the device is of a length scale comparable to a length scale between which an electron is likely to undergo some interaction, then classical parameterization is inadequate. Nanoscale transistors had this characteristic. The equivalent of electron transport for phonons is the length scale for propagation of phonons and their potential scattering. This connects sound velocity with scattering rate. If the scattering is limited, these length scales can be large. For an electron mobility of 10^6 $cm^2/V \cdot s$ such as in the high electron mobility compound semiconductor heterostructures at low temperature, this corresponds to a scattering time constant of many ps. Electrons in most semiconductors reach velocities of $\sim 10^7$ cm/s at moderate fields, and a maximum group velocity of $\sim 10^8$ cm/s, so the scattering distances can be of the order of μms. At a room temperature mobility of 10^3 to 10^4 $cm^2/V \cdot s$, such as in semiconductors or in two-dimensional systems such as graphene or carbon nanotubes, this dimension is still several nm. This order of magnitude estimation is still within our Fermi gas or Fermi liquid picture of the electron and

Fermi gas and Fermi liquid, as well as their quantum underpinnings and significance, will be discussed in detail in Chapter 4. An introductory view is to look at the gaseous state as one that is compressible. Particles—electrons for us—have plenty of space to move around and behave independently of each other. They can be brought closer together while our codification of the description of their movement remains the same. A liquid is quite incompressible, except when close to phase transition to another state. A liquid behaves quite differently from a gas. Being incompressible, particles cannot be brought together, except at the surface. The characteristics of the system can be observed and modeled largely by what happens at the liquid's surface, that is, at the Fermi energy.

Nanoscale device physics: Science and engineering fundamentals. Sandip Tiwari.
© Sandip Tiwari 2017. Published 2017 by Oxford University Press.

so should suffice.

The other end of this landscape is when quantum effects entirely determine the properties of the system. Any approach to implement quantum computation or communication is dependent on superposition, entanglement, coherence and other properties that the quantum system has. Many quite incredible features arrive from these quantum systems even if they are large in dimension—far larger than the nanometer scale. The fractional quantum Hall effect is the appearance of an entirely new phase where the collective response shows many unusual excitations with fractional quantum charge. We may no longer think of the electron as a single particle excitation interacting with rest of its environment, but of this collective ensemble as one. Topological insulators are materials where states appear that are due to composites of particles, fractional charge excitations with or without spin, and ones where coherence times could be very large. Such systems, since they preserve the quantum nature for a long time, may have relevance to the quantum-computation-oriented thoughts that we discussed in Chapter 1.

These unusual effects do not have to be of quantum origin in the sense just discussed. Placing a single electron on a capacitor raises its energy by $e^2/2C$ $J \equiv e/2C$ eV. This is the energy that must be supplied to the electron just on electrostatic considerations. A classical metal sphere in vacuum of radius r has a capacitance $4\pi\epsilon_0 r$. The electrostatic energy to place an electron, if of the order of thermal energy—$k_B T \approx 25.9$ meV at room temperature—has a radius of $e/4\pi\epsilon_0 k_B T$ with $k_B T$ in units of eV. This is about 50 nm, a sizable dimension that is reduced when the medium is a semiconductor or a common dielectric, yet still of the order of 10 nm. Such structures will show single electron sensitive phenomenon—a Coulomb blockade—at low energies. This is quite classical but shows discreteness arising from small size scale. In both this and in the ballistic transport regime, fluctuation phenomena may be suppressed because certain excitation modes are disallowed, either because of lack of energy or because another particle is not allowed in the same state because of the exclusion principle. Either way, noise is reduced.

So, beyond nanoscale transistors or the variety of phase transition devices, there are a variety of other interesting structures and devices utilizing quantum and mesoscale phenomena. We will start with a discussion of quantum computation and communication and then look at where the mesoscale provides unique properties. We then discuss classical nanoscale effects and end with devices that bring us back to how some of these examples connect to quantum computation.

The most accurate measurement of fundamental constants, such as the charge of an electron ($-e$), Planck's constant (h) and the speed of light (c), directly follow from an integral quantized Hall effect measurement, even when the sample has variety of defects and impurities, as any real world ensemble sample would, because of specific quantum many-body phenomena. Flux quantum (h/e) or resistance quantum ($h/2e^2$) directly relate two of these constants, and it is measurable in samples that are decades of nanometers in dimensions.

3.1 Quantum computation and communication

QUANTUM COMPUTATION AND COMMUNICATION bring in a number of attributes that the classical approach cannot access. Quantum mechanics is a linear theory where reversibility between microstates appears in quantum systems naturally. We will also see in this discussion that superposition allows a larger information content to stay latent in the system until one makes a measurement and extracts from the system a classical output, for example, a binary 0 or 1 state. We will also see that entanglement of two quantum systems, a superposition that cannot be expressed as a product of the two quantum system wavefunctions, appears, and this has particularly powerful consequences for computation, cryptography and secure communication.

In classical circuits. we work mostly with standard static $CMOS$ circuits—$NAND, NOR$, inverter buffers and their more advanced combinations such as $sRAM$ memory, or registers, that is, flip-flops made out of inverters and other components. Occasionally, although more so in the past, one also employed dynamic circuits. These are circuits employing pass transistors, circuits using clocks, and dynamic circuits that speed up the slower transitions of circuit switching. The circuit of Figure 3.1(a), a dynamic circuit, is a faster AND gate, which is clocked and also uses the inverter hung at the output to accelerate the transition by bootstrapping the output transitions. This is use of feedback as internal reinforcement that consumes energy gainfully for speed. With power gain, these circuits work well at the cost of high energy consumption in static and dynamic circuits, with the energy cost appearing from Boltzmann entropy and the need of determinism. Even if one employs reversible computation and avoids the moving of charge from the supply to the ground during switching, leakage paths and the Boltzmann entropic dissipation still exist. The circuit of Figure 3.1(b), our quantum billiard ball implementation discussed in Chapter 1, is one that in the ideal circumstances consumes no energy. Since quantum dynamics is a reversible process, there is no analog of standard logic primitives in quantum logic gates.

We are particularly interested in reversible logic to minimize dissipation. For reversible logic, quantum or classical, the number of logical 0s and 1s must be conserved in the input and output. This is a necessary condition for conservation of information, that is, the ability to recreate the input because no information has been lost. We have seen in the toy billiards example the ability to implement common logic functions. Viewing the billiard balls as quantum objects—

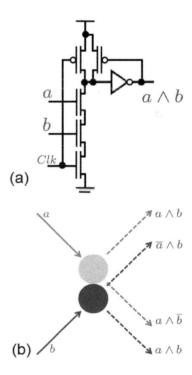

(a)

(b)

Figure 3.1: Two examples of AND gates. (a) shows a bootstrap at the output for a clocked logic to accelerate transitions in a classical AND ($a \wedge b$). (b) shows a quantum billiard ball version that has additional logical outputs.

photons interacting, for example—lets one see quantum logic gates in these examples. We look first at some of these quantum logic gates as primitives to explore reversibility, dissipation and the ability to "copy" and "observe" without generating any output—garbage— whose erasure is a cause of dissipation.

A unit wire, as drawn in Figure 3.2, is the simplest example of a single input with single output quantum gate that provides the function $f(x, t; y, t+1)$ with $x^t \mapsto y^{t+1}$. An input x at time t following operation by the unit wire gate becomes y at time $t + 1$. If the input and output are at identical positions, that is, $x = y$, then this gate is a memory. If $x \neq y$, that is, the output and input positions are different, then this is a transmission line, since $x^t = 0$ is transformed to $y^{t+1} = 0$, and likewise for $x^t = 1$ to $y^{t+1} = 1$.

A more interesting gate is the Fredkin gate, shown in Figure 3.3. The Fredkin gate executes the transformations

$$
\begin{aligned}
v &= u, \\
y_1 &= ux_1 + \bar{u}x_2, \text{ and} \\
y_2 &= \bar{u}x_1 + ux_2.
\end{aligned} \tag{3.1}
$$

u is a control signal passed on at output as v. y_1 and y_2 are conditional crossovers, if $u = 1$ than $y_1 = x_1$, and $y_2 = x_2$, but if $u = 0$, then the inputs are exchanged at the output: $y_1 = x_2$, and $y_2 = x_1$. The Fredkin gate is an example of reversible and invertible logic. If we cascade two such gates in series and let z, z_1 and z_2 be the outputs of the series stack, then $z = v = u$ for the control signal,

$$
\begin{aligned}
z_1 &= vy_1 + \bar{v}y_2 = u(ux_1 + \bar{u}x_2) + \bar{u}(\bar{u}x_1 + ux_2) \\
&= (u\bar{u} + \bar{u}u)x_2 + (uu + \overline{uu})x_1 = x_1, \text{ and} \\
z_2 &= \bar{v}y_1 + vy_2 = \bar{u}(ux_1 + \bar{u}x_2) + u(\bar{u}x_1 + ux_2) \\
&= (\bar{u}u + u\bar{u})x_1 + (\overline{uu} + uu)x_2 = x_2.
\end{aligned} \tag{3.2}
$$

The inputs are recovered. A Fredkin gate, as shown in Figure 3.3, "steps back in time" when the output is fed back into the gate. It is informationally reversible or invertible.

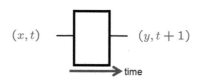

Figure 3.2: A unit wire as a single input, single output gate.

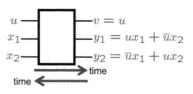

Figure 3.3: A Fredkin gate that executes a conditional crossover.

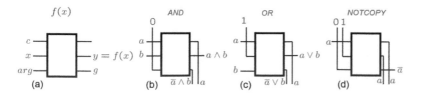

Figure 3.4: (a) A Fredkin-like gate as a general template for Boolean logic. (b) shows an *AND* gate, (c) shows an *OR* gate, and (d) shows a *NOTCOPY*, that is, a copy of the inversion of an input. These can be seen to be all implementable using the Fredkin gate.

The Fredkin gate is a generalizable reversible gate for achieving any Boolean function in invertible logic, as shown in Figure 3.4.

Figure 3.4(a) shows the generation of an output $y = f(x)$ under control c and argument arg, with the ancillary output as the garbage g. Figure 3.4(b) is a specific example of this general Fredkin-like gate showing the AND function output, Figure 3.4(c) shows the OR function as an output, and Figure 3.4(d) shows copying of the inversion of an input—the $NOTCOPY$ function.

Another gate that is an appealing primitive in quantum implementations is the Toffoli gate, which executes an XOR function. The symbol $\oplus$ is used to indicate the XOR function, as shown in Figure 3.5. A Toffoli gate outputs

$$c' = c \oplus (a \wedge b), \tag{3.3}$$

while a and b are unit wires. If $c = 1$, then the Toffoli gate behaves as a $NAND$ gate. This functional relationship also implies that, when a and b are both logical 1, then the output c' is the $NOTCOPY$ of the input c.

The combination of operations whose truth table is in Figure 3.5 is a controlled controlled NOT, or $ccNOT$, operation. $\oplus$ is a doubly controlled operation on c by a and b—a $ccNOT$. Two Toffoli gates in series result in the identity operation, that is, all the inputs are reversibly recovered.

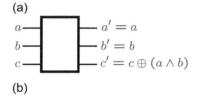

(a)

(b)

a	b	c	a'	b'	c'
0	0	0	0	0	0
0	0	1	0	0	1
0	1	0	0	1	0
0	1	1	0	1	1
1	0	0	1	0	0
1	0	1	1	0	1
1	1	0	1	1	1
1	1	1	1	1	0

Figure 3.5: (a) A Toffoli gate, with $c' = c \oplus (a \wedge b)$ as one of the Boolean logic outputs. The truth table in (b) shows that this function serves as a controlled controlled NOT ($ccNOT$) that exchanges the c bit.

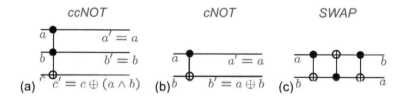

(a) $ccNOT$ (b) $cNOT$ (c) $SWAP$

Figure 3.6: (a) shows the wire representation of the $ccNOT$ gate—the Toffoli gate. (b) shows a two-wire $cNOT$ gate, which is useful for copying bits to the controlled line. (c) shows a $SWAP$ function where the Boolean signals on the lines are exchanged.

A simple pictorial representation of the Toffoli gate in a line form is shown in Figure 3.6(a). It serves as a template for implementing measuring functions needed for quantum logic operations. The output c' is the XOR or $ccNOT$ function of $c' = c \oplus (a \wedge b)$. A controlled NOT, or $cNOT$, is a two-line *in* and two-line *out* gate, where the second line is conditionally controlled by the first, as shown in Figure 3.6 (b). Here, the input (a, b) is transformed to $(a', b') = (a, a \oplus b)$. If $b = 0$, then the input $(a, 0)$ is transformed to $(a', b') = (a, a)$. With $b = 0$, the $cNOT$ makes a copy of the a bit to the controlled line. Figure 3.6(c) shows a $SWAP$ function, where the data from each line are swapped to the other line using $cNOT$ functions. The Toffoli gate, that is, the $ccNOT$ gate, is a special case of a Fredkin gate and, like the Fredkin gate, is informationally reversible. The $ccNOT$ is functionally complete, as is a $NAND$, in that any possible set of logic operations, that is, one representable in a truth table, can be imple-

The reader should convince herself that this $SWAP$ operation, as claimed, is indeed true.

mented through it. *cNOT* or *XOR* by themselves are not sufficient. Nor are *AND* and *OR*.

Both the Fredkin gate and the Toffoli gate have been shown to be reversible. But, how does one make them dissipationless? Recall our earlier insistence that dissipation occurs when information is destroyed—a precisely known bit is made random. How do we handle this dilemma that existed with Maxwell's demon? If garbage is produced by the computation, in time, as with the demon, absent infinite resources such as an infinite tape or memory, the computational operation will choke if the garbage is not erased.

The answer, due to Bennett, is to perform the operation in three stages. First, compute and keep the answer as well as the information that is no longer required, that is, the garbage, such as the various computational results written down in the bottom of the gates in Figure 3.4. Second, we copy the answer onto the tape or into the memory reversibly, that is, without dissipation. Third, we now reverse all that was done in the first step. The garbage is now removed by the program, and all the initial conditions are restored. The final state of the computer and its memory is identical to what it was at the start. So, we proceeded with the calculation reversibly from a known initial state, copied the result of interest, and then proceeded back to the precise initial state that we started with reversibly. No erasure was involved. The entire process in principle can be executed without dissipation. In this process, the dilemma encountered earlier, where Maxwell's demon with finite memory, operates a reversible engine requiring erasure of information to get to a precisely known state, from which to continue the experiment is now resolved. All we needed is the memory or tape to copy the result that we are interested in. And, presumably, the result is a finite-sized result.

These gates, in principle, can be implemented through a variety of reversible technological schemes—the quantum billiards is but one example, with correspondence to photons as the billiard balls. This billiard ball example employs particles moving at high speeds, but one could also have done it adiabatically through systems involving electrons or magnetic domains. This generality, together with these possibilities for gates, gives us a way to quantum compute.

How do the classical and the quantum approaches differ in their foundations?

While we will dwell on several aspects of this, the primary one is in the nature of representation of information. A classical bit is either a logical 1 or 0, on or off, high bias voltage or low, a black mark or not, or some other similar physical representation. These 1s and 0s are stable pointer states. A quantum bit—qbit—in contrast, is a

Perhaps this is why the infinite needs to be invoked if one postulates the existence of a supreme authority overseeing the details of the universe.

In the example of billiard ball computation, once the posed problem is tackled, we run the billiard ball computation backward while keeping the result.

A result of infinite size is an oxymoron.

superposition of states. We may write its state representation as

$$|\psi\rangle = c_0\,|0\rangle + c_1\,|1\rangle. \qquad (3.4)$$

What this equation states is that, in a two-level system composed of one eigenfunction $|0\rangle$ and another $|1\rangle$, the state function is a superposition of these two basis states. The coefficients tell us the probability of finding the system in the corresponding $|0\rangle$ or $|1\rangle$ state. It is neither $|0\rangle$ nor $|1\rangle$. It will be one of these should we make a measurement on it. And it is at that time that one gets that single classical bit of information. This is what we mean when we state that this is a two-level system that is in superposition.

Generalizing this superposition, in the classical digital computer then, the set of computational states is some set of mutually distinguishable abstract states that define the accessible configuration or phase space of the machine. These are the possible machine states. This is a stable pointer set which points to where to go next and the information that is needed to work with—this is the abstract collection of the contents of registers, memories, and the states of the variety of computational units, et cetera. The specific computational state of the classical digital machine in use at any given time represents the specific digital data and the instructions being processed at that moment within the machine. Everything is known, and the process works deterministically. In the quantum computer, the computational state is not always a pointer state. Making a measurement reduces it to a precisely known state, not a superposition or other complex subsets of it. The computational state is a composite superposition.

We differentiate the two—classical and quantum—by associating a classical bit (cbit) with the former, and a quantum bit (qbit) with the latter. Figure 3.7 shows a pictorial view of the difference between the cbit and the qbit.

The cbit provides an information of

$$H = -\sum_1^2 (1/2)\log_2(1/2) = 1\,b, \qquad (3.5)$$

with the state being either a 0 or a 1. Classically, any physical representation's presence or absence represents this bit. In Figure 3.7(a), this is represented as an up or a down. Any two-level quantum system may also be used to represent information. The state function of Equation 3.4 then represents, if a measurement were made, the possibility of finding the system in a collapsed eigenfunction state of $|0\rangle$, with probability $|c_0|^2$, and $|1\rangle$, with $|c_1|^2$. There is a probability of finding it in $|0\rangle$ as well as $|1\rangle$—it may appear as either, is a continuous normalized function of $|0\rangle$ and $|1\rangle$, and only when the measurement is made does it show its classical state. This is a qbit.

Calling the quantum bit "qbit" rather than the popular term "qubit" is a bow to David Mermin who wonders out aloud in his *Quantum Computer Science,* published by Cambridge University Press, why in this instance one should ignore the rule that a vowel must follow any "q." A personal reason is the advantage this word coinage gives to this poor Scrabble player. The accumulation of these small advantages matters in this game, which is usually played as "Scrabble for blood," aka, for winning.

(a) cbit

$$0: \qquad\qquad 1:$$

(b) qbit

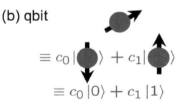

$$\equiv c_0\,|0\rangle + c_1\,|1\rangle$$

Figure 3.7: (a) shows a bit that can be in one of two states shown here as up or down. This is a cbit. Electron spin is an example of a two-level system, with (b) showing a quantum bit state as superposition from the $|0\rangle$ and $|1\rangle$ eigenfunctions that the two-level quantum system allows. This is a qbit.

It represents the possibility of appearing in either of the states at the time of measurement—not just the logical 0 or logical 1 of a classical bit. The qbit has an inherent superposition, but when measured, it too provides one of the two eigenstates constituting it. The qbit resulted in only a bit of information, but when pooled together, special cases of superposition make more powerful computation and communication capability possible.

Another way of looking at this is through the Bloch sphere shown in Figure 3.8, which geometrically shows the state space of a two-level quantum system. A Bloch sphere is of unit radius with points on the surface representing the state function of the two-level system $|\psi\rangle$. Figure 3.8 chooses a coordinate orientation so that $|0\rangle$ is the north pole of the system, and $|1\rangle$ is the south pole. The different superpositions of $|0\rangle$ and $|1\rangle$ and a few others are identified for the equatorial plane. Any state function of this quantum system, in the polar coordinates, can be represented by

$$|\psi\rangle = \cos\frac{\theta}{2}|0\rangle + \exp(i\varphi)\sin\frac{\theta}{2}|1\rangle. \tag{3.6}$$

These are all points on the Bloch sphere as superposition of the qbits.

Qbits derive their power and utility from this superposition and entanglement—a subset of states possible through superposition. Consider states of a composite of two two-level quantum systems a and b. The pure composite states of this systems are $|0\rangle_a|0\rangle_b$, $|1\rangle_a|0\rangle_b$, $|0\rangle_a|1\rangle_b$ and $|1\rangle_a|1\rangle_b$. For each of these instances of pure composite states, we can describe properties with specificity, since both a and b are *a priori* defined to be in a pure $|0\rangle$ or $|1\rangle$ eigenfunction. All possible superpositions of these product states are also allowed states of the composite of a and b. In general, the state function

$$|\Psi\rangle = c_{00}|0\rangle_a|0\rangle_b + c_{01}|0\rangle_a|1\rangle_b + c_{10}|1\rangle_a|0\rangle_b + c_{11}|1\rangle_a|1\rangle_b \tag{3.7}$$

describes this.

The interesting feature is that a superposition state of this two-system composite,

$$|\psi\rangle = \frac{1}{\sqrt{2}}\left(c_{00}|0\rangle_a|0\rangle_b + c_{11}|1\rangle_a|1\rangle_b\right), \tag{3.8}$$

is also a valid allowed state of the composite, but it has the unique property that it cannot be written as a product of state of system a and state of system b. Such states are *entangled*. It is a unique superposition of the product states that cannot be written as a product of systems of which it is composed. What this says is that while one specific case of Equation 3.7,

$$|\psi\rangle = \frac{1}{2}\left(|0\rangle_a|0\rangle_b + |0\rangle_a|1\rangle_b + |1\rangle_a|0\rangle_b + |1\rangle_a|1\rangle_b\right)$$

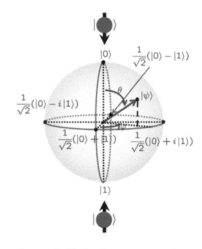

Figure 3.8: Bloch sphere to geometrically show the state space of a two-level system.

$$= \frac{1}{\sqrt{2}} (|0\rangle_a + |1\rangle_a) \otimes \frac{1}{\sqrt{2}} (|0\rangle_b + |1\rangle_b), \tag{3.9}$$

is writable as a product, Equation 3.8 is not. Equation 3.9 describes a system decomposable into its composing parts, while Equation 3.8 shows that it is possible to build a composite that is not separable in its composing parts. This unique composite has properties that are going to be interlinked—a and b are going to exhibit correlations in properties—even if its composing quantum systems are distant from each other at some later time after having been built together into a state described by the wavefunction of Equation 3.8.

In order to understand this, we now build systems involving more qbits. If we take our electron spin system of Figure 3.7(b) and expand it to 2 spins or more generally n spins, as in Figure 3.9, we may write the wavefunction of the system as

$$|\psi\rangle = c_{00} |00\rangle + c_{01} |01\rangle + c_{10} |10\rangle + c_{11} |11\rangle \tag{3.10}$$

with 2^2 of superposition states, and

$$|\psi\rangle = c_{0...00} |0...00\rangle + c_{0...01} |0...01\rangle + \cdots + c_{1...11} |1...11\rangle \tag{3.11}$$

with 2^n superposition states. We have simplified here the notation employed in Equation 3.7, and the meaning of this notation can be directly seen.

These 2-qbit and n-qbit systems of Equation 3.8 and Equation 3.9, respectively, as written here, are the quantum registers of this quantum computation. This modified Dirac notation for the various product states has a direct correspondence to the vector notation of quantum mechanics. In general, we may write any quantum bit as

$$|x\rangle = \begin{bmatrix} c_0 \\ c_1 \end{bmatrix}, \tag{3.12}$$

which is a unitary column. In this notation, our basis states are

$$|0\rangle = \begin{bmatrix} 1 \\ 0 \end{bmatrix}, \text{ and } |1\rangle = \begin{bmatrix} 0 \\ 1 \end{bmatrix}. \tag{3.13}$$

So $|0\rangle, |1\rangle$ is the basis, with $|x\rangle = c_0 |0\rangle + c_1 |1\rangle$. More than one qbit then is the result of a tensor operation, with

$$\begin{bmatrix} c_0 \\ c_1 \end{bmatrix} \otimes \begin{bmatrix} c'_0 \\ c'_1 \end{bmatrix} = \begin{bmatrix} c_0 c'_0 \\ c_0 c'_1 \\ c_1 c'_0 \\ c_1 c'_1 \end{bmatrix}, \tag{3.14}$$

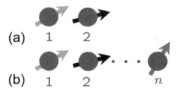

(a) 1 2

(b) 1 2 n

Figure 3.9: A quantum system with a superposition of 2 spins in (a), and n spins in (b).

and our notation identifies the basis of this two-qbit system as

$$|00\rangle = \begin{bmatrix} 1 \\ 0 \\ 0 \\ 0 \end{bmatrix}, \quad |01\rangle = \begin{bmatrix} 0 \\ 1 \\ 0 \\ 0 \end{bmatrix},$$

$$|10\rangle = \begin{bmatrix} 0 \\ 0 \\ 1 \\ 0 \end{bmatrix}, \quad \text{and} \quad |11\rangle = \begin{bmatrix} 0 \\ 0 \\ 0 \\ 1 \end{bmatrix}. \tag{3.15}$$

An n-qbit system exists in the form

$$|\Psi\rangle = \sum_{00\cdot 0}^{11\cdot 1} c_x |x\rangle, \quad \text{with} \quad 1 = \sum_x |c_x|^2. \tag{3.16}$$

n states define this n-qbit system, which gives access to 2^n different superposition states.

We can now show how a 2-qbit state,

$$|\psi\rangle = \frac{1}{\sqrt{2}} |00\rangle + \frac{1}{\sqrt{2}} |11\rangle, \tag{3.17}$$

cannot be written as a tensor product of two single qbits. Let $|x\rangle = c_0 |0\rangle + c_1 |1\rangle$, and $|y\rangle = c'_0 |0\rangle + c'_1 |1\rangle$, be the two qbits. The tensor product is

$$\begin{aligned} |x\rangle \otimes |y\rangle &= (c_0 |0\rangle + c_1 |1\rangle) \otimes (c'_0 |0\rangle + c'_1 |1\rangle) \\ &= c_0 c'_0 |00\rangle + c_0 c'_1 |01\rangle + c_1 c'_0 |10\rangle \\ &\quad + c_1 c'_1 |11\rangle. \end{aligned} \tag{3.18}$$

Comparing Equation 3.17 and Equation 3.18 forces the conditions for the amplitudes of

$$c_0 c'_1 = c_1 c'_0 = 0, \quad \text{and} \quad c_0 c'_0 = c_1 c'_1 = \frac{1}{\sqrt{2}}. \tag{3.19}$$

This is a contradiction, and therefore the two-qbit system is not a tensor product of two single qbits.

The state described by Equation 3.17 is an entangled state—a superposition state that is not a product state. It has incorporated intimately within it properties of the two qbits from which it was constructed. But, it makes it impossible to find the system collapsing either into $|x\rangle$ or $|y\rangle$, states from which it was built, upon measurement.

Any 2-level system providing a means to achieving qbits—spin 1/2 as well as photon polarization—will do just fine. And they may provide a convenient path to computation and communication. The

Qbit basis	Eigenstates
Rotational	$\left|\overset{\circ}{0}\right\rangle = \cos\theta\,\left|0\right\rangle + \cos\theta\,\left|1\right\rangle$
	$\left|\overset{\circ}{1}\right\rangle = \cos\theta\,\left|0\right\rangle - \sin\theta\,\left|1\right\rangle$
Diagonal	$\left|\nearrow\right\rangle = (1/\sqrt{2})(\left|0\right\rangle + \left|1\right\rangle)$
	$\left|\nwarrow\right\rangle = (1/\sqrt{2})(\left|0\right\rangle + \left|1\right\rangle)$
Chiral	$\left|\circlearrowleft\right\rangle = (1/\sqrt{2})(\left|0\right\rangle + i\left|1\right\rangle)$
	$\left|\circlearrowright\right\rangle = (1/\sqrt{2})(\left|0\right\rangle - i\left|1\right\rangle)$
Bell	$\left|\beta_{00}\right\rangle = (1/\sqrt{2})(\left|00\right\rangle + \left|11\right\rangle)$
	$\left|\beta_{01}\right\rangle = (1/\sqrt{2})(\left|01\right\rangle + \left|10\right\rangle)$
	$\left|\beta_{10}\right\rangle = (1/\sqrt{2})(\left|00\right\rangle - \left|11\right\rangle)$
	$\left|\beta_{11}\right\rangle = (1/\sqrt{2})(\left|01\right\rangle - \left|10\right\rangle)$

Table 3.1: Examples of different bases. The rotational, diagonal and chiral bases show entangled 1-qbit states, while the Bell basis is over entangled 2-qbit states.

plane of polarization of a linearly polarized photon, and the direction of rotation of a polarized photon, are both legitimate representations of pure basis states. When we build a 2-qbit system—a register, for example,—all we have needed is any complete orthonormal set of eigenstates to build the superposition set. The n-qbit system thus has the 2^n combinations of Equation 3.16, which is a tensor product of the n-qbit ($\left|00\ldots0\right\rangle$, $\left|00\ldots1\right\rangle$ through $\left|11\ldots1\right\rangle$) combination. Rotations of single qbits, or other unitary transforms of it, are also legitimate basis for constructing the n-qbit basis states. For 1- and 2-qbit systems, Table 3.1 shows a select set of possibilities. The Bell basis here is in a 2-qbit state.

We can now make a few remarks about the differences between the various approaches towards logical function that we have referred to in the text and what the cbit and qbit does in the logical operation.

Unitary operators satisfy $\hat{U}^T\hat{U} = 1$. Under unitary transformation of the state kets, the inner product, that is, the norm remains unchanged, that is, $\langle\alpha|\beta\rangle = \langle\alpha|\hat{U}^T\hat{U}|\beta\rangle$. Schrödinger evolution of a state is unitary.

John Bell set the Bell inequality test for a locally causal description's disagreement with predictions of quantum mechanics using this Bell basis. It establishes a clear predictive path to distinguish probability in quantum description with its entanglement properties versus the expectations from classical description. Appendix I dwells on this important foundational question.

$_{in}\diagdown^{out}$	0	1
0	0	1
1	1	0

(a)

$_{in}\diagdown^{out}$	0	1
0	0.1	0.9
1	0.9	0.1

(b)

$_{in}\diagdown^{out}$	0	1
0	$\frac{1}{2}$	$\frac{1}{2}$
1	$\frac{1}{2}$	$\frac{1}{2}$

(c)

Figure 3.10: (a) through (c) show tables of the cbit responses of NOT gates. In each case, the input is the first column, the output is the first row, and the table entries represent the matrix of probability responses connecting the output vector to the input vector. (a) is for a deterministic NOT, that is, one that is always correct, (b) is for a nondeterministic NOT, and (c) is for a completely nondeterministic NOT.

Figure 3.10 shows three examples of a NOT function using classical bits: the first is a deterministic NOT, the second is a nondeterministic or classical probabilistic NOT, and the third is a completely nondeterministic, that is, random NOT. The input state is in the first column of these tables. The output state is the first row. The matrix that maps the input column vector to the output column vector are the probability entries of this table. In an always-correct, classical deterministic *NOT*, an input logical 0 state maps to a logical 1 state with unity probability, and an input of logical 1 state maps to logical 0 with unity probability. Figure 3.10(a) shows this. Now, if there is a potential for error, as we discussed in Chapter 1, that is, the function is only probably correct, then the matrix changes. In Figure 3.10(b), an input of logical 0 results in a logical output of 1, its complement, only 90 % of time. It fails 10 % of the time. The third case shows a completely non-deterministic *NOT*. The result is truly random and results in 0 and 1 responses with an equal probability of 0.5, independently of the input.

Now, we look at the quantum *NOT* gate through the $NOT^{1/2}$ gate, for which there is no real equivalent in the classical gates. The three different examples shown in Figure 3.11, each when serially sequenced, result in *NOT* function, since $NOT = NOT^{1/2} \times NOT^{1/2}$. The difference between these examples is that, in the first two cases, shown in Figure 3.11(a) and (b), there exist phase shifts in the output state, and the last one—shown in Figure 3.11(c)—is a unitary transform. The amplitude matrix here shows that the probability corresponding to these entries in all these cases is $1/2$, not unlike the case of a completely non-deterministic *NOT*. So, if one inputs the state $|0\rangle$ to the first $NOT^{1/2}$ gate, the output state is $(1/2^{1/2})(|0\rangle - |1\rangle)$. Both $|0\rangle$ and $|1\rangle$ are possible outcomes with $1/2$ probability. This is also true for the example of Figure 3.11(c) since $|c|^2 = cc^*$. However, sequence another of a similar $NOT^{1/2}$ in series in all these cases, and you get a *NOT* function.

A completely nondeterministic, that is, random *NOT* as a logical gate is an oxymoron. It is included here to provide contrast with the quantum *NOT*. But, every failure is an opportunity! Complete randomization is a very useful random number generator. *RSA*—for Ronald L. Rivest, Adi Shamir and Leonard M. Adleman, who wrote the original code—would have been better off with the security of encryption had they not adopted the dual elliptic curve random number generator supplied by the National Security Agency (*NSA*). This opened a back door into what was in 2013 one of the more popular secure digital communication tools. And it had the stamp of approval of the National Institute of Standards and Technology (*NIST*)—an organization devoted to standards. One wonders, whose standards—individual, state, humanity, nature, . . . ? So did Edward Snowden.

A complex square matrix is unitary if its product with the conjugate transpose leads to an identity matrix. An orthonormal matrix is its analog for a real square matrix. A unitary matrix is invertible; its conjugate transpose is also unitary, and it preserves the norm.

(a) (b) (c)

Figure 3.11: Three different examples of $NOT^{1/2}$ gates, showing the amplitude matrices connecting input states to output states. The case shown in (c) is an example of a unitary $NOT^{1/2}$ gate.

This sequencing of the three examples of Figure 3.11 results in

$$\begin{bmatrix} 1/\sqrt{2} & -1/\sqrt{2} \\ 1/\sqrt{2} & 1/\sqrt{2} \end{bmatrix} \times \begin{bmatrix} 1/\sqrt{2} & -1/\sqrt{2} \\ 1/\sqrt{2} & 1/\sqrt{2} \end{bmatrix} = \begin{bmatrix} 0 & -1 \\ 1 & 0 \end{bmatrix};$$

$$\begin{bmatrix} 1/\sqrt{2} & 1/\sqrt{2} \\ -1/\sqrt{2} & 1/\sqrt{2} \end{bmatrix} \times \begin{bmatrix} 1/ & 1/\sqrt{2} \\ -1/\sqrt{2} & 1/\sqrt{2} \end{bmatrix} = \begin{bmatrix} 0 & 1 \\ -1 & 0 \end{bmatrix}; \text{ and}$$

$$\begin{bmatrix} (1+i)/2 & (1-i)/2 \\ (1-i)/2 & (1+i)/2 \end{bmatrix} \times \begin{bmatrix} (1+i)/2 & (1-i)/2 \\ (1-i)/2 & (1+i)/2 \end{bmatrix} = \begin{bmatrix} 0 & 1 \\ 1 & 0 \end{bmatrix}. \quad (3.20)$$

The negative sign in the *NOT* output of case (a) and case (b) speaks to a phase shift of π in the amplitude, leaving the probability as unity. Even though the probabilities in the case of the random deterministic *NOT* and the quantum *NOT* are the same, the former is a complete randomizer that loses all information content and is at maximum entropy, while the latter is not. This is easily seen in the sequencing examples in Equation 3.20.

The case shown in Figure 3.11(c) does transform the state $|0\rangle$ to $|1\rangle$, and $|1\rangle$ to $|0\rangle$. The matrix of *NOT* through $NOT^{1/2}$ gates that this represents is the Pauli X matrix. This Pauli X matrix is certainly a classical reversible *NOT* gate—the pure states, the north and the south poles of the Bloch sphere, that is, the computational basis states, did properly transform.

The arbitrary superposition state $|\psi\rangle = c_0 |0\rangle + c_1 |1\rangle$, which is representable by Equation 3.6, through a sequence of the case shown in Figure 3.11(c) gates—the Pauli X gate being a double operation—will evolve to $|y\rangle = c_0 |1\rangle + c_1 |0\rangle$. But, is this a quantum reversible *NOT* for qbits? A quantum *NOT* must map any arbitrary starting state $|\psi\rangle$ on the Bloch sphere to the point on the Bloch sphere projected exactly opposite through the center—the antipodal state $|\psi^\perp\rangle$. In polar coordinates, this antipodal state, geometrically, is

$$
\begin{aligned}
|\psi^\perp\rangle &= \cos\left(\frac{\pi - \theta}{2}\right)|0\rangle + \exp[i(\varphi + \pi)]\sin\left(\frac{\pi - \theta}{2}\right)|1\rangle \\
&= \sin\frac{\theta}{2}|0\rangle - \exp(i\varphi)\cos\frac{\theta}{2}|1\rangle .
\end{aligned} \tag{3.21}
$$

We need to compare it to the Pauli X matrix operation on state $|\psi\rangle$, using Equation 3.6, that is,

$$
\begin{aligned}
X|\psi\rangle &= \begin{bmatrix} 0 & 1 \\ 1 & 0 \end{bmatrix} \begin{bmatrix} \cos\theta/2 \\ \exp(i\varphi)\sin\theta/2 \end{bmatrix} \begin{bmatrix} |0\rangle \\ |1\rangle \end{bmatrix} \\
&= \begin{bmatrix} \exp(i\varphi)\sin\theta/2 \\ \cos\theta/2 \end{bmatrix} \begin{bmatrix} |0\rangle \\ |1\rangle \end{bmatrix} \\
&= \exp(i\varphi)\sin\left(\frac{\theta}{2}\right)|0\rangle + \cos\left(\frac{\theta}{2}\right)|1\rangle .
\end{aligned} \tag{3.22}
$$

A global phase difference in states is indistinguishable, so we multiply by $\exp(-i\varphi)$ to compare with our prediction of $|\psi^\perp\rangle$. We find

$$
X|\psi\rangle = \sin\left(\frac{\theta}{2}\right)|0\rangle + \exp(-i\varphi)\cos\left(\frac{\theta}{2}\right)|1\rangle \neq |\psi^\perp\rangle . \tag{3.23}
$$

Comparing this to Equation 3.21, we can see that the X gate is a classical *NOT* gate but does not operate as a general quantum *NOT* gate.

A true quantum *NOT* will give

$$
U_{NOT}|\psi\rangle = c_1^* |0\rangle - c_0^* |1\rangle \equiv |\psi^\perp\rangle . \tag{3.24}
$$

The Pauli matrices, which are unitary and Hermitian,

$$
\mathbb{I} = \begin{bmatrix} 1 & 0 \\ 0 & 1 \end{bmatrix},
$$

$$
X = \begin{bmatrix} 0 & 1 \\ 1 & 0 \end{bmatrix},
$$

$$
Y = \begin{bmatrix} 0 & -i \\ i & 0 \end{bmatrix}, \text{ and}
$$

$$
Z = \begin{bmatrix} 1 & 0 \\ 0 & -1 \end{bmatrix},
$$

suffice to specify as a weighted sum any 1-qbit Hamiltonian. These are equivalent to the Pauli operators

$$
\begin{aligned}
\mathbb{1} &= |0\rangle\langle 0| + |1\rangle\langle 1|, \\
\hat{\sigma}_x &= |0\rangle\langle 1| + |1\rangle\langle 0|, \\
\hat{\sigma}_y &= |0\rangle\langle 1| + |1\rangle\langle 0|, \text{ and} \\
\hat{\sigma}_z &= |0\rangle\langle 0| - |1\rangle\langle 1|.
\end{aligned}
$$

The last of the matrices/operators is important for the Ising interaction that we discuss in Chapter 4. The others are also important for the discussion of spin.

So, quantum gates have these very unique properties that arise from superposition and from the specific instance of superposition in the entanglement.

Several other quantum gates are of interest, and we mention these as well as ones already discussed in the matrix notation for use with the state vectors.

While classically, there is only one single-bit gate, the *NOT*, which flips the bit, or if one insists, also the *THROUGH*, which performs an identity operation, for qbits, there exist many more—all preserving the norm, that is, being unitary. Two examples are the *SHIFT*,

$$U_S = \begin{bmatrix} 1 & 0 \\ 0 & -1 \end{bmatrix}, \tag{3.25}$$

which causes a phase shift of π for $|1\rangle$, and the *Hadamard* gate *SHIFT*,

$$\mathfrak{H} = \frac{1}{\sqrt{2}} \begin{bmatrix} 1 & 1 \\ 1 & -1 \end{bmatrix}. \tag{3.26}$$

A Hadamard gate rotates the qbit sphere by $\pi/2$. Rotations and phase shifts are examples of unitary transforms, since they all preserve the norm.

In our matrix notation, we can now summarize some of the principal gates discussed earlier. The *cNOT* gate's operation is described by

$$\begin{matrix} |00\rangle \mapsto |00\rangle \\ |01\rangle \mapsto |01\rangle \\ |10\rangle \mapsto |11\rangle \\ |11\rangle \mapsto |10\rangle \end{matrix} \equiv \begin{bmatrix} 1 & 0 & 0 & 0 \\ 0 & 1 & 0 & 0 \\ 0 & 0 & 0 & 1 \\ 0 & 0 & 1 & 0 \end{bmatrix}. \tag{3.27}$$

The *SWAP* operation is described by

$$\begin{matrix} |00\rangle \mapsto |00\rangle \\ |01\rangle \mapsto |10\rangle \\ |10\rangle \mapsto |01\rangle \\ |11\rangle \mapsto |11\rangle \end{matrix} \equiv \begin{bmatrix} 1 & 0 & 0 & 0 \\ 0 & 0 & 1 & 0 \\ 0 & 1 & 0 & 0 \\ 0 & 0 & 0 & 1 \end{bmatrix}. \tag{3.28}$$

The operation of *ccNOT* or Toffoli gate operation is described by

$$\begin{matrix} |000\rangle \mapsto |000\rangle \\ |001\rangle \mapsto |001\rangle \\ |010\rangle \mapsto |010\rangle \\ |011\rangle \mapsto |011\rangle \\ |100\rangle \mapsto |100\rangle \\ |101\rangle \mapsto |101\rangle \\ |110\rangle \mapsto |111\rangle \\ |111\rangle \mapsto |110\rangle \end{matrix} \equiv \begin{bmatrix} 1 & 0 & 0 & 0 & 0 & 0 & 0 & 0 \\ 0 & 1 & 0 & 0 & 0 & 0 & 0 & 0 \\ 0 & 0 & 1 & 0 & 0 & 0 & 0 & 0 \\ 0 & 0 & 0 & 1 & 0 & 0 & 0 & 0 \\ 0 & 0 & 0 & 0 & 1 & 0 & 0 & 0 \\ 0 & 0 & 0 & 0 & 0 & 1 & 0 & 0 \\ 0 & 0 & 0 & 0 & 0 & 0 & 0 & 1 \\ 0 & 0 & 0 & 0 & 0 & 0 & 1 & 0 \end{bmatrix}. \tag{3.29}$$

The Fredkin gate—a controlled *SWAP*—in this matrix picture is described by

$$
\begin{matrix}
|000\rangle \mapsto |000\rangle \\
|001\rangle \mapsto |001\rangle \\
|010\rangle \mapsto |010\rangle \\
|011\rangle \mapsto |011\rangle \\
|100\rangle \mapsto |100\rangle \\
|101\rangle \mapsto |110\rangle \\
|110\rangle \mapsto |101\rangle \\
|111\rangle \mapsto |111\rangle
\end{matrix}
\equiv
\begin{bmatrix}
1 & 0 & 0 & 0 & 0 & 0 & 0 & 0 \\
0 & 1 & 0 & 0 & 0 & 0 & 0 & 0 \\
0 & 0 & 1 & 0 & 0 & 0 & 0 & 0 \\
0 & 0 & 0 & 1 & 0 & 0 & 0 & 0 \\
0 & 0 & 0 & 0 & 1 & 0 & 0 & 0 \\
0 & 0 & 0 & 0 & 0 & 0 & 1 & 0 \\
0 & 0 & 0 & 0 & 0 & 1 & 0 & 0 \\
0 & 0 & 0 & 0 & 0 & 0 & 0 & 1
\end{bmatrix} . \tag{3.30}
$$

All of these different gates are reversible with Toffoli, that is, *ccNOT*, and Fredkin, giving us a reversible basis. Classical gates can be simulated using reversible gates by generating them from reversible gates. A sufficient universal basis for these is the *NOT* function and the *cNOT* function. The *ccNOT* function, that is, the Toffoli gate, and the Fredkin are reversibly complete. The Hadamard gate is an extremely useful single qbit gate that maps the pure computational basis to superposition states, that is,

$$
\begin{aligned}
\hat{\mathfrak{H}}|0\rangle &= \frac{1}{\sqrt{2}}\begin{bmatrix} 1 & 1 \\ 1 & -1 \end{bmatrix}\begin{bmatrix} 1 \\ 0 \end{bmatrix} \\
&= \frac{1}{\sqrt{2}}\begin{bmatrix} 1 \\ 1 \end{bmatrix} = \frac{1}{\sqrt{2}}(|0\rangle + |1\rangle), \text{ and} \\
\hat{\mathfrak{H}}|1\rangle &= \frac{1}{\sqrt{2}}\begin{bmatrix} 1 & 1 \\ 1 & 1 \end{bmatrix}\begin{bmatrix} 0 \\ 1 \end{bmatrix} \\
&= \frac{1}{\sqrt{2}}\begin{bmatrix} 1 \\ -1 \end{bmatrix} = \frac{1}{\sqrt{2}}(|0\rangle - |1\rangle),
\end{aligned} \tag{3.31}
$$

that is, qbits.

Operating on a basis state $|x\rangle$, the Hadamard gate generates the output state $(1\sqrt{2})(|0\rangle + (-1)^x |1\rangle)$. If we take n states prepared in a pure $|0\rangle$ basis and perform the Hadamard operation on each of them independently in parallel, then the state one generates is an n-qbit superposition and is composed of 2^n eigenstates. These eigenstates are all the possible bit strings one can write using n-qbits. An 8-qbit register has $2^8 = 256$ possibilities. For example, a state $|43\rangle$ springs from the computational basis state $|00101011\rangle$. What we have just shown here is an example of a polynomial quantum operation being the equivalent of an exponential classical operation.

We end with two examples of circuits in Figure 3.12, to illustrate the consequences of this discussion. In Figure 3.12(a), the Toffoli gate (*ccNOT*) and the *cNOT* gate allow one to design a half-adder by using the approaches we have discussed. If the input $|y\rangle = 0$, then the lower two lines of the gate are the sum and the carry bit.

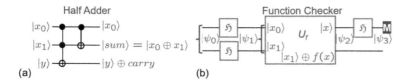

(a)

(b)

Figure 3.12: (a) illustrates a quantum half-adder composed of a $ccNOT$ and a $cNOT$, and (b) is a function checker that evaluates whether a single-variable Boolean function $f(x)$ is constant or balanced, by using the Deutsch-Jozsa algorithm.

The second example, shown in Figure 3.12(b), illustrates determining whether a one-variable Boolean function $f(x)$ is a constant, that is, whether $f(0) = f(1)$ or whether they are balanced, that is, even if $f(0) \neq f(1)$, the outcome happens with equal probability for both. This is an example of the use of the Deutsch-Jozsa algorithm. The circuit accomplishes this using Hadamard gates, and the general gate does so using the Toffoli-like XOR approach in the U_f gate, as shown in Figure 3.12(b). The following is a summary of how this operates and its ability to hold information that is not manifested until the end. The system is initialized in the qbit state $|\psi_0\rangle = |01\rangle$. The Hadamard gate creates a superposition of states for the wavefunction $|\psi_1\rangle$, with the states $|x\rangle$ and $|y\rangle$ as the input to the unitary U_f gate, which computes $f(x)$; for example, see the Fredkin-based $f(x)$ creation in Figure 3.4(a). The output $|\psi_2\rangle$ interacts with the third Hadamard gate, leading to the wavefunction $|\psi_3\rangle$. If one now performs measurement on the output of the Hadamard gate line qbit, a $|0\rangle$ measured state shows that the function is constant, and a $|1\rangle$ shows that the function is balanced. A classical algorithm for this problem will require two evaluation cycles of $f(x)$, while, in a quantum circuit, it was accomplished in one cycle because, in effect, the outcomes of $f(0)$ and $f(1)$ were present in the superposition during computation. And if one were to have a more complicated n-scale problem, the worst-case ratio would be 2^n. In the least worst-case event, one where all the first 2^{n-1} cases are 0s, then the question remains for the next case, $2^{n-1} + 1$, where if it is 0, the function is constant, and if 1, the function is balanced. This is the essence of the success of the Deutsch-Jozsa algorithm for function checking.

There are numerous such examples. In particular, Shor's algorithm for prime number factorization, as well as Grover's algorithm for search, uses similar superposition-enabled simultaneous composition and decomposition to reduce the polynomial time challenge of classical algorithms. Both of these approaches relate to the asymmetries of classical approach. Multiplication is easier, classically, and factoring is hard because of the numerous possibilities in the reverse operation of

The reader should convince herself of the following. With the input as $|0\rangle$ on both lines, the states are

$$
\begin{aligned}
|\psi_0\rangle &= |00\rangle, \\
|\psi_1\rangle &= |00\rangle - |01\rangle + \\
&\quad |10\rangle - |11\rangle, \\
|\psi_2\rangle &= |0f(0)\rangle - \left|0\overline{f(0)}\right\rangle + \\
&\quad |1f(1)\rangle - \left|1\overline{f(1)}\right\rangle, \text{ and} \\
|\psi_3\rangle &= |0f(0)\rangle - |1f(0)\rangle - \\
&\quad \left|0\overline{f(0)}\right\rangle + \left|1\overline{f(0)}\right\rangle + \\
&\quad |0f(1)\rangle - \left|1\overline{f(1)}\right\rangle - \\
&\quad \left|0\overline{f(1)}\right\rangle - \left|1\overline{f(1)}\right\rangle.
\end{aligned}
$$

This means that the output line is $|0\rangle$ if $f(0) = f(1)$, and $|1\rangle$ if $f(0) \neq f(1)$.

factoring. Shor's algorithm allows a fast determination of the period of a periodic function. This is useful in cryptography and other technical problems where one is often looking for a needle in a haystack. Grover 's algorithm gradually shifts from an initial uniform superposition to concentrating on a focused choice. The speedup comes from the probability rising quadratically while amplitude does so only linearly. Superposition leaves the rare and the not-so-rare possibilities to continue to exist as the problem is being resolved in a quantum fashion.

These different properties of the quantum operations on superposition and entanglement provide numerous new possibilities in computation. This is the theme discussed up to this point and in communication that we will discuss. Classical computation works with one input at a time, individually or in parallel. In quantum computation tackling exponentially growing number of inputs at the same time becomes possible. Classical computation is one logical step at a time, individually or in parallel. Quantum computation evolves a superposition of computation on all inputs in time. Classical computation produces one output at a time, individually or in parallel. In quantum computation, the output measurement can be rather clever, for example, joint properties of set of answers to computational problems can be extracted through the choice of observation. The measurement itself will be for a complex function rather than in cbits.

Here, we have again come back to the degrees of freedom and the thermodynamic thought process.

	Classical	Quantum
Unit	Bits: $\lvert\psi\rangle_n$ of 0 or 1	Superposition: $\lvert\psi\rangle = c_0\lvert0\rangle + c_1\lvert1\rangle + \cdots$
Capacity	Linear: N	Exponential: $\mathcal{O}(2^n)$
Computation	Serial (*BLAS**)	Parallel (superposed)
	$x \mapsto f(x)$	$\sum\lvert\psi_n\rangle \mapsto \sum f(\lvert\psi_n\rangle)$
Gates	*NAND, NOR,* ...	Qbit rotation + *cNOT*
Reversible operations	Permutations	Unitary transformations
Learning of state from bits	Yes	No
Information from bits	Direct	Measure
Information acquired	0 or 1 in sequence	$\lvert c_n\rvert^2$ for $\lvert\psi_n\rangle$
After acquisition	$\lvert\psi\rangle$ (same)	$\lvert\psi_n\rangle$ (a change)
Usage	General	Specific (factorization, search, etc.)
Issue	Thermodynamics	Loss of coherence

BLAS: Basic linear algebra subprograms.

Table 3.2: Some of the salient characteristics of classical and quantum computation.

Table 3.2 summarizes some of the basic thoughts in this classic-quantum computation milieu. It shows the differences in the basic unit of computation, the difference of linear versus exponential capacity, the serial-superposed computation itself, the difference in basic gates, how the bits are measured and the information that is acquired as a result, and examples of utility and constraints.

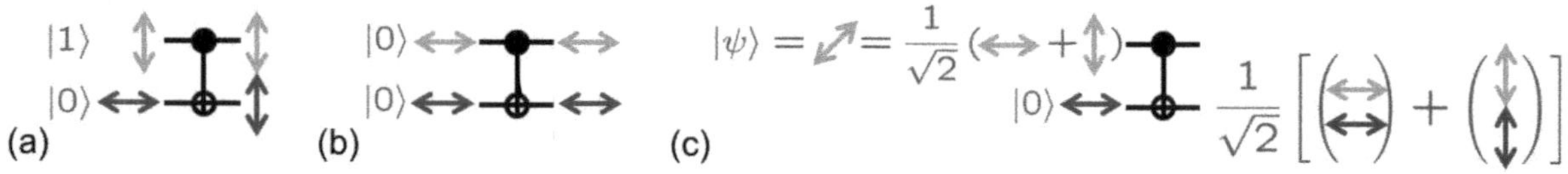

Figure 3.13: The *cNOT* response operating with pure states in (a) and (b), and an entangled control signal state in (c). The latter results in the production of an *EPR* pair.

1- and 2-qbit gates acting on qbits suffice for implementing quantum data processing. We have seen that the *cNOT* gate, that is, the 2-bit *XOR*, flips the second input if the first input is a 1, else it passes the inputs undisturbed. We show this in Figure 3.13, where the pure states are employed to operate the *cNOT* gate in the first two examples, and an entangled state operates in the third. These examples use polarization, horizontal and vertical, for example, of a photon, as the pure state. In Figure 3.13(a), if the control signal, the signal a of the *cNOT* gate of Figure 3.6(b), is the state $|1\rangle$, then the operation $b' = a \oplus b$, will convert $|0\rangle$ to $|1\rangle$. In Figure 3.13(b), since the control signal and input signal are in the same state, they pass through undisturbed. This all is directly following what we have discussed. The interesting consequence, however, of this quantum *cNOT* gate is that if the control signal is entangled, as in Figure 3.13(c) then the output of the signal line is in entangled superposition.

This entangled superposition is an *EPR* pair, so named for Einstein, Podolsky and Rosen, who raised the issue of nonlocality, such as an interaction occurring simultaneously at a distance apparently raises. If such a state were to exist, it would appear to violate commonly held notions of causality. However, we will presently see that there is no contradiction here.

Quantum superposition shows up spatially, such as in the two-slit experiment measurement in its periodicity. It can show up temporally—this is a bit more subtle—in the oscillation in time of an ensemble's quantum properties resulting from excited state amplitudes. Rabi oscillations are an example of this. Subtler still is the effect of spin-based symmetry restrictions for distinguishable multiparticle wavefunctions. Exchange interaction leading to spin alignment and magnetism is an example of this. These superpositions therefore range over size and complexity—from an elementary particle to large ensembles. The *EPR* pair, a specific subset of superposition, brings forth the notion that two particles, having interacted at some time in the past but now considerably spatially separated in our practi-

The paper, "Can quantum-mechanical description of physical reality be considered complete?" by Albert Einstein, Boris Podolsky and Nathan Rosen in Physical Review from May 1935 is iconic and a must-read for those who enjoy thought experiments. Starting with the statement that predicting a physical quantity with certainty is a sufficient condition for its reality, it queries the physical reality of quantities whose quantum operators do not commute, for example, position and momentum. The quandary it raises is whether the description of reality given by a wavefunction is complete or whether the two physical quantities with non-commuting operators do not have simultaneous reality, that is, raise questions that either relate to local and nonlocal hidden variables or the issue of nonlocality in general. Pauli, who was given to rather stinging comments when in disagreement, has been accused of being the self-styled vicar on Earth of the infallible deity. When he found that the factor of 2 was apparently in the spin of an electron, a factor that is relativistic in origin, he found it offensive, at least initially. At the time of the publication of the *EPR* paper, he is said to have remarked, "One should no more rack one's brain about the problem of whether something one cannot know anything about exists all the same, than about the ancient question of how many angels are able to sit on the point of a needle." *EPR* did not go away.

cal classical sense, are still not independent. Measurement at any time determining some property of one, say position, also instantaneously determines this same property with complete certainty for the other, no matter what the distance is between the two particles. The paradox here is that this second particle's property is determined without any measurement having been performed on it—this is non-locality—the choice of measuring position was only made after this spatial separation. Particles don't know which measurement of these two canonically conjugate variables is going to be performed. Even if both properties cannot be simultaneously measured, how does one get a completeness in measurement over two particles nonlocally? And if a nonlocal description suffices, then how do causality and special relativity get reconciled? We will see that these notions are quite reconcilable when we study the use of *EPR* entanglement.

Photon polarization provides a convenient means to tackling communications and *EPR*, and this is the approach that Figure 3.13 corresponds to. If one generates two photons in an entangled state and they follow separated paths, the entangled state as drawn in Figure 3.13(c) is a state of the whole system, which is not expressible in terms of the states of its parts. Figure 3.14 pictorially describes this. The first two entangled representations of the *EPR* pair of 2 qbits are formally equivalent, with each coupled polarization of the two particles possible with a probability of $1/2$, and this is not the same as the last coupling of states. The two photons are in a definite state of sameness of polarization, even if neither of the photons has a polarization of its own. When one is specifically measured to its horizontal or vertical polarization state, so will be the other.

$$\frac{1}{\sqrt{2}}\left[\left(\leftrightarrow\right)+\left(\updownarrow\right)\right]=\frac{1}{\sqrt{2}}\left[\left(\nwarrow\hspace{-1mm}\searrow\right)+\left(\nearrow\hspace{-1mm}\swarrow\right)\right]\neq\left(\nwarrow\hspace{-1mm}\searrow\right)\text{ or }\left(\leftrightarrow\right)\text{ or }\left(\updownarrow\right)$$

Figure 3.14: Equivalence and non-equivalence of an entangled *EPR* pair with the basis states.

We can look at the commutation properties too for this *EPR* pair. Let the separated particles have coordinates (q_1, q_2) and momenta (p_1, p_2). The commutation conditions are

$$[q_i, q_j] = 0, \quad [p_i, p_j] = 0, \quad \text{and} \quad [q_i, p_j] = -\frac{\hbar}{i}\delta_{ij} \;\; \forall\, i, j = 1, 2. \quad (3.32)$$

If we define new conjugate variables (q_1', p_1') and (q_2', p_2') through rotation in angle θ,

$$\begin{bmatrix} q_1' \\ q_2' \end{bmatrix} = \begin{bmatrix} \cos\theta & \sin\theta \\ -\sin\theta & \cos\theta \end{bmatrix} \begin{bmatrix} q_1 \\ q_2 \end{bmatrix}, \text{ and}$$

$$\begin{bmatrix} p'_1 \\ p'_2 \end{bmatrix} = \begin{bmatrix} \cos\theta & \sin\theta \\ -\sin\theta & \cos\theta \end{bmatrix} \begin{bmatrix} p_1 \\ p_2 \end{bmatrix}. \tag{3.33}$$

The commutation rules hold true for these transformed canonical conjugate coordinates. But, even if q'_1 and p'_1 cannot be precisely measured simultaneously since they do not commute, it is possible to have this two-particle system prepared in such a way that

$$\begin{aligned} q'_1 &= q_1\cos\theta + q_2\sin\theta, \text{ and} \\ p'_2 &= -p_1\sin\theta + p_2\cos\theta, \end{aligned} \tag{3.34}$$

do commute to have precise values simultaneously. One can measure either q_1 or p_1, and one should be able to precisely predict the corresponding property for the second particle.

If a precise measurement of physical property is a measure of reality, then an *EPR* pair allows measurement on one particle to simultaneously predict the second particle's corresponding property.

So if this is not instantaneous action at a distance, thus violating special relativity, how does this come about?

Quantum theory is a reversible theory—time-reversal symmetric. We can look at the measurement as follows in this quantum view. A measurement, for example, on the particle on the right of Figure 3.15 is accompanied by the propagation of information backward in time to the source, thus aligning the second particle's corresponding property. This message passing backward in time is safe from any paradoxes so long as the source cannot control the message, as is case for the *EPR* source, or the receiver disregards the message, that is, no measurement is performed.

Bell's inequality shows that locality, in the sense that a measurement on one system does not have an effect on another system sufficiently far away, thus suppressing interaction, is in contradiction with quantum mechanics. The relationship provides an approach to experiments whose outcomes will be different between theories that have "locality" built in, and those that do not. Appendix I discusses the use of the Bell states referred to in Table 3.1 to set the arguments of this foundational conundrum.

For us, the important conclusion is that cause and effect can be simultaneous yet spatially separated, that the "spooky action" was in the past instead of being in the moment with spatial separation, and that it provides a means to discriminate between theories where measurement results arise from ignorance of preexisting local properties. This subjective-objective conundrum was remarked on in Chapter 1, in the discussion of probabilities and the nature of probabilities as represented in the square of the coefficients of eigenfunctions of the state function. The latter is not the same as the classical view of prob-

$[q'_1, p'_2] = 0$, or $[q'_2, p'_1] = 0$, follows formally through these equations. Bohm, another icon along with Bell in probing the foundations, introduced the intuitive spin approach in the 1950s.

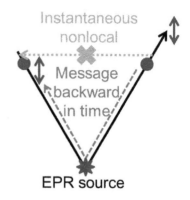

Figure 3.15: Observation on one particle of an *EPR* pair; the information can be viewed as propagating backward in time to the source. Nonlocal instantaneous action does not exist. The *EPR* pair is a doppelgänger.

Causality and the arrow of time in our daily experience is the outcome of the 2nd law of thermodynamics—the statistical evolution in an ensemble. Wheeler, Feynman and others have argued for the abandonment of microscopic causality. In between is probability, as the issue of "Whose probability?" as discussed in Chapter 1. It changes as more information becomes available. To paraphrase F. Scott Fitzgerald, we are like boats sailing against the current, pushed back ceaselessly into the past.

The de Broglie paradox is a good example of a simple thought experiment for the causality-locality-nonlocality in our classical picture. de Broglie, by making the connection between particle and wave, was instrumental in seeding the quantum revolution. This contribution, made in his Ph.D. thesis, was recognized by the Nobel committee in 1929. See the namesake paradox exercise at the end of the chapter.

ability that we learn in statistics or a signals course. While the qbit has the notion of randomness in it, there is also the idea of entanglement within it. In the *EPR* experiment, one measurement describes the outcome of another measurement elsewhere. A fair dice only has a probability in the asymptotic limit. And it has no memory of the past.

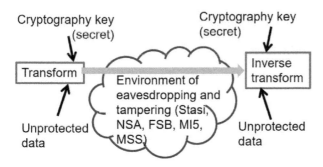

Figure 3.16: Communication so that messages may be transmitted without being intercepted.

The power of the *EPR* pair and the entanglement it represents at the quantum scale can also be seen in communications. Entanglement is a convenient way to transmit quantum information even when no quantum channel is available. And it is a powerful way to secure communications where quantum computational approaches may break the classical cryptographic code employed to prevent eavesdropping, as represented in Figure 3.16. So, we start with this protected communication problem.

Our story of communication is a parable. Our protagonists are Anarkali (*A*, not for Alice) and Balaji (*B*, not for Bob), who develop affection for each other. The environment consists of three individuals named Finch, Akbar and Terry and an organization we call the Sangh, a secret religion-centric organization—*FATS* for short, not Eve—who are the eavesdroppers. Anarkali and Balaji wish to secretly communicate. In a classical channel of communication they must work with cbits. On the other hand, they may have access to a quantum channel, for example, a narrow atom-sized opening or an entangled photon path through which they could entangle and exchange qbit interactions.

First, consider the classical exchange. The simplest cipher is to take the message of plain text $\mathcal{M}$, employ a secret key $\mathcal{K}$ that only Anarkali and Balaji know, and create a cipher message $\mathcal{M}_c$ in accord with Figure 3.16. *FATS*, the eavesdroppers, have access to $\mathcal{M}_c$ but not $\mathcal{K}$. Their ability to deconvolve the eavesdropped message depends on the ability to break the key $\mathcal{K}$. The key's integrity depends on the strength of the cipher. One may write the coding and

Ours is a pseudo-story of human love, with lessons from history. Anarkali is a Persian girl, perhaps a slave, in the 17th century in northern India and with who the prince of the Moghal empire falls in love; the prince is replaced here by Balaji, to represent the entanglement of religion. Akbar was the emperor who entombed Anarkali, so the story goes, according to William Finch and Edward Terry, two Britishers. There are no good guys here. Finch was an agent of the British East India Company— a company that vies with the Dutch East India Company for honors as an enterprise that was the most inhumane and destructive force on the people whom they conquered. The world has done worse. The genocide of Native Americans, the killings in World War 2 including the dropping of *atom bombs*, the continuing depriving of the human rights of the Roma, the slavery, including its causal effects today or slavery's modern-day equivalent of bonded labor and child labor in countries destroyed by imperialism and colonialism—these surely rank much higher.

decoding simply as $\mathcal{M}_c = \mathcal{M}_c(\mathcal{M}, \mathcal{K})$ for Anarkali transforming the message and $\mathcal{M} = \mathcal{M}(\mathcal{M}_c, \mathcal{K})$ for Balaji inverse transforming it back. In a Caesarean cipher, a primitive example of security, the secret key uses an integer I to shift the letter. There are $I = 25$ possibilities with an alphabet of 26 letters. Because this transposing, a shift, has only 25 keys, it is easy to break with access to only a few of the cipher messages. One may make this a substitution cipher in which $\mathcal{K}$ consists of mapping each letter to another unique letter in the 26-letter alphabet. Thus, there are $26! \approx 4 \times 10^{26}$ possibilities for $\mathcal{K}$. But, is this secure? No, and it is not even complicated. The asymptotic limit of the frequency of occurrence of letters is known. It is just a little more secure than the Caesarean cipher. Access to a few cipher messages $\mathcal{M}_c$ will suffice for code breaking.

This was an exercise in Chapter 1.

A perfect secret message is one where cipher $\mathcal{M}_c$ gives *FATS* no information of $\mathcal{M}$. This requires that there be no change in the classical probability in the Shannon entropy sense for any possible message. With

$$\mathfrak{P}(\mathcal{M}_i|\mathcal{M}_{cj}) = \mathfrak{P}(\mathcal{M}_i) \ \forall \, i, j, \ \text{and its implication,}$$
$$\mathfrak{P}(\mathcal{M}_{cj}|\mathcal{M}_i) = \mathfrak{P}(\mathcal{M}_{cj}) \ \forall \, i, j, \tag{3.35}$$

the probability of recovering original text is independent of having the cipher text.

But, all messages encrypted with the key $\mathcal{K}$ must lead to distinct messages so that they are distinguishable. So, the number of ciphertexts must be the same as plain texts. This implies, classically, that perfect secrecy also requires that the number of possible keys to be as great as the number of possible messages. There are different ways to accomplish this, and a common one, for the *ASCII* binary code, which has 128 symbols, is the Vernam cipher, which uses a key of N randomly chosen bits. This key $\mathcal{K}$ of N bits, known only to Anarkali and Balaji, creates a cipher by modulo 2 addition, that is, an *XOR* function. As an example, the modulo 2 addition causes

The 128 symbols of *ASCII* in a 7-bit string represent the lower case and upper case letters and the other symbols, including the space, that can be seen on the common keyboard.

$$\mathcal{M} = \ldots 001011010 \ldots, \ \text{and}$$
$$\mathcal{K} = \ldots 101110100 \ldots, \ \text{leading to}$$
$$\mathcal{M}_c = \mathcal{M} \oplus \mathcal{K} = \ldots 100101110 \ldots. \tag{3.36}$$

The problem with this is, of course, that the key needs to be communicated too. So, one needs to worry about key exchange protocols. An example in physical terms for security here is that Anarkali places the cryptographic key in a case, padlocks it, sends it to Balaji, but keeps the key with her. Balaji locks the case with another padlock and sends it back. Anarkali opens her padlock and sends it back. Balaji opens his padlock and can now receive the cryptographic key.

Of course, this example could just as well have been executed with a message and the physical case could have been a binary case—information is physical. *FATS* would not get access to the secret in this process. Three journeys were needed, and that is the cost of this security and integrity.

The path to making the cryptographic key secure is through making the path to opening this padlock or unraveling the attempt at achieving randomness in a finite set impossible or difficult. One-way functions—functions that are easy to calculate but are difficult to invert—are one example of this. The Shor algorithm of prime number factorization depends on the ease of prime number multiplication against factorization. Its objective is, given n, an integer, finding the prime numbers $p_1, \ldots, p_j$ so that $n = p_1 \times \cdots \times p_j$. We do not, as yet, know any polynomial time classical factorization algorithm. Public key cryptography is example of this approach. In this, Anarkali thinks of two large prime numbers p_1 and p_2 and makes $n = p_1 p_2$ public. Balaji talks to Anarkali by preparing a message, performing an arithmetic operation on the message with n, so that it is now deconvolvable when p_1 and p_2 are known. Since fast prime factorization is not known in polynomial time, and numbers are such that this would take days to weeks to break classically, this approach therefore achieves reasonable security for financial transactions expectations. In order to receive a message from Anarkali, the *RSA* ciphersystem—the system that *NIST* and *NSA* caused to be compromised—requires Balaji to generate a public key while keeping a private key d. So, he generates two prime numbers p_1 and p_2 and computes $n = p_1 p_2$. He then chooses an integer e so that $1 < e < (p_1 - 1)(p_2 - 1)$, with the greatest common divisor for e and $(p_1 - 1)(p_2 - 1)$ being 1. Balaji's private key d is subject to the condition $1 < d < (p_1 - 1)(p_2 - 1)$ with the constraint

$$d \times e \equiv 1 \text{ modulo } (p_1 - 1)(p_2 - 1). \qquad (3.37)$$

The decryption possibility exists because the greatest common divisor of e and $(p_1 - 1)(p_2 - 1)$ is unity.

While all these approaches make security of communication higher, all the classical communication channels employed by Anarkali and Balaji are insecure against unlimited computing power, or if $P = NP$. We discussed the nature of this and other issues of complexity in Chapter 1.

Quantum cryptographic approaches employing qbit entanglement can be secure against any quantum eavesdropping as well as unlimited computing through the secret cryptographic key.

The procedure for communication between Anarkali and Balaji now depends on two unique properties of quantum systems. The

This is again a degrees of freedom and entropic issue. Multiplication leads to one unique result, while factorization leads to many, with the one we are interested in consisting of a unique set of prime numbers.

$P = NP$ or $P \subset NP$, the P-versus-NP question, is whether a problem may be solved with reasonable computing resources in reasonable time, that is, have polynomial demands on computing, that is, belongs to P-class problems, or if it is NP, that is, has non-polynomial demands. IF $P = NP$, then there exists a shortcut in NP to a quick solution. If $P \neq NP$, then Anarkali and Balaji should be secure against the powers of unlimited classical computing.

first is the no-cloning theorem, which forbids perfect copying of an arbitrary quantum state. The argument of no cloning is as follows. In order to copy an arbitrary quantum system A in a pure state $|\psi\rangle_A$, one needs to start with another quantum system B in some initial state $|i\rangle_B$ and perform the cloning operation, for example, through $\hat{U}_c$, which transforms $|\psi\rangle_A |i\rangle_B$ to $|\psi\rangle_A |\psi\rangle_B$. Such a cloner will map computational basis states perfectly, so

$$\begin{aligned} \hat{U}_c |0\rangle |0\rangle &= |0\rangle |0\rangle, \text{ and} \\ \hat{U}_c |1\rangle |0\rangle &= |1\rangle |1\rangle, \end{aligned} \tag{3.38}$$

where the subscripts A and B have been dropped for convenience. Now consider what happens when the same operation is performed on two example superposition states. If it is an exact cloner, it will have to achieve the following:

$$\begin{aligned} \hat{U}_c \frac{1}{\sqrt{2}}(|0\rangle + |1\rangle) |0\rangle &\overset{?}{=} \frac{1}{\sqrt{2}}(|0\rangle + |1\rangle)\frac{1}{\sqrt{2}}(|0\rangle + |1\rangle), \text{ and} \\ \hat{U}_c \frac{1}{\sqrt{2}}(|0\rangle - |1\rangle) |0\rangle &\overset{?}{=} \frac{1}{\sqrt{2}}(|0\rangle - |1\rangle)\frac{1}{\sqrt{2}}(|0\rangle - |1\rangle). \end{aligned} \tag{3.39}$$

But if $\hat{U}_c$ clones $\{|0\rangle, |1\rangle\}$ correctly, then

$$\begin{aligned} \hat{U}_c \frac{1}{\sqrt{2}}(|0\rangle + |1\rangle) |0\rangle &= \frac{1}{\sqrt{2}}(|00\rangle + |11\rangle), \text{ and} \\ \hat{U}_c \frac{1}{\sqrt{2}}(|0\rangle - |1\rangle) |0\rangle &= \frac{1}{\sqrt{2}}(|00\rangle - |11\rangle). \end{aligned} \tag{3.40}$$

These are not the clones.

So, while computational basis states can be cloned, an arbitrary state cannot be. The second property that the quantum teleportation or telefacsimile depends on is the use of *EPR*. Thus, quantum cryptography provides more subtlety than the purely classical approach does, as security to any intervention by *FATS* in a prepared message between Anarkali and Balaji can be noticed.

Teleportation or quantum communication of a qbit in the state

$$|\psi\rangle = c_0 |0\rangle + c_1 |1\rangle \tag{3.41}$$

is summarized in Figure 3.17. Anarkali is in possession of $|\psi\rangle$ but doesn't know c_0 or c_1. Anarkali and Balaji share an *EPR* pair—so a pair of qbits that have been prepared in a Bell state, for example,

$$|\phi\rangle = \frac{1}{\sqrt{2}}(|00\rangle + |11\rangle). \tag{3.42}$$

Anarkali can manipulate only the first qbit of this *EPR* pair, and Balaji, only the second one.

The use of terms such as teleportation or telefacsimile, commonly used for having $|\psi\rangle$ appear at another spot, while attractive, are really not accurate. An *EPR* pair was entangled, and $|\psi\rangle$ was rebuilt elsewhere. This $|\psi\rangle$ is neither a copy nor a transported $|\psi\rangle$. It was recreated—not "teleported" nor "telefaxed."

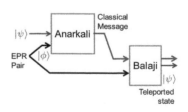

Figure 3.17: Communication between Anarkali and Balaji using an *EPR* pair.

Anarkali prepares the starting 3-qbit state of

$$|\chi\rangle = |\psi\rangle\,|\phi\rangle = \frac{1}{\sqrt{2}}\left[c_0\,|0\rangle\,(|00\rangle + |11\rangle) + c_1\,|1\rangle\,(|00\rangle + |11\rangle)\right]. \quad (3.43)$$

Anarkali has two qbits in her possession. She uses a $cNOT$ and a Hadamard operation on the first qbit, which started off with $|\psi\rangle$, in order to entangle $|\psi\rangle$ and $|\phi\rangle$:

$$
\begin{aligned}
|\breve{\chi}\rangle &= \hat{\mathfrak{H}} \times cNOT\,|\chi\rangle \\
&= \hat{\mathfrak{H}}\frac{1}{\sqrt{2}}\left[c_0\,|0\rangle\,(|00\rangle + |11\rangle) + c_1\,|1\rangle\,(|00\rangle + |11\rangle)\right] \\
&= \frac{1}{2}\left[c_0\,(|0\rangle + |1\rangle)\,(|00\rangle + |11\rangle)\right. \\
&\quad \left. + c_1\,(|0\rangle - |1\rangle)\,(|10\rangle + |01\rangle)\right].
\end{aligned}
\quad (3.44)
$$

Balaji perceives this state as

$$
\begin{aligned}
|\breve{\chi}\rangle &= \frac{1}{2}\left[|00\rangle\,(c_0\,|0\rangle + c_1\,|1\rangle) + |01\rangle\,(c_0\,|1\rangle + c_1\,|0\rangle)\right. \\
&\quad \left. + |10\rangle\,(c_0\,|0\rangle - c_1\,|1\rangle) + |11\rangle\,(c_0\,|1\rangle - c_1\,|0\rangle)\right] \\
&= \frac{1}{\sqrt{2}}\left[|00\rangle\,|\psi\rangle + |01\rangle\,X\,|\psi\rangle + |10\rangle\,Z\,|\psi\rangle + |11\rangle\,XZ\,|\psi\rangle\right] \\
&= \frac{1}{\sqrt{2}}\sum_{k=0}^{1}\sum_{l=0}^{1} X^l Z^k\,|\psi\rangle,
\end{aligned}
\quad (3.45)
$$

where X and Z are the Pauli matrices ($XZ = iY$ for Pauli matrices) that operate on Balaji's qbit. Balaji has received four transformed variations of Anarkali's message.

The encrypting and decrypting here uses the following clever procedure. Anarkali makes a measurement in the computational basis of the two qbits $(1,2)$ that she possesses. This gives four combinations for $|kl\rangle$, where k and l can be either 0 or 1 with equal probabilities. This measurement transforms the system to $|kl\rangle\,X^l Z^k\,|\psi\rangle$. Because Anarkali made a measurement, Balaji possesses a modification of the state $|\psi\rangle$. Anarkali can provide Balaji (k,l) through a classical channel subject to all the constraints of classical transmission. However, Balaji has a transformed version from which he can extract ψ by performing the reverse operation, $Z^k X^l \equiv (X^l Z^k)^{-1}$.

Two classical bits were transmitted and these were sufficient for message reconstruction because of the shared EPR pair for the qbit $|\psi\rangle$ being transported. In effect, the information was split into a classical part carried by the 2-bit (kl) message, and the quantum part, which is carried by an entanglement with an EPR pair. Cloning was avoided by destruction of $|\psi\rangle$ by Anarkali and its recreation by Balaji. The two classical bits must be received by Balaji before this recreation may take place.

So, this is not "*spukhafte Fernwirkungen.*" Security has been preserved without any spooky action at a distance.

3.2 *At the mesoscale*

PERHAPS THE MOST CLASSICAL of electrical observations is Ohm's law—the linearity of the relationship between current and voltage in a conductor, that is, $I = V/R$, where I is current, R is resistance, and V is the bias voltage difference between leads. This law states that current—the charge flux of electrons—is directly proportional to the electrostatic potential energy difference, with the linearity coefficient related to the properties of the material, and its structural form and dimensions. The thermal equivalent of this, arising from energy carried by different excitations in the material is thermal current $J_{\mathfrak{T}} = \kappa \Delta T$, where $J_{\mathfrak{T}}$ is the heat flux, κ is the thermal resistance and ΔT the temperature difference between leads.

Where may one expect a breakdown in these relationships?

The classical description is a macroscopic simplification where one finds parameters that encapsulate a lot of details underneath. Quantum-mechanically, this ensemble of atoms in a metal or a semiconductor has bandgaps, effective masses for electrons, and a Fermi energy representing the probability of $1/2$ for occupation of occupiable states; yet, it is sufficient to think of resistance in a metal, or mobility and diffusivity in a semiconductor, to describe how the charge particles move. The parameterization is sufficient for achieving an adequate description for phenomena that one may be interested in. It works because the states are much closer together in energy than energy scales of interest, and scattering is random and sufficient in number that one may view transport as a flow in the presence of friction. But, under certain conditions, for example, when the ensemble is small—as in small structures—this parameterization fails because characteristics associated with quantum effects—quantization, uncertainty, superposition, wave-particle duality, or quantum phenomena such as tunneling or even discretization of small numbers without resorting to quantum considerations—become important.

Since this characteristic of charge quantization and the wave nature of the electron is well understood from the electronics view—we employed it in our discussion of nanoscale transistors—we start by exploring magnetic fields and another important theme related to quantum foundations: what is more fundamental? Fields or potentials? In the motion of a charged particle such as an electron affected by fields in regions that the electron does not even enter in a multiply connected region. A good example for exploring this is the Aharonov-Bohm effect, which is schematically represented in a simple form in Figure 3.18. This is the magnetic analog of the two-slit experiment. An electron stream in a metal wire is split into two parts

For a detailed study of thermal and electrical transport, see S. Tiwari, "Semiconductor physics: Principles, theory and nanoscale," Electroscience 3, Oxford University Press, ISBN 978-0-19-875986-7 (forthcoming).

This view of Fermi energy, or the electrochemical potential energy, is certainly appropriate for thermal equilibrium in a metal with a large density and close spacing of states. But, what does one mean that the Fermi level is in the band gap of a non-degenerately doped semiconductor where there are no states? What we mean is that the state occupation and the occupation's distribution in the states are consistent with Fermi-Dirac occupation statistics at the specified Fermi energy even if there are no states at the energy or in its vicinity.

We also bring this subject up first in S. Tiwari, "Quantum, statistical and information mechanics: A unified introduction," Electroscience 1, Oxford University Press, ISBN 978-0-19-875985-0 (forthcoming) and S. Tiwari, "Semiconductor physics: Principles, theory and nanoscale," Electroscience 3, Oxford University Press, ISBN 978-0-19-875986-7 (forthcoming), from the viewpoint of gauge invariance while introducing the notion of the vector potential **A**. However, it is also present in locality and nonlocality, the *EPR* effect, et cetera, so it is important to understand it in the context of the mesoscale.

and recombined. We first assume a coherent structure, that is, one devoid of scattering—electrons traveling in vacuum such as in the double-slit system, but it is also realistic at nanoscale in a low scattering system. We can look upon this arrangement as a baffled solenoid or an infinite solenoid with a wire channel protected from fields, or the electrons traveling inside a torus. The wire in which the electrons travel does not itself have a field in this split part. The electrons encounter no fields.

Should there be any interference effect when the electrons following the two different paths rejoin? Experimentally, the answer is yes, there is. Superconducting loops provide very precise evidence of this. This in turn implies that there are phase shifts along the different paths that the electrons take. This observation of interference in the ring, and its implication of different phase shifts in the paths even if there is no field, points to the incompleteness of the field-based viewpoint that one traditionally takes. Thinking through the problem in potentials, so $\mathbf{A}$ for the vector potential, and φ for the scalar potential, one incorporates the nonlocal contributions—of sources elsewhere and in earlier times—locally. Stokes's and Gauss's theorems, the first for a loop, and the second for an enclosed volume, let us incorporate why infinite solenoids still need a return flux somewhere, even if the solenoid is infinite. Stokes's and Gauss's law dictate the global implications of the fields, even if this solenoid problem is not so simply connected. There are two considerations here: if one thinks in terms of fields, then one employs overt nonlocality; if one thinks in terms of potentials, then spatial and temporal nonlocality is from within. The Aharonov-Bohm effect is direct evidence of nonlocality in terms of potentials. It is this same nonlocality that appears in the *EPR* effect.

One way to view this nonlocality is as follows. The eigenstates that are the solution to the Hamiltonian of the system are all the possible results of experiments. One may build a wavepacket from the eigenstates that are localized on one side of this centrally confined magnetic field. One may, equivalently, build another wavepacket for the other side of the field. Are these packets independent? One also has the option of building a wavepacket that has a finite penetration. This wavepacket will be bent by the field. The results of all the experiments must be consistent, so the first two wavepackets must result in observations that are consistent with this last penetrating wavepacket. And the two wavepackets traveling along the two sides of the loop must also be internally consistent with all possible experiments with them. This implies that the two wavepackets are tied together. And this is reflected in a definite phase shift that is a function of the enclosed flux. This argument also implies that all space is penetrated, in the sense that what happens outside is re-

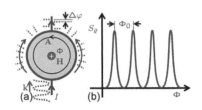

Figure 3.18: (a) shows the flow of current through a metal loop with a magnetic field enclosed inside normal to the loop. The loop itself is devoid of field. The loop could be a toroid or the solenoid creating the field could be baffled. (b) shows a schematic power spectrum of magnetoresistance in a coherent structure.

The application of Stokes's theorem to the Aharonov-Bohm experiment produces an immediate contradiction. Curl-less electromagnetic potential produces an interference between the two electron beams that enclose the non-zero flux but passes through a region free of field.

lated to what happens inside. Potentials penetrating continuously, while fields are abruptly shutting off, require one to place additional boundary conditions that can make the discontinuous change happen in fields. This is necessary to make this consistency come about without introducing ambiguity. This we have not done. The wavepacket construction cannot proceed disregarding the presence of the field, and this is what the Aharonov-Bohm effect shows; and viewing it from the potential viewpoint provides a self-consistent picture.

The phase of electron wavefunction encircling a magnetic flux Φ and a vector potential $\mathbf{A}$ associated with the magnetic field is shifted by

$$\Delta\varphi = \frac{e}{\hbar} \int \mathbf{A} \cdot d\mathbf{r}. \tag{3.46}$$

In Figure 3.18, the coherent beam following the two arms is phase shifted as shown, with the difference identified as $\Delta\varphi$. This difference periodically oscillates with the flux. The period of this flux, following Equation 3.46, is $\Phi_0 = h/e$, the flux quantum. The phase shift for changing fields and vector potential is

$$\Delta\varphi = 2\pi \frac{\Phi}{\Phi_0}, \tag{3.47}$$

where $\Phi = \int \mathbf{B} \cdot d^2\mathbf{r}$ is the flux threaded through the enclosed area. Interference contributions, being periodic in Φ_0, cause periodicity in the observed response. One sees this periodicity in the power spectrum, or in the resistance between the two ports—periodic oscillations that appear as a function of flux or applied field change.

The presence of finite scattering events such as from impurities or defects will cause phase changes, but because of the finite number of such scattering events, the resulting oscillatory structure will still be clearly discernible. This is observable in small metal rings—$\sim 100\ nm$ in diameter at low temperatures. Small rings with few scattering centers will also show persistent long-lived current at low temperatures, phenomenon which is not due to superconductivity. Since a current loop has a magnetic moment, if an external probe field exists that is misaligned with this moment, there will exist a torque on the ring. It is a small torque, but measurable. Cantilevers with such rings, another type of nanoscale object that feels torque, have been employed to measure the oscillatory cantilever response showing the effect of torque arising from persistent currents in the rings.

We have looked at classical, integral and fractional quantum Hall effects in detail in our discussion of semiconductor physics. We here start with a short summary of that discussion since it highlights several mesoscopic properties and because the fractional quantum Hall state has implications for quantum computation through forbidden transitions.

Classical, integral and fractional quantum Hall effect are discussed at length in S. Tiwari, "Semiconductor physics: Principles, theory and nanoscale," Electroscience 3, Oxford University Press, ISBN 978-0-19-875986-7 (forthcoming).

The Hall effect is the change in longitudinal and transverse resistance due to the presence of a magnetic field transverse to the plane of current flow. Resistance and potential are equivalent, with the current as a normalization constant. Figure 3.19 is an example of this measurement on a Hall sample whose properties are being observed. In the Drude classical description of free electron gas, the current density in the presence of a magnetic field $\mathbf{B} = B_z\hat{z}$ is

$$
\begin{aligned}
\mathbf{J} \;=\;& qn < \mathbf{v} > \\
=\;& \frac{q^2 n}{m^*}\left\langle \frac{\tau_{\mathbf{k}}}{1+(\omega_c\tau_{\mathbf{k}})^2}\right\rangle \mathcal{E} \\
& -\frac{q^3 n}{m^{*2}}\left\langle \frac{\tau_{\mathbf{k}}^2}{1+(\omega_c\tau_{\mathbf{k}})^2}\right\rangle (\mathcal{E}\times\mathbf{B}) \\
& +\frac{q^4 n}{m^{*3}}\left\langle \frac{\tau_{\mathbf{k}}^3}{1+(\omega_c\tau_{\mathbf{k}})^2}\right\rangle \mathbf{B}(\mathcal{E}\cdot\mathbf{B}).
\end{aligned}
\tag{3.48}
$$

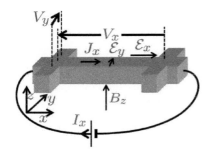

Figure 3.19: A Hall measurement where a small current flows at extremely low fields in the presence of a transverse magnetic field, with measurement of a voltage or field that is equivalent to a resistance in the plane that is transverse to the applied voltage and magnetic field. With an applied bias voltage V_x and a magnetic field of B_z, a current J_x flows and a transverse voltage V_y develops, resulting in a longitudinal resistance of ρ_{xx} and a transverse resistance of ρ_{xy}.

Here, $\omega_c = qB_z/m^*$ is the cyclotron resonance frequency, the orbiting consequence of the charged electron under the influence of the Lorentz force $q\mathbf{v}\times\mathbf{B}$, and $\tau_{\mathbf{k}}$, the momentum scattering time. The first term in the equation is the ohmic term observable as the static term in the limit $\omega_c \to 0$. The second term is the classical Hall effect, and the third term is an additional magnetoresistance contribution.

With small static magnetic fields, that is, $(\omega_c\tau_{\mathbf{k}})^2 \ll 1$, that is, a condition where during a time period of the magnetic field $2\pi/\omega_c$, numerous scattering events occur, the current density simplifies to

$$
\mathbf{J} = \frac{q^2 n}{m^*}\langle\tau_{\mathbf{k}}\rangle \mathcal{E} - \frac{q^3 n}{m^{*2}}\langle\tau_{\mathbf{k}}^2\rangle (\mathcal{E}\times\mathbf{B})
\tag{3.49}
$$

With $\mathbf{B} = B_z\hat{z}$, and $\mathcal{E} = \mathcal{E}_x\hat{x} + \mathcal{E}_y\hat{y} + \mathcal{E}_z\hat{z}$ in a general structure, if one places no conducting path in the y direction and only measures a voltage, J_y is forced to vanish, and one can relate the electric fields $\mathcal{E}_x$ and $\mathcal{E}_y$ thus obtaining the relationship

The Lorentz force is now balanced by the electric field that develops due to the accumulation and depletion of charge transversely.

$$
J_x = -\frac{qn}{B_z}\frac{\langle\tau_{\mathbf{k}}\rangle^2}{\langle\tau_{\mathbf{k}}^2\rangle}\mathcal{E}_y.
\tag{3.50}
$$

This appearance of a field in the transverse direction due to the presence of current J_x and B_z is the Hall effect, as shown in Figure 3.19. In a Hall measurement, one observes the lateral voltage (V_y) or transverse resistance, that is, $\rho_{xy} = V_y/I_x$, as well as the longitudinal voltage V_x or the resistance $\rho_{xx} = V_x/I_x$. These voltages and currents in matrix form are

$$
\begin{bmatrix} V_x \\ V_y \end{bmatrix} = S\begin{bmatrix} \rho_{xx} & \rho_{xy} \\ -\rho_{xy} & \rho_{xx} \end{bmatrix} = \begin{bmatrix} I_x \\ I_y \end{bmatrix},
\tag{3.51}
$$

where S is a normalization factor arising from the connection of voltages to fields, and currents to current densities, through the sample geometry.

At low fields, these resistances follow the linear relationship expected from ohmic law, where the conductivities are being affected by the magnetic field and the transverse field develops due to the Lorentz force effect, but scattering is unlimited. Ohm's law is still valid. Figure 3.20 shows the dependences of these resistances to the factor $\omega \tau_\mathbf{k}$.

In a low field classical Hall measurement, one is able to determine the Hall constant from the transverse measurement

$$R_H \equiv \frac{\mathcal{E}_y}{J_x B_z} = -\frac{1}{en} \frac{\langle \tau_\mathbf{k}^2 \rangle}{\langle \tau_\mathbf{k} \rangle^2}, \tag{3.52}$$

which relates carrier concentration to the scattering properties within the material. This ratio, a property of the material determined by the scattering mechanisms' characteristics, is the Hall factor

$$r_H \equiv \frac{\langle \tau_\mathbf{k}^2 \rangle}{\langle \tau_\mathbf{k} \rangle^2}, \tag{3.53}$$

knowledge of which allows one to calculate the carrier density. Since the longitudinal resistance ρ_{xx} is known, one can then determine the Hall mobility, which signifies the mobility under the imposed conditions of magnetic field and its effect on the charge carrier's traversal. The sign of the measurement is related to the charge, so it also allows one to determine the polarity of the material.

When dimensions are quantized, for example, a two-dimensional electron gas, and these have high mobility, and so limited scattering, in the Hall measurement at sufficiently high fields, one begins to observe new phenomena that are strictly quantum in origin.

Let us now assume that there is no scattering; we ask, what do the quantum-mechanical constraints dictate?

We will look at this problem in two ways—first, through a classical view into which we incorporate quantum-mechanical thoughts, that is, a mixed view, and second, directly through quantum mechanics.

Mixed view: The classical way of looking at this field effect, where scattering is absent, is to see that the electron, still a free particle in an electron gas, has a time response to the Lorentz force; this response is dictated by

$$m^* \frac{d\mathbf{v}}{dt} = (q = -e)\mathbf{v} \times \mathbf{B}. \tag{3.54}$$

So, the electron orbits around the field with the cyclotron frequency of $\omega_c = eB_z/m^*$ in a radius of $r_c = v/\omega_c$. Classically, no limits have been placed on what velocity,r momentum or position this electron

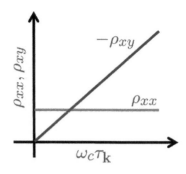

Figure 3.20: Transverse ρ_{xy} and longitudinal ρ_{xx} resistances as a function of $\omega \tau_\mathbf{k}$ in a low field Hall measurement.

may have. However, quantum-mechanically, these are prescribed through coherence—integral wavelengths of circumference of cyclotron orbit, that is,

$$2\pi r_c = \frac{nh}{p} = \frac{nh}{m^*v} \;\; \forall \, n = 1, 2, \ldots,$$

(3.55)

which relates as $v = r_c\omega_c = n\hbar\omega_c/m^*v$ for the electron's motion characteristics. The kinetic energy, by extension, is dictated to have discrete possibilities of

$$\frac{m^*v^2}{2} = n\frac{\hbar\omega_c}{2}.$$

(3.56)

The presence of magnetic field then changes the energies of the allowed states to

$$E_n - E_0 = \left(n + \frac{1}{2}\right)\hbar\omega_c = \left(n + \frac{1}{2}\right)\hbar\frac{eB_z}{m^*}.$$

(3.57)

Absent magnetic field, the two-dimensional system has quantized subbands, with the density of each subband given by $\mathscr{G}_{2D}(E) = m^*/\pi\hbar^2$. The number of states in the energy range $\hbar\omega_c$ is

$$N_0 = \hbar\omega_c\frac{m^*}{\pi\hbar^2} = \frac{2eB_z}{h}.$$

(3.58)

The states that existed within the energy band $\hbar\omega_c$, absent magnetic field, are states associated with motion in the allowed transverse plane of conduction. The presence of magnetic field does not impart energy to these electrons, since the motion and field are orthogonal— all it has done classically is to impart the movement in an orbit through a centripetal Lorentz force and placed quantum restrictions on what momentum these particles may have. The particles are restricted to the allowed energies defined by the Equation 3.57. What this new allowed energy distribution reflects is the concentration of these states within the $\hbar\omega_c$ band into the energy E_ns. This energy level is the Landau level. This description has resulted in a density of states composed of delta functions in energy as a function of magnetic field:

$$N_s(E, B) = \frac{2eB_z}{h}\sum_{n=0}^{\infty}\delta\left[E - E_0 - \left(n + \frac{1}{2}\right)\hbar\omega_c\right].$$

(3.59)

Since there has been a discretization introduced through the magnetic field, the resistance characteristics in the Hall measurement will also change. Figure 3.21 shows schematically a type of change that can occur in the resistances as a function of magnetic field, in this case, of the type that one observes in a high magnetic field for high mobility two-dimensional electron systems, which in our single particle picture can also be redrawn in terms of the cyclotron frequency.

Lev Davidovich Landau (1908–1968), the master Soviet physicist of the 20th century, appears again and again in any subject area of physics one may choose. The Landau-Lifshitz series of books is the abstract, mathematically precise, and thorough physics thinking of the pre-1960s—it is the "Old Testament." One may claim that the Feynman lectures on physics, written in a more readable and conceptual modern format, comprise the "New Testament." The former is an essential document for working out precise details, and the latter, for intuitive thinking.

High mobility, but not extremely high mobility with extreme homogeneity, as we will presently see.

This is the integral quantum Hall effect. It occurs under conditions of $\omega_c\tau_k \gg 1$ and in the presence of inhomogeneities that allow a distribution around the Landau energies to be observable. Our quantum description will clarify this.

The important observation from Figure 3.21 is the breakdown of the Ohm's law relationship, manifested as a pronounced modulation of the longitudinal resistance, which peaks as a function of field, with a chirp-like behavior. These are the Subnikov-de Haas oscillations. This resistance is

$$\rho_{xx} = \frac{1}{\sigma}\left[1 - 4\frac{\zeta}{\sinh(\zeta)}\exp\left(-\frac{\pi}{\omega_c\tau_k}\right)\cos(2\pi\nu)\right],\qquad(3.60)$$

where σ is a conductivity normalization factor at zero magnetic field, $\zeta = 2\pi^2 k_B T/\hbar\omega_c$, and $\nu = (E - E_0)/\hbar\omega_c$, written here without proof. These oscillations are observed when the single particle scattering time is longer than the period of cyclotron oscillations, that is, $\omega_c\tau_k \gg 1$, which is equivalent to $\hbar\omega_c \gg 2\pi^2 k_B T$. The plateaus in the transverse resistance occur at

$$\rho_{xy} = \frac{h}{\nu e^2}\quad\text{for }\nu = 1, 2, \ldots.\qquad(3.61)$$

The quantum conductance has now shown up again since the measurement effect is under constraints where the cyclotron orbit is under coherent conditions, with scattering absent. The most interesting aspect of this measurement is the feature that this behavior is independent of the nature of the host two-dimensional electron gas material. A unique emergent property composed of the fundamental constants h and e appears, independent of the sample material.

This mixed view has many deficiencies, the principal of which is that we arrived at Landau energy states into which the band of energy states in $\omega_c\tau_k$ states collapsed, and yet there is a band over which conduction occurs. We observed plateaus and oscillations in conduction, not delta functions! This is related to inhomogeneities that allow a spreading of states that the quantum-mechanical view effectively captures.

Quantum-mechanical view: The Hamiltonian of this problem is

$$\mathscr{H} = \frac{(\mathbf{p} + e\mathbf{A})^2}{2m^*} = \frac{1}{2m^*}\left[(p_x + eA_x)^2 + (p_y + eA_y)^2\right].\qquad(3.62)$$

When $\mathbf{B} = B_z\hat{\mathbf{z}} = \nabla \times \mathbf{A}$, then $\mathbf{A} = xB_z\hat{\mathbf{y}}$, and $\mathbf{A} = -yB_z\hat{\mathbf{x}}$, are both appropriate potentials. We make the gauge choice of $\mathbf{A} = -yB_z\hat{\mathbf{x}}$, and since $[\mathscr{H}, \mathbf{p}] = 0$, that is, they commute, we recognize p_x as a good quantum number. This means that, using $p_x = \hbar k_x$ and using spatial function separation for the linear Schrödinger equation, we find that

$$\psi(x, y) = \exp(ik_x x)\phi(y).\qquad(3.63)$$

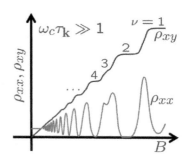

Figure 3.21: The integral quantum Hall effect: a schematic of the transverse ρ_{xy} and longitudinal ρ_{xx} resistances as a function of the magnetic field B_z in conditions of $\omega_c\tau_k \gg 1$, so in a high field Hall measurement.

Solving this equation, we obtain

$$
\begin{aligned}
\mathscr{H}\psi(x,y) &= \frac{1}{2m^*}\left[(\hbar k_x - eyB_z)^2 + p_y^2\right]\exp(ik_xx)\phi(y) \\
&= \left[\frac{e^2B_z^2}{2m^*}\left(y - \frac{\hbar k_x}{eB_z}\right)^2 + \frac{p_y^2}{2m^*}\right]\exp(ik_xx)\phi(y) \\
&= \left[\frac{1}{2}m^*\omega_c^2(y - y_0)^2 + \frac{p_y^2}{2m^*}\right]\exp(ik_xx)\phi(y) \\
&= \left[\frac{1}{2}m^*\omega_c^2\breve{y}^2 + \frac{p_y^2}{2m^*}\right]\exp(ik_xx)\phi(y), \quad\quad (3.64)
\end{aligned}
$$

with $y_0 = \hbar k/eB_z$, and $\breve{y} = y - y_0$. The cyclotron response part, with the change of coordinates to $\breve{y}$, is a harmonic oscillator–like response. The magnetic field or potential keeps the wave in a path which has energy exchange characteristics like that of a simple harmonic oscillator exchanging potential and kinetic energy. Here, it appeared due to the interaction reflected in how the momentum and the vector potential interact.

The net result is that the energies in the presence of a magnetic field are

$$
E_n = \left(n + \frac{1}{2}\right)\hbar\omega_c, \quad \text{with } n = 0,1,2,\dots. \quad\quad (3.65)
$$

The energy levels of two-dimensional electrons in a magnetic field are similar to those of a simple one-dimensional harmonic oscillator! These levels are discrete even if, absent magnetic field, they form a very closely spaced distribution whose density is $\mathscr{G}_{2D}$. They do not depend on y_0 or, equivalently, k_x. Momentum in the x coordinate has no kinetic energy. So, for any n, any k_x is allowed, with energy levels being very degenerate.

The x dependence of all Landau levels is a plane wavefunction, so the probability is a constant in this direction. The state is an "extended" state in this direction. In the mixed description, without subterfuge, all we can say is that there exist gaps over an energy range of $\hbar\omega_c$ between the allowed Landau states, which are themselves linearly varying with the magnetic field. This happens only for two-dimensional systems, not three-dimensional systems, where we still have the momentum direction aligned with the field with a distribution of energy states. The states in the x direction—the extended states with a plane wave in the x direction—in order to be localized in the y direction, are Gaussian and so satisfy the constraints specified. We will find that these Gaussian states for any k_x are centered at $y = y_0$ and have an approximate width of $\ell = (\hbar/eB_z)^{1/2}$, where ℓ is a magnetic length scale. We assume $\phi(\breve{y}) = \mathcal{A}\exp(-a\breve{y}^2)$, with $\mathcal{A}$ and a as constants, one for normalization and one for the length scale of Gaussian function, and where we have transformed the positional reference so that $y \rightarrow \breve{y}$. So, for $\phi(\breve{y})$, with this Gaussian function substitution, in this orthogonal in-plane direction, we solve

$$
\begin{aligned}
& \left(-\frac{\hbar^2}{2m^*}\frac{d^2}{d\breve{y}^2} + \frac{1}{2}m^*\omega^2\breve{y}^2\right)\phi(\breve{y}^2) \\
= \ & E\phi(\breve{y}^2) \\
= \ & \left[-\frac{\hbar^2}{2m^*}(4\breve{y}^2a^2 - 2a) + \frac{1}{2}m^*\omega^2\breve{y}^2\right]\exp(-a\breve{y}^2). \quad (3.66)
\end{aligned}
$$

For this to be consistent with the eigenenergy E, the $\breve{y}$ dependence in the coefficient must vanish. This results in $a = \sqrt{m^*\omega/2\hbar}$, which also leads to an energy contribution of $E = \hbar\omega/2$. This the lowest Landau energy level. Since this Gaussian function is localized around y_0, one can estimate that the length scale of localization is of the order of $\sqrt{a}$ for Gaussian functions. Substituting ω_c, $a^{-1/2} = 2^{1/2}\ell$, where

$$
\ell = \left(\frac{\hbar}{eB_z}\right)^{1/2}. \quad (3.67)
$$

The magnetic length scale, the length scale of localization in the quantum Hall effect, is independent of mass. At $B = 1\ T$, $\ell = 2566\ nm$. So, macroscopic objects with high mobilities are entirely capable of exhibiting the integral plateaus, the Gaussian localization occurs centered at y_0, this centering position is dependent on k_x, and k_x is the solution of coherence requirement. If the sample is rectangular with a dimension constraint of $L_x \times L_y$, then

$$
k_x \in \{k_{xj}\}, \quad \text{where } k_{xj} = \frac{2j\pi}{L_x} \ \forall\, j = 1,2,\ldots, \quad (3.68)
$$

and the extent of the allowed y_0 is limited by L_y—the sample size. $y_0 < L_y$ constrains $k_y < eBL_y/\hbar$. The maximum j is limited by this sample dimension to $j \leq L_x k_{max}/2\pi = L_x L_y/2\pi\ell^2$. This is the total number of states in each Landau level, as defined by the sample volume and field. This determines the degeneracy per unit area to $N_0 = eB_z/\hbar$. Each Landau level holds the same number of electrons. If exactly ν levels are filled, then $N_s = jN_0$. So, the resistance is

$$
\rho_{xy} = \frac{B}{ejN_0} = \frac{h}{\nu e^2}, \quad (3.69)
$$

independently of the material from which this high mobility two-dimensional electron system is constituted.

As remarked, due to the inhomogeneities of the system, the system still obeys the quantum-mechanical constraints but has a Gaussian broadening. The filling of the Hall plateau values, rather than at one specific field, arises from this broadening. Imperfections are important to the observations of the integral Hall effect plateaus. Another important consequence of inhomogeneity is that spin degeneracy is lifted, so the Zeeman doublet results when the effective g factor is non-vanishing. The spin interaction with the magnetic field has now become important, and the idealized Landau degeneracy, due to the additional term in the Hamiltonian, causes a splitting into two Gaussian bands symmetrically.

Figure 3.22 represents what this description corresponds to. The electronic states shown in Figure 3.22(a) show the Landau level picture. The positions of the harmonic oscillator potentials in the y direction are centered at k_x, which is quantized. Figure 3.22(b) shows the resulting idealized density of states with the Landau quantization of the two-dimensional electron gas, spin degenerate at $B_z = 0$.

The degeneracy, if there were no spin, is eB_z/h, and Hall plateaus exist at $h/\nu e^2$, where $\nu = 1, 2, \ldots$. With spin, each Landau level transforms to a doublet where each part has a eB/h degeneracy, so the entire Landau level degeneracy is $2eB_z/h$. When Zeeman splitting is not resolvable, for example, with too high a temperature, then Hall plateaus at $h/\nu e^2$, where $\nu = 2, 4, \ldots$, will be observed. If spin is resolved, $\nu = 1, 3, \ldots$, also appears. Zeeman splitting being linear in the field, spin is usually resolvable at high fields but not at low fields.

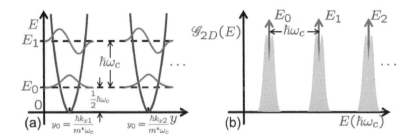

Figure 3.22: (a) shows the electronic states in a Landau level where the wave number $k_x = k_{x1}, k_{x2}, \ldots$, determines the centering of the harmonic potentials. (b) shows the idealized and real two-dimensional densities of states. The broadening of the δ function arises from inhomogeneities of the system, and spin degeneracy lifting leads to Zeeman doublets from the finite effective g factors.

This understanding of the integral quantum Hall effect has fundamentally arisen from the existence of gaps in the energy spectrum where inhomogeneity induced Gaussian broadening. The gaps come about within the single-free-electron model responding to a magnetic field in a two-dimensional electron gas. Higher quality samples, absent the homogeneities, show additional energy gaps, which are associated with fractional quantum Hall numbers. These are gaps that the single electron quantum mechanics is unable to explain. The fractional Hall plateaus shown in Figure 3.23 arise from mutual Coulomb interactions between electrons. These plateaus and associated gaps—low longitudinal Hall resistance regions—arise from many-body interactions. This behavior is that of a new collective state of a many-electron system. It possesses novel correlations— new emergent collective excitations of the quantum liquid. We will discuss this only briefly.

The fractional quantum Hall effect is the result of a dynamic interplay between the magnetic field and the electron-electron interaction in a two-dimensional electron gas. This interaction results in a stable incompressible fluid with fractionally charged excitations, because

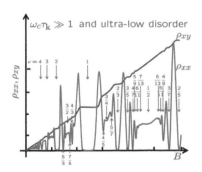

Figure 3.23: Longitudinal and transverse Hall resistance in ultra-low disorder two-dimensional electron gas Hall measurement under conditions of very high magnetic fields.

the field is such that the Landau level filling is a fraction of the form $1/n$, with n odd. This is to say that the field B_z is such that the number of electrons per unit area is $1/n$ times the number of available states per unit area of the lowest Landau level. This is the fraction that is fillable. For excitations to happen, energy gaps exist within the spectrum of the lowest Landau level as a consequence of the many-body interactions. The wavefunction of the ground state—the Laughlin wavefunction—is

$$\psi = \prod_{i<j} (z_i - z_j)^n \exp\left(-\frac{1}{4\ell^2} \sum_i |z_i|^2 \right), \qquad (3.70)$$

where $z_i = x_i + iy_i$ is a complex coordinate of the ith particle. With n as an odd integer, this wavefunction is antisymmetric under particle exchange. At the high fields of the measurement, the system becomes spin polarized, so the antisymmetry must be in the spatial part. The filling of the Landau level of this state is $1/n$.

How can many-body bound states come about in the presence of Coulombic repulsion in the presence of magnetic field? A simple illustration of this is provided for two interacting electrons in Figure 3.24. If two particles are released from the initial positions specified by the arrows, they follow orbital paths whose first two periods are shown. They remain bound. This binding happens because for each pair, there exists a competition between the Coulombic repulsion and magnetic confinement. As the particles try to escape and lower their Coulomb energy, the effective magnetic potential confines them. This is in the laboratory frame. One could have viewed this in the rotating frame, where the scalar potential makes this problem easier to tackle. The binding that results from the interplay between electric repulsion and magnetic confinement gives rise to the additional energy level structure within the lowest Landau level. So long as the interaction energy is less than the magnetic confinement energy, the system remains in a bound state. Any repulsive interaction potential is converted to kinetic energy in the retreating particles, and this energy is itself lost to magnetic potential energy. At high magnetic fields, the binding is strong, and the energy level structure corresponding to this binding has observable consequences.

The wavefunction of Equation 3.70 represents the embodiment of all these constraints. It depends simultaneously on the positions of all the particles in the system. It has stability when the fraction of filled states has values such as $1/3, 1/5, 1/7, 2/3$, et cetera, many of which are shown in Figure 3.23. These fractions are either 1 divided by an odd integer, or 1 minus such a fraction. This odd denominator arises from Pauli exclusion and because of the phase. What is also interesting about this system is that when one adds an extra electron when a

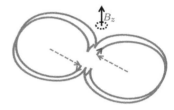

Figure 3.24: Paths of two electrons in a static out-of-plane magnetic field and interacting through Coulomb repulsion. Arrows indicate the initial position of release from rest. The electrons remain coupled to each other.

For a detailed discussion, see S. Tiwari, "Semiconductor physics: Principles, theory and nanoscale," Electroscience 3, Oxford University Press, ISBN 978-0-19-875986-7 (forthcoming).

Landau level is 1/3rd full, then the extra charge will appear in three physically separate places in the sample. A similar but inverse effect occurs when one removes an electron. These fractional charges are quasiparticles that behave much as the electron does. In this fractional quantum Hall system, the fractionally charged quasiparticles are localized in the sample, leading to the observed Hall resistance.

The real space visualization of the integral Hall effect is useful in getting insight into these phenomena. Figure 3.25 provides this perspective.

Classical or quantized—integral or fractional—all are different cases of the same effect: the Hall effect, which arises when a magnetic field exists perpendicular to a sample's surface in which current is flowing. The bias voltage parallel to the flow of the current is essentially a measure of the difference between the energy of the electrons exiting over those entering. The perpendicular voltage, which is in the direction where no current flows, is the primary effect, which is the measure of the direct consequence of the applied magnetic field. Normalizing with current flow makes the measured resistances associated with these voltages current independent. In the classical Hall effect, the Lorentz force sets up an electric field that forces a balance with the Lorentz force. The guiding center of the Lorentz force term changes with the velocity with which the charge particle moves. Small velocities have a weaker force, and the center for the rotational effect imparted is farther away. This is essential to the picture of reference frame of field analysis and easiest to tackle in the gauge transformation that we employed. But, we wish to look at this through a visual picture; therefore, we now use the laboratory reference frame.

The primary effect, classically, in the movement is the appearance of Hall voltage in the orthogonal direction, and any change in the direction of current is minor. Electrons accumulate at one edge of the sample and deplete from the other edge because of this Lorentz force. But, the total perturbation is very small. So, ρ_{xy} appears, and ρ_{xx} is still linear. This is the situation shown schematically in Figure 3.25(a) for electron movement where we have entirely ignored scattering. If the current and the force causing the current are absent, so $\mathcal{E}_x$ is zero, then the electron rotates around the magnetic field. If scattering is present, the result is an orbit that is the result of rotational motion due to the magnetic field, and scattering that may be randomizing or non-randomizing. Very close to the sample edges, there is a region where length scale is smaller than the scattering length scale, and electrons will still accumulate and deplete. Now, we ignore this scattering since we are interested in the condition of a high-enough field, where $\omega_c \tau_k \gg 1$. The higher the magnetic field, the more the

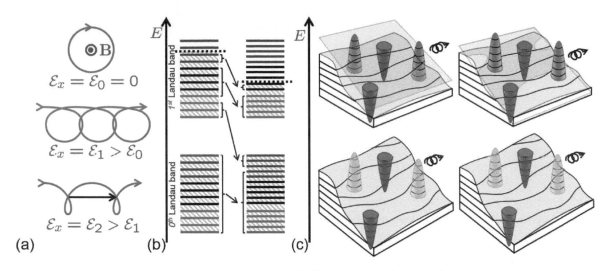

Figure 3.25: A physical picture of the integral quantum Hall effect. (a) shows the path of an electron with increasing magnetic field as the conduction-causing field is raised. (b) shows occupancy and density changes in the lowest two Landau subbands as the two-dimensional electron gas goes through a plateau to the conducting region. (c) shows the corresponding spatial picture, with localized low and high energy states and the delocalized states that cause conduction.

Lorentz force causes a deflection, the more the electrons are forced orthogonally, the more the accumulation and depletion and therefore the greater the Hall voltage and the electric field that balances the Lorentz term. And if a field is present, represented through a current flow between connected leads in the x direction, then electrons follow cycloid paths, as shown in the lower part of Figure 3.25(a).

It is the plateaus in the Hall resistance ρ_{xy} as a simple $\nu e^2/h$ relation with ν as an integer under the cyclotron path condition, as well as the disappearance of resistance ρ_{xx} in the regions of these plateaus, that comprise the remarkable part of the integral Hall effect—an effect that appears only in two-dimensional electron gas, not in three-dimensional or one-dimensional systems. With electric and magnetic fields present from the external sources, the electrons flow in a cycloid path. If the magnetic field is stronger and the electric field weaker, the electron's motion is guided strongly by magnetic field, but if the magnetic field is weaker and the electric field stronger, then the electron's motion is more linear, and the guiding center resulting from the magnetic field moves with greater drift velocity. A strong magnetic field makes the electron more localized. High magnetic field slows forward motion. Electron energy, dictated by the magnetic field, exists through the Landau energy levels the electrons occupy. So, the circular orbit of the top part of Figure 3.25(a) represents a Landau level condition that is identical for electrons semi-classically.

Now, if there is disorder, the Landau level must spread due to the disturbance caused by the disorder that breaks the coherence. The number of independent quantum states per unit area is proportional to the applied magnetic field. A disorder represented by a slightly positively charged impurity, for example, localizes the electron more, while an impurity that has a trapped electron delocalizes the moving electron more. The former state is slightly stabler and has slightly lower energy than the latter. The single Landau energy level has now become a band, as shown in the first two Landau bands in Figure 3.25(b). Here, the middle band states in each of the Landau bands are extended states, and they are straddled by the localized states, which are lower in energy and higher in energy, respectively, due to the small perturbation from disorder. The states near the bottom of the band are localized in some region of the sample. The electrons occupying these states do not leave this region. Occupation is defined by the position of the Fermi energy, and this position changes as one moves along in the direction of the applied current or voltage. These are the darkened peaks and valleys of Figure 3.25(c), which plots the real space along the horizontal axes. The electron doesn't leave these regions of peaks and valleys. The lower energy localized states are in dips of the potential landscape of Figure 3.25(c) for these same two Landau subbands. We also have the extended states, which have remained undisturbed. And we have the higher energy localized states, which that are shown as peaks. The presence of increasing magnetic fields also creates energy states. So, in Figure 3.25(b) and (c), which show the energy state and the real space picture, respectively, as one applies a higher magnetic field, an increase in the total number of states and in the states occupied, as well as a change in behavior in real space, occur.

The Hall resistance and the appearance of plateaus can now be seen with reference to Figure 3.25(b) and (c). Let us examine the case where the first Landau band is occupied in the localized high energy states, as shown on the left in Figure 3.25(b). So, if current is flowing in the Hall sample, with the Fermi energy in the subband of the upper part of the localized subband states, all the extended states and the lower energy localized states are filled. We now increase the magnetic field while changing the current flow so that the Hall voltage, the voltage between the two edges of the setup, are kept constant. We have a situation of occupation of states that is represented in Figure 3.25(b). With increasing magnetic field, the number of Landau levels increases in linear proportion. In every region of space, that is, in the representation of Figure 3.25(c), additional quantum states that have energy similar to that of neighboring states become available.

Many of these additional states will have energy that will be below the local Fermi energy. So, the energy of the occupied states will be lowered, as in the right hand side of Figure 3.25(b). Electrons from higher occupied states now occupy lower energies in the region adjacent to the Fermi energy. So, the lower position of the Landau band is occupied.

When Fermi energy stays in the subband of the high energy localized states as the magnetic field is changed, the extended Landau band remains fully occupied. Localized electrons do not travel. Changing the fraction of localized states occupied does not affect the electrical conduction properties. The current is a constant, and the Hall resistance ρ_{xx} is a constant, even as the magnetic field is being changed. What is happening in the conduction? Increasing the magnetic field slows electron forward movement, but this effect is canceled by the increase in the number of extended states that are now newly available, are occupied, and carry current. With Hall voltage constant, and current constant, ρ_{xy} is a constant.

Now, as one increases the magnetic field even more, high energy occupied localized electron states are exhausted. The Fermi energy is now in the extended states. In these, which are partially occupied or unoccupied, current flow occurs, and Hall resistance increases. This Hall resistance continues to increase with magnetic field so long as Fermi energy remains in the energy range of extended states. This is the changing resistance region between plateaus.

If the magnetic field is increased even more, all the Landau band extended states are exhausted, and the Fermi energy returns to the occupation of localized states. So long as there is at least another Landau band below this Fermi energy, the extended states of that band will be able to provide another region of changing resistance ρ_{xy}.

The ratio of Hall resistances is a ratio of integers because for any Hall voltage, the current is proportional to the number of occupied subbands of extended states. Each plateau occurs when an integral number of subbands are occupied. The change from plateau happens when extended states become occupied, that is, when Fermi energy is in their subband. We will presently see that this is equivalent to a topological argument.

The current flows without energy dissipation in the localized states—the plateaus are regions of no dissipation. Here, all the unoccupied states are localized—they do not connect to any other states as shown in the left part of Figure 3.25(c). The lowest subband is all occupied, so it too doesn't provide conduction. Energy dissipation only occurs when the Fermi energy and its changes in the Hall sample are present across the extended states as shown in Figure 3.25(c)'s

One could have followed this same argument with current externally held constant. In this latter case, we will find that the measured Hall voltage will be a constant. This holds in both cases whenever Fermi energy is in a localized state.

right side. There are now occupied and unoccupied states that are connected to each other, and the dissipation happens in this transition where the excess energy goes to the other degrees of freedom available in the system—phonon modes, for example.

The integral quantized Hall effect occurs due to an integral number of Landau levels being filled. This line of reasoning is not extendable to the fractional quantum Hall effect, where localization is absent. The fractional effect, for example, a plateau at $e^2/3h$, happens with lowest Landau band 1/3rd occupied. This requires the electron-electron interactions to be accounted for all electrons in the system, and this maps to the entire quantum-mechanical wavefunction. So, this physical picture doesn't apply to the fractional quantum effect. We now have to look at the excitation in the composite system as a collective response, so for our picture of current as one of composite fermions—composites of electron excitations in this many-body interaction. Composite fermions are bound electron states that are attached to topological features—they are variously called defects, vortices or quantized stable states. The presence of a magnetic field results in a strongly correlated quantum liquid in these conditions of Coulomb repulsion. The interaction between the composite fermions is negligible to a good approximation, making the behavior of the system one of a new composite fermion quasiparticle in a quantum liquid. Composite fermions exist in Landau-like levels in an effective magnetic field where the number of Landau levels filled is given by $\nu = \rho\phi_0/B_z$, where ρ is the particle density. The vortices here can be related through a field, $B_z^* = B_z - 2p\rho\phi_0$, where $2p$ represents the vortices bound to composite fermions. The vortices have their own geometric phases that compensate the Aharonov-Bohm phase in this magnetic field.

Composite fermions fill with a filling factor $\nu = \rho\phi_0/|B_z^*|$, and one may write

$$\nu = \frac{\nu^*}{2p\nu^* \pm 1},$$

(3.71)

where the sign change is related to the geometric phase and Aharonov-Bohm phase compensation. At very low magnetic fields, Shubnikov-de Haas oscillations appear periodic in $1/B_z^*$. The integer quantum effect occurs with integer filling of composite fermions, that is, $\nu^* = n$, where $n = 0, 1, \ldots$, corresponding to

$$\nu = \frac{n}{2pn \pm 1}.$$

(3.72)

Holes in Landau levels also produce vortices which correspond to

$$\nu = 1 - \frac{n}{2pn \pm 1}.$$

(3.73)

The reader should see S. Tiwari, "Semiconductor physics: Principles, theory and nanoscale," Electroscience 3, Oxford University Press, ISBN 978-0-19-875986-7 (forthcoming) for details of the related argument.

The reader should look up Appendix J for a discussion of the geometric phase, aka the Berry phase, for a preliminary physical understanding. Topology has a deep and intrinsic connection to the arguments because of the geometry, symmetry and other features that appear in this quantum treatment. Topology provides a basic view of the invariants and why certain relationships come about.

These mean that one will observe disappearance of ρ_{xx} at sequences of $n/(2n+1)$, $n/(2n-1)$, $n/(4n+1)$, ... This is the fractional quantum Hall effect where the first terms of each of these series are $1/3, 2/5, 3/7, \ldots$; $2/4, 3/5, 4/7, \ldots$; and $1/5, 2/9, 3/13, \ldots$; respectively.

In this discussion, we have made a topological connection between the various Hall effect observations. What do we mean by topological here? The classical Hall effect shows the Lorentz force effect on a charge particle's motion ($\rho_{xy} \propto B_z$). The quantum Hall effect shows that states can be different even if they have the same symmetry. By this, we mean that the plateaus of the integral effect are certainly different states, but the states are of the same symmetry. One way to view this is that the integral effect occurs through chiral edge states, as shown in Figure 3.26. While there is this confined two-dimensional cyclotron motion, conduction takes place and is in a plateau for a range of fields until the symmetry of the system changes again through the allowed Landau states and cyclotron motion changes. The inside of the two-dimensional system is nonconducting, while the edge states are conducting. The plateaus are quantized to incredible accuracy—parts per billion. The changes in the plateaus represent a different topological constraint on the wavefunction. The plateau represents a topological insulator—the insulation arising from the topology of the wavefunction, with insulation inside the two-dimensional region, and conduction along the edges.

We now make this connection to topology and introduce the spin Hall effect, as well as two-dimensional and three-dimensional topological insulators for their potential as media for qbit states and quantum computation.

Topology is the study of properties of spaces that do not change with continuous deformation such as stretching and bending. Two spaces that can be deformed into one another through continuous transformation are homeomorphic. Because invariants are natural to topology and appear in our broken symmetry and conservation discussions, topology appears naturally in these physical arguments here. The non-conduction is a result of the topology, which, as Figure 3.26 shows, occurs through the edge states.

The Gauss-Bonnet theorem relates the curvatures on any surface. A surface in three-dimensional space has two curvatures, κ_1 and κ_2, which we may relate through the divergences at any given point. For

$$\kappa = \kappa_1 \kappa_2 \propto \boldsymbol{\nabla}_2 \cdot \boldsymbol{\nabla}_1 - \boldsymbol{\nabla}_1 \cdot \boldsymbol{\nabla}_2, \quad (3.74)$$

the Gauss-Bonnet theorem relating the curvature on surfaces states

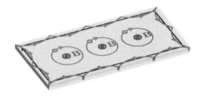

Figure 3.26: The topological view of the integral quantum Hall effect. Plateaus over a range of magnetic field arise in different states with the same symmetry but different wavefunction topologies. Chiral edge states give rise to the longitudinal resistance that show periodicity and vanishing conduction, while the transverse resistance shows plateaus.

For a brief introduction of geometric—topological—attributes and effects, with implications for our discussions, see Appendix J. A Möbius strip has only one surface and one edge. A Klein bottle is a non-orientable surface that, unlike the Möbius strip, has no boundary. A Klein bottle is similar to a sphere. Left and right cannot be consistently defined. A coffee cup is like a donut because of the hole that exists in the handle.

that

$$\int_M \kappa \, d^2r = 2\pi\chi = 2\pi(2 - 2g).$$

(3.75)

In this equation, M is the manifold—a surface is a two-dimensional manifold, where the surface of a sphere, or a planar surface, is of dimension 2—χ is an integer—a quantized number—and g is the number of handles on the object. Invariant properties appear, and note the correspondence between this relationship and Gauss's law, which we have frequently employed. The Gauss-Bonnet theorem points to the quantization. Absent handles on the surface, for example, in the case of a sphere, $\chi = 2$, in contrast, in the case of a donut, $\chi = 0$.

The Bloch wavefunction $\psi_{n\mathbf{k}}(\mathbf{r}) = u_{n\mathbf{k}}(\mathbf{r}) \exp(i\mathbf{k} \cdot \mathbf{r})$, in the quantized conductance limit, leads to the conductance

$$\frac{1}{\rho_{xy}} = \frac{e^2}{h} \times$$

$$\sum_{bands} \frac{1}{2\pi} \left(\left\langle \frac{\partial u(k)}{\partial k_x} \bigg| \frac{\partial u(k)}{\partial k_y} \right\rangle - \left\langle \frac{\partial u(k)}{\partial k_y} \bigg| \frac{\partial u(k)}{\partial k_x} \right\rangle \right).$$

(3.76)

This equation is precisely the form of the Gauss-Bonnet relationship and gives topological invariance. Since the conductance is quantized topologically according to this relationship, it changes nonlinearly in between the integers. The plateaus are therefore distinct. Two different approaches to the argument, quantum-mechanical and topological, have now let us arrive at the integral Hall quantization.

The topological argument begets the question of the role of magnetic field in the argument. Can one have a quantum Hall effect without a magnetic field, that is, in terms of our symmetry arguments, in situations that are time-reversal symmetric? The answer to this is yes. And what is particularly interesting about this answer is that it is possible even in three dimensions, thus relaxing our two-dimensional constraint. The reason is that spin-orbit coupling,

$$\mathscr{H}_{so} = \frac{\hbar}{4m^2}(\boldsymbol{\nabla}V \times \mathbf{p})\sigma,$$

(3.77)

has an energetic interaction like that of a magnetic field and is dependent on momentum. If the spin-orbit coupling is strong enough, a quantum spin Hall effect, where the strong spin-momentum coupled interaction manifests itself through a behavior similar to that of the integral Hall effect but without the necessity of the presence of a magnetic field occurs.

The quantum spin Hall effect therefore may occur with a number of heavy elements—Hg, Te, Bi, Sb, et cetera. Figure 3.27 shows the band structure of unstrained $HgTe$. The Γ_8 point is a symmetry point in a light-hole, that is, with p-like character, band. There also exists

A handle is a topological structure that can be visualized as being formed by punching two holes in the surface, attaching oppositely oriented zips on them and then zipping them together. It is called a handle because that is the shape that results. See Appendix J.

The Chern number is the curvature of the wavefunction in Hilbert space. The summation term (a summation over integrals of divergence in the $\mathbf{k}$ space) is associated with the Chern number. See Appendix J, where the geometrical connections are established from a Berry phase viewpoint.

Historically, $HgCdTe$ has been a compound with defense implications for infrared vision. With a 15 % alloy of $CdTe$, it has zero bandgap. At 20 %, it has a bandgap of 0.1 eV at 77 K; this bandgap is useful in an 8–12 μm atmospheric window.

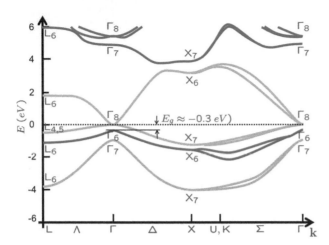

Figure 3.27: Band structure of un-strained $HgTe$ over the ± 6 eV range across major symmetry points of the Brillouin zone.

a heavy-hole band, another set of p-like states which, at the Γ point, is very close to the light-hole band. The light-hole band is inverted from the usual behavior. The s-like band shows up as Γ_6 here. This material is a semimetal with a negative bandgap, $E_g \approx -0.3$ eV. What is unique about this band structure is that there exists a band inversion due to the spin-orbit effect. The Γ_6 is below the Γ_8 point, and $E_g = E_{\Gamma_6} - E_{\Gamma_8} \approx -300$ meV. This band inversion can be manipulated, through structural effects—quantum wells, strain, et cetera, and this makes it possible to access the topological properties of $HgTe$. $CdTe$, for example, has normal Γ_6 s-like states at around 1 eV. And the heavy-hole and light-hole, that is, p-like states are pushed down to -0.5 eV. $CdTe$ has a bandgap of about 1.5 eV. In materials like $HgTe$, which has such strong spin-orbit coupling, if one now applies the external electrochemical forces of Hall measurement without the magnetic field, as schematically shown in Figure 3.28(a), a Hall voltage develops because spin and momentum are coupled. Spin accumulation and depletion take place. And if the material has high mobilities, one will see from topological arguments a behavior similar to that shown in Figure 3.26. Figure 3.28(b) shows a schematic of the edge currents that, due to strong spin-orbit coupling, flow in opposite directions. when no external magnetic field is present.

This topological insulation is different from that of a conventional insulator. Figure 3.29 shows a comparison of insulation and what happens at the surface in topological materials. Figure 3.29(a) shows a normal insulator, such as a direct bandgap semiconductor at low temperatures. The conduction band states, that is, the states that are unoccupied at low temperatures, arise from s-like states, and the filled valence band states are p-like.

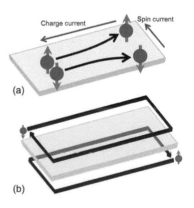

Figure 3.28: In the presence of strong spin-orbit coupling, a flow of electron charge current occurs, together with a spin current that causes spin accumulation and depletion with a Hall voltage. This is shown in (a). (b) shows that the topological quantum Hall implication of this spin response is the flow of edge currents in opposite directions, due to spin.

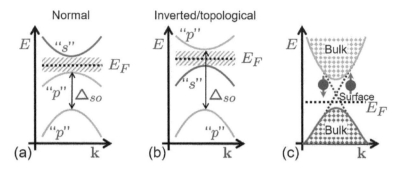

Figure 3.29: (a) shows a normal insulator, where s-like states are largely empty, and p-like states are filled. (b) shows a topological insulator, where p-like states exist above s-like states. (c) shows that, at the surface, edge states are formed where spin and momentum are coupled through their signs.

This figure also shows that, in this example, there is a strong spin-orbit coupling, the split-off band is deeper and it too arises from the p-like states. In the topological material, this picture is inverted. The Γ_6 in $HgTe$ is below the Γ_8 point, and s-like states exist below the p-like states. The spin-orbit coupling has pushed the p-like states with their strong orbital coupling higher in energy. In this case, unlike $HgTe$, where we must also tackle the semimetal negative bandgap through structural means, there is insulation in the bulk of the material because the Fermi energy is between empty and filled bands, which are separated by a large number of $k_B T$s. In Figure 3.29(c) one can view both what happens in the bulk of the two-dimensional system and at the surface. The bulk states are sufficiently separated, so there exists no conduction. But, because of the topological constraint, as with the integral quantum Hall effect, coupled one-dimensional states exist where the spin and the momentum are interlocked. The Fermi level passes through these states, and, depending on its position, different integers of states will be filled, which show up as the quantization in any measurement. We have now created the conditions shown in Figure 3.28(b). For one sign of allowed momentum, one spin is allowed. And for the symmetrically opposite momentum, only the opposite spin is allowed. This is the opposite current flow in edge states shown in Figure 3.28(b). This spin orientation and its locking to momentum is the key. The channels are topologically protected against back propagation. The protection comes from the necessity to have both spin and momentum flip for any transfer between the states. These are one-dimensional channels. We will also see that, in three-dimensional systems, the collection of protected channels form a disk of opposite momentum and opposite spin on a Dirac cone, whose points of one-dimensional intersection with an orthogonal plane through the apex are shown in the simple situation diagrammed in Figure 3.29(c).

From a physics point of view, this is insulating. From an engineering point of view, this is a very low temperature phenomena. Even a bandgap of 0.3 eV is too conducting for room temperature.

We have seen that $HgTe$ is a semimetal with bulk band inversion. Our topological argument in creating this two-dimensional spin Hall effect is the following. A normal insulator has a bandgap that is large enough that thermal transition of electrons from the valance band to the conduction band is insignificant at the temperature of use. Vacuum, just outside these normal insulators, could be viewed as consisting of states that are connected adiabatically through the surface that intervenes between them. The symmetry of the available states for the presence of electrons and similarly for positively charged holes remains the same. The bulk and the surface are adiabatically connected. Interactions on the surface are quite similar to the interactions that take place in the bulk of the material. Scattering happens, nothing uniquely new except a continuous change should be expected.

Now consider what happens with inversion. We now have a bulk insulation arising topologically. We saw in Figure 3.29(c) a way to maintain the bulk insulation but form the surface channel. This is the equivalent of the integral quantum Hall effect. The bulk of the two-dimensional layer was non-conducting, since electrons are trapped in the cyclotron orbits. At the edge of the two-dimensional region, though in a real space picture, they bounced back from the surface and continued along in a cyclotron orbit.

The simplest way to invert $HgTe$'s subband and create the quantum spin Hall effect topologically is to form the $HgTe$-based structure as a two-dimensional quantum well layer clad by a large bandgap material such as $HgCdTe$, with a sufficient molefraction of $CdTe$ so that subband edges can be suitably shifted. Well widths change subband edges, and strain too affects the bandstructure. An appropriately designed quantum well can make this $HgTe$ quantum well layer a two-dimensional topological insulator. Strain in a three-dimensional structure can turn it into a three-dimensional topological insulator.

Figure 3.30(a) shows a small thickness quantum well of $HgTe$, (b) shows one with a large thickness, and (c) shows what happens at the edge in (b). The interesting changes arise because this $HgTe/HgCdTe$ heterostructure is a type III system. A small well thickness raises the electron states and lowers the hole states sufficiently to make $HgTe$ quite like a normal semiconductor. A bandgap exists. The electron and hole energies for the lowest subband are similar to those of a normal semiconductor. $HgTe$ at this small and subcritical wel thickness behaves as an "insulator" at the low temperatures appropriate to the bandgap.

If the well is thicker, more than 6.5 nm here, however, the hole and electron subband alignment are inverted, like those of bulk $HgTe$.

The hole in the semiconductor is a quasiparticle representing the collective excitation of the filled band of electrons.

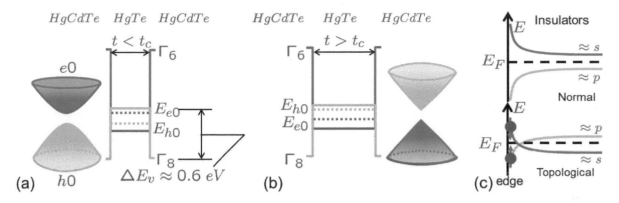

Figure 3.30: (a) shows a subcritical thickness $HgTe$ quantum well. (b) shows a supercritical thickness $HgTe$ quantum well. Both are clad with the larger bandgap $HgCdTe$ barrier. Note that the subband is inverted in (b). The valence band discontinuity in the system is ∼ 0.6 eV, and the conduction band discontinuity is significantly larger. The critical thickness for crossover is ∼ 6 nm. (c) shows the formation of the up and down spin channels at the one-dimensional edge of the well, even as the inside of the well is insulating due to the spin-orbit catalyzed spin Hall effect in the bottom half. The top half shows a normal insulator at the edge, with vacuum.

There is still a bandgap in the middle of the well, albeit with inverted bands; but, at the edges, inside the well, the electrons can conduct in a one-dimensional channel without scattering into each other because of the spin-momentum coupling argument. The bulk and the surface are now adiabatically disconnected. This is now a quantum spin Hall insulator. The edge states provide a conduction path. The two edge states that appear form two conducting one-dimensional channels with each providing e^2/h, that is, a single quantum of conductance. As in the spin Hall effect, the two spin orientations conduct in opposite directions. If this is a normal insulator, as in the top half of Figure 3.30(c), it will remain insulating throughout.

At what thickness does this crossover occur? This is shown in Figure 3.31, which should be looked at together with the bandstructure shown in Figure 3.27. The subband edge electron and hole energies are shown as a function of the quantum well thickness. The hole energies arise from the heavy hole band. A number of subbands of holes—all due to the heavy-hole band—may be of interest depending on the thickness of the quantum well, but normally one needs to concern oneself with only one of the electron subbands. The crossover from the bulk-like inverted topological structure to the normal structure occurs at $t = t_c \approx 6\ nm$ for $HgTe$. Wells thinner than t_c wells look like those of a normal semiconductor, with s-like states higher than p-like states. The energy of the two-dimensional states is a Dirac cone structure of heavy mass. At the crossover, one sees a Dirac cone—a diabolo—and then again a heavy-mass Dirac cone. Thicker

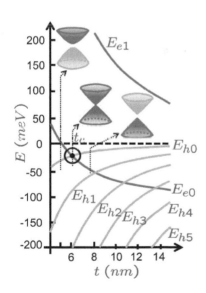

Figure 3.31: The subband edge energies for electron and hole states due to heavy holes in quantum wells of $HgTe$ clad by $HgCdTe$ as the size of the well is changed. For $t < t_c$, the quantum well is a normal insulator. At $t > t_c$, the system inverts. $t_c \approx 6\ nm$.

wells bring out the properties of the inverted/topological structure, where surface channels exist because of the integral spin Hall effect due to spin-orbit interaction or equivalently, the topological consequences of the geometric phase argument.

The topological constraints also imply that three-dimensional structures can exhibit this topological insulating behavior. These structures then are not constrained by the considerations of what we just discussed with $HgTe$. Bi_2Te_3, Bi_2Se_3, Sb_2Te_3, et cetera, are examples of three-dimensional topological insulators, where the helical edge states—the ones with the spiraling orbits—lead to this topological quantum phase transition, as shown in Figure 3.32. This arises from the non-trivial Berry's phase of such materials' topologically ordered two-dimensional gas on the surface, and we will not dwell on the related argument. The two-dimensional states on the surface are protected from each other; the electron in the conduction band state in the topological material is connected to the hole state, while the reverse happens for the valence band, and the reason for this is the non-trivial Berry's phase of such materials' topologically ordered two-dimensional gas on the surface.

The distinguishing property here is the topological property of wavefunctions, not of broken symmetry. The surface states are topologically protected. In the case of Bi_2Se_3, there are an odd number of Dirac cones in the first Brillouin zone on the surface. This number is a quarter of those in graphene. The spin and the momentum are locked; because the spin-orbit coupling is strong, the behavior is relatively robust against defects and impurities that are non-magnetic. Because of the nature of Dirac cones and the constraints on transitions, these topological insulators can achieve high mobility, and, in the presence of a magnetic field, it has novel quantum Hall effects. The strength of the spin-orbit coupling from which these properties result, also means that a three-dimensional material can have at its surface two-dimensional states through which the spin current flows and balances, so that topological insulation arises. Figure 3.32 shows a representational example of this, as an analog of the two-dimensional picture in Figure 3.28(b).

We have made the connections between the integral quantum Hall effect, the quantum spin Hall effect, and topological insulators. Is there something equivalent that we can say about the fractional quantum Hall effect? This is an important question, but we have only partial thoughts and answers. We will tackle some of these in the context of quantum computation for which this Hall effect–related foray is pertinent.

The fractional quantum Hall effect, as a many-body effect with localization absent, came with the addition or subtraction of electrons

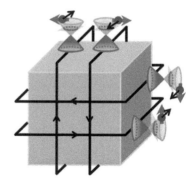

Figure 3.32: Formation of two-dimensional states on the surface of a three-dimensional topological insulator in the presence of strong spin-orbit coupling.

At least as of 2016, this story is incomplete in many significant ways—there is so much room for the excitement of discovery!

occurring with occupation in our Landau band picture of partial occupation, or, as we said, the collective response of a composite system with quantized states or vortices. This is a topological artifact. Vortices exist where the phase changes over small dimensions. The locking-in of spin and momentum causes that to happen at the surface.

Because these fractional quantum Hall states are quasiparticle excitations, there may exist excitations that are neither bosons nor fermions. Multiple important properties interlinked to the physical origins of these quasiparticle excitations are now possible. First, the quantum Hall effect is nonlocal. The spin Hall effect is also nonlocal. The conduction is non-dissipative. It takes place in the edge states that have been protected by the time-reversal symmetry. The conduction is describable via the Landauer formalism of quantized conductance. Second, the states are protected from local non-magnetic perturbations because of the locking. This is immunity to small perturbations. The third is that one can now find a topological state where excitation is neither a boson nor a fermion. This is the case with the $\nu = 5/2$ state and possibly the $\nu = 12/5$ state. The 5/2 state has a quasiparticle charge of $e/4$.

We have dwelt on these quantized Hall and topological insulators since these are systems in which the quantum processes have a direct visible pronounced effect and are very robust through the topological protection. This third property is additionally significant and potentially provides robustness against decoherence effects resulting from interactions with the environment that quantum computing must contend with.

The quasiparticle that is neither a fermion nor a boson and in which we are particularly interested is the non-Abelian quasiparticle. Different fraction-state quasiparticles can be either Abelian or non-Abelian. A non-Abelian group is a non-commutative group, that is, any group whose members, say a and b, do not commute, that is, if $[a, b] \neq 0$, is a non-Abelian group. To us, Abelianism represents a property associated with exchange of particles. Let ψ_i be a m dimensional vector, and let

$$|\psi_f\rangle = \hat{\mathscr{U}} |\psi_i\rangle = \begin{bmatrix} a_{11} & \cdots & a_{1m} \\ \vdots & \ddots & \vdots \\ a_{m1} & \cdots & a_{mm} \end{bmatrix} |\psi_i\rangle, \qquad (3.78)$$

which describes a unitary operation of the kind we have employed in our discussion of quantum computation. Since two different unitary operations do not commute, the states we work with, the qbits, which are composed of quasiparticles, remain protected. This is the property that non-Abelian excitation of the 5/2 state

A direct connection between integral and fractional quantum Hall effect is also made through parton construction. In this an argument can be made of wavefunction construction where, for example, a 1/3 fractional state is due to an $e/3$ occupying three $\nu = 1$ states.

This phase relationship, which is, topological, appears in the ferromagnetism that we discussed, so vortices exist there too. In general, such structures are called skyrmions. In two-dimensional vanilla magnetics too, vortices will appear. What these describe is a state which may not be deformed into a ground state where spin, for example, is aligned. In the quantum Hall effect, skyrmions are the lowest energy charged excitations of a quantum Hall spin magnet, and they couple charge to magnetic vorticity.

For a composite of two particles, say 1 and 2, if $\psi(2, 1) = -\psi(1, 2)$, then it is a fermion. If $\psi(2, 1) = \psi(1, 2)$, it is a boson. A non-Abelian particle is one, where under unitary operation $\hat{\mathscr{U}}$, $\psi(2, 1) = \hat{\mathscr{U}} \psi(1, 2)$, but two different unitary operations do not commute, that is, $[\hat{\mathscr{U}}_\beta, \hat{\mathscr{U}}_\alpha] \neq 0$.

brings. But, Abeleian particles—fermions and bosons, which are
particles for which $|\psi(1,2)\rangle = \pm|\psi(2,1)\rangle$—are also of interest.
Anyons are particles in two-dimensional systems with the property
$|\psi(1,2)\rangle = \exp(i\theta)|\psi(2,1)\rangle$, where θ is a real number. So, fermions
and bosons are two special cases of this general quasiparticle de-
scription, whose properties should be expected to vary continuously
between Fermi-Dirac and Bose-Einstein statistics.

For us, the consequence of topology for the formation of non-
Abelian anyons on the surface in two-dimensional and one-dimensional
systems is that it is possible to represent evolution in them through
braiding. Figure 3.33 shows a pictorial view of this description. The
unitary operation changes the state of the system shown in Figure 3.33(a) for two anyons. Only two basic moves—clockwise and
anticlockwise swaps—are needed to generate the combinations of
braidings on a set of anyons. Figure 3.33(b) shows this for three-
anyon operations for a three-particle system. If the anyon is a non-
Abelian, then the sequence of operation matters, and the end result is
different for the two exchanged sequences. The non-Abelian unitary
response represents the matrix of numbers dependent on the phase
factors that multiply the wavefunction. Since they are different for
the order of space and time, the end result is dependent on the order
and the suppression of the effect of interactions arising due the interaction with the environment. The transformations are more robust
against decoherence arising from the environment because of the
properties discussed.

Looking at Figure 3.33, one can see why these operations are associated with a braid—threads connected together to form a braid—or
a braid group. Braids are the folding of these threads, which represent evolution in space and time, of the interactions of non-Abelian
particles. A braid group represents all these operations—a series of
these non-commutative unitary operations. Topology, with the object
being smoothly deformed, distinguishes cutting, joining, knotting,
et cetera, from squashing or bending. Closed loops cannot be turned
into closed loops with a knot without cutting the string from which
the braid is made. Topological computing—quantum computation
using these topological states—uses braided strings. The braided
string represents the unitary transformation in time and space—
a world line of the movement of particles in space and time. The
evolution of a qbit then depends, because of topology or the gauge
argument, on the phase argument. The phase of the wavefunction
depends on the interaction of this unitary operation. Topologically,
the particles must follow either the clockwise path or the counterclockwise path and not cross paths—crossing paths is what happens
in knots and is the equivalent of a particle interaction. In a two-

Composite states of fermions and
bosons would make them distinguishable, so the two may not form a
composite together. A ground state of
$\nu = 1/3$ electron liquid in the fractional
quantum Hall effect is topologically
3-fold degenerate on a torus.

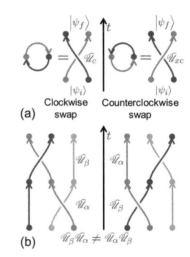

Figure 3.33: (a) shows the transforma-
tion of the state $|\psi_i\rangle$ to $|\psi_f\rangle$ under a
unitary operation of clockwise swap
and counterclockwise swap, generating
the braidings of the world line on a
set of anyons. (b) shows two unitary
operations $\hat{\mathcal{U}}_\alpha$ and $\hat{\mathcal{U}}_\beta$ on a three-anyon
system; these operations do not com-
mute.

dimensional system, this crossing is topologically disallowed, making the system relatively robust. A non-Abelian anyon manipulation is a member of the braid group. The absence of deformation of clockwise to counterclockwise operations means that certain operations are disallowed in multi-dimensional systems. Simply stated, topology suppresses this interaction. Non-Abelian particle excitations suppress interactions, making it possible for these particles to depend on the order, so time, and path, so position, of the interactions to lead the world line to the final result. These braids represent the collection of quantum operations of crossovers in an entangled path, just as they do in real life.

So, what does this computation, where the computation follows a path incorporating these thoughts, mean? A quantum computation represents the transformation of input qbits to a final state which is a quantum transformation. A braid represents a sequence of anyon transformations, so we may make the appropriate world line. We first need to create the qbits by using non-Abelian anyons, and then we perform the unitary operations on these anyons. The evolution of the qbit depends, because of topology or the gauge argument, on the phase argument. The phase of the wavefunction depends on the interaction of this unitary operation, and so on the order in which swapping takes place. The accuracy depends on the number of twists, the specific anyon species and their mapping to the qbits from which we compute.

Figure 3.34 shows four qbits made using eight anyons. The anyons are swapped, that is, braided, using the processes shown in Figure 3.33 for each step of computation on the qbits. In the end, pairs of anyons are brought together for measurement. The topology of the complete braid determines the output, which is immune to small perturbations because of the topological protection. In this example, the unitary steps performed multiple swaps and counterswaps simultaneously. The swaps could also have been performed in separate steps.

Figure 3.35 shows an implementation of the quantum *cNOT* gate using six anyons. In this case, compositionally, the qbits are triplets—the superposition consisting of two bits for input, and one bit for the condition signal. This particular implementation is sequential in the sense that only one swap or counterswap is implemented per time step, with one triplet left unmodified along the world line. It takes $87 \equiv 43$ twists for the simple set of operations shown in Figure 3.35, but these are all rows of threads at each step braided together, where two threads are braided—clockwise or anticlockwise. In the figure shown, the qbit triplets are the set of the entangled superpositions created and represented in the set where only one of

This is spoken from an upbringing where long hair and its braiding is considered as important and with the same objective as everything that the Western world does in name of fashion!

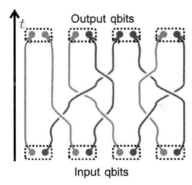

Figure 3.34: Computation with qbits using anyons. Four qbits using eight anyons are used here through three stages of braiding—unitary processes—to produce a four-qbit computational output. Here, the unitary processes simultaneously work on all the qbits.

the anyon particles of the starting pair is shown in the world line.

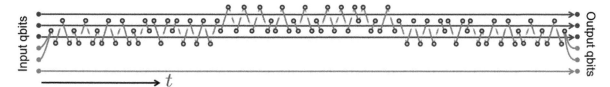

Figure 3.35: Implementation of a *cNOT* gate using anyon-based topological qbit states. Six anyons are used for the two qbits of the input and the one qbit of condition signal. Here, computation, organized using triplets composed of two anyons each, uses a unitary process that works by keeping one triplet intact and moving the anyons of the other triple around.

A schematic implementation of the fractional quantum Hall state of an Abelian 1/4 state undergoing a *NOT* operation is shown in Figure 3.36, using depleting electrodes on two-dimensional electron layers. Electrodes are employed to create islands for trapping anyons, as shown in Figure 3.36(a). The currents flow along the edges, but the positioning of the electrodes allows tunneling to take place in narrow gaps as in Figure 3.36(b). So, it becomes possible to initialize this gate by placing two anyons on one island, and then applying a bias voltage that moves it the other island. This pair of anyons is the input qbit. Measurement of programming of this initial condition is possible through passing of a current and application of a bias voltage at this spatial boundary. The qbit flip—the *NOT* operation— is induced by applying a bias voltage that causes the anyon to tunnel across to the other edge as shown in Figure 3.36(b). The passage of this anyon changes the phase relationship. The two-anyon qbit state flips to the opposite state, as shown in Figure 3.36(c).

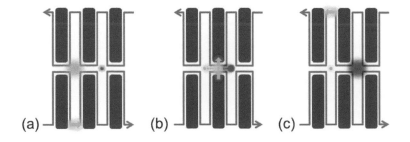

Figure 3.36: (a) shows the initialization of a *NOT* gate by placing two anyons (the two dots) on an island and trans-ferring one to a second island with bias voltage. The *NOT* operation is induced by tunneling of another anyon (another dot) between the edges, as shown in (b) causing the output in space and time to be in the flipped state shown in (c).

In all these examples, minimizing interactions with the environment, as these could result in errors, is of supreme importance—any such errors through loss of coherence require correction, which is nonlinear in the resources needed. Accuracy depends on the number of twists in the braid, which needs to grow slowly, but determining the correspondence between the specific braiding and the computation task, as of 2016, is an unsolved challenge.

Finally, we connect the end of this discussion to quantum computation. Topological and quantum computation are equivalent and capable of simulating each other. Different non-Abelian quasiparticle states, however, will have different levels of robustness. The 12/5 fractional state appears only at very low temperature, while the 5/2 state is more easily obtained. But, the 5/2 state's winding capacity is very sparse, so, with it, only two types of logic gates, an incomplete set, are possible. The 12/5 state, on the other hand, is computationally complete.

3.3   Single and many electrons in a nanoscale dot

FROM THREE-, TWO- AND ONE-DIMENSIONAL SYSTEMS, we now turn to systems small in all dimensions to explore their physical electronic behavior in structures useful as devices. These may be even $0D$ in their behavior, with $3D$ quantum-confinement effects remaining important. But, first we start with an entirely semi-classical form—a small-sized metal box, where we ignore for now the finite, but small, quantum-induced energy separation in allowed electron states, so the electrons are still in the Drude picture, with thermal energy much larger than the confinement energy spacing. We clad this box, which we will refer to as a dot, with two very thin barrier regions, as shown in Figure 3.37, and tie them together through a bias voltage source. We will worry later about the energy states due to spatial quantum confinement. The thin barrier regions permit tunneling, that is, there exists a linear resistance at small bias voltages, had one placed the contacting regions on them. Here, we happen to have a very small metal dot, one with a high density of states, intervening between the barrier regions. We will see that linear tunneling resistance will show up, but only at high bias voltages.

The current-voltage characteristics of this system are subject to the following non-ohmic constraints. The electrostatic energy needed to place an electron on a neutral dot is $e^2/2C$—this is the electrostatic energy associated with a capacitance with a charge of $-e$ due to the electron on it. The electrochemical energy source of the bias voltage V can supply this $e^2/2C$ energy so long as the voltage that develops at the dot ϕ is subject to the electrochemical energy constraint of $e\phi \leq e^2/2C$. What this inequality states is that there can be excess potential energy with the one electron present. Ideally, there is no current flow for a width in bias voltage that is twice this voltage, that is, a window of e^2/C (e/C in units of V). This is the Coulomb

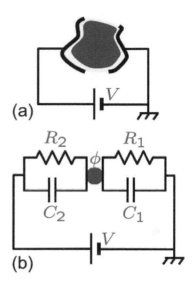

Figure 3.37: (a) shows a metal particle—a dot—with a tunneling insulator connected to an electrochemical energy source and (b) shows an equivalent circuit diagram assuming a classical metal particle—a nanodot—clad by a tunneling barrier modeled as a parallel arrangement of resistance and capacitance.

We discuss this non-ohmic constraint on a nanoscale dot geometry—a particle or fabricated planar nanosized disk in a two-dimensional electron gas—because that is where devices and structures are easily made and used. But, this capacitance argument also implies that one would see these effects, subject to constraints of localization and resistances, for any capacitor between any two electrodes. Such nonlinearities have been observed.

blockade.

Our nanometer size scale makes observation of this effect possible. First, let us look at the self-capacitance of a metal dot 10 nm in diameter. If we consider only the classical electromagnetic, perfectly conducting, metal sphere picture, then the self-capacitance C_{self}, with R denoting the sphere's radius, is

$$C_{self} = 4\pi\epsilon_0 R = 0.55 \times 10^{-18} \, F. \qquad (3.79)$$

Getting an electron onto this free nanodot will change its energy—potential—by $e^2/C_{self} \approx 0.29 \, eV$. This is $> 10 k_B T$ at room temperature. Our schematic in Figure 3.37 is a little more complex. It shows a nanosized dot with two electrodes driven by a battery. The two capacitances reflect that any electrical polarization—a continuous charge arising from the displacement of the electron charge cloud from the background ionic charge—or the discrete electron charge is coupled to multiple electrodes or other terminating surfaces. We assume that the insulator is $d = 1 \, nm$-thick SiO_2, so $\epsilon_r \approx 4$. The insulator is thin, so we will use a parallel plate capacitance. Since the insulator thickness $d \ll R$, the permittivity is four times higher. The ratio of this parallel capacitance $C \approx \epsilon_0 \epsilon_r 4\pi R^2/d$ to C_{self} is $\epsilon_r R/d$, or about 40. Even with a very partial coverage of the dot, this parallel plate capacitance will dominate, and it will scale the voltage observable through Coulomb blockade by this ratio of 40, but the result is still quite a small capacitance. In this dot, the number of electrons arising from the atoms that the dot is composed of is very large. The volume of the dot is $\sim 5 \times 10^{-19} \, cm^3$. If we assume the dot metal to be aluminum, which has a density of 2.7 g/cm^3, with its gram atomic weight of 27, the dot has $2.7 \times 5 \times 10^{-19} \times 6 \times 10^{23}/27 \approx 3 \times 10^4$ electrons. And yet, one may observe a phenomenon that is sensitive to the addition of a single electron. This shows the importance of the Coulomb electrostatic energy and its consequential magnitude at small dimensions—at nanoscale. Soon, we will see the large effects arising from it at the atomic size scale, through the Mott-Hubbard phase transition.

This argument started with the use of an energy constraint of $e^2/2C$ as the cost of placing an electron on a neutral dot. No current flows until there is enough voltage to supply the energy for the movement of an electron through the dot—or rather, placement of the electron on the dot, which it may leave if it can tunnel out to a vacant state in the drain. The applied voltage supplies this energy, subject to establishing a dot potential of $e\phi \leq e^2/2C$. So, the width of the Coulomb blockade region is the sum of the two, that is, e^2/C. For any total capacitance C of the dot, the blockade reflects the energy $e^2/2C$ of the electrostatic energy of the electron when it is on the dot,

Another way of looking at this argument: it takes $e^2/2C$ to place an electron on a neutral dot or to place an antiparticle—hole—for the electron to fit in to make the dot neutral. An electrostatic, or charge blockade range, must exist over this entire span. It is the range between two successive charge transfer events.

Very small particles do not charge up easily in the atmosphere! Regarding the use of $4\pi\epsilon_0 R$ as capacitance, a careful look is called for. One of the exercises at the end of the chapter tackles this. This problem clarifies the implications of various assumptions made in determining this simple capacitance formula. The self-capacitance of a disk of radius R is $8\epsilon_0 R$. A disk is the common form employed in single electronics. It can be formed in a two-dimensional electron gas by depleting off regions outside the confined disk.

This energy argument translates to the atomic scale too. Whether an electron localizes in the proximity of an atom or becomes an adventurous spirit by hopping around in the crystal is also subject to whether there exists enough energy for localization (Coulomb) versus for movement (kinetic). This leads to the Mott localization effect, whose Mott-Hubbard incarnation is a thermodynamic metal-insulator phase transition that we will look at in Chapter 4.

and the maximum potential energy of $\pm e^2/2C$ possible in general without the electron. So, the blockade span is the sum of these two. This blockade region can be shifted even if this blockade width is e^2/C, because the polarization can be modulated. This is also equivalent to stating that a charge gap of e^2/C arose representing the electrochemical energy change necessary between two dot states that differ by one electron. And one may shift this charge gap up and down within a limited range via polarization. Stray charge may induce a polarization charge of either polarity on the dot, causing it to shift in its electrostatic potential $e\phi$. A dot with no electron placed on it can still have electrostatic energy, even if no bias voltage is applied, because of the polarization charge. We will see this presently; but, by symmetry, it should be clear that, in an ideal environment, absent polarization charge and subject to only the fields with symmetric electrodes, the current-voltage characteristics will show a symmetric blockade region, as shown in Figure 3.38(a). A Coulomb staircase, as shown in Figure 3.38(b), is the formation of repeating steps in the current-voltage characteristics for this same circuit. This occurs under conditions of asymmetry, when one of the barriers is less conducting than the other. More than one electron may exist and transport through the structure. Each of the plateaus then corresponds to a storage of $1, 2, \ldots,$ of electrons. Since all these blockaded regions are only a function of charge and total capacitance, they suffice for the purposes of determining the total capacitance and from it the size scale of the dot. All one needs is a simple measurement of the bias blockade width, not of the $aF-fF$ capacitance.

We will include the continuously varying polarization charge as a residual term—an offset charge—of Q_{oc}. The discreteness of electrons, say n of them, where n is an integer, means that there is a total charge of $(-ne + Q_{oc})$ on the dot. Absent an applied bias voltage, that is, $V = 0$, the two capacitors C_1 and C_2, as the only electrostatic connection to the dot, represent an electrostatic charging energy of

$$U_c = \frac{(-ne + Q_{oc})^2}{2(C_1 + C_2)}, \tag{3.80}$$

where n is an integer. This energy is a minimum when n is closest to Q_{oc}/e, and $|-ne + Q_{oc}| \leq e/2$. We ask what the change in electrostatic energy is if we add 1 more electron:

$$\begin{aligned}
\Delta U_c &= \frac{[-(n+1)e + Q_{oc}]^2}{2(C_1 + C_2)} - \frac{(-ne + Q_{oc})^2}{2(C_1 + C_2)} \\
&= \frac{1}{2} \frac{e^2}{C_1 + C_2} \left(2n + 1 - 2\frac{Q_{oc}}{e} \right). \tag{3.81}
\end{aligned}$$

If there is no offset polarization and we started from a neutral dot

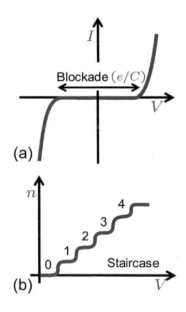

Figure 3.38: Current-voltage characteristics of an idealized symmetric Coulomb blockade, following the arrangement of the battery in Figure 3.37(a). No stray charge, and perfect symmetry, have been assumed. (b) shows the steps in number of electrons on the dot when tunneling conductance asymmetry exists in the structure and a staircase forms.

$(n = 0)$, this equation reduces to $\Delta U_c = e^2/2(C_1 + C_2)$. This is the simplest case, and the one with which we started the physical argument of this discussion. The interesting observation is that, with finite integral n, in Equation 3.81, if $-(ne + Q_{oc})/e = 1/2$, then this energy difference still vanishes. The offset charge polarization changes the extra electrochemical energy that the added electron must have in the source region, even making it vanish—despite the presence of the charge gap. For the special case of $n = 0$, this says that an offset charge of $-e/2$, makes the energy difference disappear for the placement of the additional electron. This is the blockade region of Figure 3.38(a) being shifted left by $e/2C$ V even as the blockade gap remains as e/C V. Now, consider again the case of $Q_{oc} = 0$ and we ask what the voltage applied should be so that the potential $e\phi$ of the dot would be sufficient for that increase from n to $n + 1$ of the electrons. This is the voltage partitioning for the two capacitors in series, with

$$\phi = \frac{C_2}{C_1 + C_2}V \tag{3.82}$$

as the voltage between the ground and the dot across C_1. This is the increase in charging energy of Equation 3.81, and hence it lets us determine the applied bias voltage for the electron transfer onto the dot (V_{th1}). Substituting with $Q_{oc} = 0$ but letting n be an integer that is not necessarily 0,

$$\Delta U_c = e\frac{C_2}{C_1 + C_2}V = \frac{(1 + 2n)e^2}{2(C_1 + C_2)}$$
$$\therefore \quad V_{th1} = (1 + 2n)\frac{e}{2C_2}. \tag{3.83}$$

So, if $n = 0$, and $Q_{oc} = 0$, the electron transfer requires a bias voltage of $e/2C_2$. This is the condition for transferring an electron from the ground across C_1 onto the dot. And we find that if $n = 0$, and $Q_{oc} = e/2$, then ΔU_c vanishes and correspondingly, so does the threshold for placing the electron. For the first case, one could also have asked what the threshold voltage is for transferring the electron from the dot across C_2. By symmetry, it will be $e/2C_1$. Current is determined by the flow of the electron onto and then off the dot, so the threshold voltage for conduction is the lower of these voltages, $e/2C_>$, where $C_>$ is the higher of the two capacitances C_1 and C_2. So, $Q_{oc} = e/2$ makes the threshold disappear, so that there is no blockade, though there is a shifted blockade gap, and $Q_{oc} = 0$ makes a threshold voltage appear. Steps in the Coulomb staircase arise from changing the electron count from n to $n + 1$. Equation 3.83 tells us that these steps will be incremental, the first being $e/2C_>$ determined by the larger of C_1 and C_2, with inter-step spacing of $e/(C_1 + C_2)$.

We now turn this diode-like arrangement into a transistor-like arrangement. The polarization charge will remain a natural part that we will control through the gate voltage. We now have a gate electrode, as well as the source electrode and the drain electrode between which particle charge exchange is allowed to happen, as in Coulomb blockade.

A schematic of this structure is in Figure 3.39(a), which shows the three electrodes where bias voltage is applied and a ground where all the other fields causing the offset may diverge to. So, C_π is a parasitic capacitance of all field terminations other than those of the gate (C_G), the drain (C_D) and the source (C_S). This structure could have been fabricated on a two-dimensional disk such as with a two-dimensional electron gas instead of in a three-dimensional geometry. Figure 3.39(b) shows the electrostatic energy of this dot, which is now formed perhaps through traditional fabrication technology so that all these electrodes may be formed and accessed. n is the number of excess electrons resident on the dot, and the abscissa ($\bar{n}$) plots the sum of particle and offset equivalent number of electrons. The gate does not inject or extract electrons but modulates the polarization. Any additional offset polarizations—those associated with C_π— cause a constant shift in this collection and so will be ignored in this discussion, for convenience. This shift similar to the movement of the Coulomb blockade region around the origin of Figure 3.38(a).

In a transistor, the bulk and interface states in oxides, or the dopants in the substrate depletion region, all cause a polarization effect—a continuous charge change through minute displacements of the mobile electron charge cloud from the immobile background ionic charge. Since this charge is fixed in space and time, this offset effect is a constant. Good thing too in this instance, the depletion region dopants keep the threshold voltage a constant. The gate voltage V_G also modulates this polarization charge but that can be continuous because the bias voltage can be continuous. This causes the electrons to stream in and out beyond the transistor threshold voltage through the source and drain reservoirs, thus leading to the useful current conduction between the source and the drain.

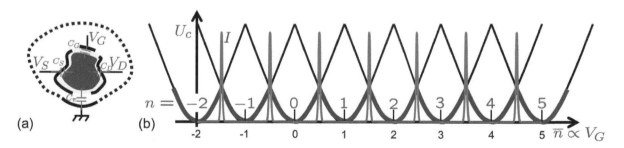

Figure 3.39: (a) shows a schematic of a single electron transistor consisting of a source electrode, a drain electrode and a gate electrode, together with a ground for the environment. (b) shows a plot of allowed energies for different counts of electrons and a residual polarization charge. The current is shown as sharp peaks where the charging energy curves cross.

Our offset charge here is being modulated by the gate voltage, so $Q_{oc} = C_G V_G$. If the number of excess electrons on the dot, being the total number of electrons minus the protons, is ne, then the charging energy is

$$U_c = \frac{(-ne + Q_{oc})^2}{2C} + K = \frac{(ne - Q_{oc})^2}{2C} + K. \qquad (3.84)$$

Here, K, a constant, is related to the choice of reference for the energy scale. For us, the important point is that $Q_{oc} = C_G V_G$ can now be continuously varied by V_G. C in this equation is the total capacitance that determines the electrostatic charging energy. Let

$$U_0 = \frac{e^2}{2C} \qquad (3.85)$$

be a constant of charging energy for this structure. With C as the total capacitance between the dot and the environment, more generally to be written as C_Σ, this U_0 is the change in electrostatic energy when the charge of e or $-e$ exists on the dot. By writing $\bar{n} = Q_{oc}/e$, we have written an equivalent electron count for the offset. This is a continuous function. So, we now have

$$U_c = U_0(n - \bar{n})^2 + K, \quad \text{and}$$

$$\bar{n} \propto n + \alpha, \quad \text{where } n \text{ is an integer, and} \quad -\frac{1}{2} < \alpha < \frac{1}{2}. \quad (3.86)$$

This is the set of solutions shown in Figure 3.39(b). The crossing of two curves allows a change in electron particle count by 1, as one electron transfers over to the next curve. This crossover point is a point of bistability, so n and $n + 1$ electrons are both allowed on the dot. This bistability and crossover become possible because the offset charge is being continuously modulated by V_G. This event repeating as a result of modulation by V_G is the basis of Coulomb blockade oscillation.

A change of charge of $\pm e$, that is, passage of a charge, means that there exists a pronounced peak in the current, as shown in Figure 3.39(b). The electron charge transfer happens as a rapid rise and fall in current with applied gate voltage and appears as a conductance oscillation.

This transfer, however, is subject to the other conditions of the system and so to the electrochemical potential of the source and the drain too. We will look at this presently to see how blockade and conduction appears in the two-dimensional gate bias voltage and drain-to-source bias voltage space. But first, we wish to complete the discussion of this simpler blockade and conduction as a function of a change in gate voltage, which effectively modulates the offset charge or, equivalently, α. Offset charge spans e under the constraint $|\alpha| < 1/2$. The charging energy is proportional to its second power, that is, the modulation over an unrestricted range of gate voltage results in

$$U_c = U_0 \alpha^2 \qquad (3.87)$$

centered at integer n. An equivalent way of looking at this is to see what happens when one increases or decreases the electron particle count by 1 on the dot:

$$U_c(n \pm 1) - U_c(n) = U_0(1 \mp 2\alpha) \geq 0, \qquad (3.88)$$

and the two energies for states with one electron difference are equal. So the condition of when a transfer is possible given by

$$U_c(n + 1) - U_c(n) = 0 \quad \therefore \quad \alpha = \frac{1}{2}. \qquad (3.89)$$

So, we have arrived at a result identical to the starting physically intuitive discussion of a one electron charging energy of $e^2/2C$.

The current peaks at the electron transfer condition and its sharpness can be viewed through the following argument. Let $U \gg k_B T$. This condition is necessary for any observable sharp rise and fall in response to the bias voltage. Let the dot have n electrons placed on it. Figure 3.40(a) shows this in a Coulomb blockaded band diagram. Placing an additional electron requires the dot's potential to be changed by e^2/C—this is a single electron charge gap of blockade in this structure. Recall our $n = 0$ discussion—this is the sum of the electrostatic energy $e^2/2C$ when an electron has been placed and the maximum of $e^2/2C$ that is the dot potential requirement for this placement. In Figure 3.40(a), no current flows since there is this charge gap and the transfer is not energetically allowed. But, if we shift electrochemical potential enough to that of Figure 3.40(b), the empty $n + 1$ state can be transferred into from the source, and, with the drain-to-source bias condition shown, which can be even vanishingly small, an empty state is available in the drain for the electron to transfer over. A single electron current flows through.

The crossing of two curves in Figure 3.39(b), the bistable point, is where n and $n + 1$ electrons may exist simultaneously. For Figure 3.40(b), if the source electrochemical potential is slightly higher, an electron below the Fermi energy in the source can still transfer over to the dot ($n \mapsto n + 1$) and from the dot over to an empty state ($n + 1 \mapsto n$) above the Fermi energy of the drain. This shows that there is going to be a width to the conductance spectrum's current signals. Energy is conserved in each of these steps—elastic tunneling—and the tunneling matrix elements are small. This implies $U_0(1 - 2\alpha) \leq |eV|$. If $U_0 = e^2/2C$ is large, this condition of finite measurable V implies that sharp maxima in conductance must be observable at $\alpha = 1/2$. This $\alpha = 1/2$ occurs when the gate voltage induces the proper offset charge, and this bias voltage is

$$V_G = \left(n + \frac{1}{2}\right)\frac{e}{C_G} \qquad (3.90)$$

for all integer n. So, the gate voltage tunes the parameter α or, equivalently, Q_{oc} for the conduction to occur. And when this parameter α is away from the $1/2$, blockade occurs.

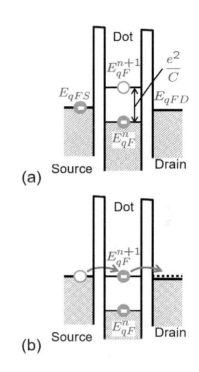

Figure 3.40: Single electron tunneling on the nanodot. A bias voltage of $-V/2$, $V/2$, and V_G are applied to the source (V_S), drain (V_D) and the gate referenced to the ground. (a) shows a blocking condition, while (b) shows a condition where the electron transfers—this is around the bistability point where n and $n + 1$ electrons are allowed energetically.

We did not consider the inelastic case. It will exist and so will resonant and sequential tunneling processes in this electron transport, as they do in a resonant tunneling diode, where single electron effects are neglected because of the large capacitances.

This discussion has assumed many constraints: very low temperatures, so no thermal tails, no broadening, and other constraints on the barrier and the structure. In addition, we have until now ignored the effect of the source and the drain potentials. We will now relax these constraints.

First, consider the scales of energy that are involved. These must be larger than the thermal energy. For Coulomb blockade, this means that the single electron charging energy scale must exceed thermal energy, that is,

$$\frac{e^2}{2C} > k_B T. \tag{3.91}$$

Since we have a nanoscale confined system and we wish to precisely see the effect on each energy level of confinement, the confinement energies must be larger than thermal energies, that is, $E_{i+1} - E_i > k_B T$. This condition is only of interest if one wants to distinguish the effect of Coulomb blockade for an electron on these precise states.

Our next condition is related to another energy consideration. That the electron passing through an energy level E_i means that it is leaking in and leaking out. This spread in time of leakage is related to spread in the energy of the eigenstate that the electron passes through. There exists an energy width of the eigenstate, and this energy width should be significantly smaller than the energy change. The electron passing through this dot is impeded in this process, under the condition where this transport is allowed by the barrier. The single electron must dwell long enough to be considered localized and yet must leak out. The lifetime of the electron in the eigenstate is inversely related to the leakage rate. This determines the energy width of the eigenstate, so $\Delta E = \hbar/\tau$, where τ, the lifetime, is related to the leakage rate Γ as $\tau = 1/\Gamma$. Therefore, $\Delta E = \hbar\Gamma$. The current flowing through the structure is $I = e\Gamma$, so we may write for energy balance in an elastic process, with a tunneling barrier of resistance R_T,

$$\Delta U = eV = eIR_T = e^2\Gamma R_T. \tag{3.92}$$

We can rewrite the eigenstate energy width in terms of this energy change as $\Delta E = \hbar\Gamma = \hbar\Delta U/e^2 R_T$. This gives us the second constraint,

$$\Delta U \gg \Delta E \quad \therefore \quad R_T \gg \frac{\hbar}{e^2}. \tag{3.93}$$

So, our first constraint establishes the observability of Coulomb electrostatic energy by requiring it to be of similar or higher magnitude as thermal energy. And the second constraint establishes that system energy change when an electron traverses through be significantly larger than the energy spread of the eigenstate, so the barriers

The energy width statement is a statement of the precision with which the eigenenergy can be measured. The uncertainty relationship describes to us its limits.

We have now a number of resistances in h/e^2 that we have seen arising from the interplay of quantum processes in our macroscopic world. The single electron effect requires a tunneling resistance of $R_T \gg \hbar/e^2 \approx 4.1\ k\Omega$. The electron resistance of a quantum conductance channel is $h/2e^2 \approx 12.8\ k\Omega$. Superconducting Cooper pair tunneling requires a barrier of $R_{scT} \gg h/4e^2 \approx 6.4\ k\Omega$. Note that, in the electron quantum conductance, we are associating the resistance with an electron of spin $+1/2$ and $-1/2$. The channel supports both. h/e^2 is associated with one electron in the conductance channel.

must have adequate tunneling resistance to keep the leakage rate and linewidth small. This is equivalent to saying that the electron on the particle must have a state of confinement and dwelling on the particle so that one may effectively say that it is on the particle. The energy and the barrier resistance property reflect this. A secondary consideration, if confinement-induced energy states are desired to be observed, is that these confinement energy separations be larger than the thermal energy—this is related to the observability of different quantized states, in addition to their charging electrostatic energy effect.

Since the polarization is an effect reflecting the electrostatic coupling to all the surrounding regions, in addition to the gate modulation that we just looked at under a constant infinitesimally small bias voltage, we must incorporate the effects of all others. At the simplest, in a three-terminal structure, this includes modulation from the drain and the source. The drain bias potential and the source bias potential also affect the polarization charge.

This means that our electrostatic charging energy Equation 3.80 needs to be modified. The offset charge is

$$Q_{oc} = -C_g V_G - C_s V_S - C_d V_D. \tag{3.94}$$

We will also write the total capacitance of charging as $C_\Sigma = C_g + C_s + C_d$. So, the charging energy equation for adding an electron, so that $n \mapsto n+1$, now becomes, assuming that the source and drain are symmetric with a capacitance of C_c,

$$
\begin{aligned}
\Delta U_c &= \frac{e^2}{2C_\Sigma}\left(2n+1+2\frac{C_g V_G + C_c V_{DS}}{e}\right) \\
&= (2n+1)\frac{e^2}{2C_\Sigma} + \frac{C_g}{C_\Sigma}e V_G + \frac{C_c}{C_\Sigma}e V_{DS}, \tag{3.95}
\end{aligned}
$$

which includes modulation by all the potentials and ignores any residual constants of polarization.

This results in Coulomb diamonds, as shown in Figure 3.41(d), where gate and drain-to-source bias potentials are varied. This is a "phase diagram" showing the various states of this system—blockaded regions and conduction regions in the bias voltage parameter space. Varying one of the bias voltages and keeping the other constant allows one to modulate the size of the blockade window, which is akin to shifting it by using the offset charge of our starting discussion. If $V_{DS} = 0$, we return to our original two-terminal situation and have the maximum blockade range in response to the gate bias voltage. But if $V_{DS} \neq 0$, the response changes, because a certain amount of the offset charge is coupled to the drain and source electrodes, thus shifting from the symmetric conditions of $V_{DS} = 0$.

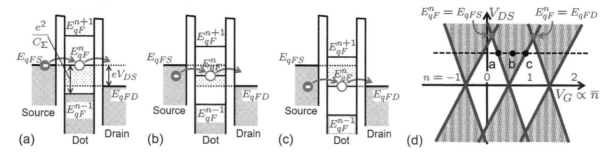

Figure 3.41: Coulomb blockade diamonds as gate and drain-to-source bias potentials are varied. At a finite non-zero drain-to-source bias voltage, conduction is possible from source to drain for a finite gate bias voltage range, as shown in (a), where the onset of conduction happens through the charging energy level for the nth electron added to the dot, (b), where conduction is still possible through a state below the quasi-Fermi energy of the source, and (c), which shows the extinction of conduction through the charging energy level for the nth electron added to the dot. (d) shows the corresponding conducting and blockade region, with a, b and c identified for the three conditions of the previous figure panels.

Figure 3.41(a) shows, for a given applied bias potential of V_{DS}, the condition of gate bias potential V_G when conduction just begins to occur through the nth electron charging energy level. Note that $e^2/2C_\Sigma$ is the single electron charging energy associated with this event and it exists as a charging energy gap above the $(n-1)$th electron's filled energy level. At the bias condition of Figure 3.41(a), electrons can begin to tunnel to E_{qF}^n from the source and onward out to empty states in the drain. As one raises the gate bias potential more, as in Figure 3.41(b), thus lowering the energy in the dot, electrons below the quasi-Fermi energy of source can still tunnel through the energy level E_{qF}^n and onward to the empty states of the drain. Finally, one reaches the condition of Figure 3.41(c), where the energy level in the dot coincides with the quasi-Fermi level of the drain, and states for tunneling out no longer exist. Current conduction is now extinguished. The application of the drain-to-source bias potential has the consequence of opening up a window of conduction and reducing the blockade window. This set of conditions for conduction corresponds to the points marked as a, b and c in the Coulomb diamonds shown in Figure 3.41(d): a is when the source quasi-Fermi level coincides with the nth electron charging energy level, b is in the region where one of the filled states of the source has electron conduction permissible, and c corresponds to the drain quasi-Fermi energy coinciding with the nth electron charging energy level. What has happened here, distinguishing it from the previous description of infinitesimally small drain-to-source bias, is the change from a bistable point to an extended region of conduction. Along the abscissa, so vanishing drain-to-source bias potential, the Coulomb blockade region exists exactly as before, causing Coulomb oscillations of current at the bistability points. But, away from this condition, the blockade width decreases, and a conduction channel exists.

The effectiveness of gate or drain-to-source bias potential in modulating the offset charge and thus the shape of the diamond is reflected through the capacitances. So, the slopes of the lines are connected to the ratio of these capacitances. If $C_g \gg C_c$, then the gate offset charge dominates, and the lines are relatively vertical, with a weak dependence of the blockade window on the drain-to-source bias potential. A reversal of this capacitance relationship collapses this window, as it should. No barrier, or a large capacitance reflect the lack of single electron dwelling on the dot and thus indicate a continuing, uninterrupted flow of current, where the effect of the gate potential is a continuous change in what occurs in a flowing charge. This is field-effect transistor.

We now include an additional wrinkle to make this analysis more practical/useful with semiconductors. We have not included, up to this point, the energy quantization arising from quantum confinement in the particle, by assuming that the density of states is very large. This works for metals and at temperatures that are not extremely low, that is, when $\Delta E \ll k_B T$, where ΔE is the difference in energy between consecutive allowed confinement eigenstates, still remains valid. In semiconductors, this assumption breaks down.

The energy—electrostatic (U_c) and potential (U_p)—for an n excess electron system is

$$U = U_c + U_p = U_0(n - \bar{n})^2 + \sum_{i=1}^{n} E_i, \qquad (3.96)$$

where E_i is the ith single particle energy level. Each E_i energy level can still accommodate a spin-up and a spin-down electron, but also requires that for the placement of the electron, the requisite single electron charging energy also be supplied. The energy cost of introducing the nth electron is

$$U(n) - U(n - 1) = E_n + 2U_0\left(n - \bar{n} - \frac{1}{2}\right). \qquad (3.97)$$

This bistability exists only when $U(n) - U(n - 1) = 0$, consequently,

$$\bar{n} = \bar{n}_n \equiv \left(n - \frac{1}{2}\right) + \frac{E_n}{2U_0}. \qquad (3.98)$$

In Coulomb blockade oscillations, the spacing between successive conductance peaks ($V_{DS} \approx 0$) then is given by

$$\bar{n}_{n+1} - \bar{n}_n = 1 + \frac{E_{n+1} - E_n}{2U_0}. \qquad (3.99)$$

In determining the threshold voltage of a field-effect transistor, well before the nanoscale limit, we have had to include the energy shifts arising from even one dimension of confinement, let alone three. One additional remark is that the addition of this single electron does not in any noticeable way perturb the energy levels that arise from spatial quantization. Within the nanoscale confined region, this energy structure arose from the interaction of thousands of spatially confined electrons and ions, as is reflected in the band structure. This assumption will only break down when one cannot employ the unit cell translational approach of the Bloch picture any more. The addition of one more electron will not change the perturbation term in the Hamiltonian of this Bloch description in any significant way, but it does matter that this system is coupled to an external world electrostatically. And this coupling insists that there be the requisite electrostatic charging energy for placement of this electron on the dot. The energy level itself is defined by the dot and its boundary conditions—quantization effects—but bringing an electron onto the dot is subject to the electrostatic charging energy resulting from the interaction of this nanoscale dot with its environment.

The spacing in the normalized parameter $\bar{n} \propto V_G$ is modified from the integral change by an additive perturbation that is half of the energy level spacing normalized to the single electron charging energy. If the energy level spacings are small, the diamonds are equal in size and equally spaced. But if the allowed energy level structure changes, such as in semiconductors, molecules or other nanoscale materials, and the energy spacings are of the order of magnitude of U_0, then one would see a variety of structures reflective of the energy levels of the system. Single electron measurements allow us to extract detailed quantum confinement and other energetic information in small dots or other small systems such as those composed of atoms with multi-state accessibility, as well as molecules and particles tethered with molecules.

Since electrons are fermions, the filling of these energy levels by electrons is also going to be constrained by the exclusion principle—each energy level may only have one up and one down spin. Spin and exclusion have very important effects in magnetism because of exchange wavefunction asymmetry. Exchange also plays an important role here in the pairing and addition of electrons under the asymmetry expectation. First, consider what should happen without exchange. Any energy level that has n electrons, where n is odd, can accommodate another electron of the opposite spin. This was unstated in the single electron discussion up to this point. This constraint for electron transfer is

$$E(n) - E(n-1) = 0, \text{ if } n \text{ is odd,} \qquad (3.100)$$

and is simultaneously subject to the electrostatic charging energy constraint. But, we may also have

$$E(n) - E(n-1) > 0, \text{ if } n \text{ is even,} \qquad (3.101)$$

which is again simultaneously subject to the electrostatic charging energy constraint. If there is an even number of electrons occupying an energy level, then the additional electron must occupy a higher energy level—an increase in energy is the consequence. The $n = 0$ case is trivial. The electron occupies the next available energy level. If $n = 2$, then states are filled up to an energy level, and the next electron must go to a higher energy level. And these energy levels, in general, are not equally spaced. So, this occupation of energy is subject to a bimodal behavior. The electrostatic charging energy is a constant, but the energetics of quantization of the electron states is number dependent. It fluctuates. For the first case, n odd, this is the simple case we studied. But, for placing an additional electron, at this point with an even number of electrons on the dot, there is a change in energy, reflected in Equation 3.99: this change in energy will fluctuate. The spacing between an odd peak and an even peak following it is constant, but the spacing between an even peak and a following odd peak varies. This is shown in Figure 3.42. This bimodal distribution is an example of Wigner-Dyson distribution originating in the spectral energy spacings of heavy atoms and resembles the eigenvalue spacing distribution of random matrices such as of large disordered systems' Hamiltonians.

The exchange energy of the dot provides another perturbation term in this interaction. Its importance will depend on the exchange integral—J—that contributes to JS^2 of energy perturbation. If all energy levels due to quantization were equally separated, then the spin of an even electron state should be 0 as a result of both $+1/2$ and $-1/2$ spins occupying that state. An odd electron state will have $S = 1/2$. In this case, following the above discussion, the level spacings are random. Now, if $E(n) - E(n-1)$ happens to be small, it is possible for the dot to have a larger spin. In an even electron state, electrons may now take advantage of this small energy spacing by aligning their spins to achieve a total spin of $S = 1$. Exchange now causes an additional alteration in the Coulomb peak spacing.

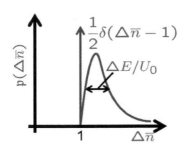

Figure 3.42: The probability distribution of successive energies of conduction. The distribution is bimodal. If there is an odd number of electrons, the energy for occupation by the next electron is separated by a constant energy shown by the delta function. If the number of electrons is even, then the next electron must be in a higher energy quantized state, and the distribution of spacing fluctuates.

Exchange integrals will appear at length in the magnetic discussion.

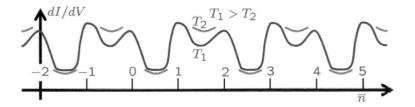

Figure 3.43: Effect of temperature in causing Coulomb blockade peaks to expand for odd-even electron number transitions in the weak coupling limit.

The next consideration of import is the coupling of leads. Our argument in determining the minimum tunneling resistance incorporated the leakage rate (Γ) in the argument that $\Delta U \gg \Delta E(= \hbar\Gamma)$. $\hbar\Gamma$ defines the width of the energy level, and this is related to the lead coupling. When the energy level separation is much larger than thermal energy, weak coupling causes Coulomb blockade peaks to expand in width as shown in Figure 3.43. A large lead coupling destroys the tunneling resistance condition $R_T \gg \hbar/e^2$, so Coulomb blockade peaks must disappear. If the lead coupling is very weak—a small width $\hbar\Gamma$—then Coulomb blockade peaks acquire width due to the reduced coupling. This is because conductance in the transition from an odd to an even number of electrons increases, resulting in a change in the conductance in the valley region. The odd-to-even transition has an increase in width, as shown in Figure 3.43. Odd electron states also have higher conductance. There is an additional effect that we will discuss, the Kondo effect, again related to spin, in how screening occurs, leading to resonant processes through virtual states. Valleys can disappear at absolute zero due to the Kondo effect.

The Kondo effect is the result of interaction between intrinsic angular momentum and the surrounding electron sea. A magnetic atom in a non-superconductive metal, a spin trapped in a quantum dot, and other systems where the interaction between electrons is strong, such as high temperature superconductors, show the consequences of this interaction. In all these cases, the electron sea of spin $1/2$ particles is responding to a local perturbation, in a process that is akin to Debye screening of ionized charge by electrons and holes but which arises from an entirely different microscopic process.

If a material undergoes a superconducting transition, at a critical temperature, the resistance vanishes. But, in a normal metal, such as copper, gold or aluminum, the resistance depends on the number of defects and increases when the defect count increases. And when the temperature is lowered, the resistance decreases, with vibrational effects decreasing asymptotically in the purest and cleanest samples. When magnetic atoms are added, near a critical temperature of transition, the Kondo temperature T_K, the resistance begins to increase. Around T_K, the magnetic moment of the impurity ion is screened by the spins of the electrons in the metal. This interaction between the magnetic ions with the conducting electrons, because of spin, causes the resistivity to increase as

$$\rho(T) \approx \rho_0 + \alpha T^2 + \beta \ln\left(\frac{E_F}{k_B T}\right) + \gamma T^5. \qquad (3.102)$$

The increase is the result, at low temperatures, of the sea of electrons with their up and down spins interacting with a single spin. We have written this equation here without proof.

In single electron tunneling, the Kondo effect—a conductance behavior change, as shown in Figure 3.44—is the result of interaction between delocalized states outside with the localized spin in the dot. Note also that here the conductance increases when the number of electrons in the dot is odd. This dot is like an impurity for the electrons of the bulk. The electrostatic charging energy (U_0) is akin to an on-site repulsion energy. The applied gate bias voltage changes the on-site electron energy (ε_0) so it can be tuned. In the Kondo effect in the dot, when the number of electrons is even, the conductance vanishes, while an odd number of electrons causes the conductance top to increase up to the quantized channel conductance of $2e^2/h$.

The reason that the Kondo effect causes an increase in conductance when there is an odd number of electrons is simplistically shown in Figure 3.45. If there is one electron occupying the dot, it exists there

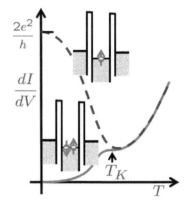

Figure 3.44: The Kondo effect during single electron tunneling. Conductance increases when an odd number of electrons populate the dot.

Anderson localization—the impurity model, which we will tackle in Chapter 4—also provides a way to examine the Kondo effect. The interplay between the on-site electron repulsion energy U and the single electron on-site energy is similar to the localization-delocalization argument. An energy width ($\hbar\Gamma$) of any energy level in the dot arises from the hybridization between the localized state of the dot and the delocalized states of the bulk.

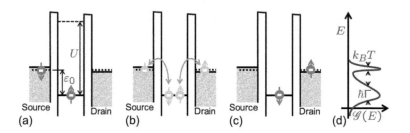

Figure 3.45: (a) shows the initial state of a dot with a single electron occupying an energy level under conditions with ε_0 localization energy—energy associated with the coupling to the electrodes and a large charge gap. (b) shows a virtual state as a consequence of uncertainty when the localized electron moves to the drain at an energy close to the Fermi energy, thus allowing an opposite spin electron to arrive in the dot from the source. (c) shows the final state reflecting passage of the dot-occupying electron to the drain while resulting in a spin flip of the incoming dot-occupying electron. In (d), this interaction is reflected in the large energy width $\hbar\Gamma$ state, and the resonance in energy close to the Fermi energy.

under conditions of a large charge gap (U) but with an energy of ε_0 reflecting its localization within the dot w.r.t. the electrons in the electrodes, for example, the source. $U \gg \varepsilon_0$. Adding an electron requires a larger energy U. And removing the electron costs the energy of ε_0. The virtual state, pictorially described in Figure 3.45(b), reflects the consequences of the uncertainty relationship in short times. A momentary tunneling out of the electron from this dot onto a forbidden state in the drain allows an opposite spin electron to transfer in from the source—with the electrons exchanging their energies in the process. The final state (c) that this process results in has the spin of the localized electron flipped while an electron has passed through leading to the increased conductance. In energy, what this reflects is an energy width $\hbar\Gamma$ for the width of the localized energy state—the leakage associated with this Kondo effect—and an additional resonance peak just above the quasi-Fermi energy in the drain. This is shown in Figure 3.45(d).

The Kondo effect, due to the isolated spin in the dot, can be coupled to a The number of the interesting mesoscale phenomena we have discussed in this chapter. One could, for example, imagine placing one of these dots in one arm of the interference ring. In this situation, the Coulomb oscillation peaks are distorted due to the Fano effect—the creation of an asymmetric lineshape due to the interference of two scattering amplitudes, one of which is in a continuum, and the other of which is due to a discrete excitation. In this interferometer, the continuous arm is the continuum, and the arm with the dot provides the discrete excitation.

The Fano effect and the Kondo effect, together in this interferometer, can cause plateaus with enhanced conductance of the even electron Coulomb blockaded regions. So, a variety of many-body interactions between a localized electron spin and the surrounding conduction electrons happen, and, in this, the first order effect is a modification of the lineshape.

We now look at what would happen if one placed two such dots in a sequence between the drain and source electrodes, as shown in Figure 3.46 schematically. Our interest is in seeing in what way, if any, our descriptive understanding should be changed. So, we will consider only the primary electrostatic charging effect of the Coulomb blockade with which we had started—the classical limit in linear formulation where the drain-to-source bias potential voltage is vanishing.

Let ϕ_1 and ϕ_2 be voltages developing on each of the dots. Now there are two gates, and the effective charge is due to the excess electrons and the polarization from the gate. We ignore the other offset charges, which do not vary, and we also ignore the quantum-mechanical nature of the discrete electrons. We will use the matrix approach. The charges are

$$\begin{aligned} Q_1 &= C_s(\phi_1 - V_S) + C_{g1}(\phi_1 - V_{G1}) + C_i(\phi_1 - \phi_2), \text{ and} \\ Q_2 &= C_d(\phi_2 - V_D) + C_{g2}(\phi_2 - V_{G2}) + C_i(\phi_2 - \phi_1), \end{aligned} \quad (3.103)$$

that is,

$$\begin{bmatrix} Q_1 + C_s V_S + C_{g1} V_{G1} \\ Q_2 + C_d V_D + C_{g2} V_{G2} \end{bmatrix} = \begin{bmatrix} C_1 & -C_i \\ -C_i & C_2 \end{bmatrix} \begin{bmatrix} \phi_1 \\ \phi_2 \end{bmatrix}, \quad (3.104)$$

with $C_{1(2)} = C_{s(d)} + C_{g1(g2)} + C_i$, where the subscripts define the choice of capacitance C_1 or C_2. So, $C_{1(2)}$ represents the sum of the capacitances tied to the dot. This relationship leads to the dot potentials

$$\begin{bmatrix} \phi_1 \\ \phi_2 \end{bmatrix} = \frac{1}{C_1 C_2 - C_i^2} \begin{bmatrix} C_2 & C_i \\ C_i & C_1 \end{bmatrix} \begin{bmatrix} Q_1 + C_s V_S + C_{g1} V_{G1} \\ Q_2 + C_d V_D + C_{g2} V_{G2} \end{bmatrix}. \quad (3.105)$$

With the potentials of the dots known, the electrostatic energies follow as before. We do this in a matrix form and consider the linear case with vanishing drain and source potentials. The electrostatic energy of the system is

$$U_c = \frac{1}{2}[\phi]^T [C][\phi], \quad (3.106)$$

and, choosing an integral number of electrons on the dots, that is, $Q_{1(2)} = -n_{1(2)}e$, we find

$$U_c(n_1, n_2) = n_1^2 U_0^1 + n_2^2 U_0^2 + 2n_1 n_2 U_0^i + \varepsilon(V_{G1}, V_{G2}), \quad (3.107)$$

with

$$U_0^1 = \frac{e^2}{2(C_1 C_2 - C_i^2)/C_2},$$

$$U_0^2 = \frac{e^2}{2(C_1 C_2 - C_i^2)/C_1},$$

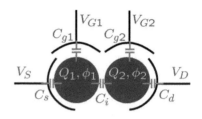

Figure 3.46: A double dot structure.

This matrix approach will become all the more important in Chapter 5, where we tackle nanoelectromechanics and the interaction of a multitude of changing capacitances and the different forms in which energy will exist in the system.

$$U_0^i = \frac{e^2}{2(C_1 C_2 - C_i^2)/C_i}, \text{ and}$$

$$\begin{aligned}
\epsilon(V_{g1}, V_{g2}) = &-\frac{1}{e}\left[C_{g1}V_{g1}(2n_1 U_0^1 + 2n_2 U_0^2)\right. \\
&\left. + C_{g2}V_{G2}(2n_1 U_0^1 + 2n_2 U_0^2)\right] \\
&+ \frac{1}{e^2}(C_{g1}^2 V_{G1}^2 U_0^1 + C_{g2}^2 V_{G2}^2 U_0^2 \\
&+ \frac{1}{2}C_{g1}V_{G1}C_{g2}V_{G2}U_0^i).
\end{aligned} \tag{3.108}$$

The normalization term of electrostatic charging energies, U_0^1, for example, is modified from before. It is a scaled electrostatic charging energy due to changes in coupling, that is, capacitances w.r.t. to that of an isolated dot. If the coupling is negligible and the two dots can be treated as being isolated, these electrostatic charging normalization energies reduce to the linear sum form of the single dot case, that is, uncoupled double dots have the charging energy

$$U_c(n_1, n_2) = \frac{(-n_1 e + C_{g1}V_{G1})^2}{2C_1} + \frac{(-n_2 e + C_{g2}V_{G2})^2}{2C_2}. \tag{3.109}$$

This is precisely the form one would expect as a simple extension of Equation 3.84. At the other end of this argument, for highly coupled dots, that is, $C_i \gg C_s, C_d, C_{g1}, C_{g2}$, the charging energy is

$$U_c(n_1, n_2) = \frac{[-(n_1 + n_2)e + C_{g1}V_{G1} + C_{g2}V_{G2}]^2}{2(\check{C}_1 + \check{C}_2)}. \tag{3.110}$$

$\check{C}_{1(2)} - C_1 - C_i$ is the capacitance of the dots to the surrounding excluding the coupling between them. This form is precisely what one would expect of charging energy for a larger dot arising from the joining of the two dots, but still subject to the same charging energy constraints. It has $n_1 + n_2$ electrons and a capacitance of $C_\Sigma = \check{C}_1 + \check{C}_2 = C_s + C_{g1} + C_d + C_{g2}$ to the surroundings.

If we now vary the gate bias voltages V_{G1} and V_{G2} and if the dots are reasonably coupled through a capacitance C_i, and a tunneling resistance subject to the conditions that we had derived for the observation of Coulomb blockade exists, we expect that n_1 and n_2, the integer charges on the dots, will remain close to each other. This is the energetically favored condition, because transfer of electrons becomes possible only when the total energy allows the energy-conserving crossover exchange to be possible.

The energy needed to add one additional electron is the way we define the electrochemical potential—we have used E_F and E_{qF} in thermal equilibrium and away from it notationally. So, $E_{qF1}(n_1, n_2)$ is the energy, the quasi-Fermi energy, that is needed for the addition of the n_1th electron to the first dot while n_2 electrons exist on the second dot. A similar definition follows for $E_{qF2}(n_1, n_2)$ for the second dot. From Equation 3.107, these quasi-Fermi energies follow as

$$
\begin{aligned}
E_{qF1}(n_1, n_2) &= U_c(n_1, n_2) - U_c(n_1 - 1, n_2) \\
&= \left(n_1 - \frac{1}{2}\right)2U_0^1 + n_2 2U_0^i \\
&\quad - \frac{1}{e}(C_{g1}V_{G1}2U_0^1 + C_{g2}V_{G2}2U_0^i), \quad \text{and} \\
E_{qF2}(n_1, n_2) &= U_c(n_1, n_2) - U_c(n_1, n_2 - 1) \\
&= \left(n_2 - \frac{1}{2}\right)2U_0^2 + n_1 2U_0^i \\
&\quad - \frac{1}{e}(C_{g1}V_{G1}2U_0^i + C_{g2}V_{G2}2U_0^2).
\end{aligned}
\tag{3.111}
$$

The interaction between the dots is reflected through $2U_0^i$, both through the electron and through the polarization of the coupling dots. If one keeps the gate voltages constant, then changing the electron count by 1 requires an energy of

$$
E_{qF1}(n_1 + 1, n_2) - E_{qF1}(n_1, n_2) = 2U_0^1 \equiv \frac{e^2}{C_1},
\tag{3.112}
$$

precisely what one would expect from our analysis of the simpler case. This is the charge gap—the electrochemical energy change in the dot needed for a change of one electron. This is the charging in energy in the classical limit, where we have ignored the spatial quantization effects of the dots. The charging for the second dot by one electron, given all other conditions constant, its quasi-Fermi energy change, by analogy, is $2U_0^2 \equiv e^2/C_2$.

What is the electrochemical energy change of the first dot when an electron is added to the second dot? Or its complement, the electrochemical energy change of the second dot when the first is populated with an additional electron, all other conditions being kept constant? This is

$$
\begin{aligned}
E_{qF1}(n_1, n_2 + 1) - E_{qF1}(n_1, n_2) &= E_{qF2}(n_1 + 1, n_2) - E_{qF2}(n_1, n_2) \\
&= 2U_0^i.
\end{aligned}
\tag{3.113}
$$

It too is precisely twice the electrostatic charging energy form that is expected.

Equation 3.111 describes the state diagram connecting electron count parameters n_1 and n_2 with the applied gate bias potential voltages V_{G1} and V_{G2} under linear conditions of $V_{DS} \approx 0$ with corresponding electrochemical potentials at drain and source as zero reference. The maximum in n_1 and n_2 that keeps the electrochemical potential of each of the dots below the zero reference is the steady state of the system. This is what we do when we draw a picture, such as that of Figure 3.40(a), of a block of the electron transfer, with n electrons in the dot. The stable charge configurations can now take different forms in the (V_{G1}, V_{G2}) phase space, as shown in Figure 3.47 for the two dot problem.

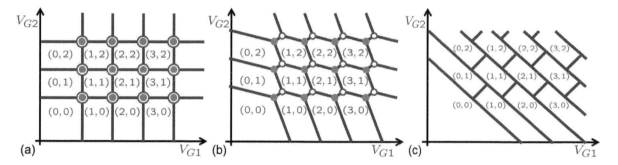

Figure 3.47: Charge stability diagrams of coupled two dot systems. The electron count (n_1, n_2) of the electrons on each dot is shown. (a) shows the case when the inter-dot coupling is small ($C_i \rightarrow 0$), (b) shows the case when C_i becomes important but $C_i < C_{1(2)}$, and (c) shows the case when the capacitances are comparable, that is, $C_i \approx C_{1(2)}$.

The case when $C_i/C_{1(2)} \rightarrow 0$, that is, when there is very weak coupling, is depicted in Figure 3.47(a), which shows the resulting transitions between electron states with individual gate bias voltages that are independent of each other. Conductance oscillations, just as for the single dot case, occur as a function of each of these gate bias voltages. The second dot in this case behaves as a drain or source electrode, except under the constraint that an integral number of electrons are stored in it. Figure 3.47(a) shows that, depending on the choice of gate bias voltage, a different number of excess or deficit electrons exist on each dot, over a range of gate bias voltages. And each one of these stable regions is connected to another one through a single electron change in storage for a range of variations. A vertex exists where, in theory, it is possible for the electron count to change by unity for both of the dots. When the ratio $C_i/C_{1(2)}$ is increased by increased coupling between the dots, hexagon-shaped stable regions appear. The slopes, like the diamonds in the single dot case, arise from the total charge dependence on both of the gate voltages applied through the capacitance ratios. The triple points are a splitting of the vertex due to the increased coupling. The increased coupling allows exchanges such as $(0,2)$ and $(1,1)$ to occur, that is, an electron exchange between the dots themselves. As this coupling increases further, one reaches a maximum that is seen in Figure 3.47(c). This is the double dot behaving like a single dot.

The triple points of the hexagon structure, because of dot-to-dot coupling, under single electron transport conditions where the bar-

Recall that $C_{1(2)} = C_{s(d)} + C_{g1(g2)} + C_i$, so $C_i/C_{1(2)} \rightarrow 1$ is the condition for a large C_i.

riers have transparency but where conditions of the structure also
define electron localization, are interesting points. The extent of
the hexagon arm is related to the coupling, and the triple points
represent points of resonance. Figure 3.48 shows the triple points
schematically as points where multiple electron configurations are
simultaneously stable. The transport between the electrodes is resonant, since all energy matching conditions are satisfied for the three
possible charge configurations. Electrons traveling in one direction
establish one of these. And its quasiparticle hole traveling in the
other direction defines the other triple point. Each of these transport
steps can be viewed as moving in a loop through each of the stable
regions around each of these points, as shown in Figure 3.48.

If one added spatial quantization and spin interactions to this
discussion, one can see that a variety of interesting effects will be
observed. We will not dwell on them. The single and double dot
examples have been given as a way to understand the energetics and
the magnitudes of the effects. We will make some remarks toward
the end of this chapter on the engineering utility of these effects in
devices.

We now return full circle to emphasize the limitations of, and the
caution one must exercise in employing, mean field approaches—the
use of means over ensemble to analyze energetics. Single electron
phenomena at nanoscale, as well as the and phase transition discussion of Chapter 4, provide an opportunity to discuss this issue—they
are examples where contradictions arise and important details get
lost when a mean taken over the contributing interacting entities is
employed to describe the effect on the action of interest, but in circumstances where local and global interactions are both important.
Unlike in phase transition, where the renormalization of fluctuations
connects the multitude of interactions, here it is an issue related to
the electrostatic energy of the charge particle—the electron—arising
from an interaction with its environment but not with itself, as it is
localized to the nanoscale dot.

Mean field averages through energy all those that contribute, including the electron, whose energetics is of interest. A simple example of this local-global detail is the energetics of an electron traveling
between two conducting electrodes in vacuum and so with no local
polarization interactions, as in Figure 3.49(a), in contrast to an electron traveling through the nanoscale dots of a conductor surrounded
by vacuum, as in Figure 3.49(b). In transport in vacuum, the energy
of the electron decreases between the plates because of Coulomb interactions. The presence of the electron between the plates induces
a positive polarization charge on the plates, consistent with a lowering of electron energy. We include only this electron-polarization

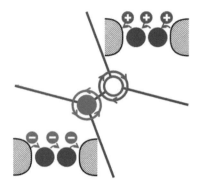

Figure 3.48: The triple points of a
hexagon structure are points of the
degeneracy of charge states where
electron or hole transport resonantly
occurs between the electrodes.

charge to determine this energy. Arriving at a mathematical description of the polarization charge, the fields and the energy of this result by solving Maxwell's equations, wel find a satisfactory answer for a limited domain—in the absence of the importance of a quantum-mechanical description in the metal, a reasonable spatial separation between the metal and the electron, static conditions and a description valid only for the region between the electron and the plate. The projection of this into an image charge form, although very useful, has a number of additional limitations. Replace the metal with an insulating dielectric, and the schematic picture for energy still remains adequate, but the description of what happens in the plate, when the metal is replaced by a dielectric, fails. But, in either case, the potential in the gap must decrease.

In the case of transport through islands surrounded by vacuum, by tunneling between them, the energy has the opposite behavior. Energy increases, as we discussed, because of the electrostatic interaction with the environment—the same reason as for the case in Figure 3.49(a) but with precisely the opposite consequence.

This brings up the question of how these energies should be calculated in general for a system consisting of conducting materials and discrete charges that we encounter, for example, even in the simple case of a parallel plate capacitor where the insulation is not absolute and discrete electrons float around, or even a p/n junction region with moving carriers together with immobile ionized dopants and clad with a quasineutral region at its edges.

As in phase transitions, where fluctuations are at every scale, here too the breakdown and limits to adequacy of mean field arise from nonlinearity. For Coulombic interaction, it is due to the energy's inverse spatial dependence. In Chapter 5, we will see the effects of nonlinearity on even a larger dimensional scale for electromechanical objects—chaos, bifurcation and the development of self-similar features. Most of the major equations that we have employed—Schrödinger's, Maxwell's, or the special case of Poisson's—are formally linear. In many examples of the use of these equations in situations with many interacting entities, the irreversibility, as we have seen, arises from the one-to-many interactions—the degrees of freedom formulation of this argument. Microscopically, we used a linear picture. In this problem of nanoscale, despite it being in microscopic reversible condition, the nonlinearity arises from the ability of the electron to influence its surroundings and for the surrounding to influence it.

In Figure 3.49(a), in the potential in a gap problem, physical changes occur when an electron enters, due to the polarization of the medium and its boundaries. The electron, however, does not *di-*

And this is the reason that the image charge approach works in calculating this energy in static conditions when restricted to describing the details only between the plate and the electron. For the limitations, see the discussion in the appendix of S. Tiwari, "Quantum, statistical and information mechanics: A unified introduction," Electroscience 1, Oxford University Press, ISBN 978-0-19-875985-0 (forthcoming).

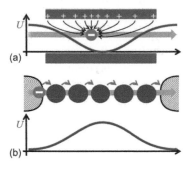

Figure 3.49: (a) shows electron energy as an electron travels in vacuum between two conductors. (b) shows electron energy as it travels through conductive islands surrounded by vacuum and polarizing electrodes.

rectly interact with itself. The energy of a point charge, if one used
our classical electromagnetic approach, will be infinite. When an
electron is assembled, a fundamental event that is beyond the scope
and interest of our discussion, it cannot possibly have needed an infinite energy. The total energy must be of the order of $m_0 c^2$, and this
energy resides within the spatial extent of the electron. And this electron's self-energy must change when it is a bound state. Quantum
field theory allows us to calculate these energies and avoid the infinities. In our range of problems, the issue of these infinities arises in a
more mundane but still very intriguing and important form.

Consider a system consisting of metallic conductors and discrete
charge, such as free electrons or ionized dopants—two metal plates
with distributed charge in a dielectric or with a vacuum between
them, such as shown in Figure 3.49. This could be a capacitor that
is leaky because of a poor insulator, or any electronic structure with
moving electrons or a transition region, so this is quite a general
problem that we see in all of electronics. The energy of the system of
interest to us, the electrostatic energy, is

$$
\begin{aligned}
U_{es} &= \frac{1}{2}\epsilon_s \int \mathcal{E}(\mathbf{r}) \cdot \mathcal{E}(\mathbf{r}) d^3\mathbf{r} \ \text{where} \ \ \mathcal{E}(\mathbf{r}) = -\nabla V(\mathbf{r}), \\
&= -\frac{1}{2}\epsilon_s \int \mathcal{E}(\mathbf{r}) \cdot \nabla V(\mathbf{r}) d^3\mathbf{r} \\
&= -\frac{1}{2}\epsilon_s \int \nabla \cdot [V(\mathbf{r})\mathcal{E}(\mathbf{r})] \, d^3\mathbf{r} \\
&\quad + \frac{1}{2}\epsilon_s \int V(\mathbf{r})\nabla \cdot \mathcal{E}(\mathbf{r}) d^3\mathbf{r} \\
&= \frac{1}{2}\sum_n V_n Q_n + \frac{1}{2} \int V(\mathbf{r})\rho(\mathbf{r}) d^3\mathbf{r}.
\end{aligned}
\tag{3.114}
$$

We have reduced the general problem of determining the electrostatic energy to two terms: one arising from the conducting bodies
as the sum of their potentials V_n and charge Q_n, and the second due
to the free electrons between the metallic conductors, as the integrated product of volume charge density and local potential, where
Poisson's equation $\nabla^2 V(\mathbf{r}) = -\rho(\mathbf{r})/\epsilon$ describes the local potential
that satisfies the boundary conditions imposed by the conductors of
V_n. A standard capacitor assumes no electron charge in the dielectric. Our use of the standard capacitance arises from the first term.
If our charge was an ionized dopant—immobile—then the second
term must be considered. This is what we employ for threshold voltage and for p/n junction's transition capacitance. If moving electron
charge is present, then again, this second term factors in determining
the electrostatic energy. This term then is associated with the diffusion capacitance in p/n junctions, bipolar transistors, and any other
structure where these moving electrons exist. Equation 3.114 can be

rewritten in various forms, for example, a form with a constant and position-dependent terms. The classic problem of a metal sphere, or a dielectric sphere in an electric field, for example, is reducible to this form. The last term then sets the implications through Gauss's law, while the first term is related to what happens on the conducting boundary conditions.

It is the second term that causes for us the problems arising from self-energy. In a non-metallic region,

$$\rho(\mathbf{r}) = \sum_i q_i\, \delta^3(\mathbf{r} - \mathbf{r}_i), \tag{3.115}$$

where q_i is the point discrete charge located at $\mathbf{r}_i = \mathbf{r}_i$. We have defined the potential as

$$V(\mathbf{r}) = \frac{1}{4\pi\epsilon} \int \sum_i \frac{q_i}{|\mathbf{r} - \mathbf{r}_i|}\, d^3\mathbf{r}. \tag{3.116}$$

At $\mathbf{r} = \mathbf{r}_i$, this term blows up. The reason we do not encounter this problem in much of electrical engineering is that we employ reference potentials. So, we only look at changes in potential energies. Even if it blows up, this referencing—subtraction—cancels the infinities out. If we are calculating absolute values, it is a problem. So, we always reference and that goes together with how we employ devices: there is always a return path to the energy source, either directly or through a common attachment to the ground—an infinite sink that is the thermodynamic reservoir for electronics. But, an additional problem is related to the averaging using the integral. The discrete charge is one that exists in limited numbers in nanoscale geometrics and it should not be included in calculating the potential if one is calculating the effect on this electron. So, we tackle this problem by modifying the potential to include only what arises as a result of other electrons. So, Equation 3.114 may be modified to the form

$$U_{es} = \frac{1}{2} \sum_n V_n Q_n + \frac{1}{2} \sum_i \int V_i(\mathbf{r})\rho_i(\mathbf{r}) d^3\mathbf{r}, \tag{3.117}$$

where $\rho_i(\mathbf{r}) = q_i \delta^3(\mathbf{r} - \mathbf{r}_i)$ as before is the charge density due to the ith charge, but

$$\nabla^2 V_i(\mathbf{r}) = -\sum_{j\neq i} \frac{\rho_j(\mathbf{r})}{\epsilon}. \tag{3.118}$$

Here, we are now excluding the ith particle, for which the energy contribution is being calculated. It sees the potential arising from all the rest. And this potential satisfies the same boundary conditions as $V(\mathbf{r})$ of Equation 3.114.

Our statement that a charge is only affected by the potential produced by other charges, can now be expanded to include in the definition of other charges the image charges on the electrodes of the

John David Jackson's *Classical Electrodynamics* is essential reading and an exercise for understanding and learning basic electromagnetism in detail. The breadth of mathematical techniques and intriguing complexities that it teaches is astounding—even in classical situations, and this is without having to worry about these self-energies.

other charges, as well as its own image charge. Its own image charge is included because it represents the interaction of the charge with the conducting boundary conditions and not with itself. Rigorously excluding any interaction of the charge with itself, now let us write the algorithm for determining the electrostatic energy as follows. The potential can be calculated as

$$V(\mathbf{r}) = V_{int}(\mathbf{r}) + V_{ext}(\mathbf{r}), \tag{3.119}$$

where the internal potential term V_{int} arises from the charge density $\rho(\mathbf{r})$ that resides internally in the system, and the external potential term V_{ext} arises from charges on the conductors and other sources outside the region where the potential is being calculated. With this modification, our electrostatic energy term can be rewritten as

$$U_{es} = \text{self-energy terms} + \frac{1}{2} \sum_{i<j} \frac{q_i q_j}{4\pi\epsilon |\mathbf{r}_i - \mathbf{r}_j|}$$
$$+ \frac{1}{2} \sum_i q_i V_{ext}(\mathbf{r}_i). \tag{3.120}$$

This form can be applied to the classical calculation for the single-electron problem. And it can be employed with quantum-mechanical detail using the wavefunction to describe charge's distribution internally, as well as the external consequences, so long as we reference and thus remove the first term, which is really not of interest in our low energy condensed matter problems.

3.4   Summary

This chapter reviewed thoughts related to the use of nanoscale phenomena and their devices, which have broad implications through their unique properties. We have been particularly interested in the device physics at the intersection of the applicability of classical and quantum-mechanical approaches—the nanoscale. Our first broad subject area of discussion was quantum computation. Starting with principles discussed in Chapter 1 and emphasizing the notion of reversibility in information and completeness in computation, we have discussed the notion of qbit—a superposition state—to be contrasted with that of cbit—the classical state—where the precise end result of a measurement, a logical 1 or a logical 0, can be foretold whether a measurement has been made or not. In a superposition state, the end result of a measurement can only be probabilistic: sometimes a logical 1 and sometimes a logical 0. The qbit state has an additional characteristic: it is a superposition state that may not be represented as a product of two states of the system from which it is constituted.

Such a state has entangled properties derived from the states of the system from which it is constituted as a continuous function. If one creates an entangled pair (*EPR*), a property (e.g., polarization) upon being determined by a measurement on one of the composing systems, causes this property of the composing systems to be also decomposable backward in time. Superposition allows computation employing qbits, absent measurement of intermediate steps, that is exponential in base 2 of the bits as opposed to the linear classical capacity. Entanglement and the *EPR* pair raise a number of issues that conflict with our classical and daily world notions which are intimately tied with locality—that cause and effect are the results of actions in the moment and space. An *EPR* pair, even if its components are separated in space in the classical sense, does not comprise two independent particles. They were connected back in time, and if, in the present moment, a property is determined for one of them, then the same property for the second particle is foretold. This is nonlocality. It does not depend on how far away that second particle is from the first, or on the speed of light. *EPR* provides a means to secure communications. Since quantum computation allows one to compute exploiting entangled forms of superposition, a number of difficult problems, including important ones of prime number factorization and of combinatorial search, may be solvable more rapidly than they are in classical computation.

We followed this discussion with an exploration of other phenomena of mesoscale. This included the Aharonov-Bohm effect, which represents both the wave nature of the electron and the nonlocality of potentials. This wave discussion, and its behavior in reduced scattering, and its relationship with quantum conductance allowed us to discuss the mesoscale aspects of the Hall effect. The integral quantum Hall effect results from the collapse of Landau states to integral states, together with the breakdown of the Ohm's law relationship. While the integral quantum Hall effect depends on inhomogeneities to bring about this collapse, when two-dimensional electron gas is exquisite, with nearly perfect characteristics of homogeneity and extremely low scattering, one observes the fractional quantum Hall effect. This is quite a new state of matter—a collective excitation where individual electron identity is lost and replaced by a collective behavior with fractional electron, Abelian and non-Abelian statistical responsse. Fermions are particles whose wavefunction is asymmetric under exchange, that is, $\psi(1,2) = -\psi(2,1)$, and bosons are particles that are symmetric under this exchange. In general, though, one may expect $\psi(1,2) = \exp(-i2\pi\phi)\psi(2,1)$, which behaves fermionically at $\phi = \pi$ and bosonically at $\phi = 0$—this is the set of Abelian particles. Non-Abelian states, such as the 5/2 state of the fractional quantum

Hall effect, are neither fermions nor bosons. Non-Abelian states are also non-commutative. This property protects them during unitary operations, so they may be braided to perform quantum computation while remaining immune to decoherence. In three-dimensional systems, this becomes possible in systems with specific spin-orbit interactions and crystal structures.

We extended this discussion to the mesoscale domain where, even when quantum effects are absent, interesting new behavior arises— single electron effects. We analyzed the importance of electrostatic charging in this, deriving the constraints under which the electron may be deemed localized and yet allowed to transport so that one may observe current and its absence under bias voltage. We emphasized here the importance of $e^2/2C_\Sigma$, the electrostatic charging energy of a nanoscale dot with a total capacitance C_Σ, and the charging energy gap—the change in electrochemical potential with an electron addition or removal, e^2/C_Σ. Current conductance oscillations and Coulomb blockade occur in simple two-terminal structures. Adding a gate electrode allows one to form a transistor where the gate bias voltage allows the modulation of conduction, or its absence. This behavior appears as a Coulomb diamond in the V_G-V_{DS} space. Since the presence of a single electron implies that a specific spin is present, spin characteristics too become important. The presence of an odd number of electrons, that is, when a specific spin is already present, or of an even number of electrons, and therefore freedom to the next electron, leads to a bimodal distribution in energy spacing. This spin property also connects to the observation of the Kondo effect arising from the interaction between a localized spin on the dot, and the delocalized electrons in the connecting leads. We ended this discussion with an extension of the electron transport in nanoscale dots to two dot systems, to emphasize the features related to the stability of the different states where an integral number of electrons stay in the dots. This allowed us to examine the flow of charge carriers in the linear region, where electrons flow from the source to the drain, and holes flow in the other direction, for specific gate bias voltage conditions.

These discussions of single and two dot systems, and our earlier discussion of phase transitions, are connected through the use of mean fields. So, we ended with a discussion of self-energy, emphasizing the changes in relationships needed to only include energetics arising from the interaction of the electron with its environment but not with itself.

3.5   *Concluding remarks and bibliographic notes*

PARTICLE-WAVE DUALITY and its multitude of manifestations—interactions of excitations between collective modes as well as between individual particles and the collective, and even the appearance of new collective condensate states where the Fermi gas or liquid picture entirely breaks down—comprise a broad topic that necessitates the use of a number of different ways to describe a bounded task of interest. Numerous device consequences follow from this—some are in science and engineering practice, and some are possibly in the distant future. But each of them, due to this particle-wave duality and its implications for the fundamental understanding of the state of matter and its use, has a very intriguing, challenging and joyful description.

Quantum computation and communication intrinsically depend on the wave nature with its quantum constraints. A gentle introduction with a physics bent, where the rough edges of this intersection of seemingly disparate disciplines of computing and quantum are dealt head-on is by David Mermin[1] in a thoughtful treatise. Williams[2] also explores this subject, including the communications and cryptography forays, in a more traditional language. For those interested in cryptography—a fascinating subject where quantum mechanics' properties seem to be particularly suitable—is tackled in depth by Bouwmeester et al.[3]. Towards the end of his life, Feynman too was interested in this subject. His lectures at Caltech are available in a very readable book form[4].

Bell states and the nature of questions that entanglement raises have occupied many of the best minds of our times. A particularly apt discussion for those intrigued by foundational questions—classical observation of a quantum system, which is the Landau-Lifshitz school of early quantum mechanics teaching or microscopic causality, et cetera—are beautifully debated by John Bell[5] himself.

On the question of the measurement, an early book by Wheeler and Zurek[6] is also a detailed exploration.

A superb discussion of the intersection of quantum and information is in the text by Peres[7].

Entanglement and the information discussion of Chapter 1 too have strong connections. How does one measure this non-latent content that entropy measures? This remains an open question, just as algorithmic entropy was a number of decades ago until elucidated by Kolmogorov. A similar question is related to the Bayesian approach to quantum mechanics, an approach which seemingly tackles some of the questions of observer-system or of mathematical time versus

[1] N. D. Mermin, "Quantum computer science: An introduction," Cambridge, ISBN-10 0-521-87658-3 (2007)

[2] C. P. Williams, "Explorations in quantum computing," Springer, ISBN 978-1-84628-886-9 (2011)

[3] D. Bouwmeester, A. Ekert and A. Zeilinger, "The physics of quantum information: Quantum cryptography, quantum teleportation and quantum computation," Springer, ISBN 3-540-66778-4 (2001)

[4] R. P. Feynman, "Feynman lectures on computation," Addision-Wesley, ISBN 0-201-48991-0 (1996)

[5] J. S. Bell, "Speakable and unspeakable in quantum mechanics," Cambridge, ISBN 0-521-81862-1 (2004)

[6] J. A. Wheeler and W. H. Zurek (eds), "Quantum theory and measurement," Princeton, ISBN 978-0691613161(1983)

[7] A. Peres, "Quantum theory: Concepts and methods," Kluwer, ISBN 0-792-33632-1 (2002)

the notion of time—present, past or future—of observers (humans or apparatus) much more clearly.

The book by Peshkin, a theorist, and Tonomura[8], an experimentalist, is a good in-depth description of the intriguing questions that the Aharonov-Bohm effect raises.

The Drude model and the classical Hall effect is discussed in nearly all of the solid-state physics and semiconductor physics texts. A very readable text for mesoscopic transport, providing an exploration of the wave-like behavior observed in the presence of a variety of interactions, is a small monograph by Imry[9]. Quantum transport and dissipation due to perturbation interactions is explored by Dittrich et al.[10]

For the integer and fractional quantum Hall effect, two very readable articles are from Thouless[11] and Halperin[12].

For an introductory exploration of the integer and the fractional quantum Hall effect, two texts, the first by Heinzel[13], and the second by Ihn[14], stand out. These are also good books for exploring a variety of mesoscopic effects and the electron transport therein.

An introductory article for exploring topological insulators is the one by Kane and Moore[15]. A more advanced discussion is by Hasan and Kane[16]. And the potential of using the non-Abelian states in quantum computation is articulated by Nayak et al.[17].

The single electron effect in nanoscale and quantum-confined structures is broadly explored in the edited text by Grabert and Devoret[18]. A more detailed discussion is to be found in the review paper by van der Wiel[19]. Ihn's book[20] looks at the Kondo effect in these single electron structures.

Röthig et al.[21] dwell in detail on the self-energy considerations, some of which we emphasized, in nanoscale structures along with those due to quantum considerations.

3.6 Exercises

1. Can a qbit be read without modifying it? **[S]**

2. The de Broglie paradox or puzzle, a twist on the Schrödinger's cat paradox, is the following. A closed box in Paris with reflecting interior walls contains a single particle. Without any attempt to locate the particle, a reflecting double partition is inserted, dividing the box into two. One of the two separated halves is sent off to Tokyo. It is indeterminate whether the box in Paris has the particle. An experiment performed in Tokyo allows one to determine if the particle is there or not. At the moment this question is resolved in Tokyo, Paris's indeterminacy is also gone. The exper-

[8] M. Peshkin and A. Tonomura, "The Aharonov-Bohm effect," Springer-Verlag, ISBN 3-540-51567-4 (1989)

[9] Y. Imry, "Introduction to mesoscopic physics," Oxford, ISBN 0-19-955269-X (2002)

[10] T. Dittrich, P. Hänggi, G.-L. Ingold, B. Kramer, G. Schön and W. Zerger, "Quantum transport and dissipation," Wiley-VCH, ISBN 3-527-29261-6 (1998)

[11] D. J. Thouless, M. Komoto, M. P. Nightingale and M den Nijs, "Quantized Hall conductance in a two-dimensional periodic potential," Physical Review Letters, **49**, 405–408 (1982)

[12] B. A. Halperin, "The quantized Hall effect," Scientific American, **254**, 40–48 (1986)

[13] T. Heinzel, "Mesoscoping electronics in solid state nanotstructures," Wiley-VCH, ISBN 978-3-527-40638-8 (2007)

[14] T. Ihn, "Semiconductor nanostructures: Quantum states and electronic transport," Oxford, ISBN 978-0-19-953442-5 (2010)

[15] C. Kane and J. Moore, "Topological insulators," Physics World, **24**, 32–35 (2011)

[16] M. Z. Hasan and C. L. Kane, "Colloquium: Topological insulators," Reviews of Modern Physics, **82**, 3045–3067 (2010)

[17] C. Nayak, S. H. Simon, A. Stern, M. Freedman and S. Das Sarma, "Non-Abelian anyons and topological quantum computation," Reviews of Modern Physics, **80**, 1083–1159 (2008)

[18] H. Grabert and M. H. Devoret (eds), "Single charge tunneling: Coulomb blockade phenomena in nanostructures," Plenum, ISBN 0-306-44229-9 (1992)

[19] W. G. van der Wiel, S. De Franceschi, J. M. Elzerman, T. Fujisawa, S. Tarucha and L. P. Kouwenhoven, "Electron transport through double quantum dots," Reviews of Modern Physics, **75**, 1–22 (2003)

[20] T. Ihn, "Semiconductor nanostructures: Quantum states and electronic transport," Oxford, ISBN 978-0-19-953442-5 (2010)

[21] C. Röthig, G. Schön and M. Vojta (eds), "CFN lectures on functional nanostructures, Volume 2: Nanoelectronics," Springer, ISBN 978-3-642-14375-5 (2011)

iment in Tokyo produced an instantaneous result in Paris. What gives? Explain.   [S]

3. If Anarkali or Balaji's parents wanted to find out what was the message exchanged between the two in the classical channel that Anarkali and Balaji are trying to protect, what do the parents have to do? Can Anarkali find out that the message has been intercepted?   [M]

4. Consider an epitaxial sheet of graphene that is on a 1000 nm-thick SiO_2 sheet, charged with a gate voltage 150 V and in a magnetic field in the range 8–12 T when measured at 1 K. What would be the (approximate) cyclotron resonance frequency? If the mobility is 10^4 $cm^2/V \cdot s$, would one expect to see well-defined quantum Hall plateaus in this region? If so, to which steps would it correspond?
    [M]

5. Which should be larger for electrons in semiconductors—the mean free path for energy relaxation, or the mean free path for momentum relaxation?   [S]

6. Consider spherical copper and $GaAs$ quantum dots that are 10 nm in diameter. Assume that they are in a SiO_2 medium and that interfaces are ideal. Calculate the Coulomb charging energy for each case. At what temperatures is the thermal energy equal to the charging energy?   [S]

Assume that the dot is spherical.

7. Zhu et al. derive the classical capacitance of a dielectric sphere with 1 and more electrons. Read it carefully. How does your result compare? Why does the Gaussian model underestimate the potential energy? What would be the implications of this reasoning if one were to consider the quantum-mechanical nature of the electron in a metal at these dimensions?   [S]

To see the Zhu et al. publication, see J Zhu, T. LaFave and R. Tsu, "Classical capacitance of few-electron dielectric spheres," Microelectronics Journal, 27, 1293–1296 (2006).

8. Let us apply a combination of the classical and the ballistic pictures in a metal—copper again. Calculate the time it takes for an electron to travel 10 nm in a nanowire of copper, assuming that the motion is ballistic. Justify why you chose the specific properties of that electron. How much would this time be in silicon going in the $\langle 100 \rangle$ direction on a (100) face for the same conditions? For this latter, has enough information been provided? If not, make a suitable assumption. How much time, an estimate, will it take for an electron to go through a 10 nm diameter single electron silicon transistor?   [S]

9. In Figure 3.50, which shows charging energy as a function of applied gate voltage in a single electron transistor structure, we

state that, at the half-integer number $\bar{n}$, which is achieved through the gate voltage, an electron can conduct through from source to drain. Explain, using a simple band diagram with the energy states in the source, the quantum dot and the drain, precisely what is happening energetically in a metal (or semiconductor) quantum dot at the $\mathscr{X}$ point in the figure to make this feasible.   [S]

10. In the presence of the Coulomb blockade effect, what is the shortest charging time constant one may obtain where the effect is directly observable? Just an estimate with reasoning will suffice.

   [S]

11. Why does an electron not feel any effects of its own charge on itself? A simple argument, please.   [S]

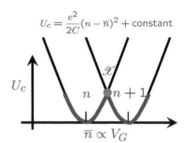

$$U_c = \frac{e^2}{2C}(n - \bar{n})^2 + \text{constant}$$

Figure 3.50: Charging energy as a function of normalized electrons on a quantum dot. n indicates the integer component of the normalized electrons count. $\bar{n}$ represents the polarization electron count that is modified through an applied bias.

4
Phase transitions and their devices

WHEN PARTS THAT ARE REASONABLY UNDERSTOOD, atoms, for example, are brought together into larger ensembles, such as a crystal, for example, new properties emerge. These properties cannot be understood or predicted as simple extrapolations of the properties of the constitutive elements—the atoms in this example. Such behavior in ensembles due to changes in the scale and the embedded complexity, pervades in nature. Bandgaps appear in a silicon crystal, which is indirect. Many compound semiconductors are direct. Some show a piezoelectric effect, such as *GaAs*, but some do not, such as silicon. Indeed, properties such as the Dirac point and relativistic electron behavior appear in carbon in its graphene crystal phase (Chapter 2): a collection of new properties, some of which we usually ascribe to the photon, such as absence of mass, and some to elementary charged particles at relativistic speeds. The appearance of these new properties is not a violation of microscopic equations of motion. They are a collective excitation response, a form of new rigid behavior from the ensemble as a whole, when scales and complexity change upon the assembling. The emergence of this new physical behavior occurs around the nanoscale, where the constitutive atomic world and the macroscopic world intersect, and it is important for understanding devices utilizing them. It is useful to have a perspective of these connections.

Some laws, such as Newton's, break down at the quantum or relativistic scale. Some laws, such as that of the conservation of energy, are more fundamental in the sense that we have not yet found a natural situation where they do not apply. This latter is an example of a fundamental principle guiding nature's behavior that manifests in the laws that we employ to explain phenomena at different scales and complexity.

What really happened at big bang is still quite unclear! Is there a negative energy?

Principles of symmetry guide most of the observations of laws of nature that we see—symmetries dictate that specific distinct pro-

Nanoscale device physics: Science and engineering fundamentals. Sandip Tiwari.
© Sandip Tiwari 2017. Published 2017 by Oxford University Press.

cesses should all proceed at the same rates. Symmetry, in this sense, is a principle of invariance. An observation does not change when we make changes in our point of view, such as via rotation or moving. Laws of nature too have symmetry. They remain the same when we change the point of view of observation in specific ways. Symmetry here is the set of ways that leave the laws unchanged. Newton's laws, for example, don't change with the changing of clocks or reference points of measurements, or with rotation of the reference frame. These Newtonian—invariance with time or rotation, among others— or Galilean—invariance with a constant velocity moving frame—or Lorentzian—the invariance of the speed of light, causing motion to shrink the length and slow down time, independent of the speed of the observer—perspectives are examples of symmetry at work.

Broken symmetry is an example of a fundamental physical principle that explains many observations of interest to us. Broken symmetry is an outcome that is not apparent in the physical state, observed even if the underlying equations that physically describe the behavior respect a symmetry. For example, equations governing the motion of a planet around the sun are spherically symmetric—a point-sun and point-planet, if they were the only objects, would be in a circular orbit. But, planets have elliptical orbits with the long axis pointed every which way. Any perturbations, even tides, cause these orbits to change. The long axis of Mercury's orbit swings at an angle of 2π about every 2250 centuries. These broken symmetry effects are ubiquitous—they can be found in particle physics and its strong and weak nuclear forces, in the current search for supersymmetry, from which these breakdowns presumably come about, and in condensed matter and its electromagnetic forces with quantum constraints, as well as in this example of the motion of heavenly bodies with gravitational forces.

We clarify this nature of symmetry and broken symmetry through a quantum example that is closer to our nanoscale interests. A stationary state must have the symmetry of its laws of motion. In quantum systems, stationary states are symmetrical or form from asymmetric states into a symmetric combination. A single particle in a parabolic potential well, such as in Figure 4.1(a), which shows a simple harmonic oscillator, has as its ground state a symmetrical eigenfunction $|\psi_0\rangle$ of eigenenergy E_0. The next excited state, E_1, results from a symmetric combination of the asymmetric eigenfunction $|\psi_1\rangle$ under the transformation $z \mapsto -z$.

We now change the energetics of this one particle problem by introducing a little perturbation that disturbs the eigenenergies of the system. Figure 4.1(b) shows a small symmetric barrier where the perturbation energy is less than the unperturbed system's $E_1 - E_0$

Philip Morrison paraphrases Leibniz thus: symmetry is related to the indiscernibility of differences. Once you walk into the hall of a Palladian building, you can't quite remember whether you turned left or right. It is the sameness of each face, edge or corner that makes an honest cube a good dice that falls with equal chance on each of its faces.

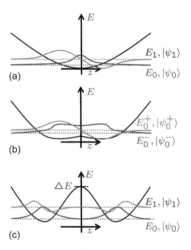

Figure 4.1: A hypothetical particle in a potential well, showing symmetry consequences in the presence of perturbation. (a) shows the two lowest energies and eigenfunctions of a symmetric potential well. (b) shows a symmetric two-well potential with a very small barrier in it together with the lowest two energies and eigenfunctions. Zero order approximation in this small perturbation makes even eigenfunctions lower in energy, and odd eigenfunctions higher. (c) shows the symmetric two-well potential with a large barrier in it, with its lowest two energies and eigenfunctions.

energy, and the width is shorter than the confinement scale. The example in Figure 4.1(a) has now become a system under excitation. Our solution must reduce to, in the zero order approximation, the unperturbed solution—a node-less, that is, even function that we had for the lowest energy in (a). The ground state is therefore even, and the excited state is an odd eigenfunction under symmetry. The particle can be found in either of the wells with equal probability. If the particle was observed in one of the wells at some instant, at a later time it may be observed in the other well. The small barrier makes the time scale of such transitions small. The ground state stationary solution is as shown in Figure 4.1(b), with equal probability at z and $-z$. Quantum processes allow the system to transition between states described by the eigenfunction $|\psi_0^-\rangle = (1/\sqrt{2})(|\psi^l\rangle + |\psi^r\rangle)$ with energy E_0^- and $|\psi_0^+\rangle = (1/\sqrt{2})(|\psi^l\rangle - |\psi^r\rangle)$ with energy E_0^+. The stationary state of Figure 4.1(b) has equal probability of being in either one of these states. It is an equal superposition of the asymmetrical states. If a large collection of such systems were observed, or a large number of observations made on the same system, it will show the particle to be equally likely in either of the wells.

Now if the barrier is very large, such as in Figure 4.1(c), with an energy $\Delta E \gtrsim E_1 - E_0$ energy of the unperturbed Figure 4.1(a) starting point, the single potential well of one-body problem has effectively become a double potential well problem. The particle will stay in whichever well it is found in for long times of the order of $\Delta E/\hbar$. These are two separate and distinguishable states, and they are asymmetrical. The stationary state of this condition is that of being on the left side or the right side and not a combination of the two. The system has achieved stability in a new configuration.

A broken symmetry has manifested itself with this larger perturbation where $\Delta E \gtrsim E_1 - E_0$. New properties will appear because of the large ``impenetrable'' barrier. Chiral molecules, molecules lacking mirror symmetry, that is, an $\mathbf{r} \nrightarrow -\mathbf{r}$ transformation, such as sugar, or lactic acid and countless others, are examples of stable molecules where the barrier between the stable configurations is large. They are stable under normal energetic perturbations. While the symmetries of the overall governing laws remain the same, the manifest symmetry is broken because of the nature of the perturbation-induced barrier. The properties of the different configurations of these molecules can be enormously different. Appendix K unifies several of the thoughts discussed here together with the mathematical approach of the description.

When a hydrogen atom binds with another hydrogen atom, a new stable system appears with less manifest symmetries while still being composed of hydrogen atoms. This composite system behaves

Chiral molecules, for example, are optically active and rotate plane-polarized light in one direction or the other. They are also biologically very active through their role in molecular recognition as a result of the asymmetric structural form. The enormity of these property differences was felt when thalidomide, whose one configuration has analgesic properties, was found to create severe birth defects when in another configuration. d-Ethambutol is used in the treatment of tuberculosis. i-Ethambutol causes blindness.

Weak nuclear forces have broken mirror symmetry, even as electromagnetic and strong nuclear forces obey mirror symmetry.

close to the description of Figure 4.1(a)—a result of being like the left or right well of Figure 4.1(c), whichever it starts in, under the large perturbation ΔE arising in the bonding.

Contrast this with the molecule ammonia. An ammonia molecule is a pyramid formed from three hydrogen atoms, with nitrogen at the apex, and it alternates between two configurations, with the nitrogen atom on either side of the hydrogen atom plane, dwelling in each for several picoseconds. It follows the description of Figure 4.1(b), with the nitrogen atom tunneling through the hydrogen molecule plane as the nitrogen atom undergoes an $\mathbf{r} \mapsto -\mathbf{r}$ transformation through the xy plane of the three hydrogen atoms. The magnitude of the perturbation, which is caused by the interaction energy, defines whether it is a description akin to that of Figure 4.1(b), a stationary state composed of two possible physical alternatives, or Figure 4.1(c), a stationary state that for all practical purposes has a new configuration.

Atoms, with their electrons and ions and their interactions in a crystal, are subject to these emergent behaviors as scales and complexities change. This is constructionism. There are of the order of 10^4–10^5 atoms in $10 \times 10 \times 10~nm^3$ of a crystal. The crystal itself has less manifest symmetry than do its constitutive atoms, that is, their ions and electrons, their nuclei and electrons, or, further down, the more reductive elementary particles and excitations of high energy physics. It will have new emergent properties.

The broken symmetry that occurs when a crystal is formed is translation.

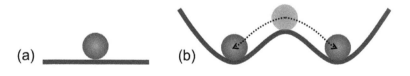

(a)   (b)

Figure 4.2: A ball placed on a flat horizontal surface in (a) stays in place through friction's restorative force. Place it in the center of raised dimple bowl as in (b), another system with its own symmetries, and it will fall into the bottom groove with the smallest of perturbations. The ensemble of ball and bowl now has less manifest symmetry.

In these examples, broken symmetry arises when the interaction energy has a deeper effect on how the collective ensemble composed of the constitutive elements responds. This is related to the energy perturbation and its relationship to the natural energies of the system. Gravitation has a stronger effect at the larger mass scale, such as in our macroscopic classical world. The ball in Figure 4.2 has manifest symmetries, such as rotational symmetries, that we can observe. On a flat surface, a ball sits in a stable position under small perturbation and aided by friction, that is, electromagnetic forces and dissipation to degrees of freedom of the surrounding environment in quantum limits, as a restoring force. So does a bowl, with a rotationally symmetric raised bottom and sides. But, a ball placed in the center of the bowl, such as shown in Figure 4.2(b), in response to a small perturbation, will drop into the bottom ring in an infinite number of positions on its perimeter. The ball and bowl system has less manifest symmetry, which results in a different collective behavior. At the nuclear scale, gravitation is weak. Parity violation, another broken symmetry, explains the observations where weak nuclear forces are important. Each of these, across a breadth of sizes and energies, shows the importance of broken symmetry.

A particularly interesting solid-state example of broken symmetry, and apropos to a discussion later on in this chapter, is of Peierls instability. Consider a linear chain of atoms, as shown in Figure 4.3(a), and a distorted form with displacement, as shown in Figure 4.3(b). The arrangement in Figure 4.3(a) has N unit cells a apart. The arrangement in Figure 4.3(b) has $N/2$ unit cells $2a$ apart. The energy band picture of the first will consist of a band in the first Brillouin zone bounded by $k = \pm\pi/a$, with states contributed by the N unit cells. The energy band of the second will consist of a band that will have its first Brillouin zone extending over $\pm\pi/(2a)$, with $N/2$ states, and the second Brillouin zone extending beyond to $\pm\pi/a$, with another $N/2$ states. Between these two bands will be a bandgap arising from the perturbation and interference from the periodicity. A symmetry broke, and this new bandgap appeared. This effect is seen in crystals, but particularly so in quasi-linear chain molecules.

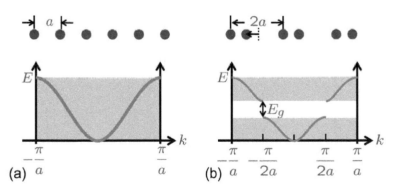

Parity, mirror symmetry and, chirality are used by different communities to describe the same phenomenon. Parity and mirror symmetry indicate invariance under spatial inversion. Chirality is the left-handedness and right-handedness of threading. A chiral molecule's mirror image cannot be superimposed on itself. When we view ourselves in the mirror, we see ourselves under the transformation left to right. But it is left to right of us with us behind the mirror, not vertically inverted. This too is an example of broken symmetry. The person behind is virtual. If real, this ensemble of two would have some different emergent properties. "Mirror mirror on the wall. Who is the fairest one of all?" so the story goes in "Snow White and the seven dwarfs." An interesting question: Why don't we also see ourself upside down in the mirror? The answer is in the transformations that our brain applies through its circuitry. What we see is our brain's interpretation of what arrives from the eye post-detection.

Figure 4.3: (a) shows a linear chain of atoms with N unit cells with atoms spaced a apart. (b) shows an arrangement where alternate atoms are closer together, and the unit cell spacing is $2a$, with $N/2$ unit cells. (a) and (b) also show the corresponding E-k diagrams.

The connection of symmetry to the physical and dynamic effect is tied through Noether's first theorem. This theorem, proved in 1915, shows that any differentiable symmetry of action of a physical system can be translated into a conservation law. Time translation symmetry leads to conservation of energy, space translation to conservation of momentum, and rotation to conservation of angular momentum. In all these, action as a time integral of the Lagrangian function leads to the system behavior through the principle of least action.

Amalie Noether, a German mathematician, was among the earliest and major contributor to the development of abstract algebra with contributions to topology. This theorem exemplifies what Eugene Wigner termed "the unreasonable effectiveness of mathematics." It is on par with any of the most significant insights from science.

See Appendix M for a summary discussion of action, and S. Tiwari, "Quantum, statistical and information mechanics: A unified introduction," Electroscience 1, Oxford University Press, ISBN 978-0-19-875985-0 (forthcoming), for more definitive arguments.

There is another interesting consequence related to broken symmetry known as the Goldstone theorem. The theorem states that a broken continuous symmetry in the presence of short-range interactions results in a collective excitation mode that has no gap, that is, the energy dispersion curve remains continuous even at zero energy. A classic example of this is acoustic phonons when continuous rotational symmetry is broken in the formation of a crystal. This collective mode at wavevector $\mathbf{q} \to 0$ is describing a slow transition from one state to another. A gapless energy spectrum also becomes important in spin waves, the equivalent of acoustic waves for spin. Goldstone's theorem in a Bose-Einstein condensate creates a specific mode that particle physicists associate with the Higgs boson. In ferroelectric transition—spontaneous electric polarization arising from translational symmetry breakdown—this can lead to launching of antiferroelectric domains. These are the soft modes of the gapless transition.

Our interest in this subject is because of it's importance to phase transitions. The phase transitions that we encounter earliest in our studies are those that water undergoes when progressing through the gaseous, liquid and solid phases. Steam, when condensed to liquid water, is less symmetric and has short-range order. The water-to-ice transition is another step in symmetry reduction, with more order through bonding. The system is not really violating the symmetry of space and time. In its collective response, under any excitation, the system finds it energetically more favorable to have a set of fixed relationships within, and responds as a whole to the stimulation. Phase transitions are examples of such singular and sharp changes in which the manifest microscopic symmetries reduce.

For the solid state, from which we make our nanoscale devices, these broken symmetries occur with major changes in the behavior of the solid state. In condensed matter, ferroelectricity, superconductivity, antiferromagnetism, et cetera, are all examples of these changes that have taken considerable theoretical effort to explain. It took several decades before the physics underlying superconductivity was understood. The physics underlying high temperature superconductivity, which was first observed in cuprates in 1986, remains a mystery in 2016.

In this chapter, we will explore these phase transitions and their use in devices. The four examples that we will dwell on in some depth are:

- ferroelectricity arising from spontaneous electric polarization,

- conductivity transitions arising from electron-electron and electron-phonon coupling,

In world history, one might even view conquests and revolutions—events with a broken symmetry, with their results in the long memory of those who suffered—as a Goldstone mode.

The English word ``ambivalent'' describes the strong contradiction arising from having strong inclinations in opposite directions. Phase transitions are just that. The difficulty in the mathematical treatment of this is that the number of interactions of all the elements within—the equations governing their behavior, the variables and the constraints—are at an Avogadro scale. So, one has to find ways to exact solutions that are manageable. This is possible, as many of the underlying behavioral effects zero out and the essentials remain. This is renormalization, which is pervasive, and its techniques are useful in many different problems that are inherently complex because of the number of interactions. The problem of satisfiability (SAT) when there are a large number of equations with a large number of variables and constraints (recall that all computable problems can be recast in this form, from our discussion in Chapter 1) has benefited immensely from this and underlies the area known as ``belief propagation.''

Broken symmetry, emergent phenomena, reductionism and constructionism are all quite loaded terms that bring out religious fervor among practitioners. Many-body theorists and particle physicists employ the terms in different ways with changing definitions and boundaries. P. Anderson, in ``More is different,'' Science, **177**, 393–396 (1972), articulates the power of constructionist thought tied to broken symmetry. But, the angst of the inter- and intra-community debate is evident through R. Peierls's ``Spontaneously broken symmetries,'' Journal of Physics A, **24** 5273 (1991) and the letters debate in Physics Today, **44**, pp. 13–15 and p. 118. All dwell on ferromagnetism—is it an example of broken symmetry or not?— as do R. Laughlin and D. Pines in ``The theory of everything,'' Proceedings of the National Academy of Sciences, **97**, 28–31 (2000). Laughlin's book, ``A different universe,'' Perseus, ISBN 0-465-03829-8 (2005), is highly recommended to inquiring minds.

- magnetism, arising from spontaneous magnetic polarization, and

- amorphous–crystalline transitions arising from structure. These changes are stable, maintain stability at the nanoscale, are manipulable by electrical and optical stimulation and therefore are interesting in the devices that we will explore.

4.1   Phase transitions

BROKEN SYMMETRY RESULTS in thermodynamic phase transitions. The change in the material's characteristics, its order, is described by a field which is the order parameter field, or order parameter, in short.

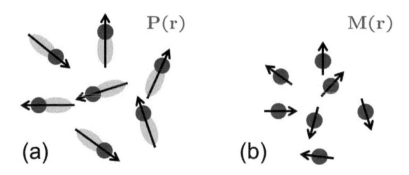

Figure 4.4: Electric and magnetic polarization as the order parameter field can be viewed as the vector fields in space, that is, $\mathbf{P}(\mathbf{r})$—a polar vector—and $\mathbf{M}(\mathbf{r})$—an axial vector. Each can change directions spatially; the magnitude locally will be quite constant, but a random distribution can result in a vanishingly small order parameter.

Spontaneous electric polarization and spontaneous magnetic polarization, that is, magnetization, as shown in Figure 4.4, are examples of these order parameters for ferroelectric and ferromagnetic materials, respectively. Spontaneous polarization is a stable polarization that appears and exists whether an external field is applied or absent. Dielectrics have polarization too. But it is not spontaneous. It appears in the presence of an external field. In its linear limit at the fields we usually employ, the displacement field $\mathbf{D} = \epsilon_0 \mathcal{E} + \mathbf{P} = \epsilon_0 \mathcal{E} + \chi \mathcal{E} = \epsilon_0 \epsilon_r \mathcal{E}$, where $\epsilon = \epsilon_0 \epsilon_r$ is the permittivity of the dielectric, and $\epsilon_r = 1 + \chi$ is the dielectric constant, where χ is the susceptibility. This polarization arises because even if atoms may be neutral, electrons' response and ions' response to forces are different, and when one makes ensembles of atoms, the interactions' response manifests itself. Silicon and its amorphous compound SiO_2 are both dielectrics and have different permittivities. Electron mass and ion mass are different, so there are frequency dependences reflecting their ability to respond to electric forces and interactions be-

Both electric and magnetic polarization are expressed in moments per unit volume. In the case of electric dipole moment, the unit is charge per area, where one length dimension cancels. The absence of a magnetic monopole, though, is circumvented for purposes of easy visualization and mathematical reduction by introducing a magnetic charge. The Système international d'unités also brings a little twist, which we will discuss later.

The silicon transistor community calls material that has a dielectric constant that is higher or lower than that of SiO_2 ($\epsilon_r = 4.2$) at a frequency of 1–10 GHz "high k" or "low k," respectively, sometimes writing these terms as "high κ" and "low κ," that is, $\epsilon_r \equiv k \equiv \kappa$.

tween them. The different natures of bonding, periodicity and other interactions mean that different dielectrics have quite different polarization responses. Even SiO_2 itself in its crystalline form of quartz is a piezoelectric—a material in which application of stress breaks the rotational symmetry. Stress causes net charge to appear on the two surfaces, with each unit cell contributing a net dipole. These order parameters describe both the state of the material and the change that took place when a phase transition occurred.

Quartz oscillators exploit this ability for frequency and time measurements and numerous other uses such as thickness measurement during deposition in vacuum systems.

Phase	Broken symmetry	Order parameter	Example
Crystal	Translation	Density, $\rho(\mathbf{r})$ $\rho(\mathbf{r})$	Si
Ferroelectric	Inversion	Polarization $\mathbf{p}(\mathbf{r})$	$PbTiO_3$
Antiferroelectric	Inversion	Staggered polarization $\mathbf{p}(\mathbf{r})$	$PbZr_{1-x}Sn_xO_3$ $0.05 < x < 0.3$
Ferromagnet	Time and spin	Magnetization $\mathbf{m}(\mathbf{r})$	Fe
Antiferromagnet	Time and spin	Staggered magnetization $\mathbf{m}(\mathbf{r})$	NiO
Superconductor	Gauge	Pair amplitude $\langle \psi(\mathbf{r})\psi(\mathbf{r}') \rangle$	Nb
Liquid crystal	Rotation	Director $d(\mathbf{r})$	$C_{34}H_{50}O_2$ Cholesteryl benzoate

Table 4.1: Examples of broken symmetries commonly employed in electronics. The liquid crystal employs molecules that are elongated. Director is the preferred orientation of the long direction of molecules. The gauge for the low temperature superconductors is the phase in the vector potential as the broken symmetry. Superfluidity—as an equivalent of superconductivity—has phase rotation as the broken symmetry, and the amplitude of the condensate as the order parameter. Hundreds, if not more, phases of matter, as well as the inhomogeneity of the universe, can be traced to different broken symmetries.

A broken symmetry caused this change in order and appearance of phase. Some examples of phases of interest in electronic devices are listed in the Table 4.1. The broken symmetry for electric polarization is inversion ($\mathbf{r} \mapsto -\mathbf{r}$) since an inversion also causes an inversion of the polarization. The time-reversal symmetry breakdown for magnetization can be understood by imagining a circularly moving charge played backward, that is, in $t \mapsto -t$. This is moving in the opposite direction, causing the magnetization to be also reversed. In both cases, a symmetry transformation of the symmetry group causes the polarization to be reversed. There are hundreds of different phases of matter arising from the different broken symmetries, and we would surely discover many more in time. Examples from the past decades include the high temperature superconductivity, the fractional quantum Hall effect and the Bose-Einstein condensate.

The thermodynamic discussion in the earlier chapters underscored the importance of variances. Fluctuations cause these variances in those examples. And we have also observed how larger errors caused by higher disorder are associated with higher temperatures in ensembles.

In the case of an ensemble of particles, and their collective behavior, one would expect ordering to appear with decreasing temperature. We now explore this through a phenomenological approach, due to Landau and subsequently expanded to many systems by others. The Landau phenomenological approach employs the energy description underlying the behavior of the ensemble to extract the order parameters and to classify the phase transitions that the different behaviors of ordering take. We posit that the broken symmetry at the transition from a state of high symmetry to a state of lower symmetry is such that any expansion of the terms contributing to energy remain valid above and below the temperature that characterizes this broken symmetry—T_c, a critical temperature. Any expression must also be invariant under symmetry operations of both phases. The state of the system in equilibrium follows from the minimum in thermodynamic potential. If the system is under applied stress, such as from pressure or mechanical, and an applied electric field $\mathcal{E}$, then, the free energy—the Gibbs free energy—is

$$
\begin{aligned}
\mathcal{G} &= U - TS - \sigma u - \mathcal{E} \cdot \mathbf{D} \\
&= \mathcal{F} - \sigma u - \mathcal{E} \cdot \mathbf{D} \\
&= \mathcal{H} - TS,
\end{aligned} \tag{4.1}
$$

where U is the internal energy, $\mathcal{F}$ is the Helmholtz free energy ($\mathcal{F} = U - TS$), $\mathcal{H}$ is the enthalpy ($\mathcal{H} = U - \sigma u - \mathcal{E} \cdot \mathbf{D}$), σ is the stress, and u the deformation. In this expression, stress and deformation—related to pressure and volume and electrostatic as a part of the electrochemical process that all electronic devices rely on—are included as energy reposed. The phase transition arises from the competition between the different forms that the energy exists in, and the system striving to achieve its lowest energy state. The order parameter comprises the physical, macroscopic and microscopic parameters that appear in the lower symmetry phase. This order parameter will depend on temperature as well as on the other variables that affect the energy of the system.

A phase transition is of nth order if there is a discontinuity in the nth derivative of the Gibbs free energy, with temperature at a specific temperature that we will call the critical temperature T_c. Second order phase transition has a discontinuity in the second derivative with temperature of Gibbs free energy. First order phase transition has this discontinuity in the first derivative with temperature. The

Recall the discussion in Chapter 1. Internal energy U is the energy required to create a system in the absence of temperature or volume changes. So, it includes the various forms—electrostatic, electromagnetic, chemical, et cetera. From the environment at temperature T, some energy transfers spontaneously cause a change in the entropy S of the system. The Helmholtz free energy $\mathcal{F} = U - TS$ is the available energy of this higher disordered state—it excludes the entropic contribution. The Gibbs free energy $\mathcal{G} = U - TS + PV$ is the energy acquired by the work PV needed for the system to occupy volume V at temperature T. Enthalpy, $\mathcal{H} = U + PV$, is the energy change associated with internal energy and the work done by the system. It is useful when a system causes thermal changes through energy release or absorption.

In the usual undergraduate formulation of this equation, pressure and volume are the terms that are represented in the mechanical deformation form here (σu), and the electrical energy term is ignored since the problems are of thermomechanics. More generally, when other forms of energy are also present, such as $\mathbf{H} \cdot \mathbf{B}$ for magnetic, they too need to be incorporated.

consequence is that the order parameter will be continuous in the second order and discontinuous in the first order transition.

The stable structure is the state which minimizes the Gibbs free energy of Equation 4.1. If the mechanical boundary is stress-free—no pressure or no mechanical stress arising from films with different properties—then $\sigma = 0$, if the external field is absent, that is, $\mathcal{E} = 0$, and if other energy contributions that are relevant but not included in the equation are also absent, then the equation reduces to the minimization of $\mathcal{G} = U - TS$.

The electric and magnetic polarization dipoles in Figure 4.4 are meant to be randomly oriented, that is, $\langle \mathbf{P} \rangle = (1/N)\sum_i \mathbf{p}_i$, and $\langle \mathbf{m} \rangle = (1/N)\sum_i \mathbf{m}_i$, are vanishingly small. In these expressions, the microscopic electric polarization is $\mathbf{p} = q'\mathbf{\Delta z}'$, the dipole resulting from a displacement of charge q' by a distance $\mathbf{\Delta z}'$, and the microscopic magnetic polarization for quantum particles is $\mathbf{m} = -g_{s,l}\mu_B m_{s,l}$, where μ_B is the Bohr magneton, $g_{s,l}$ is the g factor, and $m_{s,l}$ the spin or azimuthal/orbital quantum number. The magnetic polarization can arise from these orbital and spin sources. The macroscopic polarizations follow from this. When a ferroelectric phase forms, the polarization from a unit cell would be $P_z = (1/\Omega_0)Z^*e\Delta z'$, where Ω_0 is the unit cell volume, and $\Delta z'$ the displacement of charge Z^*e. In a dielectric or a paramagnetic material, change in temperature does not bring out any spontaneous ordering. But, in others, such as ferromagnet iron or ferroelectric $PbTiO_3$, they become finite below the critical temperature T_c. Slow spatial variations and fluctuations within, as shown in Figure 4.5, should be expected because of energetics and structure. The Landau approach approximates this. It doesn't consider the fluctuations, thus focusing on the average expectation values of the order parameters. The slow spatial variations can be incorporated within a mean description. We will discuss this later and also make remarks on the limitations of the approach. Its power is in its general capabilities for analyzing a variety of phase transitions.

What one observes during the phase transition is that the order parameter, which we will denote as $\eta(\mathbf{r})$ in general, undergoes either a discontinuous or a continuous change as shown in Figure 4.6. The ice–water transition, with density as the order parameter, is an abrupt transition. The volume increases when the ice is formed. The amorphous–crystalline phase transition of phase change memory too is an abrupt transition with density change. These abrupt transitions are first order transitions aligned with this discontinuity. Second order phase transitions are continuous in the order parameter. The discontinuity is now in the derivative of the order parameter. Ferromagnetic phase transitions are of the second order. These transitions

This classification is a vestige of history—due to Paul Ehrenfest—and is, even if originating from one of the great masters of science's flowering in the early 20th century, outdated. We now understand much more about the fluctuations and complexities of different phase transitions, which cannot be simplified to simple differentials. In phase transitions we see a myriad of complexities—weak transitions, points of multiple criticality, and many different sources of interactions contributing to energy.

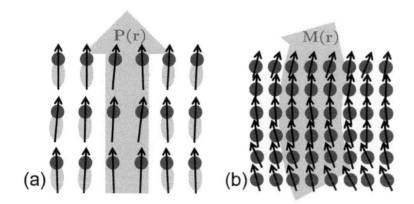

Figure 4.5: Order parameters fluctuate and may also have a slowly varying change due to the energetics of the system and inhomogeneity of the structure. (a) shows the fluctuations in a ferroelectric with its microscopic charge dipoles and (b) shows the fluctuations and a spatial variation in two dimensions for ferromagnetism.

occur at a specific temperature, T_c.

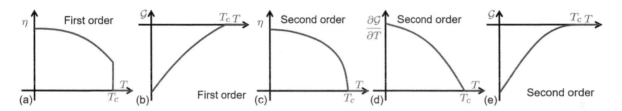

Figure 4.6: An abrupt phase transition—a first order phase transition with the order parameter η undergoing a discontinuous change at a temperature T_c—is shown in (a). (b) shows a corresponding Gibbs free energy change. (c) through (e) show characteristics in second order transitions where $\mathcal{G}$ is continuous through the critical temperature T_c. (c) shows the order parameter continuity, (d) shows the derivative of Gibbs free energy with temperature, and (e) shows the Gibbs free energy for the second order phase transition.

Whether first order or second order, the order parameter changes from a null value to a non-zero value upon the phase transition. The key ideas of the Landau phenomenological theory is that the free energy can be expressed as a power series of the relevant order parameter, that the expansion around the high symmetry phase is valid above and below the transition temperature and that the expression is invariant under symmetry operations of the phases.

Let us first consider a homogeneous and continuous phase transition, so the second order transition. Since the order parameter is small near T_c in the second order phase transition, one can expand the free energy (Gibbs, which is function of pressure (p), temperature and the order parameter) in a Taylor series:

$$\mathcal{G}(p, T) = \mathcal{G}_0 + \alpha\eta + \mathcal{A}\eta^2 + \mathcal{C}\eta^3 + \mathcal{B}\eta^4 + \cdots. \qquad (4.2)$$

For $T > T_c$, the state is disordered, that is, $\eta = 0$, and for $T < T_c$, η is finite and small. Absent external fields, this requires $\alpha = 0$.

T_c historically has been called the Curie temperature. Its origin is in magnetism, as Pierre Curie showed that magnetism disappears above a critical temperature. We will refer to this as either critical temperature, for its general validity, that is, utility in other phase transitions, or the Curie temperature, depending on context.

If the order parameter is a function of position, that is, $\eta(\mathbf{r})$, as it may be, due to surface, interface, grain boundary or other interactions, we will have to qualify our treatment a bit. This is an inhomogeneous situation. Formation of domains in ferroelectric and ferromagnetic systems, where the polarizations may be pointed in different directions, also means inhomogeneity will be important.

$\alpha \neq 0$ implies the absence of a free energy minimum since finite η existing at all temperatures would imply no minimum at $\eta = 0$ for disordered systems for $T > T_c$. The square term implies $\mathcal{A}(p, T) > 0$ for $T > T_c$ and $\mathcal{A}(p, T) < 0$ for $T < T_c$. At the critical temperature T_c, $\mathcal{A}(p, T)$, a continuous function, reverses sign. This leads to a free energy change of the type shown in Figure 4.7. For the next term, an odd power in the order, again we will assume that $\mathcal{C} = 0$. In the nature of the Taylor expansion, $\mathcal{B} > 0$. Each of these higher order terms is asymptotically smaller so that a finite free energy exists. Minimizing the free energy w.r.t. the order parameter then leads to

$$\frac{\partial \mathcal{G}}{\partial \eta} = 2\mathcal{A}\eta + 4\mathcal{B}\eta^3 = 0, \quad \text{with}$$

$$\frac{\partial^2 \mathcal{G}}{\partial \eta^2} = 2\mathcal{A} + 12\mathcal{B}\eta^2 > 0. \tag{4.3}$$

Near $T = T_c$, we expand the Gibbs free energy's second power's coefficient in the first linear term of Taylor expansion as

$$\mathcal{A}(p, T) = a(T - T_c). \tag{4.4}$$

The solution near the critical temperature is

$$\eta = 0 \quad \text{for } T > T_c, \quad \text{and}$$

$$\eta = \pm\left(\frac{1}{2}\frac{\mathcal{A}}{\mathcal{B}}\right)^{1/2} = \pm\left(\frac{1}{2}\frac{a}{\mathcal{B}}\right)^{1/2}(T_c - T)^{1/2}$$
$$\text{for } T < T_c, \tag{4.5}$$

which is the nature of the dependence shown in Figure 4.7.

If the system ordering is symmetric, odd power terms will be zero. So, polar opposite dipole polarization and magnetization are equivalent, and the free energy can only have even order terms in the order parameter field. A small $\mathcal{C}$ can still allow the minimum to exist, given the magnitude of the lower second power term, and these minima may exist and be allowed by symmetry for other situations. For now, we are assuming $\mathcal{C}$ to be vanishingly small.

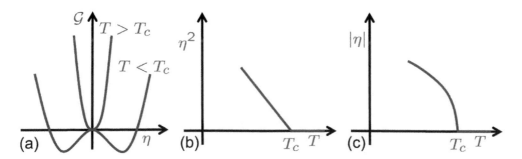

Figure 4.7: Gibbs free energy (a), the order parameter squared (b) and the order parameter for a second order homogeneous phase transition (c), showing the nature of changes above and below the critical temperature T_c.

The Gibbs free energy, which at $T > T_c$ has a minimum at $\eta = 0$, below T_c, achieves two finite order parameter fields either side of where the two new minima occur in thermal equilibrium. This similar behavior applies, at constant volume, to the Helmholtz free energy $\mathcal{F}$ if second order phase transition occurs. Since the derivative of the free energy is a linear function of η, and it disappears at $T > T_c$, the derivative of the free energy is discontinuous at the critical temperature.

This simple picture has several important implications. The entropy of the system due to order parameter change, in constant stress or pressure conditions, is

$$
\begin{aligned}
S = -\frac{\partial \mathcal{G}}{\partial T} &= S_0 \text{ for } T > T_c, \text{ and} \\
&= S_0 - a\eta^2 = S_0 - \frac{a^2}{2B}(T_c - T) \\
&\text{for } T < T_c,
\end{aligned}
\tag{4.6}
$$

where $S_0(T)$ is the entropy of the system above the critical temperature, and this behavior, derived from our linear term approximation, is valid in the vicinity of the critical temperature. The entropy is a continuous function at the critical temperature for second order transitions. Various parameters flow from this entropy. For example, the constant pressure specific heat of the material, arising from the order parameter and denoted c_p, is equal to $T \left(\partial S / \partial T\right)|_p = a^2 T / 2B$ for $T < T_c$ and vanishes above T_c. The entropy undergoes a kink at T_c while being otherwise continuous, and it causes a jump in the specific heat of $a^2 T_c / 2B$ due to order at the critical temperature. The measurement of the change in entropy near the transition provides us the information to determine the degree of freedom participating in the ordering process. The measurement of heat capacity, so from latent heat, gives us this change in entropy and the existence of this transition.

An example of a second order phase transition is ferroelectricity arising from spontaneous polarization. Figure 4.8 shows its appearance at $T < T_c$ in titanates such as $PbTiO_3$, where it is quite prevalent because of the electronic nature of Ti—its size, electrons in the d orbitals, and bonding and bonding energy with O, which surrounds it in an octahedral arrangement on cubic face centers. The figure shows an example where the movement of a stable ion, in a non-centrosymmetric crystal, leads to a polarization of $(0, 0, \pm \mathbf{P}_z)$, where z direction is chosen to align with the direction of the order parameter. The spontaneous polarization causes a small relaxation of $\eta = (\eta_{11}, \eta_{11}, \eta_{33}, 0, 0, 0)$ in the unit cell under both short- and long-range interactions. In this example of ferroelectricity, one can determine the dielectric susceptibility that quantifies the polarization. The free energy to be minimized, with variations arising from polarization changes, is

$$
\mathcal{G} = \mathcal{G}^\star - \boldsymbol{\mathcal{E}} \cdot \mathbf{D} = \mathcal{G}' + \mathcal{A}P^2 + \mathcal{B}P^4 - \epsilon_0 \boldsymbol{\mathcal{E}} \cdot \boldsymbol{\mathcal{E}} - \boldsymbol{\mathcal{E}} \cdot \mathbf{P}.
\tag{4.7}
$$

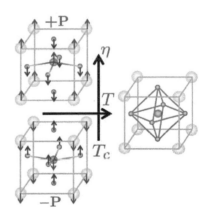

Figure 4.8: The movement of an ion, here of Ti, in an octahedral arrangement of oxygen ions, leading to ferroelectricity with two opposite spontaneous polarizations in a cell. Polarization is the order parameter. The picture is simplified since it does not show the larger crystal structure where additional short- and long-range effects will happen.

Any size-stress related changes, being weaker, are ignored in this expression. The spontaneous polarization minimizes the energy when

$$\frac{\partial \mathcal{G}}{\partial P} = 2\mathcal{A}P + 4\mathcal{B}P^3 - \mathcal{E} = 0. \tag{4.8}$$

The order parameter, spontaneous polarization, with fields aligned to polarization follows from this as

$$
\begin{aligned}
P &= 0 \text{ for } T > T_c, \text{ and} \\
P &= \pm \left(\frac{a}{2\mathcal{B}}\right)^{1/2} (T_c - T)^{1/2} \text{ for } T < T_c. \tag{4.9}
\end{aligned}
$$

The electric susceptibility χ is related as

$$\chi = \frac{1}{\epsilon_0} \frac{\partial P}{\partial E}, \tag{4.10}$$

which reduces this equation to the form

$$\epsilon_0 \chi 2(\mathcal{A} + 6\mathcal{B}P^2) - 1 = 0 \ \therefore \ \chi = \frac{1}{2\epsilon_0} \frac{1}{a(T - T_c) + 6\mathcal{B}P^2}. \tag{4.11}$$

Near the critical temperature, the dielectric susceptibility is

$$
\begin{aligned}
\frac{1}{\chi} &= 2a\epsilon_0(T - T_c) \text{ for } T > T_c, \text{ and} \\
\frac{1}{\chi} &= 4a\epsilon_0(T_c - T) \text{ for } T < T_c. \tag{4.12}
\end{aligned}
$$

The dielectric susceptibility undergoes a divergence at the critical temperature of phase transitions, as seen in Figure 4.9 for $PbTiO_3$.

We now turn to first order phase transitions. Consider a weaker transition where there exists the cubic invariant in free energy, that is, $\mathcal{C}$ is non-zero. The odd power should lead us to a sign reversal and hence a discontinuity. The relationship can be analyzed in the same way as before, that is, finding the minimum in free energy as a function of the order parameter by determining the polynomial equation where the derivative vanishes.

One can see this behavior, which is more interesting, in Figure 4.10 with $\mathcal{C} < 0$. When the temperature is high, with the second order term dominating, there is one minimum, as in the curve marked 1 in Figure 4.10(a). With a decrease in temperature, due to the coefficient $\mathcal{A}$, which was linearized in the case of a second order transition, the behavior will have changes of the type shown in Figure 4.10(a). In the curve marked 2, an additional minimum appears at the temperature T'. This is a metastable high energy state with a local energy minimum. The lowest minimum, though, is further away in temperature and in order parameter. The second order term is still positive, with a temperature greater than the critical temperature. What this says

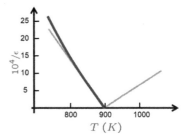

Figure 4.9: The inverse of dielectric susceptibility near the critical temperature (860–900 K) for the ferroelectric $PbTiO_3$—a second order phase transition. The theoretical estimate is shown for below and above the critical point, together with an experimental measurement.

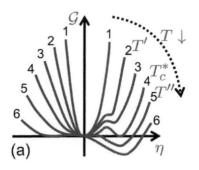
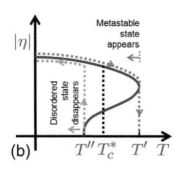

Figure 4.10: The free energy change with an order parameter in (a), and the order parameter change with temperature in (b) for a first order phase transition with a weak third order term ($C < 0$). (a) shows Gibbs free energy, and (b) shows the order parameter, which may take two oppositely signed values.

is that there is some probability that it is possible for the system to jump to this minimum and thus exhibit what appears to be a first order phase transition. At the temperature T_c^*, for the curve marked 4, there are now two minimums of identical energy—at the zero and at a finite order parameter. A small perturbation to a lower temperature makes the non-zero order parameter minimum a global minimum. The order parameter jumps. This transition therefore is of first order. But, it is a weak one in the sense that a local minimum still exists at the magnitude of zero in the order parameter. Only at further lowering of temperature to $T = T''$ does the second order term change sign and this minimum descend rapidly as the only allowed order parameter. This characteristic of order parameter is shown in Figure 4.10(b). If one cools down from a high temperature, while a minimum does appear at $T = T'$ and there are two minima at $T = T_c^*$, only at T'' does the system have the ordered state as the unique minimum. A system in an unordered state remains in the local unordered state till $T = T''$, at which temperature it jumps to an ordered state. Similarly, if one were to heat the system from a low temperature ordered state, so at a finite order parameter, that minimum only disappears at $T = T'$, at which point the system would jump, that is have a first order transition to the disordered state. T' and T'' are spinodal points. The change in order parameter shows hysteresis. The first order change takes place at two different temperatures.

These two temperatures, T' and T'', are temperatures where $\mathcal{A}$ changes sign, a local and metastable minimum appears for the first, and a global degenerate minimum appears for the latter. T' and T'' are singular points, while T_c^* is a point of degenerate energy minimum. The temperatures T' and T'' are more significant in that these singularities are unique—parameters tied to the derivative of free energy will diverge at these temperatures, but not at $T = T_c^*$. Susceptibility will diverge at these distinctly separated temperatures.

Superheating and supercooling phenomena, such as of water, are examples of this weak first order transition. It also appears in many other examples of broken symmetry.

If they can be moved by other parameters on which the free energy depends, such as pressure in this example, then the hysteresis can be modulated. We will see that electrostatic energy provides us with a tool for this in the discussion of the metal–insulator transition where we have access to the electrical potential at the boundaries of the system.

These hysteresis effects are indicative of first order phenomena. But, this combination of mean field parameter conditions is not the only route to first order phase transitions. Consider a system where the sixth power is significant, but no cubic invariance exists. Order may come about because one of the lower powers has a negative coefficient. In this case, a higher order term, such as the sixth power which has a higher positive contribution at larger η, can bring about a minimum. So, when $C = 0$ in this discussion, one will have to look carefully at the sixth power of order parameter in the coefficient D. The reason is that with $A = a(T - T_c)$, order may come about because one of the lower powers has a negative coefficient. For example, B has been considered positive heretofore. But if because of stress or the field, or other parameters on which it depends it becomes negative, then no minimum exists, with both the second and the fourth power terms negative. The term in sixth power, $D\eta^6$, will now return the energy to higher values at a larger magnitude of order parameter. This is shown in Figure 4.11 for

$$\mathcal{G} = \mathcal{G}_0 + A\eta^2 + B\eta^4 + D\eta^6, \tag{4.13}$$

with $D > 0$, and $B < 0$. At low temperatures, the first two power terms are negative, and the sixth power term brings about a minimum at a non-zero order parameter. We now have three minimums, as shown in Figure 4.11. T_c^* is a tricritical point, and the transition has now become a first order transition. This change requires that $B(p, \mathcal{E}, T)$ become negative. Thus, Equation 4.13 is a little more demanding analytically. It gives, for the appearance of order,

$$\eta = \pm \left\{ \frac{1}{3D} \left[-4B + 4\sqrt{B^2 - 3a(T - T_0)D} \right] \right\}^{1/2}. \tag{4.14}$$

Here, we have taken only the positive root inside the polynomial solution to get a real solution. This lets us see how the hysteresis and supersaturation occurs.

Figure 4.11 shows the order parameter and its behavior as the temperature is reduced in this situation. The comments regarding the uniqueness of certain temperatures of the previous weak first-order transition discussion apply here too. The unique points, Y, Λ and Γ correspond to order parameter magnitudes of $(B/3D)^{1/2}$ at Y, $(B/2D)^{1/2}$ at Λ, and $(2B/3D)^{1/2}$ at Γ, respectively .

Hysteresis means a memory of the path to the condition. The behavior of T' is arrived at as the system is cooled from heated conditions. This is similar to supersaturation. A hysteresis window opens, and when the temperature drops below T'', the system returns to a stable single state. Supersaturated solutions precipitate out into a solid from a liquid phase.

A minor algebraic trivia is that only polynomial equations in one variable of degree less than five are explicitly solvable with radicals. Fortunately, the equation for finding the minimum, following differentiation, is of second power in the square of order parameter and is therefore quite solvable for the minimum energy condition.

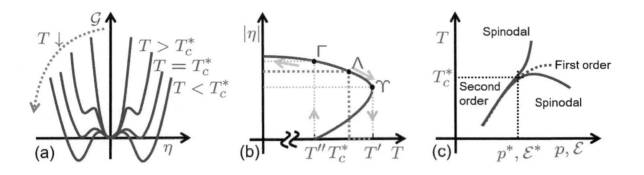

Figure 4.11: The free energy change with an order parameter when the sixth order term becomes important while the fourth order term is negative and the second order term is positive but small. Such a system can exhibit second order phase transitions, first order phase transitions with its overheating and cooling spinodal hysteresis, and a triple point, where all the phases coexist.

The critical temperature in this example occurs at

$$T_c = T_0 + \frac{1}{4}\frac{\mathcal{B}^2}{a\mathcal{D}},$$ (4.15)

and the finite order parameter, as a metastable point, appears at

$$T = T_0 + \frac{\mathcal{B}^2}{3a\mathcal{D}}.$$ (4.16)

The tricritical point, the point of triple minimum, where the second order phase transition becomes a first order one, occurs when the temperature T^* and pressure or electric field (p^* or $\mathcal{E}^*$) are such that

$$\mathcal{A}(p, \mathcal{E}, T) = \mathcal{B}(p, \mathcal{E}, T) = 0.$$ (4.17)

Symmetry breakdown resulting in phase transitions is how we started the discussion. Does the reverse, the existence of phase transition, imply symmetry breakdown? The answer to this question is no. Such a phase transition is always a first order transition. Figure 4.12 shows the spinodal and the critical temperatures, as functions of the other energy changing parameter, pressure or field. In a gas-liquid system under pressure, as the pressure is increased, the spinodal temperatures change, as does the triple minimum, but eventually all come together to a point. This point, a critical point ($p_c, T_c, \mathcal{E}_c$), has no change in symmetry across the first order transition. In principle, the transition can occur continuously between the random and the ordered phase by going around the critical point.

This phase transition without symmetry change can only happen in a first order transition. A second order transition will always have a symmetry change between the two phases. Either the minimum in

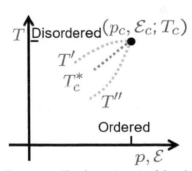

Figure 4.12: The change in spinodal and critical temperatures for order change as a function of the parameter that can affect energy of the system, for example, pressure in a gas-liquid system. The three temperatures eventually meet at a critical point of ($p_c, \mathcal{E}_c; T_c$) where there is no broken symmetry as it is the limit of phase transition.

We will also have a short discussion of quantum phase transitions. Some of these can happen without broken symmetry, for example, some of the quantum spin liquid to polarized phase of high field energy exists at the zero value of the parameter, in which case there are no other minima, that is, it is the only stable phase, or it does not exist when, at a finite order parameter, a minimum exists with a broken symmetry. There are no three minima of coexistence.

The symmetry-defying phase transition is also observed elsewhere. We will in particular look at the metal–insulator phase transition in metal oxides. Figure 4.13 shows a schematic rendering of the characteristics of V_2O_3. Point Π is a critical point.

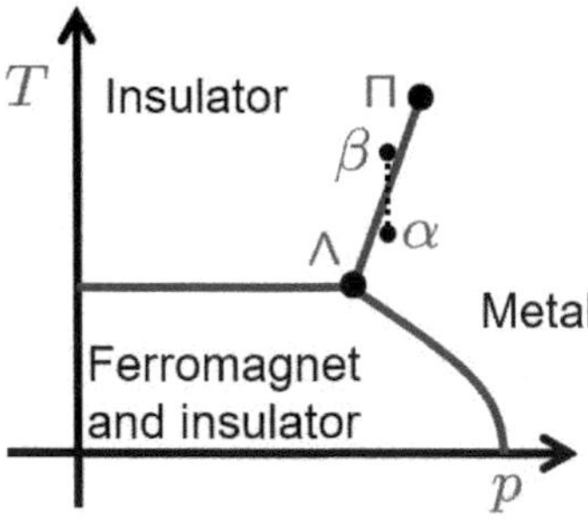

In Figure 4.13, the line $\Lambda\Pi$ is a metal–insulator transition line in the non-magnetic phases. If one is at point α and then raised the temperature to go to β, one would go from a metallic phase to an insulating phase. The metallic phase has a lower entropy than the insulating phase. This comes about because, in the metallic state, only the states around the Fermi surface fluctuate; the states below have no disorder contribution. They are all occupied. In the insulating state, all the electrons are localized to the ions, and they have therefore localized spin. Localized spin can be disordered. The informational arguments of Chapter 1 apply. The disordering from the filled Fermi sea with its ordered spin reaching close to the disordered localized spin has an entropy change of $k_B \ln 2$ per spin. So, the insulating phase has higher disorder—a disorder in the spin—of the localized electrons.

We can now compare the entropy and susceptibility behavior during the second and the first order phase transitions. The first order phase transition change is shown in Figure 4.14. The entropy change at T_0 is

$$\Delta S = \frac{\mathcal{A}\mathcal{B}}{2\mathcal{D}} = \frac{a(T - T_c)\mathcal{B}}{2\mathcal{D}}, \qquad (4.18)$$

an abrupt change takes place in it, and simultaneously, the susceptibility goes through a discontinuous change at the temperature T_0. In the second order system, these parameters are piecewise continuous.

Figure 4.13: The critical temperature versus pressure behavior of V_2O_3. The behavior shown is adapted from D. B. McWhan et al., Phys. Rev. B, 7, 1920 (1973).

The Pauli exclusion principle determines occupancy, with each electron having a unique $\mathbf{k}$ and other unique quantum numbers. The specific heat of a Fermi gas directly shows this effect.

This is an example of a metal–insulator transition due to electron correlation. It is a Mott transition. More on this later, as it can be interesting in nanoscale devices.

Figure 4.14: Discontinuous entropy change in a first order phase transition (a), and the corresponding discontinuous inverse entropy change in (b).

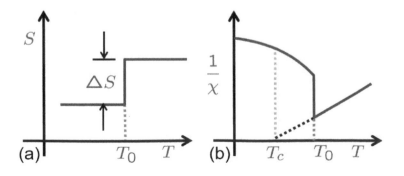

Stress and strain too, as a source and sink of work and, therefore, energy, play an important role in phase transitions and, like other energy mechanisms, couple to them. Piezoelectricity is the appearance of polarization—stable, spontaneous electric polarization—interlinked with mechanical stress. A minimum energy in ferroelectric, non-centrosymmetric structures will occur with some reconfiguration of strain. The internal energy of the system can now be expanded in this order parameter and strain. It can now be viewed to be arising from energies that are caused by the order parameter, the elastic energy arising from the strain, and a third term arising from the coupling between the order parameter and the strain, in the form

$$U(\eta, \varepsilon) = U_1(\eta) + U_2(\varepsilon) + U'(\eta, \varepsilon). \tag{4.19}$$

Now, the primary order parameter is the spontaneous polarization η, and the secondary order parameter is the strain ε. This is at least a mathematical expansion where the terms are reconfigured, but it also keys to the use of averages for these. Higher order effects such as fluctuations and other scale and perturbation effects are not included. We will not dwell on this, and keep it in mind together with other limitations of the simple phenomenological nature of Landau descriptions. It gives us a tool to incorporate a variety of energy- modifying interactions and to work with the mean.

Magnetic phase transition in a compressible structure also gives an example of the energetic coupling. Now, the magnetic order parameter of η couples to the deformation u. The Gibb's free energy can now be approximated as

$$\mathcal{G} = \mathcal{A}\eta^2 + \mathcal{B}\eta^4 + \kappa u^2 + \gamma \eta^2 u, \tag{4.20}$$

where the u^2 is the harmonic energy component of the deformation, and the term following is the anharmonic coupling between the two mechanisms, where the field effect of deformation—a linear term—is coupled to the harmonic energy term. This equation is quite solvable analytically. If the coupling γ is strong enough, this transition will go from 2nd order to 1st order. Coupling of degrees of freedom has a tendency to reduce the phase transition to that of order 1 if sufficiently strong.

But, one aspect that is important for devices and which we should look at carefully is what happens in inhomogeneous conditions, which materials at nanoscale can be subject to because of the preponderance of surfaces and interfaces. Strain, for example, has a strong size effect. Thinner films can be under strain, but if the surface is an open condition, they will relax. If the device size is small, then strain will relax multi-dimensionally. So, order, disorder and inhomogeneity may all appear in systems of interest to us. Figure 4.15 shows a pictorial view of what we are attempting to understand in this case

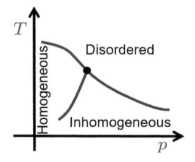

Figure 4.15: The appearance of homogeneous, inhomogeneous and disordered phases as a result of multitudes of energy mechanisms at work. This can be particularly acute at the nanoscale.

due to temperature and pressure. One example of importance is that there will be critical thicknesses below which ferroelectricity will disappear, in other materials, it may appear due to surface energy consequences.

In *PbTiO₃* particles, the critical temperature drops with a corner point at about the 30 *nm* size.

With the order parameter now a function of the position **r**, the total energy must be obtained as an integral over the volume, that is,

$$\mathcal{G} = \int_\Omega \breve{g}(\mathbf{r})\, d^3\mathbf{r} = \int_\Omega \Big[\breve{g}_0 + \breve{\alpha}\eta(\mathbf{r}) + \breve{\mathcal{A}}\eta^2(\mathbf{r}) + \breve{\mathcal{C}}\eta^3(\mathbf{r}) + \breve{\mathcal{B}}\eta^4(\mathbf{r}) + \cdots \Big] d^3\mathbf{r}, \quad (4.21)$$

where $\breve{g}(\mathbf{r})$ is the Gibbs free energy density. The other terms are also in density form and so indicated with the breve symbol (˘) on top of the Equation 4.2 coefficients that apply to a homogeneous volume.

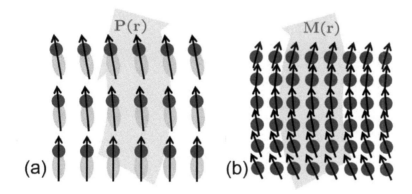

Figure 4.16: A gradually changing order parameter, in a ferroelectric system (a) and a ferromagnetic system (b), ignoring the fluctuation effects, which a mean field theory cannot reasonably handle.

The energy cost of the spatially changing order parameter $\eta(\mathbf{r})$ must be taken into account. So, look at Figure 4.4 and Figure 4.16, where a slow change takes place in an ordered two-dimensional material. Since we are interested in the ordered regime where mean field theory works, these changes will be slow, and therefore the energy penalty must be proportional to the gradient, that is, $\nabla\eta(\mathbf{r})$. Since $(\nabla\eta(\mathbf{r}))^2$ is the invariant term containing $\nabla\eta$ for a scalar η, the free energy can be rewritten in a phenomenological and general form as

$$\mathcal{G} = \int_\Omega \Big[\breve{g}_0 + \breve{\mathcal{A}}\eta^2(\mathbf{r}) + \breve{\mathcal{B}}\eta^4(\mathbf{r}) + \breve{\mathcal{N}}|\nabla\eta(\mathbf{r})|^2 + \cdots \Big] d^3\mathbf{r}. \quad (4.22)$$

We have now written the Gibbs energy in the form of a functional. This is the Ginzburg-Landau functional of the order parameter function $\eta(\mathbf{r})$. It is very useful. It is particularly important at domain walls, where these changes accelerate or are discontinuous. This term $\breve{\mathcal{N}}$ and its higher power terms, for example, $(\nabla^2[\eta(\mathbf{r})]^2)$ and so on, become important as interactions proliferate. We now allow η to

The tremendously useful principle of least action and the Lagrange equation of classical mechanics were derived using functionals in S. Tiwari, "Quantum, statistical and information mechanics: A unified introduction," Electroscience 1, Oxford University Press, ISBN 978-0-19-875985-0 (forthcoming).

be complex. We had earlier encountered this conundrum in determining order parameters and had chosen to take a real solution in Equation 4.14. To minimize the Gibbs free energy in Equation 4.22 by variational calculation we employ absolute values squared since it may be complex, and minimize by differentiating and by including only the terms through the fourth power and the inhomogeneity term; thus,

$$\mathcal{A}\eta(\mathbf{r}) + 2\mathcal{B}\eta|\eta|^2 - \mathcal{N}|\nabla\eta|^2\eta = 0. \tag{4.23}$$

This is a differential equation that is nonlinear because of the cubic power of η. We linearize near T_c as we did earlier for the coefficient in η^2. If $\mathcal{N}(p, \mathcal{E}, T)$ is negative, the equation implies that $\nabla\eta(\mathbf{r}) \neq 0$, which is inhomogeneity. An example of this is a spiral change instead of staying in one orientation. These vortices are found in magnetism. But, similar to our earlier sixth power term discussion, we should include a higher order term in the expansion when encountering an expansion's limits. In this case, this term is $\mathcal{N}(\nabla^2\eta)^2$, where $\mathcal{N} > 0$. We will now consider the Fourier components of the order parameter using the form

$$\eta(\mathbf{r}) = \eta \exp(i\mathbf{k} \cdot \mathbf{r}). \tag{4.24}$$

The rationale for this is that this makes $(\nabla\eta)^2 = |\mathbf{k}|^2\eta^2$ when we consider the first and dominant Fourier component at a wavevector of k_0 in a stable assembly. A minimum in free energy for k_0 occurs when

$$\frac{\partial}{\partial k^2}\left(\mathcal{D}k^2\eta^2 + \mathcal{N}k^4\eta^2\right) = 0$$

$$\therefore k_0^2 = -\frac{\mathcal{D}}{2\mathcal{N}} \text{ for } \mathcal{D} < 0, \text{ and } \mathcal{N} > 0. \tag{4.25}$$

What this implies is that a structure with a wavevector k_0, that is, a period of $2\pi/k_0$, forms as a primary result. η may also possibly have no primary single Fourier component. In these cases, additional overlaying structures may also appear. The periodicity of this overlaying structure is incommensurate with underlying lattice periodicity.

The change of $\mathcal{D} > 0$ to $\mathcal{D} < 0$ has a special significance, like some of the others discussed earlier. This point, the Lifshitz point, is where homogeneous order transits to inhomogeneous order. These transitions are observed in magnetically driven systems. Due to the invariance of linear derivatives, for example, energy terms of $\mathbf{M} \cdot (\nabla \times \mathbf{M})$ in magnetic systems, helicoidal inhomogeneous structures do arise.

Symmetries determine whether the phase transition is of the second order or the first order. Structural phase transitions are of the latter. Density change is discontinuous. So, amorphous–crystalline

Multiferroics, where an electric field can induce magnetism, or a magnetic field can induce ferroelectricity, can at a simple level be seen as arising from these couplings. This free energy form, however, will encounter problems with fluctuations—the randomness that comes about in all energy-driven processes—since the equation represents the interaction within a mean. We will ignore this interesting and challenging direction and its consequences for our phase transition devices, though noise and slow modes of various types will exist in this interesting collective form that is different from that of the independent free particle. In society, revolutions, with their phase transitions, happen at moments of large fluctuations.

These are superstructures in macromolecular aggregates. Self-assembly provides plenty of examples of these phase structures formed with various periodicities based on parametric constraints.

and, as we shall see, metal–insulator phase transitions due to cor-relation effects are first order phase transitions that show hysteresis behavior.

This theory of phase transitions is a mean field theory. It assumes that the interactions in the ensembles have terms that depend on large length scales interactions reducible to an effective term. The other important consequence of the use of averages, even if for only part of the interaction, is that fluctuations such as of the order param-eter are neglected. Only slow changes, such as in inhomogeneity, can be incorporated.

But, near the critical temperature T_c, fluctuations pervade. A sys-tem undergoing any symmetry breakdown is confused—short- and long-range interactions all pervade—and this would mean that mean field theory is most inexact where one is most interested.

Landau theory, however, provides us a test for the validity of mean field theory. At and near T_c, physical quantities of interest are chang-ing in profound ways, and here the scaling exponents of mean field theory, for example, that order parameter changes as $[(T_c - T)/T_c]^{1/2}$ for second order transitions, provide a test. We can determine how large the fluctuation region is in terms of the reduced temperature of $(T_c - T)/T_c$. As an example, consider fluctuation in η in an or-der state near T_c. Let $\bar{\eta}$ be the average here. We can ask, what is the chance that is of a disordered region of volume V maintains a param-eter η instead of $\bar{\eta}$? We can use Taylor expansion to determine the difference between the homogeneous and non-homogeneous states as

$$\begin{aligned} \delta\mathcal{G} &= \frac{1}{2}(\eta - \bar{\eta})^2 \frac{\partial^2 \mathcal{G}}{\partial \eta^2} \\ &= a(T - T_c)(\eta - \bar{\eta})^2 \\ &= a(T - T_c)\delta\eta^2. \end{aligned} \quad (4.26)$$

The probability of this fluctuation exists as a function of the energy, so

$$p(\delta\eta) \propto \exp\left(-V\frac{\delta\mathcal{G}}{T}\right). \quad (4.27)$$

The fluctuation amplitude, because of this distribution, then is

$$\langle |\delta\eta|^2 \rangle = \frac{T}{aV(T - T_c)}. \quad (4.28)$$

The divergence is worst near T_c. The energy associated with the fluc-tuations also changes, as it is associated with the volume V in which this parameter fluctuation is taking place. When V is small, fluctua-tions are large. So, one needs to determine the size of fluctuation to establish validity. For an inhomogeneous system, repeating this rea-soning with the Landau-Ginzburg functional for spatial fluctuations

Even with much to appreciate in this mean field approach to the analy-sis at symmetry breakdown through an expansion and in its effectiveness across many phenomena, its limitations warrant caution to avoid improper usage. To paraphrase John Ashcroft and David Mermin, it provides a grossly inadequate picture of the critical re-gion, fails to predict spin waves at low temperature and even at high tem-perature reproduces only the leading correction to Curie's law. Curie's law in magnetism has the dependence of $1/T$ in the order parameter of sus-ceptibility, which is now corrected to $1/(1 - T_c/T)$—also known as the Curie-Weiss law. This is a fair comment. In the mesoscopic discussion, we worried that a charge cannot not act on itself. A charge cannot make a unique com-mon field, because if it did, and if this field acts on charges, then it is acting on the charge itself. But, we use it in much of engineering and it only shows up as a problem at nanoscale and in effects such as Coulomb blockade. Mean field averages order parame-ters and statistical quantities but, near phase transition, the physics is in the fluctuations—the short range and the long range connections. It is, of course, quite approximate. But, it does push us forward. Richard Feynman, in his Nobel lecture, made the following anal-ogy while discussing the self-energy problem; "Like falling in love with a woman, it is only possible if you do not know much about her, so you cannot see her faults. The faults will become apparent later, but after the love is strong enough to hold you to her. So, I was held to this theory, in spite of all difficulties, by my youthful enthusi-asm." (R. P. Feynman, "Nobel lecture: The development of the space-time view of quantum electrodynamics," www.nobelprize.org/nobel_prizes/ physics/laureates/1965/feynman-lecture.html). A theory, such as Lan-dau's, needs to be appreciated for its honesty when it also makes assertions about the limits of its applicability.

leads to

$$\langle |\delta \eta_k|^2 \rangle = \frac{T/V}{aV(T - T_c) + \mathcal{D}k^2}. \qquad (4.29)$$

The real space (**r**) fluctuations and the reciprocal space (**k**) fluctuations are Fourier transforms of each other, so the correlation function speaks to the parameter decay scale.

Fluctuations, like entropy, are related to randomness and their unaccounted degrees of freedom. Therefore, this theory brings a connection between the fluctuations and dissipation. It allows us to relate the equilibrium fluctuations to out-of-equilibrium macroscopic variables.

The fluctuation-dissipation theorem states that a system's small external perturbation response is directly related to the fluctuations of the system in thermal equilibrium. The significance of this is that off-equilibrium quantities are derivable from equilibrium fluctuations.

The Gibbs free energy density $\mathcal{G}(\eta)$ of a system, as the Hamiltonian of the system, determines the partition function as

$$Z = \int \exp \left[-\frac{\mathcal{G}(\eta)}{k_B T} \right] d\eta. \qquad (4.30)$$

The order parameter η and the conjugate field $\wp$ couple, causing a change in the free energy in the system of volume V of $\eta \wp V$. The susceptibility χ, that is, the linearization of the response, is

$$\chi = \frac{d\eta}{d\wp}. \qquad (4.31)$$

The fluctuation-dissipation theorem implies that this susceptibility—an equilibrium parameter—is proportionally related to the fluctuations of the order parameter. With this conjugate field, the partition function and its implications are

$$
\begin{aligned}
Z &= \int \exp \left[-\frac{\mathcal{G}(\eta)}{k_B T} - \eta \wp V \right] d\eta, \\
\langle \eta \rangle &= \frac{1}{Z} \int \eta \exp \left[-\frac{\mathcal{G}(\eta)}{k_B T} - \eta \wp V \right] d\eta \\
&= \frac{T}{V} \frac{1}{Z} \frac{\partial Z}{\partial \wp}, \quad \text{and} \\
\langle \eta^2 \rangle &= \left(\frac{T}{V} \right)^2 \frac{1}{Z} \frac{\partial^2 Z}{\partial \wp^2}.
\end{aligned} \qquad (4.32)
$$

From these, the expectation value of fluctuations is

$$\langle |\delta \eta|^2 \rangle = \langle \eta^2 \rangle - \langle \eta \rangle^2 = \left(\frac{T}{V} \right)^2 \frac{\partial}{\partial \wp} \left[\frac{1}{Z} \frac{\partial Z}{\partial \wp} \right] = \frac{T}{V} \frac{\partial \eta}{\partial \wp}. \qquad (4.33)$$

Susceptibility and fluctuations are then related as

$$\frac{\chi}{V} = \frac{1}{T} \langle \eta^2 \rangle - \langle \eta \rangle^2, \qquad (4.34)$$

SAT—finding logical satisfiability—in the midst of large and about equal number of variables and constraints with similar number of equations has benefited tremendously from techniques developed in statistical mechanics to solve these phase transition problems. The problems have similar character. Find time and length scales in the midst of all these interactions, small and large, over space and time, and problem becomes easier. Propagation of social beliefs can be recast in this statistical mechanics format. This Landau mean-field theory is obviously inadequate for such problems.

Phase contrast microscopy, using these important ideas of transforms, is utilized across the range of wavelengths. It allows observation of this phenomenology in systems where complexity of interactions pervades. X-rays and electron beams probe the atomic scale, ultraviolet to visible and infrared are used at biological scales, and all of these, particularly far infrared, are used at cosmic scales.

In S. Tiwari, "Semiconductor physics: Principles, theory and nanoscale," Electroscience 3, Oxford University Press, ISBN 978-0-19-875986-7 (forthcoming), we work through the arguments that establish the connections between fluctuation-dissipation and the variety of responses that any system exhibits, including in noise, transport drag, and others.

Kramers-Kronig relations, which we look at in S. Tiwari, "Semiconductor physics: Principles, theory and nanoscale," Electroscience 3, Oxford University Press, ISBN 978-0-19-875986-7 (forthcoming), provide a connection.

which, for an inhomogeneous system with a wavenumber $\mathbf{q}$-dependent fluctuations, comes about as

$$\frac{\chi(\mathbf{q})}{V} = \frac{1}{T}\langle \eta_{\mathbf{q}}^2 \rangle - \langle \eta \rangle^2 = \frac{1}{T}\sum_{\mathbf{r}} \exp(i\mathbf{q}\cdot\mathbf{r})\,\langle \delta\eta(\mathbf{0})\delta\eta(\mathbf{r})\rangle. \qquad (4.35)$$

The Fourier transform of the real space correlation function of fluctuations of $\mathcal{G}(\mathbf{r})$ is the term on the right hand side. This shows how properties in equilibrium and off-equilibrium can be related.

We end this section with broader comments regarding mean field theory.

Polarization and magnetization, to a good approximation, are determined by $\mathcal{E}_{eff}$ and $\mathcal{H}_{eff}$, respectively, that is, some effective field that exists in the crystal and which arises from the internal interactions and external forces applied. The individual particles feel the average potential arising from many particles. Semiconductor crystals provide a good example for visualizing this. Figure 4.17 shows this conceptually for a conducting solid. Charged ions are located at $\mathbf{R}$. The free electron cloud surrounds these ions and is spread out in the coordinate $\mathbf{r}$. The motion of the ions is constrained by the lattice potential, while electrons move under the perturbations of interactions arising with the core ions and with each other. The interactions have a long-range interaction that appears as an average or mean field. Recall the discussion in Chapter 2 regarding discrete dopant effects and electron screening and the role of long-range and short-range interactions. So, the energetic interactions here are also being captured as a mean field. The scattering formalism employs this mean field approach. One could indeed argue that the effective mass theorem that helps us calculate the carrier behavior is a mean field result, since it captures the effects of of periodicity—both in the short range and in the long range—so long as the dimension scale is large enough to capture the internal dynamics of the crystal for transport.

4.2 Ferroelectricity and ferroelectric memories

THE APPEARANCE OF SPONTANEOUS POLARIZATION due to charge ordering, a consequence of inversion symmetry breakdown, is of special significance in electronics. The appearance of the order parameter $\eta = \pm P_s$ as two low energy states that can be switched and that can hold their state makes them useful in electronics, particularly as memories. Our discussion will largely use lead zirconium titanate (PZT) as a protoexample. But, we mean this to be a general discussion, where PZT is used to include specifics.

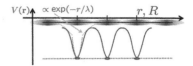

Figure 4.17: A conceptual look at a conducting solid with ions localized in a sea of electrons. The symmetry breakdown occurs with strong localization of the ions and freedom of movement for the electrons. A mean field exists from long-range interactions.

Ferroelectric phase transition has many other utilities. Having two energy minima that can be driven between the two separated but accessible states means that there are strong non-linearities. Nonlinearities mean that up and down frequency conversions can be achieved efficiently. Electroptics is one major area of such applications, but ferroelectrics are also employed in the microwave range. Ferroelectricity, as well as high permittivity, is also associated with soft phonon modes. Since phonons are broad, these nonlinearities also allow the phase transition nonlinearity to be exploited at terahertz frequency, so well below infrared frequencies—at the frequencies of acoustic phonons!

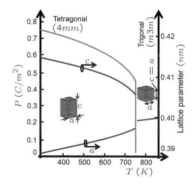

Figure 4.18: The first order phase transition, leading to the spontaneous polarization of $PbTiO_3$, is shown, together with the change in the lattice constant as the crystal undergoes a trigonal ($m3m$) to tetragonal ($4mm$) transition at a critical temperature— here, of about 763 K.

Figure 4.18 shows phase transition characteristics—the abrupt transition and the broken symmetry—when lead titanate undergoes spontaneous polarization. The material undergoes transition near 763 K. In the disordered phase, it belongs to the $m3m$ crystal group—it is trigonal—and the ordered phase is tetragonal. The ordered phase is elongated and is non-centrosymmetric, with the polarization along the c-axis. The change is of first order. It occurs with a volume change and entropy change.

We establish the two spontaneous polarization states, that is, bring the system to one or the other minimum of the energy curve, by applying an electric field across the ferroelectric. When one removes this applied electric field, the stable ferroelectric state retains the spontaneous polarization that was established. Figure 4.19 shows the two opposite polarizations that can be achieved in lead zirconium titanate ($PbZr_{0.52}Ti_{0.48}O_3$) and strontium bismuth tantalate ($SrBi_2Ta_2O_9$)—two common ferroelectrics employed in silicon technology. The two opposite polarizations correspond to the two order parameter minimum wells of the ferroelectric phase transition discussion.

There is much to discuss in these characteristics, ranging from their gradual changes to their magnitudes and what these curves really mean.

First, let us consider the implication of this polarization magnitude. Table 4.2 gives examples of some other ferroelectrics of historic or practical import.

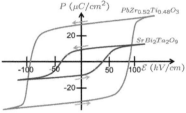

Figure 4.19: Hysteresis curves of two common ferroelectrics in electronics—lead zirconium titanate ($PbZr_{0.52}Ti_{0.48}O_3$) and strontium bismuth tantalate ($SrBi_2Ta_2O_9$). The former has more polarization, and the latter has been more reproducible under repetitive stressing.

Material	Composition	Critical temperature (K)	Spontaneous polarization ($\mu C/cm^2$)
Barium titanate	$BaTiO_3$	~303	26.0
Lead titanate	$PbTiO_3$	~860	~700
Lead zirconium titanate	$PbZr_{.52}Ti_{.48}O_3$	~640	~25
Lithium niobate	$LiNbO_3$	~1483	300
Strontium bismuth tantalate	$SrBi_2Ta_2O_9$	~760	
Yttrium manganate	$YMnO_3$	~910	
Pottasium dihyrogen phosphate	KH_2PO_4	~423	5.3
Rochelle salt	$KNaC_4O_6.4H_2$	~293	
Germanium telluride	$GeTe$	~673	

Table 4.2: Examples of ferroelectrics, with application, historic relevance and their critical temperatures. Lead zirconium titanate (PZT) and strontium bismuth tantalate (SBT) find use in electronics in memory. Lithium niobate and potassium dihydrogen phosphate (KDP) are employed in electrooptics. Rochelle salt is where ferroelectricity was discovered. Germanium telluride, or its variations such as germanium antimony telluride are ferroelectrics. In optical disks, the ferroelectric effect gives these materials superresolution since the amorphous–crystalline phase change occurs with a large dielectric constant change in the written area, resulting in superlensing.

A polarization charge of 0.25 C/m^2, or 25 $\mu C/cm^2$ corresponds to a charge density of $\sim 1.6 \times 10^{14}$ electrons/cm^2. Compare this with the maximum sheet electron density that one can obtain with a dielectric that silicon technology uses for capacitors and transistor—SiO_2 has a breakdown electric field of $10\ MV/cm$, which, by Gauss's law gives a sheet charge of $3.72\ \mu C/cm^2$ or $\sim 2.3 \times 10^{13}$ electrons/cm^2, which is about a factor of 10 smaller. One would not wish to use the oxide near breakdown, and thin oxides have soft leakage. What is attractive about this polarization, as seen in Figure 4.19, is that the polarization is accessible through electric fields of less than $100\ kV/cm$ to cause the change in state. This is again factors of 10 lower than the fields needed to get the maximum charge density through SiO_2. These curves of hysteresis show the ability to manipulate charge at high density with low fields. For example, a 20 nm-thick film can be electrically flipped, even with a $100\ kV/cm$ field, that is, with a 0.2 eV electrostatic potential change—a 0.2 V of bias voltage change. This is a small optimistic number, and the charge controllable is large.

Why does electric field change the polarization? When a sufficient potential energy is provided, in this case, electrostatic, it is an additional electrically mediated perturbation to the two-well first order phase transition system. Figure 4.20 shows the change taking place in a schematic form. Multiple domains of polarization exist, not all of which are necessarily aligned with each other. With the application of the electrostatic potential, an electrostatic energy density of $(1/2)\mathcal{E}(\mathbf{r}) \cdot \mathbf{D}(\mathbf{r})$ distributes throughout the film. The polarization, however, is inhomogeneous. Multiple domains exist, varying in size, with the polarization order parameter oriented differently. The net response is the aggregate of the local responses to the applied field, with the local response reacting to its local conditions and energetics. When a high-enough field is applied—let us consider the positive abscissa region—the Gibbs free energy minimum favors the positive spontaneous polarization minimum. As the electric field is lowered, when there is a net zero applied potential and, therefore, a low electric field across the structure, a remnant polarization of $\mathbf{P}_r$ exists. The domains are still largely in the positive minimum, even if distributed in orientations. Only at sufficiently negative potential and an electric field higher than the coercive field ($\mathcal{E}_{co}$) does the negative polarization order parameter become the lower minimum. Now, the system flips to a majority negative polarization state. Any further application of potential and field aligns the domains more strongly. The decrease of the potential and field now traces the hysteresis loop, with spontaneous polarization still negative, $-\mathbf{P}_r$, at zero field. The negative polarization minimum remains the lowest energy state until sufficient electric field is applied—the coercive field of $\mathcal{E}_{co}$—when the system flips to a majority of positive polarization domains that strongly align

Recall that, in an earlier discussion of heterogeneity, we remarked that $PbTiO_3$ particles have a corner point in polarization at close to 20 nm.

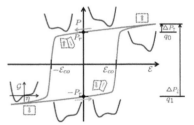

Figure 4.20: A schematic of the hysteresis loop of a ferroelectric, together with our free energy picture in the presence of the electric field at various points of the loop. Multiple domains exist, so the diagram is a cumulation over the ensemble. P_r is the remnant polarization, and $\mathcal{E}_{co}$ is the coercive field where the expectation energy of the ensemble flips.

themselves at the highest fields.

The hysteresis comes about as the system prefers to remain in its lowest energy state until the coefficient B that we discussed earlier flips sign under the influence of the electric field. The electric field acting on a charge is the equivalent of pressure in the gas-liquid phase transition. Both are a way to impart and extract energy. Note that the polarization in the stable regions does not remain flat because of macroscopic inhomogeneity. This hysteresis behavior shows the energetic changes that the system is undergoing. In any direction of the sweep, there are three distinct regions. In the sweep from negative to positive electric field, one starts with an approximately linear region where reversible domain wall motion occurs under the force of electric field. This is followed by a region of change which has a linear growth of domain and of polarization. This is the region where the polarization changes, as the field is approximately the coercive field $\mathcal{E}_{co}$. And this is followed by a part of the hysteresis curve where the domains reach near-saturation.

The usefulness of the hysteresis loop of Figure 4.20 is in the identification of the two remnant polarization states. If one takes this capacitor—a ferroelectric clad with two electrodes—and change the voltage across it to a voltage higher than that necessary for the coercive field to be applied, one can identify which one of the states of polarization—P_r or $-P_r$—the system was in before. Either a small or a large change will be observed. A small or large change in the polarization, measurable through the change in the total charge, or the current flow within a time duration of change, identifies the starting state. In our notation of Chapter 1, the q_0 state, when voltage and electric field is swept to the maximum positive, causes a change in the polarization $\Delta P_\uparrow$ available as a charge $\Delta Q_\uparrow$ in a transition held to completion. If the system is in the opposite polarization state of $-P_r$—a q_1 state—application of the full forward voltage and electric field leads to a change in the polarization of $\Delta P_\downarrow$, with the change being available as a charge $\Delta Q_\downarrow$ in a transition held to completion. $|\Delta Q_\downarrow| \gg |\Delta Q_\uparrow|$. The q_1 state gives rise to a larger current change with sign opposite to that of the q_0 state. In a circuit that we will discuss later, the passing of this current through the capacitance of an equilibrated bit line at an averaged voltage, these swings cause the voltage to swing the bit line, thus determining the state that the ferroelectric capacitor was in. This is the reading process of the state. But, it is a reading process that leads to the destruction of the stored state, since the release of the voltage after reading the process leaves the capacitor in the $+P_r$ state in this discussion. Note that we could just as well have read the capacitor by changing the voltage and the field to a more negative condition that exceeds the coercive condition of the opposite direction.

We will return to the discussion of taking advantage of this non-volatile two-state equilibrium later after discussing some of the difficulties that arise as a result of the surface, interface, structural and other interactions arising from the energy changes.

The energy stored and changed in these ferroelectric capacitors is large. The $PbZrTiO_3$ capacitor with a normalized polarization of $25 \ \mu C/cm^2$, designed to operate at silicon transistor voltages of $\sim 1 \ V$ in a $100 \times 100 \ nm^2$ areal geometry, has an energy change of $\sim QV$ in a change from $-P_r$ to P_r during one half of the scan of the loop. This means a change in energy of $\sim 2 \times 25 \times 10^{-6} \times 100^2 \times 10^{-7 \times 2} \times 1 = 50 \times 10^{-16} \ J \approx 1.2 \times 10^6 k_B T$ at room temperature. This is factors of 10 to 100 higher than dissipation in transitions, and factors of several thousands higher than the energy necessary to avoid one error during repeated switching in an ideal switch at $300 \ K$. So, a large dissipation occurs when flipping a state, and a large stored energy exists in the ferroelectric capacitor. A ferroelectric stores a large amount of energy, $(1/2) \int \mathbf{P} \cdot \mathcal{E} d^3 r$ in the polarization and $(1/2) \int \epsilon_{FE} \mathcal{E} \cdot \mathcal{E} d^3 r$ in the dielectric component. During switching, some of the energy, as when a dielectric capacitor is switched rapidly, is converted into heat and radiation—the broadband of vibrational degrees of freedom. So, energy is supplied to overcome the barrier energy separating the two states, and then the process of reaching the flipped state releases this energy. It comes out in the form of current arising from changes in the charge at the electrode. Some of this energy is dissipated in the material itself, depending, just as with dielectric capacitors, on the speed of change. When going through the full cycle around the hysteresis loop, the energy contained within the loop is dissipated. Repeated cycling and energy transfer with the structural movement causes defects and material heterogeneities, such as those at the interfaces, to increase.

The next questions we ask are, how is this polarization measured, and what are the details of its behavior within the structure? We do this by analyzing a ferroelectric capacitor. We replace the dielectric of a conventional capacitor by the ferroelectric and place two intimate contacts using a metal. Schematically, this is shown in Figure 4.21. If one places two contact metals on either side of the ferroelectric, one can measure this polarization by probing with an electrical bias. Electron accumulation and depletion at the two electrodes will form a local dipole where the polarization is terminated on an atomically thin charge layer in the metal. So, shorting the two metal electrodes has a schematic charge and field profile as shown in Figure 4.21. This is a ferroelectric capacitor. Figure 4.21(a) here represents one direction of spontaneous polarization, and Figure 4.21(b) the other.

This energy is large enough that one can utilize it gainfully by maximizing it in large capacitors at macroscopic dimensions. Strobes and flashes have utilized ferroelectric capacitors.

This the analog of Maxwell's relations of thermomechanics for the thermoelectric system.

With ideal electrodes and an ideal ferroelectric all the way to the surface/interface, there exists net charge at the metal/ferroelectric interface. In the ferroelectric, this is the net charge represented in the polarization. It is screened and thus neutralized by accumulation or separation of electron charge in the metal. The length scale of this change in the metal is short, of the order of < 1 nm, resulting from the Thomas-Fermi screening of Equation 4.53. Because of this length, there is a minute electric field under the shorted condition in the ferroelectric, with this charge dipole of limited extent at the metal/ferroelectric interface.

What happens when one places a capacitor using an applied bias to a positively or negatively polarized condition and then shorts it? If one were to switch from a branch of polarization $\mathbf{P}$ to a short, a charge density of ΔQ in units of C/cm^2 with $\Delta Q = \int \hat{\mathbf{n}} \cdot \mathbf{D} \, d^2 r = \Delta P$ will flow through the circuit. $\Delta P = P - P_r$, where P_r is the remnant polarization under the shorted conditions, and P is the polarization under the biased starting condition.

What happens in the presence of a non-ideal metal-ferroelectric interface is interesting. Figure 4.22 shows an equivalent figure of a shorted capacitor in the presence of a dielectric at the interface. Such dielectric films may come about due to material reactions to a more stable oxide encouraged by interface kinetics or due to the processes of fabrication. Now the charge density arising from spontaneous polarization in the ferroelectric and its field termination in the metal is separated by the dielectric. For the two states of polarization, now there is a net field in the ferroelectric in a direction opposite to that of the polarization.

This is a depolarizing field, which influences the hysteresis characteristics and, potentially, the long-term reliability. The magnitude of this depolarization, for a generalized metal-dielectric-ferroelectric-dielectric-metal system, can be derived as

$$\mathcal{E}_d = \frac{P}{\epsilon_0} \frac{\delta}{t}, \tag{4.36}$$

where t is film thickness, and δ an effective length scale reflective of the consequences of metal Thomas-Fermi screening, that is, the effects of the passive dielectric layer and near-electrode environment on polarization in the insulator. In the simplest idealized case of a dielectric interface layer, $\delta = t_d/\epsilon_r$, where t_d is its thickness, and ϵ_r is its relative dielectric constant. For $PbZr_xTi_{1-x}O_3$ of thickness $t = 30$ nm, with a polarization of 25 $\mu C/cm^2 \equiv 1.6 \times 10^{14}$ cm^{-2} of electron surface density and a dielectric interface of 1 nm with $\epsilon_r = 10$, this depolarizing field is about a MV/cm. If we do not short, an idealized structure will maintain its state, while a real structure will undergo kinetic effects due to the off-equilibrium reflected in

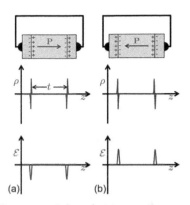

Figure 4.21: A ferroelectric capacitor, with the metal electrode directly attached to the ferroelectric. (a) and (b) show the behavior in the two order parameter states.

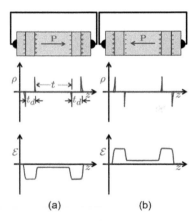

Figure 4.22: A ferroelectric capacitor, with the metal electrode attached to the ferroelectric through an intervening dielectric. (a) and (b) show the behavior in the two order parameter states. The presence of a dielectric causes the residual dielectric field in the ferroelectric as a result of a finite dielectric thickness preventing screening of the polarization charge by electrons in the metal.

In reality, most ferroelectrics have bandgaps that are finite or are similar to those of semiconductors in terms of orders of magnitude. $PbZr_xTi_{1-x}O_3$, for example, will have some surface state and semiconductor depletion effects, too. However, ferroelectrics have poor mobilities. The depletion effect have consequences for the coercive field.

this depolarization field. When a biased condition is changed to short circuit, a net charge reflective of the difference between the biased condition of polarization and starting remnant polarization will flow.

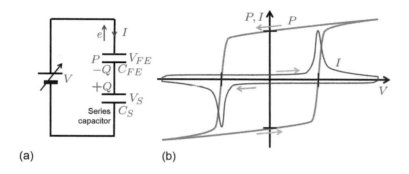

(a)   (b)

Figure 4.23: Schematic of polarization measurement of a ferroelectric capacitor by performing a bias voltage sweep with a capacitor in series.

The current that will flow and the time response represent the switching as the domains change, growing or shrinking. Switching of polarization also means that there is the crystallographic change related to the broken inversion symmetry. Figure 4.23 shows schematically the measurement of polarization-voltage and current-voltage characteristics of an idealized ferroelectric as the voltage is scanned in a loop with a dielectric capacitor in series. This measurement employs a circuit such as that of Figure 4.23(a), where a ferroelectric capacitor of capacitance C_{FE} with polarization P is placed in series with a large dielectric capacitor C_S. With the dielectric capacitances such that $C_S \gg C_{FE}$, the supply voltage $V = V_{FE} + V_S$ that is varied slowly largely appears as $V_{FE} \approx V$, with a small V_S. After the uncharged capacitors are brought together, the application of voltage $V(t)$ causes a current $I(t)$ to flow, causing a charge $Q(t)$ to be divided between the ferroelectric and the dielectric capacitors. We have

$$
\begin{aligned}
Q &= C_S V_S, \\
V &= V_{FE} + V_S \approx V_S, \\
I &= C_S \frac{dV_S}{dt} = \frac{dD_{FE}}{dt}, \text{ and} \\
Q &= \hat{n} \cdot \mathbf{D} = \epsilon_{FE} \mathcal{E}_{FE} + P.
\end{aligned} \tag{4.37}
$$

In a measurement, the instantaneous charge follows from the measurement of the series capacitor. The current in the series capacitor is $C_S dV_S/dt$. In the ferroelectric capacitor, it is the displacement current, which is the sum of the currents due to the dielectric and the polarization components, that is,

$$
I = \frac{dD_{FE}}{dt} = \epsilon_{FE} \frac{d\mathcal{E}_{FE}}{dt} + \frac{dP}{dt}, \tag{4.38}
$$

and any other leakage currents, such as the conducting particle current. If the latter is negligible because of high resistivity and barrier height, a cyclic measurement that flips the polarization to its opposite state gives the charge as an integral of current that includes the polarization. Measurement of polarization, thus, cannot be an instantaneous measurement. It requires a cyclic measurement. For any change in applied voltage in time from an initial state i to a final state f, the change in polarization is

$$\Delta P = \int_i^f I dt - \int_i^f \epsilon_{FE} d\mathcal{E}_{FE}. \tag{4.39}$$

If one performs a half loop starting at zero applied bias, one flips the polarization and the total integrated charge $Q = 2|P_r|$, which is the area enclosed in the left or right half of the current loop. When the hysteresis is symmetric, that is, there is spatial and electrode symmetry, Equation 4.38 also implies that

$$2P_r = \frac{1}{2} \left| \oint I dt \right|. \tag{4.40}$$

Multiple cycling of ferroelectrics leads to a shrinking hysteresis loop with smaller remnant polarization and also generally a change, usually an increase, in coercive fields. Figure 4.24 shows the polarization curves of a ferroelectric capacitor that consists of a deposited polycrystalline film sandwiched between two metal electrodes and which has been repeatedly cycled, as well as the changes that arise with an increasing number of cycles (N). These changes show the fatigue that ferroelectric capacitors undergo under repeated use.

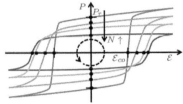

Figure 4.24: A schematic view of fatigue in poor ferroelectric capacitors. This figure shows the polarization-electric field/voltage characteristics of a polycrystalline ferroelectric clad with two metal electrodes, which here are of identical metals, and under repeated cycling. The remnant polarization and coercive field change, with the former decreasing and the latter, usually, increasing.

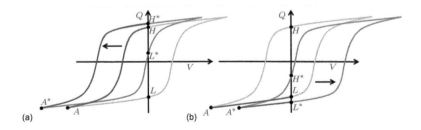

(a)

(b)

Figure 4.25 shows and describes the second phenomenon, which is commonly observed and which is known as imprint, when the capacitor is kept at one of the two spontaneous polarization states.

Exposure to elevated temperatures with data written leads to imprinting. For example, in any given state, if a high temperature is maintained, it leads to strengthening of that state and a weakening of the opposite state. This is shown in Figure 4.25. The important consequence is the strengthening of maintaining a state and a weakening

Figure 4.25: The imprint effect, as seen in the charge/polarization-voltage characteristics of a ferroelectric capacitor—a polycrystalline ferroelectric clad with two metal electrodes—when the capacitor is kept in one of its states under repeated cycling. For the remnant state H, staying in the state shifts the polarization hysteresis left, as shown in (a). For the remnant state L, repeated cycling shifts the hysteresis right, as in (b). The path $HA \mapsto H^*A^*$ in the former, with the low state shifting to L^*. For the state L, the path $AL \mapsto A^*L^*$, with the high state shifting to H^*. The opposite state can flip its polarity.

of maintaining the opposite state. Imprinting reduces the switching polarizability, as shown in the figure. This means a reduction in observable signal in the form of current or charge. The failures here arise when the level of change is such that it crosses the detection limits. Figure 4.25 shows the behavior when a high or a low state is maintained for an extended period of time at an elevated temperature. When a high state, here defined as a positive polarization charge (H), is maintained, it shifts the hysteresis curve to the left, causing H to shift to H^*. The net positive polarization increases but so does, for identical conditions, the negative polarization. The result is that, in the extreme example shown in Figure 4.25(a), any sensing scheme that depends on a polarity change for detection, fails. The same is true for the negative polarization state as shown in Figure 4.25(b). Both the fatigue and the imprint arise because every cycle of polarization reversal involves a structural change with the application of field and the energy transfer needed to overcome the barrier that separates the two low energy states. This energy is dissipated in the ferroelectric. Imprinting is when these changes emphasize certain regions in the response, and fatigue is related to loss of polarization as the structure loses its structural integrity. Ferroelectric memories need to account for these state effects during use. Their most important consequence is in the design of the cell organization and writing and interrogation mechanisms in dense arrays that still work at low voltages, low energies and high densities.

The fatigue effects arise from structural changes under the influence of the large energy changes in the system. Migration of oxygen atoms that are ionic bonded, common to nearly all the ferroelectrics, is one common suspect. This migration is usually observed in direct measurements of the atomic species of the structures, and such structural changes are associated with electronic changes that lead to trapping of electrons under repeated cycling. The fatigue effect is also coupled to the specific electrode materials and polarity because of their energy effects under cycling. Phase transitions with such structural changes also lead to other conduction transitions, such as the formation of conducting metallic filaments. Spontaneous polarization also decreases with a decrease in the separation of the operating temperature from the Curie temperature. So, a high Curie temperature is important for minimizing the consequences of imprinting and fatigue. Both these effects are important in designing circuits that can truly utilize the memory capability provided by the low fields, low voltages and large charge changes in use.

The use of ferroelectric capacitors in memory, such as in random access memory—ferroelectric random access memory (*feRAM*)—aims to utilize these properties and their constraints. As in dynamic

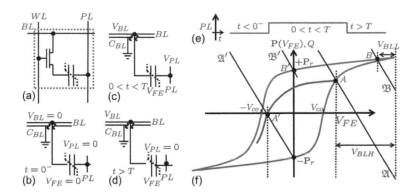

Figure 4.26: Illustration of how an access transistor with a word line (*WL*) connected to the gate, a bit line (*BL*) connected to its drain, and the source connected to a ferroelectric capacitor whose other electrode is connected to a plate line (*PL*) may be employed for writing and reading from the ferroelectric capacitor. The transistor is turned on by raising the *WL*, and we can assume here that the transistor is approximately a short connecting the bit line to one of the plates of the capacitor. A change in the plate line voltage causes a change in polarization that affects the voltage of bit line that may be measured. (a) shows the transistor and ferroelectric capacitor as a memory cell, with three lines defining the accessibility to the cell. (b) shows the initial state of a reading process where the plate line is grounded and the bit line set to zero. (c) shows the intermediate step of the reading process as the plate line is changed to a voltage V_{PL}, with bit line floating causing a change in the ferroelectric capacitor's polarization and changes in the voltage on the bit line, thus allowing the state to be ascertained. (d) shows the plate line returned to ground. (e) shows the changes in the plate line as one proceeds through the reading process. (f) shows the load lines on the hysteresis curve under the various conditions.

random access memory (*dRAM*), one may use a single transistor, single capacitor (1T1C) combination in *feRAM*. To overcome the effects arising from the shifts in the hysteresis curves when the capacitor is kept in one or the other state, or from fatigue as a result of repeated cycling, one may use other combinations. We will discuss these in detail, but we begin with understanding how the charge of a ferroelectric capacitor cell can be made available for sensing. Ultimately, we will have to find how this sensing can be part of a design that is compact, that is, employs few elements and few interconnections; is robust, that is, is insensitive to variability, cross-talk or noise; is low energy, that is, operates at low voltages and low current in short time durations for writing and reading for low energies; and is reliable, that is, will function for lengths of times desired—many years of usage.

Figure 4.26 shows a 1T1C *dRAM*-like *feRAM* cell in (a), where the ferroelectric capacitor is accessed by a transistor. The significant difference between this cell and a conventional *dRAM* cell is that, in the *feRAM* cell, the capacitor's plate, which is not tied to the transistor, is accessible by a voltage through a control line called the plate line (*PL*). The other lines that access this cell—the word line (*WL*) and the bit line (*BL*)—are similar to those in *dRAM*. So, unlike in *dRAM*, where the plate terminal is grounded, in *feRAM*, one can modulate it by voltage. This is necessary if one wants to modify or sense the direction of polarization. This spontaneous polarization, in both its polarities, exists even at zero electric field and zero voltage across the capacitor. Through Gauss's law or Maxwell's equation, the charge is related to the electric field and polarization by

$$\nabla \cdot \mathbf{D} = \rho \quad \therefore \quad \hat{n} \cdot \mathbf{D} = \sigma, \tag{4.41}$$

where σ is the charge density in Coulombs per unit area. The only way to access this polarization, which is measurable through charge

We will use Q for charge, and charge per unit area, and it is understood that one will be able to understand this from the context. The equations must be dimensionally correct. We employed this approach in our discussion of nanoscale transistors.

or its conversion to current or voltage, is through changing it. A ferroelectric capacitor holds the polarization at zero electric field and zero voltage. Measurement of the two distinct states of polarization needs a change in the voltage or the field over a bias range so that a polarization change becomes accessible. The resulting charge change then depends on the span resulting from the change in polarization and the change in the dielectric component of charge, with the latter derivable by including it in the relationship in Equation 4.41. To accomplish this, the voltage needs to be changed since that is the external electrochemical control that we have using the transistor. Since the ferroelectric capacitor is sensed or written through the bit line (BL), the voltage change must occur through the plate line PL, as is shown in Figure 4.26.

In the quiescent state, for this example, that is, $t < 0$, all the voltages are at 0 V, that is grounded, and the capacitor is in one of the two polarization states of opposite polarity. So, the bit line voltage $V_{BL} = 0$ by grounding it, even as the capacitor is polarized in one of the two states. The bit line is now floated, that is, allowed to change in potential so that it can be sensed, and then the plate line voltage is changed from 0 V to V_{PL} V. Consequently, the polarization changes, the voltage across the ferroelectric capacitor (V_{FE}) changes, and the bit line voltage V_{BL} changes. The ferroelectric capacitor at time $t = 0^-$ is in one of the two polarization states. When the plate line voltage is changed from 0 V to V_{PL} at $t = 0$ s, the ferroelectric capacitor builds a voltage across it and the bit line voltage changes. The net charge on the bit line and its connection through the transistor to the ferroelectric plate is derivable by Gauss's law. The integral across displacement over the area remains constant, since there is no applied voltage to the internal node. This is the exchange of polarization with the mobile charge with the electric field in the system composed of the ferroelectric capacitor and the bit line that has a dielectric capacitance of C_{BL}. The absence of any change in this net charge lets us determine the load line, that is, the relationship between the external and internal parameters of voltages and charge that we can visualize on the hysteresis curve of Figure 4.26(f).

The governing equations can be visualized under three specific conditions: the condition at the start ($t = 0^-$), the condition at the transition ($0^+ < t < T$) and the condition at the end ($t > T$). At $t < 0^-$, with the bit line set to 0 V, the total charge,

$$Q_{BL} = C_{BL} \times 0 - A \times P(t = 0^-), \qquad (4.42)$$

where A is the area of the ferroelectric capacitor, $P(t = 0^-) = \pm P_r$ is the initial programmed state corresponding to logical 1 and 0 or H and L, and C_{BL} is the bit line capacitance. For $0^+ < t < T$, with the

Gauss's law, or the synonymous divergence theorem, leads to

$$\int_V \nabla \cdot \mathbf{D} \, d^3r = \oint \hat{\mathbf{n}} \cdot \mathbf{D} \, d^2r,$$

where $\mathbf{D} = \epsilon_0 \mathcal{E} + \mathbf{P}$. With the surface enclosing the ferroelectric capacitor's plate and the bit line, which serves as the plate of the bit line dielectric capacitance to the rest of the environment, the exchange between the polarization and the dielectric component follows.

bit line now floating, there is a voltage buildup on the bit line and an electric field across the capacitor, corresponding to a voltage of V_{FE}. These satisfy the charge relationship

$$Q_{BL} = C_{BL} \times (V_{PL} - V_{FE}) - A \times P(V_{FE}), \qquad (4.43)$$

where the plate line voltage is a constant but $V_{BL} = V_{FL} = V_{PL} - V_{FE}$, which depends on the charge Q_{BL} and the bit line capacitance C_{BL}, changes with time. For $t > T^+$, the plate line voltage returns to $0\ V$, and

$$Q_{BL} = C_{BL} \times (-V_{FE}) - A \times P(V_{FE}). \qquad (4.44)$$

The initial polarization can be either $+P_r$ or $-P_r$, and the bit line charge does not change, that is, it only exchanges between the dielectric and the spontaneous polarization components. This allows us to draw the load lines and determine voltages.

The change from $V_{PL} = 0\ V$ to a biased plate line state, with no change in bit line charge, means

$$
\begin{aligned}
Q_{BL} &= -P(0,0^-) + C_{BL}(V_{PL} - V_{FL}) \\
&\quad - A \times P(V_{FE}) \\
\therefore\ P(V_{FE}) &= \pm P_r + \frac{C_{BL}}{A}(V_{PL} - V_{FE}) \\
&= \pm P_r + \frac{C_{BL}}{A} V_{BL}.
\end{aligned}
\qquad (4.45)
$$

This is the load line that relates the voltages, or polarizations, to the externally applied voltage V_{PL} and the ground, as shown in Figure 4.26(b–d). Various forms of this line, for different conditions, are shown in Figure 4.26(f). Of the voltage V_{PL} applied, some of it appears across the ferroelectric capacitor (V_{FL}) and some of it across the bit line (V_{BL}). The slope is $-C_{BL}/A$.

When the starting polarization state is $-P_r$ on the hysteresis curve, and the plate line voltage is applied, the load line in the steady state leads to a stable point, which in Figure 4.26(f) is represented by the intersection of the hysteresis curve with the load line marked $\mathfrak{A}$. The ferroelectric capacitor now has a polarization $P(V_{FL})$ given by this intersection point, which is marked as A on the lower hysteresis curve. The bit line voltage is V_{BLH}, a high bit line voltage, and the polarization has reversed. If the starting polarization is $+P_r$, the application of the plate line voltage will lead to the load line $\mathfrak{B}$, whose stable point at the intersection with the hysteresis curve is B, with a bit line voltage of V_{BLL}, which is a lower voltage than in the last case. The polarization remains in the same direction. The releasing of the plate line voltage takes the load line to $\mathfrak{B}'$, and the ferroelectric capacitor returns to a state of polarization of magnitude $+P_r$. So, a small bit line voltage appears, with the polarization maintained in the sense cycle. In the case $\mathfrak{A}$, however, the polarization changes sign, and when the plate line voltage returns to $0\ V$, an inner top polarization branch is traversed, with A' as the new stable point. At this point, there is a negative voltage across the ferroelectric capacitor, and a smaller positive voltage across the bit line. The state of the memory has been disturbed, and one needs to write back the negative polarization $-P_r$ to this cell. This read disturb has been, until now, unavoidable when electrically measuring ferroelectric memory to achieve a strong signal, even as the integration goal is robustness, error avoidance, low power, low energy and fast speed.

We will not discuss the write process in any detail. It is relatively straightforward. If one knows the state that needs to be programmed in, one needs to appropriately bias the bit line and the plate line and turn on the access transistor. For programming $+\mathbf{P}_r$, this means a high bit line (a positive voltage) and a plate line that is set to a low voltage such as $0\ V$. For programming the reverse, $-\mathbf{P}_r$, one sets the bit line to $0\ V$ and sets the plate line high, that is, to a positive voltage.

This is only approximately correct. The hysteresis curve as drawn is with all the voltage being applied across the ferroelectric capacitor. But with the load line $\mathfrak{B}$, only a fraction of the voltage appears across the ferroelectric, so the hysteresis loop traversed back is inside the loop as drawn. But since a bias voltage quite in excess of the coercive voltage V_{co} appears across the capacitor, the disturbance in polarization is small. This is not true for the load line $\mathfrak{A}$, where the polarization is reversed, and the amount of voltage in excess of the coercive voltage is small.

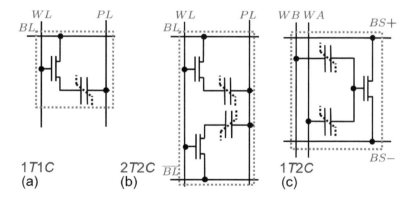

1T1C
(a)

2T2C
(b)

1T2C
(c)

Figure 4.27: Three examples of random access memory cells using ferroelectric capacitors. (a) shows a single transistor single capacitor (1T1C) cell that uses three biasing lines—a word line (WL), a bit line (BL) and a capacitor plate line (PL)—to write and read a ferroelectric capacitor cell randomly. The accuracy of sensing is improved by storing both the true value and its complement in capacitors as shown in (b), thus avoiding the imprinting effect. (c) shows an arrangement where the capacitors are back to back, with a floating connection that drives the reading transistor. (b) and (c) are arrangements that extract a high penalty in area because of their additional elements and connecting lines. (a) shows an arrangement with an additional connection to the capacitor plate, which in a *dRAM* is usually a ground.

We now explore ferroelectric cells, to understand their limitations and the issues that must be dealt with in design and exploiting this effect at the nanoscale. In particular, we are interested in dealing with the time-dependent changes, the question of sensing low voltages and compensating for various distracting effects such as interference and changes in signals that come about with scaling of dimensions and density. Figure 4.27 give three examples of representative cells. Figure 4.27(a) shows the standard cell and (b) shows a two-transistor and two-capacitor (2T2C) combination that increases the signal margin. Or, one may employ new techniques that take advantage of the uniqueness of the polarization-dielectric charge exchange as in the 1T2C approach shown in Figure 4.27(c).

In each of these examples, cells ordered in an array can be accessed to write and read individually, uniquely and in any order, that is, randomly. All these are examples of random access memory (RAM). To write and ascertain the state of any cell, one needs to employ a suitable combination of choice of bias voltages applied to the lines. Consider the simplest example of Figure 4.27(a) first, as summarized in Figure 4.28. In this figure, the circle is meant to mark the cell region as defined by the transistor, the capacitor and the connections made to them by the three lines—the word line (WL), the bit line (BL) and the capacitor's other plate line (PL)—as well as the internal connections within a cell.

The writing process is accomplished, as summarized in Table 4.3, by placing appropriate bias conditions onto the cell during the write cycle. If a logic state 1, which is equivalent to the q_1 in our notation of Chapter 1 and characterized by the $+\mathbf{P}_r$ of the ferroelectric capacitor, is to be written, the bit line (BL) is raised to high (H, a high voltage such as V_{DD}), the other plate of the capacitor is grounded by specifically grounding the plate line (PL), and the bit line is connected to the ferroelectric capacitor for the writing by raising the word line (WL) to a high that turns on the transistor. Since only the specific intersection of the WL, the BL and the PL has these conditions simultaneously appearing, only this cell is written to $1 \equiv +\mathbf{P}_r$ condition when the BL-PL voltage difference is larger than that required for the application of the coercive field $\mathcal{E}_{co}$. The rest of the cells have only some of these voltage biases, or none appearing. So, there is some disturbance, for example, a turned-on word line or a turned-on plate line will apply these voltages to all the cells in the column. However, these other cells have the bit line floating, so the entire cell's voltages rise and fall, and thus any ill effects of this disturbance are to be avoided through the design. The 0 logic state ($\equiv q_0$ in our state nomenclature), as shown in Table 4.3, is written by taking the bit line to ground (Gnd), and the plate line to high (H), that is, the exact reverse of the previous case. With the bias voltage across the capacitor reversed, the cell is now programmed to $-\mathbf{P}_r$ polarization through the application of the voltage.

Reading these states requires sensing the polarization through the charge. This is accomplished by sense amplifiers. In ferroelectric RAM, the sense amplifiers usually sense the voltage created by moving the charge on to the capacitance of the bit line. In memory, in general, depending on the speed needed of the response as well as the constraints of the memory, the sensing may either employ current or be directly voltage driven. Dynamic RAM employing a dielectric capacitor and transistor also may use the charge-to-voltage or charge-to-current conversion. Static RAM, which uses voltage-driven and

Non-volatile $NAND FLASH$ memory, a highly dense bit storage arrangement, accesses bytes—an entire word—instead of each bit individually. It is also erased in entire blocks, hence the name $FLASH$. It is not RAM. Non-volatile floating memory in NOR form is RAM.

	Logic state	Word line	Bit line	Plate line
Write	1	H	H	Gnd
Write	0	H	Gnd	H
Read	1	H	Flt	H
Read	0	H	Flt	H

Table 4.3: Write and read operation conditions for a ferroelectric memory cell consisting of a transistor in series with a ferroelectric capacitor. The word line drives the gate of the transistor. The bit line controls the writing and reading of the capacitor through the transistor. The plate line is the line connecting the other end of the capacitor. H, that is, high, means biasing to the high positive bias, V_{DD}, for example, of the circuits. Gnd is ground, that is, a bias of $0 V$, and Flt indicates that the line is left floating.

voltage-producing back-to-back amplifiers, may employ direct voltage detection or convert the voltage to current for detection. In our example, we consider the simpler case. We measure a voltage created by placing the charge on the bit line, so the voltage $V \approx \Delta Q / C_{BL}$, where C_{BL} is the bit line capacitance. The way this charge appears on the bit line is described by the last two rows of Table 4.3. If the logic state of the capacitor is 1, that is, in the $+\mathbf{P}_r$ state, and one floats the BL, applies a high voltage to the PL and turns on the access transistor via the WL, one swings the capacitor towards negative polarization. This is a large change in the charge on the bit line. If, on the other hand, the logic state is 0 and the plate line is set high, a small change in the charge will occur. So, $+\mathbf{P}_r \equiv q_1 \equiv 1$ leads to a larger signal than $-\mathbf{P}_r \equiv q_0 \equiv 0$ does. This is shown in Figure 4.28(b), together with the differences in signals one may observe because of variability.

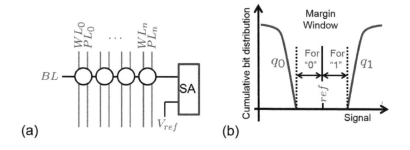

(a)   (b)

Figure 4.28: (a) is a schematic showing a ferroelectric memory device consisting of a bit line (BL), a word line (WL) and a plate line (PL) as well as a sense amplifier and a reference voltage to write and read a ferroelectric memory. (b) shows the behavior of measured signal distribution, a measurement of voltage when the charge from the ferroelectric capacitor is placed on the bit line (BL), for the 0 and 1 logic stages. The tail in this distribution is due to the variety of variational effects unfolding in the device, particularly in the ferroelectric capacitor.

In this type of ferroelectric memory, as shown in Figure 4.28(a), the sense amplifier utilizes a reference voltage V_{ref}, which is between the voltage created by a 1 state and that created by a 0 state. Figure 4.28(b) shows this as well as the variability in the measurement corresponding to the statistical distribution of the two states. Recall that fatigue, imprint and other stochastic effects will exist in this type of memory. With V_{ref} in between, the margin window for detecting q_0, which is the logic 0 state, and q_1, which is the logic 1 state, is about half of the difference between the worst-case signals. How may one generate V_{ref}? The simplest way in practice is to have two dummy cells, which are representative of the memory cells that are being written and read, set one to logic 1 and the other to 0, and take the mean of their output to create the reference to which the measured cell can be compared. This comparison is accomplished through a comparator, which is a simple back-to-back inverter operating as an amplifier driven by these signals. This $1T1C$ configuration uses three lines, so it has larger number of connections than $dRAM$.

It also occupies a larger number of lithography squares than $dRAM$ does, and it has a poorer margin, as the margin window through the reference to which a comparison is made is halved.

The other deficiency of this measurement approach is that, irrespective of which logic state, 0 or 1, the memory cell was in before reading occurred, it will always end up in one unique state, since the same high plate voltage is applied with the bit line floating in the two cases. One now needs to write back the data that was read since, for at least one of the logic states, the state will be flipped. The major attractive attribute is, of course, that this structure is non-volatile and operates at low voltages with high speeds.

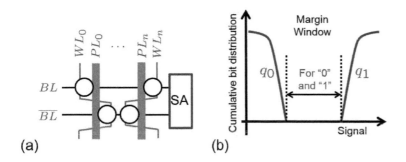

(a)

(b)

Figure 4.29: (a) is a schematic corresponding to the cell shown in Figure 4.27(b), where two transistor-capacitor combinations and a common plate are employed for memory storage. Both the bit and its complement, so 0 and 1, are stored. This requires now a BL and its complement $\overline{BL}$ connection, eliminates the reference, and opens a larger margin window, as shown in (b).

Variability due to various causes, but particularly due to imprint, fatigue and depolarization, is an important issue for ferroelectric memory. In Figure 4.27(b) and (c), we show two examples of cells that achieve improved levels of reproducibility. The operational layout and window behavior of the first is schematically shown in Figure 4.29. Since there is a common plate and doubling of the transistors and capacitors, the 1 or the 0 states in this cell are each stored in oppositely polarized capacitors, that is, one may choose $1 \equiv (+\mathbf{P}_r, -\mathbf{P}_r)$ and $0 \equiv (-\mathbf{P}_r, +\mathbf{P}_r)$. The common plate line allows one to accomplish this through the complementary $BL, \overline{BL}$ signals and through the common access transistor's word line. When a measurement is made, the entire window is measured. It is the polarity of the change that indicates the stored state, and the need for the V_{ref} of the 1T1C is avoided. The design trade-off is now the doubling of transistors and capacitors, and the need to also have signaling for $\overline{BL}$. As would be expected, these cells are more error proof and more reliable. But, all these examples demand larger area.

Conceptually, one could imagine incorporating a ferroelectric within the gate-insulator region and thus directly measuring the state through the transistor's conduction—a self-amplified self-reading device. But, a ferroelectric clad by a dielectric, as we have seen, has a depolarization field. This field then degrades the remnant polarization over time, and the structure is not non-volatile any more. We discuss Figure 4.27(c) as another example in this attempt to achieve immunity to degradation mechanisms, in this case from depolarization. Like the previous example, 2*T*2*C*, this 1*T*2*C* combination stores the data in two identical capacitors with opposite polarization to open a bigger window and eliminate common shifts in the two opposite polarizations. Depolarization fields degrade the ferroelectric storage, and this structure consisting of two capacitors tied to the gate prevents depolarization by storing opposing polarization with a shared metal electrode. Imprint and fatigue will still exist, but the window will again be large. Table 4.4 describes this cell's write and read conditins, which we will now explore to achieve physical insight into such structures.

	Logic state	WA	WB	BS−	BS+
Write	0	H	Gnd	Flt	Flt
Write	1	Gnd	H	Flt	Flt
Read	0	H	Flt	Gnd	Flt
Read	1	H	Flt	Gnd	Flt

Table 4.4: Write and read operation conditions for a ferroelectric memory cell where the transistor is only employed as a read element with a floating gate that is connected to the plates of two capacitors. This is for the cell shown in Figure 4.27(c); see that figure and Table 4.5 for an explanation of the abbreviations used.

The word lines *WA* and *WB* are connected to the two ferroelectric capacitors. If one applies a sufficiently high voltage between the two lines, as the two capacitors are in series, they will polarize in an aligned manner between the two word lines but will be asymmetric at the floating node that is connected to the transistor. We write the logic state 0 by holding *WA* high and *WB* to *Gnd*. This polarizes the two capacitors to identical magnitudes. To write logic state 1, we employ the exact opposite condition, holding *WA* to *Gnd* and *WA* high. Polarization is now in the opposite direction. Again, looking from the floating gate node, the polarizations are asymmetric but in the opposite direction. This is shown in Figure 4.30(a) and (b), in the two cells, in the polarizations indicated on the outside. If the capacitors are identical, since the polarizations are opposite, if one grounded *WA* and *WB*, or floated them, the floating gate voltage would be zero. There would be no charge on the gate electrode of the transistor or on the plates of the capacitor, and therefore no depolarization field, whether we stored 0 or 1. Note that the transistor was not connected at source (*BS*−) or drain (*BS*+) during this operation and the only electrical elements we need to look at carefully are these capacitors. The transistor only serves to read this series capacitor combination's state by probing the voltage of the common plate with suitable reading biases applied to the word line. The voltage will be defined by the word line bias voltages and the charge sharing from the transistor's capacitance load. If the voltage is low, a small or no current flows through the transistor, whose source is grounded through $BS-$. If the voltage is high, a large current flows through, and this charges the bit line $BS+$ much more. Figure 4.30 show these read operation load lines across the hysteresis of the capacitors.

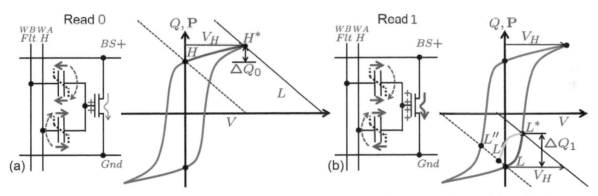

Figure 4.30: The load line behavior along the hysteresis curves during the read operation of a 1T2C ferroelectric floating gate memory cell. (a) shows the behavior when a logic state of 0 is stored in the cell, and (b) shows it for a logic state of 1.

To read, one floats the WB line, applies a voltage to WA, grounds $BS-$ and floats $BS+$. If the stored state is 0, as shown in Figure 4.30(a), the applied voltage V_H drops across the capacitor tied to WA in series with the transistor's gate dominated capacitance. This is reflected in the load line shift that is shown. A charge ΔQ_0, a small charge, is now available and shared. Depending on the transistor design, a small charge transfer happens through the channel of the transistor, and the $BS+$ voltage is modified. However, this read process of a 0 logic state leaves the polarization of the two ferroelectric capacitors in the same orientation as they were before, as shown in the left half of the figure. The trajectory HH^* shows the hysteresis arc that is traversed in the ferroelectric capacitor that was biased. The situation with reading if a logic state of 1 was stored in the ferroelectric capacitor combination is different. This is shown in Figure 4.30(b). The capacitor tied to WA now flips in polarization, if a large-enough voltage is applied, or even at low voltages traverses the steeper domain realignment region of the hysteresis curve. It shifts along the trajectory LL^*. This is a larger charge that can be distinguished from the previous case. If a very large voltage was applied to read and then released, one may traverse the entire hysteresis curve to the positive high voltage well beyond the coercive conditions and then return to L''. If a small voltage is applied, one may follow a minor hysteresis loop and return to L'. In any of these cases, one returns after reading to a condition in a quadrant of bias where polarization and electric field are both aligned. No depolarization exists. But, one must still rewrite, since the state has been disturbed.

This example is worth pondering over, since it shows that, in many of the nanoscale structures employing new phenomena, one will have to consider multiple effects together so that one may find an approach that together allows a low energy and robust operation.

We end with a discussion of how, given that one has created a small voltage on the bit line to read, does one sense. We do this with reference to the cell where both the bit and its complement are stored by employing BL and $\overline{BL}$ signals, as shown in Figure 4.27(b) . There are two significant consequences of employing this approach. One that, irrespective of the state, 1 or 0, one of the bit lines will have a smaller change in charge and the other larger so that the difference between the two will be the charge associated with the difference in polarization. This is a large span. The second is that the data is stored in the polarity. If logic 0 is one orientation of complementary storage, logic 1 is the opposite. So, the signal measured between BL and $\overline{BL}$ will be exactly reversed. One needs to measure this with respect to an appropriate reference that is in the middle of the window, irrespective of the long-term drifts in the characteristics of the storage. The simplest way to achieve this is to have a reference cell storing the H and the L state. By equalizing the bit line and its complement for this reference during a read operation, one gets the mean. Now, when reading any cell along a row by turning on its word line and causing a change in the BL to $\overline{BL}$ voltage, we can compare this voltage to the reference cell. The polarity of the change determines what state was stored in the cell. And this polarity direction change can be measured by a sense circuit consisting of back-to-back inverter.

Figure 4.31 and Figure 4.32 show examples of such approaches to sensing. This is a detail very important to any realism of a memory from nanoscale, that is, beyond the basic physics of an effect or a device, and we illustrate it using ferroelectrics. This is a little more complex as it attempts to also compensate for the longer-term drift in ferroelectric storage characteristics that we have discussed.

In Figure 4.31, the part on the left is the memory cell array, with only a 2×2 array section shown. The part on the right is the circuit that generates the reference voltage from reference cells in which H and L logic states are stored. The bit lines, and their complements, from the two sides intersect in the middle region where the sense amplifiers are, allowing one to compare the signal from the cell with that from the reference, to determine its state. This process requires

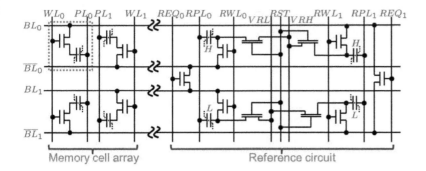

Figure 4.31: A circuit schematic showing an example of array organization and sensing that allows lower voltage operation while accommodating the imprinting, fatigue and variabilities encountered in ferroelectric memories employing transistor access to a ferroelectric capacitor to comprise the memory storage and access cell.

appropriate timing and sufficient time duration for changes to happen to near-completion in the worst-case conditions. And all the parameters—the level of the voltage that the lines are being driven with; the charge sharing between the capacitance of the bit line and that of the capacitor, as this sharing determines the level of the bit line voltage; and the interference effects that exist because of the rapidly changing signals in the lines, and the coupling between the lines, as well as the changes taking place in the ferroelectric itself—must be compensated for in the timing.

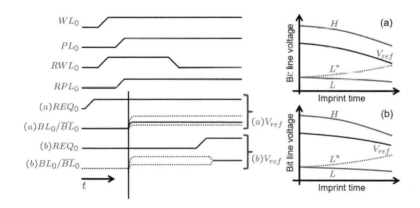

Figure 4.32: Timing diagrams for two sensing schemes. The case shown in (a) equalizes the reference cell reading of the H and L state before reading by comparing them with the reference signal. The case shown in (b) first generates the reference high and low signals before equalizing to generate the reference signal. The imprint effect of this is shown in the generated reference signal for the two cases. In the case shown in (a), the opposite state of L, L^*, which shifts up into high storage (for example, as in Figure 4.25(a)) leads to a disappearing margin after a long imprint time. In the case shown in (b), the delayed equalization means that the reference voltage is larger and the margin more robust.

The simplest case for making this comparison is that shown in Figure 4.32(a), where we only address the bits identified through the subscript 0. The read equalization signal shorts the BL and $\overline{BL}$ to equalize their voltages. A reference word line (RWL_0) turns on the word line of the reference cell for the row 0 of the memory array on which many memory cells exist, followed by the read plate signal for the reference cell (RPL_0), and the plate signal for the cell to be read (PL_0). This builds the two inputs to the sense amplifiers that need to be compared. The reference signal, because the line are equalized

and symmetric, is the mean and equalized voltage created by the
reference cell state (V_{ref}) shown in the imprint time dependence
of the case shown in Figure 4.32(a). The data H and L are being
read while the equalization is in effect, and the voltage on the bit
lines is lower because of the equalization in effect. The voltage being
measured of the cell is also below the supply voltages, due to the
voltage division between the cell plate capacitor and the capacitance
of the bit line. Because of imprinting, the voltage V_{ref}, in time, will
approach the opposite state L data. The window will narrow. In the
worst cases, it may cross. So, to avoid imprinting's consequences, one
would have to employ larger voltages and therefore also consume
more power.

One way to overcome this narrowing and crossing is to delay
equalization, shown in Figure 4.32(b), in timing and in the imprint
effect. Now the bit line voltages for the H and L of the reference cells
are allowed to build to their voltages, accentuating the H, before
equalization takes place. This, as shown in Figure 4.32(b), although
the window narrowing is still present, produces a reference voltage
(V_{ref}) that is in the middle of the worst-case window. This sensing by
a more robust generation of V_{ref} through delaying the equalization
is one of many approaches that are employed so that voltages can be
reduced and the circuits adapt to the changing characteristics of the
devices with time.

Recall our earlier discussion that
voltages applied to the ferroelectric
capacitor affect both the "same state"
and the "opposite state" retention. This
is particularly true at below coercive
conditions.

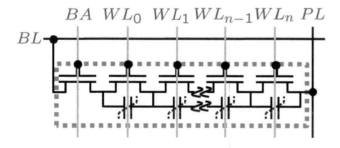

Figure 4.33: A chain ferroelectric memory cell. A capacitor can be shorted
without the loss of data encoded in the
polarization, and thus each individual bit can be measured by accessing
it through the bit access (BA) signal
applied to the access transistor to the
chain. The n-bit cell whose boundary
is shown is accessible through a common bit line (BL) and a common plate
line (PL), thus reducing the number of
connections and improving the density.

Ferroelectric capacitors have another attribute whose one consequence was employed in the 1T2C example. It is that polarization
can exist without charge, and therefore capacitors can be employed
in series. The chain ferroelectric memory device shown in Figure 4.33
exploits this property to increase density and, in particular, to reduce
the capacitance ratio between the bit line and ferroelectric capacitor
that determines the charge sharing and therefore the voltage that is
sensed on the bit line. As density and size increase, the reduction
of bit line capacitance so that the signal to be sensed remains large
is important. In chain memory, an n-bit word, for example, an 8-bit
word, is ganged and accessed through one bit line and one plate line,

with the access provided by the bit access transistor driven by the bit access signal BA. This reduces the number of connections hanging on the bit line and hence its capacitance. In order to write or read a specific bit, one accesses that bit by shorting the other bits with appropriate timing of the signals. So, in Figure 4.33, if the bit tied to the WL_{n-1} and the BL_i lines is to be written, the other memory bits stored in ferroelectric capacitors tied across transistors gated by the other word lines must be shorted. So, the word lines, other than WL_{n-1}, are raised high, turning on their transistors. The use of BL_i for the bit line, PL for the plate line, and BA for the access transistor, which now serves as the word line access for a $1T1C$ cell, provides access to the ferroelectric capacitor that WL_{n-1} accesses. Thus, this n-bit word chain is accessed by one bit line contact and one plate contact using one access transistor. These memories allow continuation of the use of low voltages and hence exploit low energies and low currents that are key characteristics of ferroelectrics. Because both the energy and current are small, achieving a high bit transfer rate in both writing and reading, a result that is particularly difficult to obtain innon-volatile memory, becomes possible.

4.3   Electron correlations and devices

THE ELECTRON, AS A CLASSICAL INDEPENDENT PARTICLE IN A COLLECTION, AS A QUANTUM PARTICLE AND AS A PARTICLE IN-TERACTING with other particles and the rest of the environment, has been employed in a multitude of ways in the solids of our discussion. The Drude model is classical and, together with the expectations of ensembles, sufficed in many situations of interest in classical electronics in semiconductors. In reduced dimensionality and reduced dimensions, as well as in degenerate conditions, it broke down, and so so we employed Fermi velocity and wave-based arguments instead. The importance of quantum effects is manifest in all these.

This breakdown of the classical physical description shows up acutely in metals. The energy of a moving electron in a metal at Fermi energy can be of the order eVs, that is, it can have a temperature of $1\ eV/k_B \approx 10^4\ K$, as opposed to the thermal energy at room temperature ($300\ K$), which is $25.9\ meV$. This is so because there are a large number of such electrons and the ones that contribute to motion are at the surface of the energy distribution. Motion description and its characteristics are not the only attributes that change in these conditions. Sommerfeld recast the nearly free electron Drude picture in metals to a middle ground that was primarily classical but with quantum-mechanical trimmings to explain the characteristics of

Arnold Sommerfeld is another doyen from the era when our understanding of the world changed from classical to relativistic and quantum mechanical—the Sommerfeld model, which is the recasting of the Drude model in light of some of quantum-mechanical considerations, is often taught. Along with Einstein and Max Born, Sommerfeld is considered one of the elders of mathematical physicists. Four of his students, and three of his postdoctoral students, received Nobel prizes in quantum physics and its intersection with relativity at high nuclear mass, as well as in high energy nuclear physics. This tree continues to propagate and does not count numerous protégés. Heisenberg, one of his students, called him the preeminent recognizer and raiser of talent. Nearly a third of the great German scientists of the first half of the 20th century were connected to him in significant ways.

specific heat as well as paramagnetism. In all these descriptions, the electrons form a Fermi gas—the solid is populated by essentially a free electron gas over much of the Brillouin zone with the electrons being very separated in a dilute condition and acting independently. Departures appear where there are degeneracies, such as through bandgaps at Brillouin zone boundaries, but, for the most part, an electron in such conditions is a nearly free electron whose momentum and kinetic energy are associated with an effective mass that incorporates the effect of the crystalline or non-crystalline environment in which it resides. However, this approach is grossly inadequate when one starts looking at a variety of properties and the more advanced examples.

When two electrons approach each other, the Coulomb interaction energy diverges. This is not a weak interaction anymore. A simple gedanken experiment suffices to illustrate the difficulties. There are n electrons per unit volume. This charge is balanced by an equal and opposite charge in the background of uniformly distributed positive charge. This is the ``jellium´´ arising from the ionic background of the matter. This electron, on average, is localized to a volume of n^{-1}. Quantum-mechanically, localizing to this volume has a kinetic energy cost, per electron, of $(\Delta p)^2/2m = (\hbar/2n^{-1/3})^2/2m = \hbar^2 n^{2/3}/8m$. When the electron is localized to a volume, the Coulomb interaction energy is $e^2/4\pi\epsilon n^{-1/3}$. This is the potential energy cost.

In a Fermi gas, with low carrier densities, these two—the potential energy from Coulomb interaction, and the kinetic energy from motion—balance when

$$\frac{\hbar^2}{8}\frac{n^{2/3}}{m} = \frac{1}{4\pi\epsilon}e^2 n^{1/3}$$

$$\therefore \; n^{1/3}\frac{4\pi\epsilon\hbar^2}{me^2} = n^{1/3}a_B^* \approx \text{a constant.} \quad (4.46)$$

$a_B^* = (\epsilon m_0/\epsilon_0 m^*)a_B$ here is the effective Bohr radius in the material. At conditions of carrier densities where one approaches this kinetic-potential competition balance, this approximation of mobile charge in a ``jellium´´ approximation will break.

Electrons screen. In situations where the average inter-electron distance is smaller than the range of interaction, for example, the screening length $\lambda_{scr} \approx (c/e^2m)^{1/2}N^{-1/6} \approx 1\ nm$ in many metals, the Fermi gas view will fail. One can imagine the short-range behavior prescribing the gas-state properties being supplanted by long-range interactions, because electron-electron interactions are spatially frequent at dimensions smaller than screening length scales and are therefore not isolated in an ensemble. The electron-electron interactions dominate, and this would change the behavior of the matter through that of the electron. For example, one can imagine that spin

This kinetic energy (T) and Coulomb interaction energy (U; not to be confused with total free energy) are two parameters whose relationship appears in many a place, since T is associated with the propensity to travel and hence provide conduction, while U specifies localization and hence the loss of conductivity. Their referential relationship will be important for conduction-insulation behavior, as many-body interactions appear in materials even more complex than the normal group *IV* and group *III–V* semiconductors.

would play a role, as quantum constraints dominate, and show different properties related to magnetization; that band behavior would change, due to long-range interactions; and that, by modifying the intensity of the interactions, one might be able to change properties in a way suitable for nanoscale devices.

The ferroelectric phase transition that we just explored is a thermodynamic phase transition in the sense that we arrived at it through internal energy and entropy competition. The breaking of the symmetry of the ground state at temperatures below the critical temperature led to an ordered lower entropy state. The entropy-energy balance is modified by temperature, stress, pressure or other electrochemical means and mechanisms for changing order.

Phase transitions also happen due to the nature of the quantum state, without recourse to broken symmetry. In this section, we started with a simple argument to bring this out. Transitions may result from the competition between the opposing behaviors of kinetic energy and potential energy. Electrons like to spread out to optimize their kinetic energy. They achieve their lowest energy by occupying the lowest available energy states—the permitted states that are spatially separated at the lowest wavevectors. But, interactions within the crystal force them to localize so that potential is minimized for stability. Electrons repel each other with a $1/|\mathbf{r}_1 - \mathbf{r}_2|$ dependence. The state of the system is defined by these competing energy components. This energy balance can also be modified by external energy exchange parameters: injection of carriers, strain, applied potentials, and interactions with other sources of energy such as photons. This phase transition, arising from quantum interactions, is what we will study now.

These situations, where interactions can be strong, are common in normal metals and other fermionic systems. Many such examples can be understood and treated as Fermi liquids consisting of weakly interacting fermions. While the electron occupies higher energy states as a fermion, following the Pauli exclusion principle, the behavior of the interacting fermions in a normal metal can be predicted via single particle excitations, that is, coupling of a quasiparticle to the rest of the fermion sea consisting of the filled states with the rest of the fermions (see Figure 4.34). This is the Fermi liquid. It is an intermediate region, where symmetry has not been broken. Strong changes in properties, however, still result.

The cause of the significant change in properties is that the scattering of quasiparticles is strongly reduced by the Pauli exclusion principle, and therefore their lifetimes are long. This argument and its applicability are based on adiabatic continuity. In the absence of broken symmetry, during the interaction, the non-interacting ground

Strongly interacting fermionic conditions can also exist. The quantum Hall effect was a consequence of strong interaction resulting in a collective response. These effects are not yet of significant import in devices, but may well become if, for example, breakthroughs happen in quantum computation.

A Fermi liquid is different from a Fermi gas in that, in the former, the fermion—the electron—is treatable as interacting with the other fermions representable by a collective behavior in a quasiparticle. In contrast, in a Fermi gas, which is a very dilute condition, the fermion is an independent particle interacting with other independent particles. A liquid, such as water in a pond, responds to a thrown stone by exhibiting collective behavior arising from the water molecules' close proximity to each other. Skipping a stone across the water shows the long-range effects of the stone's interaction with the water's collective quasiparticle response, with this particular interaction causing the stone to hop as its scattering response.

state evolves smoothly and adiabatically with increase in interaction strength.

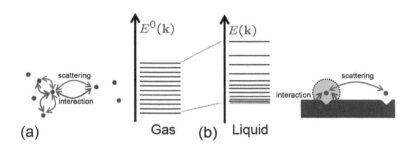

Figure 4.34 conceptually describes this evolution in this many-body interaction. Absent particle-particle interaction, the bound states of a particle confined in a quantum box evolving in a harmonic potential well change, but the number of nodes and so of quantum numbers, and the count of the states remain the same. The topological characteristics are invariant. In a metal, as the electron density increases and the screening length scale (λ_{scr}) becomes important, if the symmetry of the Hamiltonian is still reflected in the ground state, then even as the bound state energies change, the quantum states distribution remains intact, and the interaction may be interpreted as arising from a "quasiparticle" and a "quasiantiparticle." So, in Figure 4.34, the distribution function of the particle arising with the inclusion of the interaction remains the same as that of the gas, but the filled ensemble's behavior can be described as a collective response.

We discussed state evolution in topological context in Chapter 3. Fractional quantum Hall, spin Hall, and topological insulators manifest properties arising from topology and which are of interest to us in nanoscale devices.

The electron and the hole—the antiparticle of an electron—are the particles of the gas. In the liquid phase, the excitation exists only with states close to the ground state of the liquid composed of the filled states of electrons. This causes a strong change in characteristics. The scattering process is suppressed by the Pauli exclusion principle, even as the interaction becomes large. In a Fermi gas, interaction and scattering vary in similar ways, since a single independent particle description suffices at low density. In a Fermi liquid, where particles have a long lifetime due to reduced scattering, the approach using quasiparticle and quasiantiparticle is a suitable, easier and physically more intuitive means to describing the system. One need not worry about all the rest of the electrons, as their energies and their interactions are similar to those in a Fermi gas. We can just view the quasiparticle as an electron surrounded by antiquasiparticle, hole-like excitations. In the liquid, the distribution of excitations can take a well-defined shape around the surface. Because scattering is suppressed, it is stable. In contrast, in a Fermi gas, with scattering

This particle–antiparticle description is the general behavior if symmetry is maintained. He^3, a fermion, also shows this, and one may view it in terms of the quasiparticle and the quasiantiparticle, that is, the quasihole state. Many important behavioral consequences arise from this. For example electrons on the surface of He^3 form a two-dimensional state called a Wigner crystal. However, here we will only discuss effects that are known to have strong device consequences.

and interaction behaving similarly, any anisotropy will soon be lost. Physically, what this means is that important degrees of freedom in the liquid are mostly frozen out. Only electrons at the Fermi surface participate in the interaction process.

Consider what happens if an electron is added to this liquid, which resulted from the gas and therefore is composed of plane wave states. The single quasiparticle excitation implies, that, on the Fermi surface, when an electron of energy $\epsilon_\mathbf{k}$ is added, the resulting energy, which may be linearized as a first order perturbation, varies as

$$\epsilon_\mathbf{k} = \frac{\delta E(\mathbf{k})}{\delta n_\mathbf{k}} = E_F + v_F(k - k_F), \qquad (4.47)$$

where E_F is the electrochemical potential, which must be homogeneous in equilibrium as it is a thermodynamic potential. The Fermi velocity then is

$$v_F = \frac{\partial \epsilon_\mathbf{k}}{\partial k} = \frac{\hbar k_F}{m^*}. \qquad (4.48)$$

The Fermi wavevector does not change as a result of interactions.

This isotropic system is incompressible. Therefore, the effective mass m^* subsumes the lowest order effect. For us, the principal way the behavior appears in nanoscale devices is that the properties of Fermi liquids are determined by the Fermi surface of their excitations.

Why is this important? We have just set the argument that a liquid with a single quasiparticle excitation can be distinguished from a nearly free single electron.

What are the major differences in the consequences that arise from the two systems?

First, consider the single particle description. Recall that the simplest definition for a semiconductor is that it is a crystalline assembly where, at absolute zero in an ideal matter, bands are either filled or empty. Filled bands are the valence bands. Empty bands are conduction bands. A semiconductor is an example of a band insulator where temperature, doping, et cetera, allow us to modify the occupation of states and so lets us use them with a control exercised through electrical potential and charge transfer. In semiconductors, it is the interaction of the electron with the periodicity of the atoms that gives rise to the complex band structure that defines the E-$\mathbf{k}$ states available to the nearly free electron. But this band insulation due to filled and empty bands, as in Figure 4.35(a), as well as the use of it to distinguish semiconductors from metals as partially filled band materials, or semimetals as materials of overlapping bands with empty states, is not general. Transition metal oxides, halides and chalcogenides provide plenty of counterexamples. Fe is a metal with $3d^6 4s^2$, while FeO is an insulator—an antiferromagnetic one, at that—with

We can view this potential as the Helmholtz free energy change when one additional electron is added to the system, that is, $E_F = \mathcal{F}(N+1) - \mathcal{F}(N)$, where $\mathcal{F}$ is the Helmholtz free energy. $d\mathcal{F} = E_F N - S dT - p dV + \sigma d\varepsilon$, where σ is the stress force and ε is the distortion.

Strictly speaking, this incompressibility is a consequence of Luttinger's theorem, which states that the volume enclosed by a material's Fermi surface is directly proportional to the particle density. This description breaks down in a one-dimensional system. At the nanoscale, this has consequences. In these Luttinger liquids, spin and charge density waves may propagate independently. An example of this is plasmons—charge-electromagnetic excitations that propagate at a speed determined by the strength of the interaction and by the average density.

$3d^6$ as the occupation number. This band insulation property arises in quantum interference of electrons as fermions. The number of electrons per unit cell, when they have specific occupation numbers, lead to this insulation.

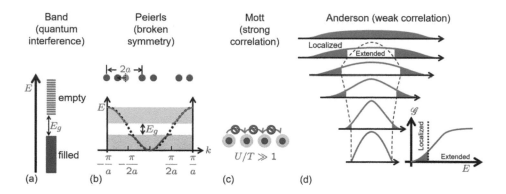

Figure 4.35: The (a) band, (b) Peierls, (c) Mott and (d) Anderson transition effects that lead to major property changes within the single electron description.

Peierls instability—a transition—shows how the metal-to-insulation change may happen when a unit cell and its occupation number is changed. We will see that the Hubbard transition is another example of this. So, more generally, the insulating characteristics, even in this single particle description, between the electron and the crystal can arise from additional effects. The Peierls transition, shown in Figure 4.35(b), causes a bandgap to appear because of electron-ion interactions with a lattice deformation. When the elastic energy that is required for a deformation can be overcome by a gain in the energy of the electrons, a new periodic potential exists. This new periodic potential causes a metal-to-insulator transition. It is a phase transition arising from a broken crystal symmetry whose order parameter is the static lattice deformation. Quantum interference as a reason for the formation of bandgaps, and occupation number, that is, numbers of particles per unit cell, and the states are both essential factors.

Now, if interactions are such in the solid that a strong repulsion between electrons exists, impeding their flow, that is, a strong interaction energy U, then for a ground state of one electron per site of a lattice, with a large repulsion for adding another electron, would mean that fluctuations—kinetic transfer—involving movement of particles from site to site have to be suppressed. This is a strong correlation–induced Mott transition, as schematically depicted in Figure 4.35(c).

Another example of a particle-induced transition is the Anderson transition, which is summarized in Figure 4.35(d)—this is a local-

We started the discussion of the chapter with the example of a thought experiment reflecting a Peierls transition. The metal–insulator transition appeared through the broken symmetry of periodicity from a to $2a$.

Our discussion of the Mott transition, and its various flavors when additional variety of energetic effects become important, will be by necessity narrow and specific to materials we wish to understand. A Mott state is insulating, and spontaneously broken symmetry may or may not be present, such as in spin antiferromagnetism. It also may show gapped or gapless neutral particle excitations, as well as presence or absence of topological order and fractioning of charge.

ization effect. Here, when an electron interacts with disorder, such as from lattice defects or impurities, it opens up a bandgap. This transition results from coherent backscattering of electron waves in a random potential. This disorder-induced metal-to-insulator transition can happen with strong perturbation in the periodicity, so a structural disorder or strong impurity scattering can cause it. The Anderson localization is a coherence effect arising from the coherence being maintained over multiple scatterings. In both the Mott transition and the Anderson transition, one may understand the insulating behavior as an energy gap for charge excitations between the Fermi energy and spatially extended states that allow conduction.

Numerous materials exists where this single particle description utterly fails and local, nonlocal, strong and weak interactions must all be considered in order to get an adequate understanding of behavior. This many-body complexity is common to materials involving elements with d and f orbital occupancy.

	3B	4B	5B	6B	7B		8B		1B	2B
4	Sc^{21}	Ti^{22}	V^{23}	Cr^{24}	Mn^{25}	Fe^{26}	Co^{27}	Ni^{28}	Cu^{29}	Zn^{30}
[Ar]	$3d^14s^2$	$3d^24s^2$	$3d^34s^2$	$3d^54s^{1*}$	$3d^54s^2$	$3d^64s^2$	$3d^74s^2$	$3d^84s^2$	$3d^{10}4s^{1*}$	$3d^{10}4s^2$
5	Y^{39}	Zr^{40}	Nb^{41}	Mo^{42}	Tc^{43}	Ru^{44}	Rh^{45}	Pd^{46}	Ag^{47}	Cd^{48}
[Kr]	$4d^15s^2$	$4d^25s^2$	$4d^45s^{1*}$	$4d^55s^{1*}$	$4d^55s^2$	$4d^75s^{1*}$	$4d^85s^{1*}$	$4d^{10*}$	$4d^{10}5s^{1*}$	$4d^{10}5s^2$
6	La^{57}	Hf^{72}	Ta^{73}	W^{74}	Re^{75}	Os^{76}	Ir^{77}	Pt^{78}	Au^{79}	Hg^{80}
[Xe]	$5d^16s^2$	$4f^{14}5d^26s^2$	$4f^{14}5d^36s^2$	$4f^{14}5d^46s^2$	$4f^{14}5d^56s^2$	$4f^{14}5d^66s^2$	$4f^{14}5d^76s^2$	$4f^{14}5d^96s^{1*}$	$4f^{14}5d^{10}6s^{1*}$	$4f^{14}5d^{10}6s^2$
7	Ac^{89}	Rf^{104}	Db^{105}	Sg^{106}	Bh^{107}	Hs^{108}				Cn^{112}
[Rn]	$6d^17s^2$	$5f^{14}6d^27s^{2\dagger}$	$5f^{14}6d^37s^{2\dagger}$	$5f^{14}6d^47s^{2\dagger}$	$5f^{14}6d^57s^{2\dagger}$	$5f^{14}6d^67s^{2\dagger}$				$5f^{14}6d^{10}7s^{2\dagger}$

Table 4.5: A table of the generally accepted group of transition metals of periods 4, 5, 6 and 7. Note that the close orbital energies of d and the s shell with a higher principal quantum number result in $4s$ and $3d$ occupancies showing unexpected changes. $4s$ is occupied before $3d$ is filled. * marks unusual, but not unexpected, changes from the usual because of the interactions. These are particularly pronounced at half and full occupancies. † indicates an expected, but not verified, electron configuration. La^{57} is often considered to be a lanthanide rather than a transition metal.

At higher atomic numbers, as d and f orbitals, which can be quite stretched out, are occupied, the state and energy become quite complex. For example, even in an atom, the energetic ordering of the electronic states, resulting from the spin and orbit interactions, becomes a veritable spectacle in the ordering of orbit filling. The interactions, captured in Hund's rules, prescribe the following: (a) the lowest electronic term corresponds to maximum **S**—this is from spin-spin interactions; (b) the lowest term also corresponds to maximum **L**—this is from the orbit-orbit interactions; and (c) the lowest term also corresponds to the largest total angular momentum **J** for shells that are more than half-full and, smallest **J** for shells that are less than half-full. This is observable in Table 4.5.

We now look at when these electron-electron interactions could become important and lead to additional new properties. Atoms, such as those of transition metals with their d and f orbitals, with large numbers of electrons will not necessarily have the same interaction effects as those in a simple semiconductor, because these orbitals have a large span and many states, and electrons have spin, and so the consequences may take different forms due to the kinetic versus potential forms of the energy and how they manifest in the many-body assembly. Transition metals, unlike normal and noble metals, have d orbital electron states that are close to the s orbital states. When the atoms are bought together, the overlap of wavefunctions results in couplings that form bands that are significantly different from those of other normal metals. In these conditions, electron-electron interactions and their correlations become important. The physical basis here is that Coulombic repulsion, which is a potential energy effect, and kinetic energy which is related to electrons moving and changing locales, that is, hopping between sites, drive the energy considerations. The dominance of one over the other determines whether electrons will localize, when potential repulsion dominates, or delocalize, when kinetics dominates.

A gap appears in the energy dispersion of many of these transition metals and their compounds due to the charge excitation spectrum. And other properties linked to this correlation and electron-electron interaction also unfold. These effects arise from electron correlations. The Wigner crystal, referred to earlier, appears because Coulomb interactions dominate and localize electrons into a periodic assembly. Insulation appears due to the electron-electron correlated interactions. Examples of these include Mott-Hubbard transitions, where no long-range order exists for local magnetic moments; Mott-Heisenberg transitions, where antiferromagnetic order appears below the Néel temperature, which is the temperature that characterizes magnetic domain stability; and Mott-Anderson transitions, which are caused by the coupling of disorder with correlations. These are all examples where correlated electron physics is necessary.

We briefly discuss the salient phenomena underlying these transitions, so that we can proceed with the discussion of nanoscale devices with them.

First, as shown in Figure 4.36, we introduce charge carriers into the crystal. This means that there is a perturbation of $\delta n(\mathbf{r}) = e\mathcal{G}(E_F)\delta U(\mathbf{r})$ in the region of the introduced particle. The switching on of this local perturbation of $\delta U(\mathbf{r})$ potential causes the electrons in the liquid to shift from the region of the particle and its vicinity so that the electrochemical potential remains homogeneous. In Figure 4.36, this is shown as the process of the change in occupation. Figure 4.36(a) shows the energy perturbation caused by the introduc-

Physicists and chemists identify elements in somewhat different ways. Chemists prefer to identify an element based on its chemical properties—the Mendeleevian approach. Physicists take a mathematical view of identity. Some elements, such as La^{57}, run afoul of this struggle. Chemists would rather it belonged to the lanthanide series. There is similarly an actinide series, though, actinium being a rare metal, it arouses less consternation.

It behooves us to use the phrase "correlated electron" and its implications with caution. However, like much else, because of popular appeal, despite us being scientists beholden to preciseness, it is a phrase far more bandied about than it should be. It appears in the technical circles of this second decade of the 21st century to be the equivalent of "paradigm" and "innovation," terms that the management crowd uses with abandon. Even in a Fermi electron gas, one can argue that the "non-interacting" electrons are highly correlated. The Pauli exclusion principle has precisely commanded the interaction for the quantum state. Our use of this adjective "correlated" in practice is to denote the importance of many-body interaction energies dominating the kinetic energy in some special way. When this happens, macroscopic properties change in a drastic way. Coulomb blockade in a quantum dot, as well as the fractional Hall effect, is also an effect arising from this strong correlation among electrons.

tion of the electron; this perturbation, which in turn, causes the mobile electrons of the Fermi liquid surface to move away, yielding the
homogeneity of the electrochemical potential shown in Figure 4.36(b).

The Coulomb perturbation is

$$\frac{1}{e}\nabla^2 \delta U(\mathbf{r}) = -\frac{\rho}{\epsilon}, \text{ or}$$

$$\frac{1}{r^2}\frac{\partial}{\partial r}r^2\frac{\partial}{\partial r}\delta U(\mathbf{r}) = -\frac{e\delta n(\mathbf{r})}{\epsilon}. \tag{4.49}$$

This equation has a solution of the form

$$\delta U(\mathbf{r}) = \frac{A}{r}\exp\left(-\frac{r}{\lambda_{TF}}\right), \tag{4.50}$$

where

$$\lambda_{TF} = \sqrt{\frac{\epsilon}{e^2 \mathcal{G}(E_F)}}. \tag{4.51}$$

λ_{TF} is the Thomas-Fermi screening length that has appeared in
Chapter 2, under the degenerate conditions when the state occu-
pancy shown in Figure 4.36 applies. For a carrier density of n, and a
Bohr radius of $a_B = 4\pi\hbar^2\epsilon_0/m_0 e^2$,

$$\lambda_{TF} \approx \frac{1}{2}\left(\frac{n}{a_B^3}\right)^{-1/6}. \tag{4.52}$$

For copper, a normal metal, $n \approx 8.5 \times 10^{22}\ cm^{-3}$, and $\lambda_{TF} = 0.055\ nm$.
In vacuum, density of states vanishes, and the perturbation caused by
an electron is $\delta U(\mathbf{r}) - e/4\pi\epsilon_0 - A$. In solids under the degenerate
isotropic conditions, $\mathcal{G}(E_F) = 3n/2E_F = (1/2\pi^2)(2m/\hbar^2)(3\pi^2 n)^{2/3}$,
as we derived earlier. These imply

$$\frac{1}{\lambda_{TF}^2} = \frac{4}{\pi}(3\pi^2)^{1/3}\frac{n^{1/3}}{a_B}. \tag{4.53}$$

For silicon under the degenerate conditions, the Bohr radius increases
by the ratio of the dielectric constant and the mass, so that the effective Bohr radius is $a_B^{Si} \approx 6.4\ nm$. At a carrier density of $10^{19}\ cm^{-3}$, a
degenerate condition, $\lambda_{TF} \approx 5.5\ nm$.

We have placed these different screening lengths under the moniker
of a general screening length λ_{scr} for the variety of conditions one encounters. For the degenerate silicon example, this screening length
scale is larger than that of copper by two decades. When the conditions are non-degenerate, that is, as in a Fermi gas, the equivalent
screening length is the Debye length

$$\frac{1}{\lambda_D^2} = \frac{e^2 n}{\epsilon k_B T}. \tag{4.54}$$

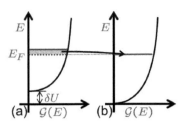

Figure 4.36: When an electron is introduced onto a weakly correlated
Fermi liquid, and local perturbation
is initiated, as shown in (a), electrons
leave the local area to maintain a constant electrochemical potential, which
is macroscopic and homogeneous
throughout the crystal in equilibrium.
The introduction of the charge creates
the energy perturbation δU, which
is shown in (a) and which causes the
carrier population to move away from
the excess charge on Fermi liquid surface, resulting in the electrochemical
potential and the state filling shown in
(b).

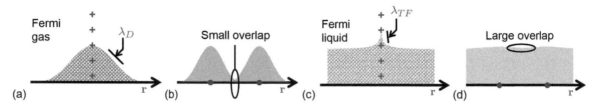

Figure 4.37: (a) shows the independent electron Fermi gas limit for screening when a charge perturbation—here, a δ-function of donors—is screened at the Debye length (λ_D) by the electron gas. The electron wavefunctions have a very small overlap and so also a low interaction energy U ,as shown in (b), with T, their kinetic energy, quite exceeding the interaction energy. If an additional donor atom is added, the donor contributes an electron that appears as an independent electron in the conduction band, as well as an ionized donor state. Beyond degenerate conditions, as in a Fermi liquid, the perturbation is screened differently, as shown in (c). It is local, with the large number of states up to the Fermi energy filled. Electrons are close together, with a significant wavefunction overlap, that is, a weak correlation exists, as schematically shown in (d). The interaction is local, but the effect is long range, spreading out at the Fermi energy, where filled and empty states coexist. Screening is at the Thomas-Fermi screening length scale $\lambda_{TF} = 2\pi/k_F$.

This is ~ 12.7 nm at 10^{17} cm^{-3} carrier density. This is another factor of 3 larger than that of degenerate silicon, which is 2 orders of magnitude larger than that in copper.

Figure 4.37 shows a pictorial description of screening in these limits of non-degenerate gas conditions and degenerate liquid conditions. In Fermi gas conditions, such as in silicon at low doping, the electron charge cloud spreads out around any charge disturbance, such as that due to donors or acceptors. The charge cloud has a small overlap, and its kinetic ability is high. A perturbation, such as from a donor charge, as shown in Figure 4.37(a) will have the electrons screen but with a spread into the distribution of states available. In degenerate conditions, highly doped silicon, for example, the disturbances are close enough for the delocalized electrons to interact with each other. The resulting interaction energy causes the states to spread, leading to the bandgap reduction. At very high electron densities, such as in metal, we need to look carefully at the energetics of whether the electron will be forced to stay confined at the site from which it became available or be allowed to spread out.

For metals, the consequences of a small screening length can be significant. At a critical electron density, the potential well with the screening field varies proportionally to $(1/r)\exp(-r/\lambda_{TF})$, with $\lambda_{TF} \propto n^{-1/6}$, and the free electron becomes localized. Hubbard's model provides a simple description that captures the competition and the effect of these forces. The Bohr length scale—which is based on the expected distance between an electron and the proton ion core in a hydrogen atom—is one end of the dimensional scale. An electron, when it moves in a crystal, overcomes the Coulombic attraction in hopping to other sites by overcoming the confining barrier. If the

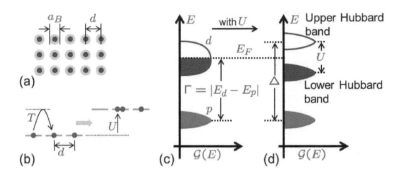

(a)

(b)

(c)

(d)

Figure 4.38: A schematic view of the Mott-Hubbard transition leading to the opening of a gap. (a) and (b) show that the localization is such that the energy needed for the movement of an electron to another locale—the adjacent site—requires a large energy U due to electron-electron correlation. a_B is the Bohr radius, and d is inter-atom spacing. The screening length is very short and hence an electron is localized to the lattice sites. This energetic condition causes, even in materials expected to have the partially filled bands shown in (c), such as for d orbitals, the creation of a gap through the formation of a top and bottom Hubbard band, as shown in (d). This gap arises due to the turning on of the energy interaction due to U in (c). Note that the p band continues to stay below the lower Hubbard band, and $U < \Delta$.

state of the system is such that most of these sites are occupied, then this electron needs considerable energy to make the motion through hopping possible. Figure 4.38(b) shows a possible hopping process. Let U be the interaction energy. This is the energy cost of two electrons, of opposite spin, to be at a site. The kinetic energy T must be sufficient as the source of interaction energy that hopping and therefore electron transport requires. This kinetic energy is proportional to a kinetic energy parameter t related to the state occupation of the electrons. t describes the kinetic hopping process—a transmittance through the overlap integral—and the Hubbard U, the interaction process. If the distance between the sites is large, that is, $d \gg a_B$, atomic overlap is small and, in turn, $T \ll U$. The ratio U/T, a relative strength, is dependent on the electron density per lattice site. If a band is half filled, a Fermi gas is formed when $U/T \ll 1$, as in Figure 4.38(c). If $U/T \gg 1$, as in Figure 4.38(d), hopping coupling and therefore conduction between sites is suppressed. A metal-to-insulator transition becomes possible with significant energy separation between upper and lower Hubbard bands.

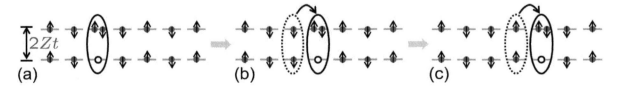

(a)

(b)

(c)

Figure 4.39: A schematic view of coherent electron movement through Hubbard bands. Occupancies can be conserved at strong coupling with double occupancy in the upper band and an antiparticle hole in the lower band as electrons move coherently. This figure shows a charge band where conduction is taking place through the quasiparticle-quasiantiparticle excitation.

Figure 4.39 conceptually describes this hopping-localization competition and transport. Given an interatomic spacing of d, within the wave picture, the energy dispersion has a bandwidth of $2Zt$,

This is, strictly speaking, in the "tight binding" approximation of independent electrons.

where Z is the count of nearest neighbors, and t is the kinetic energy parameter describing the tunneling amplitude of hopping that is dependent on d/a_B. The overlap of atomic wavefunction is small when atoms are considerably separated ($d \gg a_B$). Here, the kinetic energy $T \ll U$. Now consider half filling of bands. Every one of the lattice sites prefers to be singly occupied. In this situation, in Figure 4.39, an energy of $E^+ = E_0(N+1) - E_0(N)$ is required to add another electron. This, for $N = L$, is given by $E^+ = U - T_1/2$, where the charge excitation with energy U forms a band of width T_1, which is a function of the ratio of U with T. The upper Hubbard band thus formed is a spectrum of charge excitations for the addition of an extra electron to the half-filled ground state. This electron can be removed with an energy of $E^- = E_0(N) - E_0(N-1)$. This, for this $N = L$ condition, is $E^- = +T_2/2$, where T_2 is also a function of U/T. When $U \gg (T_1 + T_2)/2$, at half band filling, the electrochemical potential $\mu(N)$ is now discontinuous. At $N = L$, a gap of $\Delta E(N = L) = (E^+ - E^-)/2 \approx U - (T_1 + T_2)/2$, which is finite, appears. An energy gap for charge excitations of $(T_1 + T_2)/2 \approx T \ll U$ has appeared. This material has become an insulator.

The transition is definable in terms of a critical value of interaction: $U_c = (T_1 + T_2)/2$. As Figure 4.40 shows, at $U = U_c$, the Hubbard bands come together. When $U \ll U_c$, the bands overlap and the system has metallic behavior, even though Hubbard bands remain. This Mott-Hubbard transition occurs at the lower electron density because of the strong correlation there. If one places a large carrier density into the material, for example, by injecting charge, these materials will become metallic conductors. So, strong correlation exists at lower carrier densities, and the correlation weakens at higher particle density.

The energy gap that has appeared here is due to the property of the excitation. The separation shown here is a separation of excitation. It appears as a result of the nature of interaction between the electrons in a high number of electron orbitals, here d orbitals, interacting in crystal assemblies so that competition exists between localization due to interaction, and the kinetic energy of occupying progressively higher states as the electron count increases.

A Mott insulator arises in the appearance of an energy gap due localization. In Equation 4.52, what this localization-delocalization argument implies is a carrier density for the transition of $(4a_B)^{-3}$. In a semiconductor, it is the insulating-to-metallic transition at high dopings that is generally of interest. In transition metal oxides, it is the metal-to-insulator transition below a critical density that is of interest. The localization of the particle, together with the strong interparticle repulsion between the particles that impedes the flow, can be viewed,

One should note the remarkable similarity between this charge excitation argument in a periodic structure with that due to a Coulomb gap at the macroscopic size. That too is a correlated electron effect arising from single electron interaction due to energy arguments with electrostatic energy vis-à-vis thermal energy providing the necessary kinetic energy. In the periodic arrangement, when the localization happens, the energy required to move, the kinetic energy, is large. When motion is possible, it is because of the allowed excitation where the Coulomb gap, that is, the charge gap, disappears.

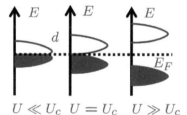

$$U \ll U_c \quad U = U_c \quad U \gg U_c$$

Figure 4.40: Evolution of Hubbard bands from conducting to insulating conditions as U is varied around the critical interaction energy U_c at the onset of the Mott-Hubbard metal–insulator transition.

The classic picture is that metals have partially filled band. But iron, with a $3d^6$ outer shell, becomes an insulator in FeO.

Several high-T_c cuprate superconductors are examples of Mott insulators arising from doped charge transfer.

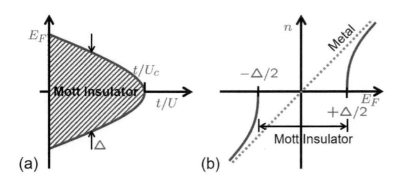

(a)

(b)

Figure 4.41: The nature and effect of kinetic versus potential energy magnitudes in the Mott transition. (a) shows the importance of a critical kinetic-to-potential ratio for the existence of the insulating phase, and the existence of the insulating phase over a range of Fermi energies. (b) shows the density change as a function of Fermi energy near the Mott insulating region created by doping, while referencing the prior metallic behavior.

as in Figure 4.38, as being a ground state of the system in which there is one particle on each lattice site and there is strong repulsion between two particles on the same site. Fluctuations involving motion of a particle from one site to another are suppressed. Conduction requires proximity in energy. The kinetic energy gain as electrons populate the periodic crystal structure occurs with the occupation of higher and higher band states. So, at any Fermi energy, the kinetic energy gain T in the periodic structure arises from the occupation of states up to the Fermi energy. The energy cost of having two carriers for example, the up and down spins in a state i at the Fermi surface, is U because of this strong localization. In $3d$ transition metal oxides, such as the monoxides of manganese, iron, cobalt or nickel, which all form Mott insulators, $U \approx 5~eV$, and $T \approx 3~eV$. The localization and the large interaction energy U result in the opening of the gap, a charge gap, and the material becoming insulating because the electron density is small and a large correlation exists.

In detail, in a Mott-Hubbard transition, a material undergoes a transformation from a metal to an insulator state as a function of the electrochemical potential's position in the symmetric form shown in Figure 4.41. A symmetric gap appears. Over a range of Fermi energies, the gap persists. As the density is changed further, by doping or by field-induced injection or carrier generation, the system goes to the metallic state, with the change in the Fermi energy. Hubbard bands can only accommodate $N/2$ electrons. Regular bands can have N electrons. And this transition from insulating to conducting occurs by the electrochemical intervention, which causes the band transformation.

For completeness, we discuss, in short, the three different forms in which this Hubbard transition manifests itself in materials. In this Mott-Hubbard transition, we did not have any ordering of the spin over a long range. In the Mott-Heisenberg transition, there exists

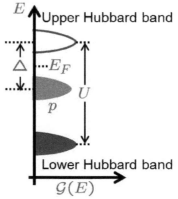

Figure 4.42: A charge transfer insulator also arises from the formation of Hubbard bands. However, unlike the Mott-Hubbard insulator, where $U < \Delta$, in a charge transfer insulator, $\Delta < U$, as shown here. The p band now intervenes between the upper and lower Hubbard bands. This figure should be compared to Figure 4.38.

long-range ordering of magnetic moment through spin in the insulating state. The antiferromagnetic coupling (J) between the electrons on neighboring sites is proportional to t^2/U. With a large U, the Hubbard gap is large; but, at low temperature, specifically, temperatures below the Néel temperature, a long-range spin order appears. The Mott-Hubbard insulator, below this temperature, now undergoes a Mott-Heisenberg insulator transition, where antiferromagnetism appears thermodynamically, that is, because of entropy. This transition can also appear in materials that are in a metallic state, where existing magnetic moments arising from the interaction energy, that is, from a paramagnetic metallic state, undergo a Mott-Heisenberg transition to an antiferromagnetic insulating state.

In a charge transfer insulator (see Figure 4.42), unlike a Mott-Hubbard insulator, the p band arising from the anions of the crystal is between the upper and lower Hubbard bands. In this insulating state, the lowest excitation arises from the p band to the upper Hubbard band. So, it is an example of a d-like quasiparticle with a p-like hole, as shown in Figure 4.39.

In the periodic table, as one goes from column 1 to 8, the interaction energy U generally increases. Alkali elements have low interaction energy and form conducting metals irrespective of the atomic size. As one goes down the columns, the interaction energy decreases. Transition metals, which are placed between these two limits, with different electrons in the band, provide examples of different insulating characteristics. Our specific interest is in transition metal oxides: perovskites and other forms, where the variety of behavior described here may appear.

We will specifically talk about magnetic ordering due to exchange energy during our discussion of magnetism and spin-based devices. For now, we take this spin ordering and its magnetic consequences as a given arising from perturbation.

This is the same Néel whose name first appeared in the discussion regarding magnetic domain stability where temperature-induced randomness, precession, and surface and bulk energies intersect. Here, $k_B T_N \ll U - T$.

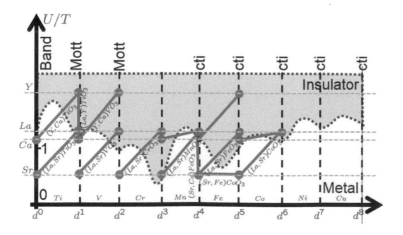

Figure 4.43: The variety of metal and insulator behavior observed in transition metal oxides. This panel shows ABO_3 compounds, where compositional change results in a metal–insulator transition. This figure has been adapted from A. Fujimori, "Electronic structure of metallic oxides: Bandgap closure and valence control," Journal of Physics and Chemistry of Solids, **53**, 1595–1602 (1992); cti, charge transfer insulation.

Figure 4.43 shows the variety if metal and insulator behaviors that occur among the transition metal oxides and their compounds, with different interaction and kinetic parameters represented as the ratio U/T. The abscissa shows different sections of constant d-orbital filling in the element C of $(A, B)CO_3$. The ordinate shows the interaction/kinetic energy ratio (U/T) arising in the A and B composition ratio of the compound. The U/T change is accompanied by a changing Δ—the p-d energy gap—as is U. La, Sr, Y and Co are common choices with the different d-filled Cs. The mixed-phase ABO_3s, where A could be Y or La—a 3+ coordination—or, Sr or Ca—a 4+ coordination—are linked through a line showing the metal-to-insulating change taking place through the composition. When moving along the abscissa, through the change in the A and B compositional ratio, one changes bandwidth through the change in the U/T ratio. Figure 4.43 shows various examples resulting from mixing. Absent oxygen, all are metallic. Present oxygen, the delocalized d electrons from transition metals strongly couple to the p orbitals of oxygen, leading to a variety of transitions as a result of the delocalization inherent to d electrons and the Coulomb interaction localizing these electrons. Starting as band insulators, d^1- and d^2-filled systems show Mott insulation, while higher d-filled systems show charge transfer insulation.

Figure 4.44 shows behavior during band filling in Mott systems, so a subset of the behaviors shown in Figure 4.43, but incorporates simpler oxides in the U/T description. $LaTiO_3$ or $LaVO_3$ are Mott insulators, as are low temperature phases of a group of oxides—V_2O_3, VO_2, NiO, et cetera—that too show metal–insulator transition. Some, such as VO_2, could very well undergo this transition due to Peierls transition, since the change takes place together with a structural transition. VO_2 is insulating in the monoclinic phase and becomes conducting in the tetragonal phase. These perovskites have electronic configurations based on d^1 or d^2. What this figure shows is the ability to cause the insulator-to-metal transition by changing charge carrier filling through the d-band filling, so by changing along the ordinate axis, with the interactions between the constitutive species at their lowest energy physical arrangements being the cause. This statement is illustrated in Figure 4.45, which puts together the insulating phase arrangements for $LaTiO_3$ and $YTiO_3$—two of the limit points in Figure 4.44. The former—$LaTiO_3$—has a bandgap of $\sim 0.2eV$, and the latter—$YTiO_3$—has a bandgap of $\sim 1.0\ eV$.

The monoclinic arrangements for $LaTiO_3$ in Figure 4.45(a)–(c) and for $YTiO_3$ in Figure 4.45(d)–(f) illustrate the d orbital interactions that occur in Mott insulation. The octahedrons are twisted, with the overlap affected in this lower energy state. Panels (c) and (f) show the front section of the unit cell, and panels (b) and (e) show a subcell

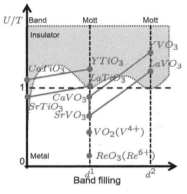

Figure 4.44: Examples of band- to-Mott transitions for the first two band fillings in mixed transition metal oxides. Changes depend on the species: the changes shown include those for Ca to Y, Sr to Ca, Sr to Y, and Sr to La, with Ti and V cores. For reference, the vanadium oxides that, along with others, show the metal–insulator transition are included. This figure has been adapted from A. Fujimori et al., "Electronic structure of Mott-Hubbard-type transition metal oxides," Journal of Electron Spectroscopy and Related Phenomena, **117**, 277–286 (2001).

There are at least three different tetragonal phases of VO_2—all conducting. That lattice distortion occurs and symmetry changes is a fact. But, it does not preclude the electron-electron interaction from being the stronger cause than the electron-lattice interaction. This makes matters somewhat complicated; hence, the many-decades debate on the cause. We will remain agnostic on this matter. Our interest is in taking advantage of the phenomenon.

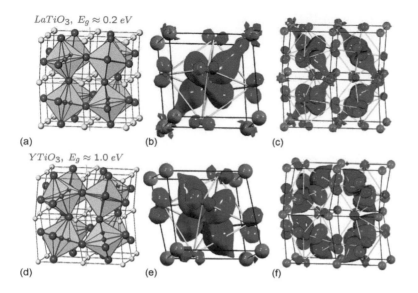

$LaTiO_3$, $E_g \approx 0.2 \ eV$

(a)   (b)   (c)

$YTiO_3$, $E_g \approx 1.0 \ eV$

(d)   (e)   (f)

Figure 4.45: The insulating–semiconducting phases of $LaTiO_3$ with $\sim 0.2 \ eV$ bandgap in (a)–(c) and of $YTiO_3$ in (d)–(f). (a) and (d) show the octahedral crystalline arrangement of the monoclinic phase. (c) and (f) show the corresponding unit cell with a representational charge density profile of the highest occupied orbitals. (b) and (e) show an expanded view to illustrate the d orbital's reach and interactions. Note the twisting that occurs over the sub-assemblies. Adapted from E. Pavarani, S. Biermann, A. Poteryaev, A. I. Lichtenstein, A. Georges and O. K. Andersen, "Mott transition and suppression of orbital fluctuations in orthorhombic $3d^1$ perovskites," Physical Review Letters, **92**, 176403 (2004).

with a representational outer d orbit probability of occupation. These orbitals twist between the different subcells, as panels (a) and (d) show more clearly in the lattice arrangement. The extent of these orbits and their occupation is central to phase and its order in these compounds.

We will use VO_2 as an example to explore metal–insulator transition in devices. As a binary oxide, it is simpler, but a reasonable stepping stone to the perovskites. VO_2's metal–insulator transition can be observed in temperature, through current injection, that is, doping, as well as in response to optical exposure—photon energy injection. Figure 4.46 shows the resistivity of thin VO_2 polycrystalline films as a function of temperature. The change is a first order phase transition with a four-orders-of-magnitude resistivity change. The insulating phase has a bandgap of ~ 0.6–$0.7 \ eV$, so it has the bandgap of a semiconductor similar to germanium or $Ga_{0.47}In_{0.53}As$. "Insulator" is a word used here for its physical antecedents—filled and empty bands at absolute zero. In the conducting phase, VO_2 shows resistivity similar to that of dirty metals. The latent heat for the transition is $\sim 5 \times 10^4 \ J/kg$.

This phase transition in VO_2 occurs with a change from monoclinic structure (see Figure 4.47(a)) to a tetragonal structural form with its distortion (see Figure 4.47(b)). The origin of the insulating phase is shown in Figure 4.47(c). Vanadium atoms have an electronic configuration of $[Ar]3d^34s^2$. It is the outer orbitals that contribute to the states around the Fermi level. The lower energy states arise from $3d$, as shown in the figure. In VO_2, V^{4+} is surrounded by an octahe-

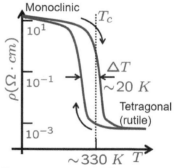

Figure 4.46: Resistivity change in a polycrystalline thin film of VO_2 undergoing a metal–insulator transition. About a four-orders-of-magnitude change in resistivity, as well as and a hysteresis window corresponding to the first order transition of $\sim 20 \ K$ is observed at around 320 K.

dron of O^{2-}. This appropriates 4 electrons—both of the s electrons and two of the d electrons—for the bonding with the $2p$ electrons of oxygen, leaving vanadium with one electron. The d state energies are also modified by the crystal interactions—one d state is lowered, two are raised, and bands are formed from the non-cubic crystalline interaction in Figure 4.47(c), shown in the middle part of the assembly of states from their atomic origin. The metallic tetragonal phase is from the bands of these states, where the remaining d electron has occupied a partially filled band. When VO_2 is in the monoclinic phase, vanadium atoms are aligned and hence dimerized, as shown in Figure 4.47(a), and Peierls distortion too occurs. This is the changed translational symmetry that we started this chapter's discussion with. The lowest states showed in Figure 4.47(c), arising from vanadium, hybridize strongly, and a bandgap opens as shown. The lower band states are now fully occupied with equal numbers of up and down spin electrons received from the dimerizing pairs of vanadium atoms.

In an engineering sense, VO_2 forms neither a good insulator nor a good metal. Good insulators have extreme resistivity. SiO_2's is $\sim 10^{16}$ $\Omega \cdot cm$ and Si_3N_4's is $\sim 10^{14}$ $\Omega \cdot cm$. With thick films, one is more concerned with defect-mediated conduction. With thin films—of the order of nms—one is concerned with tunneling. The insulating phase of VO_2 is 13 orders of magnitude off. Good conducting metals Au and Cu have resistivities of $\sim 2 \times 10^{-6}$ $\Omega \cdot cm$ and $\sim 1.7 \times 10^{-6}$ $\Omega \cdot cm$. Nichrome (Ni-Cr alloy, used in heaters) has a resistivity of $\sim 1 \times 10^{-4}$ $\Omega \cdot cm$. The metal phase of VO_2 has a resistivity about 2 times worse than that of Ti, and 5 times worse than that of V, from which it is formed.

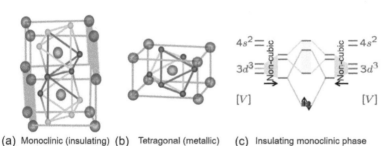

(a) Monoclinic (insulating)   (b) Tetragonal (metallic)   (c) Insulating monoclinic phase

Figure 4.47: (a) shows the salient atomic arrangement in the insulating monoclinic phase. V, the larger atom, has $4s$ and $3d$—with their 5 probability arrangements—as the outermost important orbitals to consider. (b) shows the tetragonal (also called rutile) metallic phase. (c) shows the interactions leading to insulating monoclinic phase orbital states with the lower state fully occupied.

The bands in the non-conducting (monoclinic) and the conducting (tetragonal) phases, as well as the density of states in the conducting tetragonal phase, are shown in Figure 4.48. Panels (a) and (b) are schematic drawings of the bands. In the insulating phase, the filled states below the Fermi level arise from oxygen's $2p$ states, and the lower band from vanadium's single d state. The upper bands result from the s and the other two d states. In the conducting phase, the gap disappears. Panel (c) shows a calculated density of states in the conducting phase. Here, 0 eV is the reference Fermi level, and the density of states contributions from vanadium and oxygen are separately identified. Note that, as shown in the schematic, the valence band has a large number of states originating in O^{2p}.

This figure points to an important aspect of the metal–insulator transition. In the monoclinic phase, there are two bands: the Mott-Hubbard bands—upper and lower—arising from the combined effect of strong electron correlation and Peierls instability. The bands move when the temperature is raised above the critical temperature.

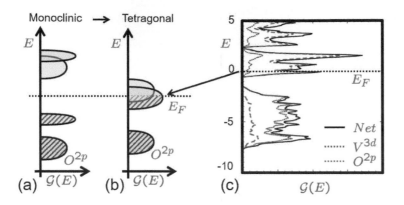

Figure 4.48: Schematic of the (a) insulating (monoclinic) and (b)conducting (tetragonal) VO_2 phases arising as a result of the interactions shown in Figure 4.47. This figure shows also the valence band arising from O^{2p} states. In the insulating state, the band primarily arising from the raised d states is shown as being above the band arising from the s states. The lower d band is the other band that arises as a result of the field interactions of the assembly and dimerization that filled it. The d band is not split in the tetragonal state. (c) shows a more accurate net density of state distribution, and the fractions arising from V^{3d} and O^{2p}.

VO_2 also undergoes this conducting transition when energy in other forms is provided to the system, or doping is created by extracting or injecting carriers.

How can such metal–insulator transitions be used to advantage in devices?

An immediate example is that if an insulator-to-metal transition takes place when sufficient carriers are injected into the structure, then one can employ a metal-clad metal-insulator transition structure for limiting the flow of power. An example of this is in radio frequency transmission and receiving where the same antenna may be employed for both processes. The transmitted signals are large, but the received signals are low. The receiving signal is amplified via sensitive low noise amplifiers. A power limiter in the receiving signal path limits the damage by constraining the power transmitted beyond the power limiter, as shown in Figure 4.49.

A metal-VO_2-metal diode shunts the transmission line along which power transmission needs to be limited. When the power is low, the diode stays in the insulating phase, as shown in Figure 4.49(b), with limited losses to the transmission. However, as the power is raised higher, the injection and extraction of carriers from the VO_2 region of the diode undergoing metal–insulator transition cause shunt conduction, and a dominant part of the input power is reflected back and dissipated within this diode. The diode will have hysteresis, since this is a first order phase transition.

If one employs VO_2 as a "semiconductor" for a field-effect transistor with this 0.6–0.7 eV bandgap, one observes rather interesting device characteristics, as shown in Figure 4.50 where the prototypical VO_2 serves as the phase transition material channel layer. One can see the normal diffusive transport turn-on of the transistor in the subthreshold region. At higher gate-to-source bias voltage and low

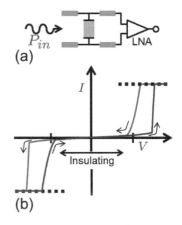

Figure 4.49: Power limiting by employing the metal–insulator transition properties in radio frequency signaling. (a) A metal-VO_2-metal diode shunts a transmission line. LNA here stands for low noise amplifier. (b) shows the current-voltage characteristics of the diode employed in the limiter.

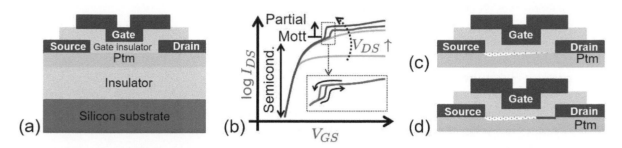

Figure 4.50: (a) shows a cross-section of a field-effect transistor employing VO_2 as a phase transition material (ptm). (b) shows the turn-on characteristics of the transistor as the bias voltages are changed. (c) and (d) show a conceptual view of the metal–insulator transition's (Mott transition) role in the observed characteristics.

drain-to-source bias voltage, one also sees the conventional onset of drift transport and the change towards linearity from the exponential dependence. But, if the drain-to-source voltage is large enough, one also sees a sudden jump in the drain current, a jump that is nearly a step. And this step has a hysteresis depending on the sweep direction, as shown in the inset in panel (b). Figure 4.50(c) and (d) show what is happening physically. When the drain-to-source bias is low as the gate-to-source bias voltage is changed, one forms an electron inversion layer at the surface, and field-effect transport occurs as in a conventional transistor. The only difference here is the characteristics of the materials—the mobility of electrons at a VO_2/gate-insulator interface, the contact resistances, et cetera. But, conventional theory fits these long-channel devices. Now, when the drain-to-source bias voltage is increased, it depletes the electrons at the drain end, but a large field exists at the drain end, through which electrons are streaming. The presence of this high field at the drain end, in the midst of this energetic electron stream, causes a doping- and field-induced transition that converts part of the channel at the drain end to being metallic. The device now has a shorter effective channel length, and the current is larger. The next question one may ask is then, why is the current not increasing? If the Mott transition happens due to the field effect, then it is at the interface and is limited by Thomas-Fermi screening ($\lambda_{TF} = \sqrt{\epsilon/e^2\mathcal{G}(E_F)}$; recall our earlier discussion), and this has a dimension scale of the order of 0.1 nm in a metal. So, the total number of conducting states is limited, and conduction occurs with low mobility, perhaps also affected by the interface. This limits the current.

In addition, because this is a first order transition, one observes the hysteresis of the inset. Our discussion of the previous example, which was shown in Figure 4.49, holds. The load line is from a true voltage source. In the presence of non linearity, the load line affects

In a voltage sweep of the gate, with the drain bias voltage constant, all w.r.t. the source, the source resistances of the applied stimulus potentials are extremely small. The load line used to make this measurement is vertical.

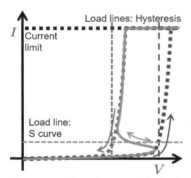

Figure 4.51: The influence of the load line affects the observed behavior of the Mott transition due to non linearity and its first order nature. A voltage source, with low source resistance, leads to hysteresis and a sharp change in current. A current source shows the "S" curve, which shows stable points in the presence of nonlinearity in the system.

what one observes. A current source would provide an *S*-like curve.
This is shown in Figure 4.51. When voltage driven, a low resistance
load line is vertical, and, depending on the sweep direction, one observes a shift in the voltage of onset of transition. If the load line is
horizontal, so observation is being made with a current source stimulus, one would observe the behavior of the onset of Mott transition at
sufficient injection and field, resulting in an increase in conductivity
and therefore a lowering of voltage, with reduced resistance, until the
change is complete The hysteresis curve provides the bounds within
which the negative resistance "*S*" curve lies. The load line as drawn
is ideal, and the device too has a bias-dependent resistance. So, in the
sweep up, and the sweep down, these resistances will limit the slope
of the sharp changes. But, the negative resistance is pronounced, and
if it is larger than the parasitics of the system, it can be exploited in
a negative resistance oscillator that could potentially reach very high
frequencies. We will presently see that the transition times are very
small and that means, with proper design-minimizing parasitics, a
high frequency oscillator a possibility.

One would expect heterogeneity in
the system, with the various granular
effects within the phase transition
material film but also because phase
transition is a change in distribution of
degrees of freedom.

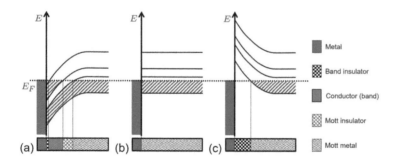

Figure 4.52: A metal/phase transition
material band picture at the interface.
The phase transition material has undergone a Mott-Hubbard transition
forming the upper and lower bands,
and it is assumed that the presence
of the interface does not disturb this
description. Three different conditions
of energy alignment in thermal equilibrium driven by the Fermi energy
of the metal and that of the Mott-
Hubbard material are shown: (a) for a
low workfunction metal, (b) when the
workfunctions match and (c) when the
workfunction of the metal is higher.
In each case, the region at the interface consists of sections with different
electronic behaviors.

Figure 4.52 shows schematically the idealized behavior at a metal
phase transition material interface in the presence of the Mott-
Hubbard transition for different workfunctions. In each of these
structures, the metal–insulator transition occurs either in an interface
region, if field mediated, or in the bulk, if induced by bulk parameters such as temperature or doping, including doping that is through
carrier injection or extraction. Devices need contacts to access them.
So, the bulk, or the interface where the metal–insulator transition exists, must be accessed by electrodes. What happens at this interface,
and how would one determine the properties? Even a simple system
such as a metal-semiconductor interface is subject to dipole charge at
the interface due to metal-semiconductor abrupt atomic change, surface reconstruction and other effects at the interface and its vicinity.
In Figure 4.52, we show an idealized behavior in thermal equilibrium,

assuming that, in a Mott-Hubbard metal–insulator transition material, where Hubbard bands have been formed so that the material is
in the metallic correlated electron conduction state, the bands persist
up to the interface with the metal. The band bending then follows
from the band and state descriptions and from Poisson's equation, as
well as from together with band occupancy.

The behavior is very intriguing, potentially pointing to remarkably different interesting manifestations that one could utilize in
devices. Assume that the Fermi level in the metal is smaller than
that in the phase transition material, as in Figure 4.52(a), and that
the behavior persists all the way to the interface. We would have in
this case, within the bulk region, the metallic conduction that arises
from the correlated electron effect of the Mott-Hubbard transition.
As we approach the interface and the bands bend, we will transit a
region where the lower Hubbard band is entirely filled and the upper
Hubbard band is empty. This is the Mott insulator region. It exists
here despite the bulk being in the metallic state! As one gets closer
to the interface, the upper Hubbard band begins to fill, and one has
a region that is conducting in the way band conductors work. We
call this band the conductor, because only a small spread in energy
in the band, unlike that of metals, is occupied. It is more like the
behavior of semiconductors that are in degenerate conditions. And,
going even closer to the interface, one now has a region where the
Fermi energy is above the highest energy state available in the upper
Hubbard band. This interface is insulating! So, a small Fermi energy in the metal, compared to that in the phase transition material,
has given rise to an electronic interface that consists of a metal-band
insulator-conductor-Mott insulator-Mott conductor—a highly unusual structure that one normally does not encounter in material interfaces. Figure 4.52(b) shows the ohmic condition when the electrons
in the metal and the electrons in the partially occupied lower Hubbard bands can freely exchange across the interface. Figure 4.52(c)
shows the electronic interface that arises when the Fermi energy in
the metal is larger than in the phase transition material. Now the
interface region is a band insulator, not unlike what happens in a
semiconductor. It shows a Schottky-like interface.

In the cases shown in Figure 4.52(a) and (c), low workfunction and
high workfunction metals create non-conducting but potentially, if
thin enough, tunneling regions at the interface. The difference between the case shown in Figure 4.52(c) and the metal-semiconductor
case is that the former will have current-voltage characteristics that
are more akin to those of a degenerate semiconductor, which too has
a partially occupied band. In the case shown in Figure 4.52(a), there
is a conducting band insulator region clad by two insulating regions

The Schottky barrier height is not
simply the difference of the workfunction of the two materials forming an
interface unlike what undergraduate
texts would have one believe. Schottky
barrier heights vary, with some correlation to workfunction. Relatively
well-behaved stable and strong-bonding
systems such as silicon demonstrate
that the barrier height is only weakly
correlated with workfunctions. *GaAs*
and other compound semiconductors
are notorious for Fermi level pinning arising from states created at
the interface—at pristine surfaces as
well as in the metal/semiconductor
system, and the barrier heights are
almost entirely unrelated to the workfunctions. Even single crystal heterojunctions of semiconductors do
not follow workfunctions, because of
the symmetry breakdown and and
other quantum-mechanical complexities. For a discussion, see S. Tiwari,
"Semiconductor physics: Principles,
theory and nanoscale," Electroscience 3,
Oxford University Press, ISBN 978-0-19-
875986-7 (forthcoming).

The workfunction, the energy required
to move an electron from the material
at its Fermi energy to the reference
vacuum, also depends on the surface
where symmetry breaks, and interactions will exist in this process of
removal. So, the workfunction also depends on the orientation of the surface.

that came about from the band properties at the interface, and there exists a conducting lower Mott-Hubbard band in both cases. Workfunctions define these conditions. Table 4.6 shows the commonly accepted workfunctions of metals—both low and high workfunctions are technologically possible.

Workfunctions					
Low	eV	Moderate	eV	High	eV
Cs	2.1	Ta	4–4.8	Co	5.0
Rb	2.3	Nb	4–4.8	Rh	5.0
Yb	2.6	Zr	4.0	Ni	5.0–5.4
Y	3.1	Al	4.0–4.3	Ir	5.0–5.7
Nd	3.2	Ti	4.3	Au	5.1–5.5
La	3.5	V	4.3	Pt	5.1–5.9
Mg	3.66	W	4.3–5.2	Pd	5.2–5.6
Tl	3.8	Mo	4.3–5.0		
Hf	3.9	Cr	4.5		
		Cu	4.5–5.1		
		Fe	4.6–4.8		
		Re	4.7		
		Ru	4.71		

Table 4.6: Workfunctions of some metals classified as "low," "moderate" and "high," qualitatively. For reference, silicon has a workfunction of 4.6–4.9 eV. It varies with direction because of what happens at the surface. Publications on VO_2 assign a workfunction of the order of 4.7–5.2 eV to the insulating phase. It increases by $\sim$0.1–0.2 eV upon transition to the metallic phase.

In any of these alignment conditions of these materials, one could alter the conditions by applying a bias voltage, where the Mott insulation and band insulation, and Mott conduction regions may appear and disappear. So, memory effects and others can arise triggered by conditions arising from the different interface conditions created, such as in the example shown in Figure 4.50, and the load line effect shown in Figure 4.51.

We discuss this alignment-induced intrinsic behavior, which is electrically driven in a simple way, in order to relate to characteristics such as those shown in Figure 4.51. So, we consider low workfunction metals, such as those in the the first two columns of Table 4.6, and high workfunction metals, such as those in the last two columns of Table 4.6. Figure 4.53 shows band bending under bias, including the thermal equilibrium condition shown in Figure 4.52(a) for a low workfunction metal in contact with a Mott-Hubbard material that has the lower band partially occupied. Both forward bias and the more interesting reverse bias conditions are shown. In thermal equilibrium, this structure has a band insulator region at the interface, with both the upper and the lower Mott-Hubbard bands occupied. If this region is thin, tunneling is certainly possible between the upper Hubbard band and the metal, and the magnitude will depend on the density of states in the Hubbard band, as well as on current continuity. Deeper in the structure, there also exists another insulating region—the Mott insulation region. This structure, in forward bias (Figure 4.53(a)), defined as the bias condition where the energy of the metal is lowered, has conduction impeded by the barriers. Increasing forward bias decreases one barrier width, that of the Mott insulator, and increases another. The reverse bias is also interesting. As progressively higher reverse bias voltage is applied, first, the interface band insulation region disappears, as in Figure 4.53(c), and there exists an upper Mott-Hubbard band conducting region at the interface separated by the Mott insulation region. Electrons in the metal can tunnel across the interface to available states in the upper Mott-Hubbard band. However, any conduction is limited by tunneling across the Mott insulation region. As the reverse bias is increased further, as shown in Figure 4.53(d), the partially filled upper Mott-Hubbard band disappears, and the Mott insulation region appears across the interface. Further increase in bias voltage, as in Figure 4.53(e), provides an ohmic conduction region at the interface. And further increase in bias voltage, as in Figure 4.53(f), results in a band insulating region at the interface.

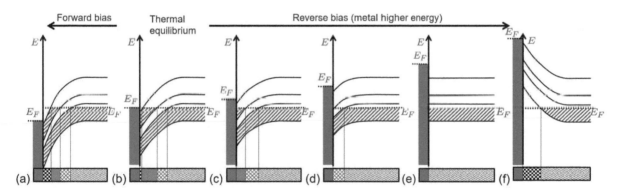

Figure 4.53: Forward-biased (a), thermal equilibrium (b) and reverse-biased (c–f) voltage band bending and conduction properties near the interface when a low metal workfunction exists at an idealized interface with a Mott-Hubbard insulator that has undergone a conducting transition. The material conduction property coding is identical to that shown in Figure 4.52.

Conduction in the structure, therefore, is highly unusual. In forward bias, conduction is impeded by band and Mott insulation. But, in reverse bias, it goes through a bias range in which conduction is ohmic at the interface. On one side of this bias voltage, conduction is impeded by Mott insulation; on the other side, it is impeded by band insulation. A filter opens for conduction in a short bias voltage range in this structure. And the structure requires a reverse bias, that is, lowering of energy of the metal electrode w.r.t. to the metal–insulator.

In other situations there is either one tunneling barrier or coupled
tunneling barriers impeding conduction.

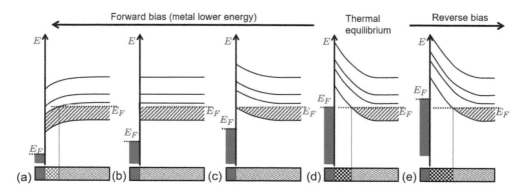

Figure 4.54: Forward-biased (a–c), thermal equilibrium (d) and reverse-biased (e) voltage band bending and conduction properties near the interface when a high metal workfunction exists at an idealized interface with a Mott-Hubbard
insulator that has undergone a conducting transition. The material conduction property coding is identical to that
shown in Figure 4.52.

Now consider the properties of the metal/Mott conductor idealized system when the workfunction of the metal is large (see Figure 4.54, for example, when using metals such as cobalt, nickel, plat-
inum or palladium. Here, it is the forward bias, that is, lowering of
the metal energy, that becomes interesting. In reverse bias, such as
that shown in Figure 4.54(e), the band insulator region that existed
at the interface stretches out further, and conduction is further suppressed. In forward bias, the conduction onset happens when the
Mott conducting region from the partially occupied lower Hubbard
band extends all the way to the interface, as shown in Figure 4.54(c).
The conduction persists when the lower Mott-Hubbard band is partially occupied, as well as being near the thermal equilibrium condition of Figure 4.54(d), as shown in Figure 4.54(b) and (c) with
electrons tunneling from this lower band. At higher forward bias,
as shown in Figure 4.54(a), this lower Hubbard band is filled adjacent
to the interface, that is, it is a Mott insulator, in our vernacular. However, electrons can still be injected into the empty states of the metal.
So, unlike the cases shown in Figure 4.53(a)–(d), where the Mott insulator has filled states in both cladding Mott-Hubbard bands, here
there are empty states of the metal for tunneling. For any forward
bias voltage beyond the limiting situation shown in Figure 4.54(c),
conduction persists between the metal and the lower Mott-Hubbard
band.

The filter opens and stays open in the forward bias region when
the metal workfunction is larger than that of the Mott-Hubbard
metal–insulator transition material.

In this idealized picture, where a metal–insulator transition has been assumed to have taken place due to temperature, low work-function metals create conditions in a range of reverse bias for conduction, and high workfunction metals create conditions beyond a threshold in forward bias for conduction.

Now consider the situation when the material has undergone the insulation transition and the lower Hubbard band is filled. We apply bias voltage. How the system behaves will depend on the conditions of transport in the interface region. If electrons can tunnel across between the metal and the Mott material, one would see the conduction. But, as in these examples studied, that conduction can be turned on and off. It is not surprising, therefore, that one sees behavior such as that shown in Figure 4.50 and Figure 4.51, and these behaviors depend on whether the sweeping is by voltage or current. A current sweep floods and extracts electrons, with a voltage responding to it. So, a condition change between Figure 4.53(e) and (f) may cause the voltage to jump up as the condition in the Mott material changes from conducting to insulating. This will cause the ``S''-type behavior shown in Figure 4.51.

The properties of the phase transition materials, of what happens within the bulk arising from the unusual physical structure and the multitude of interactions arising in transition metals with partially filled d and f orbitals surrounded by oxygen, and of what happens at the interfaces in how the Mott-Hubbard bands interact give rise to several other interesting phenomena. An important one is that two-terminal devices, in thin film form, can possibly be made conducting or insulating through application of suitable bias voltages, that is, providing a history-dependent hysteresis property, that is, a memory. This form is different from the power limiter example discussed earlier in that there is a non-volatile change in the state of conduction. We will discuss this in a subsequent section while discussing transition metal oxide–based resistive memories. Correlated electron effects may or may not play a primary role in such structures, whose conductivity may be affected by defects, vacancies and ionic movement. So, we defer this discussion to later in discussing memory-based resistance change through broken translational symmetry.

Since Mott transition can be induced by doping and field, ferroelectricity can be employed to induce it in the channel of a transistor formed using a ferroelectric gate and a metal–insulator channel. An example is a single-crystal $SrRu_xTi_{1-x}O_3$ channel, or using $LaCu_xMn_{1-x}O_3$ for the Mott conductor with $Pb_xZr_{1-x}TiO_3$ for the ferroelectric insulator in single-crystal-based systems. In the off state, there is sufficient free carrier population that the depolarization field is suppressed. This lets the ferroelectric capacitor have non-volatile

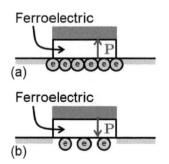

Figure 4.55: Coupling of a ferroelectric phase transition to a metal–insulator transition; here, non-volatile memory becomes possible, because residual carriers in the channel reduce the depolarization field of the ferroelectric.

storage in an integrated gain cell, where the transistor can be employed to sense. These have all been examples where correlated electron effects caused a change in property of the system that led to an electronic device use. We now turn to magnetic systems ,where the collective effects from spin and phase transition properties arising from the spin interactions can be utilized.

4.4   Spin correlations and devices

MAGNETIC PHASE TRANSITION IS A RESULT OF BROKEN SYMMETRY IN TIME. Classically, current, which is a charge flux, that is, the physical property associated with the flow of charge in space—of charge with momentum—upon reversal of time leads to a reversal in current direction and also of the magnetic field. Moving electric charges cause a magnetic field in space. This magnetic field acts on another moving charge through a Lorentz force, a force that is perpendicular to the plane of the velocity and the field and is related through a cross product. Particles also have intrinsic magnetic moment. Quantum-mechanically, this is associated with the angular momentum, that is, it can be related to the spin and orbital quantum numbers that we associate with the quantization of the angular momentum. The magnetic properties of interest of elementary particles, their atomic assembly in solid, liquid, gas and plasma forms—the matter states—arise from these intrinsic properties of the matter and the flow of current. These interactions between the particles and their ensembles at the nanoscale provide numerous interesting properties that we employ in nanoscale devices. This will be the focus of this section, and we will start, as with the discussion of ferroelectrics, with a discussion of the energetics by uniting the classical and quantum-mechanical interactions.

As for ferroelectrics, the most consequential effect of the interactions is a spontaneous polarization—magnetic here, instead of electric.

When a current I flows, in our classical view, a magnetic field **H** arises. This magnetic field is related to the current through the position-dependent current density $\mathbf{J}(\mathbf{r})$ ($I = \int \hat{\mathbf{n}} \cdot \mathbf{J} d^2 r$) by Ampère's law as

$$\mathbf{J} = \boldsymbol{\nabla} \times \mathbf{H}. \tag{4.55}$$

In detail, Stokes's theorem defines the curl of a vector, such as **H**, as

$$\lim_{S \to 0} \frac{1}{S} \oint_r \mathbf{H} \cdot d\mathbf{r} = \hat{\mathbf{n}} \cdot (\boldsymbol{\nabla} \times \mathbf{H}) \quad \therefore \quad \oint_r \mathbf{H} \cdot d\mathbf{r} = \int_S \hat{\mathbf{n}} \cdot (\boldsymbol{\nabla} \times \mathbf{H}) \, d^2 r, \tag{4.56}$$

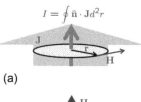

(a)

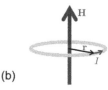

(b)

Figure 4.56: A magnetic field is created around a current (a); equivalently, a magnetic field is associated with a current loop (b). The magnetic field has chirality, and is axial, unlike the electric field, which is polar.

The relationship between the current along a linear path and the curling magnetic field in the orthogonal plane is described by the Biot-Savart law.

which lets us establish the relationship that the field at any position must satisfy through the curl of the field and hence to the current, through Ampère's law. The corollary—when an electric current flows in a loop, it generates a magnetic field **H**—also holds. Figure 4.56 summarizes this equivalence and its right-handed chirality. When a current I flows in a wire of radius R, it creates a magnetic field of magnitude $H = I/2\pi r$ for $r \geq R$. A current I flowing in a loop at radius r has at its center an out-of-plane magnetic field of $I/2r$. So, the current density, that is, charge flux, and magnetic field are related to each other through the curl operator. Magnetic field, classically, is associated with a current loop.

A spatially changing magnetic field induces a current in a conducting medium. Magnetic induction, or magnetic flux density, **B**—the magnetic analog of electric displacement **D**—is another quantity to capture the net magnetic effect arising from impressed external conditions and those arising from the effects within the material. This magnetic induction **B** is

$$\mathbf{B} = \mu_0(\mathbf{H} + \mathbf{M}), \qquad (4.57)$$

where **M**, the magnetization—magnetic polarization—is the analog of the electric spontaneous polarization **P** arising from the spontaneous phase transition effect. μ_0 is the magnetic permeability of free space, analogous to the dielectric constant of free space for electric induction. **M**, which can only exist in materials—not vacuum—can determine the maximum field that can be generated by a fully magnetized volume. For highest magnetization materials, such as Nd_2FeB_2, et cetera, this is very significant. It raises the energy density and the ability to achieve high fields in small size. Equation 4.57 also states that the symmetries of all three vector fields—**H**, **B** and **M**—are identical. They define the rotation symmetry through the curl that connects them to the electrical counterparts. They are therefore axial vectors.

These magnetic fields reverse sign with time reversal. A charge in time reversal is a current in reversed direction, creating a field of opposite sign. Electric field vectors do not change sign with time reversal; two charges separated by position vector **r** may remain static, so the field remains in the same direction even with time reversal. So, electric fields are polar vectors. Ferroelectricity is a consequence of symmetry breakdown in spatial inversion. Magnetism is a consequence of the symmetry breakdown in spin—"spin charge" can be reversed—and of time.

Magnetization and magnetic induction show many characteristics that have parallels to our observations on ferroelectrics. Magnetization, a result of spontaneous magnetic polarization, which we have

The discovery that current in a loop gives rise to an orthogonal magnetic field is due to Hans Christian Ørsted.

Earth's magnetic field, for example, is hypothesized to arise from the rotating charges of the molten core. This is a weak field arising from a small loop current. The sun with its hot plasma, has a magnetic field that is 100 times higher than that of the Earth. Locally, on the sun, such as at sun spots, it can be larger by another factor of 1000. Neutron stars have a field $\sim 10^{12}$ times higher than that of the earth. Curls do the nature's job of forcing an effect that is felt through a force that is out of plane.

We will discuss the origin and consequences of this magnetic polarization in some detail. This polarization has immense consequences for nanoscale magnetic devices, as it did for nanoscale ferroelectric devices. We have now introduced a number of different versions of magnetic fields that act to cause force with different vector and scalar operations. These different formulations of fields are, for historic reasons, tied to where they arise from. As in the electrostatic treatment, one can argue that energy is more fundamental and is the fundamental scaler. At a deeper fundamental level, the origin of these conundrums is that the energy for electrical discussions was tied to potential, and that for magnetic to kinetic causes. The kinetic energy origin of the magnetic field leads to the curls. Earlier, we have seen that the vector potential **A** was needed as the fundamental variable for magnetic quantities of interest. As an example, the magnetic flux from which the flux density is derived is related, for a single loop of a wire, for example, as $\Phi = \int_S \mathbf{B} \cdot \bar{\mathbf{n}} \, d^2r = \oint \mathbf{A} \cdot d\mathbf{r}$.

	Material	Relative permeability (μ_r)	Magnetization $\mathbf{M}$ (A/m)
	Fe	5000–200 000	$2.199/\mu_0$ @ 4.2 K
	Co	250	$1.834/\mu_0$ @ 4.2 K
	Ni	600	$0.665/\mu_0$ @ 4.2 K
	Ferrites	10-10 000	
	Permalloy (Py, Nd_2FeC_2)	100 000	

Table 4.7: Magnetization of some example materials with significant engineering use.

not yet discussed, has the following relationship with the magnetic field through the magnetic susceptibility order parameter χ:

$$\mathbf{M} = \chi\mathbf{H}. \tag{4.58}$$

Like electric polarization, magnetization can also be anisotropic, so χ in general is anisotropic. Table 4.7 lists the relative permeability and magnetization of some commonly employed magnetic materials.

There are easy directions and there are hard directions for magnetization, and this becomes important in storage media because it determines how much energy is needed to change magnetization and its direction, and thus its resistance to demagnetization. Nd_2FeB_2 magnets too have this anisotropy together with a high value for saturation magnetization giving it the capability for high storage of energy. Figure 4.57 shows the hysteresis that, like the displacement or polarization change with electric field, may be obtained in magnetic materials. When the coercive field $\mathbf{H}_{co}$ is small, it is easy to change magnetization and the magnetic induction and therefore also easy to disturb. If the remnant magnetization or magnetic induction is large, that is, large $\mathbf{M}_r$ or $\mathbf{B}_r$ remains when the magnetic field is absent, then the material is capable of large storage.

Magnetic induction follows

$$\nabla \cdot \mathbf{B} = 0, \tag{4.59}$$

another one of Maxwell's equations. This is the complement of $\nabla \cdot \mathbf{D} = \rho$ for electric induction, that is, displacement. There is no equivalent of discrete electric charge—no discrete magnetic poles. The continuous behavior can be determined through Gauss's theorem, and the definition of divergence, that is,

$$\lim_{\Omega \to 0} \frac{1}{\Omega} \oint_S \mathbf{B} \cdot \hat{n} d^2r = \nabla \cdot \mathbf{B} \quad \therefore \quad \int_S \mathbf{B} \cdot \hat{n} d^2r = \int_\Omega \nabla \cdot \mathbf{B} d^3r. \tag{4.60}$$

Since $\nabla \cdot \mathbf{B} = 0$, this equation implies that, for any volume, the magnetic induction $\mathbf{B}$ through the surfaces that enclose this volume must be zero: the total flux of magnetic induction emerging from

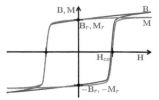

Figure 4.57: Change in magnetic induction and magnetization as a function of magnetic field in a ferromagnetic material. The hysteresis loop is of a similar nature as that due to electric polarization in ferroelectrics, except here it is from magnetic polarization.

In drawing equivalences between magnetic and electric fields, it is useful to look at these total fields as,

$$\mathbf{B} = \mu_0(\mathbf{H} + \mathbf{M}) = \mu_r\mu_0\mathbf{H}, \text{ and}$$
$$\boldsymbol{\mathcal{E}} = 1/\epsilon_0(\mathbf{D} - \mathbf{P}) = 1/(\epsilon_r\epsilon_0)\mathbf{D}.$$

The difference in the polarities of $\mathbf{M}$ and $\mathbf{P}$ and in the use of μ_r and $1/\epsilon_r$ can then be seen as arising from magnetization being reinforcing in most materials, that is, they are paramagnetic, while most dielectrics reduce the field. The fields $\mathbf{B}$ and $\boldsymbol{\mathcal{E}}$ are the total fields that act on the particle. $\mathbf{M}$ and $\mathbf{P}$ are the magnetic and electric dipole densities of the material. Dipoles in paraelectric solids reorient to the external field. Dipoles in diamagnetic solids, caused by $\mathbf{B}$, oppose it, in accord with Lenz's law, which states that an induced electromotive force gives rise to a current whose magnetic field opposes the original change in magnetic flux. This is the negative sign in Faraday's law of induction, where $\boldsymbol{\mathcal{E}} = -\partial\Phi/\partial t$, and which appears in Maxwell's equation in the form $\nabla \times \boldsymbol{\mathcal{E}} = -\partial\mathbf{B}/\partial t$. This is all a consequence of the conservation of energy. A rising kinetic energy of a charge, associated with the magnetic field, must be counteracted by a change in potential energy, which is related to the electric field.

any closed surface vanishes. It cannot escape. It must therefore loop. This equation also implies that magnetic sources inside a volume cannot change the flux of magnetic induction entering or leaving the volume through the surface that surrounds the volume. The electric field—complement of this magnetic induction—behaves distinctly differently since it terminates on charge poles.

The implication of this argument is in Figure 4.58—an example of a vertically aligned microscopic magnetic medium with no external magnetic field. At the surface, at any position, all of the magnetic vectors exist. Magnetization does not exist outside the magnetic medium and does not change the magnetic induction outside. No current is flowing inside the loop drawn. Therefore, the integral on the closed path, $\oint \mathbf{H} \cdot \mathbf{r}\ dr$, is equal to 0. The consequence of Equation 4.57 is that, inside the magnetic medium, the direction of the magnetic field $\mathbf{H}_i$, given by $\mathbf{H}_i = \mathbf{B}_i/\mu_0 - \mathbf{M}$, is opposite to that of $\mathbf{H}_x$, the magnetic field outside the magnetic medium—$\mathbf{H}_i$ is a depolarizing demagnetizing field similar to the electric field created in the case of a ferroelectric when surface polarization is not effectively neutralized. Outside, the magnetic field $\mathbf{H}_x = \mathbf{B}_x/\mu_0$ is related to $\mathbf{H}_i$ through the magnetic susceptibility of the material used for the medium. The induction inside ($\mathbf{B}_i$) is different from that outside ($\mathbf{B}_x$), but inside, as well as outside, or anywhere else, the induction entering and exiting any closed volume balances out.

In the example shown in Figures 4.58 and 4.59, the magnetic field $\mathbf{H}_x$ is a stray field that is a consequence of the magnetization of the medium. Magnetization can be non-homogeneous. It arises from properties of the material, as we will discuss. One may, though physically crude in light of the absence of isolated magnetic poles, think of this magnetization in the material as a consequence of "north" (+) and "south" (−) poles of magnetization, which exist simultaneously and balance each other as terminations of the magnetization $\mathbf{M}$. This magnetization, as simplified in Figure 4.59, is by fiat viewed as arising from the termination of magnetic polarization between two virtual opposite and equal "magnetic" charges. The stray field can then be viewed as being generated by the magnetization that is associated with this pole or magnetic charge heuristic. Since $\nabla \cdot \mathbf{B} = \nabla \cdot \mu_0(\mathbf{M} + \mathbf{H}) = 0$, the magnetic flux is continuous at the magnet/surroundings interface, and the magnetic field can be related as $\nabla \cdot \mathbf{H} = -\nabla \cdot \mathbf{M}$, continuously and at any position. The magnetic field inside the medium in this example is $\mathbf{H}_i$, and that outside it is $\mathbf{H}_x$, while the magnetic flux $\mathbf{B}_i = \mathbf{B}_x$. These relationships manifest their effects in numerous ways of import during use.

If the film is very thick or has magnetization along the plane of the film, then the condition $\oint \mathbf{H} \cdot \mathbf{dr} = 0$, where one part of the loop

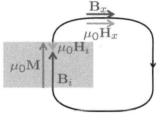

Figure 4.58: Absent externally-imposed magnetic fields, the magnetization, the magnetic field and the magnetic induction in and around a microscopic vertically magnetized medium.

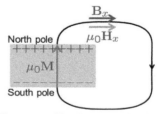

Figure 4.59: Viewing magnetization as a consequence of "north (+)" and "south (−)" pole leading to the stray magnetic field. We are assuming that the magnetization is large so that the internal magnetic field is vanishingly small.

is inside the magnetic material and the other external, implies that the demagnetization field will be small. The integral contribution inside the material across a large spatial span, balances the external part. Equivalently, the equal and opposite ``magnetic charges´´ are far apart. So, $|\mathbf{H}_i| \to 0$. If the film is thin, and magnetization is in an out-of-plane direction, then the continuity of magnetic flux implies that the external field $\mathbf{H}_x$ will be large that is $\sim \mathbf{M}$, at the interface, and, internally, a demagnetizing field that is large must also exist, since the virtual ``magnetic charges´´ are spatially close. This is similar to what happens in ferroelectrics. In thin films, such as in magnetic recording meda, we will have to worry about these demagnetization issue, since smaller film thicknesses are not bulk-like, are affected by surface energetics and therefore have paramagnetic limitations.

Closed magnetic rings have a vanishing path integral of magnetic field. Hence, no demagnetization field exists. A ring can be magnetically saturated to the maximum. If one takes this ring and forms a slit in it, the field and flux in this slit will be large, approximately the magnetization of the ring. The small magnetic domains of a thin film can then be written and read easily through such a ring with a gap.

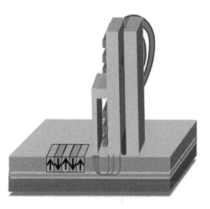

This is what is accomplished in read-write heads of magnetic disks, a part of the head drawing of which is shown in Figure 4.60. An easily saturated magnetic yoke material, a ``soft´´ material forms the yoke. A lithographically and electrochemically formed coil producea about a tesla of field in the gap for writing through the gap.

The energy in these magnetic structures is stored both internally, in the magnetized volume, and externally. The total energy storage, electric and magnetic, is

$$U = \int_{\Omega \to \infty} u \, d^3r = \int_{\Omega \to \infty} \left[\frac{1}{2} (\boldsymbol{\mathcal{E}} \cdot \mathbf{D} + \mathbf{H} \cdot \mathbf{B}) \right] d^3r. \qquad (4.61)$$

In the absence of electric fields, that is, under magnetostatic condi-

Magnetic disks were magnetized in plane for these reasons until the first decade of the 21st century. To achieve high external fields for detection at small dimensions, fields which occur simultaneously with high depolarization field in the storage medium, disk drives changed from having in-plane polarization to having out-of-plane polarization, despite this depolarization issue.

Figure 4.60: An example of a simple read-write head for perpendicular magnetic recording on a magnetic medium. The right part of the head contains the read section, which senses the magnetic field. It is a sensor whose resistance changes substantially between the two orientations of magnetic polarization that form the bit—this sensing is a magnetoresistive reading operation that has employed increasingly sensitive effects originating in different phenomena in thin film assemblies. The write part, on the left, consists of a pole created by using two yokes—a main yoke and a return yoke—that confine the high field employed for writing—the return yoke spreading out the field to prevent disturbing the region underneath, and the main yoke focusing the field for writing. High flux at the main pole, achieved through changes in areal directions with a small gap in the ring, is employed to write on the magnetic film. The read part of the head senses the orientation of the external field using the different magnetoresistive effects.

This energy storage is analogous to what happens due to electric fields in a capacitor. The energy is stored inside, but also in the stray fields outside the capacitor.

tions, in linear materials, the total magnetic energy (U_m) stored is

$$U_m = \oint_{\Omega \to \infty} \frac{1}{2} \mathbf{H} \cdot \mathbf{B} \, d^3 r$$

$$= \frac{1}{2} \mu_0 \int_i \mathbf{H}_i \cdot (\mathbf{H}_i + \mathbf{M}) \, d^3 r + \frac{1}{2} \mu_0 \int_x \mathbf{H}_x \cdot \mathbf{H}_x \, d^3 r, \quad (4.62)$$

the sum of an internal contribution where the internal field is depolarizing, and the external field that follows from the absence of divergence for the magnetic induction.

The magnetic dipole moment, viewed classically in the virtual "magnetic dipole charge" view, is summarized schematically in Figure 4.61. A point-like magnetic dipole charge of strength v is separated by the distance between the equal positive and negative magnetic charges, with the positive pole at $\mathbf{R}_2$, and the negative pole at $\mathbf{R}_1$. The magnetic field $\mathbf{H}(\mathbf{r})$ at any position $\mathbf{r}$, resulting from this magnetic dipole, is

$$\mathbf{H} = \frac{1}{4\pi\mu_0} \frac{v(\mathbf{r} - \mathbf{R}_2)}{|\mathbf{r} - \mathbf{R}_2|^3} - \frac{1}{4\pi\mu_0} \frac{v(\mathbf{r} - \mathbf{R}_1)}{|\mathbf{r} - \mathbf{R}_1|^3}. \quad (4.63)$$

The magnetic moment $\mathbf{m}$ of this magnetic dipole is an axial vector, with the dipole moment

$$\mathbf{m} = v(\mathbf{R}_2 - \mathbf{R}_1) \quad (4.64)$$

pointing from south (negative) to north (positive).

The magnetic field it creates is pointed from north (positive) to south (negative). The magnetization of the material is the density of magnetic moment, so $\mathbf{M} = \lim_{V \to 0} \mathbf{m}/V$, where V is the volume. A changing magnetization, in orientation and magnitude, is reflective of changing magnetic moment in orientation and magnitude at the microscopic scale. The field, at a distance $\mathbf{r}' = \mathbf{r} - (\mathbf{R}_2 + \mathbf{R}_1)/2$ from the center of the dipole, is

$$|\mathbf{H}| \approx \frac{1}{4\pi\mu_0} \frac{|\mathbf{m}|}{|\mathbf{r}'|^3} (1 + 4\cos^2 \varphi)^{1/2} \propto \frac{|\mathbf{m}|}{|\mathbf{r}'|^3} \text{ for } |\mathbf{r}'| \gg |\mathbf{R}_2 - \mathbf{R}_1|. \quad (4.65)$$

φ here is the polar angle w.r.t. the axis of the magnetic moment.

Closed macroscopic current loops of current I in an enclosed area S create a magnetic field identical to that of a dipole with $|\mathbf{m}| = \mu_0 I S$.

For nanoscale devices, we need to understand the nature of magnetic properties arising from electrons and atoms and their ensembles. Quantization of spin ($s = 1/2$), which through $\sqrt{s(s+1)}\hbar$ specifies the expectation value of the magnitude of spin angular momentum, and quantized values of the secondary spin quantum number ($m_s = \pm 1/2$), which through $m_s \hbar$ specifies the expectation value of the projection of spin angular momentum, are intrinsic properties of

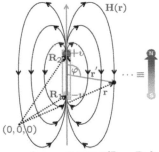

magnetic moment: $\mathbf{m} = v(\mathbf{R}_2 - \mathbf{R}_1)$

Figure 4.61: Magnetic moment $\mathbf{m}$ in terms of the virtual magnetic dipole charges $+v$ and $-v$ separated by their position $\mathbf{R}_2 - \mathbf{R}_1$, and their use in determining their contribution to the magnetic field $\mathbf{H}(\mathbf{r})$ at any position $\mathbf{r}$.

This result for dipole moment can be seen by determining the field due to the current loop and equating it to the field due to a magnetized disk occupying that area.

an electron and are associated with spin angular momentum. Protons and neutrons too are spin $1/2$ fermions and so have associated quantized spin angular momentum and its projection, which is similar to that of the electron. The orbital angular momentum of an electron in an atom is also quantized: $l = 0, 1, \ldots, (n-1)$, where n is the principal quantum number, and l is the orbital quantum number, with expectation values of angular momentum as $\sqrt{l(l+1)}\hbar$. The projection of orbital angular momentum is also quantized through the magnetic, or azimuthal, quantum number $(m_l = -l, -l+1, \ldots, l-1, l)$ as $m_l\hbar$. Magnetic energy, tied to kinetic energy, is related to these momenta. The magnetic moment is proportional to the angular momentum. Hence, electrons and atoms as ensembles with nuclear spin, electron spin and orbital quantization have magnetic properties arising from these elementary particles and their assembly, and these properties are important for the magnetic phenomena in nanoscale devices.

See S. Tiwari, "Quantum, statistical and information mechanics: A unified introduction," Electroscience 1, Oxford University Press, ISBN 978-0-19-875985-0 (forthcoming) to review. For example, briefly, the eigenfunction of an electron in an atom such as a simple hydrogen atom is $|nlm_lsm_s\rangle$. The Hamiltonian's kinetic energy operator can be written in terms of a radial part $\hat{p}_r^2$ and an angular part $\hat{L}^2$. $\hat{L}^2$ commutes with one and only one of the three components of orbital angular momentum; the components themselves do not commute with each other. By convention, we choose z as the axis of quantization for the projection, with $[\hat{L}^2, \hat{L}_z] = 0$. In the eigenfunction then, $\hat{L}^2$ quantization gives the total angular part, and $\hat{L}_z$ quantization the azimuthal or projection part. The expectations of angular momentum are $\langle \hat{L}^2 \rangle = l(l+1)\hbar^2$, and $\langle \hat{L}_z \rangle = m_l\hbar$, with specific constraints on the integers l and m_l. The same is true for the spin angular momentum also.

The magnetic moment is proportional to the angular momentum and, for historic reasons, is often written in terms of the Bohr magneton μ_B. The classical angular momentum of an electron orbiting at radius r, enclosing an area πr^2 with angular velocity ω, so that velocity $\mathbf{v} = \boldsymbol{\omega} \times \mathbf{r}$, has a magnetic moment that can be written as

$$
\begin{aligned}
\mathbf{m} &= q\mu_0 r^2 \omega/2 = -(e\mu_0/2)\mathbf{r} \times \mathbf{v} \\
&= -(e\mu_0/2m_0)m_0\mathbf{r} \times \mathbf{v} = -(e\mu_0/2m_0)L \\
&= \gamma L,
\end{aligned}
\tag{4.66}
$$

where L is the angular momentum, and $\gamma = -e\mu_0/2m_0$, the gyromagnetic ratio of an electron. The quantization of the projection of this angular momentum is in increments of $\hbar$.

A Bohr magneton is

$$
\mu_B = \frac{e\mu_0}{2m_0}\hbar = 1.17 \times 10^{-29} \; V \cdot m \cdot s,
\tag{4.67}
$$

where $\hbar$ is the incremental integral change in the angular momentum of the projections.

The electron in the eigenstate $|nlm_lsm_s\rangle$ has both a spin and an orbital magnetic moment. The expectation value of the orbital magnetic moment $\langle m_0^z \rangle$ is

$$
\langle m_0^z \rangle = -\mu_B m_l,
\tag{4.68}
$$

since m_l captures the projection. The spin magnetic moment $\langle m_s^z \rangle$ is

$$
\langle m_s^z \rangle = -2\mu_B m_s = -g\mu_B m_s,
\tag{4.69}
$$

where $m_s = \pm 1/2$ is the secondary spin quantum number. g, which is ~ 2, is called the g factor. The total magnetic moment then is the

Lamb shift, the result of the field of a moving electron or, more precisely, the fluctuations in the electric and magnetic fields, causes the energy to be perturbed so that one sees slightly different energy for $2S_{1/2}$ and $2P_{1/2}$ energy levels even in the simple hydrogen atom. This perturbation causes a fluctuation in the position of the electron, thus causing the energy shift. A small increase in the moment arises from spin—an anomalous magnetic moment increase. As a result, $g = 2.00231$ rather than 2.00000. For our magnetic discussion, this is a small effect that we will ignore.

result of spin and orbital components:

$$\langle m^z \rangle = \langle m_s^z \rangle + \langle m_o^z \rangle = -\mu_B(g_s m_s + g_l m_l), \qquad (4.70)$$

where $g_s \approx 2$, and $g_l = 1$. Since mass appears in the denominator of the Bohr magneton, the contribution of nuclear magnetic moment due to spin is scaled lower by the ratio of the electron to proton or neutron mass, a factor of $\sim 1/1836$. Nuclear magnetic moments are therefore small and will be important to our discussion only secondarily. Their primary effect is a perturbation leading to the scattering of spin—so, spin effects, such as flipping through an exchange in which angular momentum is still preserved. Magnetization of solids arises from exchange interaction. Exchange interaction is the result of collective consequences of the Pauli exclusion principle for fermions, electric Coulomb interaction, and spin coupling that causes preferred aligned arrangements with net magnetization. We will discuss this shortly. But first, we explore the behavior of this magnetic moment in simple conditions.

When a magnetic dipole of moment **m**, whether of an atom, an electron or a solid, is placed in a magnetic field, we can determine how the moment responds through the angular momentum and changes in it.

Mechanically, when a force **F** acts on a lever of length **r**, it causes a torque $\mathbf{T} = \mathbf{r} \times \mathbf{F}$ that changes the angular momentum as $d\mathbf{L}/dt$—a consequence of Newton's law. The magnetic moment, as we discussed for a current loop, arose from a circulating current in radius r, and in the quantum picture from spin, as seen in Figure 4.62. For a magnetic dipole of strength of poles $+v$ and $-v$, the field **H** exerts a force $\mathbf{F} = v\mathbf{H}$ on each of the poles, so that the net force is zero, but there still exists a torque **T**. Since $\mathbf{m} = v\mathbf{r}$, the magnetic field acting on the magnetic moment causing a torque, which is an axial vector again, follows

$$\mathbf{T} = \mathbf{m} \times \mathbf{H} = \frac{d\mathbf{L}}{dt}, \qquad (4.71)$$

absent any damping. The magnetic moment precesses, as shown in Figure 4.62. The energy of the dipole is

$$U = -\mathbf{m} \cdot \mathbf{H}. \qquad (4.72)$$

An unaligned dipole has excess energy. Classically, the relationship in Equation 4.72 follows from the fact that the force $\mathbf{F} = \pm v\mathbf{H}$ exerted on the two poles causes the dipole to align to the field by rotating the dipole through the angle between it and the field; hence, the dot product. So, for an electron, with $g \approx 2$, the energy associated with the magnetic moment is

$$U = -\mathbf{m} \cdot \mathbf{H} = -2\mu_B m_s H = \pm \mu_B H, \quad \text{where } m_s = \pm\frac{1}{2}. \qquad (4.73)$$

Exchange coupling strongly couples all electron spins, resulting in us approximating the solid ensemble as one single moment. In all these cases, the moment is an axial vector that represents the effect of current loop, and we are interested in how this behaves in a magnetic field and how the energy of this individual or ensemble moment changes through interactions.

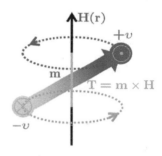

Figure 4.62: The response of a dipole **m** of the virtual magnetic charges $\pm v$ when placed in a magnetic field **H**, as a result of a torque **T** acting on it. It precesses around the axis as a result of the out-of-plane forces acting on it. The direction is opposite to that of the torque, which is out of the plane here, because γ, the gyromagnetic ratio, is negative—as is the case for an electron.

For now, we have ignored frictional forces. If none exist, angular momentum is a constant of motion, and the moment will never align. But, as in mechanics and elsewhere in the real world, where no system is truly isolated, interaction with the environment will lead to frictional forces, and consequent losses to a broadband of degrees of freedom. A compass needle aligns to the earth's magnetic field. It does so by precession of the magnetic moment about the axis of this field, which is uniform on the compass's scale length. The energy of the needle is slowly lost through damping, that is, the frictional forces to the surroundings—principally, the pivot and the air.

An electron's spin-degenerate energy level splits by $\pm\mu_B H$ in a magnetic field. This effect is observable in transitions in atoms and materials in a magnetic field. Classically, this effect is associated with the electron's moment precessing at $\omega = (e\mu_B/m_0)H$ in the two opposite orientations. The quantum-mechanical view of this follows from the time evolution via the time-dependent Schrödinger equation, or evolution under the operator $\exp(-i\mathscr{H}t/\hbar)$, where $\mathscr{H} = 2\mu_B m_s H$. The phase between the two spin states changes at the rate of $\omega = 2\mu_B H/\hbar$.

Equation 4.72 also implies that the force on the dipole, in general, can be written as

$$\mathbf{F} = \boldsymbol{\nabla}(\mathbf{m}\cdot\mathbf{H}). \tag{4.74}$$

For a dipole whose length scale is smaller than the length scale of gradient of the magnetic field, this force $\mathbf{F}$ is $\sim (\mathbf{m}\cdot\boldsymbol{\nabla})\mathbf{H}$, and it will cause the dipole to align along the direction of the maximum gradient.

The magnetic moment and the angular momentum are proportional to each other, with the constant of proportionality being the gyromagnetic ratio γ, that is, $\mathbf{m} = \gamma\mathbf{L}$. For an electron, the earlier discussion of Equation 4.66 gives $\gamma = qg\mu_0/2m_0$, which, for an electron, because of its negative charge, is negative. In the absence of damping, therefore,

$$\frac{1}{\gamma}\frac{d\mathbf{m}}{dt} = \mathbf{m}\times\mathbf{H} = \mathbf{T}. \tag{4.75}$$

The magnetic dipole precesses in a field $\mathbf{H}$ at a radial frequency of $\omega = -\gamma H$, the Larmor precession frequency, in a direction given by the left hand rule. This frequency is independent of the angle between the magnetic dipole and the field—it is the consequence of a cross product. Conservation of angular momentum means that action is balanced by reaction. So, the torque on the magnetic dipole acts with an equal reaction on the source of the magnetic field in the opposite direction.

This precession frequency $\omega = -\gamma H - e\mu_0 H/m_0 = eB/m_0$, for a flux of 1 T, corresponds to an electron spin precession time of ~ 35 ps. Nanoscale devices, where this precession is important, for example, for flipping the direction of magnetization, will have time scales of some factors higher than this. This time is sufficiently small to be of interest in many uses. Absent damping, the magnetic moment will continue to precess and not align with the field. For alignment, the damping of the precession must be through a torque that is perpendicular to the torque that causes precession. This damping torque, $\mathbf{T}_d$, with a friction coefficient ζ, is

$$\mathbf{T}_d = \zeta\mathbf{m}\times\frac{d\mathbf{m}}{dt}, \tag{4.76}$$

A magnetic resonance measurement allows us to measure the energy and the magnitude of the moment in general. Nuclear magnetic resonance under this same reasoning, but at the mass-scaled lower energy, is the foundation of the nuclear magnetic resonance imaging technique used in healthcare and elsewhere.

The Stern-Gerlach experiment is the classic experiment illustrating the magnetic consequence of spin. An atomic beam with spin is passed through a gradient in a magnetic field. The pass-through leads to a splitting of the beam into two in a direction aligned with that of the magnetic field. The resulting two beams are due to parallel and antiparallel spins, respectively.

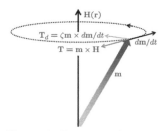

Figure 4.63: Damping and precession torque on a magnetic moment $\mathbf{m}$ in the presence of a field $\mathbf{H}$.

Damping is necessary for the compass needle to serve its purpose.

as schematically shown in Figure 4.63, with the motion towards alignment as shown in Figure 4.64. For a positive damping coefficient ζ, damping torque causes rotation towards the field's axis. Just as with mechanical friction, this damping arises in the coupling of the excess energy and the motional degree of freedom to a broadband of energy states and motional degrees of freedom in the surroundings of the precessing magnetic moment. Since it is a cross product of the moment and the moment's time change, an inversion in time causes it to also invert, unlike the precessional torque $\mathbf{m} \times \mathbf{H}$. It is dissipative because it makes the motion irreversible.

Phenomenologically again, additional terms are added to the equation of motion of the moment. One example form is the Landau-Lifshitz equation, which includes this damping term in the form

$$\frac{d\mathbf{m}}{dt} = \gamma \mathbf{m} \times \mathbf{H} + \frac{\alpha\gamma}{|\mathbf{m}|}\mathbf{m} \times (\mathbf{m} \times \mathbf{H}). \tag{4.77}$$

The next order correction is the Landau-Lifshitz-Gilbert equation:

$$\begin{aligned}(1+\alpha^2)\frac{d\mathbf{m}}{dt} &= \gamma \mathbf{m} \times \mathbf{H} + \frac{\alpha\gamma}{|\mathbf{m}|}\mathbf{m} \times (\mathbf{m} \times \mathbf{H}), \\ \text{or equivalently,} \quad \frac{d\mathbf{m}}{dt} &= \gamma \mathbf{m} \times \mathbf{H} + \frac{\alpha}{|\mathbf{m}|}\mathbf{m} \times \frac{d\mathbf{m}}{dt},\end{aligned} \tag{4.78}$$

where α is a damping parameter. These equations, because of the damping, cause the magnetic dipole to spiral in and ultimately align with the field. For a flux of 1 T, as we discussed, the electron spin precession time is ≈ 35 ps per 2π of rotation. The presence of damping causes the precession to spiral into alignment with the field, as shown in Figure 4.64, and the above equations phenomenologically describe this process. These equations may get more complicated if any of the parameters $\mathbf{m}$, ζ or α is a function of position. This may arise from several different causes: domain boundaries, thermal differences, strain effects, and a plethora of others. Domain boundaries of magnetization define regions where magnetization changes. Different directions have different effective electron mass, and that too can have consequences for the behavior of an electron.

The damping torque may also be reversed in polarity, that is, $\alpha < 0$, in which case the magnetic dipole in the steady state aligns antiparallel to the field, as in Figure 4.65. This switching of magnetization becomes possible by energy that may be supplied, for example, through a spin polarized current that interacts with the magnetic dipole. The injected electrons, when sufficiently large in numbers, may then cause the magnetization of the material to flip. This is the spin torque effect of interest in spin torque memories, since the absence of current leaves the magnetization in a state that cannot be easily switched and therefore is a state of storage. A small current,

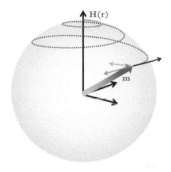

Figure 4.64: In the presence of a field and positive damping, a magnetic moment $\mathbf{m}$ precessing at the Larmor frequency eventually spirals into alignment with the field $\mathbf{H}$.

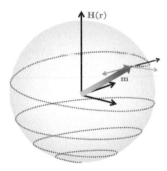

Figure 4.65: In the presence of a field and negative damping, a magnetic moment $\mathbf{m}$ precessing at the Larmor frequency eventually spirals to antiparallel alignment with the field $\mathbf{H}$.

A classical spinning top can flip from its pointed end to its broad top if it has a low center of gravity. The torque needed to cause this flip, in a direction perpendicular to the axis of the rotation, becomes feasible via an action/reaction due to a collision at its tip, which has the axial symmetry for flipping.

small enough not to cause switching, may then be used for reading if resistance is affected by the magnetic polarization.

We have now made the correspondence between the electron's magnetic moment, which arises due to its intrinsic angular momentum corresponding to the spin s and the secondary spin quantum number m_s, and the electron's orbital angular momentum arising from the orbital quantum number l and the azimuthal quantum number m_l.

This still doesn't yet explain phase transition to magnetism and the resulting characteristics, which we will exploit in devices. Our argument here will proceed as follows. Magnetism is primarily a result of the electron since the nuclear component is scaled by mass. The electron's energy has both Coulombic and magnetic components. As mentioned above in the resonance discussion, this magnetic component is significant and observable in presence of reasonably sized fields. The quantum nature of the electron as a fermion, even if only two electrons are close together, such as in a $1s^2$ configuration, let alone the $3d^{1-10}$ or $4f^{1-14}$ configuration of transition elements, will have energetic perturbations that will define the ordering of electron spin alignments because they are related to the energies. Which has the lower energy—the aligned spin or the non-aligned spin? This will determine both the short- and the long-range order in the ensemble and is related to exchange and superexchange, terms that reflect the approach by which one can visualize the creation of an asymmetric wavefunction that the electron must have as a fermion. In an ensemble, a starting approach may be to look at the lattice with the ordering of the resulting moments, as these will be related to the exchange energy. Even simple models provide good insight into this. And, from this ensemble, one should be able to see the lowest order behavior of order parameters in the magnetism. So, we will start first with a discussion of exchange interaction by looking at He—a $1s^2$ spectrum.

Helium's ground state has two electrons in the $1s^2$ configuration— a coupled state with the nucleus containing two protons. Excitation of electrons from the two electron configuration and to the two electron configuration obeys the Pauli exclusion principle for fermions, and the dipole rules of transitions arise in conservation of momentum and, of course, energy. The total angular momentum,

$$\mathbf{J} = \mathbf{L} + \mathbf{S}, \tag{4.79}$$

is reflected in the neutral helium binding energy states of two electrons with different binding energies: a $1s1s$ in $1\ ^1S_0(\sim -24.59\ eV)$, a $1s2s$ in $2\ ^1S_0\ (\sim -3.98\ eV)$, and a $1s2p$ in $2\ ^1P_1\ (\sim -3.37\ eV)$, which are the single states where the spins of the electrons are opposite; and a $1s2s$ in $2\ ^3S_1$ ($\sim -4.77\ eV$), and a $1s2p$ in $2\ ^3P_{0,1,2}$ ($\sim -3.62\ eV$), where the spins are aligned. Here, we are considering only some of the lowest energy states involving the first and the second principal quantum number. The states involving the former are "singlet" states, and states involving the latter are "triplet" states, because of their degeneracy, which we will look at in a moment. Emission versus absorption must reflect the constraints of the transition rules dictated by the conservation of energy and momentum.

Emission from the lowest energy state—the ground state $^1S_0$—can occur in any of the higher energy states, since spin exchange is possible between the nuclear and the electron components during the absorption of energy from an external source. However, emission from the ionized state to the ground state, with the requirement of conservation of angular momentum, is subject to the condition that the lowest energy state $^1S_0$ occurs with $\Delta S = 0, \Delta L = \pm 1$. Transitions with change of spin are not allowed. This means that the absorption and emission processes are very different, and they show that multiplets exist in accord with the dictates of the Pauli exclusion principle.

It is the preferred and lower energy of the triplet state, with that state's preferred spin alignment, that is essential for the magnetic moment at the atomic scale, and through short- and long-range order in the solid state. First, let us consider this atomic scale discussion of the alignment in our helium example—the simplest form with 2 electrons. The Hamiltonian part due to kinetic and Coulomb potential energy is

$$\mathscr{H}(\mathbf{r}_1, \mathbf{r}_2) = \frac{\hat{\mathbf{p}}_1^2}{2m_0} + \frac{\hat{\mathbf{p}}_2^2}{2m_0} - \frac{1}{4\pi\epsilon_0}\frac{2ee}{|\mathbf{r}_1|} - \frac{1}{4\pi\epsilon_0}\frac{2ee}{|\mathbf{r}_2|}$$
$$+ \frac{1}{4\pi\epsilon_0}\frac{ee}{|\mathbf{r}_2 - \mathbf{r}_1|}. \tag{4.80}$$

With the nucleus at the origin, the first four terms are the kinetic and Coulomb potential energies written with the nucleus at the origin, a term that we denote by $\mathscr{H}^0(\mathbf{r}_1, \mathbf{r}_2)$. The last term is the electron-electron interaction $\mathscr{H}_{ee}(\mathbf{r}_1, \mathbf{r}_2)$. The electron-electron interaction has an order of magnitude that is similar to that of the first four terms combined—the sum of the kinetic and potential energies of the two electrons treated as being independent in the central field of the nucleus. Using perturbation approach here is quite deficient. The wavefunction of the two electron system $\psi(\rho_1, \rho_2)$, where $\rho = (\mathbf{r}, m_s)$ is a generalized coordinate of position and secondary spin quantum number, with the two electrons identified by subscripting of the coordinate ρ, must be

$$\psi(\rho_1, \rho_2) = \psi_1(\rho_1)\psi_2(\rho_2) = \psi_2(\rho_1)\psi_1(\rho_2), \tag{4.81}$$

See S. Tiwari, "Quantum, statistical and information mechanics: A unified introduction," Electroscience 1, Oxford University Press, ISBN 978-0-19-875985-0 (forthcoming). for the meaning of this nomenclature. It reflects $^J L_{m_l}$, where the superscript is the maximum in units of $\hbar$ of total angular momentum, whose magnitude is $\sqrt{j(j+1)}\hbar$. The capital letter is the orbital angular momentum state, that is, s, p, d, et cetera, in capitals, and the subscript is its $m_s + m_l$.

Both the allowed and the disallowed transitions, by suppressing some of the emission processes, make atomic lasers possible. Stimulated emission occurs more efficiently through select transitions while suppressing other transitions whose lifetimes become much larger.

Heisenberg, who first attacked this problem in another classic paper, employed $\mathscr{H}^0(\mathbf{r}_1, \mathbf{r}_2) + e^2/4\pi\epsilon_0|\mathbf{r}_2|$ for the zeroth order, and $\mathscr{H}_{ee}(\mathbf{r}_1, \mathbf{r}_2) - e^2/4\pi\epsilon_0|\mathbf{r}_2|$ for perturbation. This is certainly more desirable.

the product of two wavefunctions that are functions of the generalized coordinate that must also satisfy the condition that, under exchange, the $|\psi|^2$—probability amplitude—of the composite system is invariant. This means that the wavefunction is a superposition of the form

$$\varphi(\rho_1, \rho_2) = \frac{1}{\sqrt{2}} [\psi_1(\rho_1)\psi_2(\rho_2) \pm \psi_2(\rho_1)\psi_1(\rho_2)], \qquad (4.82)$$

which is symmetric for the positive sign, that is, $\varphi_s(\rho_1, \rho_2) = \varphi_s(\rho_2, \rho_1)$, and asymmetric for the negative sign, that is, $\varphi_a(\rho_1, \rho_2) = -\varphi_a(\rho_2, \rho_1)$ under electron exchange. This system of two protons, two neutrons and two electrons is asymmetric under exchange of any particle pair, and that is the only permissible solution. The asymmetric solution can now be split between positional and spin coordinates. If position is symmetric, then spin is asymmetric, and vice versa, that is,

$$\phi_a(\rho_1, \rho_2) = \varphi_s(\mathbf{r}_1, \mathbf{r}_2)\zeta_a(s_1, s_2) \text{ or}$$
$$\varphi_a(\mathbf{r}_1, \mathbf{r}_2)\zeta_s(s_1, s_2), \text{ where} \qquad (4.83)$$

$$\varphi_s(\mathbf{r}_1, \mathbf{r}_2) = \frac{1}{\sqrt{2}} [\varphi(\mathbf{r}_1)\varphi(\mathbf{r}_2) + \varphi(\mathbf{r}_2)\varphi(\mathbf{r}_1)], \text{ with}$$

$$\zeta_a(s_1, s_2) = \frac{1}{\sqrt{2}} (|\uparrow\rangle_1|\downarrow\rangle_2 - |\downarrow\rangle_1|\uparrow\rangle_2), \text{ and}$$

$$\varphi_a(\mathbf{r}_1, \mathbf{r}_2) = \frac{1}{\sqrt{2}} [\varphi(\mathbf{r}_1)\varphi(\mathbf{r}_2) - \varphi(\mathbf{r}_2)\varphi(\mathbf{r}_1)], \text{ with}$$

$$\zeta_s(s_1, s_2) = \begin{cases} |\uparrow\rangle_1|\uparrow\rangle_2 \\ \frac{1}{\sqrt{2}} (|\uparrow\rangle_1|\downarrow\rangle_2 + |\downarrow\rangle_1|\uparrow\rangle_2) \\ |\downarrow\rangle_1|\downarrow\rangle_2 \end{cases}. \qquad (4.84)$$

The solution that is symmetric spatially and asymmetric in spin has only one unique form: it is a "singlet" state. The total spin quantum number for this state is $S = 0$, which only allows $M_s = 0$. The solution that is asymmetric spatially and symmetric in spin has three possibilities: it is a "triplet" state. The total spin quantum number for this state is $S = 1$, which allows $M_s = -1, 0, 1$.

The ground state $1s^2$ is symmetric spatially, with $|nlm\rangle = |100\rangle$. The spin part, therefore, is asymmetric under exchange, as written in Equation 4.84. This ground state's energy can be calculated under the perturbation of $\mathscr{H}_{ee}(\mathbf{r}_1, \mathbf{r}_2)$—the electron-electron Coulomb potential—in Equation 4.80. The result is $5e^2/4\pi\epsilon_0 a_B$, with the first four terms—the kinetic and Coulomb potential energy of the two electrons in the central field—of Equation 4.80 resulting in $-4e^2/4\pi\epsilon_0 a_B$. As we had noted, the second term is significant. Spin's consequence is in the wavefunction's symmetry of the spatial component and in the asymmetry of the wavefunction's spin component. It does not affect the energy of the ground state.

[3] He is a fermion. ^4He is a boson. They behave very differently in condensed conditions, specifically because of the nature of change in interaction, as they behave as two very different quantum particles subject to very different quantum statistics.

H. A. Bethe and E. E. Salpeter's "Quantum mechanics of one and two electron systems," Springer-Verlag (1947), has the definitive discussion of this problem. This symmetry and asymmetry discussion appears again and again in the sciences. A reading of the Bethe-Salpeter discussion is highly recommended.

However, this asymmetry requirement has consequences for the excited states. Since the electrons are in different orbitals (for example, $1s2s$ or $1s2p$) there are energy changes, that is, energy splitting. The spatial components of the two electrons are no longer identical, and therefore under exchange, there are differences in energy. If the spatial part is symmetric, spin state is asymmetric, and a singlet state arises whose wavefunction is the product of symmetric spatial components and asymmetric spin components. The first expression of Equation 4.84 is now restated for the excited singlet as

$$\varphi_{xs}^{S}(\mathbf{r}_1, \mathbf{r}_2) = \frac{1}{\sqrt{2}} \left[\varphi_{100}(\mathbf{r}_1)\varphi_{nlm}(\mathbf{r}_2) + \varphi_{100}(\mathbf{r}_2)\varphi_{nlm}(\mathbf{r}_1) \right]. \qquad (4.85)$$

The triplet state arises from the three different spin wavefunctions that go with the asymmetric spatial wavefunction. This triplet state's spatial function, a modification of the third expression of Equation 4.84, is

$$\varphi_{xs}^{T}(\mathbf{r}_1, \mathbf{r}_2) = \frac{1}{\sqrt{2}} \left[\varphi_{100}(\mathbf{r}_1)\varphi_{nlm}(\mathbf{r}_2) - \varphi_{100}(\mathbf{r}_2)\varphi_{nlm}(\mathbf{r}_1) \right]. \qquad (4.86)$$

For both the excited state's singlet state (Equation 4.85) and the excited state's triplet state (Equation 4.86), the central field energy, that is, that arising from the first four terms and excluding the electron-electron interaction term of the Hamiltonian, is the same; therefore,

$$
\begin{aligned}
\langle \varphi_{xs}^{S}(\mathbf{r}_1, \mathbf{r}_2) | \mathscr{H}^0 | \varphi_{xs}^{S}(\mathbf{r}_1, \mathbf{r}_2) \rangle &= \langle \varphi_{xs}^{T}(\mathbf{r}_1, \mathbf{r}_2) | \mathscr{H}^0 | \varphi_{xs}^{T}(\mathbf{r}_1, \mathbf{r}_2) \rangle \\
&= E_{100} + E_{nlm}. \qquad (4.87)
\end{aligned}
$$

But, the electron-electron interaction will be different, since the singlet's and triplet's spatial wavefunctions are very different under the $|\mathbf{r}_1 - \mathbf{r}_2|$ Coulomb potential dependence of the denominator. We have for the electron-electron correction,

$$
\begin{aligned}
E_{ee}^{S} &= \langle \varphi_{xs}^{S}(\mathbf{r}_1, \mathbf{r}_2) | \mathscr{H}_{ee} | \varphi_{xs}^{S}(\mathbf{r}_1, \mathbf{r}_2) \rangle = I + J, \\
E_{ee}^{T} &= \langle \varphi_{xs}^{T}(\mathbf{r}_1, \mathbf{r}_2) | \mathscr{H}_{ee} | \varphi_{xs}^{T}(\mathbf{r}_1, \mathbf{r}_2) \rangle = I - J, \text{ and} \\
E_{ee}^{S} - E_{ee}^{T} &= 2J. \qquad (4.88)
\end{aligned}
$$

In these expressions, I is the Coulomb integral of electrostatic repulsion energy between the spatial $|100\rangle$ and $|nlm\rangle$. J is the exchange integral quantifying the energy with exchange of the quantum states between the two electrons. The first energy relationship, for the singlet, is with antiparallel spin alignment and symmetric spatial alignment. The electrons are close to each other. The second, for the triplet, has parallel spin alignment. The electrons are in asymmetric spatial alignment and are farther from each other. Energy of the singlet state is higher because of the higher electrostatic repulsion of the electrons close to each other. The difference in energy between the two, $\Delta E = E_{ee}^S - E_{ee}^T = 2J$, is the splitting energy. The triplet state is lower in energy. This "singlet"–"triplet" splitting arose from the electron exchange, or simply "exchange interaction." J can be positive or negative, depending on the system. In this example of He, it is positive.

This central field problem shows that spin places conditions that result in a change in the wavefunctions so that the electron-electron interactions show up as an energy splitting. Two electrons, with the two spin possibilities for each electron, so four different possibilities, instead of being all degenerate in energy, are now in a singlet state, higher in energy, with a symmetric spatial wavefunction with an asymmetric spin component of wavefunction, and an asymmetric spatial wavefunction with three different symmetric spin components. When energy transitions to the ground state of ^4He from these excited states are observed, these energy differences are observable. Here, the triplet is the lower energy state, but it doesn't have to be. The sign will depend on the exchange integral.

Similar observations follow from a discussion of the hydrogen molecule. The hydrogen molecule is the simplest example of two nearest neighbors. The Hamiltonian could be considered to be composed of (a) the kinetic energy of electron 1, together with the electron's Coulomb potential energy from its interactions with the core of atom A as well as a similar term for electron 2 with the core of atom B—this is the Hamiltonian of each hydrogen atom—and (b) the Coulomb energy arising from electron-ion and electron-electron interactions not included in (a)—electron 1 with core B, electron 2 with core A, core A with core B, and electron 1 with electron 2. One could then split this Hamiltonian in a variety of ways to solve it. The bonding that follows in H_2, analyzed in the Heitler-London calculation for this two electron problem, strongly emphasizes the point of strong correlations and localization of electrons, unlike the largely free independent electron picture that sufficed for semiconductors. The covalent bond ground state is the correlated electron pair—electron 1 with electron 2—showing that the orbitals overlapping between the atoms have opposite spins. The first excited state is a triplet state, where the electrons are now in the "antibonding" two electron state. Our interest is not in this chemistry, but the spin alignment, magnetic moment and phase transition properties that ensue in larger ensembles whose simplest example for understanding is this molecule. Bonding—of molecules, and of solids, as an extension—is an example of correlation. Most molecular bonding is relatively weak in correlation. So, a mean field method, or a generally self-consistent field method, where the electron experiences a spatial average over the positions of other electrons, suffices. This is not a good approximation for magnetic systems, where this correlation is very pronounced and central.

Our interest is not really in this central field problem, but in what happens when different atoms with these extra electrons are brought together. This past discussion gives us insight and an approach to explore this non-central field problem. We can employ electron exchange over neighbors, the simplest example of which is two nearest neighbors with an electron each, assuming that the perturbation effect of the rest of the assembly is secondary. So, we employ two nearest neighbor sites.

Consider a simplified example of two atoms A and B, each with one electron, with the electrons labeled 1 and 2, respectively. The spatial wavefunctions are φ_A and φ_B, respectively. As in the example of He, the asymmetric total wavefunction φ should be composed of either a symmetric space and antisymmetric spin function (the "singlet" state) or an antisymmetric space and symmetric spin function (the "triplet" state), that is,

$$
\begin{aligned}
\varphi_s &= [\varphi_A(1)\varphi_B(2) + \varphi_A(2)\varphi_B(1)] \\
&\quad \times (|\uparrow\rangle_1|\downarrow\rangle_2 - |\uparrow\rangle_2|\downarrow\rangle_1) \text{ for the singlet, and} \\
\varphi_a &= [\varphi_A(1)\varphi_B(2) - \varphi_A(2)\varphi_B(1)](|\uparrow\rangle_1|\uparrow\rangle_2), \\
&= [\varphi_A(1)\varphi_B(2) - \varphi_A(2)\varphi_B(1)](|\downarrow\rangle_1|\downarrow\rangle_2), \\
&= [\varphi_A(1)\varphi_B(2) - \varphi_A(2)\varphi_B(1)] \\
&\quad \times (|\uparrow\rangle_1|\downarrow\rangle_2 + |\uparrow\rangle_2|\downarrow\rangle_1) \text{ for the triplet.} \quad (4.89)
\end{aligned}
$$

The energy splitting arising from the exchange of the coordinates of electrons 1 and 2 is electrostatic under the constraints of the exclusion principle. This difference, following the same notation as for He, is

$$
\begin{aligned}
2J &\equiv E_{ee}^S - E_{ee}^T \\
&= 2 \int [\varphi_A^*(1)\varphi_B^*(2)] \\
&\quad \times \mathscr{H}_{ee}(\mathbf{r}_1, \mathbf{r}_2)[\varphi_A(1)\varphi_B(2)] \, d^3r_1 \, d^3r_2, \quad (4.90)
\end{aligned}
$$

where

$$
\mathscr{H}_{ee}(\mathbf{r}_1, \mathbf{r}_2) = \frac{1}{4\pi\epsilon_0} \left(\frac{e^2}{|\mathbf{r}_1 - \mathbf{r}_2|} + \frac{e^2}{|\mathbf{R}_1 - \mathbf{R}_2|} - \frac{e^2}{|\mathbf{r}_1 - \mathbf{R}_1|} - \frac{e^2}{|\mathbf{r}_2 - \mathbf{R}_2|} \right), \quad (4.91)
$$

assuming spatial separation between the positions of center cores ($\mathbf{R}_1$ and $\mathbf{R}_2$) and the charge clouds, and between the charge clouds themselves. The first term in this Hamiltonian is the electron-electron correlation term that the spin considerations of the wavefunction stress.

Renormalization, handling the infinities that arise in our simple forays, where these denominators vanish, is a major theme in quantum and statistical mechanics, raising its head in quantum electrodynamics, phase transitions and a plethora of other directions. We discussed some of this in our self-energy foray in Chapter 3.

For the two electrons, localized as 1 and 2, or more generally i and j, and their spins, the exchange interaction energy is

$$E_{ij} = -2J_{ij}(\mathbf{S}_i \cdot \mathbf{S}_j). \tag{4.92}$$

A positive J reduces energy and favors spin alignment. If spins are aligned, it favors ferromagnetism. In terms of the total angular momentum $\mathbf{J}$, the moment resulting from the alignment is

$$\mathbf{m} = -g\mu_B\mathbf{J}. \tag{4.93}$$

If the magnetic moment mostly originates in spin, as is the case of ferromagnetic atom assemblies, then $\mathbf{J} \approx \mathbf{S}$, and

$$\mathbf{m} \approx -g\mu_B\mathbf{S}. \tag{4.94}$$

Away from cores and between them, the exchange integral drops off more rapidly than the other Coulomb terms do. This short-range interaction is important for the magnetism of solids. What shall we do when dealing with an assembly of atoms? The Heisenberg Hamiltonian addresses the preferred parallel alignment in its exchange model:

$$\mathscr{H}_{eff} = 2\mathscr{H}_0 + \mathscr{H}_C + \mathscr{H}_{xc}, \tag{4.95}$$

where the first term is the central field Hamiltonian, the second is the Coulomb term, and the final is the exchange energy term. The exchange Hamiltonian, $\mathscr{H}_{xc} \propto \mathbf{S}_1 \cdot \mathbf{S}_2$, is obtained by extension from the two spin example just discussed. For the ensemble, one may include all the exchange interactions by taking pairs. The effective Hamiltonian, the Heisenberg Hamiltonian, is therefore

$$\mathscr{H}_{eff} = -2\sum_{i<j}^{N} J_{ij}\mathbf{S}_i \cdot \mathbf{S}_j, \tag{4.96}$$

where we have used $J_{ij} = J_{ji}$, we have eliminated double counting, and J_{ij} is the exchange integral

$$J_{ij} = \int \varphi_i^*(\mathbf{r}_1)\varphi_j^*(\mathbf{r}_2)\frac{e^2}{4\pi\epsilon_0|\mathbf{r}_1 - \mathbf{r}_1|}\varphi_i(\mathbf{r}_1)\varphi_j(\mathbf{r}_2) \, d\mathbf{r}_1 \, d\mathbf{r}_2. \tag{4.97}$$

If spins are aligned, $\mathbf{S}_1 \cdot \mathbf{S}_2 = 1/4$ in the triplet, if the spins are antiparallel, $\mathbf{S}_1 \cdot \mathbf{S}_2 = -3/4$ in the singlet. If J_{ij} is positive, one obtains ferromagnetic coupling, with the triplet as the lower energy state. If J_{ij} is negative, one obtains antiferromagnetic coupling, with the singlet as the lower energy state. Note also that our $2J_{12}$ result for the two electron case follows from

$$\langle \mathscr{H}_{eff}\rangle^S - \langle \mathscr{H}_{eff}\rangle^T = -2J_{12}\left(-\frac{3}{4} - \frac{1}{4}\right) = 2J_{12}. \tag{4.98}$$

This $\mathbf{J} = \mathbf{L} + \mathbf{S}$, the total angular momentum, should not be confused with the symbol J of the exchange interaction energy magnitude. Magnetism needs this spin-orbit coupling even if the exchange interaction energy is much larger. This is because the orbital field, $\mathbf{H}_{orb} = (1/4\pi\mu_0 r^3)\mathbf{m}_l = -(e/4\pi m_0 r^3)\mathbf{L}$, can be large. Hence, we obtain the energy $\mathbf{m} \cdot \mathbf{H}_{orb}$, where $\mathbf{m}$ is strongly from spin. This energy is

$$E_{so} = -\mathbf{m}_s \cdot \mathbf{H}_{orb} = -\frac{e^2}{4\pi\epsilon_0 m_0^3 c^2 r^3}\mathbf{L} \cdot \mathbf{S}.$$

In 1928, a couple of years after the *He* study, which itself was a couple of years after the uncertainty principle, Heisenberg explored magnetism to show its quantum-mechanical origin.

This Heisenberg Hamiltonian approach took account of the spin and atomic moment coupling, since the atomic moment is included in the Hamiltonian. A general Heisenberg Hamiltonian will also include higher order terms of this perturbation expansion, as well as other fields, such as an external magnetic field. Many of the important characteristics of magnetism can be explored through simplified models where the spin contribution is included in a form that has principal components incorporated in a solvable way. Since the situation, even for such simple cases, has both short-range effects, which we emphasized here, and long-range effects, which are present in any phase transition, any approach rapidly becomes complicated. The Heisenberg model thus provides the broader phenomenological and comprehensive outline.

Its simplest reduction is the nearest neighbor Ising model

$$\mathscr{H}_{eff} = -2\sum_{i<j}^{N} J_{ij}S_i^z S_j^z = -2J\sum_{i<j}^{N} S_i S_j \tag{4.99}$$

over nearest neighbors in a two-dimensional lattice, where each atom at each site is a magnetic moment which points parallel or antiparallel to some preferred axis (z).

Figure 4.66 shows a two-dimensional square array example. Here, with $J = J_{ij}$ for the nearest neighbor interaction energy, in order to determine the thermodynamic parameters, one can write the partition function in the presence of a field H aligned with the preferred axis as

$$Z_N = \sum_S \exp\left(\frac{2J}{k_B T}\sum_{i<j}^{N} S_i S_j + \frac{H}{k_B T}\sum_i^{N} S_i\right). \tag{4.100}$$

This is a function of the coupling term J and the applied field $\mathbf{H}$.

The magnetization, on average, is

$$M = \frac{\partial}{\partial H/k_B T}\lim_{N\to\infty}\frac{1}{N}\ln Z_N\left(\frac{2J}{k_B T}, \frac{H}{k_B T}\right). \tag{4.101}$$

For a square lattice, we can expand the partition function

$$\begin{aligned}
Z_N &= \left(2\cosh\frac{H}{k_B T}\right)^N \left\{1 + 2N\frac{2J}{k_B T}\zeta^2 + N\left(\frac{2J}{k_B T}\right)^2\zeta^2\right.\\
&\quad + N\left(\frac{2J}{k_B T}\right)^2\left[(2N-7)\zeta^4 + 6\zeta^2 + 1\right]\\
&\quad \left.\mathscr{O}\left[\left(\frac{2J}{k_B T}\right)^3\right]\right\},
\end{aligned} \tag{4.102}$$

with $\zeta = \tanh H/k_B T$, and $\mathscr{O}()$ indicating the asymptotic order. This gives

$$M = \tanh\frac{H}{k_B T}\left(1 + 4\operatorname{sech}^2\frac{H}{k_B T}\left\{\frac{2J}{k_B T}\right.\right.$$

The next order term $\mathscr{H}_{eff} = -2\sum_{i<j}^{N} J_{ij}(\mathbf{S}_1\cdot\mathbf{S}_2)^2$, which is the coupling between neighbors, for example, has a $\cos^2$ dependence on the interaction angle of spin in the multi-dimensional assembly. This is observable in multilayer structures. An additional term is $-g\mu_B\mathbf{H}\cdot\sum_i\mathbf{S}_i$, to include the Zeeman term for changing energy in an external field.

The Hubbard model that we employed to understand hopping in metal–insulator transitions, and the Heisenberg approach, are good study examples for the energies of multiatom, multielectron systems. In the Hubbard model, the two competing energies are the Coulomb energy between electrons and the kinetic energy leading to hopping from one atom to another. Spin enters into this through the constraint of the exclusion principle. Hopping involves electrons of identical spin. It is delocalizing, that is, it leads to band behavior. The Coulomb energy is localizing. It keeps electrons apart, even if they are of opposite spin, by moving the electrons to separate atoms.

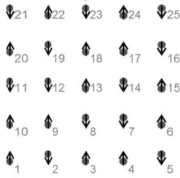

Figure 4.66: Magnetic moments in a 5×5 two-dimensional array of lattice sites for exploration of a 2D Ising model. The sequence of moments is $\{\uparrow, \downarrow, \uparrow, \uparrow, \downarrow, \ldots, \downarrow, \uparrow, \downarrow, \uparrow, \downarrow\}$. See S. Tiwari, "Quantum, statistical and information mechanics: A unified introduction," Electroscience 1, Oxford University Press, ISBN 978-0-19-875985-0 (forthcoming) for a discussion of partition functions and their use in determining probabilities for macroscopic variables of an ensemble through energy minimization. We also employed partition functions in Chapter 1, in our analysis of the entropy of information via probability.

$$+ (3 - 7\zeta^2)\left(\frac{2J}{k_B T}\right)^2 + \mathscr{O}\left[\left(\frac{2J}{k_B T}\right)^3\right]\Bigg\}\Bigg). \quad (4.103)$$

This result is more general, but we use the square lattice since it is easy to handle mathematically.

This equation is constructive. The expansion is an odd function of $H/k_B T$. If $2J/k_B T$ is small, that is, the temperature is high or the exchange integral is small, then the magnetization is of the same polarity as the field. This is marked as the high T behavior in Figure 4.67. At a field of zero, the magnetization disappears if T is sufficiently large.

If the temperature T is low or the exchange integral is large, so that $2J/k_B T$ is a dominant term in Equation 4.103, one can expand the relationship in terms of

$$\begin{aligned}\varsigma &= \lim_{T \to 0} \tanh \frac{H}{k_B T} 4 \operatorname{sech}^2\left(\frac{H}{k_B T}\frac{2J}{k_B T}\right)\\ &= \exp\left(-\frac{8J}{k_B T}\right).\end{aligned} \quad (4.104)$$

We will write the first few terms of this expansion in the low T or high J limit:

$$\begin{aligned}Z_N &= \exp\left(2N\frac{2J}{k_B T} + N\frac{H}{k_B T}\right)\left[1 + N\varsigma^2 \exp\left(-2\frac{H}{k_B T}\right)\right.\\ &\quad \left. + 2N\varsigma^3 \exp\left(-4\frac{H}{k_B T}\right) + \cdots\right]\\ &\quad + \exp\left(2N\frac{2J}{k_B T} - N\frac{H}{k_B T}\right)\left[1 + N\varsigma^2 \exp\left(2\frac{H}{k_B T}\right)\right.\\ &\quad \left. + 2N\varsigma^3 \exp\left(4\frac{H}{k_B T}\right) + \cdots\right].\end{aligned} \quad (4.105)$$

There are two sets of terms—the first due to parallel spin, and the second due to antiparallel spin. For large N, these terms in any power of ς become independent of N. If the field is positive, the first set dominates. The magnetization, using the same approach of differentiation, follows as

$$M = 1 - 2\varsigma^2 \exp\left(-2\frac{H}{k_B T}\right) - 8\varsigma^3 \exp\left(-4\frac{H}{k_B T}\right) - \cdots. \quad (4.106)$$

For small ς, that is, small positive fields, it approaches a finite nonzero value. By symmetry, for negative fields at small fields, it approaches an identical negative non-zero value. This behavior, shown as the low T curve in Figure 4.67, is that of rapid change near zero field.

This two-dimensional example, and others like it, such as on a Bethe lattice, show the general behavior pattern that these interactions, nearest neighbor and more than nearest neighbor, take. In the Ising spin array, and in the others, under the rules that we defined

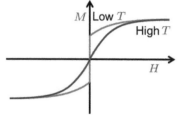

Figure 4.67: Magnetization as a function of magnetic limit in the limits of high temperature and low temperature, as described by the square lattice Ising model of Equation 4.103. At high temperatures, magnetization passes through the origin. At low temperatures or with a high exchange integral, it rapidly changes.

A Bethe lattice is an irregular array where q points connect to a core to form a lowest order shell, and further shells are built by connecting another $q - 1$ points.

for the interaction through the exchange integral, one sees a continuous and slowly varying behavior of magnetization change at high temperature, with no magnetization at zero field, and a rapid change around zero field at low temperatures. The system undergoes a very pronounced change in its behavior as it varies over a temperature range that is related to the exchange coupling term. A large positive exchange integral leads to pronounced magnetization.

We have now seen mathematically the first indication of phase transition here.

Anisotropy, hard and easy magnetization directions, arises due to these strengths and angles between interacting parameters. A general Heisenberg exchange interaction term is $J_\parallel S^z S^z + J_\perp (S^x S^x + S^y S^y)$ in both the short and the long range. If $S > 1/2$, then terms of higher order in S will also arise.

This model leads to a number of features of phase transitions that we will not dwell on in detail but which are important since they elucidate the complexity and the range of interactions that occur at phase transitions. Lars Onsager for example, showed universality through a solution of the square lattice where the exchange interaction is different in different directions. The Mermin-Wagner theorem states that, in one and two dimensions, continuous symmetries cannot be spontaneously broken at finite temperature in systems with sufficiently short-range interactions. This too has consequences for phase transitions in general, and magnetization in lower dimension systems in particular.

For us, this complexity has implications through the correlations that capture the effect of similarity—similarity of spin alignment leads to ferromagnetism, for example, the correlation between spins i and j can be defined as

$$c_{ij} = \langle S_i S_j \rangle - \langle S_i \rangle \langle S_j \rangle, \tag{4.107}$$

for translationally invariant conditions. One set is either equal to or exceeds the other. Correlation will depend on the separation between the i and the j sites. When the temperature is quite different from the critical temperature T_c, as the distance becomes large, the correlation decays to zero. So, $c(x) \propto x^{-\tau} \exp(-x/\xi)$ for $x \to \infty$, with τ as the exponent of spatial dependence, and ξ as the correlation length in the direction of interest for which this correlation is written. At the critical point, $\xi \to \infty$.

It is instructive to see properties, such as susceptibility, that are of interest, look at what happens with and without phase transition, and see how this all relates to mean field theory, from where we started. For magnetism, it is the exchange interaction that is of relevance in this discussion.

When spins are preferentially aligned with exchange interaction positive, for real three-dimensional materials, one obtains ferromagnetism. Here, one may have both hard and easy axes of alignment, that is, anisotropy, as a consequence of anisotropic exchange interaction. When the spins are anti-aligned, that is, have alternating spin polarities, we obtain antiferromagnetism, for equal up and down spin, or antiferrimagnetism, for alternating high and low spin and other cases where there is chiral dependence with gradual phase shifting.

If exchange interaction is absent, that is, atoms are distributed without any energy consequences from the exclusion principle and Coulomb energy coupling, we can look at the distribution in the presence of the field through a distribution function.

If we consider this classically, the probability of moment $\mathbf{m}$ in a collection with magnetic moment $\mathbf{M}$ is

$$\mathfrak{p}(\mathbf{M}) \propto \exp\left(-\frac{\mathbf{M} \cdot \mathbf{H}}{k_B T}\right), \tag{4.108}$$

the Boltzmann relationship, and expectations can be found through the probabilities. So

$$\langle \mathbf{M} \rangle = \frac{\int \mathbf{m} \exp(-\mathbf{M} \cdot \mathbf{H}/k_B T) \; d\Omega}{\int \exp(-\mathbf{M} \cdot \mathbf{H}/k_B T) \; d\Omega}, \tag{4.109}$$

with Ω as the elemental solid angle to cover the volume. The magnetic susceptibility for non-interacting spins, by our definition of susceptibility, is

$$\chi_0 = N\left\langle \frac{\partial \mathbf{M}}{\partial \mathbf{H}} \right\rangle = \frac{N}{T}\langle \mathbf{M}^2 \rangle = \frac{1}{3}\frac{Ng^2\mu_B^2 S(S+1)}{T} \propto \frac{1}{T}. \tag{4.110}$$

The susceptibility in the absence of exchange interactions varies as the inverse of temperature. We saw this in the behavior above the critical temperature. This is the Curie law.

The quantum case requires consideration of the spin states—that is, S^z with $M^z = g\mu_B S^z$, and $-S \le S^z \le S$. For the simplest one electron state case, that is, $S = 1/2$, the simple relationship that this provides is $M = g\mu_B \langle S \rangle$, where

$$\begin{aligned}\langle S \rangle &= \frac{1}{2}\frac{\exp(g\mu_B H/2k_B T) - \exp(-g\mu_B H/2k_B T)}{\exp(g\mu_B H/2k_B T) + \exp(-g\mu_B H/2k_B T)} \\ &= \frac{1}{2}\tanh\left(\frac{g\mu_B H}{k_B T}\right).\end{aligned} \tag{4.111}$$

If the field is small, the susceptibility is

$$\chi_0 = N\left\langle \frac{\partial \mathbf{M}}{\partial \mathbf{H}} \right\rangle = N\frac{g^2\mu_B^2}{4T}. \tag{4.112}$$

In the more general case for broader choices of S, L and the total angular momentum J, the magnetization curve is

$$M(H) = g\mu_B J \left\{ \coth\left[\left(1 + \frac{1}{2J}\right)\frac{g\mu_B J H}{k_B T}\right] - \frac{1}{2J}\coth\left(\frac{1}{2J}\frac{g\mu_B J H}{k_B T}\right)\right\}. \quad (4.113)$$

What this equation illustrates is the effect of Zeeman splitting arising from the spin and magnetic field interaction, as illustrated in Figure 4.68. Figure 4.68(a) shows that this doesn't shift this signal. At high fields, or lower temperature, the lowest energy states are preferentially filled, and the lower energy in field-magnetic coupling occurs with spin in the direction of the field. This normal Zeeman effect was first realized in the energy spectra of atoms such as hydrogen, where the total spin angular momentum is zero. Much more common, and an extension of this argument, is that, when both spin and orbital contributions are present in an atom or an assembly of atoms, the magnetic fields from both internal and external sources cause more complex energy splitting and spread, as shown in Figure 4.68(b). This latter is the anomalous Zeeman effect.

Magnetization saturates through spin alignment under quantum-mechanical constraints as it did through magnetic moment alignment under classical constraints. Equation 4.113 describes this for the saturation for the general case for a general S not restricted to 1/2.

This brief diversion can now be integrated to see the meaning of mean field and its effectiveness. The mean field is a self-consistent field that incorporates the average of the effect of the other elements of the system on each element in an ensemble—so, all other spins on the spin of interest. The self-consistency is arrived at by decoupling. So, the exchange interaction

$$\begin{aligned} J_{ij}\mathbf{S}_i \cdot \mathbf{S}_j &\equiv J_{ij}(\mathbf{S}_i \cdot \langle\mathbf{S}_j\rangle + \langle\mathbf{S}_i\rangle \cdot \mathbf{S}_j - \langle\mathbf{S}_i\rangle \cdot \langle\mathbf{S}_j\rangle) \\ &= J_{ij}(\mathbf{S}_i \cdot \langle\mathbf{S}\rangle + \langle\mathbf{S}\rangle \cdot \mathbf{S}_j - \langle\mathbf{S}\rangle \cdot \langle\mathbf{S}\rangle). \quad (4.114) \end{aligned}$$

This mean field—an internal effective field—is the effect of the surrounding, that is,

$$H_{int} = -\frac{2}{g\mu_B}\sum J_{ij}\langle S\rangle \approx -\frac{2}{g\mu_B}J'\nu\langle S\rangle, \quad (4.115)$$

where the simplification utilizes the dominance of coupling from ν nearest neighbors, and J' is an average over neighbors. This leads to

$$\langle S\rangle = \frac{1}{2}\tanh\left(-\frac{J'\nu\langle S\rangle}{T}\right)$$

$$\therefore \quad M = \mu_B \tanh\left(-\frac{J'\nu M}{2\mu_B T}\right), \quad (4.116)$$

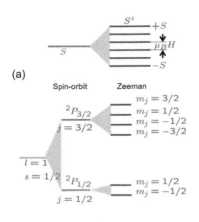

(a)

Spin-orbit   Zeeman

(b) $s = 1/2$

Figure 4.68: (a) shows the normal Zeeman splitting, which is the effect of spin-magnetic field interaction. (b) shows the field-induced consequences when both orbital and spin contributions are present. Note that this is for $l = 1$, that is, p orbitals. For d orbitals, this is more complex.

where magnetization and spin are related through the gyromagnetic ratio and the Bohr magneton. The magnetic field relationship here is in the absence of external magnetic field, but an external magnetic field can be included by an extension of this approach.

Equation 4.116, our mean field equation, is in a form where the explicit solutions cannot be written. However, Figure 4.69 shows a sketch, as for other explicit equations that we have encountered, such as for the scaling lengths of double-gate transistors (in Chapter 2), of the solutions. One solution, if the external field is zero, is at the origin, since the tanh function vanishes at the origin. But, there are also non-zero solutions for low temperatures if magnetization exists. We will define a temperature to be associated with this via

$$T_c = \frac{2}{3}\frac{J'\nu}{k_B}S(S+1),\tag{4.117}$$

for the general case. For the $S = 1/2$ case, the example shown in the figure, this reduces to $T_c = J'\nu/2k_B$. The critical temperature corresponds to the free energy picture discussed in Figure 4.7—a second order phase transition.

Dimensionality will also play a role here. Exchange interactions, both short range and long range, will have consequences based on dimensionality. For example, a one-dimensional Ising system has no phase transition—a consequence of the Mermin-Wagner theorem mentioned earlier. We show this through $S = \pm 1/2$. The partition function, for a one-dimensional N-long array, is

$$\begin{aligned} Z_N &= \sum_{S_i=\pm 1/2} \exp\left[-\frac{2J}{k_B T}(S_1 S_2 + \cdots + S_{i-1} S_i\right.\\ &\quad \left.+ S_i S_{i+1} + \cdots + S_{N-1} S_N)\right]. \end{aligned}\tag{4.118}$$

With one additional particle,

$$\begin{aligned} Z_{N+1} &= Z_N \sum_{S_N, S_{N+1}=\pm 1/2} \exp\left(-\frac{2J}{k_B T}S_N S_{N+1}\right)\\ &= 2Z_N \cosh\frac{2J}{k_B T}; \end{aligned}\tag{4.119}$$

$Z_1 = 2$, since either spin is allowed; therefore,

$$Z_N = 2^N \cosh^{N-1}\frac{2J}{k_B T}.\tag{4.120}$$

For large N, the Gibbs energy

$$\mathcal{G} = -k_B T \ln Z = -N k_B T \ln\left(2\cosh\frac{2J}{k_B T}\right),\tag{4.121}$$

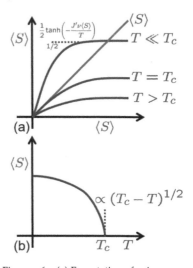

Figure 4.69: (a) Expectation of spin as a function of changing temperature for the $S = 1/2$ case, using mean field approximation to find the temperature-dependent solution of the explicit relationship. (b) shows the resulting temperature dependence of $\langle S \rangle$, that is, the proportionality to magnetization, and the existence of the critical temperature T_c where only the solution at the origin exists.

which is monotonic in temperature and vanishing at $T = 0$ K. Therefore, a one-dimensional Ising system has no phase transition at real temperatures.

If the one-dimensional Ising system has a correlation over a distance, for example, is proportional to $1/\xi \exp(-r/\xi)$, then the system does have a phase transition. The two-dimensional Ising system can be viewed in a similar way as the balancing between the lowering of energy from order and the increase of entropy from disorder. The former is favored at low temperatures, and the latter at high temperatures. The critical temperature, following the thermodynamic argument of Helmholtz energy (see Chapter 1), is $T_c \approx \Delta U / \Delta S$.

In Figure 4.70, if A is a fully ordered state in the Ising lattice, a spontaneous change state of flipped spins requires that the energy be lowered. An island of opposite spin comes about with a raising of energy by $2J$ for each coupling along the entire perimeter. So, $\Delta U = 2J\varsigma$, where ς is the perimeter length. The number of such possible configurations of size ς is 3^ς, defined by the three possibilities that each site can extend in. There are N possibilities of choosing the point, so the entropy is proportional to $\ln N 3^\varsigma$, and the free energy change is

$$\Delta U - T\Delta S = (2J - k_B T \ln 3)\varsigma - k_B T \ln N. \qquad (4.122)$$

For ς of order N, that is, when such a domain begins to appear, and for finite N, neglecting its logarithm compared to the other sizable term, a critical temperature $T_c \approx 2J/k_B \ln 3$ exists above which disorder prevails. Onsager's exact result is

$$T_c = \frac{2J}{k_B \ln(1 + \sqrt{2})}. \qquad (4.123)$$

A similar argument applies for the one-dimensional case, where the spin changes at one point, $\Delta U = 2J$, with N choices, so $\Delta U - T\Delta S = 2J - k_B T \ln N$. For $N \to \infty$, this says that $T_c \to 0$, as we had concluded.

These approximate calculations show that the exchange energy is close to $k_B T_c / 2$.

For the second order phase transition, the free energy diverges as the phase transition is approached, implying that the susceptibility will explode. According to the fluctuation-dissipation theorem, this susceptibility results from correlation of magnetization, that is,

$$\chi = \frac{\partial M}{\partial H} \propto \int \langle M(0)M(r) \rangle \, dr. \qquad (4.124)$$

In the ferromagnetic state, the magnetic moment has the inclination to align with adjacent sites, but this inclination is opposed by the inclination for an increase in entropy—a tendency to randomize spin.

The mathematical resolution of the 2D Ising problem is a tour de force that was among many contributions—the reciprocal relations of irreversible thermodynamic processes being the most recognized—from Onsager. We make only summary remarks. It is two-dimensional and three-dimensional systems that we concentrate on.

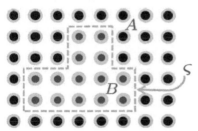

Figure 4.70: A two-dimensional Ising lattice with an enclosed island B of perimeter ς and of spin opposite to that of the surrounding region A.

Figure 4.70 reminds us of the game of Go—a territorial game played with white and black stones. These stones are metaphors for the two opposite spins attempting to capture the most territory, thus polarization. One needs to be a real expert in seeing the multitude of short- and long-range interactions. Much of the energy is expended on the perimeter. Just like in Go. The player of Go is an expert at recognizing patterns. When short- and long-range interactions exist, as in most worldly circumstances, the human being resorts to pattern recognition. This principle has been key to the success of modern artificial intelligence techniques. The Ising model that captures the essential features of this behavior is a contribution from 1920s by Lenz—a German physicist—and Ising—his student— who examined a one-dimensional line of spins—the negative case. The Ising model is really a model of models and is useful in many places—even in problems such as modeling crowded dance floors. Ising came to the United States and taught at Minot—now Minot State University—in North Dakota, and later on at Bradley University.

So, away from the critical point, there exists a correlation length ξ, where

$$\langle M(0)M(r)\rangle \quad \propto \quad \exp\left(-\frac{r}{\xi}\right) \quad \text{away from the critical point, and}$$

$$\propto \quad r^{-\gamma} \tag{4.125}$$

in the proximity of the critical point. This correlation length is a measure of the spatial extent of the alignment of spin. As a material approaches the transitioning of phase from a disordered state, the order increases, that is, ξ grows. The correlation does not change exponentially in the vicinity of the critical point–only as a power of the distance. These features show the complexity of order-disorder near the phase transition point.

We have now successfully connected broken symmetry, phase transition and mean field through a detailed and quantitative statistical-mechanical argument.

Figure 4.71: A spin wave—a magnon—arising from magnetic interactions that cause an evolution of the phase. This is a bosonic system, similar to phonons, due to exchange interactions. The precession, or equivalently the phase evolution, appears as the spin wave. Spin waves arise both in ferromagnetic and antiferromagnetic systems.

In the starting discussion of phase transitions, we had remarked on the launching of broadband long-range oscillations—a nonlinear effect that goes together with the presence of order parameter—as a consequence of the Goldstone theorem. This is the elementary excitation in such a system. In magnetism, this is manifested as spin density waves: spins, incrementally tilted, that is, phase shifted, from site to site, around the mean field, and corresponding to a boson particle wave. These are magnons, and a simple pictorial view of a magnon is shown in Figure 4.71. Magnons, that is, spin waves, exist in magnetic systems—ferromagnetic as well as antiferromagnetic systems. They are also indicators of the energy associated with magnetic interactions. One can view magnons as waves associated with the $\exp(-i\mathcal{H}'t/\hbar)$ time evolution for this energy. As for phonons, the energy due to magnetization arising from spin and orbital contributions is

Phonons—bosons—are the equivalent of spin waves for symmetry breaking of translation.

$$\langle E_{sl}\rangle = E_0 + 2\sum_q J(0)\omega_q\left(n_q + \frac{1}{2}\right), \tag{4.126}$$

and the real space dependence of the wave can be represented as

$$\langle S_n \rangle = \langle S \rangle \exp(i\mathbf{Q} \cdot \mathbf{R}_n). \qquad (4.127)$$

A traveling spin wave then is the phase-aligned spin as it moves through a material without the actual movement of the particle, just as in the case of phonons. Spin dynamics is similar to lattice dynamics—magnetic moments of spins are coupled with the exchange interaction, just as atomic masses are coupled via their springy bonds. The absolute value of magnetization changes in space, such as in anisotropic conditions.

We should mention here superexchange, a common cause of the antiferromagnetism that is prevalent among many magnetic oxides. The valence bonding in this case causes magnetic ordering to flip. Figure 4.72 shows the example of MnO, a largely ionic material with linear chain of Mn^{2+} and O^{2-} ions. The O^{2-} has a filled p orbital along the axis of bonds, with the Mn^{2+} $3d$ orbital containing five electrons. The covalent content, that is, sharing through the anti-aligned spin, means that this specific spin electron from oxygen can be shared with the d orbital of manganese. This in turn makes its opposite spin electron in the oxygen p orbital available to share with the next manganese atom in the chain. The spin has now been reversed—oxygen mediation through its valence bonding has caused antiferromagnetism through this superexchange.

The other significant way that spin enters in discussions is in insulation and conduction. We discussed the importance of this in the discussion of the Mott-Hubbard insulator state—the ground state arising from Coulomb interaction when hopping energy is higher than localization energy. But, localized electrons have the dual orientation of spin as a degree of freedom. In N sites, this is 2^N. An anti-spin alignment creates a non-zero average which lifts spin degeneracy and can create antiferromagnetic order. So, Mott-Heisenberg insulators where spin is anti-aligned also exist, as shown in Figure 4.73. Our V_2O_3 example discussed before showed such an antiferromagnetic behavior.

We have now seen a number of important effects arising from spin in an ensemble. We employed a single particle picture and the Pauli exclusion principle. The Zeeman effect, that is, the presence of a magnetic field, causes energy states to change depending on the electron's spin, as seen in Figure 4.68. Occupation of these states, under Fermi-Dirac constraints, causes a preponderance of one orientation of spin—the majority spin—and what we call Pauli spin paramagnetism. When there is no external field, exchange interaction causes a lowering of ground state energies, causing a preponderance of parallel spins.

$O^{2-} : 2p$

$Mn^{2+} : 3d$ $Mn^{2+} : 3d$

Figure 4.72: MnO where valence bonding causes magnetic ordering to flip because of shared anti-aligned spin.

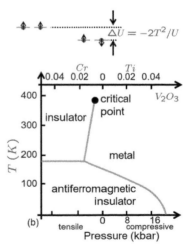

Figure 4.73: V_2O_3 as a Mott-Heisenberg insulator as a result of magnetic antiferromagnetic ordering. V_2O_3 is a classical metal–insulator system showing such a Mott-Heisenberg behavior.

This parallelization of spin becomes stronger as the electron density increases, since the magnetization $M = g\mu_B(N_\uparrow - N_\downarrow)$, where $N_{\uparrow,\downarrow}$ is the density of the alignment of spin. We used the Ising model to explore the consequences of the nature of interactions, short range and long range, in the system. But, we have not considered the periodicity of potentials, as we have in our other discussions—of semiconductors, of ferroelectrics and of metal–insulator transition materials.

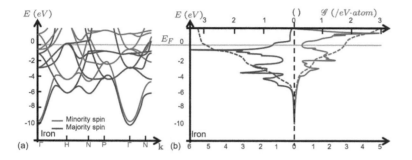

Figure 4.74: (a) shows the majority and minority spin E-$\mathbf{k}$ states for Fe (Fe^{26} : $[Ar]3d^6 4s^2$). (b) shows the majority and minority spin densities of states. normalizing them to per atom, so showing them in units of $eV^{-1} \cdot atom^{-1}$. The figure also shows an integrated density of states up to the energy. At the Fermi energy E_F, the accumulation of majority spin states exceeds its minority counterpart.

The band structure of a ferromagnetic material reflects the energy interaction consequences of the spin interactions that we just revisited. For the Bloch electron in the semiconductor, all the characteristics for spin up and spin down are identical, although through the Pauli exclusion principle, their existence must be accounted for in density of states and other places. But the density of each is the same in a volume. They respond to fields and diffuse the same way. The only place where there are effects traceable to spin's existence is when an external magnetic field is incorporated: spin-orbit coupling causes a specific band with states to appear, and the transitions can depend on the polarization. In a ferromagnetic material, this periodic Bloch electron picture will change: $E_{n,\mathbf{k}_\uparrow} \neq E_{n,\mathbf{k}_\downarrow}$. Figure 4.74(a) shows the band diagram for iron. The preferred alignment states lead to majority spins, and the higher energy, non-preferred alignment states lead to minority spin bands, even as the general natures of the momentum relationships in these bands remain rather similar. Figure 4.74(b) shows the corresponding density of states, normalizing them to per atom. Since Fe has a $3d^6$ shell, 6 is the largest number of states available per atom—this is nearly achieved for the majority band. The integrated density of states shows the cumulative effect of filled states. Near Fermi energy, one can see that the majority spin cumulative state is larger than its minority counterpart.

Materials with a large number of electrons, such as metals, because one is far up in energy in the bands, have rather complicated

and interesting-looking Fermi surfaces. The presence of majority
and minority spins, as in magnetic materials, accentuates this characteristic. Figure 4.75 shows two representative materials—copper, a
non-magnetic metal, and cobalt, a magnetic metal. Conducting electrons, that is, electrons on the Fermi surface, under a bias voltage,
move by scattering to the lower energy. In the case of copper, this is
relatively simple. The states available are along the near circular cut
of **k** states. Electrons move by scattering along these states. In cobalt,
these states are fewer for the majority spin electrons. Only some,
arising from s states, are available to move. The filled band, with its
higher spin density, leads to the ferromagnetism. The minority band
is much more complicated, with states along several of the states
available at the cut. Conduction will be dominated by these spin minority carriers that transport by scattering along these states, with the
spin preserved. But, as we will see spin flipping can also occur due to
interactions with d states and due to specific impurities that permit
the flipping. Conduction in Co is inherently different from that in Cu,
due to spin's effect on scattering.

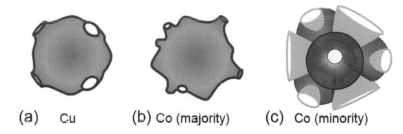

(a)   Cu   **(b)** Co (majority)   **(c)** Co (minority)

Figure 4.75: The Fermi surface of (a)
copper (Cu), a non-magnetic metal
with an equal count of the two spins,
and (b,c) cobalt (Co), a magnetic metal,
where majority (b) and minority (c)
spins have different occupancies, with
the majority band highly filled.

Now, if one looks at the density of states and the occupied states
in thermal equilibrium, one sees the situation shown in Figure 4.74(b).
The densities of states show the energy shift that was just remarked
on. We also see that there are three regions of high density of states.
These three regions are broadened regions of energy states from the
different d orbitals and their interactions. Different d orbitals in a
chain of atoms overlap differently at different wavevectors. Some
have very pronounced orientation dependence. With a Δ exchange
splitting related to spin interactions, the majority and minority spin
bands have a cumulative effect that is the separation of the top of the
majority band from the minority band. The shift here is the direct
consequence of spin polarization within the crystal. It exists without
an external field being present.

Transition metal atoms, with their successive filling of d electron
states, show s-d hybridization, since $4s$, with its higher principal

This shift is most easily seen at the
points of van Hove singularities—which
occur in the Brillouin zone at critical
points where the gradient of density
of states is not continuous, and one
observes kinks in the distribution.

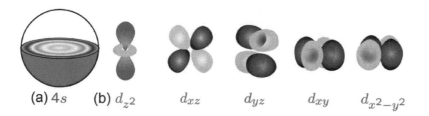

(a) $4s$ **(b)** d_{z^2} d_{xz} d_{yz} d_{xy} $d_{x^2-y^2}$

Figure 4.76: The orbital shapes of $4s$ and $3d$ orbitals whose interactions define the density of states in transition metal elements. (a) shows the probability amplitude ($|\psi|^2$) for $4s(l = 0)$, and (b) shows it for the $3d(l = 2)$ orbitals of constant probability. d_{z^2} corresponds to $m_l = 0$, d_{xz} and d_{yz} correspond to $m_l = \pm 1$, and d_{xy} and $d_{x^2-y^2}$ correspond to $m_l = \pm 2$.

quantum number, and $3d$ are energetically close, as well as have significant overlap in the valence shell. In the five d orbitals (Figure 4.76), the wavefunction disappears at the origin. Four of them form four lobes in four planes, and one forms a double-lobe with a ring. The s orbital for the $n = 4$ principle quantum number hybridizes and forms hybridized structures similar to sp^3 and sp^2, which we encountered earlier with semiconductors.

Figure 4.77 shows the relative radial probability distribution for $3d$, $4p$ and $4s$ orbitals of a $3d$ transition metal with a Z^*—an effective nuclear charge sans screening—of 10. The overlap is significant, and all this contributes to the effective Hamiltonian interaction, which in turn leads to the formation of different band states with significant spreads. Depending on the interaction within and the atom-to-atom separation, bands and bandgaps will form. And these will be different with different transition metals, where different states contribute based on the nature of orbital filling within the bonding with its atomic separation and arrangement in the solid state. So, one would expect variety in the band picture; at the very least, one will have s bands, d bands and p bands, and if f states are important, such as in the lanthanides or actinides, they too will introduce a twist in the complexity.

When bands are formed through the overlap of wavefunctions, these bands form s-d states, and the breaking of degeneracy through interaction leads to them intersecting each other in energy. An example of the resulting band picture is Figure 4.74(a) for iron, with the spin dependence of different branches of allowed states. So, the periodicity of atoms, the bonding interactions between the $3d$ electrons and the $4s$ electrons with the $2p$ electrons, leads to a band picture that can be quite complex in detail but that can also be simply looked at for instructional and learning purposes, as in Figure 4.78. The energy smearing of the levels that result from this includes magnetic interactions causing the band states to change and if magnetization exists or an external field is applied, it causes additional changes in the bands of anti-aligned and aligned spins. In ferromagnetic materials, it exists without the presence of an external field.

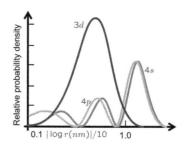

Figure 4.77: The relative radial probability densities, proportional to charge density, of $3d$, $4s$ and $4p$ orbitals, as expected from a transition metal element where $Z^* = 10$, using the Coulomb interaction.

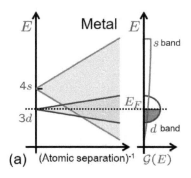 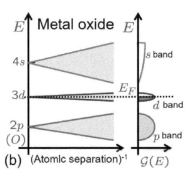

Figure 4.78: The forming of bands in an ensemble as the atoms are brought together and the $4s$ and $3d$ states interact. (a) shows the schematic behavior in a transition metal assembly, and (b) shows it for an oxide of the transition metal element, so for interactions that also involve oxygen's $2p$ states.

One can visualize this as a result of the accumulation of the perturbations and speculate what the consequences will be. Figure 4.78 shows what is likely to happen when transition metal elements are brought together, and it compares it to the situation where transition elements are brought together with oxygen. We have encountered oxygen in reference to ferroelectrics before, in perovskites. But, oxygen compounds with transitions metals are ubiquitous—the strongest ferromagnets arise in such system. In Figure 4.78(a), the case of the transition element assembly, the interactions of the $4s$ and $3d$ states, as the interatomic separation is reduced, lead to larger bands. s orbitals are radially symmetric, so the interaction is large. The $3d$ states interacting with other $3d$ states and with the $4s$ states also form a band that we identify here as the d band. The Fermi level lies within this d band. The spreading of this energy level through the interactions with oxygen is shown in Figure 4.78(b). There is again a partially filled d band, as well as the p band that arises from oxygen. The $4s$ band, higher up in energy, is now narrower in width and largely unoccupied. The d electrons can have strong magnetic contribution and this shows up both in the strong ceramic magnets as well as in the ferromagnetic interactions that lead to insulators of the Mott-Heisenberg nature.

This interpretation of the behavior of the collection of interactions is the Stoner-Wohlfarth-Slater model, or Stoner model for short. Lattice periodicity is reflected in the energy state periodicity and the width of the bands, and shifting of bands depends on whether magnetic interactions lead to lowering of energy—an anti-aligned state—or raising of energy for the aligned spin state. Recall that the magnetic interaction energy is proportional to $-\mathbf{m} \cdot \mathbf{H}$, where $\mathbf{m} = -2\mu_B\mathbf{S}$. The width of the band is inversely proportional to the lattice constant.

For the purpose of describing behavior in devices, as with the description of the parabolic picture with semiconductors, in magnetism we will employ majority and minority spins, or the spin orientations of the electrons. This only has meaning together with quantization direction—either arising from the magnetization direction from the internal interactions, or due to an external field.

So, in Figure 4.79, we show the schematic bands with energy for two example cases in $3d$-occupied non-magnetic and magnetic materials. The magnetization arising from the magnetic moment **m**, proportional to μ_B, defines the spin direction that has lower energy for a state—it is the anti-aligned direction of spin. The $3d$-dominated band has a finite width, and here is simplistically drawn as a semicircle. The $4s$-dominated states, with $n = 4$, have four anti-nodes radially with high probabilities closer to the nucleus and so are deeper in energy. These are metals, so the valence band(s) are only partly occupied. Because of magnetization and exchange interactions, one of the $3d$ bands is higher than the other for the magnetic case Figure 4.79(b). It has fewer electrons, is spin aligned and is the minority band. In the case of the non-magnetic material, the states for aligned and anti-aligned spin, as well as their occupation, are entirely symmetric, just as in the case of the conventional semiconductors. The collective shift of the two bands for the magnetic case (Figure 4.79(b)) is identified by the exchange energy Δ, and the gap between the top of the filled majority band and the top of the filled minority band is identified through the Stoner gap Δ_s.

This characterization is an approximation to capture the essentials of reality. Figure 4.80 shows example of two transition metals, one non-magnetic—copper (Cu^{29}:$[Ar]3d^{10}4s^2$)—and one magnetic—cobalt (Co^{27}:$[Ar]3d^74s^2$). The s-derived states extend across the energy, but the density of states from $3d$ and its hybridization and other interactions is large and also quite spread out. Copper is non-magnetic, since the d orbital is fully occupied, and the two spin polarities are equal in population from each atom, so the modified states in the assembly have an equal occupation of the two. Cobalt, on the other hand, with 7 of d electrons has a higher occupation of one of the spin polarity states as a consequence of lowering that spin orientation band by the exchange interaction. The aligned spin band is less occupied (minority in spin) than the anti-aligned band. Both the exchange energy Δ and the Stoner gap Δ_s are identifiable in the detailed picture here. This picture can also be related to Figure 4.74.

The carrier density in majority and minority spin bands is

$$N_{maj} = N_\downarrow = \int_{-\infty}^{E_F} \mathcal{G}_\downarrow(E)dE \text{ and}$$

$$N_{min} = N_\uparrow = \int_{-\infty}^{E_F} \mathcal{G}_\uparrow(E)dE. \tag{4.128}$$

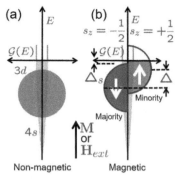

Figure 4.79: Schematic of valence bands arising from $4s$ and $3d$ states in non-magnetic and magnetic materials from the transition series. The simplified picture shows the d band as a semicircle of the two alignments of spin. The s band spans deeper in energy. Bands are partly filled. In a magnetic material, the centers of the d bands, with the oppositely aligned spins, separate by an energy Δ of exchange splitting. One of these aligned spin d bands becomes the minority band.

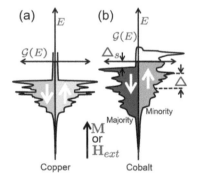

Figure 4.80: Density of states in energy for copper (Cu^{29}: $[Ar]3d^{10}4s^2$) and cobalt (Co^{27}: $[Ar]3d^74s^2$). Copper with a filled d band is non-magnetic. Cobalt with partial filling of $3d$ band is magnetic. The states arise from numerous interactions between the constitutive s, p and d states in the materials. But, one can see the $4s$- and $3s$-derived states extending across the energy depth and at Fermi energy.

The total magnetic moment arising from the spin occupancy is

$$\mathbf{m} = \mu_B(N_{maj} - N_{min}), \qquad (4.129)$$

where the Bohr magneton is determined by the electron's angular momentum.

We now have the framework for discussing devices with spin polarization leading to both magnetism and spin polarized current. First, we consider the earliest application of this in sensing small magnetic fields and storage such as in a disk drive, where the bit is written as a polarization of magnetization.

4.4.1   Magnetic storage

MAGNETIC MEDIA, SUCH AS DISK DRIVES or solid-state magnetic memory, such as magnetic RAM and spin-torque RAM, are examples of the effective use of the non-volatility of magnetization.

Magnetic media employs polarization on a flat surface—either in plane or out of plane—of a disk for storage. The data bits are written in localized regions using a local external field created using current. It is sensed using a change in current or, equivalently, resistance, due to the external magnetic field of the bit stored as magnetization. Both writing and reading is performed by a sensor probe assembly that is commonly called the read-write head. The bit's polarization is written in thin films that are continuous or potentially composed of nanoparticles. So, the field sensor on the read-write head is in close proximity to the spinning disk on which the thin film resides. The sensing, based on resistance changes caused by the film, involves careful electronic designs that can sense the minute changes in currents.

Maximizing these current changes means finding ways to maximize the resistance changes arising from the effect of magnetic field, that is, magnetoresistance. Over time, approaches have involved the use of anisotropic magnetoresistance and larger changes based on physical phenomena such as giant, colossal, et cetera, magnetoresistance in films that are incorporated on the read-write head of the disk. Writing in thin films, or assembly of particles, so changing the polarization, is subject to anisotropy, easy and hard axes of polarization, domains of orientations and their movement, energy needed and stored, competing randomizing phenomena due to environmental sources of energy, and properties of the materials themselves. We can connect our physical description up to this point to all of these.

Resistance in magnetic films is a consequence of a number of interactions above and beyond what one sees in a normal metal. We

A more precise statement, and we have skipped this diversion, is that the magnetic moments will be noninteger multiples of the Bohr magneton.

Superlatives, such as "large," "very large" or "ultra" in scales of integration, or this example in resistance with "giant," "colossal"—there is even one called "extraordinary"—or the use of "smart" with any gadget or activity with an electronic connection is an acute infectious disease among engineers and scientists. "When you catch an adjective, kill it," an admonition from Mark Twain, falls here on deaf ears. The use of acronyms to shorten a repetitive long phrase capturing a phenomenon or a quantifiable mathematical scale dimension is fine, "anisotropic" is welcome, "terascale" to definitively describe a mathematical magnitude is fine, but adjectives such as "LSI," "VLSI," "GMR," "CMR," et cetera, look pitiful in time as engineering and science marches on and one starts running out of letters and adjectives of glorification. Superlatives are a moral hazard.

discussed the consequences of the nature of cobalt's Fermi surface and spin interactions earlier. Table 4.8 shows characteristics of some illustrative materials around room temperature. Except for copper, here as a reference, the rest are magnetic. We have largely thought of magnetism in terms of $3d$ electrons and transition metals. But materials with higher orbitals partially filled, gadolinium here, for example, with $5d$ and particularly $4f$, also has a very high moment, albeit with a low Curie temperature. All have a much higher resistance by nearly factors of 5 or more than copper. The main characteristic that is different is that copper is non-magnetic. This gives a clue that the lack of states that can be occupied in d orbitals has a major influence on resistivity. The d states have the central role in the appearance of magnetism, and their interaction with s states has a similarly important role in resistance. The existence of both together causes an even more important coupling consequence— magnetoresistance in individual and coupled layers.

	Lattice		$\mathbf{m}$	T_c	ρ
			μ_B	K	$\Omega.m$
Cu^{29}	fcc	$[Ar]3d^{10}4s^{1*}$	—	—	1.7×10^{-8}
Fe^{26}	bcc	$[Ar]3d^{6}4s^{2}$	2.2	1043	9.7×10^{-8}
Co^{27}	hcp	$[Ar]3d^{7}4s^{2}$	1.7	1388	6.3×10^{-8}
Ni^{28}	fcc	$[Ar]3d^{8}4s^{2}$	0.6	631	6.8×10^{-8}
Gd^{64}	hcp	$[Xe]4f^{7}5d^{1}6s^{2}$	7.6	289	1.3×10^{-6}

Table 4.8: Some elements, magnetic and non-magnetic, of use and importance in magnetism-based devices, and a few of their properties of interest.

*In copper, the $3d$ shell is filled before $4s$.

Magnetoresistance is the change in resistance in response to a magnetic field. It may occur because of an externally applied field, since it changes the occupation of states, the transport properties of different spins as a result of changed scattering within a film and at film's interfaces, or other causes. It is the ratio of the change in resistance under application of the field versus its initial value, so,

$$\varrho_{MR} = \frac{R_H - R_0}{R_0} = \frac{\Delta R}{R_0}, \qquad (4.130)$$

where ϱ is the magnetoresistance ratio , R_H is the resistance in presence of field, and R_0 the zero-field resistance. As with the Drude picture of the free electron, as discussed in Chapter 2 and Chapter 3 , the classical Hall effect appears as a result of the Lorentz force in the presence of a field. Under the Lorentz force of $q\mathbf{v} \times \mathbf{B}$, the path of the electron deviates due to this cross force between scattering events because of cyclotron motion. The total path length of electron is longer. It encounters more scattering. On an average, the resistance of the electron ensemble in a normal metal is larger—the magnetoresistance is larger. It is a very small effect but net positive. Our clearest indication of this motion is the Hall voltage in semiconductors, where the mobility of electrons is much larger. In a metal, where the Fermi surface is considerably more complicated and the bands filled to much higher energy, the electron will travel by the path of least action—it finds a path of least dissipation. It minimizes scattering to the different parts of the Fermi surface. When a magnetic field is applied, the electron is forced to travel through a path of increased scattering. Magnetoresistance should again be positive for most conditions.

Magnetoresistance in ferromagnetic films and in combinations of films utilizing normal metals, insulators and antiferromagnetic films unfolds a variety of rich phenomena and strong effects, even at room temperature, that are very useful. Magnetoresistance as a signature of field polarity, even magnitude, is the basis for both the digital and analog detection of localized fields in magnetic domains—so for reading recorded bits in magnetic disks and tapes, in magnetic card readers, and as magnetic sensors. Ferromagnetic metals are utilized for anisotropic magnetoresistance, that is, a field direction dependence. In ferromagnetic metal multilayers, one can observe larger magnetoresistance—this is called giant magnetoresistance. In perovskite manganites, one observes even higher resistance changes—this is called colossal magnetoresistance. Each of these depends on on the inter- and intralayer effects.

We explored the d and s states in one example of a ferromagnetic material in Figure 4.79. d states have a narrow energy spread and therefore high density compared to the s states. This is conducive to ferromagnetism by increasing the difference between majority and minority state occupancy even for small exchange splitting. Small band extant means larger mass, since energy is inversely proportional and the number of states arising from d states of the atom is larger than that arising from s states. The conductivity in the Drude model is

$$\sigma = \frac{N_s e^2 \tau_s}{m_s^*} + \frac{N_d e^2 \tau_d}{m_d^*}, \qquad (4.131)$$

where $N_{s,d}$ is the number of electrons, and $\tau_{s,d}$ a scattering time constant, with the subscripts denoting the origin of the states. The scattering time constants should not be too different, so, as the effective mass of electrons is different in the two bands, the conductivity is pronouncedly due to s electrons, which are lighter. In the ferromagnet, the majority band is filled. The scattering involves s band states of both spins, and the field-aligned d states. Transitions involving spin flip are rare But, here what this implies is that only half of the s electrons near the Fermi energies can participate. Two immediate

Magnetoresistance in metal films of thickness less than the mean free path may turn negative at low temperatures if electrons are forced to take a cyclotron path that manages to avoid excessive surface scattering.

We will revisit the spin flipping effects in the transport discussion. Spin-orbit coupling makes a weak spin flip scattering possible. Another process arises from transition metal impurities which can couple spin flipping directions quite effectively.

consequences are the following: (a) above a critical temperature, an absence of magnetism implies more scattering due to the involvement of both spins, so a rapid increase in resistivity should occur; and (b) magnetic field increases spin polarization, and this reduces the s-to-d scattering. This latter will cause magnetoresistance to decrease. The spin-preserving scattering has significantly shorter scattering time than spin flipping does. So, any distribution of up and down spins caused by a perturbation will first decay at the fast time constant and then by a slow spin flipping process.

This slowness of the spin flip process means that one can model these two spin carrier densities in a two current model if one is interested in short time constant phenomena. We will return to this transport discussion shortly.

The magnetoresistance in a ferromagnet changes with the orientation of the field relative to the direction of current flow. This resistance change of a few percent arises from the spin orbit coupling. Figure 4.81 shows an illustration of the changes in resistance, with fields parallel and perpendicular to the ferromagnetic film plane. This is anisotropic magnetoresistance, and the cause of this is spin-orbit coupling. Conduction electrons arise from the outermost orbitals, not the core. But, these electrons arise from both the s orbital, such as $4s$, and from $3d$ orbitals. While the s electrons have no orbital angular momentum, the $3d$ electrons do. The spin angular momentum was the dominant cause of magnetism through exchange. The s electrons of conduction see the residual orbital angular momentum through scattering, that is, a perturbation. This is observable also in the gyromagnetic ratio that too deviates from the value of 2 that is expected for spin-only conditions. The angular momentum is pointed out of the plane of the maximum charge density, such as for $3d$. When this angular momentum is perpendicular to the momentum direction of the electron, it encounters less scattering than when it is aligned, as illustrated in Figure 4.81. This latter case has a larger cross-section for the scattering of electrons. So, the asymmetry of the $3d$ electrons appears through field-direction dependence in the scattering of the $4s$ electrons and, hence, in anisotropic magnetoresistance. The resistance with a parallel field is higher, and that with a perpendicular field lower, because the parallel field has a larger scattering cross-section from the asymmetric $3d$ electrons.

When one makes multilayers of ferromagnetic metals, two additional effects arise which cause the observable magnetoresistance ratio to increase by an order of magnitude. One is spin-dependent transport—interlayer transport has a dependence on the moment alignment. The other is interlayer exchange coupling. This latter is a consequence of the different magnetization orientations that become

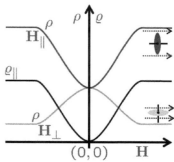

Figure 4.81: Resistivity in a ferromagnetic metal film with an applied magnetic field. $\rho_\parallel$ is for an in-plane field, and $\rho_\perp$ for an out-of-plane field. ϱ shows the magnetoresistance associated with the in-plane field condition.

A resistance change of a few percent of the anisotropic magnetoresistance was good enough for its use in reading disk drives in the period towards the end of 20th century.

possible when one makes multiple magnetic-layer structures.

When multiple thin layers of magnetic metals are interspersed with thin layers of non-magnetic metals, the magnetic field interactions between the layers, that is, the coupling, can be either ferromagnetic or antiferromagnetic. The coupling is oscillatory in the interlayer separation. If one creates a structure such that adjacent layers are antiparallel, that is, antiferromagnetic at zero applied field with interlayer exchange coupling negative, upon the application of a field, they can be forced parallel aligned, and the exchange coupling can be changed positive. This field effect causes a resistance change, since transport in the parallel and antiparallel arrangement is different because of dependence on spin. An anti-parallel structure enhances scattering for the reasons that we have discussed. The magnetoresistance ratio is higher, by an order of magnitude, than that of anisotropic magnetoresistance. Figure 4.82 shows an example of the ratio of this resistance in magnetic/non-magnetic/magnetic film structures. Additional non-magnetic/magnetic layers can also be employed and the resistances will be affected by the alignment of the moments.

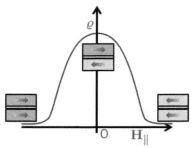

Figure 4.82: The magnetoresistance ratio as a function of field in a magnetic/non-magnetic/magnetic metal film structure designed to be antiparallel at zero field. An example would be 12 *nm* Fe/1 *nm* Cr/12 *nm* Fe. By changing the field in plane, the magnetic layer can be made parallel and the resistance ratio drops.

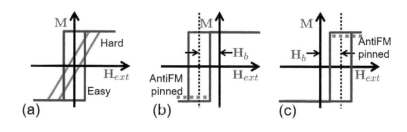

(a)   (b)   (c)

Figure 4.83: Illustration of exchange biasing by antiferromagnetic–ferromagnetic film coupling. (a) shows the normal hysteresis of easy axis polarization and the difficulty of it along the hard axis in a ferromagnetic film. (b) and (c) show the use of moment pinning by exchange bias from the antiferromagnetic (antiFM) film in either of the two directions. The shift arising from exchange bias is shown as the bias field H_b; **M** is magnetization, and H_{ext} is an externally applied magnetic field.

Exchange bias is a result of the exchange coupling that we have discussed. It is the net effect on magnetization resulting from an internal or an external field. So, when a ferromagnet is coupled to an antiferromagnet—the form where magnetic moments are organized in an antiparallel fashion—the magnetization loop can be shifted from the origin as shown in Figure 4.83, as a result of what we will call the exchange bias field, or the bias field. This bias field arises because of the antiferromagnet's magnetization. At the interface, exchange interaction will place an orientational constraint. This exchange bias allows one to maintain the field, in both direction and magnitude, in a ferromagnet layer by pinning it, making it a reference layer for coupling to another ferromagnet layer that is free to respond to interactions with external fields. This effect is therefore unidirectional. Recall that the anisotropic effect is bidirectional, that is, two opposite polarizations are the simultaneous energy minima.

Antiferromagnetism, in contrast to ferromagnetism, arises when the lowest state of the spontaneous magnetic polarization is the antiparallel organization of moments within the crystal. The crystal now is organized on two different sublattices, which are identical and of the crystal's symmetry but which are coupled antiparallelly in an ↑↓↑↓ ... order. The ensemble has no magnetization, even if unit cells have opposite magnetic moments. NiO and CoO are examples of A–A antiferromagnets where the two sublattices have same atomic species. $MnFe$ is an example of an A–B antiferromagnet where the atomic content of the two constitutive parts is different, even if the net magnetization vanishes. The critical temperature at which the transition between antiferromagnetism and paramagnetism takes place is the Néel temperature (T_N), the equivalent of the Curie temperature for the transition between ferromagnetism and paramagnetism. Table 4.9 gives examples of some common antiferromagnetic materials in use.

The coupling of the antiferromagnet layer to the ferromagnet layer exercises an energetic influence at the interface. This coupling can be programmed, and it is uniquely an interface effect. If one grows an antiferromagnet layer together with a ferromagnet layer, raises the temperature of the resulting structure above the Néel temperature, sets a field and then cools the structure, the resulting ferromagnet's hysteresis loop, instead of being symmetric in an external field, will be shifted either towards the positive or the negative direction, as shown in Figure 4.83. This procedure establishes a prescription for the orientation of the magnetization and for the interaction between the antiferromagnet and the ferromagnet film. In turn, it prescribes the exchange field $\mathbf{H}_b$. Instead of being symmetric the magnetic hysteresis curve is shifted in one direction or the other, and the shift of the mean is the exchange bias. In either of the directions, the interface of the magnetic moment of the antiferromagnetic film is pinned in one direction or the other, and its presence determines the behavior of the ferromagnetic layer. The ferromagnet layer is now pinned by the interaction at the interface through the coupling to the antiferromagnet layer.

Ferromagnetism under normal isolated conditions has uniaxial anisotropy—it can have magnetization in one direction or its opposite, since it has two equivalent easy directions. It is not unidirectional. The cause of this is magnetocrystalline anisotropy, which we will discuss when understanding storage and the changes of polarization. Briefly, since the film is an atomic crystalline arrangement with electron spin alignment, orientation should have an effect. Therefore, one would expect anisotropy. This is magnetocrystalline anisotropy. When a ferromagnet is put together with an antiferro-

	Néel temperature (K)
Cr	308
NiO	525
CoO	291
FeO	198

Table 4.9: Some commonly used antiferromagnetic materials and their Néel temperatures.

Interface effects are extremely important in nanoscale devices. Inversion layers cannot exist without interfaces. Tunneling depends on rapid-enough change at interfaces in thin materials. Loss of useful properties is related to interfaces—superparamagnetism or loss of ferroelectricity—where the energetic interactions of interfaces are dominant. Catalysis, the lowering of this energy, is a supremely useful property for a variety of chemical processes.

magnet, by growth or deposition, if no external field has been applied, irrespective of temperatures, the hysteresis loop will still be symmetric. The hysteresis will show the same uniaxial anisotropy as it would standing alone. However, the ferromagnet's easy axis may now be determined by the magnetocrystalline anisotropy of the antiferromagnet and not by its own. This depends on the several properties that determine the energy content from these sources. The existence of this anisotropy due to extrinsic causes and the exchange coupling with the antiferromagnet is reflected in this anisotropy being a uniaxial exchange anisotropy. What makes this uniaxial exchange anisotropy powerful and unique is that when the setting of the ferromagnet is accomplished with a setting field and the temperature lowered below the Néel temperature, or if the growth or deposition of the antiferromagnet is accomplished in the presence of the setting field, then we observe a unidirectional magnetic anisotropy—not uniaxial—the two field directions are not equivalent. The bias field, or the transferred exchange field, $\mathbf{H}_b$, defined in Figure 4.83, sets the preference of the magnetization. If the bias field is negative, as in Figure 4.83(b), the magnetization switches to the positive direction more easily in the presence of an external field. If the bias field is large and negative, then the magnetization direction may always stay positive, that is, aligned with any external field including its limiting case of zero field. The exact reverse of this happens when the bias field is positive, as in Figure 4.83(c). The setting field during the creation of this antiferromagnet/ferromagnet assembly pins the magnetic moment at the interface. If the antiferromagnet film is thick enough, it is no longer possible to change this interface's pinned moment, since its changing requires reversal of the antiparallel moments throughout the film. The pinned moment, a "domain wall," uniquely exists in this thin system. It contains little energy, but it determines several of this assembly's properties: not only the shift in the magnetization curve, but also an increase in the coercivity and a small increase in the saturation magnetization. This exchange bias has established a choice in the direction of the magnetization of the ferromagnetic layer. The spins at the interface are anchored substantially in the bulk of the antiferromagnet. The limiting of this takes place through domain walls. For thin-enough films, these domain walls may be absent. The fields required to reverse them then are large—orders of a tesla.

So, to summarize this anisotropy discussion, the uniaxial exchange anisotropy arises from magnetocrystalline anisotropy that an adjacent antiferromagnet may create via exchange coupling. This uniaxial anisotropy will a have preferred easy axis. On the other hand, exchange bias creates a preferred direction of magnetization by pro-

This is another subtle example where nonlinearity makes an apparently small-energy, short-range perturbation to bring about a large, long-range effect. We have encountered a number of these, for example, in all different phase transitions at critical temperature, and will encounter a number of these in the discussion of nanoscale mechanics, where nonlinearity also shows interesting chaos and bifurcation effects.

viding exchange coupling of moment at the interface between a ferromagnet and an antiferromagnet.

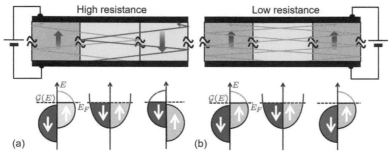

(a)   (b)

Figure 4.84: Magnetoresistance in a
magnetic/non-magnetic/magnetic
metal film, due to spin-dependent
transport. (a) shows that antiparallel
alignment causes the transport of
electrons of one of the spin polarities to
be suppressed in one of the anti-parallel
aligned metals. (b) shows that when
the moments of two magnetic layers
are parallel aligned, electrons with
one of the spin polarities can traverse
through all the films. The resistance of
the configuration in (b) is smaller than
that in (a).

This coupling of an antiferromagnet to a ferromagnet has been
tremendously useful through its consequences for giant magnetoresistance. The change in resistance arises from the difference in scattering between antiferromagnetic and ferromagnetic layers within the
spin-dependent band picture that we have discussed. A normal metal
does not show field dependence, since both parallel and antiparallel
spins are present in equal density. However, if there is a magnetic
field present, internally generated as in a ferromagnet, or externally
present, the spin polarized bands are not aligned in energy. There
are majority and minority bands, as discussed with Figures 4.79 and
4.80. This changes the electron behavior in the films since, for any
specific polarization of the electron, there may or may not be states
available. Figure 4.84 shows the reasoning behind this characteristic
and its effect on transport. When the moments of the two magnetic
layers are anti-aligned, one of the bands—the majority band of that
material—has few or any electrons available for conduction, since all
the states are largely filled. Figure 4.84(a) is a simplification of what
one sees in Figure 4.80 as employed in Figure 4.79. The normal metal
has each spin polarity in equal density, and states available at Fermi
energy for each. The majority spin electron bands, anti-aligned to
each other, are filled. On the left, the spin-down band is filled and the
magnetization up while, on the right, the reverse is true. Electrons
of either of the spins can traverse only through one of the magnetic
films and the normal metal. States are largely absent for the majority
band, which has opposite polarity in either of the anti-aligned magnetic layers. Electrons of the polarity for which states are unavailable
in the magnetic material are reflected back when incident from the
normal metal. This is a high resistance state. In Figure 4.84(b), with
aligned magnetic moments, electrons in the band of minority spin
in either of the magnetic layers are free to traverse all the films. The

Note that the spin of the majority
band is opposed to the direction of the
magnetic moment.

majority spin band conduction is suppressed because most of these states are filled. This is the low resistance state of this film. Since the condition of alignment and reversal of the moment can be achieved through an external magnetic field, the field dependence of the magnetoresistance here and the resistance through control of transition across states constrained by the field is large. This ``giant magnetoresistance´´ is much higher than the anisotropic magnetoresistance.

A giant magnetoresistance read sensor employing these principles, and in use till the early part of 20th century, used a resistance measurement of a multilayered film structure in which the field to be detected was in the plane and either parallel to the current or perpendicular to the current. Figure 4.85 shows example structures for current-in-plane and perpendicular-to-plane geometries.

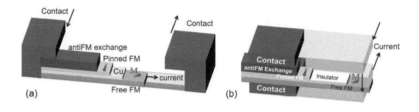

(a)   (b)

Figure 4.85: A read head traditionally consists of a two terminal device measuring current through a sensor structure that can detect weak magnetic fields. The giant magnetoresistance read sensor employs an antiferromagnet layer (antiFM) together with a pinned ferromagnet (FM) layer, with exchange, so that the magnetization of the free ferromagnet layer can rotate in response to the field and also lead to a change in resistance. (a) shows a current-in-plane geometry of a read head, and (b) shows a geometry in which the current is perpendicular, the spin valve form, that is with a tunneling barrier rather than a normal metal. The sensor is mounted on the read head assembly. These two sensing approaches employ terminal arrangements that can be brought in close proximity to the magnetic storage disk, tape, strip or other device.

The structures in Figure 4.85 employ a fixed magnetization direction in one of the ferromagnetic layers—the pinned layer. This is obtained by using exchange interaction with an antiferromagnet layer in direct contact. A free ferromagnet layer separated from the pinned layer by a conducting spacer layer then can have a weak interaction with the pinned layer, leading to an easy rotation of the magnetization in response to additional fields. The change in the relative angle of the free and pinned layers shows up as the magnetoresistance signal. The resistance changes are relatively high. One may employ a metal intermediate layer, such as in the plane geometry with low resistance. But, the alternate sensing that we will now discuss, particularly tunneling magnetoresistance, also employs two terminal devices with perpendicular current (as is also possible with conducting layers) for sensing the magnetic field.

With conduction in metallic films, resistance magnitudes are low. One would like high resistance changes such as through insulating tunneling barriers in transport, or better yet, through a metal–insulator transition. If, instead of metallic ferromagnets, a perovskite manganite whose crystal form is shown in Figure 4.86, such as $La_{0.67}Ca_{0.33}MnO_3$, is employed, very large changes in resistances can be observed. Here, the presence of field changes an insulating state to a conducting state. This is colossal magnetoresistance. Be-

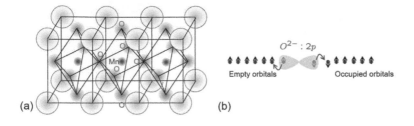

Figure 4.86: $A_x B_{1-x} CO_3$, where C is a small transition atom such as Mn, and A and B are large cations such as La or Ca, so mixed perovskites as mixtures have a magnetoresistance where the material changes conductivity dramatically since the change is associated with a transition from an insulating to a conducting phase. (a) shows the perovskite structure, and (b) shows the superexchange and double exchange that causes ferromagnetism, with the insulating-conducting transition that underlies colossal magnetoresistance.

cause of the high fields needed, which are considerably higher than what is available in high density storage media, this effect is not in use in storage media. This magnetoresistance is effectively a metal–insulator transition that is magnetic field induced. In this, it follows our earlier discussion of various phase transitions. For example, in Figure 4.72, we noted the supermagnetic exchange causing an antiferromagnetic transition, which in turn caused an insulating phase to appear. Colossal magnetoresistance arises with this superexchange and double exchange. These compounds are antiferromagnetic insulators, just like MnO_2 in the earlier discussion. But, here, the superexchange is mediated by oxygen between the filled and empty d orbitals. It leads to ferromagnetism. The reason is the presence of double exchange. Manganese, which has 5 d orbital electrons, when coupled to oxygen, can exist as Mn^{3+} or Mn^{4+}. This means both the chain Mn^{3+}–O^{2-}–Mn^{4+} and its asymmetric arrangement Mn^{4+}–O^{2-}–Mn^{3+} are possible. In the Mn^{3+} state, the manganese atom is bound to six neighboring oxygen atoms octahedrally. One d electron of Mn^{3+} is involved in each of these bonds. The orbitals mix, hybridize and form a bond differing geometrically and in length. La is triply charged, while Ca (Sr too shows this characteristic) is doubly charged. In such mixed perovskites, there is now a redistribution of electrons, and some of the Mn atoms are in the Mn^{4+} state. This is one less electron on Mn. This quasi-hole can hop between manganese sites, making the structure conductive like a metal. And this particular state of conductivity is possible under suitable field conditions, which will depend on temperature. Here, the double exchange with superexchange resulted in both ferromagnetism and metal transition. It also requires admixing. It is maximum when the Ca molefraction is about 33 %. It is instructive to look at Figure 4.43, where one can see that a number of transitions between metallic and conducting regions are possible through admixing. A perturbative field will influence which side of the transition the material resides.

To achieve higher resistance, one of the more effective means is the use of tunneling barriers that impede free flow but still cause modulation of the conductance through the spin-dependent density of states, both of the injector and of the collector. By employing insulating intervening layers, the resistance is higher, even as the density of states–dependent transport characteristics of giant magnetoresistance are preserved. This is tunneling magnetoresistance.

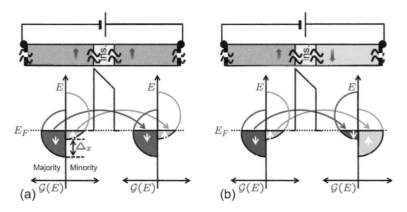

Figure 4.87: Magnetoresistance in a tunneling magnetic/insulating (Ins.)/magnetic film, due to spin-dependent transport, for an example where the majority and minority bands are both partially occupied. (a) shows tunneling with aligned moments in the magnetic films, and (b) sows it for anti-aligned moments of the magnetic films. Figure 4.85(b) shows an example of a read structure using this approach.

Figure 4.87 shows an example of such a structure, and Figure 4.85(b) showed an implementation in a read head. This genre of structures that can throttle transport are also called spin valves. When the moments of magnetic layers are aligned, tunneling occurs in this example from both the majority spin-down and the minority spin-up bands, with nearly identical densities of states if the bands are partially occupied. With anti alignment of the moments, as in Figure 4.87(b), and consequent shifts of the spin-up and spin-down bands, the number of filled states available for injection into the majority spin on the left is reduced. So tunneling is reduced. The resistance is now higher. Because of the intervening insulating layer, the resistance is high and more amenable to electronic detection.

Tunneling magnetoresistance is the preferred embodiment for use in a giant magnetoresistance read sensor head because of its very favorable resistance characteristics. Such a device, not too dissimilar from that shown in Figure 4.85(b) and with two terminals, now employs a current perpendicular to the plane geometry. Comments from the earlier in-plane discussion apply here too. The antiferromagnet layer pins a ferromagnet layer by exchange coupling. The free ferromagnet layer is separated from this pinned layer by an insulator, and small external fields can rotate its magnetization. This, in turn, affects the current through the geometry, leading to a relative change in resistance that can be much higher because the high resistance state can be made to be quite insulating, and the low resistance state quite conducting, unlike in the current-in-plane geometry, where resistances are naturally low.

For purposes of understanding stability, disturbance and manipulation of storage through polarization of the media, we should look in a little more in detail at magnetization and its axial relationships. First, consider the case of complete isotropy. All directions then are equivalent. Exchange, at low temperatures, parallelizes the spins, and this defines the magnetic moment $\mathbf{m}$, which is the product of the gyromagnetic ratio, the Bohr magneton and the sum of the spin vectors. The interaction energy of the moment with the field is the dot product $U = -\mathbf{m} \cdot \mathbf{H}$. In thermal equilibrium, the probability of any angle θ between these vectors, $\mathfrak{p}(\theta)$, is proportional to $\exp(\mathbf{m} \cdot \mathbf{H}/k_B T)$ or the exponential of the cosine of the angle. The probability-weighted average of this cosine is as follows:

$$
\begin{aligned}
\langle \cos \theta \rangle &= \frac{\int_0^{2\pi} \cos\theta \exp(mH/k_B T) \sin\theta \, d\theta \, d\phi}{\int_0^{2\pi} \exp(mH/k_B T) \sin\theta \, d\theta \, d\phi} \\
&= \frac{(\cos\theta - k_B T/mH) \exp[(mH/k_B T)\cos\theta]\big|_0^{\pi}}{\exp[(mH/k_B T)\cos\theta]\big|_0^{\pi}} \\
&= \coth(mH/k_B T) - k_B T/mH \\
&= \mathcal{L}(mH/k_B T),
\end{aligned}
\tag{4.132}
$$

which is the Langevin function. It is the projection of the unit vector of magnetization in the direction of the field.

Since the magnetization of the films arises from the atom and electron assembly, it is subject to the forces within that cause a magnetocrystalline anisotropy. Alignment, atomic contributions, exchanges and overlap are different in different directions. This means that there are easy and hard directions of magnetization in a crystal, as conceptually shown in Figure 4.88 for iron.

The saturation magnetization is harder to achieve along some directions than others. The body diagonal of the body centered cubic (BCC) unit cell's is the hard axis, with a large aligned energy that requires a larger field to flip. This is the case for iron. The edge of the cube is the easy axis, and the diagonal along the face lies in between. In materials where the unit cell is hexagonal closed packed (HCP), such as cobalt, for example, the easy axis is along the c-axis, and the hard axis lies in the basal plane. Nickel, a face-centered-cubic (FCC) assembly, has an easy axis along the body diagonal. The difference between the easy axis and the hard axis can be seen in the magnetocrystalline anisotropy energy. It is an extensive parameter. When magnetization is along the hard axis, the energy is higher. Although the limiting saturation magnetization is the same, the easy axis is easier to polarize.

As discussed in the context of Figure 4.58, the internal field due to

This relationship's immediate conclusion is that, underneath Heisenberg exchange interaction isotropy, just as in atomic systems, paramagnetism lives in the ensemble of the atoms. The distinctions are of discreteness and continuity.

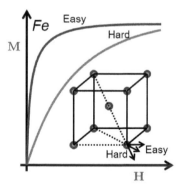

Figure 4.88: Magnetization in iron. The hard direction is along the diagonal of the BCC iron cube's unit cell, as the packing and bonding is highest there, the edge of cube is the easy direction, and the diagonal along a face is an intermediate direction. Hard here means that flipping the magnetization of that direction requires the most work. In Co, which is HCP, the easy axis is along the $\langle 0001 \rangle$ direction, which is the c-axis, and the hard axis is in the basal plane, which has a similar magnetization-magnetic field character.

the magnetization is depolarizing. This depolarizing field results in a magnetostatic energy. A horseshoe magnet can lift magnetic material. However, place an iron bar across its poles, and its external field is reduced and, equivalently, the magnetostatic energy and the internal depolarizing field. There is now much less magnetostatic energy to do the work against gravity.

Anisotropy in layered structures arises for many reasons. Magnetocrystalline anisotropy is caused by spin-orbit interaction. Shape anisotropy exists where the bulk and the surface force alignments dominate. So, as shown in Figure 4.89, a homogeneous material layer will have magnetization dominated by the magnetostatic properties, properties that we have discussed to this point and that depend on the material and the structure it has. But, we can also create artificial layered structures, such as alternating layers of magnetic and non-magnetic films, thin enough that the coupling mechanisms are physically modified by the process. This ability to define magnetism orientation and to manipulate it in layered structures is at the heart of using magnetism fruitfully in nanoscale devices.

The reduction in energy can happen through formation of domains. An illustration of a domain—a region where all the moments are aligned in the same direction—is shown in Figure 4.90(a). Here, in a "thick" magnetic film, the external field has been reduced by keeping the magnetic induction within the material and thus reducing the depolarizing field. But, it happened through the formation of domains—regions where the direction of moment changed. In Figure 4.90(a), if the film were of crystalline iron, that is, cubic, the easy axis would be the preferred direction for the domain. In cubic crystal, this arrangement lets all the domains be along the easy axis. If cubic symmetry is not present, this picture can get complicated. Domain alignment is also influenced by another property found in ferromagnetic materials: magnetostriction, whereby the process of magnetization causes minute changes in the physical dimensions of the material and thus can affect domain shape. This is the small perturbation effect on the crystal arrangement arising in magnetic energy—a short-range influence from the long-range accumulation of magnetic moment. The dimensional changes along polarization can be positive, that is, elongation, as in iron, or negative, as in nickel. So, multiple effects exist simultaneously. Minimizing $U = -\mathbf{m} \cdot \mathbf{H}$ favors either alignment or anti-alignment through the magnetocrystalline energy and remains a major driving force.

The lowest energy arises when the sum of exchange, magnetostatic, magnetocrystalline and domain wall energies are minimized. In turn, this means that smallest magnetic energy occurs when the spatial extent of the external field is small, and the flux and field

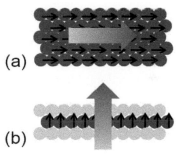

(a)

(b)

Figure 4.89: (a) shows a film of homogeneous magnetic material. In this film, if it is sufficiently thick, that is, several 10s of nms, the magnetization direction will be defined by the easy axis in the plane. (b) shows an artificial layered structure of magnetic and non-magnetic layers that are sufficiently thin, that is, fractions of nms. Now, a consequence of spin-orbit effects, difficulty in twisting the angular momentum of d orbitals of transition elements, may dominate, and the easy axis may be out of plane.

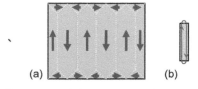

(a)   **(b)**

Figure 4.90: (a) shows an energy lowering domain arrangement in iron (BCC) in a thick-film or macroscopic form. (b) shows the possibility of lowering depolarizing fields through rapid change in small-sized stretched domains.

reside largely within the magnetized materials. This process also favors rapid change at the walls between different magnetization orientations, so in thin films, arrangements such as that shown in Figure 4.90(b) become possible. Here, only a few of the moments in the domain wall region have the non-easy axis of orientation.

Microscopically, what happens in magnetized thin films as a result of these couplings? Figure 4.91 shows the underlying nature of magnetization. Here, we show two vertically polarized main domains with the constitutive contributions within them that give the net magnetic moment. The region in between is the domain boundary along which the rotation of the magnetic polarization takes place. It is the size of this region, the energies of the domain region, and the thermalization process that will determine the stability and limits of density.

These domain walls are of interest to us since the recording of polarization, that is, how far bits have to be from each other, is related to it. Also, when one changes orientations these domain walls have to move. This change will most likely happen through paths of least energy demand. Domain walls, where the magnetic moment rotates to the new alignment, are shown in Figure 4.92 for the two polarization orientations.

Figure 4.91: Bits written in magnetic polarization on a thin film where the polarization is vertical, that is, along the thin film's thickness direction. Two regions of opposite polarization, that is, bits "0" and "1," are shown with a boundary region over which the bit changes direction.

Walls are usually classified as Bloch walls if in these walls the polarization rotates within the plane of the wall. They are Néel walls if the polarization rotates perpendicular to the plane of the wall.

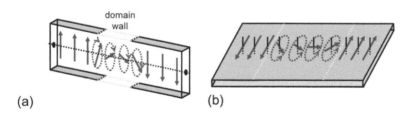

(a)   (b)

Figure 4.92: (a) shows the change in orientation of magnetization in the domain wall region, with vertical polarization. The low depolarizing condition for the domain wall here is when the moment rotates to the opposite alignment, with the rotation axis in the plane of the film. (b) shows in-plane polarization domains with a Néel domain wall. The rotation here is around an axis that is out of plane.

For vertically magnetized domains, as in Figure 4.92(a), the rotation shown has no magnetization in the plane. Magnetization extends uniformly at the angle of the twist through the film and terminates at the film boundaries. This is conducive to low energy. In Figure 4.92(b), which shows in-plane polarization domains, the rotation is around an axis that is out of plane of the film. The domain wall surface, rather than the film surface, serves as the termination of magnetic moment. This too is conducive to a reduction in magnetostatic energy, as the rotation occurs around the axis with the least magnetization and smallest field, thus keeping $-\mathbf{m} \cdot \mathbf{H}$ small. In-plane polarization would lead to a larger domain wall to accommodate enough room for rotations. Both of these polarizations have been employed in magnetic recording, with vertical polarization providing the higher density.

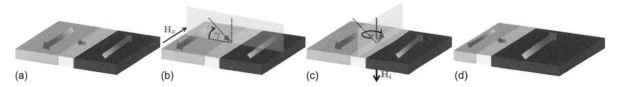

Figure 4.93: (a) shows an initial steady state of an in-plane polarized film with one grain boundary that is a Néel domain wall with a perpendicular magnetic moment. (b) shows the application of a field causing a torque on the domain wall. (c) shows the simultaneous rotation of the magnetic moment of the domain wall as a result of the demagnetizing field. This causes the aligned magnetization domain to expand, as shown in (d), and the domain wall to move.

Domain wall motion under the presence of an external field is a result of the torque that an applied field exercises on the domain. It is instructive to see how the magnetic moment and the magnetization—the moment density—change in response to the fields. Figure 4.93 shows the effect in an in-plane magnetic film due to an in-plane applied field. When an external magnetic field $\mathbf{H}_x$ is applied, as shown in Figure 4.93(b), the two domains—aligned and anti-aligned—have no torque on them, since a torque is a cross product of the field and the magnetic moment. But, there exists a torque on the magnetic moment of the domain wall. It causes the moment to rotate out of the plane of the film, and as it rotates, a demagnetizing field $\mathbf{H}_i$ builds up. $\mathbf{H}_i$ causes the domain wall moment to twist, as shown in Figure 4.93(c). Since the field is out of plane, the twisting causes the domain wall magnetic moment to rotate towards alignment with the applied field. The application of the field causes the forces to appear on the domain wall, which by orienting its magnetization makes one domain larger at the expense of the other and causes the domain wall to proceed orthogonally to the applied field. Any magnetization that is aligned or anti-aligned, or close to it, has a vanishing torque.

Films can be amorphous, polycrystalline or crystalline. So, the magnetization response behavior will depend on the nature of the film. Crystalline films will show the field direction dependence w.r.t. the crystal orientation, while amorphous films will not. In all these, a finite domain wall will still exist between oppositely polarized regions. In both the out-of-plane and the in-plane polarizations, the bulk contribution to energy is still proportional to $-\mathbf{m} \cdot \mathbf{H}$. So, one would expect a behavior with an angular dependence of the type shown in Figure 4.94 to be between the polarization and the field. This figure follows the bistable memory energy picture that we discussed in Chapter 1. These written bits are stable so long as the barrier energy between them satisfies the constraints in the discussion of errors in that chapter.

The energy required to rotate the moment of the domain away

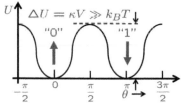

Figure 4.94: Energy as a function of the azimuthal angle from the polarization axis for hexagonal magnetic anisotropy. Both "0" and "1" correspond to regions of lowest energy; these regions are separated by an energy barrier that is related to anisotropy.

from the easy direction is the energy to change the electron spin and the electron orbital angular moments. Because of the coupling between the two, discussed earlier, the energy that needs to be overcome is the spin-orbit coupling. The d orbit, much more than the spin, is coupled to the atom, and its arrangement is also very asymmetric. The orbit's angular momentum is orthogonal to the bonding, that is, the direction of wavefunction maxima. So, the easy axis, as we have seen, is generally in the direction with least wavefunction overlap. Anisotropy reflects this energy requisite of moment alignment due to magnetocrystallinity, and it reflects this symmetry of the crystal structure. In general, one may write the anisotropy energy in terms of the angle. For hexagonal crystals with the c-axis as the easy axis, a uniaxial anisotropic energy exists with the dependence in an expansion form as

$$U = (\kappa_1 \sin^2 \theta + \kappa_2 \sin^2 \theta + \cdots)V \approx \kappa V \sin^2 \theta, \qquad (4.133)$$

which is the form shown in Figure 4.94. κ_is are the anisotropy constants, and V is the volume. For cubic Fe, this relationship will have a dependence on the three different directions of symmetry: axis, face diagonal and cube diagonal. It is written in squares and higher powers of sinusoidal functions of angles from the axis. A large anisotropy requires a larger field so that sufficient torque $\mathbf{T} = \mathbf{m} \times \mathbf{H}$ needs to be applied to cause the barrier of ΔU to be surmounted in order to progress to the next lower energy state, where the polarization is either aligned or anti-aligned to the magnetization. The field that causes this to occur, arising in the write head, must also be aligned because of the energy-moment relationship.

Anisotropy will also arise from shape and this becomes important at the nanoscale. A polycrystalline material may not have any preferred orientation and therefore no ensemble average crystalline anisotropy. Since the energetics have a dimensional dependence reflecting the accumulation of unit cell contributions, only in a spherical, truly random polycrystalline particle would one see true angular independence. For irregularly sized materials, the longest dimension will be easier to magnetize, since this direction has the lowest energy, that is, the lowest $-\mathbf{m} \cdot \mathbf{H}$.

Nanoparticles in thin films provide an avenue towards higher density. They can be synthesized as stable single crystals and potentially assembled into monolayer low defect density thin film. A principal reason is that, once the size is below a critical dimension, only one domain may exist, and it has a high saturation magnetization that is stable so long as the superparamagnetism—loss of this magnetization—that arises when thermal energy becomes important does not become a constraint. The domain wall width is the result

Equation 4.133 is written both in sine (sin) and cosine (cos) functional forms. These are just differences in reference axis. Both give an energy barrier of κV.

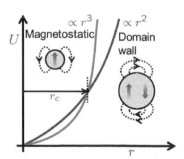

Figure 4.95: The energy size dependences of magnetostatic and domain wall energies that lead to a size below which it is favorable to have only a single domain.

of two competing mechanisms—magnetocrystalline anisotropy that favors thinness because of the easy axis alignment through rotation, and exchange energy, which favors thickness and separates the domains that are in opposite polarization. Magnetostatic energy is proportional to volume—an r^3 dependence—and domain wall energy is proportional to the wall area—an r^2 dependence. This means that the curves will cross, as schematically seen in Figure 4.95, at a critical size. Large particles, with multiple domains, because of the demagnetizing field, can be kept in a saturated state of magnetization only by application of a field that exceeds the demagnetization field. Small particles, with their single domains, are always saturated. The magnetization is then in the same direction throughout. Only the anisotropy has to be overcome, and one need not concern oneself with demagnetization or domain walls. Small particles also provide the highest polarization.

The interaction of magnetoanisotropic energy in small particles results in a number of interesting properties. In addition to being a single domain and so possessing the highest magnetization achievable, the particle is also incapable of using domain wall motion for minimizing energy as a response to the application of an external field that torques the particle magnetization. The magnetization is in the easy direction, and it is easily held there, since the rotation through hard direction is inhibited. Absent domain walls, the considerations of Figure 4.92 now do not apply. The switching must therefore occur through the hard direction back to the anti-aligned easy direction. This means that the field must be larger. The particles increase in coercivity as the critical size for single domain is reached with the reduction in dimensions. So, even for the easy direction, the coercivity becomes large, as shown in Figure 4.96(a). This increase in coercivity lends increased stability against stray fields to the stored polarization. The second appealing property is the existence of a more square hysteresis loop, as shown in Figure 4.96(b), in the easy direction. This is because the easy axis is now inhibited against easier torquing and change of orientation. If one were to measure the hysteresis curve in the hard axis direction in a single domain nanoparticle, one would not observe a hysteresis and see a largely linear response, since removal of the field would restore the magnetization to the easy direction.

Shape and magnetocrystalline anisotropy determine the magnetization of the synthesized particle along the easy direction. Since domain walls do not exist, applying an external field to flip the polarization requires the particle to respond not by domain wall motion but by polarization transition. The anisotropy energy holding the particle in the polarized state, according to Equation 4.133 and Fig-

Typical domain walls are 100 *nm*. So, below this size, one would expect single domains in nanoparticles for common transition metal magnets employed in storage.

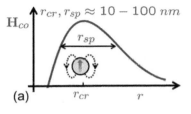

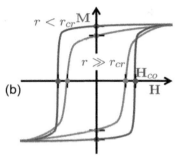

Figure 4.96: (a) shows coercivity of magnetic nanoparticles in the direction of the easy axis, with a reduction in particle size as they go from having multiple domains to having single domain, with magnetism disappearing at the smallest size, due to paramagnetism. The figure identifies r_{cr} as a critical dimension where the coercivity is the highest, and r_{sp} as a span over which high coercivity exists with a single domain. (b) shows the impact of the absence of domain walls in creating a more "ideal" square magnetization hysteresis loop in nanoparticles at $r < r_{cr}$.

The origin of the word "superparamagnetism" lies in the correspondence in the local moment behavior of paramagnetic materials. In both cases—paramagnetism, and size reduction leading to paramagnetism-like behavior in magnetic material—thermal fluctuations disturb the alignment of the magnetic moment even if their absolute magnitudes are considerably different, that is, the importance of the Bohr magneton versus the number of atoms in the nanoparticle, times the Bohr magneton.

ure 4.94, is κV, so as the volume is reduced, so is this energy. When $\kappa V \approx k_B T$, absent an external field, thermal fluctuations in polarization, flipping between the various easy polarization directions, for example, six for cubic iron, or two for hexagonal-closed-packed cobalt, will spontaneously occur, resulting in a change of states. This is superparamagnetism. Superparamagnetism—this effective loss of magnetism—is accentuated in real conditions, where the particles have size distribution and alignment distribution. The consequence with decreased anisotropy energy and a decreased particle-to-particle spacing is a rapid drop in coercivity and saturation magnetization with decreasing size in closely packed films, due to interparticle interactions, where particle magnetization now influences its neighborhood. Figure 4.97 shows this effect on the coercivity and the hysteresis loop in presence of interparticle interaction.

Superparamagnetism limits the smallest particle size where magnetization can be maintained with assured immunity from thermal fluctuations. The anisotropicity provides a barrier, our $\Delta U = \kappa V \sin^2 \theta$ of Equation 4.133, to magnetization reversal for a single isolated particle.

When one has these particles in an assembly, there is a statistical distribution of the orientations—the precession angle around the field. This is like the paramagnetic statistical distribution. The theory of this relaxation process, from Néel, employs the uniaxial anisotropy energy and fluctuations from magnetostriction, that is, the arising from strain existing with the magnetization and of vibrations. So, there is a prefactor in units of frequency in an exponential relationship of the time constant.

An alternate theory, from Brown, considers magnetization direction fluctuations as a random walk problem, that is, it is not a problem of correlation in time but of response in time to random forces. The resulting equation is in the form of the Fokker-Planck differential equation or the Kolmogorov forward equation; both of which describe the probability density function evolution in time. We skip the mathematical details of this, but its principal eigenvalue also allows the writing of a relationship very similar to that of Néel. The time constants in the Néel and the Brown arguments are related as follows:

$$\frac{1}{\tau_{N,B}} = \mathcal{A}_{N,B} \exp\left(-\frac{\kappa V}{k_B T}\right), \text{ where}$$

$$\mathcal{A}_N = \gamma H_c(|3Y\lambda + DM_s^2|)\left(\frac{V}{\pi G k_B T}\right)^{1/2}$$

is the Néel prefactor, and

$$\mathcal{A}_B = \frac{\gamma H_{co}}{2}\left(\frac{\kappa V}{k_B T \pi}\right)^{1/2} \text{ is the Brown prefactor.} \quad (4.134)$$

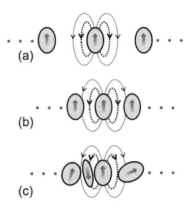

Figure 4.97: Magnetization in an assembly of nanoparticles in the single domain limit. (a) shows an idealized condition where particle uniformity is ideal, and separation is large enough to minimize interparticle energetic interactions. (b) shows decreasing particle spacing, which causes fields from one particle to influence interactions with another particle. (c) shows the more realistic condition of particle size and spacing distribution, even if in the single domain limit.

The Fokker-Planck equation describes the evolution of a distribution function in a Markov process. A Markov process is one where history is forgotten as events take place. See S. Tiwari, "Semiconductor physics: Principles, theory and nanoscale," Electroscience 3, Oxford University Press, ISBN 978-0-19-875986-7 (forthcoming) for a discussion of this in connection with noise and even the arrow of time. Kolmogorov, through his description of probability and probability distribution function evolution, has much to say about this. There are connections here to entropy and algorithmic complexity, which is a subject where Kolmogorov made his first appearance for us in this text.

The prefactor of the Néel relationship is related to Young's modulus, the magnetostriction coefficient, saturation magnetization, coercivity, the gyromagnetic ratio and the demagnetization factor that magnetostriction and the crystal vibrations connect to. Much of the literature employs the gyromagnetic precession frequency for this prefactor. The prefactor of the Brown relationship is related to a field constant and anisotropy. In these equations, Y is Young's modulus, λ is magnetostriction, D is the demagnetization factor, M_s is the saturation magnetization, $H_c = 2\kappa/M_s$, s and $\gamma = g\mu_B\hbar$.

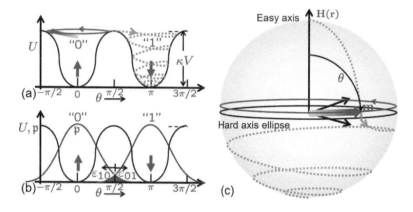

Figure 4.98: Precession-based simplistic argument of magnetization flipping. (a) shows at high energy the potential of the magnetic moment precessing over into the opposite polarization state, while (b) shows the fluctuation-induced random walk and the probability distribution of he polarization angle, leading to a finite probability of being in the opposite polarization state. The precession process is elaborated in (c) where, at high-enough energy, close to $\sim \kappa V$, that is, an energy close to the product of the anisotropy constant and the volume, the magnetic moment can pass the hard axis ellipse, and if it does, then quickly settle into a polarization that is opposite along the easy axis.

What these equations reflect is that the magnetic moment can flip if it has high-enough energy arising from the strain fluctuations due to thermal vibrations, and it can precess across the barrier—the Néel argument, as seen in Figure 4.98(a) and (c), or that it can diffuse across to the opposite state, as any system with gradients and fluctuation energy has the propensity to do. The latter example, shown in Figure 4.98(b), is similar to the Boltzmann probability of error argument that we explored in Chapter 1 for the different memory structures and for logic manipulation. Note that energy wells here are similar to our classical bistable memory's restorative form. In either case, the barrier energy is about κV.

The superparamagnetic limit is the limit where magnetism becomes unsustainable. In the Néel model, the particle is modeled as precessing over the barrier. In the superparamagnetic limit, this means that the barrier is being assumed to be nearly flat. The Brown model's random walk assumes that the magnetic moment precesses freely, behaving like a classical gas particle in motion. The one degree of freedom has an energy of $k_B T/2$. In the particle at the paramagnetic limit, magnetization changes occur through anisotropy fields generated by the temperature and strain induced by field fluctua-

At 30 nm, τ_N and τ_B are about equal, ranging from 10^{10} s^{-1} to 10^{12} s^{-1} and varying as three-half powers of size. So, at 10 nm, they would be a factor of 5 smaller. The energy in the exponential decreases as the third power of the dimension because it is proportional to volume.

tions, that is, they are asymmetric. So, at best, these approximations are useful as guidelines and for capturing the essence of the underlying phenomena. The simplicity of the equations is appealing, and the expressions in Equation 4.134, particularly the second, are employed to get the first estimates of the superparamagnetic limit for particles to be used in magnetic storage.

Domain size, domain walls and domain wall movement are all important considerations in polarization-based storage. Both the parallel spin structure of ferromagnetism, and the antiparallel spin structure of antiferromagnetism, exhibit domains. Early bubble memories employed this writing and movement to achieve archival disk memories. If we looked at a small single crystal magnetic particle, which is not constrained by its surface, we would notice the anisotropy and that different fields are needed to coerce the magnetization to change direction. If the magnetization is along the easy axis, switching it to the opposite direction by avoiding switching through the hard axis will be easier. It will require less field. So, orientations of applied field as well as the orientation of crystalline/polycrystalline material, if employed, are important.

Since domain motion consists of the formation and size changes of regions that are energetically most favored, as well as the movement of boundaries via the assimilation of existing domains and the reorientation of domains, when a field is applied, the domain whose magnetization is most closely aligned and has the strongest energetic interaction starts to grow at the expense of others. A demagnetized state is one where the domains are randomly oriented. A thin film will show a variety of these in all possible orientations and at small size scale—including in crystalline and polycrystalline films, where grain boundaries would act as only one of the sources of change in energy dynamics. When a moving domain encounters defects—sources of magnetic energy perturbations such as point defects, dislocations, grain boundaries, or substitutional and interstitial impurities—the principle of least action would imply that it will attempt to eliminate or minimize its magnetostatic energy. The defect will be a local energy minimum, and it will pin the boundary. Sufficient energy must exist through the external field to cause motion to continue. So, the domain clinging to the defect, given that energetically favorable conditions usually favor a loop to minimize demagnetization, can only free up and keep moving if sufficient field exists. This field is the coercive force. When one has a large-enough sample, the coercive field that we encountered in Figure 4.57 is the ensemble average of that magnetization loop measurement.

Magnetic disk writing and reading employs the combination of these physical principles. A disk consists of a thin film of ferromag-

This argument reflects a self-inconsistency. Both these models have underlying them assumptions that break at the conditions at which they are applied. And one can see the breakdown of these assumptions when particles with changes in exchange energy, such as those that occur when ferromagnetic films are combined with antiferromagnetic surface layers, et cetera, are measured.

The earliest ferromagnetism-based memories were based on ferrite cores. At the crosspoint of a ferrite bead, where arrays of perpendicularly strung wires were employed to write any individual bead with half current flow through each wire, a third wire strung through all the beads was employed to read any random bead that was accessed by the orthogonal crosspoint wires. The earliest Apollo missions employed this type of memory; the core memory planes used for those missions were a few inches in size and could contain up to 10 *kB* each, which was considered to be a lot of memory, at that time. Ferrites are ferrimagnetic transition metal oxides, most commonly heat-pressed ceramics from $NiO\text{-}Fe_2O_3$ powder. In bubble memories, bubbles of polarization move in a patterned permalloy of a garnet film that defines the path of movement by using a rotating magnetic field for the stepping in a maze. This inherently needs domain wall movement. These were sequential memories since a bubble had to be brought to the reading point. Bubble memories had much higher density. But, they soon fell out of favor when Winchester drives, the earliest spinning magnetic disks with read-write heads, came from *IBM*. Tape drives provide the highest storage numbers. Like disk drives, they employ a read-write head but with a very thin moving tape. However, disk drives can access data in a specific location faster than tape drives can because disk drives access data through the radial movement of the drive head across the spinning disk, whereas tapes need to wind or rewind to the specific locale.

netic material where bits can be written in the polarization. Together with this magnetizable film there are additional thin film layers introduced to protect the film from physical damage during use while spinning at a high speed very close to the head, and additional layers to improve the field intensity and its return to the head as well as for the accentuation of texture—crystallization orientation—conducive to high density robust writing and reading. The polarization depends on the properties of the film, particularly its magnetic saturation, which in turn determines how much external field must be applied through the write/read head. The smallest-sized bit that can be written is limited by paramagnetism—the stability of the bit against thermal and other energetic interactions of the environment, and the spatial and field capability of the read/write head must be able to coerce the bit in both states of polarization. So, thin films (amorphous, polycrystalline, crystalline or consisting of a planar assembly of nanoparticles) and the read-write head are part of a system design that must work with high speed spinning and an efficient and error-resilient read-write process in order to provide both speed and density at reasonable power.

Since the last decade of the 20th century, the films have tailored texturing, that is, with crystalline grains. In the first decade of the 21st century, segregated films, that is, ones where the oriented grains were separated from each other through insulating regions that were usually made of SiO_2, were employed. By the second decade of the 21st century, there arose interest in patterned media, where nanocrystals are assembled in planar films, again separated from each other by insulating regions. In all these, at least until the middle of the second decade, multiple grains or nanocrystals are employed to store the information of one bit.

The writing part of the head is essentially a very miniaturized coil capable of achieving high field and fabricated using techniques of microlithography and employing a high susceptibility material, such as a ferrite, to concentrate the field. It is placed on the head so that the direction of the field is appropriate to the polarization needed. So, for lateral or perpendicular recording, the head assembly's writing part would consist of the region creating this field and a return path. Figure 4.99 shows an example of a write coil.

A read-write head is a combination of the writing part and the reading part which are assembled together to access the track on the spinning disk. The read part of the head uses the magnetoresistance signal, that is, a voltage generated for a current driven through the film stack, to measure the state of polarization. This requires fast, low noise and low interference techniques so that the read operation can be rapid and reliable. The writing part needs to concentrate the

A density of 1 Tb/in^2 (1.6×10^{11} cm^{-2}), about the limit possible under the constraints of superparamagnetism, means an area of about 600 nm^2, per bit. At a rotational speed of 10 000 rpm a 5 $inch$-diameter disk, the speed at which the disk flies by the head at its edges is $\sim 10^4$ cm/s at a distance of ~ 5 nm away from the surface. This is about 1/3rd the speed of sound in air!

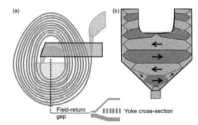

Field-return gap ||||| Yoke cross-section

Figure 4.99: (a) A miniature coil is employed to achieve the high fields of the writing part of the head. For a horizontally polarized bit, the maximum field is horizontal; for writing a vertically polarized bit, the maximum field must be vertical in both directions. (b) A simplified layout of the head assembly is shown here. The head moves on an armature, extremely small dimensions are achieved through the fabrication techniques of microelectronics, and a high current in the coil becomes possible by using a highly conducting metal, gold, for example, by electroplating with a placement that straddles a split yoke. So, the writing region has a concentration of field while a return path reduces it by spreading it out.

field and provide enough of it while allowing high density through small polarization areas and inter-bit and inter-track distances. So, the write head needs to have the ability to achieve both high fields and rapid field drop-off.

Figure 4.100 shows a cross-section of a perpendicular recording head that integrates the read and write operations; this is a more advanced version than that in Figure 4.60. The head is designed so that the field gradient it generates is high and provides a rapid cut-off of the field. The main pole employed in writing is in very close proximity to the return yoke, in order to complete the magnetic field loop; both the main pole and the return pole are made of materials with high saturation susceptibility. The pole tip is also shaped for robustness of writing while allowing higher track densities. It is elongated along the track and is narrow laterally so that adjacent tracks are not disturbed. It may also be shaped to write bit shapes that are most conducive to high density. Together, these keep the bit size small. The magnetic field from the trailing edge of the main pole is absorbed by a trailing shield, and the recording field is tilted in the direction of the track. This creates a large field gradient, reducing jitter effect at the trailing edge of the pole. Yokes, made using soft layers, isolate the head sensor from extraneous magnetic fields and noise. As the field required to write increases with increasing density, an additional trailing shield is also provided.

Reading heads exploit the magnetoresistive effect in its various forms, achieving sensitivity through the use of thin films which in turn allow narrow transition regions in the magnetic medium that is being written on. We have already discussed examples of the multilayers of such structures utilizing either the antiferromagnetic biquadratic exchange effect or weakly coupled films where one of the films is pinned by a hard ferromagnetic or antiferromagnetic layer.

The read elements for lateral and perpendicular recording are very similar. The orientation of the element and its coupling of the field determine the voltage signal generated for a constant current, which signifies through its sign the polarization of the bit, and through threshold, the robustness of the digital conversion.

The density of bit the pattern, of course, also depends on what can be achieved in the disk's films. The lateral recording media of the first decade of the 21st century employed films made from materials such as $CoCrPt$ alloys. Here, during the deposition process, $CoPt$ grains segregate, that is, form a properly oriented crystalline granular structure in the film for writing. Perpendicular recording needs a granular structure that is c-axis oriented. This granular texture is enhanced if achieved using SiO_2 walls in the writing process. Figure 4.101(a) shows an illustration of such a structure. The bit is

A bit area of 25×25 nm^2 for Tb/in^2 includes the boundary region, where the magnetization changes. Writing with a high field gradient without disturbing adjacent bits necessitates a reasonably sized boundary region. So, a bit needs to be in a magnetizable region that is $< 20 \times 20$ nm^2. For stability, > 100 k_BT anisotropic energy is needed. Even then, with the various variations, a bit error rate of 10^{-6} is likely, and it will need error correction code. If one assumes a 10 nm-thick magnetization film, or particles, the total volume of the magnetic material per bit is 2×10^{-18} cm^{-3}. For cobalt, with a covalent diameter of about 0.25 nm, this is an ensemble of 200 000 atoms. At $100k_BT$ energy, this corresponds to an anisotropy coefficient of $\sim 10^5$ J/m^3 or $\sim 0.6 \times 10^{18}$ eV/cm^3 if only one particle is employed for this writing.

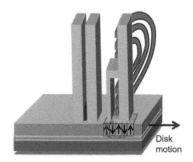

Figure 4.100: A read-write head that creates a high field and a high field gradient.

written in multiple needle-like ferromagnetic grains whose boundaries are insulating walls. This enhancement to the columnar structure becomes possible through a $Ru(0002)$ underlayer. The reason for this is that $Co(0002)$ grows preferentially on $Ru(0002)$. Underneath this structure is a continuous assistance layer that makes it possible for the field lines to concentrate and be able to return and have a low field in the return path through the use of wider yoke. So, the magnetic layers of the film, the underlying orientation-inducing layer and the continuous assistance layer make perpendicular writing possible, and the continuous film allows an increase in the magnetic interaction between grains. In these planar continuous films, the thermal stability is dependent on the stability of the grain. This necessitates that the grain be at least $\sim$6 *nm*, even if multiple grains are employed for the bit.

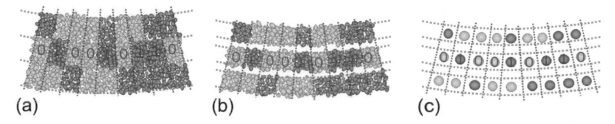

(a)   (b)   (c)

Figure 4.101: An illustration of patterned media. The darker areas are for a 1 logic state of the bit, and the lighter areas are for the 0 logic state. The middle track is identified with the logic state in the different cells along the track. The regions of magnetic material, either deposited or self-assembled, are obtained in a robust reproducible order in size and separation. (a) shows a continuous granular recording layer. (b) shows a granular layer with patterned tracks. (c) shows a bit patterned medium with high exchange–coupled granular layers. In all of these cases, multiple grains are employed per island of bit.

The rapid field change needed to limit the bit-to-bit spacing can be partly addressed through the taper designs of the head, as well as by patterning the recorded data tracks so that they are separated. This is shown in Figure 4.101(b), and modest improvements in density can be achieved using such an approach. Patterned tracks still need multiple oriented grains and a continuous assistance layer for return. Forming the track suppresses track edge noise as well as the interference that arises from adjacent tracks because of magnetization fluctuations. But, its thermal stability is still limited to the grain size, $\sim$6 *nm*.

Potentially, the largest increase in the limit to density is made possible by making the disk a patterned media surface, as shown in Figure 4.101(c). Each bit is now a single domain magnetic island formed from high exchange coupled layers, *Co-Pt*, *Co-Pd*, for example, and the location and separation of bits is maintained either by lithographically defined or self-assembled techniques. The thermal stability is now constrained by the dimension of the island and so is larger than for the previous two examples. This works together with the high coercivity and field gradients that such a patterned media structure can provide and is at a dimension smaller than the dimension where domain walls may arise. One way to achieve the writing in high coercivity material may then be by local heating during the writing process. Local heating, such as via a semiconductor laser, reduces the field required to write. But, it better not disturb what is written in the proximal region.

Let us now consider current transport in ferromagnetic material–containing device structures, examples of which were the read heads employing giant magnetoresistance or tunneling magnetoresistance. Notionally, our emphasis is on spin polarization. Current, the flux of charged particle momentum, is a response to the distribution of off-equilibrium created by change in energies. Bias voltage applied on terminals is a change in electrostatic energy. But, off-equilibrium is also caused by changes in magnetostatic energy if one has a magnetic field arising from a magnetic potential. Electrons have charge and spin. And the transitions that take place obey conservation of energy, and the momentum—linear and angular.

In discussing magnetoresistance, we referred to the two current transport model because spin flip scattering is much less common. So, the s electrons principally carry the current, and since their number of states is small, the resistance is dominated by scattering from s to d bands. The spin preservation leads to the s-d change. And the high resistance and the change in resistance is an indicator of the partially filled minority spin d band to which the smaller number of electrons available from the s band scatter to, and the clearest indicator of this is the drop in resistance due to suppression of scattering when the material undergoes the phase transition when temperature is reduced below the critical temperature T_c.

The anisotropic magnetoresistance for the two orientations of magnetization shown in Figure 4.81 arises from spin-orbit coupling. The change of magnetization direction w.r.t. the current direction causes asymmetric scattering because of the changes in density of minority spin and of magnetic moment magnitude—the s electrons see the effect of the orbital angular momentum. We showed this as a cross-sectional area representative of the dominant scattering effect due to d's orbital angular momentum.

It is possible to inject spin polarized electrons into other materials—metals, insulators, semiconductors—or to modify the properties of materials by introducing magnetic elements—for example, manganese introduced into $GaAs$ in high enough densities makes it magnetic at low temperatures. $Zn_{1-x}Cr_xTe$ is a near–room temperature magnetic semiconductor using Cr—a magnetic atom—in $ZnTe$—a group II-VI semiconductor. The spin preference of carriers recombining is now reflected in the polarization of the photon. These spin-dependent properties may also be coupled to other attributes of different structures in interesting ways. In all these, one couples unique properties—longer coherence lengths, longer ballistic mean free paths, unique dimensionality, et cetera. A few of these are worth mentioning for their interesting attributes.

The scattering of electrons—low energy electrons—in ferromagnetic conditions increases with increasing number of d states available for scattering. And these materials, in a sense, preferentially filter minority spins through this scattering-based "absorption" process. The distance that an electron of any specific spin travels between scattering, characterized by its scattering length ($\lambda_\uparrow$ or $\lambda_\downarrow$), depends on the magnetic state of the ferromagnet. Passing a current through a ferromagnet of sufficient-thickness spin polarizes the electron flux, and this polarized current is now subject to manipulation by devices with material structures that filter, probe or do other physical manipulation. We give a few examples in Figure 4.102.

Figure 4.102(a) is an example where a ferromagnet electrode injects through a semiconductor or a ferromagnetic semiconductor, of which $Ga_{1-x}Mn_xAs$ is one example, causing polarized photon output by direct recombination. While $Ga_{1-x}Mn_xAs$ has a Curie temperature below room temperature, $Zn_{1-x}Cr_xTe$ has a Curie temperature very close to room temperature, and T_cs greater than room temperature are not physically disallowed. Efficient polarized photon generation and perhaps even single photon sources with polarization programmability become potential possibilities.

Figure 4.102(b) shows an example of preferred spin injection from a ferromagnet layer into a semiconductor, where a coherence length that is long compared to that in metals is used to manipulate collection signals through magnetized electrodes. If use of a spin amplifier, low power conversion of signal, and manipulation within the spin domain are feasible, this makes possible a different way to process signals within a common volume.

Figure 4.102(c) shows an example where the two-dimensional spin polarized electron gas of a transistor is manipulated through the gate using the Rashba effect and the moving electron's field. The Rashba effect is spin-orbit interaction causing spin splitting in materials in the absence of inversion symmetry, so it also appears in structurally inversion-asymmetric semiconductors—in their inversion layers, wires, quantum-dots, et cetera—and it can be tuned by a field. With changes in spin occupation in the split energy levels, a field-effect-controlled spin transport becomes possible and observable in

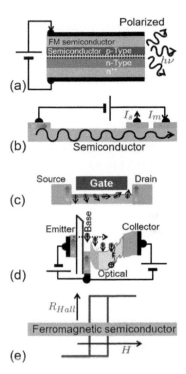

Figure 4.102: Some example device structures employing spin polarization. (a) shows a light-emitting diode with preferential photon polarization emission. (b) shows the possibility of a ferromagnet injecting preferential spin electrons into a semiconductor, where they propagate with long coherence length and show a response subject to interaction at the electrodes. (c) shows the use of a spin-polarized two-dimensional electron layer, where an electric field and the Rashba effect may cause spin orientation and hence electrode-dependent drain current. (d) shows an energy-filtering arrangement where spin polarized injection and ballistic injection through the base is detected through integrated polarization-selective optical detection, and (e) shows a ferromagnetic semiconductor, such as $Ga_{1-x}Mn_xAs$, where transition elements are incorporated in a semiconductor and thus lead to a large Hall effect; FM, ferromagnet.

the magnetoresistance. In Figure 4.102(c), this is incorporated for a spin polarized injection. Such a gate-induced structural inversion asymmetry can be introduced in a variety of structures.

Figure 4.102(d) shows a device geometry employed for energy filtering of a polarized electron beam. A thin ferromagnet base in a ferromagnet-injected geometry of a metal-base transistor is used to energy selectively inject electrons into an optically active region where the polarized electrons can be detected. This structure is an example of what is possible when using the intrinsic magnetic characteristics with gain. So, a ferromagnetic base whose magnetism orientation could be modified would give rise, in a suitably designed transistor structure, to an ability to obtain current gain where base interactions modify the number of polarized electrons collected. Finally, Figure 4.102(e) shows that a ferromagnetic semiconductor will have a strong hysteresis and hence a hysteresis in Hall resistance that could potentially underlie a memory.

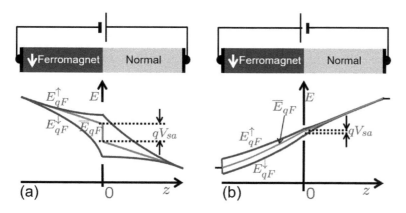

Figure 4.103: Quasi-Fermi energies during spin injection in a ferromagnet/normal metal structure under bias. (a) shows injection of electrons from the ferromagnet, and (b) shows electron injection from a normal metal. Spin-dependent effects become acute at the interface when the electrons are injected from the ferromagnet into the normal metal.

Similar to using spin bands for the two different orientations, we may describe occupation properties through a Fermi, or quasi-Fermi, energy where the occupation probability is $1/2$. This is similar to how, in Chapter 2, we used this notion for the two directions of motion in the quasi-ballistic nanowire transistor geometries to connect specific population to underlying distribution functions. If a ferromagnet and a normal metal are in contact and we apply a bias that injects electrons from the ferromagnet into the normal non-magnetic metal, the normal metal too will have spin asymmetry that has an orientation that is similar to that of the ferromagnet over a length scale—mean free paths $\lambda_\uparrow$ and $\lambda_\downarrow$—before the scattering evens the orientations out. Spin equilibration requires scattering embodied in the relaxation time constants $\tau_\uparrow$ and $\tau_\downarrow$, and therefore a distance scale. The same is true for spin randomization in time for a polarized collection of electrons that appear at the normal metal interface with the metal. In Figure 4.103 this behavior is reflected in the form of the two electrochemical potentials—the quasi-Fermi energies for the two spin orientations. The current flowing through the structure is

$$J = \frac{\sigma}{q} \nabla \overline{E}_{qF},$$

(4.135)

where $\sigma = \sigma_\uparrow + \sigma_\downarrow$ in light of limited spin flip scattering and the consequent two current approximation. The quasi-Fermi energy employed, however, needs a comment. In thermal equilibrium—no current, and zero external bias voltage—the electrochemical potential is constant. This is independent of spin orientation, as a consequence of detailed balance, and this is how we have drawn it in earlier pictures. We may write an averaged quasi-Fermi level as $\overline{E}_{qF} = (E^\uparrow_{qF} + E^\downarrow_{qF})/2$. If there is no spin preference, current continuity implies a continuous electrochemical potential between the contacts. The spin-up and spin-down electrons in the ferromagnet have different conductivities. Let $\sigma_\uparrow = \alpha\sigma$, that is, $\sigma_\downarrow = (1 - \alpha\sigma)$, where α has the physical meaning of being the fraction of conductivity due to spin-up electrons. In a normal metal, α is precisely $1/2$. Majority spin, up spin in this example, where the magnetization is down, has higher conductivity, and $\alpha > 1/2$. So, there is an accumulation of spin-up electrons. At the interface, the density of states in the materials change. The change in the average quasi-Fermi energy, reflective of the accumulation of spin up electrons, is the spin accumulation potential, qV_{sa}.

If the conductivities of up and down electrons are different, then so are the currents. The current in this example is polarized with spin up electrons. We will denote this ratio, in correspondence with α, as β. $J_\uparrow = \beta J$, so $J_\downarrow = (1 - \beta)J$. The polarization of the injected current (ϱ_J) is

$$\varrho_J = \frac{J_\uparrow - J_\downarrow}{J_\uparrow + J_\downarrow} = \beta - (1 - \beta) = 2\beta - 1,$$

(4.136)

so Equation 4.135 can be rewritten as

$$J = J_\uparrow + J_\downarrow = \frac{\sigma_\uparrow}{q}\nabla\overline{E}^\uparrow_{qF} + \frac{\sigma_\downarrow}{q}\nabla\overline{E}^\downarrow_{qF}$$

$$\therefore \quad \nabla\overline{E}_{qF} = \alpha\nabla E^\uparrow_{qF} + (1 - \alpha)\nabla E^\downarrow_{qF}.$$

(4.137)

This gives the quasi-Fermi energies as

$$\overline{E}_{qF} = \alpha E^\uparrow_{qF} + (1 - \alpha)E^\downarrow_{qF} \text{ for } z < 0 \text{ and}$$

$$= \frac{1}{2}(E^\uparrow_{qF} + E^\downarrow_{qF}) \text{ for } z > 0 \text{ with } \alpha = \frac{1}{2}.$$

(4.138)

The quasi-Fermi levels for up and down spins are continuous at the interface, since a spin-selective force doesn't exist at the interface.

This conductivity formulation assumes that the causes are independent and random, that is, that the up-spin scattering rate and down-spin scattering rate are independent of each other, and random. So, it is the same assumption that is embedded in the Mathieson's rule used for deriving the mobility as a geometric mean from the different causes in semiconductors.

The spin accumulation potential merely reflects the difference in the averages of the occupations. Accumulation of specific characteristics occurs at all interfaces. A low temperature superconducting metal on a normal metal produces an accumulation voltage because of the differences between normal electron and a superconducting pair of electrons. Superconductors at a superconductor/small gap semiconductor interface induce superconductivity in the semiconductor. These properties are all indicators that carrier-based property changes cannot be abrupt.

These relationships then give the spin accumulation voltage as

$$V_{sa} = -\frac{1}{q}\left[\alpha_\uparrow(0) - \frac{1}{2}\right]\left[E_{qF}^\uparrow(0) + E_{qF}^\downarrow(0)\right].\qquad(4.139)$$

We interpret the behavior incorporated in these equations as the existence of two channels that are similar to the forward and backward channels of nanowires, as we introduced earlier in Chapter 2, the channels here being of up and down spin, respectively. The exchange between the two channels can be characterized by a scattering relationship that is determined by unit length, as for the nanowire channels. The spin accumulation voltage reflects the changes in number of electrons in each of the channels, and, as before, we do this through the electrochemical potential—the quasi-Fermi energy for each channel.

The electron motion in this metallic conduction system, which has a large population of electrons, is dominated by drift, that is, an electric field causes scattering-dominated ohmic transport characterized by a specific conductance. However, how does the exchange between the two spin orientations of electrons occur? It occurs through scattering between states. As electrons travel in the ferromagnet, electrons polarize in a preferred orientation—through scattering, which sets the preference of orientation. As they travel in the normal metal, they randomize again through scattering. This process of the exchange between the spin channels is a diffusion process where the scattering rates depend on the state of the system, that is, ferromagnetic or normal, and where the equilibrium population too depends on the state of the system, that is, ferromagnetic or normal.

The transport of current changes thermal equilibrium densities to $n_\uparrow(z,t) - n_\downarrow(z,t)$. This is reflected in the $E_{qF}^\uparrow - E_{qF}^\downarrow$ electrochemical energy, and it changes away from $z = 0$ through spin exchange, which occurs via a diffusion process described by scattering time constants. And this happens in both the ferromagnet and the normal metal. We can write the spin diffusion equation that describes this process. The difference of the spin populations, the excess n^{xs}, is

$$
\begin{aligned}
n^{xs}(z,t) &= n_\uparrow(z,t) - n_\downarrow(z,t)\\
&= \int_{E_{qF}^\downarrow}^{E_{qF}^\uparrow} \mathcal{G}(E)dE \approx \mathcal{G}(\overline{E}_{qF})(E_{qF}^\uparrow - E_{qF}^\downarrow).\qquad(4.140)
\end{aligned}
$$

These population differences can not be very large, a few percent, perhaps a bit more, which implies a spread in energy that is much less than k_BT. The approximation in Equation 4.140 therefore is quite accurate. With a diffusive transport, the rate of change of carrier population is the diffusion coefficient times the divergence of the population. This is just Gauss's theorem applied to the scalar population density. We can write

$$\mathcal{D}\nabla^2 n^{xs}(z,t) = \frac{n^{xs}(z,t)}{\tau_{sf}} \qquad (4.141)$$

to relate the change in carriers due to the spin flipping relaxation time τ_{sf} in a volume, to the flux in and out of the volume. The diffusion here is the diffusion of electrons, of both spins, that then interact in the volume. For a Fermi velocity of v_F and a relevant electron diffusion length of $\lambda_\mathbf{k}$, the diffusion coefficient is $\mathcal{D} \approx v_F \lambda_\mathbf{k}/3$—the factor of 3 accounting for dimensionality of movement. Equation 4.140 then leads to the second order differential equation relating the quasi-Fermi energy span:

$$\mathcal{D}\nabla^2[E_{qF}(z,t)^\uparrow - E_{qF}^\downarrow(z,t)] = \frac{E_{qF}(z,t)^\uparrow - E_{qF}^\downarrow(z,t)}{\tau_{sf}}, \qquad (4.142)$$

with the steady-state solution

$$E_{qF}^\uparrow(z) - E_{qF}^\downarrow(z) = [E_{qF}^\uparrow(0) - E_{qF}^\downarrow(0)]\exp\left(-\frac{z}{\lambda_{sf}}\right). \qquad (4.143)$$

λ_{sf} is a diffusion length due to spin flip and sets the scale length over which spin populations equilibrate by diffusion. In a ferromagnet, spin flipping diffusion leads to the build-up of an excess spin population; and in a normal metal, it leads to the loss of any excess.

It is also a form of Fick's law. Fick's law is analogous to Fourier's law, which we encounter in thermal transport, in diffusion current in electron transport, and in a variety of other situations subject to random walk.

One will have to be cautious in discussing the different scale lengths of processes. Magnetism arises from polarization that happens with $3d$ electrons. Conduction of electrons is dominated by s electrons and their spin-conserved scattering.

	v_θ (m/s)	v_F (m/s)	$\tau_\mathbf{k}$ (fs)	$\mathcal{D}$ (m²/s)	$\lambda_\mathbf{k}$ (nm)	τ_{sf} (ps)	λ_{sf} (nm)
Si ($n_{2D} = 10^{16}\ m^{-2}$)	$\sim 1.5 \times 10^5$	$\sim 1.5 \times 10^5$	65	~ 0.0015	~ 10		
GaAs ($n_{2D} = 10^{16}\ m^{-2}$)	$\sim 3.25 \times 10^5$	$\sim 4 \times 10^5$	250	~ 0.02	~ 80		
Cu	$\sim 10^5$	1.8×10^6	25	0.026	44	4.7	350
Co	$\sim 10^5$	1.8×10^6	3.2	0.035	5.8	0.4	38

Table 4.10: Some characteristic, including spin-related scales, in ferromagnets, normal metals and semiconductors near room temperature.

Table 4.10 shows some of the characteristic scales for ferromagnets, normal metals and semiconductors in selected examples. Electron scattering involving spin flipping is more than two decades lower in cobalt than in semiconductors, and it is lower in copper too. In copper both scattering mechanisms are about a decade slower than in cobalt. The distance traveled before spin flip scattering is nearly a decade higher than momentum relaxation scattering. On an average, it will take hundreds of electron scattering events before one of them will cause spin flipping and polarization change of the electron. So, it takes a much longer distance for spins to change. Momentum responds faster than spin.

Consider now how spin polarization builds in a ferromagnet during injection from a normal metal, as shown Figure 4.103(b). Electrons injected from a contact into a ferromagnet may have both orientations of polarization, so long as empty states exist. Away from the contact interface, in the ferromagnet, significant spin polarization exists. At an interface with another material, any spin accumulation voltage will depend on the scaling length for polarization, as the accumulation is related to how rapidly polarization is built or lost.

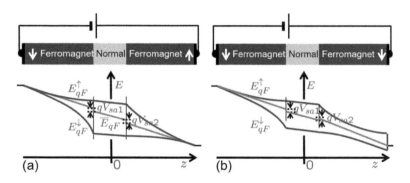

Figure 4.104: The Fermi energy change in a ferromagnet/normal metal/ferromagnet structure near the interface. The normal metal has higher conductivity than the ferromagnet, and a thickness that is much smaller than the spin flipping scale length, so $t \ll \lambda_{sf}$. (a) shows antiparallel magnetic polarization, and (b) shows parallel magnetic polarization.

The giant magnetoresistance geometries are subject to this discussion. For the two opposite cases of ferromagnet moment rotation, spin accumulation and the voltage at the interfaces have the forms shown in Figure 4.104(a). Conductance in normal metals usually being superior to that in ferromagnets, a larger potential drop occurs across the ferromagnetic regions. For a thin normal metal layer, one that is much thinner than the spin flipping length scale λ_{sf}, the spin relaxation is small. The normal metal must be of a minimum thickness to prevent exchange coupling between the two ferromagnets. Only the pinned ferromagnet has an exchange effect with the antiferromagnet. Under these conditions, if the magnetization of the ferromagnet is opposite to that of the antiferromagnet, when there is a layer of normal metal between the two ferromagnets, spin accumulation exists at both of the ferromagnet–normal metal interfaces. If both ferromagnetic layers are in the identical orientation, no accumulation exists, and the average polarized Fermi energy of the two ferromagnets is the same as that for each individual ferromagnet.

Giant magnetoresistance is a reflection of the differences between the two cases shown in Figure 4.104. It reflects the consequences of the V_{sa} spin accumulation voltages at the ferromagnet/normal metal interface. The resistance can be higher or lower, depending on the

Exchange coupling, if it exists between two ferromagnets when there is a very thin layer of normal metal between them, will force the magnetization of the second unpinned ferromagnet to rotate over a domain wall to realign with the pinned ferromagnet.

characteristics of the spin-dependent transport. Majority spin and minority spin transport characteristics affect giant magnetoresistance. In ferromagnets, parallel magnetization gives low resistance and small spin accumulation voltage. In magnetite—Fe_3O_4—because electron transport occurs via minority spins, the spin accumulation behavior reverses sign. The diffusion of the spin orientation, and the length scale λ_{sf} can be measured by injecting spin and then measuring polarization as a function of spacing.

The behavior in the normal metal layer is quite similar to what would happen if the layer had been made from a semiconductor. Metal-to-metal interfaces are quite transparent, even if the processes that take place at the interfaces due to interface rearrangements introduce interesting anomalies—we showed one such process, which is related to the existence or absence of spin-specific states, in Figure 4.84; in that case, it causes reflection. Metal/semiconductor interfaces have distinct effects that we have employed before—the barrier height being one of immense import in electronic devices. Semiconductors also have a significantly smaller carrier populations than normal metals do, and the carriers are majority or minority carriers, depending on the background doping. The spin selectivity due to reflection and the spin polarization change during transport are the basis of devices where spin-dependent properties are gainfully employed as majority and minority carriers, as we have seen in Figure 4.102.

This measurement of position-dependent spin interaction is akin to the Haynes-Shockley measurement for the majority-minority carrier interaction during minority carrier diffusion in a semiconductor.

4.4.2 Spin torque effect

THE MAGNETIC MOMENT OF AN ELECTRON, in the semiconductor and in the metal, in the presence of a field, precesses. We discussed earlier the precession of a magnetic moment in a field. For the electron, we also derived the direct relationship between the magnetic moment and spin polarization. Precession was one of the starting points for our discussion of magnetics. Equation 4.76, with a Larmor precession frequency of $\omega = -\gamma H = e\mu_0 H/m_0$, gives its magnitude and direction by the left hand rule of the response of the moment of an electron in a field H. We also wrote the Landau-Lifshitz-Gilbert equation (Equation 4.78) describing this precession in the presence of damping forces. The angular momentum of the injected spin being lost to the crystal through spin-crystal interaction is one example of a damping force. The crystal here causes this angular momentum loss through the interactions that exist from the electrons localized at lattice sites.

How do this precession and spin polarization evolve in time and space when one injects electrons into a ferromagnet? A ferromagnet has an existent magnetization due to the spin of electrons. We will call this $\mathbf{S}_F$. It is the spin moment arising from the exchange and other interactions that we have discussed. Electrons are injected into the ferromagnet and let us call this the injected spin density $\mathbf{S}_{inj}$. $\mathbf{S}_{inj}$ is the sum of all the injected spins. Each electron has a magnetic moment ($\pm g\mu_B/2$). Injected spins and existent spins don't need to be aligned, and they are certainly not in equilibrium. So $\mathbf{S}_F$, which is large, and $\mathbf{S}_{inj}$ interact, and the principal consequence of this interaction is precession: $\mathbf{S}_{inj}$ precesses.

Any interaction between the injected spin and the ferromagnet's own spin must follow conservation laws. The conservation of angular momentum tells us

$$\frac{d\mathbf{S}_{inj}}{dt} = -\frac{d\mathbf{S}_F}{dt}. \tag{4.144}$$

$\mathbf{S}_{inj}$ precesses around $\mathbf{S}_F$. The torque is the rate of change of this injected spin. As shown in Figure 4.63, this torque is

$$\mathbf{T}_{ix} = \frac{d\mathbf{S}_{inj}}{dt} = -n\mathbf{m} \times H_F = \frac{1}{2}ng\mu_B H_F \sin\theta\hat{\mathbf{t}}. \tag{4.145}$$

Here, n is the electron density; $\hat{\mathbf{t}}$ is the unit normal vector in the plane of precession, perpendicular to the background field; $\mathbf{H}_F$ is the field—a molecular Weiss field—and θ is the angle between the two spin vectors. The torque $\mathbf{T}_{ix}$ is an injection exchange torque that signifies that the injected spin is not in equilibrium with the resident spin or magnetization and therefore results in an interaction where both affect each other. The magnetic moment of the electron is aligned opposite to the spin. This torque, perpendicular to both spin vectors, is maximum when the angle is $\pi/2$. Because there exists this torque between the injected electrons' spin (or moment) and the crystal spin (or moment), the equal and opposite reaction that this denotes means that there exists also an excitation spin wave in the crystal. Even if the injected current is not polarized, scattering being spin selective in the ferromagnet, an injected spin comes about, and magnons or these spin waves that we discussed earlier exist as a collective excitation of the crystal.

Spin precession of the electron occurs at very high frequencies. The maximum torque on each electron, following Equation 4.145, is $g\mu_B H_F/2$, which is about 0.5 eV. So the precession frequency $\nu_p \approx 0.5$ $eV/\hbar = 0.75 \times 10^{15}$ Hz. The Fermi velocities in Table 4.10, ranging up to $\sim 1.8 \times 10^6$ m/s, also imply that, within a very short distances of travel, for example, $v_F/\nu_p \approx 2.4$ nm, the electron precesses through sampling the various orientations, while also undergoing the damping relaxation processes of interactions. The result is that the angular momentum of the injected spin is transferred to the medium during transit. In the diffusive regime of the exchange of polarization, the spin component perpendicular to the magnetization is lost, with the spin orienting parallel or antiparallel during transit. The transiting electrons have a random wavevector, and this samples the scattering that damps the electron spin orientation. Spin selective scattering, which we discussed earlier, strongly orients the electron spin to the preferred axes. Interfacial spin processes are particularly important since they filter by reflecting spins that are not allowed because of the absence of the corresponding states. The effect of these processes is that the angular momentum of the injected electrons, and that of the electrons in the ferromagnet, interact more than the atomic and phonon mechanisms of the crystal do, and the spins reach a steady state of polarization, as dictated by the orientation of the ferromagnet film. The process of damping, that is, energy loss to a broadband of degrees of freedom, happens through the randomness of the spin scattering processes. So, the result of spin injection is to create a number of interactions within the ferromagnet film—torques that act both on the injected stream of electrons and on the resident magnetization, enough to even flip the magnetization. This is the spin torque effect. This happens together with dissipation effects through damping, that is, relaxation processes.

The change in occupation of majority and minority bands, due to spin injection, is illustrated in Figure 4.105. The direction of magnetization of the ferromagnet is up; therefore, the crystal's spin is down. This ferromagnet is the anti-aligned ferromagnet in Figure 4.104, that is, the one into which the electrons are injected.

In Figure 4.105(a), which shows the thermal equilibrium condition, the majority and minority Fermi levels are the same. The electrons from the oppositely polarized ferromagnet injector are polarized upwards. These were majority spin electrons in the injector but are now minority spin electrons in the injected ferromagnet. In Figure 4.105(b), close to the injected interface, the minority spin electron population increases. Charge neutrality, together with its independence from spin, means that consequent with the increase in minority spin electron population is a decrease in the majority spin electron population. This constancy of the electron population density is what is implied by maintaining the average quasi-Fermi level even as the majority and minority spin quasi-Fermi energies change. These energy changes are much smaller than those from exchange splitting of the bands, by three orders of magnitude. Nevertheless, the consequence is very significant. The total angular momentum of the ferromagnet has changed, since the spin population has changed. Its magnetization has changed. The direction remains the same. If the incident spin electron flux is at a finite angle to the spin and mag-

This spin damping is the equivalent of the heat transport that takes place through phonons when energy is injected in the crystal. The crystal excitation represented in phonons, as we have discussed, is the equivalent of the spin excitation represented in magnons. Phonons can scatter spin and spin waves. However, any scattering will have to change the spin angular momentum. This is sometimes possible, such as with torsional modes of nanowires and nanotubes. But, these are only very special situations.

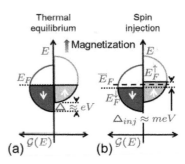

Figure 4.105: The changes in majority and minority spin electrons in a ferromagnet upon injection. (a) shows occupation of states in thermal equilibrium with a constant Fermi level. (b) shows the quasi-Fermi level for majority and minority spin upon injection of electrons.

netization of this ferromagnet, there exists torque—an equal and opposite torque pair—of which one acts on the injected spin $\mathbf{S}_{inj}$ and the opposite on the resident spin $\mathbf{S}_F$. All these changes take place without the slow spin flip process; if the spin had been collinear with the film's magnetization or its resident spin, there would be no torque, but there would still be a magnetization change, since the majority and minority population would have changed, as seen in Figure 4.105.

In the general condition, $\mathbf{S}_{inj}$ is incident at an angle relative to the quantization axis in the ferromagnetic film. The ferromagnetic film has a magnetization $\mathbf{M}_F$ determined by the majority spin that it is the axial opposite of. The resident spin $\mathbf{S}_F$ too is at an angle, and it precesses around the anisotropy axis. $\mathbf{S}_F$ is free to change orientation, subject to external forces. So, the spins interact as they precess and as spin diffusion occurs. A number of torques exist related to these magnetic moment and magnetic field interactions, together with scattering-based interactions—torques that result not only in precession but also in damping ($\mathbf{T}_d$) and antidamping ($\mathbf{T}_{nd}$). Figure 4.106 attempts to capture these.

The torque's effect is to cause the injected electron spin $\mathbf{S}_{inj}$ and moment to precess around $\mathbf{S}_F$. This is at a high frequency ($\sim 10^{15}$ Hz). $\mathbf{S}_F$ is the quantization axis, and the magnetization of the film is in the polar opposite direction. This precession, as discussed in Equation 4.145, arises from the injection exchange torque, which acts in the direction of $\mathbf{S}_{inj} \times \mathbf{S}_F$. As the injected electrons transit, diffusion occurs from majority spin through scattering. The consequence of this is a rotation of $\mathbf{S}_{inj}$ into $\mathbf{S}_F$. Equation 4.144, which represents the principle of equal and opposite reaction, tells us that, on $\mathbf{S}_F$, an opposite torque results, one that attempts to precess away from the anisotropy axis. If the ferromagnet is thick, eventually the injected electrons will align themselves to the ferromagnet's resident spin—the ferromagnet polarizes the beam. But, if the ferromagnet is thin and its magnetization and resident spin are free to rotate, we have a number of interactions to consider. $\mathbf{S}_F$, with which $\mathbf{S}_{inj}$ interacts, itself is precessing slowly around the anisotropy axis. The torque causing this precession is the cross product between the field and $\mathbf{S}_F$-based magnetic moment. Due to the crystal interactions, there is a damping torque in the direction that causes $\mathbf{S}_F$ to be anti-aligned w.r.t. the anisotropy axis. We encountered this damping term in Equation 4.76 and in modified forms in the Landau-Lifshitz equation (Equation 4.77) and the Landau-Lifshitz-Gilbert equation (Equation 4.78). Its direction is towards the axis of precession, as it arises from two cross products.

The significant new characteristic here is that the interaction be-

The injected spin angular momentum per unit volume for a current density J entering the ferromagnet is

$$\mathbf{S}_{inj} = \frac{J}{ev}\frac{\hbar}{2}\mathbf{P},$$

where J/ev is the carrier density, $\hbar/2$ the angular momentum and $\mathbf{P}$ the degree of polarization. This flux encounters spin selective scattering, similar to what we discussed for giant magnetoresistance, and exchange coupling to the local magnetization.

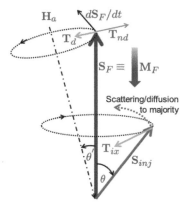

Figure 4.106: Some of the principal physical interactions occurring in a ferromagnet as the injected spin $\mathbf{S}_{inj}$ interacts with resident spin $\mathbf{S}_F$, precessing around the quantization axis defined by $\mathbf{S}_F$. $\mathbf{S}_F$ itself is precessing around the anisotropy axis if it is not aligned with it. The diffusion from minority spin to majority spin during transit is forcing alignment of $\mathbf{S}_{inj}$ to $\mathbf{S}_F$. The slow precession of $\mathbf{S}_F$ around the anisotropy axis has a damping torque due to spin-crystal interactions such as spin-orbit, but it also has a negative damping force due to the interaction between $\mathbf{S}_{inj}$ and $\mathbf{S}_F$. A large torque can cause $\mathbf{S}_F$ to flip.

tween $\mathbf{S}_{inj}$ and $\mathbf{S}_F$ causes an additional torque that is antidamping, as embodied in the relationship in Equation 4.144. This is also referred to as negative damping, as it is attempting to torque $\mathbf{S}_F$ out of the anti-alignment. This negative damping torque, the result of $\mathbf{S}_{inj}$ acting on $\mathbf{S}_F$, is forcing the magnetization of the film to reverse. This torque is shown as $\mathbf{T}_{nd}$ in Figure 4.106 and comes about because the spin of the injected electrons torques the ferromagnet's polarization. This in essence is the physical picture of the spin torque effect. It becomes significant enough to be useful at nanoscale dimensions.

We have tried to understand the physical interactions that cause this effect in at least two ways: one in terms of the forces or torques, and the other in terms of the occupation of states, which in turn is related to the angular momentum.

We can reconcile these now. The injected electrons, if fully polarized, just mean that all the carriers are majority spin in the ferromagnet and that $\mathbf{S}_{inj}$ is aligned with $\mathbf{S}_F$. The removal or absence of the minority spin electrons is what made this alignment possible. The torque that forces $\mathbf{S}_{inj}$ towards $\mathbf{S}_F$ is also perpendicular to $\mathbf{S}_F$ and in the direction of $(\mathbf{S}_{inj} \times \mathbf{S}_F) \times \mathbf{S}_F$. The equal and opposite reaction to this is the torque $\mathbf{T}_{nd}$, which is shown in Figure 4.106 and which is in the direction of $\mathbf{S}_{inj} \times (\mathbf{S}_{inj} \times \mathbf{S}_F)$, torquing $\mathbf{S}_F$ away. This torque is opposite the damping torque $\mathbf{T}_d$ discussed earlier and which is caused during the precession of $\mathbf{S}_F$ around anisotropy axis. In viewing this in angular momentum terms, recall that, when s electrons are scattered to empty d states—the d states mostly from the less-filled minority spin band—the transitions happen with a change in angular momentum. Recall our discussion of s and d electrons and conduction in anisotropic magnetoresistance; the spin-conserving transition, with the occupation of the d state, now creates a localized magnetic moment, thus causing a spin wave. Total angular momentum must be conserved. So, the spin wave excitation necessitates a change in the conduction electron spins. This spin wave is created with the flipping of conduction electron spin—the energy and angular momentum of the spin flip is converted to the spin wave excitation. Both these descriptions lead to the identical pointer to the antidamping torque that can cause magnetization to switch. This torque exists irrespective of whether $\mathbf{S}_F$ is aligned with the anisotropy field and whether the damping torque exists. Without proof, this torque has the following magnitude:

When $\mathbf{S}_F$ and the anisotropy field are aligned, the damping torque vanishes, although the antidamping torque does not.

$$\mathbf{T}_{nd} = \frac{g(\theta)\hbar}{2eS_{inj}S_F^2}\mathbf{S}_F \times (\mathbf{S}_{inj} \times \mathbf{S}_F), \quad \text{where } g(\theta) \geq 1. \qquad (4.146)$$

See See J. C. Slonczewski, "Current-driven excitation of magnetic multilayers," Journal of Magnetism and Magnetic Materials, **159**, L1–L7 (1996) for the approximations behind this derivation.

Here, $g(\theta)$ is a function of I—the current. Even if the spin densities are aligned, a small fluctuation will cause this torque to come about.

As the spin of the injected electrons changes phase, this torque is effective over a certain characteristic length scale. We end this description by emphasizing two points about these torques: the injected exchange torque ($\mathbf{T}_{ix}$) is what leads to the precession of $\mathbf{S}_{inj}$ about $\mathbf{S}_F$, and the antidamping torque ($\mathbf{T}_{nd}$) is the one that arises from the scattering of minority electrons from s to d states.

The change in the orientation of magnetization, let us call it the angle φ, caused by this spin-polarized injected charge, can be characterized as

$$\frac{\partial \varphi}{\partial t} = -\frac{|\mathbf{S}_{inj}|}{|\mathbf{S}_F|} = \frac{n_{inj}}{N_a}\omega \approx \frac{J}{eN_a}\overline{\varphi}P, \qquad (4.147)$$

where N_a is the atomic spin density and where we have employed Equation 4.4.2, which captures the injected spin density in terms of polarization.

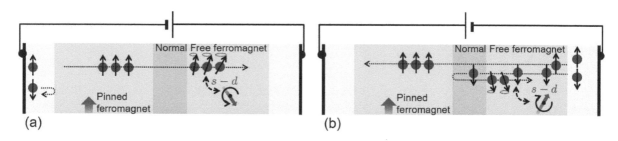

Figure 4.107: Switching of magnetization direction by spin torque. (a) shows the case when a second ferromagnet layer is anti-aligned to the pinned ferromagnet, and current flow with negatively charged electrons moving in the other direction are employed to rotate it into alignment. (b) shows how an aligned ferromagnet is anti-aligned by passing current in the opposite direction. Aligned ferromagnets represent the low resistance state, and anti-aligned ferromagnets represent the high resistance state.

Figure 4.107 summarizes the main internal events that occur in a ferromagnet/tunnel barrier/ferromagnet structure so that non-volatile resistance changes can be achieved. If the free ferromagnet is anti-aligned to the pinned ferromagnet, injection of the electrons from the pinned magnet, with the current flowing in the opposite direction causes the injected spin to apply a torque on the resident spin, that is, its resultant magnetization, as shown in Figure 4.107(a). So, the magnetic moment of the free ferromagnet responds by rotating, with an equal and opposite torque on the two interacting entities. If the ferromagnet is in alignment, the injection of these electrons does not disturb it. The pinned ferromagnet here serves to polarize the injected stream, and it is a relatively efficient process, since the minority spin carriers predominate. The majority spin electrons, as we have discussed, are reflected at entry due to a lack of available states—too few available states to transport through. The reverse of this process, shown in Figure 4.107(b), where the free ferromagnet is being torqued into alignment with the pinned ferromagnet, employs current flow in the opposite direction. The majority spin electrons of the pinned ferromagnet are reflected back and this spin orientation can interact while flowing in both directions.

In order for the magnetization flipping to be possible, the injection exchange torque must exceed the damping torque. This establishes at least one condition for the threshold in switching current. The large angles that this switching must enforce for switching of the magnetization of the material—the angle θ' shown in Figure 4.106 and which under normal circumstances of magnetization is very small—require significantly higher torque. This is because of the nonlinearity that arises due to the increasing coupling that occurs with spin waves or magnons—the excitation modes of spin within the ferromagnet.

4.4.3   Magnetic and spin-torque memories

THIS SPIN-INJECTION VALVE STRUCTURE, where non-volatile resistance changes arising from magnetoresistance can be programmed and read, is the basis for memories employing magnetism. Exchange coupling by employing an antiferromagnet to pin a ferromagnet layer and leaving another ferromagnet layer separated by a tunneling layer forms the resistance element. The free ferromagnet's orientation can be changed either through an applied field or by passing current. The former leads to magnetic memory, and if formed for random access, magnetic random access memory ($mRAM$). The latter leads to spin torque memory, and if formed for random access, spin torque random access memory ($stRAM$). Both can be read using a small current flow. Figure 4.108 shows these two types of memory in their random access memory form.

The appeal of these memory forms is that they are non-volatile and have 10s of ns write and read time, a time scale hard to achieve in a non-volatile structure—flash memories take considerably longer to write. Read process in all such structures, if through transistors, is relatively comparable in time scale—dominated by capacitances and dependent on the scale of integration and minimum linewidths of the memory bank. The principal limitation of these memories arises from the currents needed for the magnetic field or the spin torque, and the control and spatial needs. Nevertheless, non-volatility at high speed is a very desirable attribute, and magnetism-based memories are of engineering interest. The magnetoresistance element being small, these structures are usually single domain structures.

In $mRAM$ (Figure 4.108(a)), the free ferromagnet layer is switched by using a field created by passing current in the bit line (*BL*), and another line which is employed for writing and which we will call the field line (*FL*). At least in principle, achieving fields of the magnitude of an *Oersted* is needed, through passing near *m*As of total current—the sum of currents—through the wires. Reversal of polarization requires reversing the polarity of this current flow.

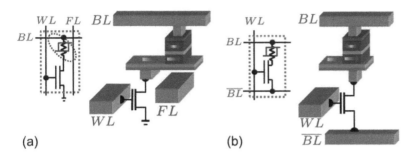

(a)   (b)

These fields have decay length scales following the Biot-Savart law. In addition, the polarization change needs to take place through the hard axis, so disturb issues, that is, effects on nearby bits are important. We will discuss this later, but, for the present, a simple description will do.

Figure 4.108: Two random access memory structures employing magnetic phenomena. (a) shows a magnetic random access memory (*mRAM*) cell, where the bit line (*BL*) and the field line (*FL*) together create a field that changes the state of polarization of a spin valve structure whose magnetoresistance changes. It is accessed via word line (*WL*), *BL* and ground connections through a transistor. (b) shows a spin-torque random access memory (*stRAM*) cell, where polarized current is employed to switch the resistance state between the *BL* and the $\overline{BL}$. The read and write operations for the cell are randomly accessed through the use of the access transistor switched by the *WL* and the other lines. The ellipse in (a) shows how the field loops from the currents in the *BL* and the *FL*. This looping is important in *mRAM*. It reinforces the field for the cell to be programmed. The circuit formulation for *stRAM* is similar to that for the 1T1C ferroelectric random access memory, except it uses a programmable resistor instead of a capacitor, and the resistor is placed on the drain side rather than the source side.

A simple description of the *mRAM*'s writing dynamics is summarized in Figure 4.109, which shows how reversal of the in-plane magnetic moment can possibly be achieved. In the initial state, the magnetic moment is closely aligned to the anisotropic easy axis of the structure, as shown in Figure 4.109(a). When a field is applied to move the magnetic moment to the hard direction, a pulse of duration *T* causes precession of the moment, as shown in Figure 4.109(b), by an angle of $\gamma = \omega T$, where ω is the Larmor precession frequency. Since an out-of-plane magnetization and, therefore, a demagnetizing field appear, as discussed in relation to Figure 4.58, the moment also precesses around this field, as shown in Figure 4.109(c). The steps shown in panels (b) and (c) of this figure occur simultaneously. Finally, in time, with damping, the moment settles along the new anisotropic easy axis.

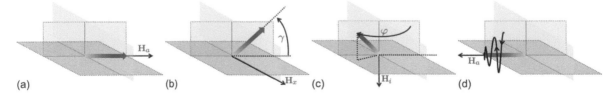

(a)   (b)   (c)   (d)

Figure 4.109: Switching of the magnetization of the free ferromagnet film towards the orientation of the pinned ferromagnet film by passage of current. The field here is constrained to being in the plane in steady state.

For reading the state of the resistance element, one employs the transistor and so uses the word line (*WL*), the bit line (*BL*) and the ground (*Gnd*) to measure changes to preset bit line signals that will be influenced by the high or low resistance of the *mRAM* resistance valve. So, the read process employs conditions that have very small fields, and small currents. But, the write process requires large-enough fields and large-enough currents so that sufficient energy is coupled to the free layer to switch orientation. The structure has four nodes that must be accessed (*BL*, *FL*, *WL* and *Gnd*), current passes in both directions in the *BL* and the *FL*, and the structure has, internally, several contact areas that require space.

In the *stRAM* (Figure 4.108(b)), the free ferromagnet layer is switched by the passage of the polarized current, as discussed for Figure 4.107. For both writing and reading, the programmable resistor element is accessed through the access transistor actuated by the word line (*WL*). The bit line (*BL*) and the bit line bar ($\overline{BL}$) are employed for writing using current of both polarities and for undisturbed reading at a smaller current. The pass transistor must be capable of these current magnitudes. The *stRAM* cell is in many ways similar to the 1*T*1*C* ferroelectric cell; the example shown is 1*T*1*R*, and the resistor is placed in a different arm of the transistor, although since the current is passed both ways, this is simply a reversal of the source and drain functions of the transistors—a change in the reference low bias voltage of the circuit.

Our discussion of the robustness of the bit storage in magnetic media, and energy in the superparamagnetic limit, dwelt quite a bit on anisotropy. The ferromagnet has both easy and hard axes for magnetization, and the preferred state is along the easy direction. Changing from one easy direction to the opposite easy direction requires passing through the hard axis or, as shown in Figure 4.98, through the hard axis ellipse. The energy barrier for change can be made to be 100s of $k_B T$ in the smallest of structures via shape anisotropy. This creates regions of stability in the presence of fields in *mRAM* structures, as shown in Figure 4.110, in different orientations; the magnetic moment realigns to the easy axis when the disturbance is removed.

Magnetic memories attempt to achieve robustness against disturbances during writing processes when the fields and currents are the largest, and in maintaining the stored information, by judicious design and by also keeping the anisotropic energy high and robust. In *mRAM*, the easy-hard axis behavior is fruitfully employed for avoiding nearest cell disturbances when currents are flowing through both the *BL* and the *FL*.

Figure 4.111 has one example of writing using phased currents through two lines to provide the field without causing errors by also writing any adjacent cells. The selected cell is identified as *s* in Figure 4.111(a), and one needs to worry most about the four nearest

In the discussion of ferroelectric memory, we dwelt in some detail on the sensing of signals and polarity, robustly and rapidly. All random access memories, including *mRAM* and *stRAM*, employ similar approaches tuned to specific manifestations of the state observables. Current sensing—flow-of-charge sensing, which is faster than voltage sensing, which needs a buildup of accumulated charge—is the preferred approach to reading for *mRAM* and *stRAM*.

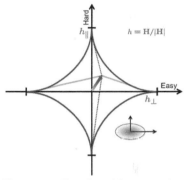

Figure 4.110: Rotation of the magnetic moment in an anisotropic ferromagnetic film. Changing the alignment of the moment from one easy direction to the other requires going through the hard axis. This implies that there is a region of stability either side of the hard axis: "astroid." Within an astroid, the magnetization returns to the easy axis upon removal of the disturbance. Flipping requires pulsing the $h_\perp$ and $h_\parallel$ sequences to rotate the magnetization past the hard axis.

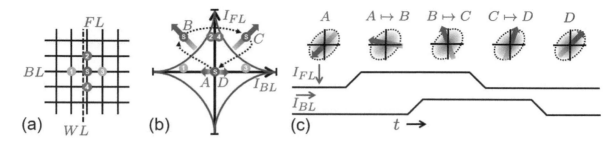

Figure 4.111: (a) shows a cell arrangement in which cell s is to be written to by passing currents in the bit line (BL) and the field line (FL), while cells 1, 2, 3 and 4 should not be disturbed. (b) shows the astroid, along with the cell's disturbance. (c) shows the timing of the changes in the currents, and hence the fields, that torque the polarization into the opposite direction in the selected cell s; I_{BL}, current in the bit line; I_{FL}, current in the field line; t, time; WL, word line.

cells, which are marked as 1, 2, 3 and 4. Let the selected cell be in the state identified as A in panels (b) and (c) of the figure. It is aligned along one of the easy axis directions, and we wish to polarize it in the opposite direction. This is represented in the timing diagram of Figure 4.111(c) as being along the easy long axis of the ellipsoid. If we turn on one of the lines used for writing, say the FL, it causes the field to act on the selected cell s and causes it to torque towards the hard axis direction as shown with the step identified as $A \mapsto B$, following the change in the current. In the steady state, the cell s now has a moment outside the "astroid." The presence of this current also disturbs the cells 2 and 3, and others along this line and in its close field proximity, but the memory's field and the anisotropy are such that these remain within the astroid. We stress cells 2 and 3 here because these cells, closest to s, will also see current that flows orthogonal to FL current and will see the largest field disturbance from it. So, if cells 2 and 3 are immune to disturbance, cells that are further away along the column will also be immune. Now, let us turn on the second current used for writing, here identified as the current in the BL line. The total field arising from the FL current and the BL current is now such that the magnetic moment has been torqued past the hard axis in the selected cell, a process identified as moving toward state C in Figure 4.111(b), and the timing region $B \mapsto C$ in Figure 4.111(c). Since the BL current is on, it also causes the cells along the bit to be disturbed more markedly. These are cells 1 and 3, and others along the row. When both these currents are on, the largest disturbance is to cells 1 through 4, which are the nearest neighbors of s. But, all these cells are designed to remain with the astroid for the programming current and field conditions. Cell s has been torqued past the hard axis. In our precession picture, the magnetic moment is now past the hard axis ellipsoid in the opposite hemisphere, the current in the FL is now turned off, that is, only the field due to the other write line due to the BL exists, and the magnetic moment is damped towards the opposite easy axis, that is, the $C \mapsto D$ change in alignment proceeds. When the BL current too is shut off, no external field exists, and, in time, the moment aligns, as shown for state D. By time phasing the two currents, it is possible to step the magnetic moment through the hard axis towards the opposite easy axis, and the choices of currents, anisotropy and the resulting fields is such that the disturbance of cells 1 through 4, which see the largest fields, is kept within the stability ``astroid''. The $mRAM$ programming to the opposite direction is quite similar: one needs to apply currents of the opposite polarity. One should also note that if a specific bit has to be written in an opposite polarity, one needs to read it first to assure the correct final state.

The field H for $I = 10$ mA of current, at the edge of a 100 nm-diameter wire, is $H = I/2\pi r = 10^5/\pi$ A/m. This field is ~ 3.2 kOe, or a magnetic induction of 40 mT. The precession occurs at approximately a GHz. So, as long as a few precession cycles are taken to flip magnetization, the switching response time can be in several ns and the reading time of arrays in 10–100 ns. This is approximately what $mRAM$s do. One can also check on the estimate of the field need. Cobalt has an anisotropy constant of 45×10^4 J/m^3. So if one assumes a ferromagnetic film of thickness 50 nm, and an area footprint of 30×30 nm^2 dimension, the energy is $\sim 2 \times 10^{-17}$ J, or ~ 4800 $k_B T$. This is still substantially larger than the $\sim 200 k_B T$ we calculated in Chapter 1, as that corresponds to a 17 nm thick and a 10×10 nm^2 footprint. But fields and currents still need to be sufficient for causing change in reasonable time. The non-scaling of current—limited by a current density of 10^6 A/cm^2, due to electromigration in wires—makes this a difficult practical and fundamental task.

The $stRAM$ cell is simpler than the $mRAM$ cell. The writing process involves bidirectional current flow, as we have already discussed, for programming the non-volatile magnetoresistance, and it can be accessed by using the word line to turn on the access transistor. The BL and the $\overline{BL}$ allow currents in both directions. Sensing is accomplished at small-enough currents to leave the state of the magnetoresistance spin valve unchanged. An $mRAM$ cell needs a high-enough sum current resulting in a high-enough local magnetic field to change the orientation of the magnetic moment past the hard axis. The spin torque requires a high-enough current in order to torque the free layer. Use of fields, the limiting issue for $mRAM$, is now avoided. We should be particularly interested in the currents and the timing dynamics to understand the power and speed characterization of $stRAM$.

Our starting point for this is Equation 4.147. Looking at Figure 4.112, one sees the three conditions of relevance, and the dynamics at work. Current density determines the spin polarized electron density that torques the magnetization. So, minimizing current requires minimizing the area of the structure. Let the critical current be I_c. If the current flow through the structure, I, is less than I_c, the magnetic moment returns towards the easy axis after the current is stopped. If $I \gtrsim I_c$, the moment changes slowly by precessing across the hard axis ellipsoid. If $I \gg I_c$, it does so rapidly. The precession frequency, as mentioned in our discussion of Figure 4.63, is $\omega = \gamma H = e\mu_0 H/m_0 = eB/m_0$. For a flux of 1 T, this corresponds to an electron spin precession time of ~ 35 ps. In frequency, this corresponds to 28 MHz/mT, or 2.8 GHz/kOe.

For the case when the magnetization is confined to a planar film with a dipolar shape anisotropy, that is, with a thickness smaller than the size of the device, the trajectory of magnetization is restricted to rotating in the plane, as shown in Figure 4.113. The writing dynamics builds up precession and then causes the switching past the plane, where the damping then helps with the reversal. Because of thermal energy, the initial starting point of the writing process may be at a distribution of angles θ_0. The write time (t_w) under these conditions is

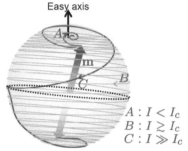

Figure 4.112: The response of the magnetic moment as it precesses under the influence of spin torque from injected current, at three different current conditions: when the injected current is below the critical current density, when it is approximately equal to the critical current density, and when it is significantly higher than the critical current density.

$$t_w \approx \frac{2}{\alpha \gamma M_s} \frac{I_c}{I - I_c} \ln \frac{\pi}{2\theta_0}, \qquad (4.148)$$

where I_c is the switching threshold. For fast switching, this means a reasonable current overdrive is required so that the ratio $I_c/(I - I_c)$ is much smaller than 1. The initial distribution of angles, due to thermal excitation, also means that there is a spread in the distribution of switching times. The spread in the initial distribution of angles also causes an increased distribution of the switching times because of the $(\alpha \gamma/|\mathbf{m}|)\mathbf{m} \times (\mathbf{m} \times \mathbf{H})$ double cross product term, which is related to the interaction between $\mathbf{S}_{inj}$ and $\mathbf{S}_F$, with $\mathbf{S}_F$ precessing around the anisotropy axis. A change in this initial condition means different initial torquing and different starting conditions for switching. This causes the random distribution in the switching time to be a function of the excess current drive. The closer the injection current is to the critical current, the larger is the spread in the distribution of the switching times. This behavior, although an issue in assuring bounds on switching time within energy constraints, also is potentially a means of generating noise for other uses.

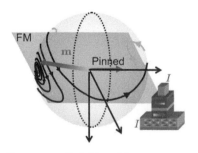

Figure 4.113: Switching of the magnetization of the free ferromagnet film towards the orientation of the pinned ferromagnet film by passage of current, with the magnetization confined to the film plane.

4.5   *Memories and storage from broken translational symmetry*

RESISTANCE CHANGES THAT ARE "NON-VOLATILE," similar to what
floating gate or floating charge storage memories accomplish, are also
possible through atomic, ionic, and electronic migrations. They can
have a broken translational symmetry, but they also rely on random
walk—suppressing or enhancing phenomena employable in data
storage. So, often the manifestations are a combination of the bistable
form and the random walk form, where the barrier in between, as
discussed in Chapter 1 and such as that shown in Figure 1.21, begins
to look more like the one shown in Figure 1.22, where random walk
at the spatial interface prevails. This is akin to what happens in a
dynamic random access memory using a capacitor and a transistor.
The random walk diffusion of electrons in the transistor controls the
time scale of stability. In optical storage, the change in property being
exploited occurs over a relatively large region, on the optical wavelength scale. In resistive memories, it occurs over a smaller length
scale—the dimensional and diffusional scale of atomic species, with
an interface between two phases that have translational differences,
or through phase changes over an entire region, such as in the metal–
insulator transitions discussed earlier. And these changes can be
brought about by energy provided through optical or electrical and
possibly magnetic means. The time scale of usefulness is controlled
by the length scale of the changes, and the dynamics.

Broken translational symmetry is important for memory through
the formation of these phases and their interface. The phases determine the conductivity, reflectivity or other properties that can be
measured and programmed. We encountered the metal–insulator
phase transition earlier. It arises from a diverse set of causes—the
Hubbard and the Peierls mechanisms being two prominent examples.
But, broken translational symmetry also occurs between crystalline
and amorphous forms. This crystalline–amorphous transition occurs in a variety of situations—in optical storage media or in phase
change memories, and to these examples can be added the metal–
insulator phase transitions that we have discussed. The movement
of atomic species can be particularly acute, with consequential optical property and electronic transport property changes in softer
materials such as chalcogenides. Local energy input brings about the
crystalline–amorphous transformations. Examples of devices using
broken translational symmetry include the various forms of optical
storage disks employing germanium antimony telluride ($Ge_xSb_{1-x}Te$,
shortened to GST), and others of this genre in various admixed forms
or other soft materials. Semiconductor lasers are used for localized

heating for crystalline–amorphous unidirectional or bidirectional transformation and for reading through optical reflectance in small, 100s of *nm* bit sizes. An electrical energy version of this is conversion by passing current. This is what happens in phase change memory. Property change through localized material change is also possible in transition metal oxides, where we have seen that properties transform tremendously because of orbital spread, the number of quantum states associated with the outermost orbits, and interactions with the local environment. Nickel, titanium, cobalt, hafnium, manganese, ruthenium, iron, copper, and others all form oxides in which one can form filaments that are conduction paths, by converting the insulating material in those regions into a conducting one; after repeated conversion back and forth one may only partially convert such regions by random walk. The random walk underlying this metastabilty may then be an important determinant of the stability of this valve that toggles between conduction and insulation. These are all examples where local translational symmetry breakdown has been employed to achieve property changes leading to random walk memory. The movement of species—ions and carriers—provides the kinetically slow conditions that allow ``non-volatility'' to be achieved on the time scale length of interest in devices.

This use of the consequences of translational symmetry breakdown through optical and electrical energy changes—in optical disks, in phase change memories, and in other electrochemical memories—is the focus of this section.

4.5.1 *Optical memories*

THE DIFFERENT FORMS OF OPTICAL DISKS used for music, video and data storage employ the property changes representative of the order parameters that can be quickly measured, written and erased.

Optical disks can be read only (*ROM*), recordable once (*r*) or rewritable (*rw*).

Read-only optical memory, such as that used in a read-only compact disk (which is called a *cdROM*), achieves binary discrimination through differences in reflectivity in a metal film, usually aluminum. Reflective areas—called lands—have aluminum present, and non-reflective areas—called pits—are areas from which aluminum has been removed. The data layer, which contains pits and lands, is complemented by another, thin, reflecting layer. The light reflected from a pit is phase shifted with respect to that from a land area. The consequent modulation of the intensity of the light as it is reflected from different areas in the optical disk is measured by the detector, which

Mott-Hubbard behavior is only one example of transformation of properties because of state changes as a result of energetic interactions in local environment. Some transformations lead to ferromagnetism, some to anti-ferromagnetism, and some to neither. Giant magnetoresistance in complex oxides such as $La_{2/3}Ca_{1/3}MnO_3$, and high-T_c superconductivity in cuprates ($La_{1-x}Sr_xCuO_4$, $YBa_2Cu_3O_7$ or $Bi_2Sr_2SaCu_2O_8$), are other examples of these property changes. Our focus in this section is on local change rather than on bulk change.

Optical disks arose at Philips in the 1970s from an an early project that explored optical storage employing lasers—it was for recording analog signal such as audio, but the storage itself was encoded and modulated binarily. The use of an optical pick-up head, the electronics and mechanics of the servo system, and the creation of disk masters became the basis of the compact disk (*CD*) systems that appeared half a decade later, from a Philips and Sony joint effort. Like the Winchester and the hard disks that followed from *IBM* efforts, optical disk technologies continue to evolve many decades later in both read-only and rewritable formats employing different media, coding and read-write wavelengths. Recordings in an optical disk are on tracks that form a continuous spiral. In contrast, a hard disk's tracks are in concentric circles.

itself has a complex design aimed at achieving resolution, speed and alignment.

Recordable-once disks employ or have employed a variety of mechanisms for recording data, including hole burning, bilayer alloying, surface and interface roughening, thin film agglomeration, phase change, and bleaching. Forms of compact disks and digital video disks (*DVD*s) have employed organic dyes spun onto the disk as part of the multiple layers that are used in recording and which include a dielectric layer and a metal layer. During recording, heating the dye bleaches it. The degraded dye, with its changed optical properties and changes in the interface, results in a phase shift and intensity modulation of the light reflected from the metal layer, with the dielectric layer also affecting the reflectivity. The differences in the optical contrast of the unbleached and the bleached areas, which correspond to lands and pits, respectively, allow the detector to sense the recording. In Blue-ray disks (*BD*), the recordable-once disks employ inorganic systems, for example, silicidation of *Cu-Si* through laser heating, resulting in changes in reflectivity.

Rewritable read-write disks employ phase change materials, harnessing the changes in reflectivity that arise from broken translational symmetry, and can be repeatedly altered and cycled.

In all these disks, the data, correction coded, that is, in a form where errors can be corrected, is encoded further with "reference" markers useful for various registrations needed for timestamps and other information for data recovery, before being written along the spiraling track of a disk. In summary, in all optical disks, the approach used for writing bits involves the use of marks and spaces for representing the data, a correction code, and techniques that allow sensing of the reflected signal at high speed. The differences in the optical contrast of marks and spaces result in the intensity of the reflected light being amplitude modulated, with the encoded patterns produced in this way having characteristics that are compatible with the optical channel being used to read or write while the disk spins at a constant speed. This means reducing transitions; hence, the coded patterns that contain the bit data can be long and short pits.

*cdROM*s are manufactured with the use of masters and so can be replicated cheaply and rapidly. In contrast, rewritable disks contain soft chalcogenides that can undergo repeated changes between two or more order parameters, with these changes arising from the broken translational symmetry triggered by optical energy. We will discuss *GST* as the classical prototype of the material used in rewritable disks.

The major part of the optical disk, as shown in Figure 4.114 for a dual-layer rewritable form, is the optical medium. It may be in one or

*cdROM*s, and vinyl records before them, are examples of the use of embossing, which is a technique that goes back to early paper printing, and before that, in the printing on cloth, metal and terracotta objects. *ROM* disks specifically employ injection molding and photo-polymerization. For the curious, a perfectly safe experiment is to place a waste *cdROM* inside a plastic sealed bag and then putting it through the electromagnetic energy interaction that takes place in a home microwave oven (2.45 *GHz*), for ten seconds. Watch the fireworks! These arise as the eddy losses cause the low temperature metal to heat up, melt, and then gasify, with charged and hot radiating aluminum nanoparticles making a spectacle. The bag here is to prevent the contaminants from making the microwave oven unhealthy and unusable for your fellow users.

GST is employed in a variety of compositions, as an admixture of germanium telluride (*GeTe*) and antimony telluride (*Sb$_2$Te$_3$*), together with other materials, or even by themselves. New materials, employing similar principles, continue to be synthesized and employed for archival storage. The use of shorter wavelengths, using smaller pits through near-field techniques, and writing and reading approaches that allow multidimensional storage are all examples of advances made in the pursuit of higher densities and speed.

more data layers on a disk-shaped substrate and capped with a cover. The laser then access the data layers through an optical assembly. This optical reading and writing focuses on the data layer and then defocuses away, for example, through transparent material. Some of this transparent material serves to protect the data layer, and some of it enhances reflection, to make the read-write process robust. High absorption, low refraction metals are employed on the side of the disk furthest from the laser. With dielectric layers enhancing interference, improved reading is achieved from the amorphous registration marks. They also enable the removal of heat and so are useful in the quenching process for amorphization. In structures with multiple data layers, semi-transparent films of conducting oxides and nitrides, such as In_xSn_yO, HfN, et cetera, between the data layers, provide cooling while allowing access to the data layers.

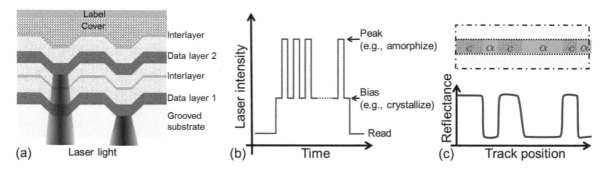

Figure 4.114: A schematic view of a read-write storage disk employing phase change. In the example in (a), two data layers are shown together with interlayers that provide thermal conduction with transparency. Data is written in the form of regions of modified optical properties, along the grooves of the substrate in the data layer film. A vertical cavity laser, a detector and multiple optical elements are employed for focusing energy, for reading and for registration. (b) shows power modulation in a film that amorphizes or crystallizes as a function of the optical power and duration. (c) shows, along a track, a reflected signal that needs to be detected. Crystallized films are polycrystalline.

The writing and reading process also requires careful registration and ``reference´´ marks that are employed by the photodetector assembly. The changes in optical properties arise from the use of two different power levels and time durations, causing the material to adopt two different phase states. Here, high power results in amorphization, and low power results in crystallization into a polycrystalline form, as shown in Figure 4.114(b), resulting in reflected signals of the form shown in Figure 4.114(c), along the track. In far-field optical disk recording, since diffraction is a primary constraint for size of writing, the wavelength of the laser is an important parameter. The dimension perpendicular to this size—the thickness of the medium through which the laser accesses this data—is a secondary constraint. Increasing density usually is reflected in the reduction in the transparent thickness through which the lasers access the optical medium of the disks. For some of the common disk formats, Table 4.11 shows some of the characteristics tied to the nanoscale dimension. Particularly notable is the dimension of the spot and the thickness of the layer through which the focusing needs to be achieved in far field at any given numerical aperture (NA) to increase disk capacity. From this, one can envision newer forms that will reduce spot dimensions by using near-field approaches.

	λ (nm)	Numerical aperture	Thickness (mm)	Track pitch (nm)	Channel bit length (nm)	Bit density (Gb/cm^2)	Spot size (nm)	Capacity (GB)	Transfer rate (Mb/s)
Compact disk (CD)	780	0.45/0.5	1.2	1600	277	~0.06	900	0.65	4.3
Digital video disk (DVD)	650	0.60/0.65	0.6	740	133	~0.43	550	4.7	11
Blue-ray disk (BD)	405	0.85	0.1	320	74.5	~2.28	238	25.0	36

Table 4.11: Some characteristic parameters employed in common disk formats. While the thickness of the transparent region accessed by the laser changes, with technology improving constantly for density, the disks are maintained at constant thickness for compatibility between generational change—the dummy substrate or cover accommodates this change. NA is the numerical aperture that defines the diffraction limit.

Because of its ability to undergo such changes in measurable optical and electrical properties, GST, which is stable in many compositions and in admixed forms with other compounds, has been used in optical disks and in phase change electrical memory. Figure 4.115 shows a simple pictorial description of differences, with respect to two properties, between the amorphous and the crystalline forms of GST. The former has higher symmetry than the latter. The changes take place at a reasonably low temperature, a little higher than $400\ K$, which is achievable with laser-based heating and cooling, as in digital disks, by passing current or via other local heating methods, as in phase change–based electronic memories. The rate of cooling determines the resulting state: a fast cooling quenches and causes amorphization to prevail, while a slow cooling allows crystallization to take place. The region of conversion of energy determines the localized heating and size that affects the properties. In disks, this size scale is determined by the extent of optical energy absorption. In electrical structures, it is determined by the region with highest electrical losses—losses that happen predominantly at interfaces. The nature of nucleation and growth determines the time scale that is characteristic of the writing times.

The changes arising from the interactions within the film, partic-

The compound $Ge_2Sb_2Te_5$ and its Ag- and In-doped compositions are among the favored materials, but there are many others. Our interest is in the underlying non-equilibrium thermodynamics, and we will largely use GST as an example that is relevant to the electrically changed material useful for memories.

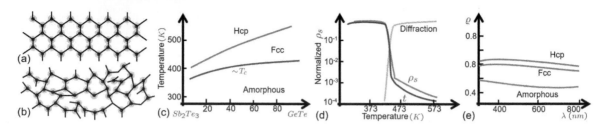

Figure 4.115: Germanium antimony telluride can be exchanged between the amorphous and the crystalline phases, with substantial changes in short-range order and with pronounced changes in properties, two of which are shown here. (a) and (b) show the phase translational symmetry of the crystalline phase and the amorphous phase, respectively. (c) shows the different phases formed as a thin film of $(GeTe)_x(Sb_2Te_3)_{1-x}$ undergoes transformation at low temperatures. *HCP* here stands for hexagonal close packed, and *FCC* stands for face-centered cubic, with T_c identifying where the amorphous-to-crystalline transition occurs under slow conditions. As the film is thin, interface conditions are important, even as the temperatures of transformation remain low. (d) shows the change in resistivity of the film during this transformation; the crystallinity of the film may be viewed through X-ray diffraction. Scales for layer thickness and diffraction intensity are not shown. (e) The reflectivity (ϱ) of $Ge_2Sb_2Te_5$, as a function of wavelengths in the visible range.

ularly those occurring during nucleation and during the growth of phases, are secondarily affected by conditions of deposition. Film porosity and additional minute impurities or residual elements—hydrogen and inert gases from the sputter deposition ambient—all affect the transition kinetics of changes between phases, even if dominated by the equilibrium phase diagram. Figure 4.115(c) shows that there is an intermediate, inter-penetrating, rock-salt-like, face-centered-cubic phase (*FCC*), as well as a hexagonal-close-packed (*HCP*) phase at higher temperatures, with transition temperatures increasing as the *GeTe* mole fraction increases. The property changes are substantial—nearly four orders of magnitude in the sheet resistance of the film, as shown in Figure 4.115(d), with the resistance being high in the amorphous phase and low in the crystalline phase. The change in the reflection coefficient, as shown in Figure 4.115(e), is also large. Amorphous films have higher real component of index of refraction than crystalline films do. This means that amorphous films reflect back less than crystalline films do.

Table 4.12 shows the average changes that occur at different compositions for the crystalline and the amorphous phases. Films of thicknesses ranging from 5 *nm* to 30 *nm* are used to achieve optimized contrast, recordability and stability. The change in reflectivity through the phase change is the basis for both recordable and rewritable disks, with the marking effects, which are in the form of different phases brought about by the laser and its properties, representing the state and hence the data. As shown in Figure 4.114, when a region storing a bit is amorphized by a large incident laser

Reflectivity, $\varrho = (n+1)/(n-1)$, where n is the real part of the refractive index. The imaginary part k corresponds to the absorption processes.

energy in a short pulse period, so by rapid heating and cooling, it forms one state. An intermediate optical intensity over a longer time period gives energy for crystallization through atomic motion, without rapid quenching, leading to the crystalline and complementary binary state. The read operation of the state of such a programmed element employs low energy which senses the reflected signal to determine one of the two possible states. This makes rewritable storage possible—high energy heating and rapid cooling for amorphization, lower energy heating and slow cooling for crystallization, and minimal energy for the read operation. Writing creates amorphized low-reflectance regions. Erasing employs crystallization.

	Structure	Crystallization temperature (K)	Lattice constant (nm)	n	k	ρ ($\Omega \cdot cm$)	U (eV)
GeTe	FCC		0.280, 0.313	6.1	0.2		3
	α		0.260	3.6	1.5		
$Ge_1Sb_2Te_4$	FCC	431	0.288	3.6	4.2		1.82
	α		0.264				
	HCP						
	α			4.1	2.1		
$Ge_2Sb_2Te_5$	FCC	442	0.283, 0.315	4.4	4.3		2.23
	HCP		0.422, 1.717	3.9	4.3		3.4–3.8
	α		0.315	4.1–4.5	2.3–2.1		
$Ge_4Sb_1Te_5$	FCC			3.2	3.9		
	α			3.8	1.3		
$Ge_1Sb_4Te_7$		423				1.52	
$Ge_1Sb_2Te_4$		431	0.288, 0.264				
$Ag_1Sb_1Te_2$	FCC			2.7	3.5		
	α			3.2	2.8		
$Au_1Sb_1Te_2$	FCC			3.1	4.2		
	α			3.3	3.2		

Table 4.12: The average Ge-Te bond lengths for the crystalline (c) and the amorphous states (α) of phase change materials with useful optical contrast (Ge-Sb-Te chalcogenide admixtures), the real and the imaginary parts of their refractive indexes at 650 mm, and the activation energies for the transitions. FCC here denotes the rock salt form—two interpenetrating FCC crystals in a checkerboard form, unlike the zinc blend form, which has tetrahedral coordination. Where appropriate, different lengths are given for the different axes, for example, for hexagonal close-packed (HCP) forms.

In both the optical and the electronic embodiment of phase change, the extent of the change, the energy needed, its kinetics—nucleation and growth—are the important physical effects that we should understand. For this, one needs to understand the non-equilibrium processes at the interfaces and of growth, and the process of crystallization. The crystallization temperature is often close to the glass temperature (T_g). The glass transition temperature is lower than the melting temperature (T_m) and is a function of the rate of heating or cooling. A slower cooling allows the glass transition to occur at lower temperatures, as the system has time to reach the metastability of the second order transition. Slow cooling then increases the crystallization probability. If the lowering of temperature is rapid, then the reduction in cooling time results in metastability being lost, and the amorphous state results. The structural relaxation of this state is very slow. This is the reason we have called this memory form a random

Recall the discussion of phase transitions. The melting temperature T_m is a symmetry breakdown condition the volume changes suddenly. However, as a second order transition, the liquid phase can be undercooled. It becomes more viscous as temperature is lowered. Ultimately, a solid forms—the temperature at which this occurs is the glass temperature, at which solid properties are acquired by the undercooled liquid.

walk memory. Figure 4.116 shows this temperature and time behavior for writing and erasing in phase change materials. Amorphization employs melting and quick quenching, while crystallization requires raising the temperature to above the glass temperature more gradually in time, with a slow decay, so that polycrystallization can take place through nucleation and growth.

Table 4.12 can now be understood as containing only a subset of properties for useful phase change materials. Writability requires glassy properties, as it relates to melting, together with good optical absorption in the amorphized state. Long-term storage requires a stable amorphous phase, so a high activation energy and a high crystallization temperature. Readability requires contrast, which Table 4.12 addresses. Erasability requires rapid recrystallization. This, in turn, means that materials with simpler crystal structures with less atomic movements and interactions, and low viscosity, will be more effective. Cyclability requires stability of the entire assembly and therefore low stress and the conflicting need for low melting temperature. We now look at the processes that lead to these properties. Central to this is the process by which crystallization takes places when the phase change material undergoes time-prescribed temperature changes.

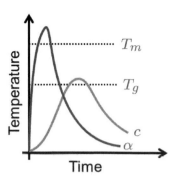

Figure 4.116: Temperature-time changes, high and fast, medium and slow, and their peak temperature positions characterizing the amorphization (α; writing) and crystallizing (c; erasure) of a phase change material. T_m and T_g are melting temperature and glass temperature, respectively.

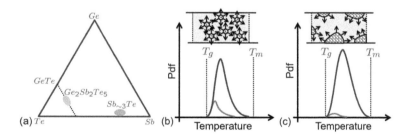

Figure 4.117: A schematic description of the nucleation- or growth-dominated dynamics of the crystallization of amorphous marks in the tracks of phase change material–utilizing optical disks. (a) shows the Ge-Sb-Te phase triangle with two compositions: $Ge_2Sb_2Te_5$, in which crystallization occurs via a nucleation-dominated mechanism, and: $Sb_{\sim 3}Te$, in which crystallization occurs via a growth dominated mechanism. The nucleation and growth curves for these two mechanisms are shown in (b) and (c), respectively, as probability densities. In (b), which shows a $Ge_2Sb_2Te_5$-like process, nucleation of sites for crystalline growth is the rate-limiting step. In (c), which shows an $Sb_{\sim 3}Te$-like process, growth of the boundaries is the rate-limiting step; Pdf; probability density function, T_g, glass temperature, T_m, melting temperature.

Crystallization of an amorphous or glassy region can occur from nucleation within the bulk, in which case several nucleation sites may initiate the process, or at sites at the surface or interface. The former is a homogeneous process starting from a compositionally homogeneous material. It is nucleation dominated. Growth occurs in the interior of a parent phase—amorphous, molten or glassy. If it occurs from nucleation at heterogeneous surfaces, it is a heterogeneous process. Different compositions of the material may show these differing mechanisms, as shown in Figure 4.117 for the amorphous-to-crystalline conversion for the GST example. Nucleation-dominated growth is limited by the number of nucleation sites that form, which then grow over a limited distance to create the polycrystalline structure. The size of the crystallites is small, and their density is high, because of the large number of nucleation sites they arise from. the growth-dominated process arises from nucleation at heterogeneous interfaces, for example, the channels, or the interface with the poly-crystalline regions of the layer in which the amorphous region may reside. With a low nucleation rate, the growth is limited by the movement of the crystallization front.

4.5.2   Homogeneous nucleation and growth

LET US CONSIDER THE HOMOGENEOUS PROCESS first. The formation of the crystalline regions thermodynamically can be viewed as a result of statistical atomic movement resulting in the crystallites. A rudimentary model of the formation of a crystallite would relate the distribution of sizes to the free energies through a Maxwell-Boltzmann relationship. This energy will change with size, since it will have a bulk contribution and an interface contribution. The size-dependent equilibrium distribution of crystallites of a dimension r can be written as

$$\langle N(r) \rangle = \overline{N} \exp\left(-\frac{\Delta U_{xlr}(r)}{k_B T}\right), \qquad (4.149)$$

where $\Delta U_{xlr}(r)$ is the energy barrier of cluster formation, being the sum of the bulk and surface contributions:

$$\Delta U_{xlr}(r) = -\Delta U_{lc,\Omega}\Omega(r) + \varsigma_S S(r). \qquad (4.150)$$

Here, $\Delta U_{lc,\Omega}(r)$ is the Gibbs free energy density difference between the two phases, where Ω is the volume, ς_S is the interface free energy density, and $S(r)$ is the surface area. At T_m—the melting temperature—the energy barrier to cluster formation is zero; below it, it is positive, making the clustering process favorable. Undercooling causes it to increase. As crystallization proceeds, the cluster free energy density difference changes because the volume contribution changes with the third power of length scale, and area as the second power. Consider a spherical cluster whose behavior is schematically shown in Figure 4.118. Volume dependence ($\Omega = 4\pi r^3/4$) and area dependence ($4\pi r^2$) imply that a maximum in the cluster difference free energy density occurs at $\partial \Delta U_{xlr}(r)/\partial r = 0$, that is, at

This model is a simple description to capture the zeroth order essentials of the process. Since these are alloys, strictly, in equilibrium, one needs to interpret these changes in terms of the liquidus temperatures that one sees in the phase diagrams.

$$r_c = \frac{2\varsigma}{\Delta U_{lc,\Omega}}, \quad \text{with}$$

$$\Delta U_c = \frac{16\pi\varsigma^3}{3\Delta U_{lc,\Omega}^2}. \qquad (4.151)$$

There exist a critical radius $r_c(T)$ and a critical difference free energy density $\Delta U_c(T)$, which is the amount of energy expended in forming a crystallite of size r_c. If $r < r_c$, the crystalline clusters—assumed spherical—will spontaneously disappear in equilibrium, because their existence is energetically unfavorable. If $r > r_c$, increasing crystallite size is energetically favorable, and they will expand. It is the existence of this activation barrier that prevents undercooling, that is $T < T_m$, from causing immediate crystallization. Nucleation in this simple description causes the formation of clusters with $r > r_c$ and which then can proceed to grow.

This equilibrium description can now be improved to include kinetics, that is, the evolution of the nucleation and growth process. We again approach this in a simple way to get physical insight. The cluster distribution for $r < r_c$ is slated for disappearance. The cluster distribution at $r = r_c$, if it acquires one more atom, is beyond the barrier and grows. It becomes a nucleation site and leads to kinetic growth. We therefore employ $\langle N_c(r_c) \rangle$, which is the expected number of clusters at a given radius, to determine the nucleation rate $\langle I \rangle$ as

$$\langle I \rangle = k S_c \overline{N} \exp\left(-\frac{\Delta U_c(r_c)}{k_B T}\right), \qquad (4.152)$$

where k is the number of atoms arriving at the cluster surface per unit time, and S_c is the size of cluster surface. In an equilibrium model, clusters also have a probability of losing atoms, so clusters with $r > r_c$ can also become smaller with a certain probability, even if that probability is smaller than the probability of becoming larger. Indeed, if we argue that there is a first order correction resulting from increasing and decreasing being of equal probability at $r = r_c$, then, at $r = r_c$, one should start with a nucleation density of $N'(r_c) = \langle N_c(r_c) \rangle / 2$. Assuming a spherical cluster, the nucleation rate is

$$\langle I' \rangle = k S_c \overline{N} \frac{1}{n_c} \sqrt{\frac{\Delta U_c(r_c)}{3\pi k_B T}} \exp\left(-\frac{\Delta U_c(r_c)}{k_B T}\right), \qquad (4.153)$$

where n_c is the number of atoms in the critically sized cluster.

These relationships show features that are similar to those that we have observed in energetic processes—they contain an exponential term related to the constraining activation energy of barriers, and a prefactor related to the fluxes arising from the processes due to the energy in the constitutive elements that cause the changes in occupation. So, prefactors may differ somewhat, such as they did with the magnetic particle relationships and the switching probability relationships, but an exponential term is the major contributor to the magnitude, through the term's strong temperature dependence. In comparing Equation 4.152, which has an equilibrium approximation based on nucleating clusters at $r < r_c$, to Equation 4.153, which has a first order corrected size distribution, the additional term in Equation 4.153 is related to the weakly temperature-dependent prefactor

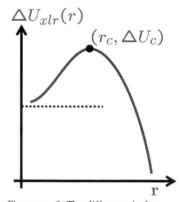

Figure 4.118: The difference in free energy of cluster formation, $\Delta U_{xlr}(r)$, that is, the reversible work energy at the liquid-solid transition at a temperature $T < T_m$.

We are implying through the use of the distribution $\langle N_c(r_c) \rangle$ that there is a cut-off at r_c—this is the Volmer-Weber model, which employs the equilibrium model as the starting point. The Becker-Döring model makes the cluster distribution disappear gradually beyond r_c. It is a steady-state model.

and the number of atoms in the cluster. For most of the materials of interest, this is one to two orders of magnitude, unlike the exponential.

How may one determine the arrival rate of atoms—k—through which we have prescribed the crystallization kinetics? If growth at the interface is dominated by diffusive hops between the solid-liquid interface, it is limited by interface diffusion. This is a short-range motion at the interface onto a suitable vacant site on the surface for occupation and subsequent clustering. It is not the long-range motion of constitutive atoms, with a gradient of their density, between the interface and somewhere farther away in the liquid—the classical definition of diffusion-dominated growth. The contrasting mechanism is when crystalline clustering occurs just because there is enough energy available thermally for the atomic arrangement to come about. This is limited by collision processes. The rates are

$$k = \frac{6D}{\lambda^2} \text{ for interface diffusion, and}$$

$$k = \frac{c}{\lambda} \text{ for collision-limited processes.} \qquad (4.154)$$

Here, λ is a scale length characterizing interatomic distance, D is the diffusivity of the amorphous or liquid phase from which the crystallites are formed, and c is the speed of sound, which characterizes the thermal motion.

Both these limits of crystallite growth have an activation energy where ΔU_c is a pronounced component, with a slightly larger modification for the interface-diffusion case. Interface diffusion is an energy barrier–limited process. It therefore often follows an Arrhenius activation energy relationship, and hence the rate I or I' is modified by a change in the activation energy, $\Delta U_c' = E_D + \Delta U_c$, where E_D is the diffusion activation energy. In growth from liquid, D is relatable to the liquid shear viscosity η as $D = k_B T / 3\pi\eta\lambda$. This viscosity-diffusivity relationship holds well for undercooled liquids.

We can now see why the glass-to-melting-point temperature difference is important in growth kinetics, and the critical role that viscosity plays in it, as shown in Figure 4.119. Close to the glass transition temperature T_g, the viscosity strongly increases as a result of the solidification from the liquid. $T > T_m$ is a stable equilibrium region where the liquid exists. The temperature region $T_m > T > T_g$ is a region of metastable equilibrium in an undercooled liquid. And the temperature region $T_g > T$ is one where the amorphous state persists as an unstable state—it is a non-equilibrium state that the material has been frozen in. If infinite time existed to achieve metastability, the viscosity would follow the equilibrium curve shown. The glass temperature, as discussed earlier, depends on the rate of cooling because

The liquid shear movement-diffusion relationship is often referred to as the Stokes-Einstein relationship. Its origin is the same as that of drift and diffusion in electron flow in semiconductors. They are both specific instances of the Einstein relationship characterizing Brownian motion, where a large particle undergoes fast molecular scattering interactions with the smaller particles surrounding it.

the dynamics is being changed, typically at viscosities of 10^{13} *poise*. The faster the cooling the process is, the higher is the glass temperature. Fast cooling, therefore, allows one to rapidly move into the frozen non-equilibrated amorphous state on the right hand side of the curve. The arrows in this figure show how the rates of cooling change the nature of the viscosity with temperature. In an interface-diffusion-limited crystallization process, the nucleation rate vanishes at the melting point T_m, since $\Delta U_c(T_m) \to \infty$, with the barrier disappearing. This follows from Equation 4.151. Therefore, one observes a rate dependence that has a strong maximum between T_g and T_m. In a collision-limited kinetic growth process, this is not observed.

4.5.3 *Heterogeneous nucleation and growth*

WALLS, IMPURITIES, AND VARIOUS OTHER DEFECTS AT INTERFACES are also sources of nucleation—an extrinsic process that is heterogeneous and subject to the specific conditions of the sample. If one assumes isotropy of phases, then a crystalline cluster interfaced with a liquid or amorphous phase is also isotropic, that is, the cluster grows spherically from a flat surface. This simplified model is shown in Figure 4.120, with a diagram of a cross-section of a crystallite growing at the interface of a wall with an amorphous or liquid phase after nucleation at the wall. The crystallite is in the shape of a truncated sphere of radius r. The positioning of the spherical curvature arises in the energy densities at the liquid-crystal (lc) interface, the liquid-wall (lw) interface and the crystal-wall (cw) interface.

If θ is the wetting angle, then the exposed fraction $f(\theta)$ of the sphere is

$$f(\theta) = \frac{1}{4}(2 + \cos\theta)(1 - \cos\theta)^2. \tag{4.155}$$

If $\varsigma_{cw} - \varsigma_{lw} < \varsigma_{lc}$, then heterogeneous nucleation is preferred to homogeneous nucleation. The reversible energy recovered in heterogeneous nucleation is related to that of homogeneous nucleation as follows:

$$\Delta U_c^{het} = f(\theta)\Delta U_c^{hom} \quad \therefore f(\theta) = \frac{16\pi\varsigma^3}{3\Delta U_{lc,\Omega}^2}, \tag{4.156}$$

with the critical radius of the crystallite remaining unchanged. Absence of wetting, that is, no heterogeneous nucleation, reduces this expression to that for homogeneous nucleation. The kinetics of growth now follows as for the homogeneous case.

We have looked at these kinetics since this nucleation is important for sizes and the energy-power-time characteristics for causing phase transition. This is important not only for energy introduced optically but also for energy introduced electrically, where, instead of being

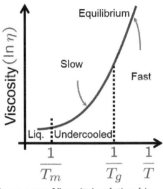

Figure 4.119: Viscosity's relationship with inverse temperature, for glasses; Liq., liquid.

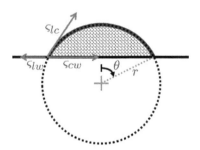

Figure 4.120: A cross-section of a nucleated and growing crystal at the interface between a wall and an amorphous or liquid phase. The growth happens in the presence of interface energies corresponding to the liquid/amorphous-crystal (lc), liquid/amorphous-wall (lw) and crystal/wall (cw) interfaces.

concerned with reflection, one is interested in direct current sensing for ascertaining the memory state through its resistance. We discuss this now along with other memory forms where resistance change in a two-terminal structure is employed.

4.5.4   Resistive memories

WHETHER DUE TO ELECTRON CORRELATION or due to changes in the state of material through other means—a change in composition through valence changes and ionic movement, oxidation-reduction or amorphous–crystalline transitions or others—any significant change that occurs in resistance is an effective means of implementing memories that are electrically controlled and thus potentially suitable in integrated circuits. Often multiple effects can be present simultaneously, and can be difficult to distinguish between them, particularly since interfaces with changing electronic properties—Schottky, or ohmic or recombining—and chemical properties—sources of oxidation or reduction—as well as defects in all their manifestations may all contribute. For example, in thin films, with an applied bias voltage and ion and electron flow, conductive filaments may form, leaving at the contact interface a very small region whose properties may become programmable. This region itself too may be filamentary. Even the soft breakdown of SiO_2 is an example of a percolative filament—one that may be modified by the bipolarity of annealing and consequent ionic movement. A catastrophic breakdown is a one-time programmable memory. These are useful too. A common form is a fuse on an integrated circuit; which when blown off, the fuse isolates off a section, such as a defective column, of a memory array.

We will collect these devices, independent of the causation, under the moniker of resistive memory. So, these include electron correlation structures, which are based on the Mott-Hubbard or other correlation-mediated forms of phase transition; amorphous–crystalline transition structures, where symmetry change occurs through local heating and cooling with a crystalline–amorphous transition in continuous, granular or partial form, with the latter two forms resulting in the creation of interface regions; structures based on interface effects; and structures based on the use of filaments, including forms where filaments are coupled to a region that may change its resistive behavior due to a variety of reasons not included in this list. In all these cases, the films where the resistance change occurs are thin, and the essential phenomena of programmable resistance may arise in an even thinner region adjacent to a contact or an interface. If such a resistance change is programmable, the traditional

Soft breakdown is discussed at length in S. Tiwari, "Device physics: Fundamentals of electronics and optoelectronics," Electroscience 2, Oxford University Press, ISBN 978-0-19-875984-3 (forthcoming). The model of breakdown include defect-mediated conducting channel formation, which is a consequence of percolation as defect density increases.

technique of transistor- or diode-based access becomes applicable, as shown in Figure 4.121.

In its simplest form, with only one polarity needed for the resistance programming, one of the electrodes—the PL or the BL—can be made a common ground, as in dRAM, where capacitor charge storage, instead of this resistance, is employed. If both polarity voltages are needed to achieve the programming, then this structure becomes the equivalent of the single transistor ferroelectric memory. Instead of a field-effect transistor, one may employ a bipolar transistor, which too provides isolation and individual access. These are all random access through the individual access, so we will call these resistive random access memory (rRAM). Or, one may employ a diode in series to achieve isolation between memory sites, and if the windows of operation for programming and reading are large, no isolation may be needed. In these last two cases, the assembly becomes a cross-bar lacking more or less in disturb immunity and local selectivity, and possibly amplification, but achieving higher density. What form one may use depends on the characteristics achievable. A single polarity operation is preferable, since it provides circuit and density simplification. A small current operation is useful, since it allows scaling of the field-effect transistor. But, most of all, the structures must have reliability as well as a reproducible, very narrow distribution of programmed states at low energy of operation, with enough speed for high densities to be possible.

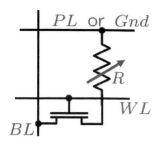

Figure 4.121: A programmable resistive element as a random access memory. If only one polarity of voltage is needed for functioning, one of the electrodes—the bit line (BL) or the plate line (PL)—may be kept as a common ground. The transistor can also be a bipolar transistor, a diode instead of the transistor can also provide access, or one may attempt to use this resistance programmability in a cross-bar without the random access.

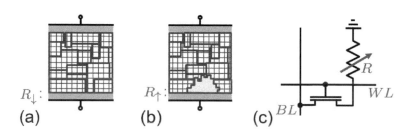

Figure 4.122: An amorphous–crystalline transition–based modulatable resistance memory—phase change random access memory (pcRAM). The presence of an amorphous region, as shown in (b), causes resistance to be high; in contrast, crystallinity; as shown in (a), causes the resistance to be low; BL, bit line; R, resistance; $R_\downarrow$, high resistance state; $R_\uparrow$, low resistance state; WL, write line (also, word line).

The amorphization–crystallization transition that we just discussed is an example that can be achieved through current-driven Joule heating. This is an example of thermochemical conversion. Since heating drives this transition at the electrode where the fields are the highest, as well as seeding sites, it is polarity insensitive. This makes the random access device shown in Figure 4.121 the simplest one—it is like a dRAM with one electrode at a common ground potential. Figure 4.122 shows this. Low temperature-melting chalcogenides, such as $Ge_2Sb_2Te_5$, and other mixed and stoichiometric forms that we encountered for optical memories are common materials for such phase change memories. Phase change random access memories (*pcRAM*s) are to us a subset of resistive random access memories (*rRAM*s).

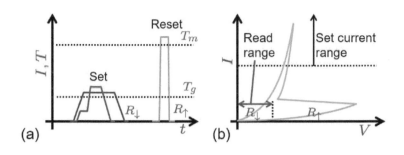

(a)

(b)

Figure 4.123: (a) shows the current pulsing for forming the low resistance $R_\downarrow$ crystalline clusters state ("set") and the high resistance amorphous $R_\uparrow$ state ("reset"). Here, an alternative to the single-pulse technique is shown for the slower crystallization cycle, where a low field thermal ordering of seed is followed by the crystallization for the "set" cycle. T_g approximately indicates a temperature necessary for crystallization, that is, near glass temperature, and T_m is the temperature necessary for melting the phase change material. (b) shows the current-voltage characteristics in the two states. Reading takes place in a low voltage range. There exists a threshold for snap-back for the amorphous high resistance state.

The behavior of these materials is similar electrically to that of the optical memories but with additional consequences arising from the electrode interactions, electron transport, heating and cooling dynamics, and the consequent kinetics. Figure 4.122 illustrates this by comparing the effects of insulating amorphous and conducting crystalline regions on resistance. The process of forming the crystalline, low resistance state $R_\downarrow$—a "set" state which has one or a collection of clusters—or the amorphous, high resistance state $R_\uparrow$—a "reset" state—occurs through current pulsing, as schematically shown in Figure 4.123. This random-accessed control of the resistive element can occur through a transistor—field effect or bipolar—or via a diode, both of which must be capable of supplying the necessary current. The larger current, that is, the one for the reset operation, is of the order of 100s of μAs. The input electrical energy, like the absorbed laser energy in optical recording, melts a localized volume in the highest field region, so at the electrode. Crystalline cluster formation and amorphization occur near the electrode, where defects and interface effects cause seeding. The cooling rate determines whether the material is quenched into an amorphous state, so by a rapid heating pulse of high current that is rapidly shut off, or into a crystalline state, so by a slow heating pulse of lower current and a slow cooling accomplished by turning off the current at a slower rate. This is essentially how, in Figure 4.123, the crystalline region with $R_\downarrow$, and the amorphous region with $R_\uparrow$, are achieved.

The crystallization speed has to be considerably slower than the amorphization speed, and this is an important determinant of the electronic speed of operation of the structure. A time constant of a *ns* implies a crystallization speed that is only a factor of 1000 slower than sound velocity! Also, attempts at increasing speed by limiting the crystallization region or its quality have consequences for the stability of material. Non-volatility, which is the attractive attribute of phase change memories also requires high stability. As with optical memories, nucleation—in this case, of small crystallites—and their growth need to be controlled to have control of the crystallization process. Growth-driven behavior is more uniform, as it progresses more evenly than nucleation dominated behavior, which has a spread of sites initiating the process. This is to say that when the final state is nucleation-dominated, the distribution of resistances has a larger tail. While, electrically, it is preferable that both of the ``set´´ and ``reset´´ steps be accomplished through a simple electrical pulse, it is also possible to use a small electric field with Joule heating to induce initial thermal ordering to seed fast nucleation growth via a higher energy electrical pulse. This approach would allow stability of the amorphous phase, as it controls the cluster size distribution and hence the variance of the states. This is the example of the more complex pulsing cycle shown in Figure 4.123.

Many of the comments made for the optical case also apply to the electrical situation, so we will not repeat them. Phase change memories are interesting electrical analogs of optical storage. But, one will have to worry about all the consequences of their random walk, which is beholden to numerous Poisson statistics consequences.

Two other resistance changes that we have already discussed are the metal–insulator change due to the Mott-Hubbard transition, and that due to the Peierls transition, in Section 4.3. Figure 4.124 illustrates conducting–non-conducting change across a structure and in an interface region. Materials that undergo these changes, with VO_2 being a prominent example, do so throughout, due to correlated electrons, structural distortion or both. Figure 4.49 and Figure 4.50 showed two examples of devices using a metal–insulator transition occurring throughout an extended region in VO_2. To be used in this form, the hysteresis window must come about under operating conditions at the temperature of operation and be bias modulatable—so this is not as straightforward as a $1T1R$ circuit approach. Figures 4.52, 4.53 and 4.54 illustrated what happens at the interface when barriers influence the occupation of the lower band of a Mott-Hubbard transition. This situation's characteristics will depend on the workfunction of the electrode. Achieving two states of conductivity in these structures may require either both polarities to be available or near $0\ V$ for one state, and a slightly higher bias voltage for the other state, as the figures illustrated for the interface-dominated conductivity.

But, there are many other ways that resistance may be changed,

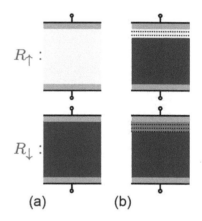

Figure 4.124: (a) shows a structure that undergoes a conducting–non-conducting transition, such as a metal–insulator transition, throughout the material—VO_2, for example—and (b) shows a structure with an interfacial region, where resistance change occurs through correlated electrons.

because many interactions may be enhanced through structural changes and through how external energetic interactions are applied to induce local changes. Changing the potential causes electrons and ions to move. The momentum and energy of the electrons may be exchanged with those of the crystal, breaking bonds and creating vacancies and interstitial ions, and these vacancies and ions can move. Interfaces change the energetics, so the behavior of electrons at interfaces would change and will be reflected in barrier heights, in recombination through interface traps, and even in the linearity of an interface with ohmic behavior. All this involves the electrons, the crystal interface, defects of various types, and atomic motion. These processes, whether barrier mediated or random walked, will be subject to thermal energetics. Electrochemical and thermochemical energy interactions and their combinations will affect response. A < 1.8 nm-thin SiO_2 gate oxide—a material that is normally a robust insulator in a transistor—is subject to soft breakdown—the formation of percolation paths between the gate and the channel when sufficient defect generation has occurred. Larger geometries have drift in operational characteristics due to the motion of hydrogen and boron—two easily moved species—at and near the interface. So, even a material with very strong bonds may lead to resistance change that can be modulated. Both these examples of SiO_2 can be annealed by temperature and even with the application of bipolar voltage pulses to enhance ionic movement. If these motional effects on conductivity are reproducible and reliable over time and have a narrow distribution with sufficient separation, then they too are potentially usable as programmable resistance for memories.

With all these changes in materials and electrodes possible, a fair collection of ways of affecting resistance changes exist, and we will discuss a few salient examples of these. But, before doing this, we illustrate a few of the current-voltage characteristics that are observed in the "set" ($R_\uparrow \mapsto R_\downarrow$) and "reset" ($R_\downarrow \mapsto R_\uparrow$) process that we encountered in Figure 4.123. Three examples of this process are shown in Figure 4.125.

Figure 4.125(a) shows an entirely symmetric transition in a symmetric material structure. The characteristics and the changes caused by the application of stimulus are symmetric in both directions. The earliest examples of this electrically programmable reversible resistance change were observed in low temperature semiconductor glasses such as As-Te-I and As-Te-Br and their mixtures with Cu—materials where T_g, the glass temperature, is below ~ 500 K and which are also chalcogenides. Figure 4.125(b) is a more definitive representation of the observations in a wider group of structures—not just semiconducting glasses—of recent times. One may change

Electromigration in circuit wiring often leads to creation of voids, which are seeded at various boundaries because the electron wind causes atomic movement.

See S. Tiwari, "Device physics: Fundamentals of electronics and optoelectronics," Electroscience 2, Oxford University Press, ISBN 978-0-19-875984-3 (forthcoming) for a discussion of soft breakdown and negative- and positive bias temperature instabilities—important limitations to the scaling of conventional silicon oxide field-effect transistors.

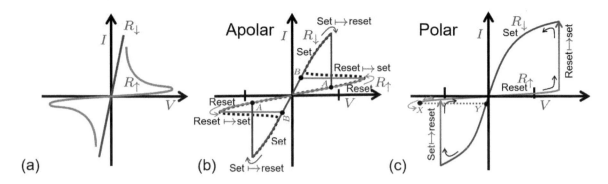

Figure 4.125: (a) shows an apolar (polarity-insensitive) transition between the $R_\uparrow$ and the $R_\downarrow$ states. This is an entirely symmetric effect in bias voltage in a symmetric structure. (b) shows another example of apolar switching behavior. Any asymmetry in voltage here may be caused by the nature of the interfaces and temperatures. It shows in more detail the entirely symmetric behavior of (a). (c) shows an example of polar transition. Here, each of the transitions happens for only one type of polarity. In each case, the read operation may be achieved by using a small current or bias voltage.

the high resistance ($R_\uparrow$—"reset") state to an ohmic-conducting low resistance ($R_\downarrow$—"set") state via a snap-back, with appropriate pulsing and timing of the electrical pulses. The behavior is quite similar to that observed in the other chalcogenides that we have discussed for use in memory elements. These electrical characteristics are polarity insensitive—they are agnostic w.r.t. polarity—and comprise what we will call apolar switching behavior. Here, the effect is the same irrespective of the polarity of the applied voltage. While in chalcogenides the transition occurs through amorphous–crystalline changes which arise from heating and cooling, due to thermoelectric conversion, and which give the transition its apolar switching characteristics, it doesn't have to be due to thermal causes, as we shall see. Figure 4.125(c) shows $R_\uparrow \mapsto R_\downarrow$ and $R_\downarrow \mapsto R_\uparrow$ changes occurring at opposite polarities—these transitions are polarity sensitive. This kind of switching, which the literature often refers to as unipolar switching, we will call polar switching. This polar form of transition can be observed in a variety of combinations of materials and electrodes.

Since we have already discussed the low temperature semiconductor glasses—chalcogenides—as apolar examples caused by temperature-time's independence from the polarity of the bias voltage, we will start with examples of polar switching and see that quite a few different physical processes can cause these electrically-driven transitions.

A well-understood example of polar switching is that of an electrolyte with electrochemically active electrodes made of, Ag or Cu, for example, both of which have relatively mobile species. The

The literature uses a variety of terms for polarity-insensitive set-reset transitions—bipolar being a common one. We will use the term "apolar switching" to specifically connote polarity-insensitive transitions. For the polarity-sensitive case, unipolar transition is a term that is commonly used. However, the term "bipolar" indicates the presence of polar opposites—from a specific electronic device based on the use of two different carrier types, to human personality frailties—rather than an insensitivity to polarity. "Apolar" is a clear term indicating agnosticism to polarity. The opposite of "apolar" is "polar," which means "dependent on polarity," a concept that "bipolar" does not capture.

electrolyte—which, in the absence of filaments, acts as an insulator between the electrodes—could be a metal oxide or even a chalcogenide. In such an assembly, ionic transport and redox reactions take place simultaneously, the end result of which is the formation of filaments, or their dissolution, as shown in Figure 4.126. The filament arises when the electric field resulting from the application of bias voltage causes ions (Ag^+ ions , or Cu^+ and Cu^{2+} ions) to move from the cathode to the anode; the beginning of this process is shown in Figure 4.126(a), where the device is still in a high resistance $R_\uparrow$ state. Figure 4.126(b) shows the completion of filament formation, which causes the $R_\downarrow$ state (``set'') to appear.

The polar transition that occurs during this process is shown in Figure 4.125(c). A $Pt/SiO_2/Ag$ assembly—consisting of an inert electrode and together with an active electrode, with a thin SiO_2 layer between them—will also give rise to a polar transition. In this case, the $R_\uparrow \mapsto R_\downarrow$ (``reset''$\mapsto$``set'') transition occurs with the Ag electrode as the cathode, and the Pt electrode as the anode, with the Ag^+ ions migrating and accumulating to extend a filament that then stretches from the anode to the cathode. The cathode is the active electrode in these polar transitions. A reversal of the bias potential reverses this. As well as Pt, Pt-Ir and W can be used for the inert electrode. However, if conducting materials other than these are used for the inert electrode, a polar transition may not necessarily be obtained, depending on the specific characteristics of the material used. Each of these three ``inert'' anode electrodes—Pt, Ir and W—are permeable to oxygen. So, the mechanisms underlying the polar transition in this resistive assembly are complex, as oxidation, reduction, and electrochemical and thermochemical forces each play a role. The reason the reduction-oxidation reaction is important in this case is that, because of it, there is the motion of ions, not just of electrons. Thus, the interelectrode medium is better described as an electrolyte, rather than as a conductor or an insulator. The situation shown in Figure 4.126(a), the removal of Ag^+ ions from the cathode, shows an oxidation reaction accompanied by the dissolution of the electrode—the removal of silver. The steps in the formation of the filament are charge transfer, the diffusion of Ag^+ ions, and the nucleation of the filament that then propagates. Any of these steps could be a rate-limiting step, depending on the electro-thermo-chemical properties of the assembly. Accumulation of ion charges causes chemical potential gradients. An electromotive force will also be generated by the Nernst potential. In addition, because of because of the nanometer-sized dimensions, and the formation of regions with high curvature, such as the filament, the free energy of the curved regions, relative to the flat regions, will be different. The electromotive force in the structure is thus an accu-

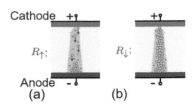

Figure 4.126: A resistive memory based on electrolytic processes—an electrolytic memory. (a) is a high resistance state ($R_\uparrow$) in which the presence of bias voltage causes the mobile electrical ions (Ag^+, or Cu^+ and Cu^{2+}) to move towards the anode, initiating filament formation. The low resistance state ($R_\downarrow$) happens when filament formation is complete, causing the resistance to drop ($R_\downarrow$). Application of the opposite electric field causes the filament to break and $R_\uparrow$ to be restored.

A reduction-oxidation, or redox reaction, is a chemical reaction where oxidation states changes. An electron must transfer between species for the reaction to happen. An oxidation reaction must be accompanied by a reduction reaction. $Ag \rightarrow Ag^+ + e^-$ is an oxidation reaction. A reverse of this reaction, or $Ag^+ + OH^- \rightarrow AgOH$ is a reduction reaction.

The Nernst potential is $\Delta U = k_B T \ln n_2/n_1$, the electrical potential between regions of two different ion densities (n_2 and n_1). It is simply a potential, following the Boltzmann distribution, that arises when a concentration gradient exists. The electric potential is higher in the region with a lower concentration, since the Boltzmann distribution has $\exp(-U/k_B T)$ dependence.

mulation of the chemical potentials arising in the concentration differences and the applied electric potential. The presence or absence of OH^- or other oxygen-based ions affects the electromotive force, since it changes the chemical potential in the system, just as the ability or inability to generate Ag^+ ions does. So, the Nernst potential, the diffusion potential, the Gibbs-Thomson potential and the applied electrostatic potential drive the creation of Ag^+ ions, their transport and their preferential accumulation on highly curved surfaces, thus leading to filament formation.

In general, in this electrochemically active system consisting of electrode/electrolyte/inert electrode, three mechanisms may be rate limiting for the switching behavior:

1. As just discussed, one mechanism is electrochemical redox, with an electrochemically active electrode metal and an interelectrode medium—the electrolyte—that is conducive to cation transport. In the example just discussed, during the $R_\uparrow \mapsto R_\downarrow$ transition, Ag^+ ions are generated because of a favorable Nernst potential. The ions transport because of the electromotive force and preferentially deposit on the filament due to the Gibbs-Thomson effect. A $Cu^{++}/SiO_2/Pt$ structure exhibits similar behavior. This electrochemical redox and ionic transport and deposition-based resistance transition is polarity sensitive. This transition is polar.

2. Valence change may occur as a rate-limiting step when reduction at the anode becomes rate limiting. The presence of OH^-, for example, improves this rate. Metal oxide electrolytes often have valence change as a rate-limiting step. An initial larger voltage step is necessary to make the structure switch. This step is an electroforming step. It is an electrochemical oxidation step that releases oxygen from the anode. Since anion transport, as in the previous mechanism, is electromotive mediated, this valence-change rate-limiting process too is polarity sensitive. Structures using TiO_2, Ta_2O_5 or $SrTiO_3$ as an electrolyte and with electrodes of Pt, Ti, or Ta, which are oxygen permeable, appear to have valence-controlled switching. This transition is polar, but it does not have to be filamentary.

3. If thermochemical redox dominates the electrochemical process, that is, if thermal energy drives stoichiometry change through redox and through diffusion and conductivity changes, then the switching behavior will be temperature sensitive because of thermal activation but will be insensitive to the polarity of the bias voltage. This transition will be apolar. Joule heating will be necessary for this. $Pt/NiO/Pt$ structures are examples of thermochemical switching devices, and the transitions in them are apolar.

The lower energy at edges, corners and tips is due to the Gibbs-Thomson effect. Increasing curvature affects vapor pressure and chemical potential. Single-crystal growth proceeds via edges because of the lower energy barrier there. The melting point and the freezing point are also lowered by the Gibbs-Thomson effect. It is also why beer bubbles form from the dissolved gases on the asperities of a mug, with frosted frozen mugs being particularly effective for this.

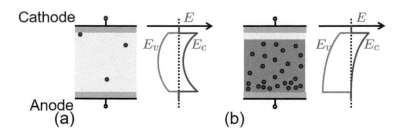

Cathode

Anode
(a)

(b)

Figure 4.127: (a) shows schematically a $Pt/TiO_2/Pt$ structure, together with a conduction and valence band diagram showing the barrier to conduction at the interface. When a preforming bias voltage—a voltage larger than that used in operation—is applied, oxygen vacancies—which are donor-like—are created, making TiO_2 n-type, and these vacancies drift to the anode under the forming bias. This diffusion reduces the barrier at the anode, as shown in (b).

In the second mechanism—the valence change—the necessity of electroforming indicates the necessity of oxygen or hydroxyl anion generation. Electroforming generates vacancies of oxygen in the electrolyte. In TiO_2 and $SrTiO_3$, oxygen bubbles are observed at the cathode, as well as oxygen vacancies at the anodes. This suggests O^{2-} migration to the cathode, and vacancy migration to the anode. So, the oxygen vacancies, denoted by V_O^{2+} in the redox equation

$$O_O \leftrightharpoons \frac{1}{2}O_2 + V_O^{2+} + 2e^-, \qquad (4.157)$$

representing the creation and annihilation of vacancies, drift towards the anode and accumulate. Figure 4.127 describes the consequences. Figure 4.127(a) shows a representative, for example, $Pt/TiO_2/Pt$ structure before preforming. There exist injection barriers for electrons, and TiO_2 may have some defects such as doubly positively charged oxygen vacancies. When a large voltage is applied to cause additional V_O^{2+} vacancies, TiO_2 becomes n-type. The application of this forming voltage also drives oxygen vacancies towards the anode—the lower potential electrode. Accumulation of the vacancies at the anode also freezes them, since the field is now lowered. The n-type region now extends further to the anode and the barrier height is lowered, so the anode contact may even become ohmic. These are structures where the transport at the interfaces—rectifying, ohmic conducting, or recombining—makes a difference in observed resistance, but with areal scaling. $SrTiO_3$, unlike TiO_2, is p-type, and, with electroforming scavenging the oxygen vacancies at the interface, the depletion region is now extended at the interface. Instead of being more conducting, the structure becomes more resistive. So, the behavior seen when $SrTiO_3$ is used as the electrolyte is the opposite of that seen when TiO_2 is used. In fairly oxygen-deficient TiO_2 films, the presence of a large number of vacancies results in the formation of filaments, which do not arise from the movement of ions between electrodes but instead through the local induction of phase change: TiO_2 orders into the conductive Ti_4O_7 phase when the concentration of oxygen vacancies exceeds a critical density.

NiO-based resistive structures are unique because of their apolar behavior. *NiO* is structurally quite stable, but its passage of high local current density, as well as the time constants of the switching, indicates that it can form current filaments. The formation of these current filaments depends on elevated temperatures that come about from Joule heating in the small regions through which the current flows. Thermal energy being the major cause of the accelerated molecular movement that leads to the formation and dissolution of the filaments, the state-change behavior of these resistance structures is apolar.

Finally, one should also stress that the presence of filaments with correlated electron effects—such as the Mott-Hubbard transition—may also happen in these structures, since many of these oxides are transition metal oxides where the correlated phase transition effects are known to take place. In these situations, defect movement and ionic movement now couple in a small interface region at the tip of the filament with correlated electron effects, as shown in Figure 4.128.

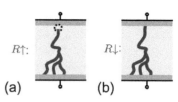

Figure 4.128: Combining filament formation with correlated electron effects as a possible means programmable resistance. (a) shows a high resistance state ($R_\uparrow$), where the filament does not completely span the region between the electrodes. (b) shows the low resistance state ($R_\downarrow$) that forms via the conductive transition of the small interface region at the tip of the filament by a correlated electron effect such as the Mott-Hubbard transition.

4.6   Summary

IN THIS CHAPTER, WE HAVE TRAVERSED a rather broad expanse of condensed matter phenomena and the devices that use them— nearly a book's worth. Multiple stable forms of phase transitions exist, and there also exist multiple electrical, optical and magnetic means to their manipulation. This makes their effective use in solid-state memories possible. How these stable states come about and the characteristics of these transitions are rich in detail.

Our argument started with establishing the origin of phase transitions— the appearance of new order and properties, which we called out in the order parameters. We reflected on how the symmetry of an action connects to conservation laws. Broken symmetry in translation, rotation, mirroring and other attributes leads to the appearance of order parameters. Spontaneous electric polarization is the result of broken inversion symmetry. If one inverts position ($\mathbf{r} \mapsto -\mathbf{r}$), polarization too inverts. Spontaneous magnetization—magnetic polarization—is the consequence of broken time-reversal symmetry. A moving charge under $t \rightarrow -t$ causes magnetization to be reversed. This appearance of an order parameter is a stable property—a new phase—that is potentially of use, since it is energetically favorable.

We explored this energy theme of thermodynamic phase transitions through a discussion of Gibbs free energy, as well as the other free energies under their specific constraints, through their thermodynamic foundations. During a phase transition, depending on the energy-constitutive relationship, one observes the change in order parameter taking many different forms. We refer to these as the 1st, 2nd, ..., nth orders of phase transition, to show that it is the nth derivative of Gibbs free energy that is discontinuous. Irrespective of this derivative change, a non-zero order parameter appears upon thermodynamic phase transition. The Landau mean field theory posits that the response of the assembly can be viewed through an average over the statistical causes. One includes the interaction represented by each contributing term as arising through a form where the rest of the ensemble is represented through an average field obtained over the entire ensemble. While this does not reasonably address the fluctuations—short- and long-range interactions are central to phase transitions—it suffices for many of the consequences that we wish to utilize in the stable region. We employed the Landau theory to explore the various features that can be seen when the Gibbs free energy is expanded in power law over the various energy-contributing interactions. Superheating, supercooling, hysteresis, critical points where multiple phases may be present simultaneously, et cetera, are all interesting observations that one can extract from the Landau formalism. Phase transition occurs with entropy change.

Multiple consequences of these phase transition phenomena are of interest to us: (a) the appearance of ferroelectricity, with its useful consequences for electronic memory; (b) correlated electron effects, which appear in unusual forms in various phase transitions and have useful consequences in memories and other devices; (c) spin and ferromagnetism, with their use in storage media and magnetic memories; (d) amorphous–crystalline transitions and their use in optical storage; and (e) resistance changes that can be electrically programmed through a collection of different effects, including those arising from phase transition, and their use in resistive memory.

Ferroelectricity is the result of the appearance of spontaneous polarization. Polarization charges as large as 25 $\mu C/cm^2$—nearly 16 times the highest electron concentration supportable by SiO_2 at breakdown—appear in materials such as $PbZr_{0.52}Ti_{0.48}O_3$, and coercive fields of $\sim 100 \ kV/cm$ are sufficient to switch them. The charge must be terminated by the electrode without there being much of a potential drop, so there cannot be an interfacial region between the ferroelectric and the conducting electrode. Otherwise, it will cause a significant depolarizing field, which will lead to degradation in time. Other degradation effects, such as fatigue manifested as decreasing remnant polarization with cycling, or imprinting, which shifts characteristics, too can happen. Ferroelectric memories attempt to overcome these effects through different designs to achieve robustness. The cell designs range from $1T1C$ which is similar to a $dRAM$, to multiple transistors with multiple capacitors. Our discussion here emphasized the variety of considerations that enter into designing high density circuits so that sufficient margin, high density, high speed and low energy characteristics can be achieved in large arrays. This was to illustrate that nanoscale's dimension-related manifestations lead to a challenge in design that must account for the ensemble characteristics of any effect that is to be utilized.

The nearly free electron situation that occurs when the density of the carriers interacting with a crystal is low can lead to a number of electron-ion interactions with interesting consequences. When energy needed to distort a crystal is exceeded by the energy available from the electrons, a new periodic potential can result. Under suitable conditions, this may lead to metal–insulator transitions. The symmetry broken here is that of the crystal, and the order parameter is the lattice deformation. Other, similar transitions also exist that are useful in devices.

Electron correlation has its origin in the collective interaction of electrons as quantum particles in a crystal with a periodic array of atoms. The Fermi gas and the Fermi liquid description of the electron ensemble shows the large changes that result from the electron–crystal interactions. In a liquid, single particle interaction—a quasi-particle coupling to a fermion sea—becomes weaker, by the Pauli exclusion principle. In the liquid, the interaction occurs locally and close to the surface of the filled states.

We discussed at length the Mott-Hubbard transition and charge transfer transition, both of which can be metal–insulator transitions, as examples of the energy competition that underlies the variety of phase transition phenomena in condensed matter. These are examples where the broken symmetry was not necessary. The phase transition arose because of quantum-mechanical consequences. From a device view, these are interesting for what they lead to at interfaces and for their use in traditional electronic devices such as memories and transistor-like elements. VO_2 was a particular example of interest to us.

Spin correlation leads to magnetism, which too is of interest in nanoscale devices. After exploring the underlying physical descriptions of spin and of magnetic moment, we discussed the origin of magnetism to understand the connection of spin correlation here. We employed the Ising toy model as well as statistical approaches, not only to explore magnetism but also to emphasize the nature of exchange interaction. These also gave us a description through which we could show correspondence to the earlier Gibbs energy–based phase transition description. Spin, interacting spins in crystal assemblies, and the majority and minority spin states in magnetic materials

And now one can appreciate the argument between Anderson and Peierls on ferromagnetism's phase transition origins.

have numerous effects. V_2O_3, for example, can be a ferromagnetic insulator or a metal with a transition temperature around 150 K.

Using this physical description, our discussion of magnetic storage first looked at magnetoresistance, which is the change in resistance in the presence of a magnetic field. This magnetoresistance effect—which is small in a thin film between two different orientation of the magnetic field—can be made larger in multi-film structures through suitable design where the resistance effect exploits the transport dependence on the alignment of spin vis-à-vis the magnetization of the film. Exchange bias too may be employed here to choose the magnetization of a ferromagnetic layer and another coupled layer that may be allowed to change direction. These effects on transport through spin-magnetization interactions led us to looking at structures where electron transport is across layers rather than along layers. We looked at a number of different "valve" structures where the ratio of resistance arising from two different polarizations of the field being detected can be made large. Again, exchange bias, with insulating films between the magnetic film structures, was employed to see how the read heads of a disk drive work in detecting the minuscule fields of disk drive bits.

Our discussion of the magnetization of thin films, which can be useful for storing data, looked at the roles of domains and domain wall motion, the energetics of these structures and, finally what happens in them when one makes them extremely small. This is the condition of superparamagnetic limit—a size scale where magnetism becomes unsustainable. Fluctuations of the programmed field polarization become large.

Spin polarization can also be incorporated in other materials—materials that may have long coherence lengths, large mean free paths or unique dimensionality properties. This discussion, in particular, looked at the behavior of ferromagnet/semiconductor junctions, where spin injection and polarization can be sustained over fair size scales.

Spin and magnetization at the nanoscale are also of use in non-volatile memories. We discussed two forms: magnetic random access memory, and spin torque random access memory. The first uses magnetic fields to cause the magnetization in a multimagnetic film stack to rotate and passes current through the stack to determine its state. The second employs spin polarized current flow to write data by polarizing the free magnetic film. In each of these cases, the structure is random access and uses a transistor as the isolation device. To understand the dynamics of these devices, we looked at torques and at the entirety of the electromagnetomechanical interaction that takes place when spin polarized current passes through a magnetized film, and various torques and damping effects come into play. Together, this led us to a preliminary understanding of the time scales, critical currents, and other considerations for stability in these non-volatile memories.

We have also explored broken translational symmetry—its extreme case of the amorphous–crystal transition—to see how it is employed in optical storage. This discussion largely restricted itself to chalcogenides—$Ge_xSb_{1-x}Te$ being a prominent example. This discussion had a larger materials science emphasis, where kinetics and nucleation, together with the optical and structural design of the disk, provide the ability to write small bits that can be easily optically read. The nature of nucleation and the growth process, the relevance of equilibrium phase diagrams, and the approach to the crystalline or the amorphous phase through temperature-time evolution were essential elements of this discussion.

Chalcogenides' amorphous–crystalline transition can also be extended to electrical structures, where the phase change can be assessed through a measurement of resistance. These are examples of resistive random access memory devices. They are temperature driven and exhibit apolar behavior. $rRAM$ can also be achieved through other means. Electrolytic cells are forms where mobile ions—Ag^+ or Cu^{2+} ions—move under bias through an insulating film, forming a filament bridge that results in lower resistance. This is an example of polar transition—a polarity-dependent transition. Reduction and oxidation occur at the two electrodes of the electrochemical cell. In general, even a simple structure such as this may have multiple mechanisms underlying the transition, depending on the electrode, the electrolyte, the nature of the defects and the other species available, given that redox is essential.

Transition metal oxides and perovskites, when combined with various types of electrodes, also show a programmable resistance capability. This transition behavior is apolar in some cases. In each of these, the cause of the programmable resistance depends on the specifics of the electrodes and the material, since the creation of vacancies, other defects or other charged species, as well as the nature of the doping of the insulating medium, matters. In addition, some materials form filaments, while others may have an areal effect where modulation of conduction behavior at the entirety of the interface matters.

4.7 Concluding remarks and bibliographic notes

PHASE TRANSITIONS AS STABLE STATES with specific property changes that can be harnessed in devices comprise a fair expanse of solid-state territory. We started with a discussion of the nature of the broken symmetry that brings most of these transitions to come about and then looked at their various manifestations that are useful in devices.

Symmetry, in a human view, is a harmony of proportions. The geometric appearance of this in translation, rotation, mirroring, et cetera, is reflected in the development of group theory in the early 20th century. Group theory became an essential tool in the discovery and practice of quantum mechanics. Hermann Weyl and Eugene Wigner brought these mathematical tools to elucidating quantum mechanics. Symmetry, in this route, shows up in nature and the physical world in multitudinous ways. A short monograph by Weyl[1] harmonizes the art and science beautifully. This 1952 writing explores the connections between paintings, such as Michelangelo's "Creation of Adam" from the Sistine Chapel, and architecture, for example, that of the Moors, and other Islamic architecture, to crystals and their properties.

The appearance of interesting properties that do not have the same apparent symmetry as the underlying laws is what we have called broken symmetry. A very incisive discussion of broken symmetry can be found in the Dirac memorial lecture given by Rudolf Peierls in 1992. Most of the arguments are in the detailed paper given here as reference[2]. Altmann[3] provides a beautiful introduction linking Ørsted and electromagnetics, Hamilton and quaternions, and Peierls instability of linear chains in linear molecules. A very fruitful understanding of symmetry in sciences can be gained from this short monograph.

Symmetry arguments are often central to many relationships that one derives. For example, the Poynting vector ($\mathcal{E} \times \mathbf{H}$) follows as a result of electrical-magnetic symmetry in energy. Symmetries and their breaking are patterns of nature that give it its richness—from particle and high energy physics, to molecules, to us and the environment in which we dwell. Noether, who placed the connection between symmetry and physical and dynamic effect on mathematical footing, is increasingly recognized for her incisive contributions in recent times. One of her earliest fans was Einstein. A recent book[4] is a particularly delightful reading, as it builds the arguments progressively in very clear chapters. In the old mathematics building at Göttingen, one can see her picture in the company of Gauss , Riemann, Klein, Hilbert, Weyl, Courant, Minkowski, and other mathematics greats, in the corridors. In the first half of the 20th century, the incredible interplay of physics and mathematics fostered at Göttingen—with participants spanning the alphabet from Born to Wigner—was the greatest collective intellectual contribution of human history. To sidestep the

[1] H. Weyl, "Symmetry," Princeton, ISBN-13 978-0691080451 (1952)

[2] R. Peierls, "Spontaneously broken symmetries," Journal of Physics A, **24**, 5271–5281 (1991).
[3] S. L. Altmann, "Icons and symmetries," Clarendon, ISBN 0-19-855599-7 (1992)

Multiplication of a vector by i is a rotation by $\pi/2$. What then would this look like for a complex number? It is a quaternion with three imaginary units.

[4] D. E. Neuenschwander, "Emmy Noether's wonderful theorem," Johns Hopkins, ISBN 978-0801896934 (2010)

discrimination against women, Noether taught courses listed under Hilbert's name. Her own research proceeded independently, without the discrimination practiced in teaching.

For an exposition, at an advanced graduate level, of broken symmetry and phase transitions, see Piers Coleman's book[5]. Khomskii tackles the connection between excitations and order in phase transition in a considerate way that can be understood by engineers[6]. An intuitive understanding from the Ising toy model, as well as from renormalization, is in Gitterman and Halpern[7]. A word of caution, though. Even though the title speaks to modern applications, it is intellectual flights in physics, and not engineering. The text is an excellent discussion of phase transitions and renormalization in a mathematically intuitive way. Another text, by Atland and Simons[8], is also very appealing for its discussion of excitations, among other topics.

Mean field theory—the Landau school's approach—and its application to phase transitions can be found in many of the solid-state physics texts. Nishimori's[9] book is interesting in how it connects the Ising discussion to the traditional mean field. It also is a good source for seeing the connection to information processing.

Ferroelectricity, in books, appears mostly as a classical study. From a device perspective, the writing seems to be largely reports. The book by Rabe et al.[10] does tackle the middle space between science and engineering well in some of the chapters. For those interested in this subject historically, the book by Jaynes, dated 1953[11], is an interesting read. Jaynes later went on to contribute immensely to the understanding of Bayesian approaches and the principle of maximum entropy that was discussed in Chapter 1.

Electron correlation, and its multifaceted appearance in condensed matter—as described by Peierls, Mott, Hubbard and Anderson—are treated quite intuitively and with rigorous mathematical support at an advanced level in the text by Gebhard[12].

A classical treatment of materials and their ferromagnetism and antiferromagnetism may be found in the text by Chikazumi [13]. This is a good introduction to ferromagnetism from a materials science perspective.

A more modern treatment, without the classical baggage, can be found in the books by Majlis[14] and White[15]. These excellent books make a good connection to the device-oriented implications of this topic.

An understanding of magnetic domains—their physical character and energetics—is quite important to any use of the phenomenon in devices. A long discussion of this topic can be found in the book by Hubert and Schäfer[16].

[5] P. Coleman, "Introduction to many-body physics," Cambridge, ISBN-13 978-052186886 (2015)

[6] D. I. Khomskii, "Basic aspects of the quantum theory of solids," Cambridge, ISBN 978-0-521-83521-3 (2010)

[7] M. Gitterman and V. Halpern, "Phase transitions: A brief account with modern applications," World Scientific, ISBN 981-238-903-2 (2004)

[8] A. Altland and B. D. Simons, "Condensed matter field theory," Cambridge, ISBN-13 978-0521769754 (2010)

[9] H. Nishimori, "Statistical physics of spin glasses and information processing: An introduction," Clarendon, ISBN 978-019-850-941-7 (2001)

[10] K. Rabe, Ch. H. Ahn, J.-M. Triscone, "Physics of ferroelectricity: A modern perspective," Springer, ISBN 978-3-540-34590-9 (2007)

[11] E. T. Jaynes, "Ferroelectricity," Princeton, ISBN 1114812307 (1953)

[12] F. Gebhard, "The Mott metal–insulator transition: Models and methods," Springer, ISBN-13 978-3540614814 (2000)

[13] S. Chikazumi, "Physics of ferromagnetism," Oxford, ISBN 0-19-851776-9 (1997)

[14] N. Majlis, "The quantum theory of magnetism," World Scientific, ISBN 981-02-4018-X (2000)

[15] R. M. White, "Quantum theory of magnetism," Springer, ISBN-10 3-540-65116-0 (2007)

[16] A. Hubert and R. Schäfer, "Magnetic domains: The analysis of magnetic microstructure," Springer, ISBN 978-3-540-64108-7 (1998)

For a more recent understanding of electrons (few or not-so-few) and their spin in bulk, in thin films, and transporting both in plane and out of plane, as well as their variety of superlative-qualified magnetoresistance effects, see the edited volume from Maekawa[17].

Spin waves and the collective excitations, sometimes referred to as magnonics, are to be found in two well-written books—those of Rössler[18] and Demokritov et al.[19] Spin dynamics—its stability, noise, and the more advanced subject of chaos and related properties such as bifurcation, which we will discuss in connection with electrome- chanics, are tackled by Hillebrands and Thiaville[20]. A very compre- hensive treatment of spin and its practical use in devices as well as characterization is the excellent text by Stöhr and Siegmann[21].

Optical energy–induced structural transitions, and their conse- quences, are of immense commercial import. However, there are very few books devoted to this subject. The non-equilibrium kinet- ics makes it more of a practical technological direction of pursuit. One book, edited by Raoux and Wuttig[22], does look at the crystalline changes and their consequences, with an emphasis on chalcogenides.

For the electrical modulation of chalcogenides, transition metal oxides, perovskites and others, and their use in resistive random ac- cess memories, the story is still quite incomplete. A review paper by Terao et al.[23] may be of interest to the reader, as it connects cor- related electron themes with resistive memory systems. For a more mainstream view of the subject and its complexity and dependence on materials, structures, environment and intentional and uninten- tional trace impurity incorporation, see Waser[24] and Magyari-Köpe et al.[25]. This subject is of considerable current interest. One would ex- pect that more comprehensive and recent review papers will continue to appear.

4.8   Exercises

1. Can ferroelectricity exist in a centrosymmetric system? Reason why in brief. Quasicrystals exist. These are crystals with periodic arrangements that are quasiperiodic, such as the two-dimensional form shown in Figure 4.129, have rotational symmetries that are forbidden and a structure that can be reduced to a finite num- ber of repeating units. Can quasicrystals be ferroelectric? Briefly explain.   **[M]**

2. Consider a system of two neutral atoms separated by a fixed dis- tance a, with each atom of polarizability α. Find the relation be- tween these two parameters for such a system to be ferroelectric.

[M]

[17] S. Maekawa (ed.), "Concepts in spin electronics," Oxford, ISBN-13 978-0198568216 (2006)

[18] U. Rössler, "Solid state theory: An introduction," Physica-Verlag, ISBN 978-3-540-92761-7 (2009)

[19] S. O. Demokritov and A. N. Slavin, "Mangnonics: From fundamentals to applications," Springer, ISBN 978-3-642-30246-6 (2013)

[20] B. Hillebrands and A. Thiaville, "Spin dynamics in confined magnetic structures, III," Springer, ISBN 978-3-540-20108-3 (2006)

[21] J. Stöhr and H. C. Siegmann, "Magnetism: From fundamental to nanoscale dynamics," Springer, ISBN-10 3-540-30282-4 (2006)

[22] S. Raoux and M. Wuttig (eds), "Phase change materials: Science and applica- tions," Springer, ISBN 978-0-387-84873-0 (2009)

[23] M. Terao, T. Morikawa and T. Ohta, "Electrical phase-change memory: fundamentals and state of art," Japan. Journal of Applied Physics, **48**, 080001- 14 (2009)

[24] R. Waser, "Resistive non-volatile memory devices," Microelectronic Engineering, **86**, 1925–1928 (2009)

[25] B. Magyari-Köpe, S. G. Park, H.-D. Lee and Y. Nishi, "First principles cal- culations of oxygen vacancy-ordering effects in resistance change memory materials incorporating binary tran- sition oxides," Journal of Materials Science, **47**, 7498–7514 (2012)

3. For a one-dimensional chain of atoms of polarizability α and separation a, show that the array can polarize spontaneously if $\alpha \geq a^3/4\sum n^{-3}$, summed over positive integers. **[M]**

4. A material is nonlinear for a specific property if its response does not satisfy superposition. For example, for the stimuli x and y, with the response $f(x)$ or $f(y)$, $f(x+y) \neq f(x) + f(y)$. Does a material that has undergone phase transition, ferroelectric or ferromagnetic, have a linear or a nonlinear response for small perturbation signals? **[S]**

5. In the chapter, we have a discussion of second order phase transition conditions, where it is the second derivative that undergoes a discrete change. In a first order phase transition, it is the first derivative that changes stepwise. In both cases, the free energy $\mathcal{G}$ is continuous. Use the Landau-Ginzburg equation to derive the order parameter at T_c, and find T_c in terms of the Landau-Ginzburg equation coefficients. **[M]**

6. We can write the Landau-Ginzburg equation for free energy density in terms of the of order parameter P in the form

$$\mathcal{G} = \frac{1}{2}aP^2 + \frac{1}{4}bP^4 + \frac{1}{6}cP^6 + \cdots,$$

where $a = a'(T - \theta_c)$ and is temperature dependent. Let us assume that the rest of the coefficients are constants. Since the order parameter for ferroelectricity, polarization, can only point in a symmetry direction, we are writing it as a scalar.

- With $b > 0$, and $c = 0$, determine the form for equilibrium $P(T)$.

- Now, incorporate into the free energy the energy arising from the polarization coupling to the electric field $\mathcal{E}$. From this, derive the dielectric susceptibility $\chi = \partial P/\partial \mathcal{E}$ above and below the critical temperature.

- Sketch $P(T)$, $\chi(T)$, and $\chi^{-1}(T)$.

- Take the case of $b < 0$, and $c > 0$. How do the behaviors of $P(T)$, $\chi(T)$, and $\chi^{-1}(T)$ change? **[A]**

7. Consider a piezoelectric crystal. The elastic strain is linearly coupled to polarization. $P = \alpha\varepsilon$, where ε is the strain, and α is a constant of the linear relationship. This material phase transitions to a ferroelectric phase on cooling. Write a simple scalar relationship (no tensor complexities) that should be added to the free energy to incorporate the related energy contribution. **[M]**

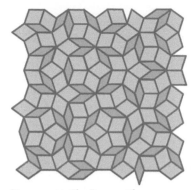

Figure 4.129: The Penrose tile as an example of a two-dimensional quasicrystal. Three-dimensional quasicrystals with any rotational symmetry are possible, and many have been seen— naturally and synthetically.

The story goes that Landau, whose Landau-Lifshitz book series are the bibles and geetas for those with physics tendencies, had a grading system for scientists based on a base 10 logarithmic scale. A class that was lower by 1 had achievements that were ten times lower than those in the class above. Einstein was in the $1/2$ class. Bohr, Schrödinger, Heisenberg, Dirac and Fermi were in the 1st class, that is, Einstein achieved about $\sqrt{10}$ times more than that group. After this mean field theory equation, Landau assigned himself to the 1.5 class. He was a hedgehog and not a fox.

8. Why are depolarization fields absent from a metal-clad ferro-electric structure but present when an insulating film is present between the metal and the ferroelectric? Is this statement strictly true or only a good approximation? Explain in brief using a band diagram. **[S]**

9. Do ferroelectrics always have to be insulating and ferromagnets conductive? **[S]**

10. A ferroelectric capacitor used in transistor-accessed memory has a remnant polarization of 20 $\mu C/cm^2$. What is the electron density that can be accessed? How much voltage will appear if all this charge was transferred onto a bit line of 0.5 pF capacitance from a ferroelectric capacitor that had an area of $500 \times 500\ nm^2$? **[S]**

11. Why is ferromagnetism not observed in copper even though it has electrons in the d orbital? **[S]**

12. Nickel—Ni^{28}: $1s^2\,2s^2 2p^6\,3s^2\,2p^6\,3d^8\,4s^2$, according to chemistry textbooks—is ferromagnetic. Nickel's electronic configuration is also written as $1s^2\,2s^2 2p^6\,3s^2\,2p^6\,3d^9\,4s^1$, according to quite a bit of research literature. Will both these configurations be magnetic? Reason in brief. **[S]**

The first orbital occupation description fits with the Madelung picture that chemists love. The $4s^2$ is occupied here after the $3d^8$. Calculations show that the second results in the lower energy states for the collection.

13. Consider a simple example of the Ising model. Let the average magnetization for a large collection N of spins $S_i = \pm 1$ follow

$$m = \frac{1}{N}\left|\sum_{i=1}^{N} S_i\right|.$$

An external magnetic field h breaks the symmetry of phase transition, causing two opposite polarizations of alignment. One form of the mean field equation at temperature T—recall our Equation 4.104—is

$$h = T \tanh^{-1}(m) - J\nu m,$$

where ν is related to the effective number of neighbors, and $J > 0$ the coupling strength that we have derived.

- Analyze the solutions of this equation to obtain the m^* values that satisfy it.

- If $h = 0$, find T_c, the critical temperature of phase transition. **[M]**

14. Ferromagnetism is observed in the second half of the 3rd transi-tion series but not in the first half. Explore and explain why. **[S]**

15. Should ferromagnetism have a depolarization issue? **[S]**

16. Let us calculate the amount of stored energy barrier (in magnetic form) and estimate the time constants for flipping between aligned and anti-aligned states in spin torque structures and in $mRAM$ structures. Let us choose a structure which is near the superparamagnetic limit (30×30 nm^2). We choose a 3 nm-thick cobalt film as the free layer, and an efficient, very thin tunneling barrier that intervenes between it and the fixed layer. The effective magnetic moment per atom is $1.72\mu_B$. Assume the fixed layer to be also cobalt, but $10\times$ thicker and separated by an infinitesimally small barrier. You can make any additional assumptions, with an explanation, to justify the calculation.

- Estimate the energy barrier between the two equilibrium states assuming single domain and entirely bulk-dominated behavior. Express your answer in $k_B T$s.

- What is the precession frequency?

- Assume that we manage to apply an effective field that 9s $10\times$ higher by passing current through the write lines. Estimate the time to flip.

- Estimate what current this corresponds to if similar conditions are recreated in a spin torque memory. **[A]**

17. If one wished for about a 10 ns response time from a magnetization flip, estimate the effective field that is needed. How much energy is then stored in a $10 \times 10 \times 10$ nm^3 volume? And how much current is needed through a 10×10 nm^2 cross-section wire nearby to cause flipping? **[M]**

18. What is the disturb process in $mRAM$, and how are the easy and hard axes of magnetization taken advantage of in reducing its consequences? Briefly, please. **[S]**

This question is non-trivial and assumes much. Do your best but state your approximations and justify them. The purpose of this question is to get an order of magnitude understanding of the parameters.

5
Electromechanics and its devices

COUPLING MECHANICAL FORCES WITH OTHER FORCES at the microscale and the nanoscale allows the creation of devices that can perform extremely precise measurements, as well as devices for everyday use. Examples where the first type of device is used include scanning probe techniques, where a vibrating probe with an electrical, magnetic or optical tip, or a tip with some combination of these properties, makes ultrasensitive measurements possible. Examples of the second type of device include inertial sensors for orientation sensing in mobile displays or for head protection in disk drives. Probe techniques let one get down to single electron, single spin, single photon and single phonon sensitivity, in addition to allowing extremely precise measurements of forces near uncertainty limits by using optical coupling. Inertial sensing allows one to perform chemical sensing near single molecule limits, or global positioning, due to precision frequency control. The nanoscale is a dimensional region where continuum mechanics and discrete mechanics intersect—continuum mechanics in the sense that one may look at any object's behavior as being describable by continuous functions in space, and discrete mechanics in the sense that discreteness at the atomic size scale and thus the position dependence of properties must be considered. An example to illustrate the importance of this discreteness is a carbon nanotube strung between two contacts, with an electric force applied through a gate separated from the tube. While, it may suffice to describe the tube itself through Young's modulus and other, similar macroscale properties—isotropic and anisotropic—at the clamping points, distortion, generation of defects, et cetera, will be dominated by behavior that must include atomic scale dynamics. A comprehensive description, while keeping the calculation manageable, will require a multiscale multiphysics where both atomic scale and continuum scale descriptions need to be incorporated and coupled through an adequate description of boundaries.

Inertial frames of reference undergo constant rectilinear motion with respect to one another. An accelerometer—an inertial sensor—can then detect acceleration. Galilean and Lorentz transformation allows one to convert measurements in one of these inertial reference frames to those of another: Galilean for Newtonian conditions, and Lorentz for relativistic conditions.

We will pursue the optical coupling aspects of mechanics in some detail in Chapter 6

Modeling the behavior of a rope hanging from two points is at many levels a difficult problem exemplary of the interweaving of complexity, simplicity and emergence. I have spent considerable free time trying to find the catenary equation for this system without using differential calculus. And even in this simple example, the continuum mechanics properties have been simplified to the bare minimum: Young's modulus $Y \to \infty$, and Poisson's ratio $\nu \to 0$.

This chapter discusses the continuous-to-discrete spectrum of the properties of electromechanic interactions, and their use in devices. Since many of the devices use energy coupling across forms, particularly the coupling of electrical and mechanical energy, we will employ a few different techniques to tackle the problem in order to extract characteristics of interest. Interaction also has statistical characteristics. Forces that arise from discreteness, such as impulsive forces, appear as shot noise. A historically important example of this impulse manifestation is Brownian motion. Objects undergoing Brownian motion, when observed with sufficient resolution, show rapid reorientation of direction followed by slow movement—fluctuation, a noise in a signal such as velocity, momentum or power flux and which rides along with the slow averaged signal that one measures. The macroscopic response is slow even if the fluctuation events are fast. An energy gradient causes slow movement in this Brownian motion in the presence of short and rapid events. This is an example of fast and slow forces at work together. So, there are important temporal attributes essential to understanding the devices in their limits or the limits of the measurement itself. The mathematical treatments reflect the choices one needs to make to gain insight while employing the simplest of techniques that will suffice. Another important attribute arising from these couplings is the ability of the forces to extend the system into the nonlinear region of response—nonlinearities are pervasive in electromechanics. They result in effects such as hysteresis, bifurcation and chaos. We therefore discuss the necessary physical description for the coupling and then look at its manifestation in interesting devices.

There also exists thermal noise in mechanics, just as in electronics—displacement fluctuations due to thermal energy in the thermodynamic spread of an ensemble.

5.1 Mechanical response

WE START WITH THE CLASSICAL DESCRIPTION of the response of a beam in elastic conditions. Figure 5.1 shows a beam under a distributed load, that is, a force per unit length arising from an external source. L is the length of the beam, w its width, and t its thickness. Weight arising from gravity, a hydrostatic force arising from a fluid pressure, or a Coulombic force, for example, cause a pressure of force per unit length of $p(\zeta)$. The beam deflects, and the moment of this force and deflection is $\mathfrak{M}$. In general conditions, such as non-uniform loads or non-uniform beams, this deflection can consist of bending (an angular deformation), shearing (a slip), translation (a uniform Cartesian shift) and rotation (a uniform angular movement around an axis).

We consider the case of a beam that has no longitudinal forces.

By elastic, we mean that when a stress is removed, the material returns to its undeformed state and that the deformation under stress is a linear response.

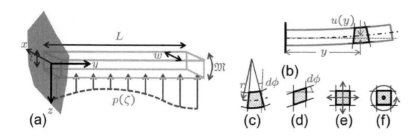

Figure 5.1: Response of a beam clamped at one end under a distributed load is shown in (a). In general, a beam undergoing deflection as shown (b), can bend (c), shear (d), translate (e) and rotate (f), depending on the external forces it is subjected to within elastic response conditions. The dot-dashed lines show the neutral axis.

The centerline then is a neutral axis with strain of opposite polarity on either side. If the load is uniform, the strain is reflected in bending. Since the force exists across the beam, we employ a moment reflecting the leverage of the position-dependent force. The beam has a compressive and tensile strain on either side of the neutral axis. The moment causes this strain, and the strain is the reaction pushing the beam back to maintain equilibrium. We employ this moment of inertia to reflect this position dependence in the inertial response of the beam.

The strain in a rectangular beam in the coordinate system of Figure 5.1(a) is $\varepsilon = z d\phi / dy = z/r$. This follows from the balance of longitudinal forces, so beam length along the neutral axis is a constant. The strain causes a reaction for restoration. The moment around the neutral axis is the integrated product of the lever of stress ($\sigma = \varepsilon Y$, where σ is the stress, and Y is Young's modulus). The moment of inertia for a rectangular beam is $I = \int_S z^2 dS = wt^3/12$, where dS is the elemental cross-section of the beam. The moment of the beam is

$$\mathfrak{M} = \int_S Y \varepsilon z \, dS = \frac{YI}{r}. \tag{5.1}$$

The radius of curvature r changes with position. Farther away, there is a larger accumulated moment of the force, and the displacement $u(y)$ is large. For this elastic—linear response—case, as a function of the position, the radius of curvature is related to the displacement as

$$\frac{1}{r} = \frac{\partial^2 u}{\partial y^2}, \tag{5.2}$$

for small deflection, and, hence, the moment as

$$\mathfrak{M} = YI \frac{\partial^2 u}{\partial y^2}. \tag{5.3}$$

Consider the general case shown in Figure 5.1(a), without the uniform force approximation. Under conditions of limited displacement, so that higher order terms in the Cartesian geometric representation are ignored, if the position-dependent force per unit length, a

A moment is a leverage. A force acting at a distance has a moment. In general, any product effect is a moment. For examples, see Chapter 2 and S. Tiwari, "Semiconductor physics: Principles, theory and nanoscale," Electroscience 3, Oxford University Press, ISBN 978-0-19-875986-7 (forthcoming) for the moment equations for density, current, and energy transport for electrons in the kinetic picture. The electrostatic potential arises as a moment from the electric field in $\psi = -\int \mathcal{E} \cdot d\mathbf{r}$.

Readers not versed in the continuum theory of elasticity should look at Appendix L for a summary of many of the mechanical engineering texts that treat this classical subject.

A more precise solution for deflection is

$$\frac{1}{r} = \frac{\partial^2 u/\partial y^2}{\left[1 + (\partial u/\partial y)^2\right]^{3/2}}.$$

pressure, is $p(\zeta)$, the position-dependent moment is

$$\mathfrak{M}(y) = \int_y^L p(\zeta)(\zeta - y)d\zeta \qquad (5.4)$$

where the coordinates are along the beam, with the origin at the fixed point ($y = 0$, which coincides with the origin of ζ). This equation gives the leverage at any given position y due to the force per unit length of $p(\zeta)$. This represents the force exerted on an infinitesimal section of the beam, multiplied by its distance from the clamping point. The shearing force at any position y is the result of the accumulation of the force beyond that position—the differential, that is,

$$Q(y) = \int_y^L p(\zeta)d\zeta = \frac{\partial \mathfrak{M}(y)}{\partial y}. \qquad (5.5)$$

The forces beyond the position y cause the shearing at y. The force per unit length then is the second derivative:

$$p(y) = \frac{\partial^2 \mathfrak{M}(y)}{\partial y^2}. \qquad (5.6)$$

If the beam is stationary, then the moment $\mathfrak{M} = YI\partial^2 u/\partial y^2$. This is what we derived when discussing the case of uniform force. The force per unit length is then related as

$$p(y) = YI\frac{\partial^4 u}{\partial y^4}. \qquad (5.7)$$

When the beam is in motion, the inertial force reacts to the force per unit length with this positional dependence. The vertical displacement, a translational movement, is related to $p(y)$ through the force law, so $p(y)$ can be written as

$$p(y) = \rho A\frac{\partial^2 u}{\partial t^2} \qquad (5.8)$$

through time-dependence.

These forces—the applied force and the strain from within—and incorporating the time-dependence of the response can be gathered together in the form

$$p = \rho A\frac{\partial^2 u}{\partial t^2} + YI\frac{\partial^4 u}{\partial y^4}. \qquad (5.9)$$

This equation can be expanded to include any other forcing or damping functions that may exist—electromagnetic, mechanical or other. This partial differential equation couples time and space in form to which the eigenmode analysis can be applied. So, we have now reduced the problem of a moving beam under forces to one of eigenmode analysis.

The wave equation, such as in electromagnetics, is second derivative in time and in space. The equation is separable in eigenmode analysis through separation of the independent space- and time-dependent parts. The spatial equation is the Helmholtz equation, whose solution is the spatial function. The time-dependent part leads to the harmonic function. The product of the two is the eigenfunction solution of the wave equation.

In the case of a fixed cantilever beam, at the fixed end ($y = 0$), the shearing force and the moment also vanish. So, the displacement and its first derivative vanish at $y = 0$. Similarly, the second and third derivatives of displacement vanish at $y = L$. The short argument that establishes this boundary condition is that, beyond $y = L$, the various forces vanish. This means that the second and third derivatives must too, since the moment vanishes.

This was a rather convoluted way to arrive at the equation for the force. The force-balancing approach works, but it easily gets out of hand, particularly regarding any intuition and questions on what coordinate system to employ. How would one handle rotations and translations happening simultaneously? One answer is to use energy principles, and scalers, through Hamilton's principle and the Lagrangian method. This gives us an elegant way to tackle this by resorting to the principle of least action. The kinetic energy is

$$T = \int_0^L \frac{1}{2}\rho S \left(\frac{\partial u}{\partial t}\right)^2 dy. \tag{5.10}$$

The potential energy is

$$U = \int_0^L \int_{-t/2}^{t/2} \frac{1}{2} Y e^2 w \, dz \, dy = \int_0^L \frac{1}{2} YI \left(\frac{\partial^2 u}{\partial y^2}\right)^2 dy. \tag{5.11}$$

With the work done by external forces, $\delta W = \int_0^L p \, \delta u \, dy$, Hamilton's principle leads to

$$\int_{t_1}^{t_2} (\delta W + \delta T - \delta U) \, dt = 0, \tag{5.12}$$

which states that the sum of the work exerted by external forces and the difference between the changes in kinetic and potential energies change, over time, for infinitely small displacement δu around the actual displacement u vanishes so long as either this perturbation or its positional dependence, that is, either δu or $\partial(\delta u)/\partial y$ vanishes.

For the beam problem, the principle implies that

$$0 = \int_{t_1}^{t_2} \left[\int_0^L \left(-\rho A \frac{\partial^4 u}{\partial y^4} - YI \frac{\partial^4 u}{\partial y^4} + p\right) \delta u \, dy \right.$$
$$\left. + YI \frac{\partial^3 u}{\partial y^3}(L)\delta u(L) - YI \frac{\partial^2 u}{\partial y^2}(L)\delta\left(\frac{\partial u}{\partial y}(L)\right) \right] dt. \tag{5.13}$$

Since the perturbation in the displacement at any position, including at $y = L$, and the perturbation in its derivative at $y = L$ are arbitrary, it must follow, from each of the associated terms in this equation, that

$$p = \rho A \frac{\partial^4 u}{\partial y^4} + YI \frac{\partial^4 u}{\partial y^4} \quad \text{for } 0 < y < L,$$

We will establish this equation by a straightforward energy argument when employing the Lagrangian approach to solving coupled problems. The boundary conditions of derivatives will then naturally follow.

The physical behavior of action is discussed at length in S. Tiwari, "Quantum, statistical and information mechanics: A unified introduction," Electroscience 1, Oxford University Press, ISBN 978-0-19-875985-0 (forthcoming) w.r.t. quantum, information and statistical mechanics, with the early thoughts from the classical treatments. A detailed discussion of Lagrangian and Hamiltonian is available there. A short treatment of principle of the virtual work, d'Alembert's principle, Hamilton's principle, and the use of energy scalers in Lagrangian is in Appendix M.

δ here is an operator characterizing difference with attributes akin to that of infinitesimally small change of differentiation. It commutes with the differentiation operator.

$$\frac{\partial^2 u}{\partial y^2} = 0 \quad \text{for } y = L, \text{ and}$$

$$\frac{\partial^3 u}{\partial y^3} = 0 \quad \text{for } y = L. \tag{5.14}$$

These are the equations of motion, absent damping, and the boundary conditions at $y = L$. At $y = 0$, the displacement vanishes, as does its first differential with position—the inclination. In time, the initial condition consists of a displacement u and a velocity $\partial u / \partial t$. The boundary conditions suffice for solving the fourth order differential equation.

One could have arrived at this same equation by using the Lagrangian method, which we will utilize in problems later on. As remarked, these are all different methods utilizing action in different forms and are formally equivalent. For the Lagrangian $\mathscr{L} \equiv T - U$, we may utilize the two canonical conjugate coordinates—position u and velocity $\dot{u}$—to write the equation of motion as

$$\frac{d}{dt}\left(\frac{\partial \mathscr{L}}{\partial \dot{u}_i}\right) - \frac{\partial \mathscr{L}}{\partial u_i} = F_i, \tag{5.15}$$

where i denotes a differential section of the beam; assembly of all the sections forms the entire beam.

This equation may be solved through eigenmode analysis. Here, we take the case of free vibration; later on, we will revisit the problem for damping and forced vibrations. A harmonic force causes a harmonic response in this linear response description. We separate the response function into a time-dependent part and a space-dependent part. The separation of variables is possible since, in the steady state, the time response of displacement at any position y has to have the same harmonic form in time. This implies that the solution function is a product of a space-dependent part and a time-dependent part. So, we wish to find the position-dependent part ($\mathcal{Y}$) and time-dependent part ($\mathcal{T}$) of

$$u(y, t) = \mathcal{Y}(y)\mathcal{T}(t). \tag{5.16}$$

The governing equation, Equation 5.14, then becomes

$$\frac{YI}{\rho A}\frac{d^4\mathcal{Y}}{dy^4} = -\frac{1}{\mathcal{T}}\frac{d^2\mathcal{T}}{dt^2} = \omega^2; \tag{5.17}$$

since neither of the dependences varies with position or time, these can be split into

$$\frac{d^4\mathcal{Y}}{dy^4} - \frac{\rho A}{YI}\omega^2 = 0, \quad \text{or} \quad \frac{d^4\mathcal{Y}}{dy^4} - k^4\mathcal{Y} = 0, \text{ and}$$

$$\frac{d^2\mathcal{T}}{dt^2} + \omega^2\mathcal{T} = 0, \tag{5.18}$$

These solution techniques are employed, with discussion, in S. Tiwari, "Quantum, statistical and information mechanics: A unified introduction," Electroscience 1, Oxford University Press, ISBN 978-0-19-875985-0 (forthcoming), where they were used to find eigenfunction solutions of quantum-mechanical problems.

where ω is the angular frequency of the harmonic force, and $k^2 = \omega/(YI/\rho A)^{1/2} = \omega/a$, with $a = (YI/\rho A)^{1/2}$ as a parametric constant. This solution for the homogeneous equation shows characteristics that are similar to those of the solution for the electromagnetic wave propagation equation, although the dependences are in a different order, and the parameters are quite different. The beam oscillates back and forth, exchanging the potential energy, which in the spring is associated with quantum-mechanical bonding, with the kinetic energy associated with the vibrational motion. The ansatz function then is

$$
\begin{aligned}
u &= \mathcal{Y}\mathcal{T} \\
&= (C_1 \cos ky + C_2 \sin ky + C_3 \cosh ky + C_4 \sinh ky) \\
&\quad \times (A_1' \cos \omega t + B_1' \sin \omega t).
\end{aligned} \tag{5.19}
$$

The boundary conditions then imply that $C_1 + C_3 = 0$, and $C_2 + C_4 = 0$, because of fixed reference position and vanishing inclination at $y = 0$. The vanishing second and third derivatives of displacement at $y = L$, which we derived as a boundary condition due to the absence of moment from the shearing force at that position, then lead to the set

$$
\begin{aligned}
C_1(-\cos kL - \cosh kL) + C_2(-\sin kL - \sinh kL) &= 0, \text{ and} \\
C_1(\sin kL - \sinh kL) + C_2(-\cos kL - \cosh kL) &= 0, \tag{5.20}
\end{aligned}
$$

whose solution exists iff

$$
\begin{aligned}
0 &= (\cos kL + \cosh kL)^2 + \sin^2 kL - \sinh^2 kL \\
&= \cos kL \cosh kL + 1. \tag{5.21}
\end{aligned}
$$

An infinite number of solutions exist. The solutions have the form $kL = \lambda_i$, where $\lambda_1 = 1.875, \lambda_2 = 4.694, \ldots$, where the higher order terms are well approximated by $\lambda_i = 2(i-1)\pi/2 \; \forall i \geq 3$, with less than a percent of error. We have now found the various eigenmodes of the positional waves that exist in the cantilever beam as it oscillates up and down. The natural frequency of each mode (ω_i) is different:

$$
\omega_i = \frac{\lambda_i^2}{L^2} \left(\frac{YI}{\rho A}\right)^{1/2}. \tag{5.22}
$$

The frequency of any mode increases as beam is made smaller in length or thicker, because of the $1/L^2$ and I/A dependence. The amplitude can now be determined through the C-coefficients:

$$
C_2 = -\frac{\cos \lambda_i + \cosh \lambda_i}{\sin \lambda_i + \sinh \lambda_i} C_1 = \alpha_i C_1. \tag{5.23}
$$

C_3 and C_4 follow, since they are related to C_1 and C_2 through the boundary conditions at $y = 0$.

The displacement then is

$$u(y,t) = \mathcal{Y}\mathcal{T} = \sum_{i=1}^{\infty} C_1 \mathcal{Y}_i \mathcal{T}_i$$

$$= \sum_{i=1}^{\infty} [\cos k_i y - \cosh k_i y - \alpha_i (\sin k_i y - \sinh k_i y)]$$
$$\times (A_i \cos \omega_i t + B_i \sin \omega_i t) \tag{5.24}$$

with $A_i = A'_i C_i$, and $B_i = B'_i C_i$, determined by the initialization conditions of motion. The displacement and speed at $t = 0$ uniquely determine the infinite A_i and B_i sequences since the initial condition of displacement is a function of position infinitely spread out across the beam. The displacement in time, as a function of position, is the sum of the eigenmode expansion of the infinite series of Equation 5.24. For example, if the initial condition was precisely one corresponding to $A_1 = 1$, where all the rest—B_1, A_i, B_i $\forall i \geq 2$—were zero, then the initial condition for displacement must be precisely $u(y,0) = \cos k_1 y - \cosh k_1 y - \alpha_1 (\sin k_1 y - \sinh k_1 y)$.

Since the solution is in terms of a basis set that is orthogonal, the eigenfunctions are orthogonal. We can show this rigorously by drawing on Equation 5.18, which the basis eigenfunctions $\mathcal{Y}_i$ of function $\mathcal{Y}$ must satisfy, first rewriting it in the reduced form

$$\frac{d^4 \mathcal{Y}_i}{dy^4} - k_i^4 \omega^2 = 0, \text{ and}$$

$$\frac{d^4 \mathcal{Y}_j}{dy^4} - k_j^4 \omega^2 = 0. \tag{5.25}$$

The standard technique is to multiply by the other basis eigenfunction and then subtract and integrate over space. So,

$$(k_i^4 - k_j^4) \int_0^L \mathcal{Y}_i \mathcal{Y}_j dy$$

$$= \int_0^L \left(\mathcal{Y}_j \frac{d^4 \mathcal{Y}_i}{dy^4} - \mathcal{Y}_i \frac{d^4 \mathcal{Y}_j}{dy^4} \right) dy$$

$$= \left(\mathcal{Y}_j \frac{d^3 \mathcal{Y}_i}{dy^3} - \mathcal{Y}_i \frac{d^3 \mathcal{Y}_j}{dy^3} - \frac{d\mathcal{Y}_j}{dy} \frac{d^2 \mathcal{Y}_i}{dy^2} + \frac{d\mathcal{Y}_i}{dy} \frac{d^2 \mathcal{Y}_j}{dy^2} \right) \Big|_0^L$$

$$= 0 \tag{5.26}$$

because the boundary conditions are that $\mathcal{Y}_i$ and $\mathcal{Y}_j$ and their first derivatives with position y vanish at $y = 0$ and that their second and third derivatives vanish at $y = L$. This requires that

$$\int_0^L \mathcal{Y}_i \mathcal{Y}_j dy = 0 \;\; \forall i \neq j, \tag{5.27}$$

This type of approach is used extensively in S. Tiwari, "Quantum, statistical and information mechanics: A unified introduction," Electroscience 1, Oxford University Press, ISBN 978-0-19-875985-0 (forthcoming) for normalization, for tackling issues of infinities arising from degeneracy and for finding eigenenergies. Here, we are not using conjugates, since the functions are real.

since $k_i \neq k_j$. In turn, the first part of Equation 5.25, when multiplied by $\mathcal{Y}_j$, and integrated over the length, also leads to

$$\int_0^L \mathcal{Y}_j \frac{d^4 \mathcal{Y}_i}{dy^4}\, dy = 0 \quad \forall i \neq j. \tag{5.28}$$

Ergo, the eigenmode functions are orthogonal. We have here established the technique to describe the position dependence and time dependence of the beam response as an eigenfunction response. We will look at example solutions when we return to this subject for forced and damped conditions.

Resonators employing circular plates are also of interest as banks of frequency-selective filters in wireless communications. So, we will analyze this problem to show the features that appear from symmetries and dimensionality. We consider only the homogeneous undamped case shown in Figure 5.2: a thin plate of density ρ, diameter D, thickness t and radius R. In this case,

$$\frac{\partial^2 u}{\partial t^2} + \nabla^4 u = 0. \tag{5.29}$$

The boundary conditions in polar coordinates, if the plate is clamped at its edges, are

$$u(\varrho, \theta; t)\big|_{\varrho=1} = 0, \text{ and}$$

$$\frac{\partial u}{\partial \varrho}(\varrho, \theta; t)\bigg|_{\varrho=1} = 0, \tag{5.30}$$

where we have written the conditions in a normalized radial unit, $\varrho - r/R$, where R is the radius of the disk. If the disk is anchored at the center, these boundary conditions change appropriately to $\varrho = r/R = 0$. We employ the traditional separation of variables technique, given the nature of the derivative dependences of the governing equation. First, we separate time,

$$u(\varrho; t) = z(\varrho, \theta) \exp(-i\omega t). \tag{5.31}$$

The spatial equation then is

$$\nabla^4 z - \omega^2 z = 0. \tag{5.32}$$

This form of equation immediately suggests that $z(\varrho, \theta) = F(\varrho, \theta) + G(\varrho, \theta)$, which must satisfy

$$\nabla^2 F - \omega F = \nabla^2 G + \omega G = 0. \tag{5.33}$$

Now, we separate the spatial variables. Take $G(\varrho, \theta) = G_\theta(\theta) G_\varrho(\varrho)$, the product of the angular and the radial parts. The solution must satisfy

$$\frac{d^2 G_\theta}{d\theta^2} + \lambda^2 G_\theta = 0, \tag{5.34}$$

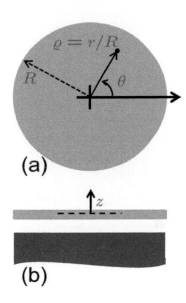

(a)

(b)

Figure 5.2: A circular plate anchored either along the perimeter or at the center to the substrate.

Perimeter-clamped plates are ubiquitous in drums. A plate anchored at the center is also known as a wineglass resonator.

and, therefore,

$$\frac{d^2 G_\varrho}{d\varrho^2} + \frac{1}{\varrho}\frac{dG_\varrho}{d\varrho} + \left(\omega - \frac{\lambda^2}{\varrho^2}\right)G_\varrho = 0. \qquad (5.35)$$

$$G_\theta(\theta) = a' \sin \lambda\theta + b' \cos \lambda\theta \qquad (5.36)$$

is a good solution, and there must be phase matching in angular displacement, that is, $G_\theta(\theta) = G_\theta(\theta + 2\pi)$. The constraint is that $\lambda = 1, 2, \ldots$, that is, positive integers. So, the angular part of this solution has the form

$$G_{\theta n} = a'_n \sin n\theta + b'_n \cos n\theta, \quad \text{where } n = 1, 2, \ldots. \qquad (5.37)$$

The radial dependence equation, recast by substituting for λ, has Bessel functions as the solutions:

$$G_{\varrho n} = \alpha_n J_n(\sqrt{\omega}\varrho) + \beta_n Y_n(\sqrt{\omega}\varrho). \qquad (5.38)$$

$Y_n(\sqrt{\omega}\varrho)$ is unphysical since it is unbounded, so we only need to consider the first term; so we have

$$G(r, \theta) = J_n(\sqrt{\omega}\varrho)(a_n \sin n\theta + b_n \cos n\theta), \qquad (5.39)$$

where $a_n = a'_n \alpha_n$, and $b_n = b'_n \alpha_n$. Solution for $F(\varrho, \theta)$, with its negative sign equation, forms a solution in Bessel functions with an imaginary argument and is

$$F(\varrho, \theta) = I_n(\sqrt{\omega}\varrho)(c_n \sin n\theta + d_n \cos n\theta). \qquad (5.40)$$

Combining Equations 5.39 and 5.39, we obtain the spatial solution for edge clamping,

$$\begin{aligned} z(\varrho, \theta) &= \left[a_n J_n(\sqrt{\omega}\varrho) + c_n I_n(\sqrt{\omega}\varrho)\right] \sin n\theta \\ &\quad + \left[b_n J_n(\sqrt{\omega}\varrho) + d_n I_n(\sqrt{\omega}\varrho)\right] \cos n\theta, \qquad (5.41) \end{aligned}$$

to which we can now apply our boundary conditions. This lets us find the allowed frequencies, that is, the values of $\omega_n^\star$, that satisfy this equation. These constraints appear through the determinant of the two boundary condition equations, for clamping on the edge, as

$$J_n(\sqrt{\omega})I'_n(\sqrt{\omega}) - J'_n(\sqrt{\omega})I_n(\sqrt{\omega}) = 0. \qquad (5.42)$$

The solution to the constraints of the problem with this flexing must now be determined numerically. The natural frequencies of this plate are

$$\omega_n = \frac{\omega_n^\star}{L^2}\left(\frac{D}{\rho t}\right)^{1/2}, \qquad (5.43)$$

where D is related to flexural rigidity of the plate, and L is the radius—the characteristic length of this system.

The natural frequency of the plate being a function of size means that an array of such disks, all programmed for different frequencies, are potentially useful as filters at frequencies of wireless communications.

We can convince ourselves of the usefulness of beams too with simple estimation. Resonance frequency relates inversely to the square root of mass. The effective mass of a beam being proportional to the order of the resonance mode, the higher order modes have less energy than the fundamental one. The cantilever beam has a mass $m = \rho Lwt$, the product of density, length, width and thickness. Its elastic constant along the direction of thickness, responsible for the restorative force is, $k_s = 12YI/L^3$, as determined earlier. The moment of inertia here is proportional to wt^3. In a restorative system such as this, the natural frequency of vibration will be related to the spring constant and mass as $\omega_0 = \sqrt{k_s/m}$. Reducing dimensions reduces mass and increases resonance frequency. If one were to place a mass Δm at the tip of the cantilever, the modified mass $m = m_0 + \Delta m = m_0(1 + \Delta m/m_0)$ would change the vibrational frequency to $\omega = \omega_0/\sqrt{1 + \Delta m/m_0}$. With a suitable choice of parameters then, this cantilever becomes an ultrasensitive mass detector. The sensitivity

$$S = \frac{d\omega}{dm} = -\frac{1}{2}\frac{\omega_0/m_0}{(1 + \Delta m/m_0)^{3/2}} \propto [L]^{-4} \qquad (5.44)$$

shows that the mass detection limit can be increased through the frequency measurement in the inverse fourth power of length.

	Energy	Force
Electrostatic ($\mathcal{E} < \mathcal{E}_{br}$)	$U_{es} = \int_\Omega (1/2)\boldsymbol{\mathcal{E}} \cdot \mathbf{D} d^3\mathbf{r} \ \sim [L]^3$	$\mathbf{F}_{es} = -\boldsymbol{\nabla} U_{es} \ \sim [L]^2$
Magnetostatic ($B = nI/\mu_0 L \ \sim [L]$)	$U_{mag} = \int_\Omega (1/2)\mathbf{B} \cdot \mathbf{H} d^3\mathbf{r} \ \sim [L]^5$	$\mathbf{F}_{mag} = -\boldsymbol{\nabla} U_{mag} \ \sim [L]^4$
Thermostatic	Nonconservative, $\int_\Omega d(TS) \sim C\rho\Omega\Delta T$	$\mathbf{F}_{ts} \sim [L]^3$
Friction (atomically smooth interface)	Nonconservative, $\int \mu_f R_\perp dr$	$\mathbf{F}_f = -\mu_f R_\perp \hat{\mathbf{r}} \sim [L]^2$

Table 5.1: Dimensional dependence of energy and forces. The electrostatic field $\mathcal{E}$ must be smaller than the breakdown field $\mathcal{E}_{br}$. The magnetic field is proportional to the current (I) and to the number of turns of a coil (n) and varies inversely with the permeability μ_0 and the length of the coil, L. If current density is kept constant then the magnetostatic energy and forces follow as summarized. If heat dissipation, flow and temperature changes are significant, one must also consider a temperature gradient that varies inversely with L. Forces associated with thermal and frictional forms of energy are nonconservative. They couple a broadband of excitations. C is specific heat capacity, μ_f is the coefficient of friction, and $R_\perp$ is the reaction force at the surface in contact.

This argument shows the importance of length and of the other parameters that determine the spring constant and the inertia of the structure. Size matters. Scaling size reduces effective mass. Vibrational amplitude will also reduce with size. Since electromechanic devices utilize multidomain coupling, it is pertinent to look at this size scaling, where the energy and forces may have dependence on length $[L]$, area $[L]^2$ and volume $[L]^3$. Table 5.1 shows the length scale dependence of different energy and forces.

Frictional forces arise from quantum-mechanical exclusion and the transfer of energy to a broadband of degrees of freedom—energy loss paths such as the various vibrational modes—when objects are brought in contact. In macroscopic objects that are not atomically smooth, contact occurs only in a small area. At a minimum, three atomic scale regions suffice. The frictional force is independent of area. The difference between static and dynamic friction arises because when the objects are in motion, they are separated further away from each other than when they are static, that is, in the lowest energy ground state. The effective contact area between them is now reduced, and thus the dissipative force is reduced. When surfaces are atomically smooth then adhesion is strong, friction is large and force is proportional to the physical area. Small and atomically smooth objects have a frictional force that varies as $[L]^2$.

An estimate for the mechanical strength of the beam can be obtained from the dynamic equation, Equation 5.9. In static conditions, $\partial^4 u / \partial y^4 = p/YI$, which relates the bending to Young's modulus, the moment of inertia, and the load per unit length. This load per unit length is $p = g\rho wt \sim [L]^2$, and the moment of inertia is $I = wt^3/12$, independent of length, so the bending has a dimensional dependence of $u \sim [L]^2$. This goes together with the beam's resonance frequency, which has the dependence $\omega_0 \sim [L]^{-1}$. The sensitivity to mass, Equation 5.44, then has a $[L]^{-4}$ dependence.

It is this sensitivity to forces, arising as inertial response, that is of immense interest in precision measurements of orientation, acceleration, et cetera, and in fundamental measurements through the nanoscale. In Figure 5.3, the simplest circuit model is shown for the cantilever as an inertial sensor. Here, the displacement of the mass m is in the cantilever assembly's reference frame, and the cantilever assembly is moving with an acceleration a in the laboratory reference frame.

The force equation is

$$F = m\ddot{y} + \gamma\dot{y} + k_s y = -ma, \qquad (5.45)$$

where γ is the damping coefficient. For a harmonic acceleration at a frequency ω, this leads to

$$-\omega^2 my + i\omega\gamma y + k_s y = -ma. \qquad (5.46)$$

Wafer bonding relies on the strong adhesion between atomically smooth surfaces by converting the physical bonding to chemical bonding. Worthwhile questions to ponder are, why is there a difference between static and dynamic friction, and exactly what is this force called friction? Friction has much in common with heat. They are both tied to coupling a broadband of excitations. S. Tiwari, "Quantum, statistical and information mechanics: A unified introduction," Electroscience 1, Oxford University Press, ISBN 978-0-19-875985-0 (forthcoming) tackles these questions through perturbation and the coupling of states from the quantum point of view.

Shorter strings have a higher frequency. This is reflected in the auditory sensation of pitch.

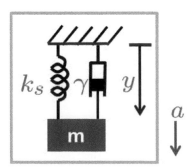

Figure 5.3: A circuit model of an inertial sensor consisting of a mass m with a displacement y in the reference frame of the sensor, under an acceleration a of the sensor, and subject to restorative forces characterized by the spring constant k_s, and damping forces characterized by the damping constant γ.

The acceleration in the laboratory reference frame can be deduced in numerous ways—displacement of the proof mass, stress in the spring such as in the cantilever body, et cetera. Consider the proof mass displacement in these harmonic conditions:

$$y = -\frac{m/k_s}{1 + i\omega\gamma/k_s - \omega^2 m/k_s}a = \frac{1/\omega_0^2}{1 + i\omega\gamma/k_s - \omega^2/\omega_0^2}a. \qquad (5.47)$$

If the acceleration is constant or slowly varying in time, then $\omega \ll \omega_0$, that is, the frequency of the forcing signal is very different from the natural frequency of the inertial sensor; then

$$y = -\frac{m}{k_s}a \quad \therefore \quad S = \frac{dy}{da} = \frac{m}{k_s} = \frac{1}{\omega_0^2}. \qquad (5.48)$$

As the spring constant $k_s \sim [L]$, and mass $m \sim [L]^3$, the sensitivity S of displacement to acceleration will vary as $[L]^2$. The sensitivity of the measurement is higher by a second power of the displacement— the inertial sensor has increased sensitivity to the acceleration.

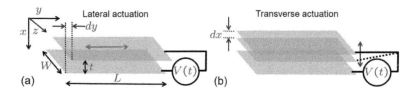

Figure 5.4: A laterally operating actuator in (a), and a transversely operating actuator in (b).

Actuators, the complements of sensors, derive their utility from the energy and the generated forces. So, the discussion of energy and forces in Table 5.1 is particularly apropos. Figure 5.4 shows an idealization of an electrostatic actuator with lateral and transverse motion, under an applied bias voltage of $V(t)$, and with an air gap between the plates. The actuator itself consists of two conducting plates of width W, and length L, and which are separated by a distance t, and it has a bias voltage of $V(t)$. For lateral actuation, the force is

$$\begin{aligned} F_y &= -\frac{\partial U_{es}}{\partial y} - \frac{\partial}{\partial y}\int_\Omega \frac{1}{2}\epsilon_0\mathcal{E}^2 xyz\,dx\,dy\,dz \\ &= \frac{1}{2}\epsilon_0\mathcal{E}^2 Wt = \frac{1}{2}\epsilon_0 V^2\frac{W}{t}, \end{aligned} \qquad (5.49)$$

and for transverse actuation, the force, is

$$\begin{aligned} F_x &= -\frac{\partial U_{es}}{\partial x} - \frac{\partial}{\partial x}\int_\Omega \frac{1}{2}\epsilon_0\mathcal{E}^2 xyz\,dx\,dy\,dz \\ &= \frac{1}{2}\epsilon_0\mathcal{E}^2 WL = \frac{1}{2}\epsilon_0 V^2\frac{WL}{t^2}. \end{aligned} \qquad (5.50)$$

For the same dimensions and an energy density defined by the electric field $\mathcal{E} = V/t$, the force transversely is high—proportional to

the product of energy density and the width and length—compared
to the force longitudinally, which is the product of energy density,
width and plate separation. Plate separation t is usually significantly
smaller than the length L of plates. The transverse motion is subject
to higher stress and squeezing, due to the displacement t and consequent dissipative losses. This also results in lower quality factor
Q. The force for the lateral motion is more insensitive to lateral displacement, while for that transverse motion is strongly sensitive to
the separation.

Actuation can also be triggered thermally, and, although slow as
it is subject to thermal time constant, a thermal actuator can unleash
a large energy due to higher energy density in thermal forms. Figure 5.5 shows a thermal actuator in a cantilever form consisting of a
bimaterial strip of different expansion coefficients.

Thermal expansion causes a change in length of $\Delta L = L\alpha\Delta T$
for a change ΔT in temperature in a material of thermal expansion
coefficient α. The change in elastic energy resulting from this change
in length is $\Delta U_\theta = -k_s \Delta L^2/2$. This results in the force

$$F_y = -\frac{\partial U_\theta}{\partial y} = k_s(\alpha\Delta T)^2 L. \tag{5.51}$$

The force varies as the second power of temperature differential
and expansion coefficient and can be large. The other characteristic
of thermal systems is the large energy density that can be stored.

		Maximum energy (J/m^3)	
Electrostatic	$\mathcal{E}_{br} \approx 3 \times 10^8 \ V/m$	$U_{es} = (1/2)\epsilon_0\mathcal{E}_{br}^2 \approx 4 \times 10^5$
Magnetostatic	$M_s = 2.5 \ T \ (Fe\text{-}Co)$	$U_{ms} = M_s^2/2\mu_0 \approx 2.5 \times 10^6$
Thermal	$\Delta T \approx 350 \ K$	$U_\theta = c\rho\Delta T \approx 5.8 \times 10^8$
Mechanical	$\varepsilon_{max} = 0.045,$	$U_{mech} = (1/2)Y\varepsilon^2 \approx 1.5 \times 10^8$
(silicon 30 nm)	$Y = 150 \ GPa$	

Table 5.2 shows the nearly three orders of magnitude difference in
stored energy between electrostatic and thermal assemblies. So, if
speed is not a consideration, thermal systems are a possibility that
should be considered.

A comb drive is a common form of electrostatic actuator employed
in inertial measurement systems where both transverse and lateral
actuation exist. It is useful to look at its sensitivity using the simplistic analysis that we have employed so far. Figure 5.6 shows a comb
drive with a few of the parameters relevant to practical structures.
Comb drives are either balanced arrangements when a straight-line

The quality factor Q as a measure of
losses versus storage of energy appears
throughout this book series. Engineers
often look upon it as the ratio of energy
lost in a cycle versus the average stored
energy. If there is any damping—in
a beam, via the surrounding fluid,
anchoring leakage, the surface, for
example, due to defects, or bulk, that
is, nonlinearities, or thermoelastic
effects—the irreversible heat flow across
thickness of beam, will reduce it.

An ink jet printer head is a well-
employed example of thermal actuation. Although desktop printing is
now mostly by xerographic transfer,
3D printing, printed electronics, et
cetera, use inkjet-like printing based on
thermal actuation to precision transfer a
variety of material.

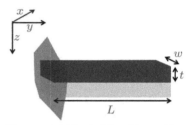

Figure 5.5: A bimaterial cantilever strip
as a thermal actuator.

Table 5.2: Maximum electrostatic energy
density and thermal energy density
in useful conditions in nanoscale
geometries. For thermal energy density,
the material is assumed to be silicon
with $c \approx 700 \ J/kg \cdot K$, and $\rho \approx$
$2.33 \times 10^3 \ kg/m^3$. The strongest $Fe\text{-}Co$
magnets have a remnant strength of
about 2.5 T. ε is strain. Nanowires of
silicon have been measured to have a
maximum strain of 4.5 %.

The comb drive is named so after its
visual similarity to the namesake used
in our daily life.

movement is desired or round when a rotation is desired. The specific example shown here is linear. It consists of two interlocked combs, one static and one moving, with conducting fingers.

Let y_0 be the unbiased transverse gap between the fingers, and L_0 the finger length, which $x_0 = L$ overlaps. In this comb drive arrangement, two primary capacitances—C_1 and C_2—represent the coupling of electrostatic forces. And let us assume fringing capacitance, C_π. If N_r is the number of fingers in the moving part, the primary capacitance for a thickness t in the depth direction of the figure is

$$C_1(y) = N_r \left(\frac{\epsilon_0 L t}{y_0 + y} + C_\pi \right), \text{ and}$$

$$C_2(y) = N_r \left(\frac{\epsilon_0 L t}{y_0 - y} + C_\pi \right). \quad (5.52)$$

Were the arrangement symmetric, that is, $y = 0$, the capacitance would be $C = C_\pi + 2N_r\epsilon_0 L t/y_0$. Any differences in alignment show up as a differential change in the capacitances. The same is true for applied voltages. So, if one applies a voltage V_y to the moving part, and a differential voltage $2V$ between the static fingers, there is a net force on the moving part. To the first order, the electrostatic force, the change in energy and the displacement are related as

$$F_{el} = -\frac{1}{2}\frac{C_0}{y_0}\left[(V - V_y)^2 - (V + V_y)^2\right] \approx \frac{2}{y_0}C_0 V V_y, \quad (5.53)$$

where $C_0 = 2N_r L t/y_0$ is the primary capacitance. The sensitivity of the capacitance, measurable through charge, is therefore related as

$$S = \frac{\partial C}{\partial x} \approx 2N_r \frac{\epsilon_0 t}{y_0}. \quad (5.54)$$

For an average gap, $y_0 = 1 \ \mu m$, an overlap length $L = 50 \ \mu m$, a depth $t = 2 \ \mu m$, and a repeating number of $N_r = 50$, the capacitance $C \approx 9.2 \ fF$, and the sensitivity $S \approx 0.4 \ fF/\mu m$. A change of a fF is relatively easy to measure, and a force that can cause a μm of movement relatively easy to apply in an unloaded comb drive. One can derive the spring constant effect arising from the electric force that modifies the mechanical spring constant of the comb drive. $k_{el} = dF/dy = -2C_0 V V_y/y_0^2$, since two opposite forces exist and the spring constant is altered by this magnitude. Therefore, the resonance frequency of the structure changes to

$$\omega_0 = \sqrt{\frac{k}{m}} \approx \omega_{0,mech}\sqrt{1 + \frac{k_{el}}{m}}. \quad (5.55)$$

As a result of the application of a bias voltage V, the displacement is x.

Balanced here means that one may place an even number of rows where movement reinforces in one direction and balances in the orthogonal direction.

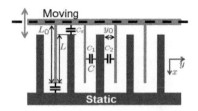

Figure 5.6: An electrostatic linear comb drive, with half of a section shown. The dashed line shows the line of symmetry.

This analysis is quite approximate—quasistatic and in its one-dimensional approximation. Motion will be both in plane and out of plane, due to the multi-dimensionality of the geometry. A qualitative view of this is shown in Figure 5.7. At the natural frequency of the structure, the resonance frequency $f_0 = 2\pi\omega_0$, the response is large rising by more than an order of magnitude above the low frequency response, and, beyond it, the response rapidly goes out of phase. The out-of-plane response is small, depending on the parasitic out-of-plane effects.

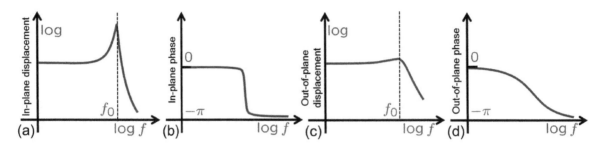

Figure 5.7: The in-plane and out-of-plane displacement and phase in response to a driving signal for an electrostatic linear comb drive. The response peaks near the natural frequency of the structure with a decades or more increase above the low frequency response. Beyond the natural frequency, the response rapidly changes to out of phase. Silicon comb drives typically have quality factors of a few 10s or more.

5.2   Coupled analysis

WE NOW DEMONSTRATE THE APPLICABILITY OF ENERGY-BASED MATHEMATICAL TECHNIQUES to solving this problem coupling electrical and mechanical energy forms. We do this for the simple structural form shown in Figure 5.4(b): a movable plate changing the size of the gap in response to bias voltage V. We consider a conservative exchange in energy between the mechanical and the electrical forms. Let Q be the charge on the capacitor; V, the voltage between the plates; displacement, x; and F, the mechanical force to keep the plate in place by opposing the electrostatic attraction. Conditions are assumed to be ideal—massless, without stiffness or damping, and with a pure one-dimensional capacitance. We take the displacement x and the charge Q as the independent coordinates, so $V = V(x, Q)$, and $F = F(x, Q)$. The boundary condition includes that, when Q vanishes F too vanishes, that is,

$$F(x, 0) = 0. \qquad (5.56)$$

To get to coordinates (x, Q), the power delivered to the structure must be the sum of the electrical power ($VI = VdQ/dt$) and the mechanical power (Fdx/dt), so, the net work on the capacitor during the time interval dt is

$$dW = VIdt + F\frac{dx}{dt}dt = VdQ + Fdx. \qquad (5.57)$$

The total stored energy $W(x, Q)$ of the capacitor is partly electrical and partly mechanical, and if we know this energy function, the force and the potential follow as

$$F = \frac{\partial W}{\partial x}, \text{ and}$$
$$V = \frac{\partial W}{\partial Q}. \qquad (5.58)$$

The complementary energy function using the Legendre transformation is

$$W^\star(x, V) = VQ - W(x, Q). \qquad (5.59)$$

See Appendix M for a discussion of complementary energy functions— functions that sum up to a constant.

The total differential then is

$$dW^\star = Q\,dV + V\,dQ - \frac{\partial W}{\partial x}\,dx - \frac{\partial W}{\partial Q}\,dQ. \qquad (5.60)$$

It follows then from Equation 5.58 that

$$Q = \frac{\partial W^\star}{\partial V}, \text{ and}$$
$$F = -\frac{\partial W^\star}{\partial x}. \qquad (5.61)$$

This ideal capacitor is a linear electrical element, so one can explicitly write the state functions $W(x, Q)$ and $W^\star(x, V)$. The linear electrical property is reflected in $V = Q/C(x)$, where $C(x)$ is the capacitance corresponding to position x of the moving plate.

Equation 5.57 shows us the way to find the different energies of interest. In our generalized coordinates, the path $(0, 0)$ to $(x, 0)$ to (x, Q) allows us to maintain $F = 0$, and $V = 0$, over the initial segment and then build electrical energy with no change in position over the second segment. From Equation 5.57 for this path,

$$\begin{aligned}
W(x, Q) &= \int_{x=0,Q=0}^{x=x,Q=0} dW + \int_{x=x,Q=0}^{x=x,Q=Q} dW \\
&= \int_0^Q VdQ = \int_0^Q \frac{Q}{C(x)}dQ = \frac{Q^2}{2C(x)}. \qquad (5.62)
\end{aligned}$$

Legendre transformation and the linear electrical constitutive relation of $V = Q/C(x)$ allows us to write the coenergy as

$$W^\star = \frac{1}{2}C(x)Q^2. \qquad (5.63)$$

Knowing $W(x, Q)$ and $W^\star(x, V)$, we may write the mechanical force necessary to balance electrostatic force in choice of coordinates. From work,

$$F = \frac{\partial W}{\partial x} = -\frac{Q^2}{2C^2}\frac{dC(x)}{dx}, \text{ and}$$
$$V = \frac{\partial W}{\partial Q} = \frac{Q}{C(x)}. \tag{5.64}$$

From coenergy,

$$F = -\frac{\partial W^\star}{\partial x} = -\frac{Q^2}{2}\frac{dC(x)}{dx}, \text{ and}$$
$$Q = \frac{\partial W^\star}{\partial V} = VC(x). \tag{5.65}$$

This Legendre transformation–based approach lets us, by choosing independent coordinates and writing work equations, find parameters of interest conveniently from scalars, unlike the vector-based approach that requires a yeoman's work.

Similar to an electric actuator, one can visualize a magnetic actuator and sensor where current flowing in a loop causes a magnetic field, or a magnetic field causes a current flow. For an electromagnet placed in an external magnetic field, which is out of plane of the current, counter forces will generate a torque to align magnetic fields—the external, and that generated by the current. Figure 5.8(a) shows a conceptual example where the current I flowing through a structure of dimension L will cause a moment that is proportional to L^2IH. The magnetic field used to actuate the diaphragm vibration in speakers in audio systems is an example of such usage. The piezoelectric effect—mechanical-electric coupling—converts applied voltage to crystal deformation that is employed for relatively rapid actuation. One can employ this deformation for actuation such as opening and closing of a valve that controls flow, as shown in Figure 5.8(b). This could also be done by thermal means, using expansion and contraction. So, actuators based on a variety of effects that can exert a force exist.

These examples, which illustrate the coupling of energy that we associate with actuators as a mechanical form, are manifestations of the exchange of coupled energy within forms—electromagnetic, mechanical or gravitational—of the macroscopic environment that we are interested in. The mechanical form is the classical view of the manifestation under quantum-mechanical constraints. The spring constant (k_s), for example, is the linear, that is, elastic term of a response arising from the spatial dependence of quantum-mechanical considerations of energy lowering and increase. To tackle these exchanges, one should work with approaches that are scaler and that

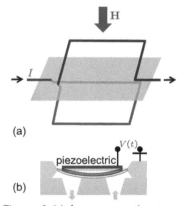

Figure 5.8: (a) shows a magnetic actuator/sensor where the magnetic field from the current of the element interacts with an external magnetic field, or, equivalently, two magnetic fields interact. (b) shows the opening and closing of a flow path via voltage-controlled piezoelectric actuation.

use the energy as the underlying consideration from which the vector fields can be extracted.

A major deficiency of this discussion to this point is that in each of the examples coupling electrical, magnetic or other forms of energy with mechanical energy, we employed an *ad hoc* approach suitable for only quite approximate low order estimates. Forces are vectors in which the choice of coordinate systems and reference frames determines the complexity and tractability of the calculation, particularly when more than one interaction needs to be accounted.

The principle of virtual work states that, in order for a system of particles to be in mechanical equilibrium, virtual work done on all the particles must vanish, that is,

$$\sum_1^N \mathbf{F}_n \cdot d\mathbf{r}_n = 0. \tag{5.66}$$

A classic problem of mechanics is a small block sliding on a larger triangular block, which itself is sliding on a surface—all under gravitational forces. While the problem is solvable using Cartesian coordinates and a moving reference frame, the approach is very unwieldy. The Lagrangian approach tackles it swiftly and decisively.

There are significant limitations here. If it is a large collection of particles forming the macroscopic assembly, any integral of external forces determined by working through displacement of the center of mass point, $\sum \left(\int \mathbf{F}_{i,ext} \cdot d\mathbf{r}_{CM} \right)$, is not the real work done. Forces have not been been multiplied by individual displacements but rather by the net force on the center of mass. Any kinetic energy associated with the motion that results will not be in terms of the velocity of the center of mass, that is, $\Delta \left(Mv_{CM}^2/2 \right)$ is not the change in kinetic energy in general, for example, work that causes an object to rotate without a center-of-mass motion. The real work is $\sum \left(\int \mathbf{F}_{i,ext} \cdot d\mathbf{r}_i \right)$. The virtual work principle applies only to static conditions; it cannot take into account perturbations arising from motion, for example, energy lost to unaccounted degrees of freedom represented in friction.

Hamilton's principle—the principle of least action—and the invariant action functional through its Lagrangian provide a natural way of tackling these problems via the Euler-Lagrange equation.

Consider the simplified electromechanical actuator shown in Figure 5.9, where there is a single degree of freedom of movement for one of the plates of a capacitor across which a bias voltage $V(t)$ can be applied. This moving plate responds with a spring constant of k_s. The choice of y as an independent coordinate is obvious. For the second coordinate, we have a choice to make—we can choose either the voltage V or the charge Q on the plates. Let us first choose V as the general coordinate with the objective to determine position y and charge Q.

In the sliding block problem, the small block warms up—another loss of energy. The Lagrangian approach can account for this.

The reader is referred to the discussion of symmetry, conservation, Hamiltonians and Lagrangians in S. Tiwari, "Quantum, statistical and information mechanics: A unified introduction," Electroscience 1, Oxford University Press, ISBN 978-0-19-875985-0 (forthcoming) for a more detailed discussion. The principle of least action, Hamilton's principle, d'Alembert's principle, Gauss's principle of least constraint and Hertz's principle of least curvature are all related to each other.

The potential energy stored arises from the mechanical stored energy and the electrostatic stored energy.

$$U = U_s + U_{es} \;\; = \;\; \frac{1}{2}k_s y^2 + \frac{1}{2}CV^2$$

$$= \frac{1}{2} \begin{bmatrix} y \\ V \end{bmatrix}^T \begin{bmatrix} k_s & 0 \\ 0 & C \end{bmatrix} \begin{bmatrix} y \\ V \end{bmatrix}$$

$$= \frac{1}{2} k_s y^2 + \frac{1}{2} \frac{\epsilon A}{t_0 - y} V^2. \qquad (5.67)$$

From the calculus of variations, we may write the following for the condition of minimization of the Lagrangian:

$$\delta \mathscr{L} = \delta U - V \delta Q = 0, \qquad (5.68)$$

which implies

$$k_s y\, \delta y + \frac{1}{2} \frac{\epsilon A}{(t_0 - y)^2} V^2\, \delta y - \frac{\epsilon A}{(t_0 - y)^2} V^2\, \delta y = 0$$

$$\therefore\ k_s y - \frac{1}{2} \frac{\epsilon A}{(t_0 - y)^2} V^2 = 0. \qquad (5.69)$$

Here, the first term is the elastic stress force (F_s), and the second is the balancing electrostatic force (F_{es}). In matrix notation, the dynamics of the actuator in terms of the independent coordinates of position and voltage are described by

$$\begin{bmatrix} k_s & 0 \\ 0 & C \end{bmatrix} \begin{bmatrix} y \\ V \end{bmatrix} = \begin{bmatrix} \frac{1}{2} \frac{\epsilon A}{(t_0 - y)^2} V^2 \\ Q \end{bmatrix}. \qquad (5.70)$$

We could have chosen charge Q as the independent coordinate that determines the electrostatic energy. It complements y, which determines the mechanical energy. In the absence of bias voltage $V(t)$, we choose mechanical equilibrium at the plate separation of t_0. This position is the reference for displacement. The capacitance is $C = \epsilon A / (t_0 - y)$. The electrostatic force for any displacement from equilibrium is

$$F_{es} = -\frac{\partial U_{es}}{\partial y} = \frac{\partial (QV)}{\partial y} = \frac{\partial}{\partial y} \left(\frac{\epsilon A}{t_0 - y} V^2 \right) = \frac{1}{2} \frac{Q^2}{\epsilon A}, \qquad (5.71)$$

balanced by the spring force $F_s = k_s y$. The balance equations can be written as

$$\begin{bmatrix} k_s & 0 \\ 0 & \frac{t_0 - y}{\epsilon A} \end{bmatrix} \begin{bmatrix} y \\ Q \end{bmatrix} = \begin{bmatrix} \frac{1}{2} \frac{Q^2}{\epsilon A} \\ V \end{bmatrix}, \qquad (5.72)$$

which have position and charge as the independent variables. This simply states the balance of mechanical and electrical forces for the plate, and the balance between the forces from charge on the plate and those from the applied bias voltage. Either approach suffices; we have written the total energy and, from it, derived the forces in our choice of independent coordinates.

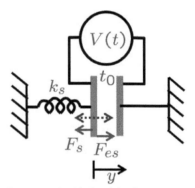

Figure 5.9: An idealized, lossless one-dimensional electrostatic actuator.

Now we look at the implication of these equations. The force balance equation is

$$F_s - F_{es} = k_s y - \frac{1}{2} \frac{\epsilon A}{(t_0 - y)^2} V^2 = 0. \tag{5.73}$$

This is an equation in the third algebraic degree of y—a nonlinear equation. The degree and the nonlinearity have physical manifestations in the actuator's response. The former means three solutions—either three real solutions, or one that we will see is stable, and two that are unstable equilibrium configurations that are disallowed because of the boundary conditions of the physical system. The nonlinearity, as we will discuss later, leads to chaos in the response.

The energy of interest in the system is

$$\begin{aligned} U &= \frac{1}{2} \begin{bmatrix} y \\ V \end{bmatrix}^T \begin{bmatrix} k_s & 0 \\ 0 & C \end{bmatrix} \begin{bmatrix} y \\ V \end{bmatrix} \\ &= \frac{1}{2} k_s y^2 + \frac{1}{2} C V^2 = \frac{1}{2} k_s y^2 + \frac{1}{2} \frac{\epsilon A}{|t_0 - y|} V^2. \end{aligned} \tag{5.74}$$

The energy minima occur at

$$\frac{\partial U}{\partial y} = 0 = k_s y - \frac{1}{2} \frac{\epsilon A}{(t_0 - y)^2} V^2 \; \forall \; y < t_0, \tag{5.75}$$

and

$$\frac{\partial U}{\partial y} = 0 = k_s y + \frac{1}{2} \frac{\epsilon A}{(t_0 - y)^2} V^2 \; \forall \; y > t_0. \tag{5.76}$$

For $y < t_0$, we find that there are two solutions. But, $y > t_0$ is an unphysical circumstance. The moving plate cannot get past the other plate, which is located at $y = t_0$. The moving plate *pulls in* and stops at $y = t_0$.

This third degree equation leads to a set of conditions, subject to applied voltage and the rest of the parameters of the system, where there are energy minima at positions, or the nonlinearity causes the plate to be pulled into contact with the other plate. The pull-in voltage V_π is the bias voltage where a small perturbation snaps the moving plate into contact with the static plate even when the moving plate is distant from the static plate.

This is the onset of instability.

The stable solution of Equation 5.75 exists for $y < t_0$ at bias voltages of

$$V = \sqrt{\frac{2 k_s}{\epsilon A}} \left[y(t_0 - y)^2 \right]^{1/2} \; \forall \; y < t_0. \tag{5.77}$$

This bias voltage solution implies that, just at the onset of pull-in,

$$\frac{\partial V}{\partial y} \bigg|_{V=V_\pi} = 0 = \sqrt{\frac{2 k_s}{\epsilon A}} \frac{1}{2} \frac{1}{\sqrt{y}(t_0 - y)} \left[(t_0 - y)^2 + y 2(t_0 - y) \right]. \tag{5.78}$$

So, the pull-in happens when

$$y = \frac{t_0}{3}, \quad \text{with } V = V_\pi = \sqrt{\frac{8k_s t_0^3}{27\epsilon A}}. \tag{5.79}$$

As the voltage is increased, and the bias voltage begins to bring the plates closer due to Coulombic attraction, at $y = t_0/3$, and the corresponding bias voltage V_π, the plates snap together.

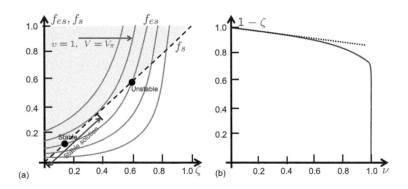

Figure 5.10: The dimensionless relationship of the forces, with displacement showing a stable solution at small displacement voltages $v < 1$, together with an unstable counterpart, the onset of pull-in at $v = 1$, and the region of no solution beyond.

These results are conveniently visualized in a dimensionless form. Using $\zeta = y/t_0$ for spatial coordinates, and $v = V/V_\pi$ for bias voltage, the governing equation is

$$\zeta - \frac{4}{27} \frac{v^2}{(1-\zeta)^2} = 0 \tag{5.80}$$

in dimensionless form, with a dimensionless elastic force of $f_s = \zeta = F_s/k_s t_0$, and a dimensionless electrostatic force of

$$f_{es} = 4v^2/27(1-\zeta)^2 = F_{es}/k_s t_0. \tag{5.81}$$

Figure 5.10(a) shows this solution with varying parameters. For voltages where $v > 1$, that is, $V > V_\pi$, two stable solutions exist where the elastic force and electrostatic forces can balance with the plates still apart. At $v = 1$, that is, $V = V_\pi$, we have two stable solutions and the pull-in solution. Any small perturbation causes this pull-in and brings the plates into contact. For $v < 1$, while an algebraic solution exists, the physical structure doesn't allow it. Figure 5.10(b) shows this change as the voltage is varied. The nonlinearity of the equation is reflected in the nonlinearity of separation as increasing voltage brings the plates together. It arises from the relationship in energy that we have derived,

$$U = U_s + U_{es} = \frac{1}{2}k_s y^2 + \frac{1}{2}CV^2$$

$$= \frac{1}{2}\begin{bmatrix} y \\ V \end{bmatrix}^T \begin{bmatrix} k_s & 0 \\ 0 & C \end{bmatrix} \begin{bmatrix} y \\ V \end{bmatrix}$$

$$= \frac{1}{2}k_s y^2 + \frac{1}{2}\frac{\epsilon A}{t_0 - y}V^2, \tag{5.82}$$

where the last term is a nonlinear term in position and continuously increases with decreasing spacing. It causes the snapping of pull-in and the chaotic effects that we look at later.

We now take a more complex example of usage employing the coenergy approach: a microphone based on capacitive response—so, a mechanical-electrical coupling, but now including more realism through the use of circuit elements. Our model for this microphone, where the diaphragm has a mass m and is coupled with a resistor-inductor to a voltage source, is shown in Figure 5.11. The moving plate of the diaphragm forms a variable capacitor and is mounted, that is, attached to an anchor, modeled here with a spring constant k_s and the dissipation constant γ'. In equilibrium, a charge Q_0 on the capacitor produces an attractive electrostatic force F_{e0} that is balanced by the elastic force of the spring. We employ x_0 and x_1 as the two equilibrium gaps. Any excitation around equilibrium causes oscillation about this equilibrium. The capacitance seen by the electrical network is $C(x) = \epsilon A/(x_0 - x)$.

The electrostatic force between the plates at equilibrium is

$$F_{e0} = -\left.\frac{\partial W_e}{\partial x}\right|_0 = -\frac{\partial}{\partial x}\left[\frac{Q^2}{2C(x)}\right]_0 = \frac{Q_0^2}{2\epsilon A}, \tag{5.83}$$

using W_e to denote the electrostatic energy. This force is balanced by the mechanical force $k_s x_1$. We now use the generalized coordinates (x, Q). Current, a dependent parameter, is $I = \dot{Q}$. To determine the Lagrangian, the different kinetic and potential energies are

$$T^\star = \frac{1}{2}m\dot{x}^2,$$

$$V = \frac{1}{2}k_s(x + x_1)^2,$$

$$W_m^\star = \frac{1}{2}L\dot{q}^2, \text{ and}$$

$$W_e = \frac{1}{2}\frac{(Q_0 + q)^2}{C} = \frac{x_0 - x}{2\epsilon A}(Q_0 + q)^2, \tag{5.84}$$

where $W_m^\star$ is the mechanical coenergy, and q is the excess charge, so that $Q = Q_0 + q$. The structure has dissipation and so has non-conservative energy and forces. We write these as

$$D = \frac{1}{2}\gamma'\dot{x}^2 + \frac{1}{2}R\dot{q}^2, \tag{5.85}$$

and the non-conservative work is

$$\delta W_{nc} = V(t)\,\delta q + F\,\delta x. \tag{5.86}$$

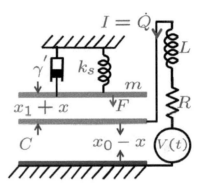

Figure 5.11: A capacitive diaphragm-based speaker. An inductor-resistor drives the moving plate, forming an *LCR* resonator. The equilibrium positions of the plate are x_0 and x_1 from the two other plates, one of which is fixed.

The Lagrangian ($\mathscr{L}$) is

$$\mathscr{L} = \frac{1}{2}m\dot{x}^2 + \frac{1}{2}L\dot{q}^2 - \frac{1}{2}k_s(x+x_1)^2 - \frac{x_0-x}{2\epsilon A}(Q_0+q)^2, \qquad (5.87)$$

leading to

$$\frac{\partial \mathscr{L}}{\partial \dot{x}} = m\dot{x}; \quad \frac{\partial \mathscr{L}}{\partial x} = -k_s(x+x_1) + \frac{(Q_0+q)^2}{2\epsilon A}; \quad \frac{\partial D}{\partial \dot{x}} = \gamma'\dot{x};$$

$$\frac{\partial \mathscr{L}}{\partial \dot{q}} = L\dot{q}; \quad \frac{\partial \mathscr{L}}{\partial q} = -\frac{x_0-x}{\epsilon A}(Q_0+q); \quad \text{and} \quad \frac{\partial D}{\partial \dot{q}} = R\dot{q}. \quad (5.88)$$

The two Lagrange equations in the two generalized coordinates then can be written as

$$F = m\ddot{x} + \gamma'\dot{x} + k_s(x+x_1) - \frac{(Q_0+q)^2}{2\epsilon A}, \quad \text{and}$$

$$V = L\ddot{q} + R\dot{q} + \frac{x_0-x}{\epsilon A}(Q_0+q), \qquad (5.89)$$

with equilibrium setting the boundary conditions where x, $\dot{x}$, $\ddot{x}$, F, q, $\dot{q}$ and $\ddot{q}$ all vanish. So, of the two Lagrange equations, at equilibrium, the first reduces the balance of elastic and electrostatic forces that we have already found, and the second gives

$$V_0 = \frac{x_0 Q_0}{\epsilon A}. \qquad (5.90)$$

If small perturbations are assumed for charge and position, then we can take first order terms for the product terms, that is,

$$(Q_0+q)^2 \approx Q_0^2 + 2Q_0 q, \quad \text{and}$$

$$(x_0-x)(Q_0+q) \approx x_0 Q_0 - xQ_0 + x_0 q. \qquad (5.91)$$

So, off equilibrium, the Lagrange equations reduce to

$$F = m\ddot{x} + \gamma'\dot{x} + k_s x - \frac{Q_0 q}{\epsilon A}, \quad \text{and}$$

$$V = L\ddot{q} + R\dot{q} - \frac{Q_0 x}{\epsilon A} + \frac{x_0 q}{\epsilon A}. \qquad (5.92)$$

The third term in this equation is the ratio of the positional disturbance and the equilibrium gap, since $x_0 = \epsilon A/C_0$, and the fourth term is the equivalent for charge. Let $I_0 = Q_0/\epsilon A$; then, the effect of our small perturbation can be represented by

$$F = m\ddot{x} + \gamma'\dot{x} + k_s x - I_0 q, \quad \text{and}$$

$$V = L\ddot{q} + R\dot{q} + \frac{Q_0}{C_0} - I_0 x. \qquad (5.93)$$

The transfer function—the relationship between acoustic force on the diaphragm and the voltage across the resistor—follows directly from this.

We went through this exercise to stress that, by using the approach of Lagrange, relatively complex energy couplings can be analyzed in a straightforward procedure. Nanoscale systems are used for detecting vibrations, measuring different forces, providing stability and often working at measurement capability limits prescribed by noise and the measurement approach. The Lagrangian methodology gives a means of analyzing it to the first order analytically. So, nanosystems used for detecting vibrations, providing stability, measuring different forces—a myriad of diverse transduction mechanisms—all can be analyzed to the first order analytically.

It is pertinent to remark on motion, since our discussion has been entirely quasistatic.

The primary effect here is due to the motion in a gas or fluid environment. It is subject to fluctuation-dissipation. In the case of a beam, there are fast phenomena, through atoms and molecules interacting with the beam randomly, and there is the slow response of the beam to the applied forces. The beam response feels the friction of the fast response—they are connected. We will dwell on the connections between fast and slow phenomena later, as they determine the limits for precision measurements, as well as the characteristics, such as power spectra, through which we may measure. But, for now, we need to understand the response of the beam in this fluidic environment, to determine the speed of the response in a framework where we may ignore the fluctuations and capture the effect in the friction caused by the fluid. The motion of the beam will be affected. The Bernoulli effect is precisely the complement of what we are interested in—fluid flow around a shape such as a wing.

To understand, we must look at the conservation equations of fluid flow under conditions commonly encountered. There are five primary variables: the position $\mathbf{r}$, the time t, the energy reflected in the entropy per unit mass s, the fluid's mass density ρ, and the viscosity η, which reflects the sheer stress response—a form of fluid friction. The relationships

Fluctuation-dissipation, in Brownian motion, in electron motion, or in any environment where rapid changes, or noise, couple together with the system's response, is dealt with detail in S. Tiwari, "Semiconductor physics: Principles, theory and nanoscale," Electroscience 3, Oxford University Press, ISBN 978-0-19-875986-7 (forthcoming), where we are largely concerned with the connection of noise to system properties, to linear response, and thus to Kramers-Kronig relationships.

$$
\begin{aligned}
\frac{\partial \rho}{\partial t} + \boldsymbol{\nabla} \cdot \rho \mathbf{v} &= 0, \\
\rho T \frac{\partial s}{\partial t} - \kappa \boldsymbol{\nabla}^2 T &= \Sigma > 0, \\
\rho \left(\frac{\partial}{\partial t} + \mathbf{v} \cdot \boldsymbol{\nabla} \right) v &= -\boldsymbol{\nabla} p + \eta \boldsymbol{\nabla}^2 v, \\
p &= p(\rho, s), \text{ and} \\
\Sigma &= \frac{\kappa}{T} (\boldsymbol{\nabla} T)^2 + \eta \left(\frac{\partial v_i}{\partial r_j} \right)^2,
\end{aligned}
\tag{5.94}
$$

where $\rho(r,t)$ is mass density, $v(r,t)$ is the velocity, $s(r,t)$ is the entropy per unit mass, $p(\rho, s)$ is the pressure, η is the viscosity, and κ is the thermal coefficient, provide a continuum description when the disturbances have scale lengths longer than the mean free paths of the fluid molecules. The first of these describes conservation of mass, the second, the conservation of energy, and the third, the conservation of momentum. The viscosity η is friction, and these equations are the conservation laws in fluids. These are equivalent to the conservation equations employed for electron flow in semiconductors in the continuum approximation—the moment equations derived from Equation 2.74. Viscosity here abstracts the collective molecular fluctuation effects in a single parameter. It is the friction.

The form we commonly use and call the Navier-Stokes equation references only one of these equations—the third equation, which is valid for incompressible fluid when there are no other bodily forces, such as electrostatic or gravitational forces, present.

More precisely, and more generally, this equation—the Navier-Stokes equation—may be written as

$$\rho\left(\frac{\partial}{\partial t} + \mathbf{v} \cdot \boldsymbol{\nabla}\right)v = -\boldsymbol{\nabla}p + \boldsymbol{\nabla} \cdot \mathbb{T} + \mathbf{f}, \qquad (5.95)$$

where $\mathbb{T}$ is the deviatoric total stress—of order 2, reflecting the deviations from the mean normal stress tensor—and $\mathbf{f}$ is the body force per unit volume—gravity, electrostatic or any other. Absent this force $\mathbf{f}$, and assuming an incompressible fluid, this equation reduces to the third equation of the group of Equations 5.94. The left hand side of the equation is the contribution of fluid mass—from changes in velocity and its accumulation—so a divergence. The right hand side has the force and the consequence of changes in the velocity coordinate of the phase space. A moving plate or beam will be subject to forces due to the gas or liquid environment it is in during the movement. This is the complement of Bernouilli's principle—a change of reference frame. And these effects are well described for classical motion in the continuum description. Navier-Stokes equations are nonlinear and partial. So, additional nonlinearities arise due to the interaction between the environment and the moving plate.

The resistance that the fluid environment places against the motion of the cantilever beam considered earlier in the time-dependent beam motion problem represented in the Equation 5.9 now comes under an applied force $F = F_0 \exp(i\omega t)$:

$$\rho A \frac{\partial^2 u}{\partial t^2} + \frac{\beta}{2w} \frac{\partial u}{\partial t} + YI \frac{\partial^4 u}{\partial y^4} = F = F_0 \exp i\omega t, \qquad (5.96)$$

where

$$\beta = 3\pi\eta w + \frac{3}{4}\pi w^2 \sqrt{2\rho\eta\omega}. \qquad (5.97)$$

The Navier-Stokes equation, named after Claude-Louis Navier and George Gabriel Stokes, like the second law of thermodynamics, is phenomenologically valid, but we, at least as yet, know no way to derive it in a suitable approximation from first principles. It is Newton's law for fluid flow, but Newton's law can be derived from quantum mechanics using Ehrenfest's theorem as well as from the principle of least action applied under classical constraints.

β here is the real part of $1/d_{11}$, where d_{11} is the strain tensor component. We have written this resistance proportional to the instantaneous velocity of the vibrating beam without proof. F_0 is the amplitude of the force, and as a reminder, ρ is the mass density of the beam, w is the width, t is the thickness, I is the inertia, and Y is its Young's modulus. We are only considering the real part of the fluid resistance. An additional time-dependent displacement term has appeared in the equation, and together with it, another nonlinearity due to fluidic damping. This equation describes the time-dependent evolution more accurately—the first term is the inertial component, the second is the damping, and the third is the inertial force of the beam itself. For example, if a harmonic external force is impressed, the response will have additional frequency components in the response. One can analyze this through the multiple eigenmode analysis we employed in the non-damped, non-forced example earlier. The displacement eigenfunction $\mathcal{Y}_n$s still satisfies Equation 5.27—the orthogonality of function—and Equation 5.28—the orthogonality with the fourth order derivative. An additional property of the eigenfunction is

$$\int \mathcal{Y}_i \frac{d^4 \mathcal{Y}_i}{dy^4}\, dy = \omega_i^2 \int \mathcal{Y}_i^2\, dy. \tag{5.98}$$

These eigenfunctions form the orthogonal basis set from which we can construct the real solution through a linear combination. Let

$$u = \sum_{i=1}^{\infty} u_n(t)\mathcal{Y}_n \tag{5.99}$$

be the solution. Equation 5.96, multiplied by $\mathcal{Y}_n$ and integrated over the length then gives the time-dependent equation for harmonic forcing function as

$$m_n \ddot{u}_n + c_n \dot{u}_n + k_{sn} u_n = F_n \exp(i\omega t) \ \text{ for } \ n = 1, 2, \ldots. \tag{5.100}$$

Here, we have simplified the appearance of the equation by using dots and double dots for time derivatives. The other parameters of the equation are given by

$$m_n = \rho w t \int_0^L \mathcal{Y}_n^2\, dy; \ c_n = \frac{\beta m_n}{2\rho w t^2}; \ k_{sn} = m_n \omega_n^2; \ \text{and}$$

$$F_n = \int_0^L F \mathcal{Y}_n\, dy. \tag{5.101}$$

m_n is a mass term related to the eigenmode whose length effect is included in the integral with position, c_n is related to how the beam damps in the eigenmodes, k_{sn} is the elastic energy component of the eigenmode, and F_n is the force amplitude resulting from integration across the beam at the frequency ω. The nth mode is a result of the

These could have been made orthonormal, in which case, much of the quantum-mechanical perturbation techniques would transfer over. However, it is important to stress that this wave equation is quite different from—and of a higher order than—the Schrödinger equation.

coupling between the applied force, damping and the beam. This is a second order equation, quite solvable, an equation of damped response encountered in the damped systems, for example, the response of an RLC network. The solution is $u_n = G_n(\omega)F_n \exp(i\omega t)$, with

$$
\begin{aligned}
G_n(\omega) &= \frac{1}{m_n\omega^2 + ic_n\omega + k_{sn}} \\
&= \frac{1}{k_{sn}} \frac{1}{1 - (\omega/\omega_n)^2 + 2i\zeta_n\omega/\omega_n},
\end{aligned} \tag{5.102}
$$

where $\zeta_n = c_n/2m_n\omega_n$. The forced displacement of the beam in the presence of fluidic damping and a harmonic force is

$$
u = \exp(i\omega t) \sum_{i=1}^{\infty} F_i G_i(\omega)\mathcal{Y}_i(y). \tag{5.103}
$$

This is the complete solution for damped and forced conditions employing eigenmode techniques.

Now let us assume that the impressed force is at a frequency close a mode, the nth mode, that is, $\omega \approx \omega_n$. The nth term of the expansion, the natural frequency of the nth mode being closest in frequency of the impressed force, will be the dominant term, since the coefficient $\zeta_n \ll 1$, and $\omega \approx \omega_n$. So, the displacement $u \approx F_n G_n \mathcal{Y}_n(y) \exp(i\omega t)$.

The mode equation 5.100 has the same form as the equation for a single degree of freedom under force vibrations. It is an eigenmode nonlinear partial differential equation where one must find a sufficient number of the terms to get an accurate estimate. The beam responds as if there is only a single degree of freedom—that of the nth mode, which is closest to the impressed frequency. The damping in the system can then be understood within this single degree of freedom. Consider the real harmonic force, the in-phase component of $\exp(i\omega t)$, $F = F_n \sin \omega_n t$. The displacement of the mass m behaves like the case shown in Figure 5.3—a mass vibrating under force, with kinetic and potential energy exchange under damping. The solution is

$$
u(t) = \frac{1}{2\zeta_n} \frac{F_n}{k_{sn}} \cos \omega_n t. \tag{5.104}
$$

The peak kinetic energy of the eigenmode is

$$
T_{max} = \frac{1}{2} m_n \left(\omega_n \frac{F_n}{8\zeta_n^2 k_n} \right)^2 = \frac{F_n^2}{128\zeta_n^4 k_n} = \frac{1}{2} \frac{1}{k_n} \left(\frac{F_n}{8\zeta_n^2} \right)^2. \tag{5.105}
$$

The total external work in a cycle is

$$
\begin{aligned}
\Delta W &= \oint c_n \dot{u} du = \oint c_n \dot{u}^2 dt \\
&= c_n \int_0^{2\pi/\omega_n c_n} \left(\frac{\omega_n F_n}{2\zeta_n k_n} \sin \omega_n t \right)^2 dt = \frac{\pi F_n^2}{c_n \omega_n}. \tag{5.106}
\end{aligned}
$$

The beam attempting to response to the force, is following it as close at it can, under constraints of damping. One can also see this after the analysis in the mathematical result.

Therefore, from the energy dissipated per cycle, the quality factor $\mathcal{Q}$ is related as

$$\frac{1}{\mathcal{Q}} = \frac{\Delta W}{W} = 4\pi\zeta_n. \tag{5.107}$$

There is a relative energy loss of 4π times ζ_n—related to the eigen-mode amplitude coefficient and frequency and inversely related to the eigenmode mass, so to the properties of the beam.

Now consider the response when an excitation is shut off and the vibration decays due to damping. Assume that the beam's free vibration decays from a starting condition consistent with the nth eigenfunction's mode. Equation 5.103, starting from this unique self-consistent initial starting condition of the nth eigenmode chosen for simple mathematical form, then has the solution

$$u = u_0 \exp\left(-\frac{t}{\tau}\right)\cos\omega_d t, \tag{5.108}$$

where $\omega_d = \omega_n(1 - \zeta_n^2)^{1/2}$, and $\tau = 1/\omega_n\zeta_n$, and where $\zeta_n \ll 1$ reflects the decay time scale.

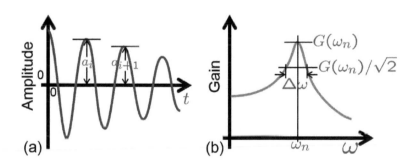

(a)

(b)

Figure 5.12: The response of a cantilever in the presence of damping. (a) shows the time dependence when a constant force is applied, and (b) shows the spectral response.

Figure 5.12 shows the response that Equation 5.100 indicates. Figure 5.12(a) shows it for the decay of the nth eigenfunction starting from a pure nth mode initial condition. Figure 5.12(b) shows the response for a harmonic force at the frequency closest to ω_n. The time-dependent solution has a progressively decreasing amplitude in time. The progressive amplitude ratio is $\exp(-t/\tau) \approx 1 + 2\pi\zeta_n$. If a cantilever is employed in controlling a position and one wishes to damp motion, one would make the factor ζ_n as large as possible. If then one were to apply a static force F_n, the static deflection of the beam would be F_n/k_{sn}, and the ratio of the resonant amplitude to this static deflection would be $2\zeta_n$. The frequency response would have resonance at a frequency close to ω_n, an energy half-width, that is, with an amplitude of $1/\sqrt{2}$ of $2\zeta_n$. The quality factor $\mathcal{Q} = \omega_n/\Delta\omega$, and $\zeta_n = 1/2\mathcal{Q}$. A high $\mathcal{Q}$ means less loss per cycle as well as a more accurate measurement of ω_n, since the peak is sharp. This, in turn, means a more accurate measurement of forces.

The use of high Q systems is therefore essential for accurate measurements with cantilevers.

The damping arises not just from the ambient in which the cantilever operates, but also from the internal frictional forces of the beam—the anharmonicity of the mechanical response, due to bulk and surface effects, as well as the anchoring of the structure. This beam effect can become important for long beams in materials such as silicon. Depending on the geometry, the anchoring makes the quality factor of losses at the anchor very significant for even small dimension variations. Another important point regarding this analysis is that, in general, we are attempting to solve a wave equation. It is significantly more complex than the electromagnetic wave case. There is dissipation, nonlinearity and these higher order terms. This means that only very specific highly circumscribed problems will have analytic solutions. The eigenmode analysis is a good example of this analytical solution approach. The three-dimensional nature of the problems is an additional wrinkle. And the complexity is also connected to the multiphysics and multiscale nature of the problem— multiple interactions are simultaneously present, more than one dimensional scale is important and therefore more than one method of analysis, for example, quantum and classical, will be important. For now, we restrict ourselves to realistic problems that can be subjected to classical analysis. We should be able to understand and extract important conclusions drawing on our classical analysis with quantum insights.

We now turn to the use of cantilever beams in precision measurements using this classical-quantum mix. These moving cantilever probe techniques are particularly apropos w.r.t. quantum measurements, that is, ultrasensitive measurement in energy limits such as of single quantum modes. Such measurements, for example, single-electron, single-photon or single-phonon measurements, or precise ``wavefunction'' measurements at a surface through charge, must necessarily be at the intersection of the classical with the quantum. We do this by first understanding the meaning and limitations of the classical description and then looking at fluctuation-dissipation in a different way than we did in our semiconductor physics discussion. We will start with understanding when equipartition of energy breaks down.

The principle of equipartition of energy is quite useful in analyzing the properties of ensembles of independent classical particles that are not subject to exclusion constraints. The Maxwell-Boltzmann distribution function and the coordinates for the position $\mathbf{q}_i$ and mo-

An example of loss sensitivity is powerfully shown in a disk resonator—a disk attached to a substrate through a small post. Damping and energy losses can occur through this small post and may become particularly efficient when there is good matching between the positional wave of the disk and the transmission modes of the pedestal. Qs can change by factors of hundreds for even fractional percentage changes in thickness.

For a discussion of the physical meaning and the manifestation of connection between fast and slow, system and noise, see S. Tiwari, ``Semiconductor physics: Principles, theory and nanoscale,'' Electroscience 3, Oxford University Press, ISBN 978-0-19-875986-7 (forthcoming).

mentum $\mathbf{p}_i$ suffice for description. Appendix D summarizes a short derivation of the expectation energy associated with each degree of freedom of motion. It is $k_B T/2$. So, in three-dimensional assemblies, each particle on an average will be expected to have $3k_B T/2$ of kinetic energy. The constraint is that there are no exclusion restrictions on position and momentum, except those that arise naturally through the thermodynamic constraints in statistics.

The first order effect that one may consider beyond this classical approximation is that of any non-continuous effect—a discretization effect such as quantum-mechanical constraints. For the classical picture to be still applicable, the discretization should be very small compared to this energy scale, that is, $\Delta E_i \ll k_B T/2$. At the nanometer scale, confinement in any or all three dimensions in a semiconductor may break this rule. Low temperatures may also break this rule.

The allowed wavevector, or equivalently, momentum and energy spacings are very small, and the classical description is quite valid in non-degenerate conditions. If the material is degenerate, states below and around the Fermi energy are predominantly occupied, and the non-interacting free ranging of the classical approximation breaks down. Certain states are not randomly accessible anymore because of the interaction and occupation that Fermi-Dirac statistics represents. Even in moving atoms, there is a size scale where this description would break down.

We can now tie this discussion of energy with Brownian motion, which underlies the damping term. Brownian motion, the mechanism that is one of the causes of damping for the cantilever, is a good example of thermal motion and noise for the cantilever beam. Figure 5.13 shows a set of fictitious particles, for example, inert gas molecules, in a classical distribution and undergoing Brownian motion, that is, scattering with each other. Because of the three degrees of freedom, the energy for each particle is

$$U = T = \sum_{i=x,y,z} \frac{\mathbf{p}_i^2}{2m} = \frac{1}{2}mv_x^2 + \frac{1}{2}mv_y^2 + \frac{1}{2}mv_z^2, \tag{5.109}$$

and, by equipartition of energy,

$$\left\langle \frac{1}{2}mv_x^2 \right\rangle = \left\langle \frac{1}{2}mv_y^2 \right\rangle = \left\langle \frac{1}{2}mv_z^2 \right\rangle = \frac{1}{2}k_B T. \tag{5.110}$$

At what size scale will the Brownian motion be observable? The Brownian motion is observable when the energy associated with a single particle in its volume $\mathfrak{V}$ exclusion zone becomes comparable to the thermal energy constraint, that is, for a density ρ, where

$$\frac{1}{2}\rho\mathfrak{V}\langle v^2 \rangle \approx \frac{3}{2}k_B T, \tag{5.111}$$

Metals have a large effective mass for the electrons, and these energy states can be close together, for example, much closer than room temperature thermal energy. But, this rule of energy separation between states as they are being filled can still be broken because of classical, that is, non-quantum, effects. Placing a single electron on a nanoscale object, one with aFs of capacitance, requires a large electrostatic Coulomb energy—on the scale of thermal energy. This single electron behavior depends on the classical electrostatic coupling of this nanometer object with its surroundings.

We will see that some forms of noise, for example, atomic vibrations, come close to being classical in the approximate limit. Examples include the Debye and the Einstein theories for the heat capacity, electrons in non-degenerate conditions, et cetera, come close to being describable classically in a range of temperatures where being subject to Bose-Einstein or Fermi-Dirac distribution among allowed states reduces to Maxwell-Boltzmann distribution.

that is,

$$\mathfrak{V} \approx \frac{k_B T}{\rho \langle v^2 \rangle} \approx \frac{0.0259 \times 1.6 \times 10^{-19} \ J}{10^3 \ kg/m^3 \times (300 \ m/s)^2}$$

$$\approx 4.6 \times 10^{-15} \ m^3 \quad \therefore \ \ell \approx 17 \ \mu m. \qquad (5.112)$$

The Brownian fluctuations represent a form of scale breakdown of the classical continuum description. When one's resolution of observation is made more precise, discreteness becomes observable. This size-scale discretization in the quantum-classical span is relatable by the inequality $\Delta q \Delta p \geq \hbar$, or $\Delta E \Delta t \geq \hbar$, of Heisenberg uncertainty. At the large size scale, the quantum uncertainty, a fluctuation, appears washed away. Where does the classical-quantum cross-over take place and become significant? Consider a classical description of molecules as objects with a mean separation $\bar{r}$ and a mean momentum $\bar{p}$. Heisenberg uncertainty here suggests the condition $\bar{r}\bar{p} \gg \hbar$ for the classical description to be applicable. The de Broglie wavelength of the molecule is $\bar{\lambda} = h/\bar{p} = 2\pi\hbar/\bar{p}$. So, the usefulness of the classical description is restricted to those cases where $\bar{r} \gg \bar{\lambda}$. Only when the molecular separation is significantly larger than the de Broglie wavelength of the molecule would a classical description suffice. Where the classical description, that is, an average behavior as a good description of an ensemble—with uniformity of description of the property in space and time—fails is when the size scale or the time scale over which the classical property is measured is limited. This is when one must account for uncertainty. In these situations, the ensemble description is not over sufficiently large numbers for it to be accurate enough to be useful.

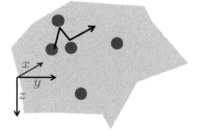

Figure 5.13: Brownian motion of particles in an ensemble.

Noise is the time fluctuation describing the spread in measurements in the conjugate coordinate—momentum. Brownian motion is the momentum fluctuation that is observable in space through a spread of measurements in its conjugate coordinate—time.

The intermediate domain of these scale relations is at small dimensions when ($\bar{r} \approx \bar{\lambda}$). Discreteness is still a good description, but the volume is still small enough that the ensemble is also small. This domain is interesting and occurs at the range that one often encounters in measurements at their limits. This is the domain between quantum and classical in the midst of a small ensemble. Here one sees all forms of noise, such as shot noise, Brownian motion, and other phenomena at the intersection of quantum with classical.

Finally, when one gets to the limit $\bar{r} \ll \bar{\lambda}$, one must describe the molecules quantum-mechanically, that is, through the wavefunction $|\psi\rangle$, from which suitable properties may be extracted.

Some order-of-magnitude calculations about this region at the interface of classical and quantum-mechanical approaches are helpful to make a connection to the natural world. The volume of the region associated with molecules is $V = \bar{r}^3 N$, where N is the number of molecules. This means $\bar{r} = (V/N)^{1/3}$. Equipartition of energy, valid in the classical conditions, implies $\bar{p}^2/2m = 3k_B T/2$, that is, $\bar{p} = (3mk_B T)^{1/2}$. This results in the following requirement for the de Broglie length in this particle or molecule gas

$$\bar{\lambda} = \frac{h}{(3mk_B T)^{1/2}} \quad \therefore \quad \left(\frac{V}{N}\right)^{1/3} \gg \frac{h}{(3mk_B T)^{1/2}}, \qquad (5.113)$$

for classical approximations to be valid. This holds true generally—for particles ranging from molecules and atoms to electrons, photons and others that we often think of only quantum-mechanically.

Let us consider the constraints of this classical-quantum boundary for both molecules in air and electrons in solids. Air is predominantly nitrogen, which has the molecular mass $m = 2.324 \times 10^{-27}\ kg$. At the temperature $T = 300\ K$ and one atmosphere pressure, the mass of an ensemble is 14 $g/mole$, that is, where a mole contains Avogadro's number, that is, 6.023×10^{23}, molecules. So, $N/V = \bar{p}/k_B T \approx 2.5 \times 10^{19}\ molecules/cm^3$. This corresponds to a mean spacing of $\bar{r} \approx 3.4 \times 10^{-9}\ m$, and a de Broglie wavelength of $\bar{\lambda} \approx 1 \times 10^{-10}\ m$. So, for air, which consists mostly of nitrogen, $\bar{r} \gg \bar{\lambda}$, and the classical description is quite acceptable under these conditions of temperature and pressure. But, this is not always true. Low temperatures or high pressures will conflict with this constraint.

Now consider the electron gas in a metal such as of the alkali group—Li, Na, K, et cetera. The outermost orbital electron, one per atom, is potentially available for conduction. And let us consider the extrema of this condition, where all are available for conduction in the crystal. This corresponds approximately to $\bar{r} \approx 2 \times 10^{-10}\ m$. The electron mass is $m = 9.1 \times 10^{31}\ kg$, corresponding to a de Broglie wavelength of $\bar{\lambda} \approx 5 \times 10^{-9}\ m$. $\bar{r} \ll \bar{\lambda}$, and the classical approximation is not appropriate for modeling such an ensemble. On the other hand, if one considers a non-degenerate electron ensemble in silicon, say at a density of $1 \times 10^{16}\ cm^{-3}$, then $\bar{r} \approx 4.5 \times 10^{-8}\ m$. Here $\bar{r} \gg \bar{\lambda}$, and the classical approach will provide quite valid estimations.

When one is in this small-dimensional limit of observation, we can directly observe the fluctuations instead of just the dissipation effect such as the drag. In Figure 5.14, which is a refinement of Figure 5.13, a microsystem $\mathfrak{S}$ of mass m interacts with its environment $\mathfrak{R}$, which is at temperature T. Our interest is in finding not only the energetics of these interactions, and their consequences in system properties, such as the ones that fluctuation-dissipation theorem points to, but also specifically the consequences for the ultimate limits of measurements that the characteristics of these interactions portend. This

Or one may even get a new quantum state of matter. A Bose-Einstein condensate formed from an ensemble of atoms is an entirely new state of matter—a quantum state which acts as a new collective whole with an entirely new set of properties. It is far more than just a breakdown of classical approximation.

Yet, raise this carrier density in a three-dimensional semiconductor system, such as through doping, confine the electrons in a two-dimensional electron gas, or lower the temperature of a very low scattering two-dimensional gas, such as in a very high mobility system, and one would see breakdown of the classical approximation. One would see quantum Hall or fractional quantum Hall effects which can only exist in quantum-constrained conditions with additional restrictions.

would entail understanding the consequences for a nanoscale system, where the energetics arise from gains due to applied forces or losses due to dissipation in time or in its reciprocal space form—frequency, that is, in the power spectral characteristics. We will find that this power spectrum contains much of the information important to us in defining limits of operation of devices such as the resolution limit of measurement or the finest scale of actuation.

The response of mass m is subject to two forces—the externally applied force, and any force that represents the interaction with the environment. The first is a slowly varying force. The second is a rapidly varying force due to the fluctuating interactions with the environment. We write the slowly varying force as $\mathcal{F}$, and the rapidly varying force as $\mathfrak{F}$:

$$m\frac{dv}{dt} = F(t) = \mathcal{F}(t) + \mathfrak{F}(t). \tag{5.114}$$

Consider molecules moving at speeds of sound velocity, so $\sim 10^3 \ m/s$. A travel distance of about the size of simple unit cell of a solid, $\sim 0.5 \ nm$, occurs in about $0.5 \ ps$. This time is a very small time. Another way of saying this is that the correlation time τ^* of the rapidly varying force $\mathfrak{F}$ representing the fluctuation interactions is very small. The slow system response takes place on a time scale much larger than the time scales over which any fluctuating rapidly varying force event has an effect that can be directly correlated to that event.

Consider the case when there is no externally applied force in the system outlined in Figure 5.14. Absent external force, the averaged velocity also vanishes.

If we consider time scales $\tau \gg \tau^*$,

$$
\begin{aligned}
m\left[v(t+\tau) - v(t)\right] &= \mathcal{F}(t)\tau + \int_t^{t+\tau} \mathfrak{F}(t') \, dt' \\
&= \mathcal{F}(t)\tau + \int_t^{t+\tau} \langle \mathfrak{F}(t') \rangle \, dt'. \tag{5.115}
\end{aligned}
$$

When time scales of interest are larger than the correlation times τ^*, one may use an averaged force to capture the effect of the rapidly varying force. It is this rapidly varying force acting in time that causes a system to move towards thermal equilibrium, absent external forces. Present external force, it will also be this force that will cause the system to reach a steady state.

We can now mathematically describe these fluctuation correlations. The change in the system $\mathfrak{S}$ after time $\tau \gg \tau^*$ is describable by Boltzmann statistics. The accessible states are determined by the reservoir $\mathfrak{R}$. The most probable microstates are the ones with the highest degeneracy, as determined through this exchange with the reservoir. The energy change from E' to $\Delta E'$ for $\mathfrak{S}$, after time τ', is

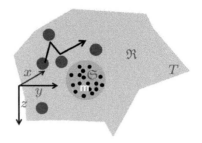

Figure 5.14: System $\mathfrak{S}$ of mass m interacting with the environment $\mathfrak{R}$ at temperature T in the scale limit where fluctuations, such as Brownian motion, whose effect is felt in dissipation, can be observed.

A correlation time is a measure of the time over which one would feel in select properties the direct consequences of the fluctuation events through the energetics they couple. At times much longer than the correlation time, the effect will appear as an average that is dissipative. See S. Tiwari, "Semiconductor physics: Principles, theory and nanoscale," Electroscience 3, Oxford University Press, ISBN 978-0-19-875986-7 (forthcoming) for a detailed discussion.

An electron in a high electric field in a semiconductor loses its energy through rapid interactions with the crystal—particularly via optical phonon emission. This energy loss in scattering too is an example of when an electron ensemble in a long sample in high electric fields reaches steady-state saturated velocity because of fluctuations, which predominantly arise in phonon scattering.

then described by

$$\frac{\mathfrak{p}_\sigma(t+\tau')}{\mathfrak{p}_\sigma^0(t)} = \frac{\Omega(E' + \Delta E')}{\Omega(E')} = \exp\left(\frac{\Delta E'}{k_B T}\right). \qquad (5.116)$$

Here, Ω represents the number of most probable microstates, and $\mathfrak{p}_\sigma$ represents the probability of any property associated with this likely collection of microstates. This probability is related to the energy change exponentially. When the time elapsed is large enough, that is, significantly greater than the scattering time, the system $\mathfrak{S}$ is likely to be in the equally likely collection of all accessible states—the most probable microstate collection Ω is determined by the statistics of the reservoir, to which the system $\mathfrak{S}$ is a minor perturbation involving the exchanges taking place in the fluctuation interactions.

How does this system change as the reservoir itself evolves?

We know, with small perturbations in energy,

$$\mathfrak{p}_\sigma(t+\tau') = \mathfrak{p}_\sigma^0(t) \exp\left(\frac{\Delta E'}{k_B T}\right) \approx \mathfrak{p}_\sigma^0(t)\left(1 + \frac{\Delta E'}{k_B T}\right). \qquad (5.117)$$

Here, the $\underline{0}$ superscript indicates thermal equilibrium. This description of probability change in terms of the time of the system describes the evolution of any property that is commensurate with these probabilities. So, the expectation of the rapidly varying force is

$$\begin{aligned}
\langle \mathfrak{F} \rangle &= \sum_\sigma \mathfrak{p}_\sigma(t+\tau')\mathfrak{F}_\sigma \\
&\approx \sum_\sigma \mathfrak{p}_\sigma^0(t)\left(1 + \frac{\Delta E'}{k_B T}\right)\mathfrak{F}_\sigma = \left\langle \left(1 + \frac{\Delta E'}{k_B T}\right)\mathfrak{F}_v \right\rangle_0.
\end{aligned} \qquad (5.118)$$

Computation of the expectation with equilibrium probabilities means that, at thermal equilibrium, $\langle \mathfrak{F} \rangle = 0$. Therefore, for a small perturbation away from equilibrium,

$$\langle \mathfrak{F} \rangle = \frac{\langle \mathfrak{F}\Delta E' \rangle_0}{k_B T}. \qquad (5.119)$$

All these relationships are good approximations for times that are significantly larger than correlation times, that is, $\tau \gg \tau^*$. These are time scales where the strong time correlations of rapidly time-varying interactions are averaged out.

We can use these relationships to calculate the magnitudes of the effects:

$$\Delta E' = -\int_t^{t'} v(t'')\mathfrak{F}(t'')\, dt'' \approx -v(t)\int_t^{t'} \mathfrak{F}(t'')\, dt''. \qquad (5.120)$$

Since the velocity changes slowly over the correlation times τ^* of interest, that is, it is perturbed and is observable, but the averages are changing slowly,

$$
\begin{aligned}
\langle \mathfrak{F}(t') \rangle &= -\frac{1}{k_B T} \langle \mathfrak{F}(t') v(t) \int_t^{t'} \mathfrak{F}(t'') dt'' \rangle_0 \\
&= -\frac{1}{k_B T} \overline{v}(t) \int_t^{t'} \langle \mathfrak{F}(t') F(t'') \rangle_0 \, dt''. \quad (5.121)
\end{aligned}
$$

We substitute $s = t'' - t'$, and we get the response to force in the time interval τ as

$$
\begin{aligned}
m \langle v(t+\tau) - v(t) \rangle &= \mathcal{F}(t)\tau + \int_t^{t+\tau} \langle \mathfrak{F}(t') \rangle \, dt' \\
&= \mathcal{F}(t)\tau - \frac{1}{k_B T} \overline{v}(t) \\
&\quad \times \int_t^{t+\tau} dt' \int_{t-t'}^0 \langle \mathfrak{F}(t') \mathfrak{F}(t'+s) \rangle_0 \, ds.
\end{aligned}
$$

$$(5.122)$$

We now define a correlation function $\mathcal{K}(s)$, where s is the dummy time separation over which this correlation is defined, as

$$
\mathcal{K}(s) = \langle \mathfrak{F}(t') \mathfrak{F}(t'+s) \rangle_0 = \langle \mathfrak{F}(t') \mathfrak{F}(t'') \rangle_0. \quad (5.123)
$$

The term inside the integrand arising from the rapidly varying force's time consequence in correlation is finite positive. It is a dissipative term. It causes the return to thermal equilibrium when the slowly varying forces are null. It causes, at thermal equilibrium, the average velocity to vanish. When a slowly varying force is present, it helps establish a steady-state response. We note, at $s = 0$, the following condition of vanishing time difference in the correlation:

$$
\begin{aligned}
\mathcal{K}(0) &= \langle \mathfrak{F}(t') \mathfrak{F}(t') \rangle_0 \\
&= \langle \mathfrak{F}^2(t') \rangle_0 \geq 0. \quad (5.124)
\end{aligned}
$$

This correlation function $\mathcal{K}(s)$ is a measure of dispersion—how rapidly the correspondence between fast force and its effect changes over time. Since, over a long time duration, the average of the rapidly changing force vanishes, that is, $\langle \mathfrak{F} \rangle = 0$, the rapidly varying force must lose all correlation with effect as a result of the accumulation of a large number of uncorrelated scattering events. So,

$$
\lim_{s \to \infty} \mathcal{K}(s) \to \langle \mathfrak{F}(t) \rangle \langle \mathfrak{F}(t+s) \rangle = 0. \quad (5.125)
$$

$\mathcal{K}(s)$ also satisfies one additional condition that follows from the expectation of the square of the forces being positive:

$$
\begin{aligned}
\because \quad \langle [\mathfrak{F}(t) \pm \mathfrak{F}(t+s)]^2 \rangle &\geq 0 \\
\therefore \quad \langle \mathfrak{F}^2(t) + \mathfrak{F}^2(t+s) \pm 2\mathfrak{F}(t)\mathfrak{F}(t+s) \rangle &\geq 0 \\
\therefore \quad \langle \mathfrak{F}^2(t) \rangle + \langle \mathfrak{F}^2(t+s) \rangle \pm \langle 2\mathfrak{F}(t)\mathfrak{F}(t+s) \rangle &\geq 0 \\
\therefore \quad 2\mathcal{K}(0) \pm 2\mathcal{K}(s) &\geq 0,
\end{aligned}
$$

that is, there is a bound on the correlation function defined by $-\mathcal{K}(0) \leq \mathcal{K}(s) \leq \mathcal{K}(0)$. Since $\mathcal{K}(s)$ is independent of time, being only a function of s, one may shift it arbitrarily. Using a new time, $t_1 = t - s$,

$$
\begin{aligned}
\mathcal{K}(s) &= \langle \mathfrak{F}(t)\mathfrak{F}(t+s) \rangle = \langle \mathfrak{F}(t_1)\mathfrak{F}(t_1+s) \rangle \\
&= \langle \mathfrak{F}(t-s)\mathfrak{F}(t) \rangle = \langle \mathfrak{F}(t)\mathfrak{F}(t-s) \rangle \\
&= \mathcal{K}(-s).
\end{aligned} \tag{5.126}
$$

This establishes that correlations are symmetric in time.

Figure 5.15 is a sketch of this correlation function with the separation time. It is highest at vanishing separation, that is, close to the time scale of τ^*, the correlation time, and it rapidly decays beyond that.

We can now evaluate the response of the system $\mathfrak{S}$ of mass m of Figure 5.14, as written down in Equation 5.122. We have

$$
\begin{aligned}
m\langle v(t+\tau) - v(t) \rangle &= \mathcal{F}(t)\tau - \frac{1}{k_B T}\overline{v}(t) \\
&\times \int_t^{t+\tau} dt' \int_{t-t'}^0 \langle \mathfrak{F}(t')\mathfrak{F}(t'+s) \rangle_0 \, ds.
\end{aligned} \tag{5.127}
$$

The response of the mass is the accumulation from a slowly varying force and a rapidly varying force. Over a small time interval τ, the response to the slowly varying force may be approximated by a constant force. The response to a rapidly varying force can be accounted for through a double integral. The rapidly varying force changes the energy—dissipating it. The average velocity is the average distance over time over which this fluctuating force causes a correlated effect. A small time duration means that the force is more correlated, with the effect,and, hence, the effect is more pronounced. A large time means that this rapidly varying force is less correlated with a motional effect—it leads to impulses in different directions—so that the average effect will reduce. It is the correlation of force with the motional effect acting over a distance that causes the energy change, so the magnitude of energy change w.r.t. the thermal energy $k_B T$ is also an important ratio. The average response to the rapidly fluctuating forces is embedded in this correlation function, which appears as a double integral in Equation 5.127. The meaning of this double integral is simply that the effect of the force must be evaluated over the extent of the time interval of interest, while taking into account that, within this time interval, the rapidly occurring events lead to a correlation time dependence—the farther apart in time, the smaller the effect.

Correlation measures the correspondence between the expectations arising from causal forces separated in time. So, the expectations are causally connected through the magnitude and not the sign of separation. Either of the two forces could have acted first.

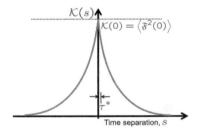

Figure 5.15: The correlation function's dependence on separation in time. It peaks at short times, that is, times close to the time scale of τ^*, the correlation time, and decays, vanishing in the limit of $s \to \infty$, where the average of the rapidly varying force must also vanish.

Figure 5.16 shows a geometric view of this integration. For any time separation τ, one must first include the effects between t and $t + \tau$, one of the intervalas over which the double integral is being evaluated. The double integral integrates a section in the (t', s) space, and we can simplify:

$$
\begin{aligned}
m \langle v(t + \tau) - v(t) \rangle &= \mathcal{F}(t)\tau - \frac{1}{k_B T}\overline{v}(t) \int_t^{t+\tau} dt' \int_{t-t'}^0 \mathcal{K}(s)\,ds \\
&= \mathcal{F}(t)\tau - \frac{1}{k_B T}\overline{v}(t) \int_{-\tau}^0 ds \int_{t-s}^{t+\tau} \mathcal{K}(s)\,dt' \\
&= \mathcal{F}(t)\tau - \frac{1}{k_B T}\overline{v}(t) \int_{-\tau}^0 (\tau + s)\mathcal{K}(s)\,ds.
\end{aligned}
$$

$$(5.128)$$

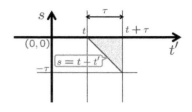

Figure 5.16: A geometric outline of the double integration of the correlation function in order to evaluate the consequences for motion.

For time differences $\tau \gg \tau^*$, we have noted that $\mathcal{K}(s) \to 0$ so long as $|s| \gg \tau^*$, that is, $s \to \infty$. The double integral can be simplified to a single integral, and the term is finite positive, so

$$
\begin{aligned}
m \langle v(t + \tau) - v(t) \rangle &= \mathcal{F}(t)\tau - \frac{1}{k_B T}\overline{v}(t) \int_{-\tau}^0 (\tau + s)\mathcal{K}(s)\,ds \\
&= \mathcal{F}(t)\tau - \frac{1}{k_B T}\overline{v}(t)\tau \int_{-\infty}^0 \mathcal{K}(s)\,ds \\
&= \mathcal{F}(t)\tau - \frac{1}{k_B T}\overline{v}(t)\frac{\tau}{2} \int_{-\infty}^{\infty} \mathcal{K}(s)\,ds \\
&= \mathcal{F}(t)\tau - \frac{1}{k_B T}\overline{v}(t)\frac{\tau}{2} \int_{-\infty}^{\infty} \langle \mathfrak{F}(0)\mathfrak{F}(s) \rangle_0\,ds,
\end{aligned}
$$

$$(5.129)$$

where we used the symmetry of $\mathcal{K}(s)$. Since the mean velocity evolves slowly,

$$
\frac{d\overline{v}}{dt} = \frac{\langle v(t + \tau) \rangle - \langle v(t) \rangle}{\tau} = \frac{\langle v(t + \tau) - v(t) \rangle}{\tau}. \qquad (5.130)
$$

Therefore, we may write,

$$
\begin{aligned}
m\frac{d\overline{v}(t)}{dt} &= \mathcal{F}(t) - \left[\frac{1}{2k_B T} \int_{-\infty}^{\infty} \langle \mathfrak{F}(0)\mathfrak{F}(s) \rangle_0\,ds \right] \overline{v}(t) \\
&= \mathcal{F}(t) - \Gamma\overline{v}(t).
\end{aligned}
$$

$$(5.131)$$

The Brownian process causes dissipation and damping, and one may quantify it with a damping parameter:

$$
\Gamma = \frac{1}{2k_B T} \int_{-\infty}^{\infty} \langle \mathfrak{F}(0)\mathfrak{F}(s) \rangle_0\,ds. \qquad (5.132)
$$

We have now found one source of the damping in mechanical movement, in the fluctuations and the drag effect from the gaseous fluid environment. Other sources exist too, for example, the anharmonicity of the elasticity of the movement, sometimes referred to as

Zener internal damping. We will look at this towards the conclusion of this chapter. All these are the result of the first order correction arising from dissipation. So, in general, this damping correction then results in an oscillator that is a nonlinear harmonic oscillator. The simplest form that one may write its force equation is the form that we have used a number of times:

$$\ddot{u} + \Gamma\dot{u} + k_s u = F(t) \tag{5.133}$$

$F(t)$, the force here, may have both a slowly varying component and a rapidly varying component. If it only consists of a slowly varying component, then in this equation, one may treat, $\mathcal{F} = F(t) - k_s u$. On the other hand, if it is only fast, then $\mathcal{F} = -k_s u$ and $\Gamma\dot{u} - \mathfrak{F}(t)$ produces the damping. The implications of Equation 5.132 can now be evaluated. If the source of noise is entirely uncorrelated, that is, white in the limit of very rapidly varying forces, then

$$\begin{aligned} \langle \mathfrak{F}(t)\mathfrak{F}(t') \rangle &= 2k_B T\Gamma\delta(t - t'), \text{ and} \\ \mathcal{K}(s) &= 2k_B T\Gamma\delta(s). \end{aligned} \tag{5.134}$$

The Fourier transform of this, that is, in the reciprocal space frequency coordinate, is

$$S(\omega) = \frac{1}{2\pi}\int_{-\infty}^{\infty} \mathcal{K}(s)\exp(i\omega s)\,ds = \frac{k_B T\Gamma}{\pi}. \tag{5.135}$$

This is the power spectral density in the double-sided form, that is, over the radial frequency band defined as $-\infty \le \omega \le \infty$. It is the power contained per unit frequency in the fluctuations that gave rise to the damping expressed through the parameter Γ. This force fluctuation, because of the correlations, will have a spectral dependence that is included in our analysis. The damping factor, arising as it does through the correlation function, includes it. This spectrum of thermal force fluctuations assumed to be the displacements in u is

$$\begin{aligned} S_u(\omega) &= \frac{1}{\left(\omega_0^2 - \omega^2\right)^2 + \omega_0^2\omega^2/Q^2}\frac{k_B T\Gamma}{\pi m^2} \\ &= \frac{1}{\left(\omega_0^2 - \omega^2\right)^2 + \Gamma^2\omega^2/m^2}\frac{k_B T\Gamma}{\pi m^2}. \end{aligned} \tag{5.136}$$

This is the same form as that of Equation 5.102 which we derived under a damping term using eigenmode analysis. There, we derived it for what happens in displacement of the beam. This spectral distribution means that we can determine the mean square displacement fluctuations where this interaction manifests itself. The spectral density

$$S_u(\omega) = \lim_{T \to \infty} \frac{u(\omega)u^*(\omega)}{T} \tag{5.137}$$

In an experimental apparatus, one measures over real frequencies, so the spectrum is single sided, that is, $0 \le \omega \le \infty$. But, from a Fourier point of view, it is more convenient and natural to perform analysis through a double-sided spectrum. This often leads to confusion, for example, in the Kramers-Kronig treatment. Suffice it to say that the double-sided spectrum and the single-sided spectrum are related:

$$\int_{-\infty}^{\infty} S_u(\omega)\,d\omega = \frac{1}{2\pi}\int_0^{\infty} S_u'(\omega)\,d\omega,$$

where $S_u(\omega)$ is double sided, and $S_u'(\omega)$ is single sided. The $1/2\pi$ factor came directly from the integral of the Fourier transform relationship to the function in time domain. Normalization leads to the factor $1/(2\pi)^{1/2}$ that one also sees. Because of symmetry, $S_u(\omega) = S_u'(\omega)/4\pi$. The roundtrip $1/2\pi$ factor in Fourier transforms is partitioned differently by physicists and electrical engineers—like i and j for imaginary unit. I favor $1/(2\pi)^{1/2}$'s symmetry of assignment to both transform integrals instead of an arbitrary $1/2\pi$ and 1. I favor i for imaginary, even if there is nothing imaginary about it. I prefer not to use j for imaginary numbers, to avoid confusion with current density.

Note here the correspondence with the Nyquist expression, used in S. Tiwari, "Semiconductor physics: Principles, theory and nanoscale," Electroscience 3, Oxford University Press, ISBN 978-0-19-875986-7 (forthcoming) when we discuss fluctuation-dissipation and tie it to thermal noise. The damping factor Γ corresponds to the resistance R. R arose from fluctuation effects from electrons subject to a Brownian-like phenomena. The damping factor in this case is the resistance to the mechanical motion of the beam.

Readers seeing the fluctuation-dissipation theorem in S. Tiwari, "Semiconductor physics: Principles, theory and nanoscale," Electroscience 3, Oxford University Press, ISBN 978-0-19-875986-7 (forthcoming) will note the correspondence of mean square fluctuations with thermal noise in semiconductors. The thermal noise can be viewed in the two coordinates—voltage and current—utilized there.

represents, for any frequency, the magnitude of the conjugate product in the infinite time limit at that frequency. So, at a resonant frequency, this density will be large, and, away from the resonant frequency and at times larger than correlation times, it will decrease.

The displacement in time, $u(t)$, or in the reciprocal space, $u(\omega)$, of a beam at any position y, is, respectively,

$$u(t) = \frac{1}{\sqrt{2\pi}} \int_{-\infty}^{\infty} u(\omega) \exp(i\omega t)\, d\omega, \text{ and}$$

$$u(\omega) = \frac{1}{\sqrt{2\pi}} \int_{-\infty}^{\infty} u(t) \exp(-i\omega t)\, dt. \tag{5.138}$$

The mean square displacement fluctuation is

$$\begin{aligned}
\langle u^2 \rangle &= \lim_{T \to \infty} \frac{1}{T} \int_{-T/2}^{T/2} u(t) u^*(t)\, dt \\
&= \lim_{T \to \infty} \frac{1}{T} \frac{1}{2\pi} \int_{-T/2}^{T/2} \int_{-\infty}^{\infty} \int_{-\infty}^{\infty} u(\omega) u^*(\omega') \\
&\quad \times \exp\left[i(\omega - \omega')t\right]\, d\omega\, d\omega'\, dt \\
&= \lim_{T \to \infty} \frac{1}{T} \int_{-\infty}^{\infty} u(\omega) u^*(\omega)\, d\omega \\
&= \int_{-\infty}^{\infty} \lim_{T \to \infty} \frac{u(\omega) u^*(\omega)}{T}\, d\omega \\
&= \int_{-\infty}^{\infty} S_u(\omega)\, d\omega. \tag{5.139}
\end{aligned}$$

How does the correlation function relate to the spectral density and displacement fluctuations? We can write

$$\begin{aligned}
\mathcal{K}(t) &= \int_{-\infty}^{\infty} S_u(\omega) \exp(i\omega t)\, d\omega \\
&= \frac{1}{2\pi} \lim_{T \to \infty} \frac{1}{T} \int_{-\infty}^{\infty} d\omega \int_{-T/2}^{T/2} dt' \int_{-T/2}^{T/2} u(t') u^*(t'') \\
&\quad \times \exp\left[i\omega(t - t' - t'')\right]\, dt'' \\
&= \lim_{T \to \infty} \frac{1}{T} \int_{-T/2}^{T/2} dt' \int_{-T/2}^{T/2} u(t') u^*(t'') \delta(t - t' - t'')\, dt'' \\
&= \lim_{T \to \infty} \frac{1}{T} \int_{-T/2}^{T/2} u(t + t') u^*(t'')\, dt' \\
&= \lim_{T \to \infty} \int_{-T/2}^{T/2} \frac{u(t + \tau) u^*(\tau)}{T}\, d\tau \\
&= \lim_{T \to \infty} \int_{-T/2}^{T/2} \frac{u(\tau) u^*(t + \tau)}{T}\, d\tau. \tag{5.140}
\end{aligned}$$

So, $\mathcal{K}(0) = \langle u^2 \rangle$. The peak in the correlation function, the one at vanishing displacement in time, is also the mean square displacement.

Having connected the response in these classical conditions by employing classical statistics, we can determine limits. Since $k_B T/2$ is the energy associated with this specific displacement,

$$\frac{1}{2}k_BT = \frac{1}{2}k_s\langle u^2\rangle = \frac{1}{2}k_s\left[\int_{-\infty}^{\infty} S_u(\omega)\,d\omega\right]$$

$$= \frac{1}{2}k_s\left[\int_{-\infty}^{\infty} \frac{k_B T\Gamma}{\pi m^2} \frac{d\omega}{(\omega_0-\omega)^2 - \Gamma^2\omega^2/m^2}\right]d\omega$$

$$= \frac{1}{2}m\omega_0^2 \frac{k_B T\Gamma}{\pi m^2}\int_{-\infty}^{\infty} \frac{d\omega}{(\omega_0-\omega)^2 - \Gamma^2\omega^2/m^2}\,d\omega. \quad (5.141)$$

This is generally valid so long as the classical approximations of continuum and distribution function are valid. If we consider the condition of strong coupling, that is, frequencies that are very near resonance frequency, then the integral can be evaluated. It is $\pi m/\omega_0^2\Gamma$. So,

$$\frac{1}{2}m\omega_0^2 \frac{k_B T\Gamma}{\pi m^2}\frac{\pi m}{\omega_0^2\Gamma} = \frac{1}{2}k_B T. \quad (5.142)$$

The importance of this relationship and the mathematical evaluation of this connection between spectral density, correlation fluctuations and displacements is that if one were interested in making very precise measurements, that is, measurements where the effect of any force coupling is being measured through its energy exchange, then the measurement would have to be near resonance, where the spectral signature can be directly measured. And, at this condition of measurement near resonance, one also knows the fluctuation effect that is being coupled through the equipartition of energy.

What is the minimum detectable limit under these conditions? If we write frequency in Hz instead of radial units, $S_F(\nu) - 4k_B T\Gamma$. The units are in N^2/Hz. If one measures with bandwidth B, that is, accounts for the energetics across this bandwidth, then the minimum detectable force is $F_{min} = (4k_B T\Gamma B)^{1/2}$. We have $\Gamma = m\omega_0/Q = k_s/\omega_0 Q$; $k_s = m\omega_0^2$; and $\omega_0 = 2\pi\nu_0$. As a function of frequency, the damping term is

$$\Gamma = \frac{k_s}{\omega Q} = \frac{k_s}{2\pi f Q}, \quad (5.143)$$

and the minimum force measurable is

$$F_{min} = \sqrt{\frac{4k_s k_B T B}{\omega_0 Q}} = 2(k_s T B)^{1/2}\left(\frac{k_s}{Q}\right)^{1/2}\left(\frac{1}{\omega_0}\right)^{1/2}. \quad (5.144)$$

To improve on the limit of measurement, one must minimize bandwidth and dissipation through a high quality factor—this reduces the energy of fluctuations coupled in–and one must also minimize temperature, thus reducing the thermal energy.

We now look at these cantilevers in use in devices under practical considerations. So, first we evaluate this limit force for our classical cantilever, with inertia $I = wt^3/12$, area $A = wt$, and a mass

This expression of minimum measurable force has a precise correspondence with Nyquist noise for electrons in conductors. Fluctuation noise places limits onto the measurements that can be performed using the system.

related to density ρ and length L as $m = \rho wtL$, which accounts for the moment. Earlier, we looked at the response of cantilever beams. The out-of-plane transverse vibration frequency—a resonant radial frequency—of the structure is

$$\omega_0 = \sqrt{\frac{YI}{\rho A}} \left(\frac{1.875}{L}\right)^2. \tag{5.145}$$

Consider silicon; Figure 5.17 shows a pictorial display of silicon's resonance frequency dependence at dimensions where the classical approximation is still valid. Frequencies in the MHz to GHz range are possible in structures. For GHz, one would need to increase thickness and reduce length, producing a stiffer beam that can resonate higher. Increasing the length of the beam without changing its thickness will reduce the resonance frequency, as longer wavelength modes are supported. The lowest measurable force from this cantilever, since $k_s = Ywt^3/L^3$, is

$$
\begin{aligned}
F_{min} &= 2\sqrt{k_BTB} \left(\frac{\rho wtL}{4Q}\right)^{1/2} \left(\frac{Ywt^3}{12\rho wt}\right)^{1/4} \frac{1.875}{L} \\
&\approx 2\sqrt{k_BTB}(Y\rho)^{1/4}w^{1/2}L^{-1/2}Q^{-1/2} \times \frac{1}{2} \\
&= 2\sqrt{k_BTB}\left[(Y\rho)^{1/2}\frac{wt^2}{LQ}\frac{1}{4}\right]^{1/2}. \tag{5.146}
\end{aligned}
$$

In this expression, the factor within square brackets is the damping factor Γ.

When one applies a harmonic force, that is, one forms a forced harmonic oscillator, such as that of the inertial sensor shown in Figure 5.3,

$$m\ddot{u} + \Gamma\dot{u} + k_s u = F(t) = F_0(\omega)\exp(i\omega t), \tag{5.147}$$

the response is

$$u(t) = \frac{1}{\sqrt{2\pi}} \int_{-\infty}^{\infty} u_0(\omega)\exp(i\omega t)\,d\omega, \tag{5.148}$$

where

$$u_0(\omega) = \frac{1}{(\omega_0^2 - \omega^2) + i\omega_0\omega/Q} \frac{F_0(\omega)}{m}. \tag{5.149}$$

The spectral density for displacement or for force from this follows as

$$
\begin{aligned}
S_u(\omega) &= \lim_{T\to\infty} \frac{u_0(\omega)u_0^*(\omega)}{T} \\
&= \lim_{T\to\infty} \frac{1}{T}\left|\frac{1}{\sqrt{2\pi}}\int_{-\infty}^{\infty} u_0(t)\exp(-i\omega t)\,dt\right|^2 \\
&= \frac{1}{(\omega_0^2 - \omega^2)^2 + \omega_0^2\omega^2/Q}\frac{S_F(\omega)}{m^2}. \tag{5.150}
\end{aligned}
$$

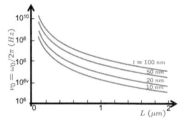

Figure 5.17: Resonance frequencies of an idealized silicon beam as a function of length, assuming a Young's modulus of 150 GPa a density of 2329 kg/m^3, and no losses (infinite Q). Beams used in low mass detection, such as in biological measurements, operate under very lossy conditions. In this idealization, since I/A, the ratio of inertia to cross-section area is independent of width, and, therefore, so is the resonance frequency. For comparison, SiC has a Young's modulus of 450 GPa, and SiO_2 has a Young's modulus of 50–70 GPa.

The damped forced oscillator is also a filter that is particularly adept at extracting the signals near the resonance frequency. It allows measurement and extraction of signals of interest through their force and displacement effect near the natural resonance frequencies of the system.

We have tackled the damping arising from the ambient environment but not from the anharmonicity of the beam. Here, we take the case of thermoelasticity. Thermoelastic damping is intrinsic to the material and caused by the flow of heat irreversibly across the thickness of a resonator. Since the oscillating beam is undergoing deformation, there is an elastic field—the stresses and strains of the structure. This field couples to the temperature fields. Such a damping can be considerably pronounced in thick and long resonators. This inelasticity is the Zener effect. In the elastic limit of an isotropic material, we related stress to strain through Young's modulus, that is, $\sigma = Y\varepsilon$. Zener inelasticity is the time-dependent damping arising from within the material's response, that is, a damping arising from within. With the Zener term,

$$\sigma = T_\varepsilon \frac{d\sigma}{dt} = Y_0\left(\varepsilon + T_\varepsilon \frac{d\varepsilon}{dt}\right), \tag{5.151}$$

where Y_0 is the zero order term used here in normalization, and T_ε is the Zener inelasticity coefficient. If we write the stress and strain in harmonic forms, that is,

$$\begin{aligned}
\sigma &= \sigma_0 \exp(-i\omega t), \text{ and} \\
\varepsilon &= \varepsilon_0 \exp(-i\omega t),
\end{aligned} \tag{5.152}$$

respectively, one can write the amplitude ratio as

$$\frac{\sigma_0}{\varepsilon_0} = Y_0 \frac{1-T}{1-T_\varepsilon} = Y(\omega). \tag{5.153}$$

The stress-strain response can then be determined to be

$$\begin{aligned}
Y(\omega) &= \left(\frac{1+\omega^2 T^2}{1+\omega^2 T_\varepsilon^2} - \frac{i\omega T}{1+\omega^2 T_\varepsilon^2}\Delta\right)Y_0 \\
&= Y_{eff}(\omega)\left(1 - \frac{i\omega T}{1+\omega^2 T_\varepsilon^2}\Delta\right) \\
&= Y_{eff}(\omega)\left(1 - \frac{1}{Q}\right),
\end{aligned} \tag{5.154}$$

where

$$\begin{aligned}
Y_{eff}(\omega) &= \left(\frac{1+\omega^2 T^2}{1+\omega^2 T_\varepsilon^2}\right)Y_R, \\
\Delta &= \frac{T}{T_\varepsilon} - 1, \\
\text{and } \frac{1}{Q} &= \frac{\omega T}{1+\omega^2 T^2}\Delta. \tag{5.155}
\end{aligned}$$

Magnetic resonance force microscopy, and others, employ this approach to select the force effect to be measured through the couplings employed and then make a system that is adapted to the conditions. Single spins have been measured using such a technique employing magnetic resonance.

We can now determine the time-dependent response of this Zener inelastic beam. The fourth order equation, our Euler-Bernoulli equation, for motion in the x direction at position y, is

$$\frac{\partial^4 u_x(y)}{\partial y^4} + \rho A \frac{\partial^2 u_x(y)}{\partial t^2} = 0. \tag{5.156}$$

For a free beam, this becomes

$$Y I_z \frac{\partial^4 u_x(y)}{\partial y^4} + \frac{\rho A}{Y_{eff}\,(1 - 1/Q)\,I} \omega^2 u_x(y) = 0. \tag{5.157}$$

With a harmonic force $F_0(\omega)\exp(i\omega t)$ and a harmonic displacement response of $u_{0x}(\omega)\exp(i\omega t)$, the eigenmode equation is

$$Y I \sum_{N=1}^{\infty} a_N \frac{\partial^4 u_{0x}(y)}{\partial y^4} - \rho A \omega^2 \sum_{N=1}^{\infty} a_N u_{0x} = F_0(\omega). \tag{5.158}$$

When Q is large, the dissipation is limited, and the nth eigenmode's frequency is modified to

$$\omega'_N = \sqrt{\frac{Y_{eff} I}{\rho A}} \beta_N^2 \left(1 + \frac{1}{2Q}\right) = \omega_n \left(1 + \frac{1}{2Q}\right). \tag{5.159}$$

It increases in frequency inversely with the quality factor. The displacement in terms of the eigenmode expansion is

$$u_{0x}(\omega, t) = \sum_{N=1}^{\infty} a_N u_{0xN} \exp(i\omega t). \tag{5.160}$$

Employing the product and integration together with the condition of orthogonality,

$$(\omega_N'^2 - \omega^2) a_N = \frac{1}{\rho A L^3} \int_0^L u_{0xN} F_0(\omega)\,dy. \tag{5.161}$$

So, the coefficient of displacement expansions can be written as

$$a_1 = \frac{1}{\rho A L^3} \frac{1}{\omega_1^2 - \omega^2 - i\omega_1^2/Q} \int_0^L u_{0x1} F_0(\omega)\,dy \tag{5.162}$$

for the first term, and so on; together, this characterizes the response in the presence of Zener inelasticity.

We have now worked through the eigenmode analysis under the conditions of Zener damping. The applied force, its magnitude, and how far it is from the natural frequencies of the mechanical system lets us determine the amplitude coefficients of oscillations under small-signal conditions. The quality factor is again an important term in determining the frequency shifts in the response, as well as the amplitude of the resulting oscillations. The damping here, to a second order, is different from the constant damping factor utilized earlier, and we found a way to handle this through an effective mean theory.

Examples up to this point have concentrated on a beam or plate vibrating—transverse resonators. Electromechanical phenomena also underlie surface and bulk acoustic wave resonance. In such structures, acoustic excitation employs piezoelectric materials, since piezoelectricity couples applied fields to mechanical displacement.

5.3   Acoustic waves

THE ABILITY TO EXCITE TIME-DEPENDENT DISPLACEMENT—even periodic—means that there is now the ability to excite acoustic mechanical modes. In structures formed with thin layers of a piezo-electric on a bulk substrate, one achieves excitation far from any fundamental acoustic mode, that is, of pure modes. An acoustic wave transfers energy from an excitation source for transmission to an elastic medium. Thus, the propagation velocities are determined by the mechanical properties of the material—silicon, for example, being a common microfabricated material. Piezoelectric properties can change this propagation substantially through the local dipole fields that are also atomic in origin, just as the acoustic motion is. The wave on a surface, akin to the wave in a water pond, is the surface acoustic wave (*SAW*). The oscillation motion and propagation in the bulk of the material is the bulk acoustic wave (*BAW*).

One may therefore question the invocation of resonance in these conditions.

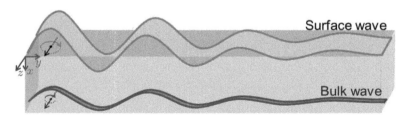

Figure 5.18: Wave propagation on a surface and in the bulk. The particles, that is, atoms maintain a mean position and move around it both in plane and out of plane as the wave travels longitudinally. Surface motion is more pronounced than the motion in the bulk.

Figure 5.18 shows a conceptual drawing of this propagation in an isotropic medium. At the surface, the atomic motion is more pronounced, and in the bulk, less so—a very compressed ellipsis with its long axis along the direction of propagation. The decay length scale is of the order of the wavelength of propagation. If the structure is small, transverse modes—shear mode waves—also exist, propagating energy in both directions. These shear mode resonances are at longer wavelengths than those for the longitudinal waves.

These are Rayleigh waves.

If one places boundary conditions, such as those shown in Figure 5.19, through the excitation electrodes, then the region in between acts as an acoustic cavity that supports waves at λ and its even fractions, like an electromagnetic cavity. The cavity stores energy through constructive interference. Acoustic mode resonators are possible in semiconductors—Si and others. When made using piezoelectric materials, such as thin films of AlN, ZnO, $LiNbO_3$, $PbZrO_3$ or the other piezoelectrics that we discussed in Chapter 4, a large reduction in size, as in SAW and BAW devices, becomes possible. In a longitudinal mode resonator propagating in the surface plane, the excitation is perpendicular to the surface. So excitation is along the direction of strongest piezoelectricity, for example, the c-axis of AlN, which has a wurtzite crystalline structure and actuation in the plane.

The fundamental frequency of the resonator, the lowest frequency supportable, is determined by the fitting of the smallest wave, a half-wave between the electrodes, or a full wave in the pitch. For a speed of sound of c, this is $f_0 = c/\lambda$. AlN has a longitudinal sound velocity of $\sim 10^4\ m/s$, so at a pitch of about $10\ \mu m = 10^4\ nm$, the fundamental frequency is $1\ GHz$. This makes resonators and frequency selection act as filters, where the frequency is determined by the pitch of the lithography of the structure. Bulk resonators use a similar approach across the depth of the structure and are so thick, but the voltages are applied in the most efficient direction for piezoelectric effect. The depth of the structure causes the frequencies to be lower and, as before, the applied signal and acoustic propagation are orthogonal. Since bulk approaches have higher coupling, bulk-like excitation in a surface-oriented structure has been used in devices with thin film features. Multiple BAR structures operating simultaneously on a common substrate couple the excited modes, so a preferred approach is the use of film bulk acoustic resonators ($fBAR$), which employ acoustical isolation through air gaps below films employed as bulk resonators. An $fBAR$ structure has electrodes across the thickness of a film but a gap below, so that elastic propagation can be suppressed.

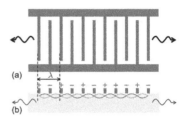

Figure 5.19: Lateral excitation using a surface acoustic wave resonator. By applying the harmonic bias voltage signal between two interdigitated electrodes in (a) on the surface, an acoustic wave is excited in the piezoelectric layer (b), which is shown in cross-section.

The shear waves in AlN have a velocity of $6.3 \times 10^3\ m/s$, that is, about a third smaller than longitudinal velocity.

5.4 Consequences of nonlinearity

ONE CONSEQUENCE OF NONLINEARITY that we have already seen is the pull-in effect. But, nonlinearity manifests itself in a rich set of ways—chaotic behavior—a variety of changes in response characteristics including limit cycles where frequency components of force response change rapidly. So the structure can behave in what appears to be a reasonable and simple-to-describe fashion and then suddenly jump to a very unexpected behavior, which is usually one not conducive to the kind of predictable and feedback-controllable behavior that we desire. However, although such chaotic behavior is complex and seemingly unpredictable, it is not random.

A nonlinear system is one whose time evolution is nonlinear. These are systems whose summarizing equations for the dynamical variables of properties of interest are nonlinear. Describing the behavior of a system requires evolution equations, parameters describing the system, and initial conditions.

We will look at our earlier examples to emphasize the nonlinearity consequences. Our force equation with damping, for a simple point mass, for example, is

$$m\ddot{u} + \gamma\dot{u} + k_s u = F_0 \cos \omega t \qquad (5.163)$$

and has the transfer function

$$|T| = \frac{F_0/m}{\left[(\omega - \omega_0)^2 + (\omega\omega_0/\mathcal{Q})^2\right]^{1/2}}. \qquad (5.164)$$

This equation shows a resonant peaking in frequency, at ω_0, of amplitude $\mathcal{Q}F_0/k_s$, together with a low frequency response of F_0/k_s. In our prior discussion of response, under somewhat different constraints, our solutions took the form of Figure 5.7 for the in-plane and the out-of-plane responses of a comb drive, and the eigenmode solution of Figure 5.12 for a cantilever. Now we introduce damping. In the coordinate system where u is the displacement, we write it as a nonlinearity in the spring constant, so let the spring have anharmonicity of higher order:

$$m\ddot{u} + \gamma\dot{u} + k_s u + k_{s2}u^2 + k_{s3}u^3 + \cdots = F_0 \cos \omega t. \qquad (5.165)$$

To simplify, but still considering the nonlinearity up to the third power term, we set $\gamma = 0$, so, for example, no fluidic damping. But, nonlinear effects of the amplitude of oscillations are included.

We start with the homogeneous equation, to see how the system behaves without the forcing function. We transform variables, rewriting the equation up to the third power—a nonlinear term—as

$$\breve{\omega}_0^2 \frac{\partial^2 u}{\partial \varphi^2} + \omega_0^2 u + \varepsilon\alpha u^2 + \varepsilon^2\beta u^3 = 0, \qquad (5.166)$$

where $\varphi = \breve{\omega}_0 t$; $k_{s2} = \varepsilon\alpha m$; and $k_{s3} = \varepsilon^2\beta m$. This transformation now makes the power of ε consistent with the power of anharmonicity. Perturbation powers are now consistent, where

$$\begin{aligned}
\breve{\omega}_0 &= \omega_0 + \varepsilon\omega_1 + \varepsilon^2\omega_2, \quad \text{and} \\
u &= u_0 + \varepsilon u_1 + \varepsilon^2 u_2,
\end{aligned} \qquad (5.167)$$

Stanislaw Ulam, the inventor of the Monte Carlo approach, is reported to have remarked "calling the subject nonlinear dynamics is like calling zoology 'nonelephant studies'." We largely study systems in their linear limit, but this is a very small subset. Ulam, a mathematician, had an elevated sense of humor with a love of tautological jokes. "A mother gives her son two ties. When the son visits wearing one of them, she comments, 'So you didn't like the other one?'" Another, from his father—the family was from Lwov (now Lviv) on the Ukraine-Poland border—is a conversation between two businessmen, who are also friends, meeting each other on a train in a region where Krakow and Saint Petersburg were frequent destinations. The first asks "Where are you going?" The second answers "I am going to Krakow," to which the first takes offense and says vehemently, "You tell me you are going to Krakow so that I will think that you are going to Saint Petersburg, but you are really going to Krakow. Why did you lie to me?"

The eigenmode solution describes spatial and time dependence completely. This is a point mass.

become ways of ordering perturbations. Substituting these into Equation 5.166,

$$0 = \left[\omega_0^2 \frac{\partial^2 u_0}{\partial \varphi^2} + \omega_0^2 u_0\right] + \varepsilon\left[\omega_0^2 \frac{\partial^2 u_1}{\partial \varphi^2} + \omega_0^2 u_1 + \alpha u_0^2 + 2\omega_0\omega_1 \frac{\partial^2 u_0}{\partial \varphi^2}\right]$$

$$+\varepsilon^2\left[\omega_0^2 \frac{\partial^2 u_2}{\partial \varphi^2} + \omega_0^2 u_2 + 2\alpha u_0 u_1 + \beta u_0^3\right.$$

$$\left.+(\omega_1^2 + 2\omega_0\omega_2)\frac{\partial^2 u_0}{\partial \varphi^2} + 2\omega_0\omega_1 \frac{\partial^2 u_1}{\partial \varphi^2}\right] + \mathcal{O}(\varepsilon^3). \tag{5.168}$$

This can be true in general iff each of the bracketed terms vanishes. The first of these cases occur with the harmonic resonator solution of $u_0 = U_0 \cos \varphi$. Using this, the second term becomes

$$\omega_0^2 \frac{\partial^2 u_1}{\partial \varphi^2} + \omega_0^2 u_1 = 2\omega_0\omega_1 U_0 \cos \varphi - \frac{1}{2}\alpha U_0^2(1 + \cos 2\varphi). \tag{5.169}$$

Since this is a homogeneous equation, there exists no energy input, so the first term on the right must vanish, else the first perturbation in the amplitude of displacement will continue to rise. So, $\omega_1 = 0$. The harmonic displacement term from where this came, $(k_{s2}u^2/2$, does not cause a perturbation in frequency. But, it does in the amplitude. We may solve the equation with $\omega_1 = 0$:

$$u_1 = -\frac{\alpha}{2\omega_0^2}U_0^2 + \frac{\alpha}{6\omega_0^2}U_0^2 \cos 2\varphi. \tag{5.170}$$

The first order correction in displacement is a static shift and an additional component at twice the frequency, since $\varphi = 2\omega_0 t$. These two results from the first two vanishing terms of Equation 5.168 can now be fed into the last term:

$$\omega_0^2 \frac{\partial^2 u_2}{\partial \varphi^2} + \omega_0^2 u_2 = -\left[\left(\frac{3}{4}\beta - \frac{5}{6}\frac{\alpha^2}{\omega_0^2}\right)U_0^3 - 2\omega_0\omega_2 U_0\right]\cos \varphi$$

$$-\left(\frac{1}{4}\beta + \frac{1}{6}\frac{\alpha^2}{\omega_0^2}\right)U_0^3 \cos 3\varphi. \tag{5.171}$$

Using similar arguments,

$$\omega_2 = \left(\frac{3}{8}\frac{\beta}{\omega_0} - \frac{5}{12}\frac{\alpha^2}{\omega_0^3}\right)U_0^2, \tag{5.172}$$

with substitution in Equation 5.171 leading to

$$U_2 = \left(\frac{1}{16}\frac{\beta}{\omega_0^2} + \frac{1}{24}\frac{\alpha^2}{\omega_0^4}\right)U_0^3 \cos \varphi$$

$$+\left(\frac{1}{32}\frac{\beta}{\omega_0^2} + \frac{1}{48}\frac{\alpha^2}{\omega_0^4}\right)U_0^3 \cos 3\varphi. \tag{5.173}$$

Two additional perturbations terms arise from the u^3 dependence of anharmonicity—one at the resonance frequency, and one at the third harmonic. This last is an additional term in odd order. It is now a source of interference.

From this homogeneous equation analysis, we conclude that the spring anharmonicity leads to a change in resonance frequency, whose first order effect is a shift in resonance frequency to

$$\breve{\omega}_0 = \omega_0 + \varepsilon^2 \omega_2 = \omega_0 + \left(\frac{3}{8} \frac{k_{s2}}{k_s} - \frac{5}{12} \frac{k_{s1}^2}{k_s^2} \right) \omega_0 U_0^2 = \omega_0 + \varsigma U_0^2, \quad (5.174)$$

where we have introduced the parameter ς to denote a nonlinearity ratio factor of the system. This is a major effect. The other additional consequence is the perturbation in the amplitude of oscillation. This resonance frequency may shift down or up depending on the sign and magnitude of the nonlinearities of mechanical stiffness terms. It decreases if $k_{s2} < 0$, but if $k_{s2} > 0$, then it will increase.

What is the major consequence of a forcing function? One can estimate this using the homogeneous solution of the nonlinearity's effect, and this gives us a simpler, understandable way for estimating. For example, in the homogeneous solution, one would expect it to be a shift in the frequency to the form Equation 5.174. So, the amplitude near the resonance is a change from Equation 5.164 to the form

$$U_0 = \frac{F_0/m}{\left[(\omega^2 - \breve{\omega}_0^2)^2 + (\omega \breve{\omega}_0 / Q)^2 \right]^{1/2}}. \quad (5.175)$$

This method will only work up to a certain point.

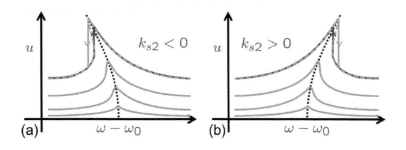

Figure 5.20: Forced excitation of a nonlinear cantilever system. The sign of the second order mechanical nonlinearity (k_{s2}) determines the direction of resonance shift. At sufficiently high excitation, bifurcation resulting in hysteresis becomes possible.

Figure 5.20 shows a schematic of the response. Recall the pull-in behavior of Figure 5.10. It arose as a direct consequence of nonlinearity. The plate pulled in into contact with the static plate even when it was farther away. Nonlinearities cause a pronounced discontinuous effect when driven strongly enough. The response shows a hysteresis. As the forcing function increases, regions come about that are no longer single-valued functions.

A bifurcation is a sudden, qualitatively different behavior of the system, resulting from a small change in a parameter. For example, a doubly clamped beam compressed by the clamp will first shrink and then, with a very small change of force, buckle at the bifurcation point. This sudden jump from one resonating curve to another, portrayed as hysteresis due to its different position depending on how the approach happens, is a bifurcation. Bifurcation can be local in the sense that a crossing of a threshold of some parameter results in a change in a local stability property or other invariant properties. Global bifurcation arises from the intersection of a large set of invariants. One can estimate this bifurcation point. With $\Delta\omega = \omega - \omega_0$,

$$
\begin{aligned}
(\omega_0 + \Delta\omega)^2 - \breve{\omega}_0^2 &= (\omega_0 + \Delta\omega + \breve{\omega}_0) \approx 2\omega_0(\Delta\omega - \varsigma U_0^2), \quad \text{and} \\
(\omega_0 + \Delta\omega)\breve{\omega} &\approx \omega_0^2.
\end{aligned}
\tag{5.176}
$$

Equation 5.175 can be rewritten as

$$
F_0^2/m^2 = 4\omega_0^2 \left[(\Delta\omega_0 - \varsigma U_0^2)^2 + \omega_0^2/4Q^2 \right] U_0^2.
\tag{5.177}
$$

Bifurcation occurs when the amplitude suddenly changes with a shift in the resonant frequency in response to system parameter changes. So, we determine $\partial U_0/\partial \Delta\omega_0$ and see where it explodes, that is, where the denominator vanishes, as shown in Figure 5.21. This leads to

$$
U_0^2 = \frac{1}{6}\frac{1}{Q^2\varsigma^2}\left[4Q^2\Delta\omega_0\varsigma \pm \left(4Q^2\varsigma^2\omega_0^2 \right)^{1/2} \right].
\tag{5.178}
$$

A bifurcation point is single valued. So,

$$
\Delta\omega_0 = \pm\frac{\sqrt{3}}{2/Q}\omega_0,
\tag{5.179}
$$

arising from the ς's opposite signs. This gives the bifurcation point as

$$
U_b = \left(\frac{\omega_0}{\sqrt{3}|\varsigma|Q} \right)^{1/2}.
\tag{5.180}
$$

One can also see that, at resonance, the response has an amplitude larger than at the bifurcation point:

$$
U_c = \left(\frac{4\omega_0}{3\sqrt{3}|\varsigma|Q} \right)^{1/2}.
\tag{5.181}
$$

A higher quality nonlinear resonator has a lower amplitude and bifurcation point. Nonlinear effects establish the range over which a resonator will have usefulness bereft of hysteresis. A response is schematically drawn in Figure 5.21, using our solution approach. The

The term "bifurcation" is due to Henri Poincaré who also pointed out the different types of such bifurcations in continuous systems. The specific one we encounter often in our electromechanical systems is the Hopf bifurcation. It is a local bifurcation. Stability is lost at a fixed point of the dynamical system. It arises from eigenvalues that are complex conjugates crossing the imaginary axis.

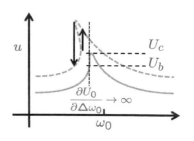

Figure 5.21: Onset of bifurcation when the amplitude-frequency curve becomes vertical. Beyond it, hysteresis appears in the frequency response characteristics.

amount of power needed to reach this limit is the product of stored energy and ω_0/Q, which is the fraction that is lost every cycle. This is $\omega_0 k_s U_c^2/2Q$.

This example serves as a good starting point for exploring the variety of interesting characteristics that nonlinear systems undergo. Chaos is a time-aperiodic behavior, that is, not exactly repeating and therefore appearing as apparently random or "noisy." But, this chaotic response, strictly speaking, is non-random, since it is in the response of our explicitly written, deterministic, time-evolution equation, whose parameters and initial conditions are defined. So, it is not the result of any rounding errors but deterministic evolution catalyzed by nonlinearity. Classical systems ranging from pendulums to planetary systems show it. Quantum systems, for example, lasers, show it. Biological systems, for example, a beating heart, show it. It is this ubiquity, its complementarity in fractals, and the nature of its universality, where similar ratios cross disciplines, that made this subject a very interesting multidisciplinary area in the last decades of the 20th century. We should emphasize that all chaotic systems are nonlinear, but not all nonlinear systems are chaotic.

Easily understandable examples are circuits made with diodes, using a diode's nonlinearity. An example with an inductor is shown in Figure 5.22. A sinusoidal and static signal forces this circuit, which contains a diode and whose energy is stored in an inductor. The diode is the nonlinear element here. Recall how the diode's response is reflected in the conduction through it. When one forward biases it, it passes current, a current that is exponentially dependent on the bias voltage drop across it. And it passes very little current in reverse bias, a reverse saturation current that we will assume to be zero. The forward current, internally in the diode, is sustained by charge storage, and this charge distribution has a gradient. The drift and diffusion current within the diode sustains the forward current. So, there is a storage of charge associated with the forward current. When a voltage across the diode is flipped from forward to reverse bias, this excess charge still needs to come out. So, a current still continues to flow for a short time—the reverse-recovery time. This time depends on the current that was flowing through before, which corresponds to the charge storage that existed in the diode. The inductor in this circuit breaks the tight coupling that exists between current and potential differences, since it introduces storage of energy when current flows. We apply a very small sinusoidal voltage v signal added to the static voltage signal V. The sinusoidal voltage measures the response, but the static voltage drives the nonlinearity. The diode and the inductor store and exchange kinetic and potential energy, much like the way that potential and kinetic energy are stored and exchanged in a

At the start of the century, Poincaré with his limit cycles, and others, had laid the mathematical foundations through the study of nonlinear equations. The late-20th century rejuvenation started with an innocuously titled paper "Deterministic nonperiodic flow" in a journal that occupies the dark corners of stacks in the library—Journal of Atmospheric Sciences. The coordinates are Volume 20, pages 130 to 141, in the year 1963. It employed a very simple approximation of the Navier-Stokes equation for thermal energy diffusion, but one that had nonlinear dynamics in it. From this simplicity came some of the development and understanding of complexity science, which in many instances is not so complex and in which, often, noise or degrees of freedom have an insignificant or nonexistent role. The author of the paper, Edward Lorentz, an atmospheric scientist on the faculty at MIT, is recalled as a self-effacing, quiet and gentle scientist, a A model we may well wish to take lessons from in these times.

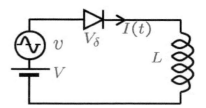

Figure 5.22: A circuit with a diode as the nonlinear element, and an inductor as the energy storage medium, to demonstrate the effects of nonlinearity in a simple model.

mechanical beam.

Using this circuit, we now show the first consequence of non-linearity—the sudden changes that we have called bifurcations.

When the period of oscillation is close to the reverse-recovery time, the nonlinear effects of switching on and switching off are dominant. At low voltages, so when $V + v$ is below the diode turn-on voltage, currents are low, and the diode is essentially off. At the turn-on voltage, one would expect the half-wave rectification. During the positive part of the sinusoid source cycle, the diode conducts, and during the negative part, it turns off. The current is very small, the amount of charge stored in the diode is small, and one just sees the clipping of the sinusoidal signals. This and response under other conditions of signal voltage, that is, $V + v$, are shown in Figure 5.23. One expects and sees the clipped manifestation of the sinusoidal forcing function in (a). The response has the same periodicity as the input voltage. But, as the sinusoidal voltage is increased, following incremental changes as shown in (b), there comes a point when it suddenly jumps to a period that is twice that of the applied signal as in Figure 5.23(c). A bifurcation has occurred. The period doubling occurred because, before bifurcation, there was sufficient time for the diode charge to be drained off, that is, the diode could shut off. But, with just enough extra applied bias voltage signal, the diode could not shut off before a positive applied signal voltage arrived again. However, with the inductor in the circuit, changes in current are not instantaneous and reverse current must first stop before going to the positive cycle. There is therefore less forward current. And now, the diode can actually shut off in the reverse cycle. This is roughly the reason for a doubling of the period at the first bifurcation point. Increase the voltage further and there is again a sudden change—period doubling to period-4 and again at a higher voltage parameter to period-8, and so on.

As the voltage is increased further, past further period doublings, one gets to a point where the sequence of peaks becomes erratic—this is chaos. One does need to ascertain that this response is not the result of noise, or any other effect. The logic of the current and charge nonlinearity in time affecting the response is shown in Figure 5.24 (a), which shows the current and the voltage of a period doubled instant. Panels (b) and (c) in this figure show the chaotic diode voltage response in the form of aperiodicity. One would also follow the paths that the parameters take, in order to see the sudden and discontinuous change with small change that arises in bifurcation. But, one also observes divergence in nearby trajectories at the onset of chaos. For any small change of initial condition, a very different response, here in the form of peaks that are not at all periodic,

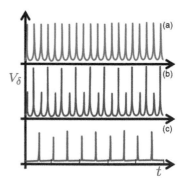

Figure 5.23: The output response for different stimulation conditions for the diode circuit in Figure 5.22. (a) shows a clipping response, (b) shows a change in response as the static bias is increased to a point where the amount of charge stored the in diode is sufficient to affect the amplitude by step recovery, and (c) shows the onset of frequency doubling because of diode nonlinearity and inductor-aided reaction to current changes. Adapted from R. C. Hilborn, "Chaos and nonlinear dynamics," Oxford, ISBN 0-19-567173-2 (2004).

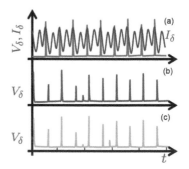

Figure 5.24: (a) shows the voltage and current during period doubling response. (b) and (c) show the chaotic response at a higher forcing function with erratic new peaks appearing. The two different traces are for very minutely perturbed initial conditions so that the first peak is in synchronization, but aperiodic peaks also appear. Adapted from R. C. Hilborn, "Chaos and nonlinear dynamics," Oxford, ISBN 0-19-567173-2 (2004).

appears. As one increases the forcing function further, chaos may disappear, and later on, appear again.

One can show the richness of this behavior through a bifurcation diagram such as in Figure 5.25, which shows bias voltage as the parameter, and the peak current response signal. We start with the period-2 response, a response that has two magnitudes. When a bifurcation occurs, leading to a period-4 response, the response signal shows four peak amplitudes, and these change in magnitude as the parameter is increased. But, in this figure, at a certain voltage, one sees chaos for a significant range of the parameter, before it disappears and one also sees a period-3 response, which bifurcates to a period-6 response and then chaos, before returning to the period-1 response. We could have chosen a slightly different sinusoidal signal, and then the bifurcation diagram may have been considerably different than this one across the same static bias voltage change.

Periodic doubling is but one route to the onset of chaos. For example, nonlinear functions, such as iterated maps, show similar features. The fundamental nature of chaos is best represented by its universality—different functions end up in convergence parameter in bifurcation diagrams, as the period doubling does. This is illustrated by showing the bifurcation for the two functions

$$x_{n+1} = Ax_n(1 - x_n) \tag{5.182}$$

and

$$x_{n+1} = B \sin \pi x_n. \tag{5.183}$$

Both are nonlinear, and their bifurcation diagram are shown in Figure 5.26. Both show features such as period doubling in the march to chaos. What is striking is the geometric convergence ratio. The ratio of differences of parameter values at which successive doubling happens is approximately constant for all the splittings and, in the limit, reaches a constant. So, in Figure 5.26,

$$\delta_n = \frac{A_n - A_{n-1}}{A_{n+1} - A_n}, \tag{5.184}$$

where the A_n are the parameter values where bifurcation appears, remains approximately constant, and, in the limit,

$$\delta \equiv \lim_{n \to \infty} \delta_n = 4.66920161\ldots. \tag{5.185}$$

This is the "Feigenbaum δ." Any iterated map function that is parabolic near its maximum, as well as a few other properties not relevant to our interest here, will have this same convergence ratio as the order of bifurcation ratio goes to the limit of ∞.

So, a diversity of behavior appears in this nonlinear deterministic calculation, including chaotic behaviors, starting from the precisely

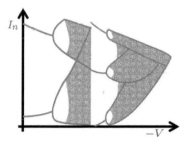

Figure 5.25: A bifurcation diagram, with applied voltage as the parameter for the nonlinear diode circuit. Adapted from R. C. Hilborn, "Chaos and nonlinear dynamics," Oxford, ISBN 0-19-567173-2 (2004).

Self-similarity properties, such as of the Mandelbrot set, the Julia set and the Fatou set are examples of a form of universalism. For a Fatou set of a function, the values have the property that all nearby values behave similarly under repeated iteration of the function. For a Juilia set of a function, the values have the property that an arbitrarily small perturbation can cause drastic changes in the sequence of iterated function values. The Mandelbrot set is the set of complex numbers iterated from the function $f_c(z) = z^2 + c$, starting with $z = 0$, for which the sequence remains remains bounded in absolute value. The constancy of certain features, independent of the scale size of underlying parameters, represents an important truth—a physical principle at work. The Julia set and the Fatou sets come from the time of the beginning of complex dynamics studies. The behavior of a Fatou set is regular. The Julia set, its complementary set, on the other hand, is chaotic. The Mandelbrot set is the earliest representative of fractals, where universality again appears.

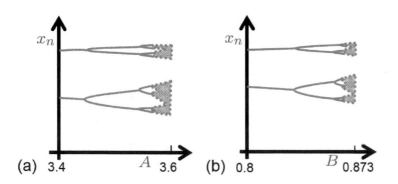

(a) 3.4 A 3.6   (b) 0.8 B 0.873

stated rules of the equations, the values of the parameters, and the
initial conditions. Mathematically stated, this is true. But, in real
systems, whether experimental or a theoretical model, there is always
some imprecision in specifying initial conditions—in real systems,
from noise itself, and in simulations, from rounding off errors or just
the imprecision of numerical implementations. This means that the
behavior does become unpredictable in the chaotic system. But, it is
not due to noise, which has its origins in randomness.

The example we explored was that of an electronic circuit, because
of its simplicity. But, we could have looked at chaos in a mechanical
system, where energies are stored and released and where nonlinearities exist with time dependences. Figure 5.27 shows an example
corresponding to the electrical example we have studied. This is a
simplified but different form of a comb drive that we looked at while
discussing Figure 5.6. We drive the moving plate with a static voltage, and we apply a sinusoidal signal on one of the static plates—an
input plate. As a result, the moving plate responds, and the output
signal is picked up in the form of a voltage across a resistor R that
is between the other static plate—an output plate—and the ground.
We will outline the underlying mathematical formulation in order
to explore the resulting behavior that brings out several interesting
properties.

The force on the moving plate is

$$F_e = \frac{1}{2}\frac{C_0}{(d-u)^2}(V + v_0 \sin \omega t)^2 - \frac{1}{2}\frac{C_0}{(d+u)^2}V^2, \qquad (5.186)$$

where C_0 is the capacitance in unforced conditions with the movable
plate a distance d apart from the static plates. We use a third power
force term for the nonlinearity, so

$$F_e = m\ddot{u} + \gamma\dot{u} + k_s u + k_{s3}u^3. \qquad (5.187)$$

The chaotic system is indeterminable,
even if it is a deterministic system. So,
while electron thermal noise may only
be suppressed by the fundamentals,
for example, in a ballistic channel, the
presence of one electron precludes
that of another with identical quantum
numbers, thus suppressing the classical
randomness of the fluctuation through
another counter-correlation arising from
Pauli exclusion. In principle, chaos can
be limited and controlled if so desired
because its manifest description is also
amenable to control through feedback
and other techniques.

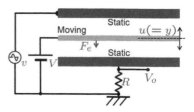

Figure 5.27: An approximation for the
analysis of a comb drive–like structure.
Two plates are static, while one is
allowed to move under the application
of a static and sinusoidal electrical
stimulation.

This is a dimensionally compressed equation where the mass is now an effective mass, and the damping and elastic constants are lumped constants. We make the equation dimension-free, as in the starting analysis of nonlinearity. The normalizations are $\varphi = \omega_0 t$; $\Omega = \omega/\omega_0$; $\eta = u/d$; $\mu = \gamma/m\omega_0$; $\alpha = k_s/m\omega_0^2$; $\beta = k_{s3}d^2/m\omega_0^2$; $\varkappa = C_0 V^2/2md^3\omega_0^2$; and $Y = 2\varkappa v_0/V$, with $\omega_0 = (k_s/m)^{1/2}$. With a minuscule sinusoidal signal, the dimension-free form is

$$\ddot{\eta} + \mu\dot{\eta} + \alpha\eta + \beta\eta^3 = \varkappa\left[\frac{1}{(1-\eta)^2} - \frac{1}{(1+\eta)^2}\right] + \frac{1}{(1-\eta)^2}Y\sin\Omega\varphi, \quad (5.188)$$

where the derivative is w.r.t. φ.

This equation embodying the response corresponds to a bias potential, referenced to the no-displacement, that is, $u(= y) = 0$, or $\eta = 0$, condition as

$$V(\eta) = \frac{1}{2}\alpha\eta^2 + \frac{1}{4}\beta\eta^4 - \varkappa\left[\frac{1}{(1-\eta)} + \frac{1}{(1+\eta)}\right] + 2\varkappa. \quad (5.189)$$

This is a nonlinear equation. Its solutions depend on applied bias potentials. When none is applied, there is one unique degenerate solution at the equilibrium point. But, as the applied bias voltage is changed, so do the number of equilibrium points and their positions. Figure 5.28 shows these under a few different conditions. At bias voltages corresponding to $\varkappa = 0.75$, no equilibrium point appears in the range shown here. Recall our pull-in discussion of Figure 5.10. This resonator has become unstable and likely has been pulled in to one of the stationary plates. At the very smallest of bias voltages, or just the sinusoidal signal with no static bias voltage applied, there exists the region of equilibrium at the center, and two unstable saddle points beyond. The resonator operates close to its free oscillation characteristics. As the bias voltage is increased, the center equilibrium point loses its stability, it becomes a saddle point, and two new low energy points emerge symmetrically on either side.

The Melnikov method is a technique for analyzing the stability of the center for time-periodic perturbations, and we can apply it here. The method provides analytical insight into stability, instability, the appearance of chaos, et cetera, in nonlinear systems. Equation 5.188 may be written for the phase space form as

$$\dot{\eta} = \zeta, \text{ and}$$

$$\dot{\zeta} = -\alpha\eta - \beta\eta^3 + \varkappa\left[\frac{1}{(1-\eta)^2} + \frac{1}{(1+\eta)^2}\right]$$
$$+ \varepsilon\left[-\overline{\mu}\dot{\eta} + \frac{\overline{Y}}{(1-\eta)^2}\sin\Omega\varphi\right], \quad (5.190)$$

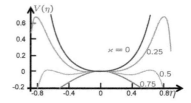

Figure 5.28: Potential as a function of dimensionless plate displacement from equilibrium for the moving plate comb drive.

Appendix N is recommended reading for those wishing to understand the rudiments of system analysis and their portrayal in phase space. The notions of stable points, saddle points, closed orbits, limit cycles, et cetera, are convenient ways of looking at and understanding behavior in a way that complements the energy-based view we have largely taken.

V. K. Melnikov's, "On the stability of the center for time periodic perturbations," Transactions of the Moscow Mathematical Society, **12**, 1–57 (1963) provides a classic technique that is useful in showing the existence of chaotic orbits in dynamical systems. It dates to the same time as Lorenz's work. The Melnikov technique is now applied across the spectrum of physical systems where nonlinear dynamics leads to chaos.

with $\overline{\mu} = \mu/\varepsilon$, and $\overline{Y} = Y/\varepsilon$, that is, both μ and Y are of order $\mathscr{O}(\varepsilon)$, with ε quite small. These conditions are satisfied by a high quality factor $\mathcal{Q}$ and a sinusoidal voltage that is a small fraction of the bias voltage. A Melnikov function, which we employ without discussion, is

$$M(\varphi_0) = \int_{-\infty}^{\infty} \left\{ \zeta_0 \left[-\overline{\mu}\zeta_0 + \frac{\overline{Y}}{(1-\eta_0)^2} \sin\Omega(\varphi - \varphi_0) \right] \right\} d\varphi. \quad (5.191)$$

This function is designed to be proportional to the perturbation of the distance between stable unstable fixed points of the homoclinic and heteroclinic orbits. (η_0, ζ_0) is the unperturbed trajectory, which follows

See Appendix N for a discussion of trajectories and fixed points in phase portrait and of orbits.

$$\zeta_0 = \frac{d\eta_0}{dt} = \pm[2V(\eta_s) - V(\eta_0)]^{1/2}, \quad (5.192)$$

defined through the saddle point.

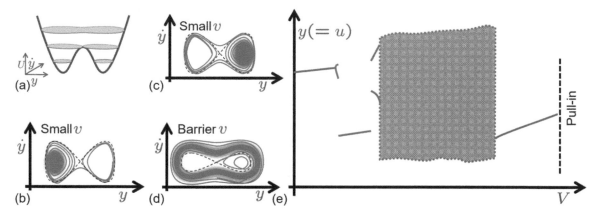

Figure 5.29: The two-well representation of the energy and phase trajectories of a moving beam between two plates. The energy picture as a function of position and velocity is shown in (a). (b) through (d) show responses for three different sinusoidal signals. For a small drive, oscillations starting around the left fixed point and finally settling on the right is shown in (b); the opposite of this is shown in (c). In (d) it is in both under conditions of increased energy supplied to the system. The bifurcation diagram of this moving plate electromechanical system with nonlinearity modulated by the bias voltage is shown in (e). The system undergoes bifurcation with branches of periodic motion followed by chaos, then a return to periodic motion and finally pull-in.

The solutions of these equations require approximations, for example, the second of Equation 5.190 may be expanded in a Taylor series form and substituted in the Melnikov function, which is thus reduced to the simpler form

$$M(\varphi_0) = \overline{\mu}I_1 + \overline{Y}I_2, \quad (5.193)$$

with

$$I_1 = -\int_{-\infty}^{\infty} \zeta_0^2(\varphi)\, d\varphi, \quad \text{and}$$

$$I_2 = \int_{-\infty}^{\infty} \zeta_0(\varphi) \frac{\sin \Omega(\varphi + \varphi_0)}{[1 - \eta_0(\varphi)]^2} \, d\varphi. \qquad (5.194)$$

The Melnikov analysis defines a threshold curve that predicts the different regions of behavior, for example, of chaos above it. It also allows one to see analytically the presence of periodic orbits, so long as sufficient accuracy from the Taylor expansion is included. Absent use of this technique, one can perform a numerical simulation with sufficient accuracy, though such a technique will not give obvious insights into the contribution of the different nonlinear connections of the energy terms.

We summarize here observations on this system that show the effect of nonlinearity in beam dynamics. Figure 5.29(a)–(d) shows the orbits at small driving perturbations in this double-well moving plate system for a chosen set of parameters. This double-well system is discussed more comprehensively in Appendix N. At low perturbations, the system remains near one of the two fixed points. Upon an increase in perturbation that makes the transfer between the fixed points easier, the phase trajectory of the oscillations appears in both regions surrounding the fixed points, with about equal likelihood. The bifurcation diagram of this system formed by slowly increasing the sinusoidal signal is shown in Figure 5.29(e). One can observe that periodic motion around one of the stable points exists at the small voltages, but, as the sinusoidal voltage is increased, chaotic behavior comes about until, suddenly, at even higher sinusoidal voltage, the periodic response returns. Finally, pull-in happens. This example exhibits nearly all the features of a nonlinear system that we have discussed to this point.

We have now seen the richness of phenomena arising from the nonlinearity in these systems. Much of this analysis revolved around resonance, and a narrow band around the resonance frequencies. This is natural for two reasons—the appropriateness of continuum treatment at the size scale of these problems and the properties of material, both of which make the wave approach and its eigenfunction solutions appropriate. Operating near these eigenmode frequencies made it possible to use them in frequency selection and force detection.

These oscillations beget a few comments. We have only considered one particular type of oscillation in a beam or plate vibrating—transverse vibrations. But beams undergo torsional vibrations, and the energies in these two different types of modes can couple. Take a beam clamped at both ends, and excite it laterally. It undergoes damped vibrations not unlike what we have described. Excite it to higher amplitude, and it will have bending vibrations. Take a standing microscale or a nanoscale beam with some mass at its end. Excite

Resonance and this coupling of slow and fast forces can be dangerous too. The then Citibank Tower in New York City was found by a student, Diane Hartley, a senior at Princeton, to be susceptible when excitation came in the form of wind at an angle to the building, that is, torsional conditions. The use of bolted bracing, which is more susceptible to slip due to shear, instead of welded joints, as well as the use of tuned masses for damping the building's movements, made this building dangerous—subject to collapse due to conditions that were likely about once in 15 to 20 years. The building had to be retrofitted after occupation in 1978. Bridges have collapsed, and modern bridges have had to be refitted—the Millennium Bridge in London, in the year 2000, went wobbly when the first crowds walked on it. In this case, it was a reinforcing effect arising from people's stepping, as it is a natural walking reaction to be in phase with the bridge. Stochastic resonance—energy coupling by random to natural resonance—is what happened here. The white noise frequencies at resonance couple and amplify, making signal stronger, while not amplifying the rest and is useful in ultrasensitive measurement. The bandpass of the resonance becomes of use, as we have seen.

it, and it will show these modes, and given enough energy, even possibly buckle. The coupling of modes and exchange of energy means that there will be bifurcations—the coupling to bending vibrations is a Hopf bifurcation and shows up as a beating phenomenon. Hysteresis, chaos, et cetera, all can appear, and, in general, this response behavior can be quite complicated.

The oscillator is an important element in systems—essential to frequency-based approaches of measurement or communication. See Appendix O for a discussion of oscillators and their appearance in basic physics, properties of materials, and in devices. The beam under force conditions that we have discussed is an example of a Duffing oscillator. A Duffing oscillator models the behavior of a double-well system, Figure 5.29(a) is an example of a Duffing oscillator in forced conditions. If started with a certain energy, so a certain amplitude, and then left to itself, a Duffing oscillator gradually loses energy and amplitude, due to damping, and finally comes to rest in a well. The period of oscillation depends on the amplitude. When harmonically forced, a large-amplitude response occurs when the frequency is close to the natural frequency of the oscillator. Since the natural frequency is a function of the amplitude, the response occurs with a change in the natural oscillation frequency of the system. This change in shape of the response curve of the system with amplitude may show the hysteresis that we discussed, depending on whether the external force is increasing the frequency through the response region or decreasing it. This is directly a result of the nonlinearity. The other consequence of nonlinearity is the chaotic behavior of different period cycles near the attractors. The nonlinear examples that we tackled were Duffing oscillator-like.

A van der Pol oscillator is another type of oscillator where limit cycles of the periodic time-dependent behavior appear spontaneously. The oscillation amplitude in a van der Pol oscillator increases with excitation, when excitation is small, but saturate at larger excitation. This is because the damping increases at a higher rate with excitation and hence places a limit. The consequence is that, in a van der Pol oscillator, the nonlinear damping factor causes the phase space trajectory of the oscillator to approach the limit cycle as $t \to \infty$. The fixed point in the system is now a repeller.

5.5 Caveats: Continuum to nanoscale

WE HAVE EMPLOYED A CONTINUUM APPROACH to the mechanical description up to this point—all properties, such as Young's modulus, stress and strain, et cetera, are continuously distributed and

Duffing employed this simple model to describe the forced vibrations of machinery in 1918. He employed the force

$$F = k_s y - \alpha y^3$$

or, equivalently, the energy

$$U = -\frac{1}{2} k_s y^2 + \frac{1}{4} \alpha y^4$$

for a two-well system. α is the anharmonicity—a softening if $\alpha > 0$, and a hardening if $\alpha < 0$.

Balthasar van der Pol employed a nonlinear damping to describe the nonlinear saturating dynamics of an electronic oscillator employing a triode. These dynamics are described by an equation that is very similar to that fir our linear damped simple harmonic oscillator:

$$\ddot{q} + \gamma \dot{q} + \omega^2 q = 0,$$

where q is charge, γ a damping factor, and ω frequency of oscillation of charge, absent damping. γ is a function of charge and causes saturation of charge q, the excitation here. This phenomenon of saturation exists in all active devices because, ultimately, there is a limit to the supply of energy into any active system.

definable throughout the medium. They may arise from phenomena at the atomic scale and from atomic bonding—matter is fundamentally discontinuous, yet we employ a continuous description that ignores any consequences of the specifics of this phenomenon on the local description. There are limits to this use of classical mechanics, since the fields and other characteristics we employ are a continuum approximation. If one has a planar, single-atom-thick sheet such as of carbon in its graphene phase, and we bend it, the picture distinguishing two surfaces and compressive stress and tensile stress shown in Figure 5.1 loses meaning. What is compression or tension in a single-atom-thick film? A tube with a hole punched in its wall has fracture or bending properties affected by the hole. A carbon nanotube with one carbon atom plucked from its wall will have its properties also affected by this hole. But, an adequate description of the former, drawing on a continuum description, will fail in the latter, where the local atomic scale interactions are now perturbed, and the mechanical properties of the nanotube will change in a very different way. The short-range interactions now matter, and a description that only utilizes the long-range description is inadequate. In this there is a direct correspondence between our discussion of stochastic effects in electronics at the nanoscale and in mechanics at the nanoscale. A long nanotube clamped at one end, for example, such as in Figure 5.30, will be subject to short-range constraints at the clamped end, while, further away along the tube, a continuum description and all these eigenmode analysis, et cetera, may be quite adequate, depending on the characteristics one is interested in.

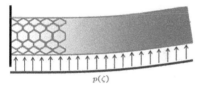

$p(\zeta)$

Figure 5.30: A nanotube as a cantilever beam. The nature of the locality of forces and interactions will result at least in significant anchor effects that will not be adequately described by a continuum approach.

One way to assess a scale length here would be to compare the dimensional scale of the characteristic—spatial frequency of vibration, for example—to the scale of the perturbation. Thick, wide and long beams have anchor losses where leakage and propagation take place over larger length scales. In a very narrow beam, such as a nanotube, this region is much much smaller than the eigenmode wavelength. What this will mean, as in the adiabatic barrier versus abrupt barrier discussion of electron transport, is that the adiabatic approximation breaks down. Changes are now at the atomic scale, and that matters. One interesting aspect of the breakdown, however, is that while the quantum description in charge transport gives little room for approximations for quantum-dominated effects to be force-fitted into a classical picture leading to our mesoscale, nanoscale and phase transition discussion, the approximations for mechanical effects, where the particles are localized, do.

For example, the bonding of atoms, arising from the spatial sharing of electrons, can be adequately described by a potential energy which fits into the classical description. For example, the Lennard

Jones potential for molecule-molecule bonding,

$$U = 4E \left[\left(\frac{\sigma}{r} \right)^6 - \left(\frac{\sigma}{r} \right)^{12} \right],$$ (5.195)

where E and σ are energy and dimensional parameters, respectively, and r is a radial coordinate, works pretty well for interaction with water molecules. So, a scanning probe system, such as a scanning tunneling microprobe, a magnetic resonance microprobe or an electric field microprobe, all with a tip very close to an atomic surface, can be modeled reasonably accurately to the first order (see Figure 5.31).

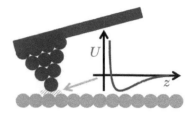

Figure 5.31: A scanning probe close to an atomic surface under the energy-lowering influence of interatomic forces.

Another limitation to continuum analysis is due to the statistical effects when the ensemble becomes small. We saw the consequence of this in the observability of Brownian motion, and the consequences of correlations with slowly and rapidly varying forces. There are two interesting offshoots of this. The first is related to our past discussion. Air at room temperature and pressure has about 10^{19} $molecules/cm^3$, so a cube of 1 $\mu m \equiv 1000$ nm has about 10^6 air molecules. We can use damping factors, Navier-Stokes equations, and other continuum descriptions, so long as we also continue to look at the fluctuations. But, what if we have a 10 nm-sized volume? This is the size scale for the vibration at the tip, where a mass is being measured to high resolution: a molecule or molecules attached to the tip. But the surrounding has, on average, 1 molecule. So, while the beam is vibrating, say, at 10 MHz is $2 \times 10 \times 10^{-7} \times 10 \times 10^6 = 20$ cm/s, which is much slower than the thermal velocity, and hence is averaging the noise, the statistical fluctuations of the sample size are not. An ensemble of N has a variance of $N^{1/2}$. Averaging long measurements will take an incredibly long time, since the rate of convergence is very slow, resulting in errors corresponding to this variance.

Another way to look at this problem is to consider a fluid-fluid interface, as shown in Figure 5.32. With ς as the surface tension, and R_1 and R_2 as the curvature radii at the fluidic interface, the stress tensors are related through

The Navier-Stokes equation is at the heart of atmospheric sciences and aero- and hydrodynamics. One physical way of looking at this is the following: particles are, at a high rate, coming in contact with other particles. If the volume is large, the ratio of the volume of the exclusion zone to the total volume is adequately small, then incompressibility is a good description, and Navier-Stokes suffices. But, make the volume small, and it is now a molecular flow regime—a mesoscale for this fluidic condition.

$$\hat{n} \cdot (S^{\underline{1}} - S^{\underline{2}}) = -\varsigma \left(\frac{1}{R_1} + \frac{1}{R_2} \right) \hat{n}.$$ (5.196)

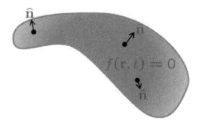

Figure 5.32: An interface described by a function $f(\mathbf{r}, t) = 0$.

Here, the Ss are the stress tensors. For stationary fluid, stress is normal—hydrostatic—and this is transformed to the pressure difference

$$p_1 - p_2 = \varsigma \left(\frac{1}{R_1} + \frac{1}{R_2} \right) \hat{n}.$$ (5.197)

An interface's position is obtained by the condition that the fluid at the boundary is stationary—the kinematic condition. This means that the interface $f(\mathbf{r}, t)$ satisfies, for each of velocity vectors $\mathbf{v}^i$,

$$\frac{\partial f}{\partial t} + \mathbf{v}^i \cdot \boldsymbol{\nabla} f = 0. \tag{5.198}$$

This is the no-slip boundary condition for the Navier-Stokes equation. Now, what happens when the mean free path of the fluid molecule is comparable or larger than the system size? The Knudsen number is one of several parameters defined in fluid mechanics and which are useful for describing the common characteristics of the fluid behavior associated with that scale. The Knudsen number is the equivalent of the volume-exclusion-to-system-volume ratio that we have looked at and is defined as $\mathfrak{K} = \overline{\lambda}/\ell$, so it is a cube root of the volume exclusion ratio. When the Knudsen number $\mathfrak{K} < 10^{-4}$, a no-slip boundary is a good approximation. Higher than this, and slipping along the interface becomes pronounced, and one force-fits an approximation that is pretty accurate. If a fluid wall, say, aligns the with x-axis and moves with velocity v_w along the x-axis, then

$$v_1 - v_2 = \frac{2 - \alpha}{\alpha} \frac{\mathfrak{K}}{1 - b\mathfrak{K}} \frac{\partial v_1}{\partial y}. \tag{5.199}$$

Here, α is a forced-fitting parameter called the accommodation coefficient, and b is a slip coefficient. The Knudsen number is thus an essential number for understanding the limit of the solid-liquid and liquid-liquid applicability of continuum mechanics inscribed in the Navier-Stokes equation.

Fluid viscosity is another parameter that we connect to this interface phenomenon tied to drag—our friction in fluid. The ratio of inertial forces to viscous forces is the Reynold's number $\mathfrak{R}$. An object of dimensional scale ℓ moving in a medium of viscosity η and density ρ at a velocity v has $\mathfrak{R} = av\rho/\eta = av/\nu$. $\nu = \eta/\rho$ is the kinematic viscosity, which is 10^{-2} cm/s for water. The ratio η^2/ρ has the units of force parameterizing the drag. If an object has a Reynold's number of 1, then this force will effectively drag the object. A small Reynold's number means that the inertial force needed for moving an object is small. As an object gets smaller, the drag effect reduces, and so does Reynold's number. A human swimming in a pool has an $\mathfrak{R}$ of 10^4; a fish in fish tank, of 10^2; and an *E. coli* bacterium which moves at speeds of the order of 30 $\mu m/s$ has an $\mathfrak{R}$ of 10^{-4} or less. Inertia plays little role in these conditions. Take away the force, and the object with a low Reynold's number almost immediately stops. Inertia and the prior velocity are irrelevant.

So, how does an object move in fluid environment using internal action? An object needs more than one degree of freedom in configuration space to be able to direct motion that is not a loop. An oar needs to be rotated around its axis, taken up and out of the water,

There may be a large range of viscosities for liquids, but it doesn't get much lower than that of water.

or undergo some other additional degree of freedom in order for a boat to move ahead. If not, so that there is just a forward and reverse motion of the oar in the water, the boat will oscillate back and forth. Human hands and legs break this symmetry during swimming. The flagellar motor or other synthase motors, such as *ATP* synthase, which is shown in Figure 5.33, do it for microbes. In a flagellar motor, it is the corkscrew-like oar of the motor that allows motion—a straight shaft will not do. In the *ATP* motors, it is the slightly off-axis shaft. In both of these, the motors have incredible speeds—100s to 1000s of revolutions per minute—and energy conversion efficiencies of more than 50%. A human produces nearly 20 *kg* of *ATP* every day through an energy-efficient, reversible cycle necessary for all these different chemo-electro-mechanical systems that are necessary for the body to function. The strong coupling that permits efficient energy conversion is crucial for these biological systems.

This is to say that $\rho \partial \mathbf{v}/\partial t$ and $\rho(\mathbf{v} \cdot \boldsymbol{\nabla})\mathbf{v}$ can be neglected. Edward Purcell has a very elucidating article on this subject, ``Life at low Reynold's number,'' *American Journal of Physics*, **45**, 5 (1977). Humans can swim well in water. But, try doing it in molasses. The Great Molasses Flood of 1919 is a tragicomedic incident when a hastily-built poor quality tank in the North End of Boston collapsed on January 15, 1919. 21 people and several horses died because the life of a meter-dimensional object is unbearable at this low Reynold's number.

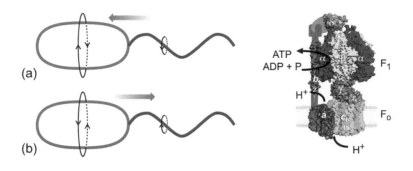

Figure 5.33: (a) and (b) show the forward and backward motion, respectively, of a flagellar motor such as that of *Escherichia coli* (*E. coli*). In (a) the filament bundle moving counterclockwise causes the cell body to roll forward clockwise. (c) shows the chemical-to-mechanical conversion motor of *ATP* synthase. Each motor employs an intermediate gated ion transport, H^+, Na^+, K^+ or Ca^+, for example, for mediating the energy conversion. The structure here shows the various protein units that deform mechanically under chemical potential change and thus cause the armature and the arm on the left and connected through b_2 to rotate. The drawing shows the $ADP + P \rightarrow ATP$ cycle—a cycle that is reversible.

So, while the continuum picture is adequate at much of the microscale and larger in such fluidic problems, it is not at very small scales. One needs to exercise adequate caution when using the approaches that we have developed because the nanoscale is a region where, many times, continuum modeling will be inadequate, and then one needs to utilize approaches that are more rigorous and appropriate to the scale of interest. For tackling the mechanical problems, one may proceed from continuum models based on bulk materials properties at 1000s of *nm* to continuum models that incorporate nanoscale material properties, such as the surface effects at 10s of *nm*, to more quantum-mechanically accurate approaches, such as molecular dynamics or tight binding, which are more *ab initio* and fundamentally more rigorous. These will be the equivalent, for tackling electronics problems, of the use of classical conductor models at large dimensions, such as in power transmission, semi-classical Drude models, such as in large dimension semiconductor devices, and quantum-mechanical and other rigorous models at the smallest scale for electronics.

5.6   Summary

This chapter focused on mechanics and the coupled behavior in environments where electrical forces are also important. We stressed the different approaches of analysis to bring out a number of interesting attributes of the behavior that are of import to devices. Lagrangian and Hamiltonian approaches, the use of conjugate variables, and energy in its kinetic and potential forms give us powerful tools for analysis in conditions where conservative and non-conservative forces exist. At its simplest, one could explore how moving plates and beams become useful as sensors and actuators. In many of these oscillatory modes, eigenmode analysis showed us the spatial and temporal dependence—in beams and plates. Inertial mass sensors and gyroscopes, et cetera, all rely on these approaches. The energy density and sensitivity analysis showed the tremendous capabilities that one can obtain. We extended this analysis in the presence of non-conservative components such as drag to understand how fast and slow forces behave. Correlation and its manifestation in spectral power density gave a powerful approach to then see how one might get the best sensitivity in structures where the mechanical resonance may be utilized together with electrical behavior. These are the forms important for measurements where one attempts to reach the quantum limits—measurements of single electron charge or phonons. One other interesting aspect of these resonances is the stochastic coupling of energy between fast and slow. Biology employs such stochastic motors to utilize energies of the order of $100 \ k_B Ts$,. This is the energy scale of many of the biological transduction processes.

All these systems also exhibit a variety of effects arising from nonlinearity. We emphasized bifurcation and chaos and utilized the phase portrait for observing the variety that unfolds. Even simple classical systems, in presence of nonlinearity, exhibit a variety of complex behaviors. Quantum systems do too. While we did not look at fluidic systems in depth, much of what we have described for mechanical elements in gaseous environment has an equivalent in the liquid environment, albeit with more complexity. Compressible and incompressible conditions will behave differently. Hydrodynamics at extremes can become unpredictable. These all manifest nonlinearity-induced behaviors such as chaos, limit cycles, hysteresis, et cetera, which are observable in simple Duffing systems. We have not yet discussed the coupling of mechanics with electromagnetic optical forces. In Chapter 6, we will dwell on this subject, since it provides a powerful means of obtaining uniquely sensitive measurements and generating interactions that can be gainfully employed in signal generation and manipulation.

5.7   Concluding remarks and bibliographic notes

MICROSYSTEMS ARE PERVASIVE in our daily life at this point, whether in mobile instruments in the form of gyroscopes, or in the car, as an accelerometer. But, these mechanical-electronic interactions and optical interactions, which we will discuss in the next chapter, are just as essential as signal measurement and control mechanisms across many domains where one of them by itself would not suffice or where these energy-coupling mechanism would provide a more sensitive or in other ways more appropriate approach.

Early classical mechanics, born of Kepler's laws, which were based on Tycho Brahe's as well as his own observations, rapidly progressed to the Lagrange and Hamilton approaches The Euler-Lagrange equation caste motion in terms of the Lagrangian, where the difference between the kinetic and the potential energy of the system is expressed using position coordinates and their derivatives. Hamilton introduced the action S as an integral of the Lagrangian in time, so that motion becomes a stationary point of action—an invariant. Equivalently, the Hamiltonian and the Hamilton equations give motion in time. This Lagrangian-derived approach of action also holds in electromagnetism, in a more complicated form from which Maxwell's equations follow. In Feynman's path formulation of quantum mechanics, the probability of an event is the modulus length squared of a complex number—the probability amplitude. This amplitude is obtained by adding together all the contributions of all paths in configuration space, with the contribution of a path proportional to $\exp(iS/\hbar)$, where S is again the action. Lagrangian and action are incredibly powerful.

Several exemplary texts—traditional and modern—exist, given the importance of this approach. For mechanics, an exemplar is by Hauser[1], but numerous others exist written for a mechanical engineering audience. A classic text for understanding the theory of elasticity is by Timoshenko[2] and was first published in 1934. A down-to-earth description of beam response may be found in the compendium by Pilkey[3].

For electromechanical microscale systems, a standard undergraduate text, one of the earliest ones, is by Senturia [4]. It is from the viewpoint of engineering and is quite comprehensive. A more advanced treatment for the various movements, sensitivities and the scaling considerations may be found in the text by Pelesko and Bernstein[5].

The methods we have used here were published by Lagrange published the methods we have used here in 1788. He seems to have gone by a fair number of different names. He was born Giuseppe Lodovica Lagrangia, or Giuseppe Luigi Lagrangia, in Turin. Lagrange followed Euler at the Prussian Academy of Sciences in Berlin, when Euler fell out with King Frederick of Prussia, who apparently was looking for "sophistication," and moved back to Saint Petersburg, having been invited back by Queen Catherine. Lagrange was a major figure in the adoption of decimal systems in France during the revolution, a matter that the United States still struggles with. William Rowan Hamilton of Dublin recast the Lagrange approach in 1833.

[1] W. Hauser, "Introduction to the principles of mechanics," Addison-Wesley, ISBN-13 978-0201028126 (1965)

[2] S. P. Timoshenko, "Theory of elasticity," Tata McGraw-Hill, ISBN 0-07-070122-9 (2010)

[3] W. D. Pilkey, "Formulas for stress, strain and structural matrices," John Wiley, ISBN 0-471-03221-2 (2005)

[4] S. D. Senturia, "Microsystem design," Kluwer, ISBN 0-306-47601-0 (2002)

[5] J. A. Pelesko and D. H. Bernstein, "Modeling of *MEMS* and *NEMS*," Chapman & Hall, ISBN 0-387-97173-4 (2003)

Preumont provides an advanced and comprehensive treatment of dynamics in electromechanical systems, using Lagrangians[6]. This book also discusses piezoelectric systems. The intricacies and uses of piezoactuation as well as acoustic interactions are tackled by the compilation edited by Safari and Akdŏgan[7].

Nonlinear aspects of movement are discussed in several texts. Particularly appropriate for the mechanics of beams is the book by Younis, which is devoted specifically to microelectromechanical systems (MEMS)[8].

Nonlinearity, stochasticity and chaos have been important themes in the scientific, theoretical, and applied mechanics communities for much longer than the area of interest to us: microscale and nanoscale electromechanics. For stochastic resonance and its many manifestations the review paper by Gammaitoni et al. is highly recommended[9].

A very readable, intuitive and comprehensive discussion of chaos exists in a number of texts. Acheson[10] provides a very intuitive introduction to chaos. Hilborn[11] provides a more advanced and comprehensive treatment, including the fractal aspects of chaos. This text is a very readable and rich text for an introductory but detailed analytical discussion of nonlinearity and chaos. Strogatz's[12] is a textbook replete with insights and examples.

Chaos also occurs in non-classical systems. The adventurous will not be disappointed by Gutzwiller's exposition[13]. It is a very well-thought-through and cogently written honest discussion by one of the great gentleman physicists of the 20th century. Those interested in a further exploration of Melnikov functions may wish to consider the book by Han and Yu[14].

The mechanics of a fluidic environment, given its importance to biotechnology, has a number of book offerings. For fluidics, Abgrall and Nguyen[15] provide a good treatment of scale and interface effects. Electrophoresis and magnetophoresis, the motion of objects in fields in a fluidic environment, are important biological techniques. Jones[16] discusses them comprehensively.

[6] A. Preumont, "Mechatronics: Dynamics of electromechanical and piezoelectric systems," Springer, ISBN 1-4020-4695-2 (2006)

[7] A. Safari and E. K Akdŏgan, "Piezoelectric and acoustic materials for transducer applications," Springer, ISBN: 978-0-387-76538-9 (2008)

[8] M. I. Younis, "MEMS: Linear and nonlinear statics and dynamics," Springer, ISBN 978-1-4419-6019-1 (2011)

[9] L. Gammaitoni, P. Hanggi, P. Jung and F. Marchesoni, "Stochastic resonance," Reviews of Modern Physics, 70, 223–287 (1998)

[10] D. Acheson, "Chaos: An introduction to dynamics," Oxford, ISBN 0 19 850257 5 (1997)

[11] R. C. Hilborn, "Chaos and nonlinear dynamics," Oxford, ISBN 0-19-567173-2 (2004)

[12] S. H. Strogatz, "Nonlinear dynamics and chaos," Perseus, ISBN 0-201-54344-3 (1994)

[13] M. C. Gutzwiller, "Chaos in classical and quantum mechanic," Springer-Verlag, ISBN 1-58488-306-5 (1990)

[14] M. Han and P. Yu, "Normal forms, Melnikov functions and bifurcations of limit cycles," Springer, ISBN 978-1-4471-2917-2 (2012)

[15] P. Abgrall and N. T. Nguyen, "Nanofluidics," Artech, ISBN 978-1-59693-350-7 (2009)

[16] T. B. Jones "Electromechanics of particles," Cambridge, ISBN 978-0-521-43196-5 (1995)

5.8 Exercises

1. Calculate the moment of inertia of a beam of thickness t and width w as well as that of an I-shaped beam such as a railroad track, where the thickness of the top and bottom sections of extent w is Δt, and the center element is Δw thick, as shown in Figure 5.34. Find the dependence of the inertia on the cross-sectional area and plot it in a suitable form to point out the optimization points where the weight of the beam can be reduced substantially while

sacrificing a smaller reduction in inertia. This points to why *I*-shaped beams are so ubiquitous. **[S]**

2. The shortest curve connecting two points in a plane is a straight line. Show that the variational form of this minimization is to minimize the integral $\int_{x_1}^{x_2} \sqrt{1 + (dy/dx)^2}\, dx$. Here, $y(x)$ connects (x_1, y_1) and (x_2, y_2), which are the two end points. Use the Euler-Lagrange equation to prove that a straight line is the minimizing curve. **[S]**

3. If silicon fractures under an axial stress of $\sim 10^9\ N/m^2$, find the maximum length of a vertical silicon beam that does not exceed the fracture stress under its own gravitational load. **[S]**

4. Calculate the magnetic energy in a toroidal solenoid whose $L = 0.2\ nH$ and which has a current of $1\ mA$ flowing through it. What electronics-compatible capacitor design would store similar energy? Assume that the capacitor is made out of SiO_2. Give the plate area, insulator thickness and necessary operating bias voltage. **[S]**

5. A bimetallic strip is 1 *mm* long and composed of two materials with thermal expansion coefficients of $\alpha_1 = 2.5 \times 10^{-6}\ K^{-1}$ and $\alpha_2 = 5.0 \times 10^{-6}\ K^{-1}$, respectively. The beam is 10 *μm* thick. Find the maximum deflection starting from none as designed at 300 *K* for temperature excursions of 100 *K*, 500 *K* and 1000 *K*. **[S]**

6. *ZnO* is piezoelectric and can be suitably deposited on a silicon cantilever. Design a cantilever with a 2000 *nm*-thick *ZnO* integrated with two electrodes above it so that the free end may deflect by $\pi/6$. Find the length and thickness of the cantilever, and the voltage needed for the deflection that is suitable in a microsystem. **[M]**

7. When silicon is oxidized at high temperatures, we may assume that the SiO_2-*Si* system is stress-free at the high temperature. When cooled, however, stress develops.

   • Estimate the thermal strain when silicon is oxidized at a high temperature—say 1275 *K*—creating a stress-free film and then cooled to 300 *K*.

   • Estimate the thermal strain when a silicon wire d_0 in diameter is oxidized and reduced to a diameter of d_c for the core and diameter d_f for the outside oxide. **[M]**

8. A cantilever of mass m_c has a point proof mass m placed at its free end, as shown in Figure 5.35. If the Young's modulus is Y, determine the resonance frequency of the structure. **[M]**

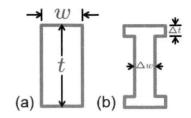

Figure 5.34: (a) and (b) show cross-sections of two beams with different size parameters. (a) is a rectangular beam, and (b) is an *I*-beam.

Oxidation of a wire causes a higher stress because of the reentrant geometry even at high temperatures. The consequence is that the oxidation rate asymptotically vanishes as the stress builds up and further oxidation of silicon is prevented. It is possible to make crystalline nanoscale silicon wires this way. With another increase in dimensional confinement, it becomes possible to make silicon nanoscale dots.

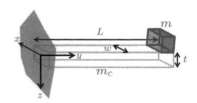

Figure 5.35: A point mass m placed at the end of a beam composed of a material of Young's modulus Y and mass m, with t, w and L as size parameters.

9. If the point proof mass is placed with its center of mass a displacement of Δy beyond the beam of length L of the previous problem, assuming that the cantilever is massless, show that the resonant frequency of the structure can be approximated by

$$f_r' = \frac{1}{2\pi}\sqrt{\frac{Ywt^3}{12ml^3}\frac{\mu^2 + 6\mu + 2}{8\mu^4 + 14\mu^3 + (21/2)\mu^2 + 4\mu + (2/3)}},$$

where $\mu = \Delta y/L$. **[A]**

10. This previous problem's relationship states that the resonance frequency of a point mass at the end of a massless cantilever is

$$f_r = \frac{1}{2\pi}\sqrt{\frac{Ywt^3}{4L^3m}}.$$

If one were to attempt to use such a resonance for mechanical-to-electrical energy conversion, for example, by a charge on such a mass oscillating between two plates, the resonance frequency needs to be close to the system's mechanical resonance. Which mechanical form in daily living frequencies may allow a practical mechanical-to-electrical conversion? Examples are the oscillations of walking, a traveling bus or car, et cetera. Is there an issue of frequency mismatch here? **[M]**

11. For a system subject to $\dot{y} = y^{1/3}$, show that the initial boundary condition of $y(t = 0) = 0$ does not have a unique solution. Why is this so? **[S]**

12. Plot the potential energy for $\dot{y} = y - y^3$ and show the different equilibrium points of the system. **[S]**

13. For a square and stiff plate supported by four cantilevers, as shown in Figure 5.36, determine the effective stiffness and the pull-in voltage V_π between this assembly and a planar electrode in parallel with the stiff plate a distance d away. Assume that the Young's modulus is Y. Estimate V_π for an effective and nominal microscale geometry. **[M]**

14. Make an equivalent circuit and write the equations to determine the response of the system shown in Figure 5.37. **[M]**

15. In a classical system subject to scattering from its environment, the system sensitivity to forces improves at lower temperature. Briefly explain why. And also why does the spectral response decay away from a peak? What is this peak due to? **[S]**

16. The interaction of the tip and sample in an atomic force microscope may be modeled using the approximate lumped parameter

Figure 5.36: A stiff plate held by four cantilevers and electrostatically actuated by a parallel plate that is not shown but is a distance d away.

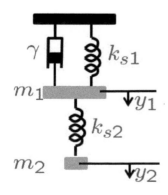

Figure 5.37: A partially damped system of two masses.

equation

$$m\ddot{x} + k_s x = \frac{Dk_s\sigma^6}{20(\ell + x)^8} - \frac{Dk_s}{(\ell + x)^2},$$

where m and k_s are the mass and the stiffness constant of the cantilever, ℓ is the tip-to-surface separation, $D = AR/6k_s$, with A being a Hamaker constant related to the van der Waals forces of this geometry, R is the radius of the contact tip, and σ is a molecular-scale dimension (≈ 0.03 nm).

- Rewrite the interaction equation in a dimensionless form to draw out the tip forces.

- What are the dimensionless equilibrium solutions? What kind of conditions of stability and bifurcations may come about?

- What is the potential energy as a function of dimensionless parameters?   [A]

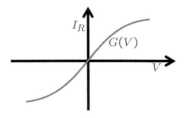

Figure 5.38: A nonlinear response of $I_R = G(V)$ from a resistor.

17. In a series RC circuit being charged from a voltage source, if the resistor is nonlinear, that is, $I_R = G(V)$, where I_R is current through the resistor, and $G(V)$ is as sketched in Figure 5.38, derive the circuit equations, and identify the fixed points of the response. What are the implications of the nonlinearity, and how does it affect the stability?   [S]

18. The Allee effect is the observation that the effective growth rate of some species is at its maximum at some intermediate population, that is, $\dot{n}/n$ peaks at some intermediate n with n as the population. Too small a population, and finding mates is hard. Too large a population, and the food and resources become scarce.

- Show that

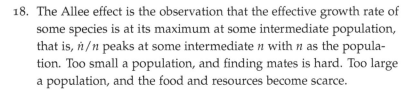

$$\frac{\dot{n}}{n} = r - c_1(n - c_2)^2,$$

under constraints on r, c_1 and c_2, is a model for the Allee effect.

- What are the fixed points of the system and the nature of their stability? Stability is discussed in the Appendix discussion of phase space portraiture.

- Comment on the form of $n(t)$.   [S]

19. Phase has appeared inextricably in nearly all discussions throughout the electromechanical system response analysis. Phase has information that is critical to analysis. A simple example to show this is a problem concerning a relationship in time. Two runners A and B are running at a constant speed around a circular track. A takes T_A time to complete the circle, and B takes T_B time. Let $T_B > T_A$. If A and B start at the same time, how long does it take for A to overtake B once?   [S]

The phase method of thinking through this problem is the most intuitive. There is an old von Neumann story that Ulam relates. von Neumann was asked the question on time and distance traveled when two cyclists separated by a distance L get going—at constant speeds different from each other's and towards each other—with two birds that too launch off from their shoulders at much faster and different speeds. The birds fly to each other and back, touching the shoulder of their host bicyclists, and repeating. When the bicyclists reach each other, how much distance did the birds travel? von Neumann thinks for a short time and gives the answer. When asked how he found it, he said he formed a series. "Is there another way?"

20. Show that a system with

$$\dot{x} = -2\cos x - \cos y, \quad \text{and}$$
$$\dot{y} = -2\cos y - \cos x$$

is reversible but not conservative. Show the phase portrait.   [M]

21. Consider the Duffing equation $\ddot{y} + y + \varepsilon y^3 = 0$, where we have introduced a nonlinearity parameter ε.

- Show that there exists a center at the origin that is nonlinear for $\varepsilon > 0$.

- Show that if $\varepsilon < 0$, trajectories near the origin are closed.

- What happens to trajectories farther from the origin?   [M]

22. We consider the movement of a sphere in air under normal temperature and pressure conditions.

- What is the approximate size of a sphere moving at a reasonable speed for a transition from turbulent flow, that is, chaotic flow, to laminar flow, that is, with parallel continuity at boundaries?

- If the dynamic viscosity varies with pressure as $\eta = \eta_0 p / p_0$, estimate the pressure for a sphere of radius 1000 nm moving at 0.01 cm/s.

- Can you estimate the spacing between air molecules at this pressure?

- Is the description of these changes consistent?   [A]

23. Because mass varies at the cube of length, inertia is usually considered unimportant to microscale motion in fluids. Take a system consisting of a small ball attached with a string to a motor that is putting it through rotational motion.

This is a model that teaches us about the conditions under which biological motors operate.

- Under the constraint of a limiting tension per unit cross-section of the tethering string, derive the scaling relationship of the rotational frequency of the ball.

- Is the inertia important in this system?   [M]

24. In two square pads of hook and loop fasteners, let L be the size length, and let ℓ^2 be the area on the pad that each hook and loop pair occupy. Let $F_0 = \kappa \ell^n$ be the force required for separation of a hook and loop pair.

Velcro is the more prominently known example of these fasteners.

- Show that the force for separating the pads is given by

$$F = \lambda^{n-2} \kappa L^n,$$

with $\ell = \lambda L$.

- Argue that the magnitude of this exponent $n \to 2$.
- Is a microscale or nanoscale hook and loop fastener effective?

[S]

25. A hollow sphere of radius R when pushed into water feels a restoring force equal to the weight of the displaced water, according to the Archimedes' principle. Ignore friction, and determine the frequency of oscillation. [S]

26. The dynamics of a spring-mass-damping system (a plate under electric force, nonlinear spring and damping γ due to the squeezing environment or Zener causes) can be described by the force equation

$$F = k_{s1}u + k_{s2}u^2 + k_{s3}u^3.$$

If the system is initially pulled a distance x_0 away from equilibrium, derive a dimensionless equation of motion. [S]

27. Take the chapter's example of single degree of freedom parallel plate capacitor of mass m actuated by a bias $V = \overline{V} + \hat{V} = \overline{V} + \hat{V} \cos \omega t$ (a superposition of static and harmonic voltage bias). Write the dimensionless equation of dynamics, and extract and comment on parameters. [M]

28. Consider a van der Pol oscillator, with its nonlinearity, described by

$$\ddot{x} + 2\alpha(x^2 - 1)\dot{x} + x = 0.$$

Analyze its stability and bifurcation for $\alpha > 0$. [M]

6

Electromagnetic-matter interactions and devices

ELECTROMAGNETIC-MATTER INTERACTIONS IN CONDENSED MAT-
TER should occur somewhere near the length scale of the photon
extending out to a distance until it is extinguished as a result of the
interactions. A photon, if in the X-ray range, is at a nanometer-scale
wavelength and has quite important interactions—wave based—that
we use in characterization of all forms of materials using the wavelength scale resolution. But, even a photon with a much, much longer
wavelength than those of X-rays, say, in the visible or infrared range,
when interacting with matter—a large population of electrons in a
metal, or the electron-ion assembly that is an organic molecule—also
is an environment of energetic coupling of excitation modes. Oscillations of the population of electrons or excited states of the molecule
arise through interactions where the nanometer spatial scale is still
important. Here, we view these as coupled interactions where we
introduce quasi-particles such as plasmons. Other interactions with
the mechanical motion are described through other quasiparticles—
polaritons, for example, which represent electromagnetic interaction with dipoles. Absorption, as well as radiation, which can be
spontaneous and random in all directions, stimulated/catalyzed by
photons and therefore direction selective, and these interaction processes with excitons will be constrained by selection rules arising
from conservation laws. All these interactions certainly will have
nanoscale features. Operating at the nanoscale will also bring about
non-linearities and changes in interaction parameters due to boundaries that structures have, for example, at metal–semiconductor interfaces, where disparate electron energy and populations couple. All
of these have features that will depend on size. So, electromagnetic
radiation interactions—optical interactions being a subset of these, in
our common usage—present numerous effects of nanoscale import
and of device utility.

We will look at some important and consequential representa-

These long-range excitations can also be
viewed as Goldstone theorem at work.

Nanoscale device physics: Science and engineering fundamentals. Sandip Tiwari.
© Sandip Tiwari 2017. Published 2017 by Oxford University Press.

tive examples. First, we discuss the Casimir-Polder effect, or Casimir effect, for short as an example of electromagnetic interaction that appears in all the different themes we have discussed up to this point—mechanic, electronic, chemical—since it is quantum mechanical in origin. We will discuss the light-matter interactions within the bulk and at the interfaces of the condensed matter state that we have been interested in throughout this text and focus around the differences in inorganic and organic forms. In particular, we will look at these for their utility in devices such as solar cells/photodiodes—forms of optical-to-electrical energy conversion device as well as sensors that are detectors and as light emitters—forms of actuators. In solar cells, our theme will be quite specific—the thermodynamic relationships represented by the Shockley-Queisser limit, that is, constrained by detailed balance and off equilibrium when photons interact with matter, or how photon excitation and energy exchange take place in organic materials. For light emission, with organics we will employ the complement of light detection in light emission, and for inorganic materials, we will use the quantum cascade laser as a representative nanoscale example. In particular, these will integrate for us localized electron-electromagnetic interactions represented by coupled quasiparticles such as plasmons—Coulombic and resonant energy exchange through dipole interactions—or the formation of excitons in molecular and organic structures; resonant transmission, such as that represented by superlattices; and the light conversion strength vis-à-vis other energetic effects, such as through phonons, in all these systems. This is the range of interactions and their use in devices that is the focus of this chapter, where even if the wavelength of the photon is long—in micrometers and longer, for infrared and microwave wavelengths—the consequences of interesting effects that take place at the nanometer scale make numerous useful optoelectronic devices possible.

6.1 The Casimir-Polder effect

How do forces behave at the near-atomic scale?

So, not what happens at the nuclear scale, but between two close metallic objects, or dielectric objects, or even two molecules—the spatial scale where both the quantum-mechanical and the classical views need to be reconciled. One may look at the ideal gas law, $PV = nRT$, where P is pressure, V is volume, n is the molefraction, R is the ideal gas constant, and T is temperature, for one as a macroscopic kinetic result from non-interacting, vanishing-sized atoms and molecules. If we do precise measurements, one finds that there need to be corrections to this. Even by the year 1873, Johannes van der Waals, not yet knowing the nature of the particles at the smallest scale or that there may be sticky or repelling forces at work or a size to these particles, showed that the relation $(P + a/V^2)(V - b) = nRT$ was a more precise description. a and b here are two constants for the specific gas or combinations of it chosen. An early success of quantum theory occurred when Fritz London showed the origin of van der Waals's correction to nonpolar gases was due to electric dipole fluctuations arising from charged constituents. He accomplished this by incorporating the fluctuations in the interaction term of the Hamiltonian and then showing that this interaction of the dipole moments of two molecules decays with a $1/r^6$ dependence. This is the electrostatic interaction at a distance. We will see that this is extremely important for the long-distance energy transfer between donor and acceptor molecules in organic electronics. This energy transfer mechanism is the Förster transfer, and we will discuss it when we explore energy conversion and transport in organic materials. But, a very important thought underlies this dipole interaction. The interaction energy itself has a $1/r$ dependence, but because of the dipole of a polarized molecule, this behavior gets positionally sensitive the closer one gets to the molecules, and the closer the molecules get to each other. For such an interaction, first order perturbation theory will imply that $\langle \mathscr{H}_i \rangle = 0$ due to the randomness of the orientations. The second order term, arising as

Not important to the argument here, but because of the dipole, the terms in Equation 6.1 arise in the form $(\mathbf{p} \cdot \mathbf{r})_1 (\mathbf{p} \cdot \mathbf{r})_2 / r^5$, which has $1/r$ Coulombic energy dependence and is reconcilable with the $1/r^6$ dependence.

$$V' = \sum_{m \neq 0} \frac{\langle 0 | \mathscr{H}_i^2 | m \rangle \langle m | \mathscr{H}_i^2 | 0 \rangle}{E_0 - E_m} \propto \frac{1}{r^6}, \tag{6.1}$$

is the van der Waals interaction. It is an effect at a short distance, calculated here as arising from the polarization of molecules in the quantum-mechanical limit. This result assumes that the distance scale $r \ll \lambda$, the characteristic wavelength of polarizability. Otherwise, at a long distance, there will be, in a low temperature limit, an additional $1/r$ dependence arising from retardation.

Fluctuating dipoles are also fluctuating electric fields. A ground state of a simple harmonic oscillator, that is, our prototype of a stable quantum system with two canonic variables, has an energy $\hbar \omega_0 / 2$—the zero point energy—arising from the non-determinism of the two canonic variables simultaneously—the Heisenberg uncertainty or, correspondingly, the fluctuating exchange of energy between kinetic and potential form. So, in its simplest form of argument, consider two parallel conducting plates with a separation of a and shorted together. $\langle \mathcal{E} \rangle = 0$. But, the expectation value of the square of the fields $\langle \mathcal{E}^2 \rangle \neq 0$, and so it is true for $\langle \mathbf{H}^2 \rangle \neq 0$. This means that

As discussed in S. Tiwari, "Quantum, statistical and information mechanics: A unified introduction," Electroscience 1, Oxford University Press, ISBN 978-0-19-875985-0 (forthcoming), and emphasized in Edward Purcell's classic text, *Electricity and Magnetism* of the Berkeley physics course series, the speed of light places constraint on energy propagation—this is what we represent through the light cone of the allowable trajectory.

expectation of the energy is not equal to zero:

$$\langle U \rangle = \frac{1}{2} \int_{\Omega} \left(\frac{1}{2} \mathcal{E} \cdot \mathbf{D} + \frac{1}{2} \mathbf{B} \cdot \mathbf{H} \right) d^3 \mathbf{r} \neq 0. \tag{6.2}$$

If there exists an energy, then there exists a measurable force. This is the Casimir-Polder force F_C and the effect that this represents is the Casimir-Polder effect, or the Casimir effect, for short.

The Casimir effect is the existence of forces on bounding surfaces when one confines a finite volume of space. One can visualize this, through a pictorial argument representing the mathematical one just discussed, using the allowed electromagnetic modes representation shown in Figure 6.1. A low frequency cut-off of the allowed modes is essential to the design and use of waveguides and for the specific modes that are allowed by cavities. Because of the conductivity of metal, the field is suppressed at the metal interface, and therefore only specific wavelengths below a cut-off, or frequencies above a cut-off, are allowed, following the electromagnetic constraints of Maxwell. In free space, that is, the region external to the confined space, all solutions of the wave equations are allowed. Between the two bounding surfaces, only certain modes subject to the following simplifications from Maxwell's equations are allowed. The equations prescribe

$$\begin{aligned}
\nabla^2 \mathcal{E}^2(\mathbf{r}) + \frac{\mu \epsilon \omega^2}{c^2} \mathcal{E}(\mathbf{r}) &= 0, \\
\nabla \cdot \mathcal{E}(\mathbf{r}) &= 0, \\
\nabla^2 \mathbf{H}^2(\mathbf{r}) + \frac{\mu \epsilon \omega^2}{c^2} \mathbf{H}(\mathbf{r}) &= 0, \text{ and} \\
\nabla \cdot \mathbf{H}(\mathbf{r}) &= 0. \tag{6.3}
\end{aligned}$$

And these are subject to the conducting boundary conditions. With perfectly conducting plates, these are

$$\begin{aligned}
\mathbf{D} \cdot \hat{n} &= \sigma, \\
\mathbf{B} \cdot \hat{n} &= 0, \\
\hat{n} \times \mathcal{E} &= 0, \text{ and} \\
\hat{n} \times \mathbf{H} &= \mathbf{J}_s, \tag{6.4}
\end{aligned}$$

where σ is the charge density, and $\mathbf{J}_s$ is the surface current density. When the interface is with perfectly conducting metal plates, this constrains the fields in the medium at the interface with the plates to $\mathcal{E} = 0$, $\mathbf{D} = 0$, $\mathbf{B} = 0$, and $\mathbf{H} = 0$.

This determines the eigenfrequencies of the field between the plates as

$$\omega_{klm}(a) = \pi c \left(\frac{l^2}{L^2} + \frac{m^2}{a^2} + \frac{n^2}{L^2} \right)^{1/2}, \tag{6.5}$$

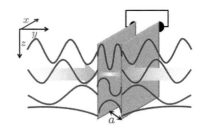

Figure 6.1: Two conducting parallel plates brought together. Vacuum fluctuations pervade outside with all modes allowed, but are limited by the electromagnetic mode constraints from the geometry between the plates.

where L is the dimension of the "semi-infinite" plates that are a apart, and l, m and n are integers. The integer m determines the the modes whose lowest energy excitations are shown in Figure 6.1. So, the zero point energy between the plates is

"Semi-infinite" in the sense that we will not worry about the details of the edges. Details away from the edges are subject to different symmetries that are also easier to analyze.

$$U_0(a) = 2 \sum_{l,m,n} \frac{1}{2} (\hbar \omega_{lmn}) = \sum_{l,m,n} \pi \hbar c \left(\frac{l^2}{L^2} + \frac{m^2}{a^2} + \frac{n^2}{L^2} \right)^{1/2}. \quad (6.6)$$

The energy difference between conducting plates, with a distance a in y and infinitely apart, is

$$
\begin{aligned}
U(a) &= U_0(a) - U_0(\infty) \\
&= \frac{\hbar c}{\pi} \sum_n \left[\int_0^\infty dk_x \int_0^\infty dk_z \int_0^\infty \left(k_x^2 + \pi^2 \frac{m^2}{a^2} + k_z^2 \right)^{1/2} dk_y \right. \\
&\quad \left. - \int_0^\infty dk_x \int_0^\infty dk_z \int_0^\infty \left(k_x^2 + k_y^2 + k_z^2 \right)^{1/2} dk_y \right] \\
&= -\frac{\pi^2 \hbar c}{720 a^3} L^2, \quad (6.7)
\end{aligned}
$$

derived by excluding high energy photons, for example, those smaller than atomic separation. Since there is a cut-off of low frequency photons, this energy is negative. Plates closer together have lower energy than those that are farther apart. The force per unit area between parallel conducting plates then is

$$F_C = -\frac{\partial U(a)}{\partial a} = -\frac{\pi^2 \hbar c}{240} \frac{1}{a^4}, \quad (6.8)$$

a 4th power dependence. The force between a conducting sphere of radius R and a plate is

$$F_C = -\frac{\partial U(a)}{\partial a} = -\frac{\pi^3 R \hbar c}{360} \frac{1}{a^3}, \quad (6.9)$$

a 3rd power weaker dependence. Less confinement of the fluctuation energy of the system leads to less force.

What is remarkable about these expressions is the electron's role in the conducting boundary condition, even though itself does not appear in the expression. The Casimir force is like a depletion force of the electromagnetic field. It appears when fields are excluded, which of course depends on the properties of the interfaces. Polarizable interfaces, such as metals, affect this exclusion, and the force is strong and attractive. It can also be a repelling force with dielectrics, as we will presently see. This Casimir force can be very significant. Two conducting plates that are 1 cm on a side and are a μm apart have a force of 100 nN. This is the weight of a particle 0.5 mm in size, so of the order of magnitude of gravitational force on a plate. If the

This integral summation is due to Evgeny Lifshitz of Landau and Lifshitz, who provided a complementary view in 1954 to that of Hendrik Casimir and Dirk Polder, from 1948, whose first paper focuses on the long-range correction with which we started the discussion, so quite an entirely different physical view. Casimir was at Philips's Eindhoven research laboratory, an example of a collection of brilliant minds in a non-academic establishment. The industrial interest arose from the nature of interactions among colloidal particles in liquids—some agglomerate, and some don't. Casimir attributes the zero point argument, which he espoused later, to a line of inquiry spurred by Bohr. Lifshitz's calculation is very complex, but it includes temperature's effect, and it shows attractive and repulsive consequences. Vitaly Ginzburg, another of the Russian doyens, is quoted as saying, "His calculations were so cumbersome that they were not even reproduced in the relevant Landau and Lifshitz volume, where, as a rule, all important calculations were given." Julian Schwinger too approached this problem, using his Green's function source theory. We all need our own zero point energies—self-drive—if we want to be a force for change.

plate spacing is reduced by a factor of 100, the plates will be pulled together even at an atmospheric pressure.

This behavior of the Casimir force is intimately connected to the boundary conditions and to microscopic properties of the materials. Change the boundary conditions, and the nature of this interaction will change—even changing sign. These boundaries, irrespective of the origin, change the coupling of the fluctuations existent in the field modes. And this leads to a measurable force.

The geometry and boundary conditions of problem will define the redistribution and whether forces will be attractive or repulsive. We illustrate this behavior by replacing the conducting plates with dielectric plates. Now, our boundary conditions must be changed, and the modes, internal and external, will penetrate and extend out and be subject to both the properties of the dielectric plates and the boundary conditions, as shown in Figure 6.2. In particular, Equation 6.4 is replaced by

$$D_n = D_n^{dp}; \quad B_n = B_n^{dp}; \quad \mathcal{E}_t = \mathcal{E}_t^{dp}; \quad \text{and} \quad H_t = H_t^{dp}, \tag{6.10}$$

with the dielectric constant a function of frequency, that is, $\epsilon(\omega)$. The absence of high polarization at the interface implies wave penetration in the dielectric plates, with a changing dielectric response with frequency. The modes are not confined anymore, spaces are connected, are subject to the dielectric response with frequency, and the boundary conditions define the coupling of the allowed modes of zero point energy fluctuations. This spread of the modes, and their penetration and spread across the surrounding space, can mean, under suitable conditions, that the electromagnetic pressure between the plates is now greater than that outside the plates. The plates will now repel. This argument states that because of the change in boundary conditions, the number of modes hasn't changed inside the ``confined´´ volume between the dielectric plates, nor is there any difference between the number of modes in this ``internal space´´ and that in the ``external space´´ outside the plates, but the modes themselves have shifted in frequency. The conducting plates caused standing waves, each of which is the equivalent of a harmonic oscillator. The frequency shift can be seen as an adiabatic shift arising through the movement of plates from having a large separation between them to having a small separation between them. The redistribution of modes causes there to be a difference between the pressure of the vacuum fields between the plates and that of the fields external to the plates.

The relationships in this reformed dielectric plate problem are now in a tensor form, and one may write the Maxwell stress tensor for the electromagnetic pressure including all the source terms for forces,

In here, we can see the origins of Casimir's interest in this problem. Particles may attract or repel, depending on the specifics of the interaction. This applies to colloidal behavior in its own way, the genesis of Casimir's interest in this problem before it took a life of its own, as many curious phenomena that science attempts to explain tend to do. As an aside in this vein, this effect pervades down to the domain of string theories where extra dimensions curl into lesser. In chromodynamics, these boundaries are topological, connecting different phases of vacuum.

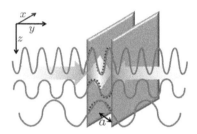

Figure 6.2: The Casimir force situation in the zero point energy picture, with two dielectric plates separated by a small separation of a. The internal space and the external space are connected, and modes are spread out subject to the boundaries and the existence conditions in regions with different microscopic properties. The dotted lines between the plates show the lowest energy internal modes of the previous conducting plates example.

with the matrix elements as

$$\sigma_{ij} = \epsilon_0 \mathcal{E}_i \mathcal{E}_j + \mu_0 H_i H_j - \frac{1}{2}\left(\epsilon_0 \mathcal{E}^2 + \mu_0 H^2\right)\delta_{ij}, \qquad (6.11)$$

where (i, j) are the orthogonal projections. The Poynting vector is a special case of this stress tensor, illustrating the propagation of energy. The fluctuation-dissipation theorem provides the correlation relationship for vector potential as

$$\langle A_i(\omega, \mathbf{r}), A_j(\omega, \mathbf{r}')\rangle = c \tanh \frac{\hbar\omega}{2k_B T} \Im[\mathcal{D}_{ij}(\omega, \mathbf{r}, \mathbf{r}')], \qquad (6.12)$$

where $\mathcal{D}_{ij}$ is the dielectric function. This leads to the pressure result

$$\langle \sigma_{yy}\rangle = \frac{k_B T}{\pi c^3}\sum_n \omega_n^3 \int_1^\infty p^2 \left\{\left[\frac{(s_1 + p)(s_2 + p)}{(s_1 - p)(s_2 - p)}\exp\left(-\frac{2p\omega_n a}{c}\right) - 1\right]^{-1}\right.$$
$$\left. + \left[\frac{(s_1 + p\epsilon_1)(s_2 + p\epsilon_2)}{(s_1 - p\epsilon_1)(s_2 - p\epsilon_2)}\exp\left(-\frac{2p\omega_n a}{c}\right) - 1\right]^{-1}\right\} dp, \quad (6.13)$$

where $\omega_n = 2\pi n k_B T / \hbar$, and $s_{1(2)} = (\epsilon_{1(2)} - 1 + p^2)^{1/2}$, with the frequencies ω_n as the discrete solutions of the Kramers-Kronig principal
value solutions from the complex plane, that is,

$2\pi k_B T/\hbar$ at room temperature corresponds to about 10^{15} Hz—terahertz.

$$\epsilon_{1(2)}(i\omega) = 1 + \frac{2}{\pi}\int_0^\infty \frac{\zeta\Im[\epsilon_{1(2)}(\zeta)]}{\omega^2 + \zeta^2}\,d\zeta. \qquad (6.14)$$

When the temperature effects can be neglected, Equation 6.13 can be
simplified, with L as the plates' size, both for the short-range and for
the long-range limit. These are, respectively,

$$\lim_{a\to 0}\langle\sigma_{yy}\rangle = \frac{\hbar}{16\pi^2 a^3}\int_0^\infty \int_0^\infty x^2 \left[\frac{\epsilon_1(i\zeta) + 1}{\epsilon_1(i\zeta) - 1}\frac{\epsilon_2(i\zeta) + 1}{\epsilon_2(i\zeta) - 1}\exp x - 1\right]^{-1} dx\,d\zeta$$

$$= \frac{L^2}{6\pi a^3}, \text{ and}$$

$$\lim_{a\to\infty}\langle\sigma_{yy}\rangle = \frac{\hbar c}{16\pi^2 a^3}\int_0^\infty \int_0^\infty \frac{x^3}{\zeta^2}\left[\frac{\epsilon_1(0) + 1}{\epsilon_1(0) - 1}\frac{\epsilon_2(0) + 1}{\epsilon_2(0) - 1}\exp x - 1\right]^{-1} dx\,d\zeta$$

$$= -\frac{\pi^2 \hbar c}{240 a^4} \text{ for } \epsilon(0) \to \infty. \qquad (6.15)$$

The latter is the case of a perfect conductor. The Casimir result is
a limit case of Lifshitz's general case. The power of these relationships is that the forces can be derived from the frequency-dependent
dielectric properties of the interacting bodies and that this can be
accomplished entirely separately.

A further generalization of this is
work by Dzyaloshinkskii, Lifshitz and
Pitaevskii, as it includes general possibilities of the microscopic properties,
including conditions such as liquid
capillarity along walls.

　　The electromagnetic modes causing this Casimir force, in these examples, are evanescent modes of the material. Thus, this evanescence
reflects the interaction of the material with its surrounding modes.
In the perfect conductor example, the boundary condition reflects

both the perfect termination of the fields, and an infinite imaginary component of the dielectric constant. In a dielectric, both the real and the imaginary part affect the field and, therefore, the mode. The temperature dependence of the force follows from the change in the properties of the material.

The roughness of surfaces, as well as their reflectivity, affects these forces. The former increases the force, and the latter decreases it. Perfect reflections, connected through the dielectric properties, generate the largest force. Surfaces reflect poorly at short wavelengths. Reflectivity improves with decreasing roughness. So, the Casimir force increases with decreasing surface roughness. So, precise measurements of the Casimir forces require very precise temperature and roughness control. Modified forms of Equation 6.13 provide these.

We conclude these discussions with two simplified models, to reinforce the importance of the fluctuations. Vacuum fluctuations explain van der Waals forces, inhibition of spontaneous emission, and vacuum Rabi oscillations.

We will discuss this fluctuation and coupling interaction, using an electrical model and a hydrodynamic model.

Consider Figure 6.3, which shows long-range and short-range interaction for two identical interacting atoms that are a distance a apart. We look upon these as simple spherically symmetric harmonic oscillators modeled as LC networks.

In Figure 6.3(a), the interaction is the result of the polarization $\mathbf{p}$ acting at a distance a—a long range interaction—and, inFigure 6.3(b), the interaction is capacitively coupled through the capacitance C_i to model behavior at a very short range. So, the mode changes that we pointed to, as prototypical examples of nature's interactions, occur here in Figure 6.3(a) through the effect felt by each of the atomic dipoles due to the other. The field due to a dipole is

$$\mathcal{E} = \frac{1}{4\pi\epsilon_0}\frac{\mathbf{p}}{a^3} \qquad (6.16)$$

in the centered orthogonal plane. The energy of the system is $U = (n+1/2)\hbar\omega_0$, where $\omega_0 = 1/(LC)^{1/2}$. The oscillators are in the ground state, $n = 0$, with $U = U(0) = 2 \times \hbar\omega_0/2 = \hbar\omega_0$ for the energy of the system. When these oscillators are brought closer together but are still in the long-range limit, the interaction due to their dipole fields—this reflects charge on the plates that is moving back and forth—breaks degeneracy. This leads to an energy perturbation $V = \mathbf{p} \cdot \mathcal{E}$, that is, $V = p\mathcal{E} = p^2/4\pi\epsilon_0 a^3$. The breaking of degeneracy leads to a splitting of the degenerate ω_0 frequency to

$$\omega_\pm = \omega_0\sqrt{1 \pm \kappa}, \qquad (6.17)$$

See the very readable discussions in D. Kleppner's Reference Frame column of Physics Today, **43**, No. 10, 9–11 (1990) and S. K. Lamoreaux's feature article of Physics Today, **40**, No. 2, 40–45 (2007).

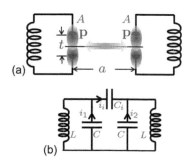

Figure 6.3: Fluctuating polarization of two atoms as (a) coupled simple harmonic oscillators modeled as dipoles coupling in the long-range limit, and (b) coupling through a capacitor in the short-range limit.

where $\kappa \propto 1/a^3$. The interaction energy is

$$
\begin{aligned}
\Delta U &= U - U(0) \\
&= \frac{1}{2}\hbar\omega_0 \left(\sqrt{1+\kappa} - 1\right) + \frac{1}{2}\hbar\omega_0 \left(\sqrt{1-\kappa} - 1\right) \\
&\approx \frac{1}{2}\hbar\omega_0 \left[-\frac{\kappa^2}{8} - \frac{\kappa^2}{8}\right] = -\frac{1}{8}\hbar\omega_0\kappa^2 \\
&\equiv -\frac{1}{8}\hbar\omega_0\frac{\check{\alpha}^2}{a^6},
\end{aligned}
\tag{6.18}
$$

where the Taylor expansion of the square root used the third term to incorporate the most significant term of the perturbation expansion. Here, $\check{\alpha}$ is the polarizability of the capacitor, accounting for the dipole moment arising from a separation in distance t of fractional charge of the capacitor. This expression has the van der Waals 6th power dependence.

The short-range interaction, at its simplest, can be modeled via a coupling capacitance (C_i), as shown in Figure 6.3(b). Let (q_1, i_1) and (q_2, i_2) be the charge and currents through the capacitors, respectively, with the current i_i showing the coupled exchange perturbation of energy through C_i. This parameterization gives

$$
\frac{q_1}{C} = L\frac{d}{dt}(i_1 - i_i), \text{ and}
$$

$$
\frac{q_2}{C} = L\frac{d}{dt}(i_2 - i_i),
$$

$$
\therefore \quad \frac{d^2}{dt^2}q_+ + \frac{1}{LC}q_+ = 0, \text{ and}
$$

$$
\frac{d^2}{dt^2}q_- + \frac{1}{L(C+2C_i)}q_+ = 0,
\tag{6.19}
$$

where $q_+ = q_1 + q_2$, and $q_- = q_1 - q_2$. We have found a second order differential equation, where the charge oscillations have the two stable frequency solutions

$$
\omega_+ = \frac{1}{(LC)^{1/2}}, \text{ and}
$$

$$
\omega_- = \frac{1}{[L(C+C_i)]^{1/2}}.
\tag{6.20}
$$

The energy lowering is

$$
\begin{aligned}
\Delta U &= \frac{1}{2}\hbar\omega_+ + \frac{1}{2}\hbar\omega_- - 2 \times \frac{1}{2}\hbar\omega_0 \\
&= \frac{1}{2}\hbar \left[\frac{1}{(L(C+2C_i))^{1/2}} - \frac{1}{(LC)^{1/2}}\right] \\
&= \frac{1}{2}\hbar\omega_0\frac{C_i}{C} \propto \frac{1}{a}.
\end{aligned}
\tag{6.21}
$$

This has a $1/a$ dependence in energy and therefore $1/a^2$ dependence in force, as one would expect for the classical Coulomb limit. Given our earlier comment in Chapter 5 on self-energy and this discussion being about being spatially near a quantum-mechanical limit, this relationship should be viewed with serious skepticism. It should be expected to fail in the shortest dimension and to work only where classical considerations suffice, for example, when using plates with moving charges.

A discussion of this dipole energetics in the long-range limit is very instructive. First, it shows that the correlations between the dipoles of these two atoms—they could have been molecules as well—are just as well described by the vacuum energy. We used electric dipoles here. If we had employed magnetic dipoles, for which the strength is reduced by $\alpha = e^2/4\pi\epsilon_0\hbar c = e^2 c\mu_0/2h = 1/137$, the fine structure constant, one would have obtained

$$
\begin{aligned}
\Delta U &= U - U_0 \\
&= \frac{1}{2}\hbar\omega_0\left(\frac{1}{\sqrt{1+\kappa}} - 1\right) + \frac{1}{2}\hbar\omega_0\left(\frac{1}{\sqrt{1-\kappa}} - 1\right) \\
&\approx \frac{1}{8}\hbar\omega_0\frac{\alpha^2}{a^6},
\end{aligned}
\tag{6.22}
$$

an increase, that is, repulsion. This repulsion would not happen between atoms, due to the fine structure argument, but might in an interaction between an atom and a cavity, where the magnetic exchange is specifically coupled. If one had employed the next excited state of the system, so, for $n = 1$, there would have been effects from the first order term, $\Delta U_\pm = \pm\hbar\omega\alpha/a^3$, for the two modes: an increase for one, and a decrease for the other. This is the dispersion in resonance. A first order effect can be large and allows long-range molecular states to have interactions in a collection of atoms. If one places a two-level system in a resonant cavity and limits the number of excitations to just 1, this cavity interaction with the two-level atom would be another example of coupled harmonic oscillators where lowering of energy—attraction—and raising of energy—repulsion—will take place. An excited atom introduced into the cavity will exchange energy with the cavity. A vacuum Rabi oscillation, oscillating at the rate of $\kappa\omega_0 = \Delta U/\hbar$, will result. If a cavity is off-tuned, the atom excited away from its initial state will take longer to return to that state than if the cavity is on-tuned. This is the equivalent of suppression of spontaneous emission.

There exists a classical analogy, not related to the zero point energy but related to radiation pressure arising from the change in modes—a force that we will call the Causseé force, to recognize its early 19th century description. The situation is summarized in Figure 6.4. If two ships exist close together in the midst of a long wave-length swell, even in the open ocean and with their sails unfurled, they will feel the radiation pressure from the ocean waves.

A classical electromagnetic radiation pressure can be understood as arising from the interaction between current—flowing electrons—in a radiation field. If $\mathbf{J}$ is a flowing current, an electron stream, for example, interacting with an electromagnetic wave, the Lorentz force is $\mathbf{J} \times \mathbf{B}$ per unit length. The normalized work by electric field is $\mathbf{J} \cdot \mathcal{E}$, and since the two fields are related through speed of light, the radiation force is $F = \mathbf{J} \cdot \mathcal{E}/c$. This is a classical force for the speed c of the flow of energy in the wave interacting with the classical system. Another example, from a quantum-mechanical view, is the force due to photon flux, such as for a sailboat in the outer space:

$$F = \frac{dp}{dt} = \Delta p \phi_{photon} = 2\hbar k \frac{P}{\hbar \omega} = F \frac{2P}{c}, \qquad (6.23)$$

where ϕ_{photon} is the photon flux, and P is the energy flux—the Poynting vector. This is a broadband action and reaction, even if small.

A ship at sea is another classical system, and the wave in this case is the wave on the sea's surface, with a long wavelength, as shown in Figure 6.4. This transverse wave has a motion in a waveform of $u(y - ct)$ in the displacement and coordinate notation that we have employed. The power exchange occurs through the dot product, so the slope of the wave displaces positionally. This slope displacement $\partial u/\partial y$ is itself a traveling wave. The ship is a harmonic oscillator that oscillates back and forth around its central axis longitudinal plane—an oscillating string whose two end points are interacting with the ocean wave whose slope displacement is a propagating wave. The positional change of the slope displacement means that the string—our ship—has a torque on it, and torque times angular velocity gives us the work per unit length, that is,

$$F = T \frac{\partial}{\partial y} \left[\frac{\partial u}{\partial y} \right] \quad \therefore \quad W = T \frac{\partial}{\partial t} \frac{\partial u}{\partial y} = T \frac{\partial}{\partial y} \frac{\partial u}{\partial y} \frac{\partial y}{\partial t} = Fc, \qquad (6.24)$$

an equation identical to that used in the classical electrical analogy. This is because the wave shape is invariant between the hydrody-namic and the electrical conditions.

The ship's oscillation—rolling—is the harmonic oscillator, absorb-ing and re-radiating energy. For each action, there is an equal and opposite effect, ignoring the dissipative forces. The wave is forcing the ships leeward in Figure 6.4. Since the ocean wavelength is long compared to the spacing between the ships, the ships are fairly in phase. So, the reaction waves emanating from the far side of each of the ships are in phase and propagate out to the sea. But, the reac-tion waves between the ships are quite out of phase and will cancel.

See S. L. Boersma, "A maritime anal-ogy of the Casimir effect," American J. Physics, **64** (5) 539–541 (1996) for a delightful discussion of this. P. C. Causseé, in l'*Album du Marin*, Char-pentier (1836), describes the ships as being pulled together even if no wind is present, so long as a there is a reasonable swell.

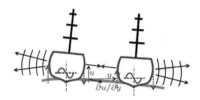

Figure 6.4: Two ships in close quarters on a swell. Their reactions towards each other cancel because of a very small phase mismatch, while those away from each other do not, causing the ships to be pulled together by the net effect of force from the ocean wave.

LightSail is an oft-delayed oft-promised solar sailboat—an umbrella structure—that would be made of $5.5 \times 5.5~m^2$ of 4600 nm-thick Mylar and would be capable of sailing in outer space, carrying two cameras, accelerometers, a telemetry system, attitude control instrumentation and battery-charging solar arrays. *NASA* keeps promising the launch, but so far it has left the amateur astronomer community disap-pointed. I came to science as a toddler flying in space atop a "Sputnik" while going to sleep on a terrace in hot Indian nights. Freeman Dyson explored in Project Orion the possibility of nuclear pulse propulsion—the use of short, high thrusts from nuclear explosions—for interplanetary travel. This project was a non-starter, especially after the intercontinental ballistic missile treaty.

Therefore, no radiated energy exists between the ships. Just as in Casimir condition, there is less energy between the ships than outside the ships, so the ships are pushed towards each other by the external radiation pressure.

6.2   Optomechanics

We have now seen that reducing dimensions to the nanoscale makes possible ultrasensitive measurements using mechanical coupling, as in Chapter 5, and also in turn uncovering fundamental quantum-mechanical phenomenon such as the Casimir-Polder effect. We extend this energy-coupling approach to explore the subject of quantum optomechanics—the coupling of cavities to mechanical resonance—where quantum control of mechanical and its converse—reaction—can be achieved. These are all examples of coupling mechanical energy to other forms of energy, including electromagnetic form of energy, through resonance.

The important consequence for us, from a device perspective of these approaches is the ability to make measurements in the quantum limit and to couple the effects between different energy forms at select frequencies in a narrow range limited by linewidth. Reducing fluctuations—noise—means that most such systems must reduce the mechanical or electrical fluctuation noise which we call thermal, or Nyquist or Johnson noise, must be minimized, that is, the expectation of the number of phonons $n \approx k_B T / \hbar \omega_q < 1$, that is, the number of such mechanical modes is kept to a minimum.

When one removes heat and phonons by heat sinking, one removes energy over the broadband, that is, the phonons spread over the entire range of acoustic and optical energy, as well as over the entire Brillouin zone. Once one has reduced the temperature sufficiently, it is the lowest energy modes with lowest energy that are occupied.

If such a mechanical object—a mechanical mirror—were placed in an electromagnetic cavity, each of the cavity mode would cause a kick per photon to be enhanced, that is, provide an impetus proportional to the force that we calculated for the flux. The thermal modes can be removed using the resulting momentum exchange via this radiation pressure—the radiation field now extracts work from the mechanical system. In this, one is employing a dynamic back-action with photons acting on the mechanical mirror through radiation pressure. The use of photons in these conditions is particularly apropos—photons have large energy compared to the energy in mechanical motion, so thermal effects of occupation are minute, and one needs to be concerned only with quantum noise—shot noise

To avoid the destruction that would be caused, if one wishes to pull the ships apart, one would not want to work against the swell's forces for that is very large. The more energy-efficient method would be to tow them away longitudinally.

Gravitational wave detectors employing large optical interferometers, for example, the Laser Interferometer Gravitational-Wave Observatory (*LIGO*), which aims at achieving displacement sensitivity below 10^{-19} $m \cdot Hz^{-1/2}$—a thousandth of a proton radius—were the earliest uses of this ultrasensitivity.

In this narrow range lies the difference between quantum and classical. Heat may be removed using a broadband of phonons such as from a typical heat sink. Once the number of phonon excitations is small, one may use a narrowband, with the added advantage of reduced, quantum-limited noise.

We discuss the thermal noise from a fluctuation-dissipation and correlation perspective in S. Tiwari, "Device physics: Fundamentals of electronics and optoelectronics," Electroscience 2, Oxford University Press, ISBN 978-0-19-875984-3 (forthcoming). A short discussion of noise in various forms, and a physical perspective, is in Appendix F arising from the discrete nature of the photon interaction events with the mirror. These cause fluctuations in the mirror response, and this is a quantum back-action. Dynamic back-action is useful, as we will see, in making precision displacement measurements and in extracting heat to bring the system to its mechanical ground state. Quantum back-action—a randomly arriving photon shot–induced fluctuation in the mirror—is an irritant. Figure 6.5 summarizes this interaction between photon flux represented as power along the abscissa and the displacement sensitivity S_{yy} along the ordinate. As the photon power increases, a detector used for measuring the photons from the cavity has increased sensitivity, but simultaneously there is an increase in the noise from the back-action. For the system itself, there is also a floor on the measurement, established by the quantum uncertainty constraint.

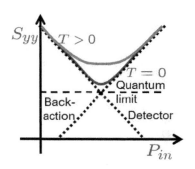

Figure 6.5: Fluctuation power as a function of input optical power for a nanoscale mechanical resonator coupled in a cavity. Limits for measurment are set by the detector's noise-constrained limit, the photon induced fluctuations from back-action, and the interaction with phonons. Quantum uncertainty places the floor within these constraints.

The mechanical element, such as the mirror of a cavity, that is allowed to harmonically oscillate, interacts with the radiation field. For a single object, for example, a mirror, with this harmonic mechanical motion, Figure 6.6 shows an optomechanical arrangement—a high quality factor ($\mathcal{Q}$) Fabry-Pérot cavity with one fixed mirror that will illustrate displacement measurement, and heating and cooling. Figure 6.6(b) shows the cavity gain and its lasing characteristics when the mechanical mirror is fixed ($\omega_q = 0$).

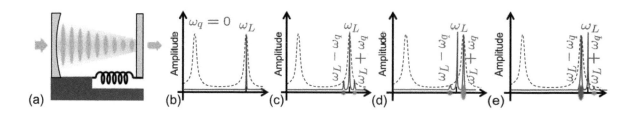

(a)   (b)   (c)   (d)   (e)

Figure 6.6: Optical-mechanical cooling and heating through coupling between an electromagnetic cavity and a mechanical oscillator, which is represented here as a mirror. (a) shows the cavity interferometer arrangement. (b) shows cavity lasing with a fixed mirror. (c) shows sum and difference frequency generation of the photon frequency, with a pump photon beam tuned to the cavity resonance. (d) shows the shifting of photons to lower frequency; cavity resonance at the sum frequency leads to increased energy at the higher frequency of the photon beam, which, in turn, cools the mechanical oscillation. (e) shows the reverse of (d) by pumping with photons of frequency higher than the cavity resonance frequency.

An optical cavity's resonance is distinct from that of atomic resonance. It is broadband. Any Fabry-Pérot cavity built with mirrors has a large number of resonances, and the cavity's energy capabilities do not saturate for conditions of interest. The radiation stays confined mostly in the cavity and leaks out at a very low rate. This is to say that the cavity has a very high quality factor $\mathcal{Q}$ and a low leakage rate Γ, which goes together with a sharp linewidth. The intensity within the cavity will be high, and with it the optical power-dependent quantum back-action and the interactions.

Figure 6.6(c) shows the interaction with the mechanical mode of energy $\hbar\omega_q$, with the mechanical mirror as a simple harmonic oscillator with the cavity precisely tuned. The consequence is the creation of two sidebands similar to the Stokes and anti-Stokes peaks of the Raman scattering process, or amplitude modulation by the mechanical motion. The reason for this dynamic back-action is that the modulation of the cavity length is reflected, through the cavity gain curve, in an amplitude modulation of the cavity's laser response. The laser frequency peak at ω_L acquires two sidebands that are displaced by the mechanical frequency ω_q. One may also view this as a phase modulation of the optical beam by the mechanical mirror. Mirror motion changes the cavity's length and hence its resonance frequency by $\omega_{cav} dy/L$, where dy is the slowly changing displacement of the mirror, and L the cavity length. With a cavity of high finesse, this motion effect is amplified through every round trip, so a phase change results even as the reflected amplitude is only minutely perturbed, due to the very small displacement. The spectral density obtained using Fourier analysis of the measured response gives the mechanical response on which the spectral response depends and from this the characteristics of the cavity—mechanical resonance frequency, quality factor, and temperature—may be determined.

This creation of sidebands is useful since it makes it possible to couple optical and mechanical energy. If one detunes the cavity and shifts the cavity gain curve, the sideband amplitudes become asymmetric, because the density of states of the cavity has been changed. In Figure 6.6(d), the higher frequency $\omega_L + \omega_q$ has a higher amplitude. Each photon here has extracted an energy from a phonon of the mirror. The cavity thus cools the mirror. Similarly, when the cavity is detuned, as in in Figure 6.6(e), the lower frequency photons at $\omega_L - \omega_q$ have a higher amplitude, since they have provided an energy of $\hbar\omega_q$ to the mirror. This cavity, depending on its tuning, has the ability to heat or cool the mirror by coupling phonon energy out or in using photons.

A cavity establishes conditions that sustain only certain modes and with that the specific characteristics that the allowed photon modes may have. The number of modes may be large but localized within the cavity, as in the normal mirror-based Fabry-Pérot cavity; it may be just a single mode; or it may be a propagating wave. With the phonon and the photon needing a quantized description for the interaction, the most direct way to tackle this is through field-quantization. The optomechanical Hamiltonian with the interaction

Recall the discussion of leakage rate Γ and linewidth in relation to quantum dots and single electron effects. The confinement in a box is that of the electron wave, and the leakage leads to the current. The linewidth is inversely related to the leakage rate.

Finesse is the mean number of reflections a photon undergoes before escaping. It is the product of the quality factor (Q) and the spectral range ($\Delta\nu = c/2L$) divided by the optical frequency ν.

The gravitational wave observatory *LIGO* achieves its $10^{-19} \ m/Hz^{-1/2}$ sensitivity by using this approach in an interferometer.

can be written as

$$\hat{\mathscr{H}} = \hbar\omega_{cav}(y)\hat{a}^\dagger\hat{a} + \hbar\omega_q(y)\hat{b}^\dagger\hat{b} + other, \qquad (6.25)$$

where the first term is the optical part of the cavity, the second is the mechanical, non-interacting part, and *other* represents the dissipation and driving terms of the system. If one Taylor expands this,

$$\omega_{cav}(y) = \omega_{cav}(0) + \frac{d\omega_{cav}}{dy}y + \frac{1}{2}\frac{d^2\omega_{cav}}{dy^2}y^2 + \mathcal{O}(y^3). \qquad (6.26)$$

For an optomechanical coupling strength of $g = y_\delta d\omega_{cav}/dy$, where y_δ represents zero point positional fluctuations, the linear optomechanical Hamiltonian is

$$\hat{\mathscr{H}}^l = \hbar\omega_{cav}\hat{a}^\dagger\hat{a} + \hbar\omega_q(y)\hat{b}^\dagger\hat{b} + \hbar g\hat{a}^\dagger\hat{a}(\hat{b}^\dagger + \hat{b}) + other. \qquad (6.27)$$

Removal of thermal energy via a single mechanical mode increases the radiation energy. This is a process proportional to the occupation factor (n) of the mechanical modes. The addition of energy is via the spontaneous process, which is proportional to n, and the stimulated process, which is proportional to $n + 1$. Detuning the cavity allows one to enhance the upper sideband. The extracted, higher energy, up-converted photons obtain this additional energy from the mechanical element, thus reducing their motional energy. In principle, it is now possible to cool the mechanical mode to its ground state, with the entire system's measurements limited only by all the sources of noise, including the limits placed by the quality of the optics. This technique is an example of sideband cooling. It is useful in nanoscale systems such as the oscillating probes discussed in Chapter 5, in coupled optomechanic bandgap structures, superconducting assemblies and others.

Cooling, the measurement of displacement, and the exploration of optical nonlinearities via this technique are in the classical regime, that is, one with weak broadband coupling. There is also a strong coupling regime, where one may measure the shot noise of radiation pressure, or generate squeezed light, as illustrated in Figure 6.7. Here, one makes the position of the photon more accurate at the expense of its energy spread. The figure illustrates the use of laser light to measure the fluctuations in the position of a mechanical resonator, as before but at a measurement rate that is comparable to its resonance frequency and greater than its thermal decoherence rate. This condition causes positional squeezing of the light. One may also squeeze the light in momentum. The membrane reflectivity determines the coupling strength. Cavity frequency $\omega(y)$ has its maximum when the mechanical element—a transparent membrane—is at a node of the cavity. A positive slope with displacement y implies

See Appendix P for a short summary of the use of ladder operators such as the creation and annihilation operators and their properties. The appendix also discusses the connections between vector potential, the harmonic oscillator modes and the field's properties. S. Tiwari, "Quantum, statistical and information mechanics: A unified introduction," Electroscience 1, Oxford University Press, ISBN 978-0-19-875985-0 (forthcoming) develops this for broader application to the electrosciences.

An analogous technique useful in quantum information processing is the preparation of quantum ground states of an ion ensemble by cooling the ions, that is, reducing their motional energy by resonant coupling with the electromagnetic source. The cavity, in general may take many forms, superconductive microwave cavities being one of the preferences for optomechanics, where $n \ll 1$ become possible by operating mechanical resonators at mK of temperature.

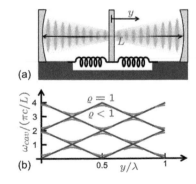

Figure 6.7: The generation of squeezed light by placement of a harmonically oscillating mechanical element between the mirrors of a Fabry-Pérot cavity. (a) shows a transparent mechanical element between two fixed mirrors. (b) shows the change in the cavity frequency which, for fixed mirrors and absent a mechanical element, is normalizable to $\pi c/L$—the roundtrip reinforced frequency. Dispersion caused by partially reflecting mirrors is compared to that from perfect reflection.

that light energy is mostly stored on the right side of the cavity, with radiation pressure acting on the left. The complementary situation occurs with a negative slope. The mechanical element, thus, is the means for exchange through optomechanical coupling, and the force on the mechanical element is also largest near the cavity's resonant condition. The observable change makes a highly sensitive force measurement possible as a quantum back-action by a photon. The mechanical membrane responds to quantum fluctuations of light, altering transmission. It causes the single mode to be squeezed. The diffraction limit is broken.

This interaction, at low temperatures, will cause cooling, reducing the nanomechanical resonator's fluctuations. These fluctuations occur under Brownian motion, with a noise distribution around the eigen-frequencies of the Fabry-Pérot resonant frequency of $\pi c/L$. As photons circulate between the mirrors, the nanomechanical resonator's fluctuations scatter the photons, and each one of these scattering events takes a finite time τ during which the photons are circulating before a new equilibrium is reached. This finite time τ causes a delay in the back-action on the resonator, since the radiation pressure changes. This retarded back-action has the effect of optical viscous damping—akin to that for particles of Brownian motion from molecular collisions—and in turn of cooling of the mechanical resonator. This extracted energy is carried away by the photons escaping the cavity.

For this Fabry-Pérot cavity, the two mirrors define the boundaries, but many modes are supported between them. An analogy to this situation is that of the electron wave picture: the bandstructure of a confined quantum box versus that for electron waves spread out over the whole crystal. In these systems, to achieve the useful back-action property, one had to bring the system to conditions of low thermal effects and length scale as well as reduce the energy for the nanomechanical resonator to the point where one could uncover the quantum-mechanical interaction effects.

Another use of these optomechanic effects is approaching the fundamental limit of the mechanical element's quantum behavior—its ground state—and because one has quantum control as well as access to a large number of modes, one may use the single mode squeezing techniques discussed earlier to more than one mode and thus achieve entanglement of multiple modes. Here, for example, the response of the mechanical element to one wavelength alters the response to the second.

Cavities can also be built using periodic structures instead of two mirrors as just discussed. Periodic arrays do this for propagating modes by forcing periodic boundary conditions that in turn define

The electron's quantized properties lead to a variety of electron-phonon interactions, such as conventional randomizing scattering—a classical analog of interaction—and, in a quantum well dominated by dipolar selection rules, which lead to polarization effects as well as a variety of bottlenecks such as of phonons—these are quantum-driven interactions.

Note that photon numbers commute, that is, $[\mathscr{H}, \hat{a}^\dagger \hat{a}] = 0$, but phonon numbers do not, that is, $[\mathscr{H}, \hat{b}^\dagger \hat{b}] \neq 0$.

the properties of the mode. This is true for photons, and it is also applicable to phonons for dimensions of guides that are on the scale of the phonon's mean free path. One may then view the phonons as a propagating quasiparticle under no or limited scattering, just as for a photon. The phonon mean free path is the constraining length scale for a meaningful discussion, unlike that for the photon, whose interactions are quite weak.

As a simple starting example, consider the two-mass-and-spring model for a one-dimensional chain, a model often employed to introduce the notion of phonons.

This is shown in Figure 6.8(a), with the horizontal axis as a normalized wavevector. There exist triply-degenerate optical and acoustic branches. The higher energy optical branch arises from the out-of-phase displacements and a rapid change in time. The acoustic branch arises from the small phase displacement change differencesy. There exists in this structure a phonon bandgap Δ_g. The low frequency "acoustic" branch here represents in-phase excitation, dominantly of the large masses. The bandgap separates these modes from those of the small masses that comprise the high frequency "optical" branch. Increasing the difference between the large and the small mass creates a wider gap.

In S. Tiwari, "Semiconductor physics: Principles, theory and nanoscale," Electroscience 3, Oxford University Press, ISBN 978-0-19-875986-7 (forthcoming), we discuss periodic structures in detail, concentrating on the photon bandstructure that can be built with one-, two- and three-dimensional periodic arrays of changing indexes and their properties: bandgaps; guided propagation, including slow waves; intensity buildup at defects; et cetera. The reader should refer to this discussion to see its correspondence with that for mechanical periodic structures. Mechanical structures are more complex, since the "masses" and the links are both assemblies in their own right, and properties arise from their togetherness and interactions.

N uncoupled atoms have $3N$ spatial degrees of freedom—3 of translation for each atom. N coupled atoms also have $3N$ degrees of freedom. Of these 3 are of translation and 3 are of rotation for entire assembly. So, $3N - 6$ is the number of phonon modes represented in the dispersion relationship. These can be longitudinal or transverse, and some will be optical, and some acoustic.

Figure 6.8: (a) shows the two-mass-and-spring model of allowed energy (frequency) as a function of wavevector for a one-dimensional chain. (b) shows an analog where finite-sized, three-dimensional, nanoscale mechanical mass elements—blocks—are bridged to each other in a one-dimensional chain.

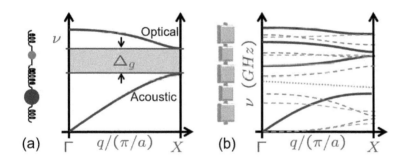

(a) Γ $q/(\pi/a)$ X (b) Γ $q/(\pi/a)$ X

If one builds a quasi-analogous structure, as in Figure 6.8(b), with three-dimensional small nanoscale mechanical masses, for example, blocks of silicon, that are coupled together, one would expect to obtain analogous characteristics for the vibrational modes of this assembly. This is a one-dimensional phononic crystal. The characteristics of the structure will be defined by the properties of the medium as well as by the thickness, the lateral dimensions, the bridge gap, the periodicity which subsumes the longitudinal dimension, and the elastic properties of the material and its surfaces. Since the stiffness of beams is constrained by the weakest parts, the links make the beam more floppy, that is, oscillate at a lower frequency. This quasi-one-dimensional chain's low frequency branches are reduced by making the links small in width so that the masses are connected very loosely. The mass elements dominate the behavior of the high frequency branches. Since these are weakly linked, the resonance behavior of these masses determines the high frequency branches, which are made relatively wavevector independent through the weak coupling between them. The vibrational modes will be of various symmetries—in phase and out of phase in the three-different Cartesian directions—and these modes are shown through the schematically drawn dashed lines. Again, there will exist vibrational bandgaps—the phononic bandgap or a bandgap in mechanical modes—through the lowering of low frequency characteristics and separating them from the flatter, high frequency characteristics.

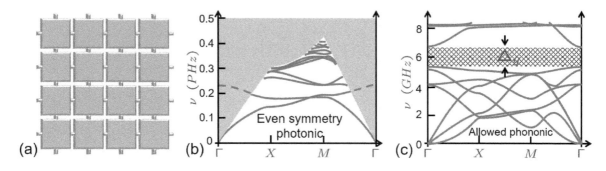

Figure 6.9: (a) shows a two-dimensional crystal of silicon, with (b) showing the structure's even symmetry photon modes and (c) showing the structure's allowed phonon modes. The structure has a thickness of 220 nm, a lattice constant of 500 nm and a link width of 100 nm.

These structures will become more complex as one turns them into two-dimensional and three-dimensional assemblies. However, a peek at this brings out an important characteristic: it is significantly harder to achieve a photonic bandgap than a phononic bandgap. Figure 6.9 shows the modes allowed for a silicon-based structure, for photons in (b) and for phonons in (c). The photon bandstructure is for even symmetry modes for propagation in plane. Square lattices such as this have low symmetry, and propagation changes substantially in different directions. So, the X and M points, which are high symmetry points at the Brillouin zone boundaries, have propagation properties that are very different from those of Γ. In vacuum, with this two-dimensional symmetry, light can propagate out of plane, but phonons cannot. The in-plane propagation is guided to specific ω-$\mathbf{k}$, as described by the lines with evanescence outside. The rest of the region, a region referred to as being outside the light line, consists of continuum and leaky guided resonant modes, indicated here as extended dotted lines. Thus, for photons, strictly, there is no photon bandgap. In specific directions, one may call it a psuedo-bandgap, since certain frequency or energy ranges are not allowed. But, this is only true for light propagation in only that specific direction, not universally, and scattering or other interactions will cause leakage. These comments hold for transverse electric (*TE*) modes, transverse magnetic (*TM*) modes, with their odd symmetry, do not even have the pseudo-guided-mode bandgap. The phonon bandstructure, however, does have a bandgap, as shown in Figure 6.9(c), because of the reasons already discussed.

Thus, creating a true photon bandgap is quite difficult. A two-dimensional structure with higher symmetry, for example, a triangular mass connected as a snowflake, does achieve a bandgap for *TE* modes, so still with constraints. The reason for this is this symmetry and the dependence of photon transmission to the slot gap between masses—the constraint through the dielectric periodic properties. Phonon transmission on the other hand depends on the size of the link. The existence of two of these tunable parameters—which are relatively independent—and symmetry allows one to achieve simultaneous control over both light and sound and thus their interactions. One may design structures with photon confinement with high Q in millions within a phononic bandgap crystal, even as one structure is at *PHz* (optical frequencies) and the other is at *GHz* (mechanical frequencies)—a seven orders of magnitude difference. We can therefore go back to the discussion of cooling and energy extraction from mechanical motion and see how one may achieve such extraction through the use of cavity-like structures: periodic waveguides, with photon and phonon confinement.

Figure 6.10 shows an example of how this optomechanical interaction can be used to reach the detection limits in the ground state of zero point motion. Light can be coupled evanescently from a fiber positioned above the mechanical waveguide—a perforated beam that is held at its edges and floats in space above the substrate from which it is created. Quantum process dictates that phonon absorption be proportional to $\langle n \rangle$ and that emission be proportional to $\langle n \rangle + 1$ of the modes of the mechanical oscillator. This is inherently asymmetric. The asymmetry of the motional sideband generated in the coupling of Figure 6.10 in this cold limit can be measured by using the tremendously narrower linewidth of the optical cavity compared to that of the mechanical frequency. In Figure 6.10, both optical and acoustic waves are localized. The cavity is provided two optical resonances—one for cooling by the fundamental mode, and one for reading out the mechanical movement. The fundamental light frequency mode of the cavity is in 100s of *THz*, corresponding to close to about 100 *nm* removed from the infrared free space wavelength of 1.54 μm of fibers.

Figure 6.10: A fiber in close proximity to an optomechanic waveguide. With a nanoscale spatial separation, light can couple to the waveguide structure, where photons and phonons are confined.

A common and convenient approach to this is using silicon-on-insulator substrates, forming the perforated beam with its anchoring in silicon and removing the SiO_2 underneath.

In atomic systems, this is in the Stokes and anti-Stokes sidebands of the fluorescence and absorption spectrum of the atom, due to motion.

Therefore, this 1.54 μm wavelength, as a second order mode of the cavity, can be used for measurement in a nanoscale assembly.

The mechanical modes of the beam to which these optical modes connect are also interesting. They appear in several forms. Figure 6.11 is a description of the photon and phonon modes of a suitably designed optomechanic periodic cavity with light coupled evanescently. The optical modes of the cavity are defined by the periodicity of the structure's dielectric characteristics, where the gaps are important. Figure 6.11(a) and (b) show the first and second order modes of the evanescently coupled light being confined in the cavity. The phonon modes are determined by the periodicity, particularly the floppy linkages—the edge connectors here—between the dielectric mass regions. These dimensions are of the order of 100s of nm or less. And we have remarked that this structural form at these dimensions can have phonon bandgaps quite easily. The mechanical modes can be both transverse and longitudinal. The high frequency transverse mode is one where the cavity expands and contracts over short periodic distances, that is, performs an out-of-phase motion, as the wave travels. We don't show the in-phase mode, which would be a very low energy mode. This is shown in Figure 6.11(c) as a "breathing" model. Two longitudinal modes are shown in Figure 6.11(d) and (e): the "accordion," where the beam elements stretch asymmetrically outward around the node, and the "pinch," where the beam elements stretch inward to constrict the node.

The breathing mode is a suitable mechanical mode for use in photon-phonon interaction. Let ω_r be the reading frequency. The Hamiltonian for this coupled condition is

$$\mathscr{H} = \hbar(\omega_r + g_r \hat{z}/z_0)\hat{a}^\dagger \hat{a} + \hbar(\omega_{cav} + g_{cav}\hat{z}/z_0)\hat{c}^\dagger \hat{c} + \hbar\omega_q \hat{b}^\dagger \hat{b}, \quad (6.28)$$

where g_r and g_{cav} are the coupling rates—the gain-dissipation coefficients—and z_0 is the zero point normalization $\hat{z} \equiv z_0(\hat{b}^\dagger + \hat{b})$ of the breathing mode operator. This equation corresponds directly to Equation 6.27 for the Fabry-Pérot optomechanic cooling discussion. The one significant difference between the two approaches is that the cooling there is due to the coupling of the oscillator to the high Q electromagnetic resonance resulting in back-action. Here, this cooling is continuous through the mode occupation difference for absorption and emission and so is similar to the Raman processes employed in atom cooling.

Figure 6.12 shows the salient procedural approach underlying this approach to optomechanic cooling. (a) shows the absorption ($\langle n \rangle$) and emission ($\langle n \rangle + 1$) displacement power spectrum of the mechanical guide. When a laser at frequency $\omega_L = \omega_{cav} - \omega_q$ is tuned for the cooling resonance, a photon population of $\langle n_{cav} \rangle$ is created at this

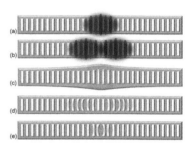

Figure 6.11: Light wave and mechanical wave interaction in a suitably designed optomechanic waveguide. (a) and (b) show the first and second order optical mode. (c) shows the symmetric transverse mechanical mode, which is often referred to as "breathing" mode. (d) shows a symmetric longitudinal mode that is sometimes referred to as an "accordion" mode, and (e) shows the asymmetric version that is sometimes referred to as a "pinch" mode.

laser frequency ω_L. The mechanical motion interaction causes this cooling laser light to scatter into Stokes and anti-Stokes sidebands. The bands are created at $\omega_L \pm \omega_q$, that is, at $\omega_{cav} - 2\omega_q$ (Stokes) and at ω_{cav} (anti-Stokes), which is the photon cavity resonance frequency. Since the cavity gain width is much broader, this up conversion gets strongly enhanced and the mechanical mode cooled—an example of the back action cooling consequences of motion. A reading laser at ω_{Lr} can measure by scanning as represented in Figure 6.12(b) describing this process. This part of the figure shows the relative position of ω_L to that of the optical cavity mode—down-shifted by ω_q. The corresponding ω_{Lr} can be swept across the Stokes band positioned at $\omega_{cav} - 2\omega_q$ for reading out. Figure 6.12(c) shows the corresponding narrative for the reading of the down-converted Stokes sideband, and (d) shows it for the up-converted anti-Stokes motional sideband. So, with ω_{Lr}, in (c), the linewidth of the readout cavity is broad, with the change in ω_{Lr} shifting the interaction between the optical and the mechanical energy form positioned at $\omega_{Lr} - \omega_q$. The mechanical resonant linewidth being much smaller, the reading laser provides a means to measure the down-converted sideband. The same happens for up-conversion, as shown in Figure 6.12(d).

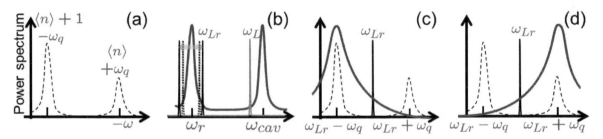

Figure 6.12: (a) shows the power spectrum for displacement of a simple harmonic oscillator—the mechanical resonator. (b) shows the frequency response and positioning at ω_r of the read out cavity, ω_{Lr} is the reading laser's frequency, ω_L is the cooling laser's frequency, and ω_{cav} is the photonic cavity's fundamental mode. ω_L, lower by ω_q from the photon cavity resonance frequency, causes cooling. (c) shows the relative position due to the read-out laser's mechanical interaction as it is swept through the Stokes motional sideband. (d) shows the same from the anti-Stokes motional sideband.

This discussion exemplifies the strength of photon-phonon coupling. We have remarked in the Fabry-Pérot discussion the potential of using this for generating entanglement.

One interesting implication in the periodic structure is that the phonons are at GHz and the photons in free space are in infrared ($1.54\ \mu m \equiv \approx 200\ THz$) at the optical communication wavelength.

Photons derive much of their applications appeal from the fact that they can travel long distances with large information content. This results from the photon's weak interaction with the environment, wide transmission bandwidth and relative immunity to thermal noise. Phonons have, particularly due to their bosonic character, many properties that are similar to that of the photon. Since they are excitations of masses around a mean position, interactions are strong, and the bandwidth and distance scale of information transmission much smaller. But, optomechanics makes possible delaying and storing phonons by resonantly imparting the continuously lost energy. For acoustic phonons, this is possible using a photon energy such as microwave—electronic or superconducting or by other means. Photons have resonances that are short lived (nss) while phonon resonances exceed 100s of μss. Photon filters are broadband (GHz) while phonon filters, for example, those based on surface acoustic waves, are narrow (of the order of 100 kHz). Phonons, with their lower energy and speed, can coherently interact with GHz electromagnetics as well as with qbits, such as those from superconductivity-based devices and circuits. So, photon-phonon coupling is a means of utilizing the desirable property of each of these forms.

The ability to couple photons and phonons makes photon-phonon coupled systems possible—and it is an ability that will be of interest at GHz, a frequency at which considerable analog and analog-digital processing takes place.

A simple example of these photon-phonon systems is one that can transform the signal in photons to one in phonons and vice versa. The latter process is simple. An adiabatically moving mirror with a GHz-to-THz scale difference changes the resonant frequency. Optomechanical waveguides, suitably designed, can also do this. An optical waveguide, so one with periodic control for optics, can be coupled to a phonon waveguide, for example, one that intersects orthogonally, and this will cause a coupling at the intersection. These are simplified forms of the fiber-optomechanic waveguide interacting, modulating and cooling systems that we have now discussed.

6.3   Interactions in particle beams

Photonic confinement and manipulation through programmed periodicity and defects has many other usage possibilities because of the high intensity, size confinement or the avoidance of diffraction limit.

An optical tweezer, an important example, is a very conventional microscopic means to hold and confine nanoscale and smaller particles in three-dimensional space. Figure 6.13 shows a schematic description of this technique of optical trapping that is employed in the tweezer. Consider a TEM_{00} Gaussian beam near its focus. A par-

Although we did not discuss this, this ability to couple allows one to cool mechanical movement by microwave coupling also. The infrared wavelength is not unique; it happens to be a dimensional scale that couples well to the mechanical length scale.

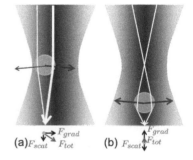

(a) F_{scat} F_{grad} F_{tot} (b) F_{scat} F_{grad} F_{tot}

Figure 6.13: An optical trap functions by employing gradient force and conservation of momentum to move nanoscale objects towards higher intensity. The light intensity achieved through the optical design is shown as a background. (a) shows that a refractive object will feel a force towards the higher intensity as a result of the photon refracting away. (b) shows that the forces also bring the object towards the focus.

ticle that is displaced off the focus, as in Figure 6.13(a), is influenced by multiple photon flux and material interaction–induced forces. The bending of the photon beam towards the higher refraction region within the object, because of the conservation of momentum, is balanced by an opposite momentum change in the object. There also exists a force F_{scat} due to the net effect of scattering through the object. The photon flux being higher in the higher intensity region, a force F_{grad} exists nudging the particle towards the peak intensity of the beam. An object that has both refracting surfaces perpendicular to the beam will have a negligible force so long as there is insignificant absorption in the object. The consequence of these two forces—the photon kick and the gradient force—is a net force of F_{tot}. A particle centered in the intensity maximum region but displaced off the focus also sees a net force, as shown in Figure 6.13(b). The gradient force in this instance pushes the object towards the focus as a consequence of the net change in angle due to refraction of the beam in the object—again with the conservation of momentum during any interaction.

This is an optical trap acting as an optical tweezer through the use of gradient force.

Another technique that has been of tremendous import through the use of small dimensions is that of confocal microscopy—a technique that allows one to move the focal plane and build three-dimensional images of translucent objects.

The principle of this approach is schematically shown in Figure 6.14. Using a laser as a monochromatic high intensity point light source, a beam splitter is utilized to collect the reflected light in another point detection unit—a narrow opening backed by a detector. Light that is reflected from the focal plane is collected, and light reflected from outside the focal plane is rejected. By scanning the laser point source, one can obtain an image in the focal plane and by moving the object in the focal plane along the optical axis, one can construct a three-dimensional image of the object being imaged. Use of this point source and point detection allows one to circumvent the limit of the wide field, where out-of-focal-plane light limits the resolution and constrains the contrast of a conventional microscope.

These are two examples of the use of optical-mechanical interactions to improve resolution statically.

Another significant application of this optical-mechanical interaction in small dimension limits is the laser cooling of ions and, by doing it in small dimension systems, employing it for achieving precision atomic clocks. The principle of atom cooling is the same as for our discussion of optomechanical cooling for the larger-sized resonator. An atom in an optical field is again viewable as a resonator—a simple harmonic oscillator, as shown in Figure 6.15, interacting

With optical beads that are functionalized, for example, attached to a DNA strand, this approach allows one to execute quite high precision measurements and positional control in molecular fluidic systems. This is useful in a realistic living environment for biological studies at the molecular scale—this approach has led to the pN force measurements of the biological motors that do much of the transduction work in our bodies and make the bacteria move.

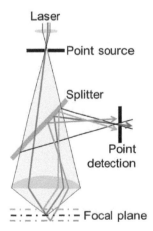

Figure 6.14: A confocal microscope. The laser acts a point source, forming a focal plane on the sample. Light from the out-of-focus areas is excluded via point detection.

with local field. When the applied frequency of the cooling laser is below the resonant frequency, that is, $\omega < \omega_0$, the cooling occurs, as it does with the mechanical resonator, because the sideband at $\omega + \omega_q$ is aligned with the optical resonance frequency ω_0. So, the power in the lower sideband is smaller. The atom cools.

This sideband, or equivalently a lower and a higher energy interaction, is applicable in many situations where this picture holds—it represents the generalized coordinate response when the two degrees of freedom interact. The significance of the implication is that whenever there is a resonance, a sharply defined peak, the interaction response will be at sum and difference frequencies, and since the gain is frequency dependent, the two sideband frequencies and their underlying physical behavior will respond differently.

Atomic clocks, frequency standards, miniaturized spectroscopy, magnetic resonance instrumentation, slowing of light, nonlinear optics with single photons, et cetera, all become possible as a result of this. The miniaturized spectroscopy for a number of applications becomes possible by the precise frequency standards that timing makes possible. Magnetic resonance becomes possible because of the magnetic field dependences that become measurable. Slowing of light is related to the giant refractive index and the nonlinear effect that a Bose-Einstein condensate using atom cooling makes possible. And nonlinearity and single photon measurements become possible due to the collective precision and large state changes made possible using this interaction approach. We use a compact atomic clock as an example of the nanoscale device usage of this phenomenon.

6.3.1   Atomic clocks

Consider a three-level system such as the Λ configuration of Figure 6.16(a) where the transitions $|b\rangle \leftrightarrow |a\rangle$, and $|c\rangle \leftrightarrow |a\rangle$, can be controlled using optical fields at precise wavelength through lasers. Such a three-level system is achievable through a number of alkali atoms that have melting point close to room temperatures and become easily available as atoms in vapor—cesium (Cs) and rubidium (Rb) are two examples. The optical wavelength is near infrared and achievable through semiconductor lasers such as those using vertical cavities. This makes a very compact source possible. The difference between two transitions is in the microwave range—10 $GHz \equiv 40$ μeV, compared to $\sim eV$ of the optical range. This allows a modulation at microwave frequency to control the occupation of the $|a\rangle$, $|b\rangle$ and $|c\rangle$ states, with the optical fields present. These states arise from spin interaction—an electron spin flip relative to the nuclear spin leads to the fine structure splitting, and the hyperfine structure with much

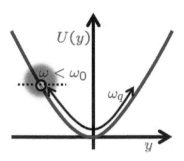

Figure 6.15: Cooling of an ion by a laser light at $\omega < \omega_0$, the cavity resonance frequency. The ion in the field can be viewed as a simple harmonic oscillator of frequency ω_q. If If $\omega > \omega_q$, the radiation will heat the ion.

High precision and accuracy of timing is quite in use in everyday life. The global positioning satellites orbit about 20, 200 km up. A receiver determines its position by using the time and position transmitted by multiple satellites. The time for the signal to travel from the satellite is $\approx 2 \times 10^7 m \div 3 \times 10^8$ $m/s = 0.66 \times 10^{-1}$ s. Civilian usage in USA is obfuscated currently to about 2 m accuracy, which we can enhance by averaging but which corresponds to a timing accuracy of $\sim 0.66 \times 10^{-8}$ s. The GPS satellites need some rather precise timing sources and are capable of much more than this accuracy.

smaller energy shifts arises from interactions within from internally generated fields. In Cs: $[Xe]6s^1$, the fine structure arises from the total electronic angular momentum $J = 1/2$ state of $6S_{1/2}$, where $l = 0$, and $s = 1/2$, and from the $J = 3/2$ state of $6P_{3/2}$, where $l = 1$, and $s = 1/2$. The nuclear interaction with this, so for $F = I + J$, the total angular momentum, where I is the nuclear term, leads to a hyper-fine interaction, which is shown here as the splitting on the right of Figure 6.16(a) and which permits the microwave interaction. Since the ground state has zero orbital angular momentum, it is relatively insensitive to magnetic fields.

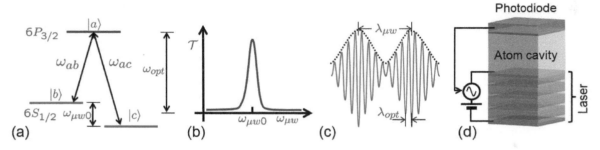

Figure 6.16: Light and microwave interaction in a three-level atom in Λ configuration. Example is of Cs. (a) shows the optical (ω_{ab} and ω_{ac}) and microwave transitions ($\omega_{\mu w0}$). (b) shows the transmission transparency in a narrow window at the difference frequency because of the non-interacting "dark" state. (c) shows the microwave modulation of the optical signal, and (d) shows a heterogeneously integrated assembly combining the cavity, a laser and a photodiode to implement the clock.

Atomic clocks such as those using hydrogen—atomic fountain clocks—require a microwave cavity, since they are based on a double resonance in the microwave and the optical frequency ranges, where the microwave field is necessary in order to couple with hydrogen. The miniaturized clock, by orchestrating optical–microwave interaction, is not constrained by this—these structures are simply based on a small volume, with the hyperfine species in a buffer gas, an integrated laser such as the vertical cavity that can be modulated by the radio-frequency signal, and a detector.

It is instructive to look at this microwave interaction matching through the following quantum-mechanical argument, in order to understand the behavior of the 3-level-system atomic clock. If an atom can be in a superposition of two atomic states, $|a\rangle$ and $|b\rangle$, then

$$|\Psi(t)\rangle = c_a(t) \exp(-i\omega_a t) |a\rangle + c_b(t) \exp(-i\omega_b t) |b\rangle \qquad (6.29)$$

tells us the probability of finding the atom in either state, with the probability being proportional to the square of the coefficient. When a perturbation $\mathcal{H}'$ is applied to cause an interaction in this arrangement through an electromagnetic field $\mathcal{E} = \mathcal{E}_0 \exp(i\omega t)$, with ω close to the resonance of light, that is, $\omega \approx \omega_{ab} = \omega_a - \omega_b$, the system responds at multitudes of frequencies. The fast response is at $\omega + \omega_{ab}$. We worry about the slow changes of response that our instrumentation can measure, and this happens on a time scale proportional $1/(\omega - \omega_{ab})$. For the non-zero interaction, the Hamiltonian matrix element is

$$\langle a|\mathcal{H}'|b \rangle = \hbar\Omega \exp(-i\omega t), \tag{6.30}$$

and its complement is $\langle b|\mathcal{H}'|a \rangle = \hbar\Omega \exp(i\omega t)$. The strength of Ω is proportional to the field amplitude, that is,

$$\Omega = \frac{\langle a|-ez|b \rangle \, \mathcal{E}_0}{2\hbar}, \tag{6.31}$$

in terms of the dipole moment. This tells us that the time evolution of the state coefficients follows

$$
\begin{aligned}
i\dot{c}_a &= \Omega \exp[i(\omega - \omega_{ab})t]\,c_b, \quad \text{and} \\
i\dot{c}_b &= \Omega \exp[-i(\omega - \omega_{ab})t]\,c_a.
\end{aligned}
\tag{6.32}
$$

This describes the Rabi oscillation for this system under conditions of stimulated transitions between two states, with Ω as the Rabi frequency of oscillation when the frequencies precisely match. With off matching, the atomic population cycles at $[(\omega - \omega_{ab})^2 + \Omega^2]^{1/2}$. This states that the atoms absorb and emit photons repeatedly, cycling through, and no energy is lost. If one included finite lifetime of the excited state, then spontaneous decay in random directions also happens, and this energy is lost even in the idealized situation.

Now, we expand this to a three-level system, that is,

$$
\begin{aligned}
|\Psi(t)\rangle &= c_a(t)\exp(-i\omega_a t)|a\rangle + c_b(t)\exp(-i\omega_b t)|b\rangle \\
&\quad + c_c(t)\exp(-i\omega_c t)|c\rangle.
\end{aligned}
\tag{6.33}
$$

The lifetimes of the ground states $|b\rangle$ and $|c\rangle$ are long, and $|a\rangle$ spontaneously decays—we assume at decay rate of Γ_a. With two fields at frequencies ω_1 and ω_2 coupling state $|a\rangle$ to $|b\rangle$ and $|c\rangle$, respectively, we limit our discussion to two interaction Hamiltonians,

$$
\begin{aligned}
\langle a|\mathcal{H}'|b\rangle &= \hbar\Omega_1 \exp(-i\omega_1 t), \quad \text{and} \\
\langle a|\mathcal{H}'|c\rangle &= \hbar\Omega_2 \exp(-i\omega_2 t),
\end{aligned}
\tag{6.34}
$$

with the time evolution of state coefficients as

$$
\begin{aligned}
i\dot{c}_a &= \Omega_1 \exp[i(\omega_1 - \omega_{ab})t]c_b + \Omega_2 \exp[i(\omega_2 - \omega_{ac})t]c_c, \\
i\dot{c}_b &= \Omega_1 \exp[-i(\omega_1 - \omega_{ab})t]c_a, \quad \text{and} \\
i\dot{c}_c &= \Omega_2 \exp[-i(\omega_2 - \omega_{ac})t]c_a.
\end{aligned}
\tag{6.35}
$$

State $|a\rangle$ couples to both $|b\rangle$ and $|c\rangle$, but states $|b\rangle$ and $|c\rangle$ couple only to $|a\rangle$.

The spontaneous emission losses are proportional to $|c_a|^2$. If an atom can be prevented from being excited, no energy will be dissipated and light propagates without absorption. This requires $\dot{c}_a = 0$ for all times. No atoms from either ground states are excited at any time, including in the presence of light field. So, the steady-state population of $|a\rangle$ must vanish, that is, $c_a(t) = 0$. A little manipulation of Equations 6.35 leads to the condition

$$\Omega_1 c_b = -\Omega_2 c_c \exp\{i\left[(\omega_2 - \omega_{ac}) - (\omega_1 - \omega_{ab})\right]t\}. \qquad (6.36)$$

By making the phases of laser fields constant, we have the constraint

$$(\omega_2 - \omega_{ac}) - (\omega_1 - \omega_{ab}) = 0 \quad \therefore \ \omega_2 - \omega_1 = \omega_{bc}. \qquad (6.37)$$

Transparency exists with $\dot{c}_a = 0$, $c_a = 0$ with this two-photon resonance when the applied optical frequency matches the hyperfine energy. When this matching is precise, that is, $\Omega_1 c_b = -\Omega_2 c_c$, then the state function reduces to

$$
\begin{aligned}
|\Psi(t)\rangle &= |d\rangle \\
&= \frac{1}{\Omega_1^2 + \Omega_2^2} \\
&\quad \times \left[\Omega_2 \exp(-i\omega_b t)|b\rangle - \Omega_1 \exp(-i\omega_c t)|c\rangle\right]. \quad (6.38)
\end{aligned}
$$

$|d\rangle$ is a state that never excites to $|a\rangle$. It is a "dark" state, where the atomic population is in a sense trapped in the two lowest ground states.

This shows that when starting from a condition where the population in $|b\rangle$ and $|c\rangle$ is about equal but not a coherent superposition, application of the light field causes excitation to $|a\rangle$, and spontaneous decay either to the "dark" state or to its orthogonal "bright" state. "Dark" state ceases to interact. "Bright"state continues to, but, over time, through excitation and spontaneous decay, the population of "dark" state continues to increase until transparency comes about. The "dark" state is a coherent state. Laser light therefore needs to maintain the relative phase for the condition of transparency to persist. This simple picture essentially underlies the detection of transparency in the miniature atomic clock system shown in Figure 6.16(b)–(c).

The coherent interaction between fields and atoms, the ability to do this simultaneously with two optical fields of long-lived states, and matching to the difference frequency through the microwave permits this atomic excitation as well as its monitoring. When the frequency difference between the two optical fields matches the splitting between two hyperfine sublevels of the atomic ground state, one has prepared a coherent superposition of the two—a long-lived non-interacting "dark" state. This state exists only in a very narrow range of the differential frequencies between the optical fields that are accessed through microwaves. When the system is prepared in this coherent population trapping state, a narrow transmission peak exists at the resonance, as seen in Figure 6.16(b). This is an electromagnetically induced transparency. By modulating the lasing signal with a microwave signal, and locking the microwave oscillator to this resonance by using a detector, as schematically drawn in Figure 6.16(c) and (d), one achieves precision matching to this resonance, and, by extension one achieves a precise timing. The current modulation of the laser makes possible a number of sidebands of the laser's output spectrum. A modulation frequency near the ground state hyperfine splitting makes the two frequency components of the optical spectrum of the laser resonate with the two optical transitions that couple to $|a\rangle$. This is the coherence of the hyperfine split ground state to the excited state of this system, and, when this occurs, the transparency increases to where one can lock, through the detection of the microwave frequency, for the greatest transparency.

For Cs, the optical wavelengths is near $\approx 852\ nm$. Accuracies exceeding well over 10^{-10} per day—several orders of magnitude worse than those of the fountain clocks but still very useful—become possible.

6.3.2   Miniature accelerators

The microwave range of radio frequency is also employed in another major set of devices with particle-electromagnetics interaction: accelerators for health diagnostics and treatment and for fundamental high energy particle studies of physics. Interestingly, there is an important nanoscale aspect of this, even though the current particle accelerators are by and large. The charged particle beam in all these apparatuses is accelerated by a periodic radio frequency signal, whose field causes the particles to accelerate. So, the spacing and timing in the design must keep the bunched particles of the beam to encounter the accelerating field when they pass through the radio frequency region.

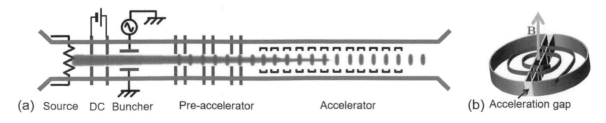

(a) Source   DC   Buncher   Pre-accelerator   Accelerator   (b) Acceleration gap

Figure 6.17: A linear accelerator for electrons (a), and a cyclotron (b).

Figure 6.17 shows two examples that encapsulate the basic idea behind these designs. Linear accelerators employ the principle that repeated application of small voltages in time-varying fields can be used to accelerate particles with repeated application of force. Cyclotrons align the time-varying field to the revolution of charged particle in magnetic field. Modern high energy machines employ combinations of both of these, exploiting their individual favorable characteristics. In a linear accelerator, a low energy charged particle such as an electron is first accelerated by a direct voltage applied before a uniform stream of these electrons is bunched by passing them through a radio frequency cavity that has a longitudinal field. This bunches these electrons spatially and in time—those at lower field accelerating towards higher field, and those beyond the high field seeing less acceleration. This requires an appropriate phasing for the alignment of the field in time and in the segments for acceleration of the particles passing through. These bunches are now accelerated in the linear acceleration part where the energy is again increased during the passage through individual segments with $\Delta U = \int \mathcal{E} \cdot d\mathbf{r}$. So, if a microwave frequency of $2.856\ GHz \equiv \lambda = 0.105\ m$ is used, as in the Stanford collider, a large structure naturally results. A linear accelerator can achieve a field of the order of $50\ MeV/m$. The bunches are far apart in space and in time. These limits arise from the frequency that determines the rate at which the accelerating fields can be applied, and the design of instrumentation to make sure that the bunches can remain focused. Figure 6.17(b) shows a cyclotron, an example of one of the earliest accelerators, that is composed of two "D"-shaped segments separated by a gap across which the radio frequency electric field is applied. The source, usually ions, is confined by an out of plane magnetic field through the Lorentz force. With a field B, balancing of forces ($mv^2/r = qvB$) gives the radius of orbit as $r = mv/qB$ and therefore the time for one revolution as $T = 2\pi m/qB$, or a cyclotron frequency of $\omega = qB/m$. If the particle is made resonant with a time-varying electric field applied in the gap, the particle will be accelerated and its radius increased. The important characteristic of these resonant circular accelerators is the synchronization between oscillating acceleration fields and the revolution frequency of particles.

So, how do the nanoscale and optics fit into this discussion? Instead of a microwave GHz frequency, lasers operate at fractions of PHz. So, if one could imagine a pulsing laser structure, a pulse for which is shown in Figure 6.18, a much larger bunching density of electrons, and an estimated electron population in the thousands, will result. A $\lambda = 2\ \mu m$ corresponds to a cycle time of $6.6 \times 10^{-15}\ s$. The bunches will thus be sub-ps in length. Semiconductor lasers are capable of 100s of W and GHz repetition rates, and banks of semiconduc-

The Large Hadron Collider—a storage ring collider—of $CERN$ has dual rings that are 8.6 km in diameter, that is, 17 km in length, cross at four places and are capable of generating two 7 TeV proton beams, as of the year 2012. A linear particle accelerator and several synchrotrons are used to raise the protons to 450 GeV energy before they are fed to the main ring. The Stanford linear accelerator is 3 km long capable of 50 GeV electrons and positrons in the year 2012. Positron emission tomography (PET), used for mapping tumors in the brain, usually employs positron emission from F^{18} that is incorporated into fluorodeoxyglucose (18F-FDG), which is stored by the brain. An emitted positron is annihilated nearly immediately by an electron. This produces two γ-ray photons traveling in opposite directions, thus localizing the event. The radioactive F^{18} has a short lifetime and is made locally using a cyclotron—a few ms in diameter—by an accelerated 10 MeV deuteron colliding with neon.

As of 2012, the giant Large Hadron Collider, with its protons moving at near relativistic speeds—off by 9 × $10^{-9}c$ from the speed of light—makes 11 000 revolutions per second. There are 2808 bunches of ~0.1 teraprotons, 25 ns apart. The bunch collision rate is 40 MHz.

Synchrotrons employ varying magnetic fields and radio frequencies so that the orbit radius is kept constant.

Synchrotrons employ varying magnetic field and radio frequency so that the orbit radius is kept constant.

tor lasers can generate many kW at hundreds of Hz repetition rate. Solid-state lasers are capable of about 10 μJ in a 100 fs micropulse, which is 10 GV/m field and this corresponds to about 10 kW/m of energy gain with a 10^4 Hz of pulse repetition.

One can therefore imagine fairly attractive and useful energy gains over short distances in an all-electro-optical linear accelerator. This is the attraction of the use of nanoscale features and a schematic description is shown in Figure 6.19(a). The laser light is coupled through a grating to the electron beam, and the grating acts much as it does with radiofrequency cavities. A grating structure designed so that the ridges are in conformity with the wavelength of light, as shown in Figure 6.19(b), and a gap so that the bunches accelerate when they traverse between the ridges will synchronously reinforce the increasing of energy. As carriers reach relativistic speeds, the structures will also have to change in order to accommodate the related features.

For both the large dimensions and the small dimensions, in the accelerators, an essential challenge is imparting the energy—synchronously, as in through phase matching and making sure that the charged bunch does not spread out. The number of protons in a bunch of the Large Hadron Collider is 1.15×10^{11} in a length of 0.3 m. At the frequency of 400 MHz employed, the bunches are separated by 7.5 m. So, there is a phase variation of about $0.3/7.5 \approx 4\%$ in the beam.

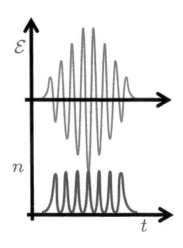

Figure 6.18: The field of a laser pulse (a) and the corresponding electron bunching response (b).

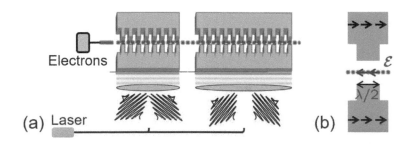

(a)

(b)

Figure 6.19: An optical accelerator with a laterally coupled laser beam front through a periodic grating channel is shown in (a). Appropriate phase delay is needed between the sections. (b) shows the features of the grating where the field inside causes the electrons to accelerate, through the sizing of the wavelength.

The large population requires focusing so that the repulsion doesn't lead to charge spreading. A common way of doing this is through magnet quadrupole, a simple example of which is shown in Figure 6.20(a): a vertically focusing quadrupole magnet. The transverse direction to this is defocusing. Quadrupoles achieve focusing using a compound lens by placing two in series—one that focuses and one that defocuses, as schematically drawn in Figure 6.20(b). The

One can now see why superconducting magnets are so important to the high energy, high density accelerators. Nature limits focusing.

combined focal length of a compound lens is

$$\frac{1}{f} = \frac{1}{f_1} + \frac{1}{f_2} - \frac{d}{f_1 f_2}, \tag{6.39}$$

which reduces to a focal length of f^2/d for $f_1 = -f_2 = f$. This is a net focusing effect. Since a magnetic field can be viewed as a gradient in scalar potential, use of a dipole causes bending through a linear dependence, use of a quadrupole causes focusing through a quadrature, and sextuoples are used for chromatic correction, that is, of momentum, through a cubic term.

A few additional points need to be repeated. First, for the Lorentz force on a charged particle ($\mathbf{F} = q\mathcal{E} + q\mathbf{v} \times \mathbf{B}$), with an increase in its velocity, the magnetic deflection is stronger than the electrostatic deflection. Electrostatic quadrupoles are used in low energy—keV— ion beam systems for lithography. Also, with the magnetic field, with force being perpendicular to the field and the velocity, the field does no work—particles cannot gain kinetic energy as a result of it, and electric fields must be used to accelerate. Magnetic fields are only employable for bending. The second is that Maxwell's equations being linear, and the fact that $\nabla \cdot \mathbf{B} = 0$, no linear combination of arrangement of magnets can focus a charged beam. In an optical accelerator, with a 4% phase spread, as in the collider, the size of this electron bunch is $0.04\lambda \approx 80\ nm$ at a 2 μm wavelengthand is shorter still at visible wavelengths. At an electron population of about 5000 in a bunch, this is an electron density of $\sim 10^{19}\ cm^{-3}$ for a spherical cloud. The Debye screening length at these magnitudes is 0.38 nm, which is a manageable number. And one may focus using techniques similar to that of the larger accelerators using microfabricated magnets.

Accelerating charge particles radiate electromagnetic energy. So, synchrotrons with bending charge beams—an acceleration—radiate light. In linear systems, this is accomplished by employing wiggler magnets—short, bending magnets with alternating fields causing the trajectory of the high velocity particle beam to repeatedly bend, resulting in the radiation. Low field wigglers are also called undulators. These make the compact, ultra-short wavelength, coherent and incoherent high brightness light sources employed in X-ray and other short light wavelength imaging studies with atomic precision, as well as in radiation therapies. Undulators—wigglers of low energy—can be microfabricated with nano- to microscale magnets consistent with the dimensions of the rest of the system.

With this discussion of particle interaction with electromagnetics, we now turn to interactions between collection of atoms and electromagnetic radiation. This collection of atoms may be a few atoms

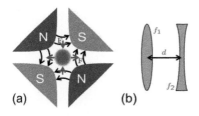

(a)   **(b)**

Figure 6.20: (a) shows a vertically focusing quadrupole. (b) shows the formation of compound lens with two quadrupole magnet in series to obtain a net focusing effect.

One will need nonlinearity for self-focusing. We saw this for optical media in Chapter 4. Plasma-based nonlinearity can also cause self-focusing and is utilized in Tokomak-like devices.

See S. Tiwari, "Quantum, statistical and information mechanics: A unified introduction," Electroscience 1, Oxford University Press, ISBN 978-0-19-875985-0 (forthcoming) for a detailed discussion. An accelerating charge particle has a Poynting vector $\mathbf{S}$—power flow per unit area—that depends on acceleration but not on velocity. This is reflected in the Larmor relationship. For velocities much smaller than c, the Larmor relationship for the energy radiated by a non-relativistic point charge is $P = q^2 a^2 / 6\pi\epsilon_0 c^2$, where a is the acceleration. It is the constraint that the speed of light be a constant that dictates radiation by an accelerating particle. If a charge suddenly appears at a point, its field effect propagates with a constant speed of c. But, its sudden appearance is an acceleration that results in its presence being felt at time t through the field in a region of length ct. There exists a wavefront of this propagation. A simple explanation, due to Purcell, is to look at a charged particle rapidly decelerating in time τ to zero velocity with the constraint of constant speed of light. In the laboratory reference frame, there exists a shell region of $c\tau$, where the field effect of the particle must change, with the shell radius defined by speed of light and the time after the stopping of the particle. There exists a transverse field in this shell region representing the radiation propagation outward.

together, such as in a nanoparticle, and the interaction may occur at surfaces or in bulk. In each of these cases, the interaction is through the electrons, which, being of low mass compared to the nucleus, have the strongest field-induced displacement.

6.4   Plasmonics

AN ELECTRON CHARGE CLOUD IN AN ELECTROMAGNETIC FIELD will respond, just as a single electron charge would, to the field and its time variations. So, a metal or a semiconductor electron ensemble will have an excitation response under the forcing and dynamic constraints. It will respond to the time variations as a collective excitation. This is a plasmonic response. The behavior of the excitation is represented by a quasiparticle—a plasmon—that models the excitation in detail, in the same way as the photon, a hole, an exciton or a phonon does. Briefly, the argument is the following: for an electric field $\mathcal{E} = e\mathcal{E}_0 \exp(-i\omega t)$, in the lowest order, we may view the charge cloud as responding through a collective displacement $\mathbf{z}$ that is reflected in a polarization of $\mathbf{P}' = -en\mathbf{z}$. In the Drude picture of electrons independent of each other, the response to the force under damping due to scattering may be described by the quasistatic response

$$m^*\ddot{\mathbf{z}} + \frac{m^*}{\tau_{\mathbf{k}}}\dot{\mathbf{z}} = -e\mathcal{E}_0 \exp(-i\omega t). \quad (6.40)$$

$\tau_{\mathbf{k}}$ characterizes the randomizing momentum scattering process causing damping through fluctuations. The collective response and forcing directions are aligned, that is, the direction $\hat{\mathbf{z}}$ aligns with the direction of field, $\boldsymbol{\mathcal{E}} = \mathcal{E}\hat{\mathbf{z}}$, with a solution in the form

$$\mathbf{z} = \frac{-e\mathcal{E}}{m^*(-\omega^2 - i\omega/\tau_{\mathbf{k}})}. \quad (6.41)$$

The polarizability of $\mathbf{P}' = -en\mathbf{z}$ can be associated with the charge cloud as in Figure 6.21.

Let $\mathbf{P}(\omega = \infty)$ and $\epsilon_r(\omega = \infty)$ be the polarization and the dielectric constant, respectively, at infinite frequency. These represent a background effect once all time-dependent effects are extinguished—a reference for us. Now, we include into this the effect of the polarizability of the electron plasma—the plasmonic response. The displacement relationship at any frequency ω is

$$\epsilon_r \epsilon_0 \boldsymbol{\mathcal{E}} = \epsilon_0 \boldsymbol{\mathcal{E}} + \mathbf{P} = \epsilon_0 \boldsymbol{\mathcal{E}} + \mathbf{P}_\infty + \mathbf{P}'. \quad (6.42)$$

With the background constraint as $\epsilon_0 \boldsymbol{\mathcal{E}} + \mathbf{P}(\infty) = \epsilon_r(\infty)\epsilon_0 \boldsymbol{\mathcal{E}}$, one can

The properties of plasmons and their coupled response when interacting with other particles are important in the discussion of scattering and reflectivity in S. Tiwari, "Semiconductor physics: Principles, theory and nanoscale," Electroscience 3, Oxford University Press, ISBN 978-0-19-875986-7 (forthcoming).

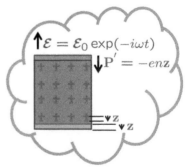

Figure 6.21: In the presence of an excitation field $\mathcal{E}$, the charge in the quasineutral material responds with displacement. Mobile electrons displace more than heavy ions. The effect is the plasma's polarization by $\mathbf{P}' = -en\mathbf{z}$.

extract the dielectric constant in presence of plasmonic response as

$$\epsilon_r(\omega) = \epsilon_r(\infty) - \frac{ne^2}{\epsilon_0 m^* \omega(\omega + i/\tau_\mathbf{k})}$$

$$= \epsilon_r(\infty)\left[1 - \frac{\omega_p^2}{\omega(\omega + i/\tau_\mathbf{k})}\right], \tag{6.43}$$

where

$$\omega_p = \sqrt{\frac{ne^2}{\epsilon_r(\infty)\epsilon_0 m^*}}. \tag{6.44}$$

ω_p is the radial plasmon frequency—a volume plasmon frequency. The larger the carrier density or the lower the mass, the higher is the plasma's polarizability and its effect on changing the dielectric constant at frequencies of interest in the visible range, the infrared range, the microwave range, et cetera. The plasma also undergoes internal scattering, which is represented here in $\tau_\mathbf{k}$, so frequency related to both ω_p and $1/\tau_\mathbf{k}$ determines this dielectric response of the plasma. We also note that Equation 6.43, as a damped response, can also be framed in the physical form

$$\epsilon(\omega) = \epsilon_r(\infty)\epsilon_0 \left(1 - \frac{\omega_p^2}{\omega^2 + i\omega/\Gamma}\right), \tag{6.45}$$

where $1/\Gamma$ is a damping frequency that corresponds to the $1/\tau_\mathbf{k}$ term employed. The dielectric function has a real and imaginary term, that is,

$$\epsilon(\omega) = \epsilon^r + i\epsilon^i, \tag{6.46}$$

where the frequency dependence arises in

$$\frac{\epsilon^r}{\epsilon_r(\infty)\epsilon_0} = 1 - \frac{\omega_p^2 \tau_\mathbf{k}^2}{1 + \omega^2 \tau_\mathbf{k}^2}, \text{ and}$$

$$\frac{\epsilon^i}{\epsilon_r(\infty)\epsilon_0} = \frac{\omega_p^2 \tau_\mathbf{k}}{\omega(1 + \omega^2 \tau_\mathbf{k}^2)}. \tag{6.47}$$

The complex dielectric function $\epsilon_r(\omega)$ describes this electromagnetic response.

A moderately doped semiconductor, $n \approx 10^{18}$ cm^{-3}, $m^* \approx 0.26m_0$—we have chosen silicon here—and $\epsilon_r(\infty) \approx 12$, has a plasma frequency $\nu_p = \omega_p/2\pi \approx 5 \times 10^{12}$ Hz. This excitation has an energy of ~ 20 meV, in the range of phonon frequencies, and well below the visible optical frequency. A $\tau_\mathbf{k} \approx 10^{-12}$ s also means that $\omega_p\tau_\mathbf{k} \gg 1$. The plasma response is significantly faster than the response time of scattering dynamics. In a metal, with $n \approx 10^{22}$ cm^{-3} and an electron mass of $m^* \approx m_0$, the plasma frequency is $\nu_p \approx 2.5 \times 10^{14}$ Hz. Visible light, red to violet, extends over 4.3×10^{14}–7.5×10^{14} Hz. So, metals retain their metallic character below plasma frequencies, and yet

So, one implication is that in visible light, most semiconductors a few absorption lengths thick with moderate bandgap and $\epsilon_r(\infty) \approx 12$ look gray. They do not show any doping response to the naked eye.

these frequencies are in a very useful range where plasmonic effects are significant and potentially of use.

Metals, at low frequencies and in the visible range, are reflective. The electron plasma responds and prevents propagation. They are good conductors. As the visible range is approached, electromagnetic waves can penetrate more and dissipate more. If the frequency is increased even further, metals become more dielectric-like, with dissipative propagation. If electrons in a metal exhibit quite free electron–like behavior, in Na, for example, the metal exhibits transparency in ultraviolet. Au or Ag, however, dissipate, due to band absorption.

Even at microwave frequency, a low frequency compared to optical frequencies, plasma's effects exist, for example, in the skin depth of the penetration of the fields into the metal. With $\omega \ll \tau_\mathbf{k}^{-1}$, and $\epsilon^i \gg \epsilon^r$, the complex refractive index, due to the imaginary term, is

$$n^i = \left(\frac{\epsilon^i}{2}\right)^{1/2} = \left(\frac{\tau_\mathbf{k}\omega_p^2}{2\omega}\right)^{1/2}. \qquad (6.48)$$

This imaginary term causes absorption with the coefficient

$$\alpha = \left(\frac{2\omega_p^2 \tau_\mathbf{k}\omega}{c^2}\right)^{1/2} = \sqrt{2\sigma_0 \omega \mu_0}, \qquad (6.49)$$

where $\sigma_0 = ne^2 \tau_\mathbf{k}/m$ is the static conductivity. In a metal, the field decays as $\exp(-iz/\delta)$, where δ is the skin depth, and x is the orientation perpendicular to the surface:

$$\delta = \frac{2}{\alpha} = \frac{c}{n^i \omega} = \left(\frac{2}{\alpha_0 \omega \mu_0}\right)^{1/2}. \qquad (6.50)$$

Equation 6.43's one implication is that the subtraction term comes close to unity at plasma frequencies. For the high frequencies with $\omega \gg 1/\tau_\mathbf{k}$ of use, one may write a plasmonic consequence of dielectric response of

$$\epsilon_r(\omega) \approx \epsilon_r(\infty)\left(1 - \frac{\omega_p^2}{\omega^2}\right). \qquad (6.51)$$

The dielectric constant vanishes at the plasma frequency $\epsilon_r(\omega_p) = 0$ and changes sign on either side. Recall that these oscillations are in the same direction as the field.

In order to understand the frequency and spatial behavior of plasmonic effects, in small structures, or along interfaces, we will now look at the reciprocal time, that is, frequency-dependent form of Maxwell's relationships. We will consider only local, linear and isotropic conditions such as in our starting discussion for plasma response. These frequency-dependent forms of Maxwell's equations

This alignment implies that the excitation of the free electron system by an electromagnetic wave is in a similar vein as the phonon polariton or the exciton polariton arising from mechanical oscillations.

are:

$$
\begin{aligned}
\nabla \times \mathcal{E}(\mathbf{r},\omega) &= i\omega \mathbf{B}(\mathbf{r},\omega), \\
\nabla \times \mathbf{H}(\mathbf{r},\omega) &= -i\omega \mathbf{D}(\mathbf{r},\omega) + \mathbf{J}(\mathbf{r},\omega), \\
\nabla \times \mathbf{D}(\mathbf{r},\omega) &= \rho(\mathbf{r},\omega), \text{ and} \\
\nabla \cdot \mathbf{B}(\mathbf{r},\omega) &= 0.
\end{aligned}
\tag{6.52}
$$

One may also perform this Fourier transform to map the real space to reciprocal space. For wave propagation, we are generally interested in the spatial dependence with time-harmonic fields. For local, linear and isotropic fields, the constitutive relationships in the spatial and frequency coordinates are

$$
\begin{aligned}
\mathbf{D}(\mathbf{r},\omega) &= \epsilon_0 \epsilon_r(\mathbf{r},\omega)\mathcal{E}(\mathbf{r},\omega), \\
\mathbf{B}(\mathbf{r},\omega) &= \mu_0 \mu_r(\mathbf{r},\omega)\mathbf{H}(\mathbf{r},\omega), \text{ and} \\
\mathbf{J}_c(\mathbf{r},\omega) &= \sigma(\mathbf{r},\omega)\mathcal{E}(\mathbf{r},\omega),
\end{aligned}
\tag{6.53}
$$

where $\mathbf{J}_c(\mathbf{r},\omega)$ is the induced conductivity current arising from particle motion within the medium. The total current is

$$
\mathbf{J}(\mathbf{r},\omega) = \mathbf{J}_c(\mathbf{r},\omega) + \mathbf{J}_s(\mathbf{r},\omega),
\tag{6.54}
$$

where the second term is any sourcing current density. The dielectric constant and the permeability are related through the susceptibility, that is, $\epsilon_r(\mathbf{r},\omega) = 1 + \chi_e(\mathbf{r},\omega)$, and $\mu_r(\mathbf{r},\omega) = 1 + \chi_m(\mathbf{r},\omega)$. Implicit here are the polarization constitutive relationships

$$
\begin{aligned}
\mathbf{P}(\mathbf{r},\omega) &= \epsilon_0 \chi_e(\mathbf{r},\omega)\mathcal{E}(\mathbf{r},\omega), \text{ and} \\
\mathbf{M}(\mathbf{r},\omega) &= \epsilon_0 \chi_m(\mathbf{r},\omega)\mathcal{E}(\mathbf{r},\omega).
\end{aligned}
$$

The time-harmonic field equation 6.52 and the current equation 6.54 by substitution lead to the wave equation in space and frequency of

$$
\nabla \times \frac{1}{\mu_r(\mathbf{r},\omega)}\nabla \times \mathcal{E}(\mathbf{r},\omega) - \frac{\omega^2}{c^2}\left[\epsilon_r(\mathbf{r},\omega) + i\frac{\sigma(\mathbf{r},\omega)}{\omega\epsilon_0}\right]\mathcal{E}(\mathbf{r},\omega)
$$
$$
= i\omega\mu_0 \mathbf{J}_s(\mathbf{r},\omega),
\tag{6.55}
$$

which is the analog of the Helmholtz equation for wave propagation in free space for our conductive, locally responsive, homogeneous material. For non-magnetic conditions ($\mu_r(\mathbf{r},\omega) = 1$) and absent current sourcing ($\mathbf{J}_s(\mathbf{r},\omega) = 0$), this reduces to the form

$$
\nabla \times \nabla \times \mathcal{E}(\mathbf{r},\omega) - \frac{\omega^2}{c^2}\epsilon(\mathbf{r},\omega)\mathcal{E}(\mathbf{r},\omega) = 0,
\tag{6.56}
$$

where

$$
\epsilon(\mathbf{r},\omega) = \epsilon_r(\mathbf{r},\omega) + i\frac{\sigma(\mathbf{r},\omega)}{\omega\epsilon_0},
\tag{6.57}
$$

See the discussion in S. Tiwari, "Semiconductor physics: Principles, theory and nanoscale," Electroscience 3, Oxford University Press, ISBN 978-0-19-875986-7 (forthcoming) for the connections through Fourier transforms of the response through causality, fluctuation-dissipation, Green's function approaches and the Kramers-Kronig relationship. Following our Green's function discussion, displacement and current for the causal homogeneous situation is

$$
\mathbf{D}(\mathbf{r},t) = \epsilon_0 \int \epsilon(\mathbf{r}-\mathbf{r}',t-t')\mathcal{E}(\mathbf{r}',t')\,dt'\,d\mathbf{r}'
$$

and

$$
\mathbf{J}(\mathbf{r},t) = \int \sigma(\mathbf{r}-\mathbf{r}',t-t')\mathcal{E}(\mathbf{r}',t')\,dt'\,d\mathbf{r}'.
$$

Conductivity arises from charge particle motion. Displacement current also gives rise to the current, but it is a current due to time-dependent changes in electric fields and polarization and supported by accumulation and depletion of charges at external boundaries, that is, away from the medium. This is displacement, since it is associated with $\partial \mathbf{D}/\partial t$.

a complex dielectric function that also includes ohmic losses from the out-of-phase effect through the curl connection of the fields. The complementary derivation for magnetic field leads to

$$\nabla \times \frac{1}{\epsilon(\mathbf{r}, \omega)} \nabla \times \mathbf{H}(\mathbf{r}, \omega) - \frac{\omega^2}{c^2} \mathbf{H}(\mathbf{r}, \omega) = 0. \qquad (6.58)$$

These two equations (Equations 6.56 and 6.58) for a non-magnetic material with local, linear and isotropic characteristics and with homogeneity, that is, spatial independence, become, respectively,

$$\left[\nabla^2 + \epsilon(\mathbf{r}) \frac{\omega^2}{c^2} \right] \mathcal{E}(\mathbf{r}, \omega) = 0, \text{ and}$$

$$\left[\nabla^2 + \epsilon(\mathbf{r}) \frac{\omega^2}{c^2} \right] \mathbf{H}(\mathbf{r}, \omega) = 0. \qquad (6.59)$$

This is a Helmholtz form where the dielectric function is complex. Plane waves are solutions of this form, except that there will exist dispersion, that is, $\omega = \omega(\mathbf{k})$ following from

$$\epsilon(\omega) = \frac{c^2 k^2}{\omega^2}. \qquad (6.60)$$

We now have the mathematical framework in place to understand the consequences of plasmons—collective charge particle excitation—in the presence of electromagnetic stimuli.

First, we will consider the non-propagating example where the materials we consider are small. For example, what happens with localized plasmons, for example, with small—10s of nm-sized—particles placed in an electromagnetic field, as in Figure 6.22? Being confined, any coupled interaction will not propagate. The field polarizes the particle. Take electromagnetic fields of interest to usage—$GHz \equiv 30 \ cm$, $THz \equiv 300 \ \mu m$, and $PHz \equiv 300 \ nm$, so up to and beyond visible frequencies—all have wavelengths much longer than the size of the particle of interest. If the particle permits collective charge excitation, it will have plasmonic effects. The effects are due to localized plasmons, constrained by the dimensions and with characteristics considerably different from those from volume plasmons.

To understand the interaction effect in a metal, we first look at the case of a dielectric particle's polarization. In Figure 6.22, let the particle be a sphere of radius a, of a dielectric function ϵ surrounded by a host medium of dielectric function ϵ_h. Let $\mathcal{E}_{inc}$ be the externally applied electric field, and let the frequency ω be such that we may consider this to be an electrostatic quasistatic condition.

The presence of the dielectric distorts the fields locally—it has different constraining electromagnetic properties than the surrounding host. At the surface of the dielectric, normal displacement and tangential electric fields are continuous. This boundary condition and

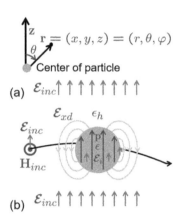

(a)

(b)

Figure 6.22: A particle in a transverse electromagnetic field propagating horizontally. The electric field is along the vertical direction causing polarization of the particle vertically. ϵ characterizes the dielectric response of the particle, and it is embedded in the host medium characterized by ϵ_h. The origin is chosen to be at the center of the particle.

the electromagnetic constraint in the particle and outside leads to a picture, as in Figure 6.22, where one may view the particle as having a dipole moment $\mathbf{p}$, an internal field $\mathcal{E}_i$ and, because of the dipole, a field $\mathcal{E}_{xd}$ external to the dielectric in the host. The external field is the sum of the applied field ($\mathcal{E}_{inc}$) and that generated by the dipole ($\mathcal{E}_{xd}$).

Scalar potentials give us a convenient route for solving this field problem. Since no charge exists, $\nabla^2\phi(\mathbf{r}) = 0$, where $\phi(\mathbf{r})$ is the potential. In spherical coordinates (r, θ, φ), absent any azimuthal (φ) dependence, we may write the solution in Legendre polynomials: $r^l P_l(\cos\theta)$ and $P_l(\cos\theta)/r^{l+1}$, where $l = 0, 1, \dots, \infty$. In order to have solutions that do not explode, the former is allowed inside the sphere, and the latter outside. As a scalar, in the presence of the field $\mathcal{E}_0\hat{\mathbf{z}}$, with $\mathbf{z} = r\cos\theta$, the solution is the sum of ϕ_i—the internal potential—and ϕ_x—the external potential:

$$\phi(\mathbf{r}) = \phi_i(r,\theta) + \phi_x(r,\theta)$$

$$= \sum_{l=0}^{\infty} A_l r^l P_l(\cos\theta)$$

$$+ \left[-\mathcal{E}_0 r P_1(\cos\theta) + \sum_{l=0}^{\infty} B_l \frac{P_l(\cos\theta)}{r^{l+1}} \right]. \quad (6.61)$$

Legendre polynomials are given by

$$P_l(x) = \frac{1}{2^l} \sum_{i=0}^{l} [{}^l C_i]^2 (x-1)^{l-i}(x+1)^i.$$

Some examples are $P_0(x) = 1$, $P_1(x) = x$, $P_2(x) = (1/2)(3x^2 - 1)$, et cetera. They are orthogonal with the l^2 inner product over the interval $[-1, 1]$ and give the integrated inner product of $2/(2l+1)$ with themselves. Because of the orthonormality and Laplace applicability, Legendre polynomials find use in gravitational and Coulomb potential problems over continuous distributions.

Continuity of the normal displacement and the tangential electric field at the dielectric/host interface implies

$$-\epsilon \frac{\partial}{\partial r}\phi_i(r,\theta)\Big|_{r=a} = -\epsilon_h \frac{\partial}{\partial \theta}\phi_x(r,\theta)\Big|_{r=a}, \quad \text{and}$$

$$-\frac{1}{r}\frac{\partial}{\partial \theta}\phi_i(r,\theta)\Big|_{r=a} = -\frac{1}{r}\frac{\partial}{\partial \theta}\phi_x(r,\theta)\Big|_{r=a}. \quad (6.62)$$

The lowest Legendre polynomial, being the constant 1, has a vanishing derivative.

The coefficients follow from the boundary conditions. Equation 6.62 prescribes

$$\epsilon \sum_{l=1}^{\infty} l A_l a^{l-1} P_l(\cos\theta) = -\epsilon_h \mathcal{E}_0 P_1(\cos\theta)$$

$$-\epsilon_h \sum_{l=1}^{\infty} (l+1) B_l \frac{P_l(\cos\theta)}{a^{l+2}}, \quad \text{and}$$

$$\sum_{l=1}^{\infty} A_l a^l \frac{d}{d\theta} P_l(\cos\theta) = -\mathcal{E}_0 \sum_{l=1}^{\infty} B_l \frac{1}{a^{l+1}} \frac{d}{d\theta} P_l(\cos\theta). \quad (6.63)$$

The coefficients can now be related through these. Some of these are

$$B_0 = 0,$$

$$\epsilon A_1 = -\epsilon_h \left(\mathcal{E}_0 + 2\frac{B_1}{a^3} \right),$$

$$\epsilon n A_n a^{2n+1} \;=\; -\epsilon_h (n+1) B_n \;\; \text{for} \;\; n \geq 2,$$

$$A_1 a \;=\; -\mathcal{E}_0 a + \frac{B_1}{a^2}, \;\; \text{and}$$

$$A_n a^{2n+1} \;=\; B_n \;\; \text{for} \;\; n \geq 2. \tag{6.64}$$

For this to be valid, $A_n = B_n = 0 \;\; \forall n \neq 1$. Only the $l = 1$ term is non-zero and satisfies

$$\begin{bmatrix} a^3 & -1 \\ a^3 \epsilon/\epsilon_h & 2 \end{bmatrix} \begin{bmatrix} A_1 \\ B_1 \end{bmatrix} = -\mathcal{E}_0 a^3 \mathbb{I}. \tag{6.65}$$

The two coefficients are

$$A_1 \;=\; -\frac{3}{2 + \epsilon/\epsilon_h} \mathcal{E}_0, \;\; \text{and}$$

$$B_1 \;=\; \frac{\epsilon - \epsilon_h}{\epsilon + 2\epsilon_h} a^3 \mathcal{E}_0, \tag{6.66}$$

with the potentials as

$$\begin{aligned} \phi_i(r,\theta) \;&=\; -\frac{3}{2 + \epsilon/\epsilon_h} \mathcal{E}_0 r P_1(\cos\theta) \\ &=\; -\frac{3}{2 + \epsilon/\epsilon_h} \mathcal{E}_0 r \cos\theta, \;\; \text{and} \\ \phi_x(r,\theta) \;&=\; -\mathcal{E}_0 r P_1(\cos\theta) + \frac{\epsilon - \epsilon_h}{\epsilon + 2\epsilon_h} \mathcal{E}_0 \frac{a^3}{r^2} P_1(\cos\theta) \\ &=\; -\mathcal{E}_0 r \cos\theta + \frac{\epsilon - \epsilon_h}{\epsilon + 2\epsilon_h} \mathcal{E}_0 \frac{a^3}{r^2} \cos\theta. \end{aligned} \tag{6.67}$$

If the dielectric sphere is absent, that is, the host dielectric extends everywhere, $\epsilon = \epsilon_h$, the two potentials match at $r = a$, and the second term of the second equation vanishes. This second term is the result external to the dielectric sphere by the presence of the sphere in the incident electric field. It is the consequence of the dipole moment of the dielectric sphere.

The polarizability $\underline{\alpha}$ of this particle is defined by $\mathbf{p} = \underline{\alpha}\mathcal{E}$. For any dipole of moment $\mathbf{p}$, the potential in space is

$$\phi_d(\mathbf{r}) = \frac{1}{4\pi\epsilon_0\epsilon_h} \frac{\mathbf{p}\cdot\mathbf{r}}{r^3} = \frac{1}{4\pi\epsilon_0\epsilon_h} \frac{1}{r^3} pr\cos\theta', \tag{6.68}$$

θ' being the internal angle. Our dielectric sphere's external effect must match the classical dipole result. The two viewpoints must reconcile. We make the correspondence by forcing equality, that is,

$$\phi_x^d(r,\theta) = \frac{\epsilon - \epsilon_h}{\epsilon + 2\epsilon_h} \mathcal{E}_0 \frac{a^3}{r^2} \cos\theta = \frac{1}{4\pi\epsilon_0\epsilon_h} \frac{pr\cos\theta'}{r^3}. \tag{6.69}$$

The angles must coincide, that is, $\mathbf{p} = p\hat{\mathbf{z}}$ must be aligned with $\mathcal{E}_0$, and we obtain for this particle

$$\mathbf{p} \;=\; 4\pi a^3 \epsilon_0 \epsilon_h \frac{\epsilon - \epsilon_h}{\epsilon + 2\epsilon_h} \mathcal{E}_0 = \underline{\alpha}\mathcal{E}_0$$

$$\therefore \;\; \underline{\alpha} \;=\; 4\pi a^3 \epsilon_0 \epsilon_h \frac{\epsilon - \epsilon_h}{\epsilon + 2\epsilon_h}. \tag{6.70}$$

Polarizability arises from the mismatch of electromagnetic properties and manifests itself when a field is present. Polarizability is proportional to the volume. Polarizability is a function of the geometry and the permittivity of the particle. Atoms have polarizability: electrons shift ever so minutely away from the nucleus in the presence of a field with the external and internal forces balancing. Molecules have polarizability. Water's polarizability, each water molecule in the field of the rest of the ensemble, aligns to occupy less volume. Intramolecular polarization is needed to understand the conformational dependence of electrostatics, and intermolecular polarization is important for a condensed phase's properties. These interactions are essential for electron-hole recombination in light-emitting diodes where inorganic centers create photons or for the recombination of photons to create electron-hole pairs that can be separated for photovoltaics or for spectroscopy, that is, the characterization used and understanding of the processes that happen during the interaction. These are all interactions where polarization, and if mobile charge is present, plasmons appear.

Equation 6.51 which describes the consequence of volume plasmon, and Equation 6.70, for the polarization of a sphere, also tell us the change that occurs in electromagnetic interactions as a result of a change in dielectric response. For example, the denominator of Equation 6.70 vanishes when

$$\Re\left[\epsilon + 2\epsilon_h\right] = 0, \qquad (6.71)$$

the Fröhlich condition. This is an electrostatic resonance—a dipole resonance. Induced field from the sphere blows up under limits placed by the imaginary part of the sphere's dielectric function. The presence of plasmonic resonance, the changes in the real and imaginary part of dielectric response at resonance, is clear through Equation 6.51.

That this dielectric function can be negative is not unusual. Many metals show negative permittivity in the optical and infrared spectra. The dielectric response of gold, for example, has a real part that goes negative at below violet frequencies and across the visible spectrum, and an imaginary part that resonates in the 400–600 nm range, as shown in Figure 6.23. So, if the sphere is composed of metal, and ϵ is bounded by $[-2\epsilon_h, -\epsilon_h]$, then the polarizability will have a sign that is precisely opposite to that of a dielectric sphere. So, a dielectric sphere coated with a metal film, as in Figure 6.24, suitably designed, may have polarization effects that cancel, making the object disappear. An invisibility cloak has been placed on it.

In practice, there are a number of constraints to this "cloaking". In Figure 6.24, a simple metal film will not really do. With a collection

This analysis is relevant not only to electromagnetic propagation, so communications in general as applied to antennas, et cetera, but also to nanobiotechnology, where techniques of nanotechnology are applied in fluidic environments of biology, exploiting physical stimulation. An applied field can move objects in a fluid environment even if they carry no charge. This is dielectric electrophoresis. The dipole moment of dielectric sphere has a pole at $\epsilon = -2\epsilon_h$, which, following Equation 6.45, implies a resonance feature at $\omega_r = \omega_p/\sqrt{3}$. Light will detect particles that are otherwise not observable due to diffraction. The reader would also appreciate a quick self-exploration of what would happen if one had a spherical cavity in a dielectric ambient. This approach still applies.

The index of refraction for most optical material at X-ray wavelengths is smaller than unity! A concave—not convex—lens weakly focuses parallel X-ray light.

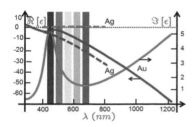

Figure 6.23: The real and imaginary part of the dielectric function of gold, around the visible optical spectrum range.

of metallic nanospheres on the dielectric, it is possible to create a
design so that, over a range of wavelengths, the polarization can
be made to largely cancel. One has to tune the plasmonic shell that
surrounds the object, and the tuning has finite bandwidth.

In other frequency ranges, for example, the microwave range, it
is possible to create structures on the material—resonators of many
varieties—that locally and in aggregate cancel the polarization. These
structures are all examples of creation of metamaterials where object's observable properties have been altered, even as the object's
physical description is still traditional. All of them are changing the
effective dielectric and permeability functions, and, thus, the interaction with the electromagnetic stimulus.

So, the plasmonic interactions of the particle, its real and imaginary dielectric functions leading to scattering and absorption arising
from polarizability, have these frequency-dependent effects. For example, nanosized gold particles in a glass matrix lead to red coloring.
The visible wavelengths in the glass are reduced to a few 100s of nm
(λ_0/n). If the gold particles are a few 10s of nm in size, Figure 6.25
serves as a schematic representation of the scales of dimensions of
the near-visible light optical wave and the particles, as well as the
effect of the fields. The gold particles polarize and, because of their
complex dielectric function, cause both scattering and absorption—
absorption because of the imaginary part of the dielectric function.
Without proof, the interaction is related as follows:

$$C_{ext} = \frac{24\pi^2 a^3 \epsilon_h^{3/2}}{\lambda_0} \frac{\epsilon^i}{(\epsilon^r + 2\epsilon_h) + (\epsilon^i)^2}, \quad (6.72)$$

where the dielectric function is at wavelength λ_0. Gold, because of its
resonances, emphasizes the reddish color in glass.

Since geometry is intrinsic to polarizability, and materials can be
anisotropic, spatial and orientational generalizations are intrinsic to
these approaches. For the sphere, a three-dimensional object, this
volume-normalized polarizability, that is, $\alpha/(V = 4\pi a^3/3)$ is proportional to $3(\epsilon/\epsilon_h - 1)/(\epsilon/\epsilon_h + 2)$. Dimensionally, these are more
generally related as $\nu(\epsilon/\epsilon_h - 1)/(\epsilon/\epsilon_h + \nu - 1)$, with ν as the dimensionality. For a circle, this will be $2(\epsilon/\epsilon_h - 1)/(\epsilon/\epsilon_h + 1)$. The
complement of this, where the permittivities are inverted, that is,
$\epsilon \mapsto 1/\epsilon$, then gives precisely the negative of the polarization. When
a sphere with core of ϵ_2 and size a_2 has a shell of ϵ_1 and size a_1, then
the polarization of this geometric form is

$$\frac{\alpha}{V} = 3\frac{(\epsilon_1/\epsilon_h-1)(\epsilon_2/\epsilon_h+2\epsilon_1/\epsilon_h)+(a_2^3/a_1^3)(2\epsilon_1/\epsilon_h+1)(\epsilon_2/\epsilon_h-\epsilon_1/\epsilon_h)}{(\epsilon_1/\epsilon_h+2)(\epsilon_2/\epsilon_h+\epsilon_1/\epsilon_h)+2(a_2^3/a_1^3)(2\epsilon_1/\epsilon_h-1)(\epsilon_2/\epsilon_h-\epsilon_1/\epsilon_h)}, \quad (6.73)$$

an equation in which volume ratios of the two regions (a_2^3/a_1^3) enter

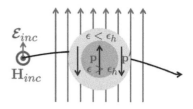

Figure 6.24: A dielectric sphere coated
with metal so that polarizations cancel,
"cloaking" the sphere.

Metamaterials employ both permittivity's negative values, as well as
permeability's, which we have not
discussed but are complementary. At
microwave wavelengths, it is possible
to create a collection of designed resonators that lead to new observable
properties from the material ensemble.
At visible wavelengths, the dimensions
shrink sufficiently so that, at least at
the moment, the freedom to design
artificial resonators is restricted, and
one needs to employ the resonances of
the material itself.

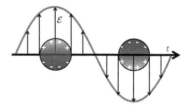

Figure 6.25: A plasmonic particle in
an electromagnetic field. The particle
polarizes, has a complex dielectric
function, and scatters and absorbs light.

naturally. The case of a circle with a shell of Figure 6.24 follows from this as an example where net polarization can be made to vanish.

We now discuss this reduction of dimensionality for the case of plasmons at the two-dimensional boundary of solids. Charge fluctuations of the free electrons in thin metals or in two-dimensional electron gases are examples where two-dimensional plasmons —confined to screening lengths—should be expected. We will call these surface plasmons. Figure 6.26 draws an interface between two materials of dielectric constants ϵ_1 and ϵ_2, the second of which supports nearly free electrons, that is, an electron plasma. One dimension, x here, is confined so that the fields disappear away from the interface, but longitudinal and transversal fields will exist and propagate laterally with dissipation. Electromagnetic interaction at the metal surface is sensitive to the surface. Smooth surfaces reflect well, whereas rough surfaces absorb more.

Metals reflect well if sufficiently smooth. Some energy is lost in the metal. This dissipation is in a region confined near the surface laterally defined by the propagation properties.

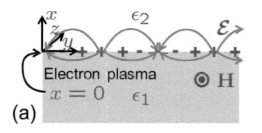
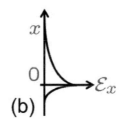

(a) (b)

Figure 6.26: Surface plasmon at the boundary of a dielectric and a free electron material, such as a metal or an inversion layer that supports electron plasma. (a) shows fields at the interface, and (b) shows the magnitude of an x-directed electric field at the interface.

$x = 0$ defines the interface boundary. With the constraints of extinction away from $x = 0$ in the x direction, and by aligning the propagation to the y direction, the field solution for Maxwell's equations, with subscripts identifying region 1—the conducting medium—and region 2—the second medium, which in general can also sustain plasmons—the fields are of the form

$$\text{for } x > 0 \ \ H_2 = (0,0,H_{z2}) \exp\left[i(k_{x2}x + k_{y2}y - \omega t)\right], \text{ and}$$
$$\mathcal{E}_2 = (\mathcal{E}_{x2}, \mathcal{E}_{y2}, 0) \exp\left[i(k_{x2}x + k_{y2}y - \omega t)\right];$$
$$\text{for } x < 0 \ \ H_1 = (0,0,H_{z1}) \exp\left[i(-k_{x1}x + k_{y1}y - \omega t)\right], \text{ and}$$
$$\mathcal{E}_1 = (\mathcal{E}_{x1}, \mathcal{E}_{y1}, 0) \exp\left[i(-k_{x1}x + k_{y1}y - \omega t)\right]. \quad (6.74)$$

Since material 1 is dissipative with $\epsilon_1 = \epsilon_1^r + i\epsilon_1^i$, we will find that the imaginary part of the dielectric function extinguishes the electromagnetic fields in material 1, but because it also forces boundary condition with material 2, it will also force extinguishing of the x-oriented electric field in material 2. The field confinement and absence of propagation in the x direction will result in k_{x1} and k_{x2} being imaginary, that is, $k_{x1} = i\kappa_{x1}$, and $k_{x2} = i\kappa_{x2}$, and hence these are dissipative.

The Maxwell equations for this problem state the following for both materials ($i = 1, 2$):

$$\begin{aligned}
\nabla \times \mathbf{H}_i &= \frac{\partial \mathbf{D}_i}{\partial t} = \epsilon_0 \epsilon_i \frac{\partial \boldsymbol{\mathcal{E}}_i}{\partial t}, \\
\nabla \times \boldsymbol{\mathcal{E}}_i &= -\frac{\partial \mathbf{B}_i}{\partial t} = \mu_0 \mu_i \frac{\partial \mathbf{H}_i}{\partial t}, \\
\nabla \cdot \mathbf{D}_i &= 0 = \nabla \cdot \epsilon_i \boldsymbol{\mathcal{E}}_i, \text{ and} \\
\nabla \cdot \mathbf{B}_i &= 0 = \nabla \cdot \mu_i \mathbf{H}_i,
\end{aligned} \tag{6.75}$$

subject to the continuity boundary conditions that follow from above:

$$\begin{aligned}
\mathcal{E}_{y1} &= \mathcal{E}_{y2}, \quad \text{and} \quad \epsilon_1 \mathcal{E}_{x1} = \epsilon_2 \mathcal{E}_{x2}, \quad \text{for electric fields, and} \\
H_{z1} &= H_{z2}, \quad \text{and} \quad \mu_1 H_{x1} = \mu_2 H_{x2}, \quad \text{for magnetic fields.}
\end{aligned} \tag{6.76}$$

The boundary condition on fields along the surface is implicit in

$$k_{y1} = k_{y2} = k_y. \tag{6.77}$$

The curl relationship for the magnetic field implies, for the y-oriented component,

$$\begin{aligned}
\frac{\partial H_{zi}}{\partial x} &= -\epsilon_0 \epsilon_i \mathcal{E}_{yi} \omega \text{ for } i = 1, 2, \\
\therefore k_{x1} H_{z1} &= \epsilon_0 \epsilon_1 \omega \mathcal{E}_{y1}, \text{ and} \\
k_{x2} H_{z2} &= -\epsilon_0 \epsilon_2 \omega \mathcal{E}_{y2}.
\end{aligned} \tag{6.78}$$

The field continuity of Equation 6.76 and this equation places the following relationship between the fields:

$$\begin{aligned}
H_{z1} - H_{z2} &= 0, \text{ and} \\
\frac{k_{x1}}{\epsilon_1} H_{z1} + \frac{k_{x2}}{\epsilon_2} H_{z2} &= 0.
\end{aligned} \tag{6.79}$$

A solution exists iff the determinant vanishes, that is,

$$\frac{k_{x1}}{\epsilon_1} + \frac{k_{x2}}{\epsilon_2} = 0. \tag{6.80}$$

This establishes the dispersion relationship for the propagating excitation of a surface plasmon coupled to electromagnetic excitation—a surface plasmon polariton. Equation 6.78 and the curl relations of Maxwell also let us write the dispersion forms

$$\begin{aligned}
k_y^2 + k_{x1}^2 &= \epsilon_1 \left(\frac{\omega}{c}\right)^{!2}, \text{ and} \\
k_y^2 + k_{x2}^2 &= \epsilon_2 \left(\frac{\omega}{c}\right)^{!2},
\end{aligned} \tag{6.81}$$

which together with Equation 6.80 give the dispersion relationship for the surface plasmon polariton as

$$k_y = k_{sp} = \frac{2\pi}{\lambda_{sp}} = \frac{\omega}{c}\left(\frac{\epsilon_1\epsilon_2}{\epsilon_1+\epsilon_2}\right)^{1/2} = k_0\left(\frac{\epsilon_1\epsilon_2}{\epsilon_1+\epsilon_2}\right)^{1/2} = n_{sp}k_0. \quad (6.82)$$

Here, k_0 is the free space optical wavevector, and n_{sp} is the effective index for the surface plasmon polariton.

The propagation solution that we have found can be summarized by the following. The magnetic field is described by

$$\begin{aligned}
\mathbf{H_2} &= H_{z2}\exp\left[i(k_{sp}y + k_{x2}x - \omega t)\right]\hat{\mathbf{z}}, \quad \text{and} \\
\mathbf{H_1} &= H_{z1}\exp\left[i(k_{sp}y + k_{x1}x - \omega t)\right]\hat{\mathbf{z}},
\end{aligned}$$

with $\epsilon_1\left(\frac{\omega}{c}\right)^2 = k_{sp}^2 + k_{x1}^2 \;\;\therefore\;\; k_{x1}^2 = \epsilon_1\left(\frac{\omega}{c}\right)^2 - k_{sp}^2,$ and

$$\epsilon_2\left(\frac{\omega}{c}\right)^2 = k_{sp}^2 + k_{x2}^2 \;\;\therefore\;\; k_{x2}^2 = \epsilon_2\left(\frac{\omega}{c}\right)^2 - k_{sp}^2, \quad (6.83)$$

where the surface plasmon polariton wavevector is given by Equation 6.82. The x- and y-directed electric fields follow from the Maxwell relationships. And when the x-directed wavevector is imaginary because of the suitable combination of dielectric functions, the x-directed field is localized, keeping the surface plasmon polariton propagating in the interface region with dissipation.

Surface plasmon polaritons exist only with *TM* polarization. *TE* polarization does not support surface modes. This is a direct consequence of the existence of the charges between which the vector electric field can exist on the surface. This wave, therefore, may also be viewed as a charge density wave.

Figure 6.27 shows some characteristic features of plasmons and plasmon polaritons under different conditions of dimensionality. We did not discuss the one-dimensional case, but the excitational interactions will, of course, occur there too. The volume plasmon energies are large. The interaction of light with the surface plasmon lifts the degeneracy. This leads to the surface plasmon polariton. The figure here shows this where the light cone intersects with the plasmon mode. If medium 2 is free space ($\epsilon_2 = 1$), and medium 1 a metal (ϵ_1 is complex with a negative real part and $|\epsilon_1| > 1$, for example, gold, as in Figure 6.23), then k_{x1} and k_{x2} are complex with a large imaginary part ($k_{x1} \approx i\kappa_{x1}$ and $k_{x2} \approx i\kappa_{x2}$), and $k_y > \omega/c$. The wave intensity extinguishes away from the interface, and the wavelength—the propagating surface plasmon polariton has a wavelength $\lambda_{sp} = 2\pi/k_y$—shrinks since the wavevector is stretched. So, at very low wavevector, the surface plasmon polariton has $\lambda_{sp} \approx \lambda$, the wavelength of light. But, at a large wavevector, $\lambda_{sp} \ll \lambda$. The extinction depths, related to κ, follow from the third relationship of

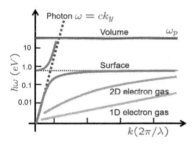

Figure 6.27: Energy versus wavevector with plasmonic conditions. The 2D and 1D electron gases are conducting semiconductor interfaces formed with inversion layers and so contain a significantly smaller electron density than metal does, in addition to the dimensional constraints.

Equation 6.83 but also directly follow from Figure 6.27. At large k_{sp}, in the region farther from the light cone, k_{x2}^2 and k_{x1}^2 are therefore imaginary numbers. Since the ϵ_2 has been chosen to be that of a dielectric, a non-dissipative material, the energy is contained there to a larger depth. When the carrier density is small, such as in electron gases of semiconductors, the plasma frequencies and energies are lower. A one-dimensional electron gas has plasmon energies that are below 10s of meV.

An important consequence of this dispersion characteristic is that with the dissipation, metal boundaries support frequencies that would otherwise not be supported because of these changes in characteristics brought about by the electron ensemble interaction. A gap between two metal plates will support lower frequencies due to plasmonic propagation. This is because of the shrinking of wavelength in a dielectric gap. Figure 6.28 shows the E-k_{sp} and E-λ_{sp} dispersion of thin silver films on SiO_2. While the thicker film reaches the expected surface plasmon characteristics of the semi-infinite medium conditions that we have derived, as one goes to thinner films, one can see that lower energies—optical signal energies—can be accessed and with low wavelengths. The dependence also shows that for silver, the confinement of field on the surface is at a depth of the order of 20 nm. That the wavelengths are small also permits higher resolution. So, plasmonic modes can support optical signals and smaller dimensions. Since the energy is concentrated in a smaller volume, gain, modulation and nonlinearities can also be introduced. When one places two metal films in close proximity, these propagation characteristics can be further improved. Figure 6.29 shows the dispersion behavior for propagation between two layers of silver. A smaller gap has a stronger plasmonic response extending out to higher surface plasmon polariton wavevector—a smaller wavelength. Also, seen here is a single silver film characteristic, together with silver's resonance response, which is near 5×10^{15} Hz.

This argument and the ability to propagate even at this small metal film thickness of 5 nm also lead to surface plasmon guiding in nanowires—an extra dimension of confinement—because, as Figure 6.30 shows, TM modes can still be supported. This gives the ability to connect to reduced dimensions—nm-scale dimensions with sub-wavelength confinement and slow propagation modes—longer in wavevector and shorter in wavelength, as is important for resolution and interaction scales. The ability to guide electromagnetic waves into these small dimensions is an important ability that plasmonics provides in the investigation and utilization of nanoscale devices and their properties.

We conclude this discussion of plasmon effects with a few exam-

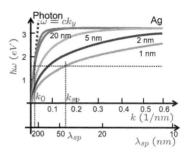

Figure 6.28: E-k_{sp} and E-λ_{sp} characteristics of thin silver films on SiO_2. For reference, the light line of the free photon shows the corresponding wavevector and wavelength; the wavevector and wavelength for the surface plasmon polariton at the optical communication energy of 1.54 eV are also shown.

A perfect metal boundary condition, such as when we do waveguide analysis and prescribe a cut-off frequency, will now break.

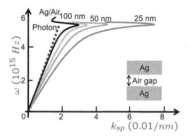

Figure 6.29: Dispersion characteristics for transmission confined by two silver layers with an airgap at gaps, of 25 nm, 50 nm and 100 nm. For comparison, the optical line and characteristics of a single metal film of silver are also shown.

Figure 6.30: Propagating surface plasmons are transverse magnetic (TM) because electric fields are polar vectors. In a nanowire, a TM mode surface plasmon with its shrunk wavelength, propagates along the wire.

ples of its use taking advantage of the energy concentration, high intensity, and wavelength shrinking. Being confined to surfaces, the surface plasmon provides, in particular, near-field interrogation and action capabilities. Uses of these principles are drawn in Figure 6.31.

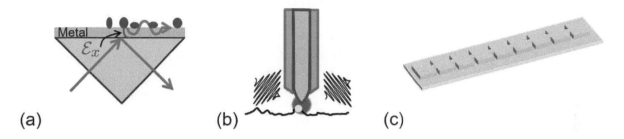

Figure 6.31: Three examples of using plasmonic effects at nanoscale. (a) shows generation of a surface wave that interacts with small objects locally and has a shorter wavelength. (b) shows a nanoscale scanning probe that has light-carrying capability through plasmonics, is also capable of carrying an electrical signal, and therefore can act as an electrical nanoscale coaxial probe that interacts with small objects locally in the presence of other optical beams. (c) shows an example of Bragg scattering in plasmonic guides that allow distributed feedback and coupling to optical waveguided structures below.

Figure 6.31(a) shows an example of a technique that provides higher resolution in measurements and is one way of generating surface plasmon polaritons. A prism lens is used to couple light to a metal film that is made thin enough to generate surface plasmons for propagation while providing sufficient field at the other interface for interactions. In a fair-sized areal region related to the extinction of the surface wave, a few μms or larger, strong field-enhanced interaction will take place on the surface. So, one may make observations on objects that have resonant effects, multi-photon interactions, et cetera, with higher resolution in this localized dimension. This approach also points to device implications. If a stronger interaction occurs at the shorter wavelength, such as a generation or recombination process, then the corresponding device will improve in this region of interaction. Metal-based grating structures on small bandgap semiconductors, for example, generate terahertz more efficiently. Terahertz detection becomes more efficient when this plasmon effect is incorporated and the carriers removed through these same metal films. In thin film solar cells, this plasmonic enhancement helps with scavenging light that would otherwise be lost to photocarrier generation by enhancing the interaction, for example, by bringing longer wavelength light to shorter wavelengths that can be captured by the material.

Figure 6.31(b) shows a scanning probe that employs the nanoscale features in the probing assembly to confine and improve interactions in a very small region. This is unlike the previous case, where energy spreads out in the surface propagation volume. The probe may contain more than one signaling mechanism. Instead of just the optical signal, one may also bring in other signals—static voltages, for example—and then add additional optical sources to flood the interaction region. Second harmonic generation, Raman enhancement, et cetera, all become possible, where the probe and its plasmonic field-enhancement also localizes the region to as small as molecular dimensions and, as a result of the interaction, provides a suitable antenna capability for the signal to be extracted. Nanotubes can be ``optically'' observed and characterized through near-field Raman techniques using this approach.

Finally, the example in Figure 6.31(c) is to point out that this plasmonic capability is also important to large-sized objects where metal for charge carriers and optical regions for waveguiding and gain exist. Usually, one keeps these away from each other since, in sources, such as lasers, the metal and plasmonic effects' major consequence is increased losses. But, one can visualize this distributed feedback structure where different order modes are coupled and change the radiating modes using surface propagation modes. For example, a higher order distributed feedback guide shifts the light line to a guided mode of a higher effective index. Bragg scattering then can couple this higher effective index mode to a low index radiating mode that end fires—without the large longitudinal and transverse distortions that usually occur in semiconductor-based guided wave structures.

6.5 Optoelectronic energy exchange in inorganic and organic semiconductors

THIS ELECTRON, DIPOLE AND ELECTROMAGNETIC INTERACTION is also central to light absorption and emission as practiced in photovoltaics and light emission—examples of energy conversion. We have confined ourselves to inorganic semiconductors up to this point, so we will start with a discussion of organic and organic-inorganic assemblies, where light absorption and emission is of increasing use. We will discuss here the processes of energy transfer in small assemblies—molecules and nanocrystals—and the processes of transport, as well as their implications for devices that absorb light using such materials, and draw some thermodynamic constraints on energy conversion efficiency from the solar spectrum due to the nature of blackbody radiation and that of the absorption processes.

Since the electromagnetic interaction with matter occurs through the response of charge, its excitation response, either as isolated

Absorption, emission, spontaneous and stimulated processes in inorganic semiconductors, and blackbody radiation are discussed at length in S. Tiwari, ``Semiconductor physics: Principles, theory and nanoscale,'' Electroscience 3, Oxford University Press, ISBN 978-0-19-875986-7 (forthcoming). The reader is encouraged to refresh this discussion.

charge or through dipoles, appears in processes such as absorption and emission. In crystalline inorganic semiconductors, one discusses this in terms of the oscillator strength—a parameter that quantifies the dipole matrix element of interaction with the fields for the material.

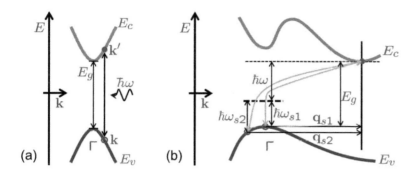

Figure 6.32: Absorption in direct semiconductors without involvement of phonons is shown in (a) and with emission of phonons in indirect semiconductors, in (b). Electrons and holes involved in the process are shown to be at the bandedge but don't have to be. Phonon-mediated absorption is less likely than direct absorption in direct semiconductors. The reverse holds for indirect semiconductors.

In direct bandgap materials, as in Figure 6.32(a), this absorption process can be direct, that is, between electron and hole states, while conserving energy and momentum. In indirect semiconductors, as in Figure 6.32(b), this absorption process is by necessity predominantly indirect where the momentum change occurs through the absorption of a phonon. Phonon wavevector $\mathbf{q}_{s1}$ and $\mathbf{q}_{s2}$ represent here two example phonons with energies $\hbar\omega_{s1}$ and $\hbar\omega_{s2}$, respectively, that make momentum and energy conservation possible in the absorption process. The indirect band-to-band process has the minimum energy conditions of $\hbar\omega > E_g - \hbar\omega_s$. The analysis of indirect transitions employs second order perturbation theory. Virtual processes, that is, processes of low probability but allowed due to uncertainty, must be considered. These virtual processes may occur through either the conduction band or the valence band, as shown in Figure 6.33.

But, in each of these cases, the electron-hole generation during absorption or annihilation for light generation must couple the electromagnetic radiation to the crystal. In indirect semiconductors, as a result of this phonon-assisted process, recombination tails much more gradually below the bandgap energy than in direct bandgap materials.

These direct and indirect recombinations can be seen through the photon's interaction with the electron that has polarizability since a field displaces it. Its presence in the crystal brings the collective effects through the crystal's assembly and periodicity and energetic constraints. This can be seen through the following summary argument.

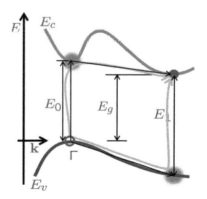

Figure 6.33: Absorption in an indirect semiconductor at the bandedge, showing two limit cases involving conduction and valence band. The virtual states are identified through fuzzy regions in the two bands.

The Hamiltonian of an electron of wavevector $\mathbf{k}$ interacting with a photon of vector potential $\mathbf{A} = A\hat{\mathbf{a}}$, an interaction made possible because of charge, is

$$
\begin{aligned}
\mathscr{H} &= \frac{1}{2m^*}(\hbar\mathbf{k} - q\mathbf{A})^2 = \frac{1}{2m^*}\left(\hbar^2 k^2 - \hbar q \mathbf{k}\cdot\mathbf{A} - \hbar q\mathbf{A}\cdot\mathbf{k} + q^2\mathbf{A}^2\right)\\
&= \frac{1}{2m^*}\left(-\hbar^2\nabla_{\mathbf{r}}^2 + i2q\hbar\mathbf{A}\cdot\nabla_{\mathbf{r}} + q^2\mathbf{A}\right)\\
&\approx -\frac{\hbar^2}{2m^*}\nabla_{\mathbf{r}}^2 + \frac{iq\hbar}{m^*}\mathbf{A}\cdot\nabla_{\mathbf{r}} = \mathscr{H}_0 + \mathscr{H}',
\end{aligned}
\tag{6.84}
$$

since $q^2\mathbf{A}$ is significant only at high intensity and nonlinear conditions. The perturbation term coupling parabolic bands in a crystal then can be written as

$$
\begin{aligned}
\mathscr{H}'_{\mathbf{k}'\mathbf{k}} &= \frac{iq\hbar A}{2m^* N}\int_\Omega \psi_{\mathbf{k}'}^* \exp(i\mathbf{q}\cdot\mathbf{r})(\hat{\mathbf{a}}\cdot\nabla_{\mathbf{r}})\psi_{\mathbf{k}}\,d\mathbf{r}\\
&= \frac{iq\hbar A}{2m^*}\int_{\Omega_0} u_{\mathbf{k}'}^*\left[\hat{\mathbf{a}}\cdot\nabla_{\mathbf{r}} u_{\mathbf{k}} + i\left(\hat{\mathbf{a}}\cdot\mathbf{k}\right)u_{\mathbf{k}}\right]\,d\mathbf{r},
\end{aligned}
\tag{6.85}
$$

where $\psi_{\mathbf{k}} = \exp(i\mathbf{k}\cdot\mathbf{r})u_{\mathbf{k}}(\mathbf{r})$ is the electron Bloch function, and the wavevector, that is, momentum, has been conserved. The first term of the integral represents the allowed transitions, and the second, the so-called forbidden transitions—these are where momentum matching must occur through phonons. The crystal momentum matrix element

$$
\mathbf{p}_{\mathbf{k}'\mathbf{k}} = -i\hbar\int_{\Omega_0} u_{\mathbf{k}'}^*\nabla_{\mathbf{r}} u_{\mathbf{k}}\,d\mathbf{r}
\tag{6.86}
$$

reduces the allowed perturbation term to the form

$$
\mathscr{H}'_{\mathbf{k}'\mathbf{k}} = -\frac{qA}{2m^*}(\hat{\mathbf{a}}\cdot\mathbf{p}_{\mathbf{k}'\mathbf{k}})\,d\mathbf{r}.
\tag{6.87}
$$

This determines the transition rate when the occupation statistics are included, and one can write recombination rates in terms of the reduced mass $m_r^* = m_e^* m_h^*/(m_e^* + m_e^*)$ in terms of the effective masses of the electron (m_e^*) and the hole (m_h^*):

$$
\begin{aligned}
r &= \frac{q^2 A^2 \omega f}{16\pi^2 m^*} f_0(1-f_0)\int_{\Omega_{\mathbf{k}}}\delta\left(E_g + \frac{\hbar^2 k^2}{2m_r^*} - \hbar\omega\right)4\pi k^2\,dk\\
&= \frac{q^2 A^2 \omega f (2m_r^*)^{3/2}}{8\pi\hbar^3 m^*} f_0(1-f_0)(\hbar\omega - E_g)^{1/2}
\end{aligned}
\tag{6.88}
$$

for the allowed, and

$$
r = \frac{q^2 A^2 f'(2m_r^*)^{5/2}}{12\pi m^{*2}\hbar^4} f_0(1-f_0)(\hbar\omega - E_g)^{3/2}
\tag{6.89}
$$

for the forbidden. f—the oscillator strength—is

$$
f = \frac{2(\hat{\mathbf{a}}\cdot\mathbf{p}_{\mathbf{k}'\mathbf{k}})^2}{\hbar m^* \omega} \approx 1 + \frac{m^*}{m_h^*}
\tag{6.90}
$$

and a measure of the number of oscillators with the frequency ω. The transition rate being directly proportional to f is the historic reason for calling it the oscillator strength. It represents the efficiency of coupling of the electromagnetic wave with the electron's polarizability, which responds to the changing field strength and is, therefore, also related to the dielectric function's behavior. When one confines dimensions, such as in quantum wells or quantum wires, the momentum matching condition being related to the conduction and valence states' specifics—heavy holes and light holes have different total angular momenta because of orbital angular momentums—polarizing of light becomes possible through the specifics of transitions that are allowed and those that are not. So, in quantum well structures, it becomes possible to polarize using these coupling processes.

What distinguishes organic semiconductors from inorganic semiconductors is the length at which the energy interaction occurs because of the nature of chemical bonding within the molecule and its separation from another molecule. Conjugated molecules, that is, ones with p orbitals with delocalized electrons, then specially exhibit interesting transport properties. Figure 6.34 shows a simple picture of the highest occupied molecular orbital ($HOMO$) and the lowest unoccupied molecular orbital ($LUMO$) subject to intramolecular and intermolecular energy constraints. The length scale of interaction is also important to the existence and the nature of electron or bound electron-hole pairs—excitons—which become more preponderant in small and large molecule semiconductors because of strong Coulomb attraction. In a light-emitting diode, the electroluminescence arises from these electrons, holes or excitons recombining, and these are a result from injection of electrons or holes from metal electrodes. Anodes—drains—are low workfunction metals that couple to the highest occupied molecular orbital levels ($HOMO$ levels) where holes can easily enter and exit. Cathodes—sources—are high workfunction metals that couple to the lowest unoccupied orbital levels ($LUMO$ levels). Exciton decay, or electron-hole recombination as in most semiconductors, produces the photons. In a solar cell, a photovoltaic device, the reverse process occurs, but if excitons are preponderant, the challenge is to cause a separation and then the efficient removal of the electrons and holes.

Exciton dissociation requires a strong carrier accepting species, for example, a strong acceptor. C_{60} is a strong acceptor capable of 6 electrons but has poor solubility, which limits its density. So, others, such as $PCBM\ C_{61}$ (full form: phenyl-C_{61}-butyric acid methyl ester), are employed. Because of the nature of the diffusive transport of exciton, the need of this electron-hole separation, and once the electron and hole are separated, the limits to the distance they can travel in most

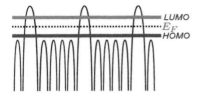

Figure 6.34: Closely spaced organic molecules, particularly the conjugated types involving connected p orbitals, have states arising in close spacing with small barriers in intramolecular bonding and larger barriers in intermolecular bonding. This makes carrier transport possible. The figure simplistically shows the highest occupied orbital ($HOMO$) and the lowest unoccupied orbital ($LUMO$) due to the native interactions.

Transport in conducting organic semiconductors, examples of crystalline forms such as carbon nanotubes or graphene being the exception, are poorly understood and modeled, like amorphous and polycrystalline semiconductors. In both cases, the statistical distribution makes the modeling a parameter fit. One is reminded of a pithy von Neumann comment, "With four parameters I can model an elephant. With a fifth, I can make it wiggle its trunk." Excitons, being net neutral, diffuse. Excitons exist. Measurements show their signatures in energy localization. But, there is an uneasy partial truth in a friend's comment that if you don't understand something, blame it on excitons.

In quantum wells and in several semiconductors, particularly the small bandgap II-VI compounds, excitons will be seen to be important.

organic semiconductors that are poor conductors with large amounts
of scattering, interfaces become important, and the devices are, by
necessity, very thin.

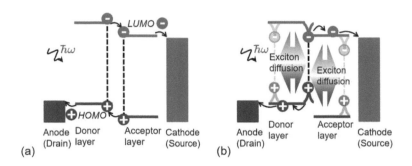

(a)

(b)

Figure 6.35: Optoelectronic interaction
in organic semiconducting films and
interfaces may be dominated by free
carriers—electrons or holes, or by
excitons. Organic solar cells may be
inorganic-solar cell-like, that is, involve
electron-hole generation and their
separation by intrinsic and applied
fields through transport as in (a), where
the generated electrons are collected
by the cathode, following transport
through the acceptor transport layer
and holes by the anode following
transport through hole transport layer.
In (b) exciton behavior dominates.
Excitons are formed from the electron-
hole pairs created and diffuse. At
the interface, the exciton breaks into
an electron and a hole, and these are
removed through the acceptor and
donor layer.

Figure 6.35 shows two quite different ways by which photovoltaic
conversion occurs in organic semiconductor structures. Each employs the quite different characteristics of different organics and their
interfaces.

Hole transport, which is often quite poor (a mobility of 10^{-3}–
10^{-9} $cm^2/V \cdot s$), and is common to small organic molecules. We
will not discuss these—there are many, and many continue to be
invented because of improvements in characteristics desired, but a
few examples include N, N$'$-bis-(3-tolyl)-NN$'$-diphenyl-1, 1$'$-biphenyl-
4, 4$'$-diamine (TPD), Tris-(8-hydroxyquinonline) aluminum (Alq_3—
an aluminum salt) and polyvinyl acrylic (CH_2-CH-OH_n—a chain
molecule) and its variants. Poor hole transport and competing non-
radiative recombination means that such layers must be thin.

Electron transport is usually better than hole transport—reaching
up to 10^{-1} $cm^2/V \cdot s$ in mobility. Again, many examples exist and
continue to be refined. Examples of conjugated compounds include
polyacetylene (PA), polythiophene (PT), poly(3-alkyl) thiophene
($P3AT$), polypyrrole (PPy), polyethylene dioxythiophene ($PEDOT$)
and poly (para-phenylene vinylene) (PPV)—a plethora of acronyms.

Figure 6.35(a) shows a structure which is based on electron and
hole transport and usually made with the small molecules that provide the transport medium for either carrier. The former is a donor
transport layer, and the latter is an acceptor transport layer—a set of
terms that have their origin in chemistry terminologies. In the solar
cell, with its thin layer, as shown in Figure 6.35(a) the electron-hole
pair is generated near the interface region, and the fields cause the
electrons and holes to be separated and collected by the electrodes
of the suitable workfunction. This process is essentially the same as
that of conventional inorganic solar cells where the *p-n* or *p-i-n* junc-

tion regions have fields that separate the carriers. In the other organic assemblies, usually made of large molecules, as in Figure 6.35(b), it is preferable for the electrons and holes generated to relax to a favorable lower energy exciton state. This neutral bound pair can diffuse, as shown in Figure 6.35(b), through the donor and acceptor layers until it encounters a suitable interface where the exciton can break apart. The interface is also conducive to separating the charge species. These separated electrons and holes are removed by transport through the acceptor and the donor transport layers. So, the generation of electron-hole pairs that rapidly form an exciton, the diffusion of the exciton, the breaking of the exciton by the interface region, the charge separation before recombining of carriers or reformation of excitons, and the field-mediated transport of electrons and holes determine the characteristics of the operation of the solar cell.

Figure 6.36 shows the reverse of the operation of the solar cell for light emission—a light emission diode. Electrons and holes are injected, and as they transport away from the injection interface, the population of excitons increases. The light emission layer is particularly amenable to the formation of excitons and their annihilation through the generation of photons.

In each of these structures, the thicknesses of the layers need to be kept small. The length scale over which excitons transport and decay to light is about 10–200 nm, a dimension that is also of relevance to solar cells. This is because the process of interest, for example, the removal of carriers from excitons, is interface specific; a competing process, such as the non-radiative recombination of carriers, is sufficiently high that one needs to keep thickness small; or parasitic effects, such as due to the resistance of films, become excessively high because of the poor transport properties.

One organic device structure that has been particularly of interest is the Grä̈zel cell, schematically drawn in Figure 6.37. Interfaces for the separation of the charge carrier are again important to the workings of the cell, which uses a light-sensitive dye. Light absorbed by the light-sensitive dyes, for example, Ru-bpy (Tris(bipyridine)ruthenium(II) chloride) or triscarboxy-ruthenium terpyridine, et cetera, cause an excitation of the molecule—to the S^* state, in this example. The exciton in the dye-doped absorber layer diffuses, as in our previous examples. The electron, as before, can be removed with an appropriate workfunction material by splitting the exciton. TiO_2 is commonly used for collecting the electrons. Dyes are small, so this electron removal process can be quite efficient. The electron is supplied back to the dye in its ground state, or equivalently the hole generated is removed from the dye through reduction-oxidation, or redox, for short, using an appropriate medi-

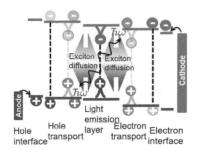

Figure 6.36: An organic light-emitting diode where electrons and holes that are injected form excitons as they transport through electron and hole transport layers, as well as in the light emission layer, where they efficiently recombine.

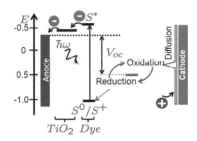

Figure 6.37: The Grä̈zel cell, consisting of a light-sensitive dye with the electron of the excited state being removed by an appropriate workfunction material such as TiO_2 and the charge being supplied through a redox reaction connected by diffusion to the cathode.

Dyes are optically active light-sensitive molecules. Like it or not, we all have imbibed Red dye 2, which was a common food coloring till the 1970s, when it was banned in the United States, as a potential carcinogen.

ator in a conducting electrolyte, which can be a polymer. The redox process coupled to the dye and the titania completes the electron transport circuit, making a solar cell possible. In such a cell, the maximum voltage that can be generated is determined by TiO_2's quasi-Fermi energy and the redox potential of the electrolyte. This gives an open-circuit voltage $V_{oc} \approx 0.7\ V$. Short-circuit current densities of 20 mA/cm^2 are about 60 % of those achievable from silicon solar cells. Since the dye is the active species, it must exist over a large surface area so that much of the incident light may be absorbed and the titania needs to conduct the electrons away. This is accomplished by making a matrix of titania nanoparticles coated with the dye placed on an inexpensive conducting electrode—the anode—which may be a metal. Interconnected titania particles conduct electricity, the dye is chemisorbed on it, and the porous regions are filled with a p-type conducting media for the hole transport. A number of materials have the appropriate workfunction for the electron extraction, ZnO, SnO_2, Nb_2O_3, et cetera, being other examples.

This excitation when the photon is absorbed or generated, and the energy transfer process involved in this donor-acceptor medium is, of course, very different from that of a crystalline semiconductor. It is also very central to the size-scale arguments that we have made, so through multiple ways, it is very central to energy conversion devices. We therefore emphasize the nanoscale features of it.

There are two principal processes through which molecules transfer energy. One is through the exchange of electrons. This requires orbital overlap, and because it is based on particle exchange, has short-range interactions. This is the Dexter process. It is effective over distances of 0.5–2.0 nm. The other is an electromagnetic interaction—a dipole-dipole Coulombic exchange of energy, which because of the field exchange, has long-range interaction. This is the Förster process. It is effective over distances of about 3.0–10.0 nm.

Since TiO_2 and ZnO are common paint solids, the Gräzel cell is based on particularly inexpensive and common material. However, dyes are notoriously unstable, and this has been an impediment to the adoption of this approach.

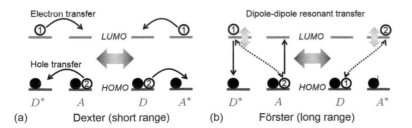

(a) Dexter (short range) (b) Förster (long range)

Figure 6.38: Electron exchange underlying the Dexter energy transfer process in (a). (b) shows the dipole-dipole Coulombic interaction of the Förster resonant energy transfer process; *HOMO*, highest occupied molecular orbital levels; *LUMO*, lowest unoccupied molecular orbital levels.

Figure 6.38 summarily describes the short-range (Dexter) and the long-range (Förster) energy transfer processes between the donor and acceptor molecules. First, consider the short-range Dexter process, shown in Figure 6.38(a). A donor molecule is excited ($D \mapsto D^*$) by incident radiation causing an electron to rise up to a higher energy state orbital. The figure shows this as the electron marked 1 rising to the *LUMO* level from a *HOMO* level. The frontier orbitals, because they interact most strongly with the surroundings at low energies, for example, in the photoreaction process, provide the lowest energy path for interaction. The electron of this excited donor transfers to the neighboring acceptor's *LUMO* level, while an electron from the acceptor's *HOMO* level transfers back to the donor level. This is the electron exchange. In the semiconductor-like description of this process, one could look at this process as the creation, in the donor, of an electron-hole pair that is transferred to the acceptor. An electron transferring from the acceptor's *LUMO*—valence band-like state— is equivalent to the "quasiparticle" hole transferring in the other direction. The right hand side of Figure 6.38(a) describes the complementary process, where an acceptor is excited ($A \mapsto A^*$), and the electron from the *LUMO* level of the excited acceptor transfers to the *LUMO* level of the donor while the electron from the *HOMO* level of the donor transfers to the acceptor. The 1- and 2-labeled interacting electrons exchange position between the acceptor and the donor, in either case. This is a tunneling process modulated by the barrier and the distance where the wavefunction decay is exponential. Because the composing eigenfunctions have an imaginary wavevector ($k(\mathbf{r}) = i\kappa(\mathbf{r})$) or, equivalently, a wavefunction $\psi(r) \propto \exp(-r/\lambda)$, where $\lambda = 2\pi/\kappa$ is a length scale, the interaction strength must reduce as its square, that is, $\mathcal{H}'_{DA} \propto\sim \exp(-2r/\lambda)$. This limits the short-range interaction to about 2 *nm*, with the short end limited to the bonding dimension length scale (~ 0.6 *nm*). This is the connection of the dimensional scale of the van der Waals contact to that of the overlap of the orbitals.

The Förster process, shown in Figure 6.38(b), is effective at much longer length scale and involves dipole-dipole interaction, that is, dipolar fields or radiation fields—the same mechanism as in antennas or in the radiation losses of our accelerating particle discussion— except that, in this instance, there is no radiation out to the environment but a radiation that is well coupled. The radiation couples the electronic excitation and, through it, causes a non-radiative transfer of energy. The dipolar fields interact through their overlap across space even as the electron remains localized to the same molecule, unlike the case in the Dexter process where it transfers. The excited states, of the donor (D^*) and the acceptor (A^*), shown on the left and the right, respectively, in Figure 6.38(b), are states in an oscillating field. The polarization of the molecule, because of the transfer of electron from the *HOMO* level to the *LUMO* level, is rapidly oscillating in time. It is a transmitting antenna that is radiating. So, if the excited donor state matches in frequency to the acceptor state, energy transfer may take place, and if it does this—a resonant process—an electron from the *HOMO* level of the acceptor is bumped up to the *LUMO* level. The donor has transferred energy to the acceptor, and, in this process, the excited electron drops back down to the *LUMO* level. So, the resonant energy transfer involves the change in occupation of electrons between levels of the same molecule: one loses energy while the other gains energy.

The picture of the energy exchange of $D^* + A \mapsto D + A^*$, where the state $|D^*\rangle$ arose from the absorption of a photon, as shown in in Figure 6.38(b), is described by the energy and configuration representation in Figure 6.39(a).

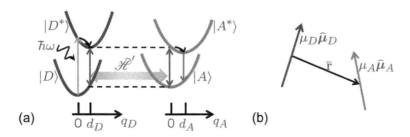

(a)   (b)

Figure 6.39: (a) shows the Förster resonant energy transfer process represented in a configuration diagram. (b) shows the molecular dipoles that interact radiatively but cause a non-radiative transfer of energy.

Förster transfer requires that the electromagnetic interaction— optically induced electronic coherence—of the donor resonate with the electronic energy gap of the acceptor. The strength of this interaction, that is, the magnitude of the transition matrix, is proportional to the transition dipole interaction, which relates to the alignment and separation of dipoles. A pictorial representation of this is shown in Figure 6.39(b). Figure 6.39(a) shows the following process at work. Light is absorbed by the donor at equilibrium energy gap. This, of course, requires the appropriate electromagnetic-matter coupling based on polarization, similar to what we have discussed for electrons and for semiconductor crystals. This state's harmonic oscillator-like picture is shown in the configuration diagram picture of $|D^*\rangle$ as a function of the configuration coordinate q_D. This rapid vibrational state relaxes in times of the order of *ps* with donor coherent oscillation at the energy gap of the donor. Resonance, that is, energy matching of the electromagnetic coupling with the ground state energy gap of the acceptor to its excited state transition, allows the excited donor to lose its energy to the acceptor. The acceptor is now in state $|A^*\rangle$, centered on the configuration coordinate d_A. In this excited state, the acceptor fluorescence is spectrally shifted from donor fluorescence. The energy matching between the excited donor and the ground state of the acceptor is what caused the transition. But, the proximity of the acceptor to the donor also matters. This can be seen in Figure 6.39(b). The Hamiltonian of the system is $\mathscr{H} = \mathscr{H}_0 + \mathscr{H}'$, where

$$\mathscr{H}_0 = |D^*A\rangle \mathscr{H}_D \langle D^*A| + |A^*D\rangle \mathscr{H}_A \langle A^*D|, \text{ and}$$
$$\mathscr{H}' = \frac{3(\mu_A \cdot \hat{r})(\mu_D \cdot \hat{r}) - \mu_A \cdot \mu_D}{r^3}. \tag{6.91}$$

This implies that the transition rate for the Förster process is

$$\mathcal{T}_{ET} = \frac{1}{\tau_D}\left(\frac{R_0}{r}\right)^6, \tag{6.92}$$

where τ_D is the fluorescence time of the donor—it characterizes donor vibrational-electromagnetic features and is thus related to the matrix elements whose explicit calculation, which is based on the golden rule and the electronic and nuclear configuration of the wavefunction, we have skirted. R_0 is a critical transfer distance where the energy transfer is equal to the rate of fluorescence. If the molecules are very far apart, all the energy will be lost through fluorescence rather than this Förster transfer.

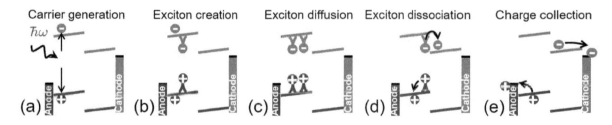

Figure 6.40: (a)–(e) show some of the significant steps in the photoelectric conversion in an organic solar cell based on exciton diffusion and splitting at the interface. (a) shows the creation of an electron-hole pair in the donor layer. (b) shows a fast relaxation to excitons. (c) shows diffusion of excitons. (d) shows the splitting of the exciton, and the charge transfer at the donor-acceptor layer interface, and (e) shows the charge movement and collection.

We collect the thoughts of this discussion together with a description of efficiency and scales of various processes that compose a realistic solar cell based on organic materials(see Figure 6.40).

Absorption length is the scale length over which light intensity drops by $1/e$. For visible light, in silicon, which is an indirect semiconductor, this is $\sim 15~\mu m$; for $GaAs$, which is a direct semiconductor, this is $\sim 1.5\mu m$. For organic semiconductors, good materials have an efficiency of absorption of greater than 50 % and an absorption region that is quite narrow, and the excitation time for electron-hole pair creation is in fss. These electron-hole pairs relax to the excitonic bound state in sub-ps time. The binding energy of these excitons is fractions of eV (0.1–1.0 eV). These neutral excitons diffuse. There is a larger population at the incident cross-section than deeper in the junction. Typical diffusion time constants are of the order of nss for the sub-100 nm-thick films, with exciton diffusion lengths of 5–20 nm. Many of these excitons are lost during diffusion, so the efficiency of this diffusion process is quite small ($\sim$ 10 %) and only a fraction make it to the donor/acceptor interface layer. This interface is efficient—the exciton dissociating in fractions of ps— with the electrons and holes created being collected by the cathode and anode with a high efficiency. The sum efficiency of the process, its ohmic losses, the short-circuit current density and open-circuit voltages, et cetera, are all the aggregated result of the static, dynamic, geometric transport and the radiative properties of the materials and their junctions.

An optically active organic semiconductor and its optoelectronic use requires the properties of good optical activity—this is tied to these energy mechanisms, good transport properties of the carriers and of excitons, as well as the ability to split and form excitons so that the non-radiative processes can be limited. By necessity, as we have emphasized, these structures must be thin, with traversal limited, as the optically active region is also limited in size.

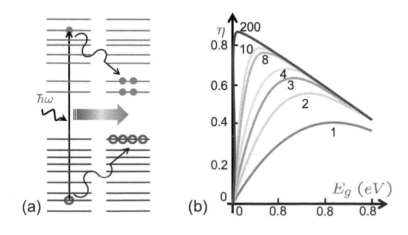

Figure 6.41: In nanocrystals of small bandgap materials with strong electron-hole binding, absorption of a photon can lead to an electron-hole pair in a higher quantized state—a hot state—leading to multiple electron-hole pair generation, as shown in (a). Varying energy photons lead to the generation of multiple such excitons that, if extracted, allow higher efficiency of conversion, as shown in (b) since excess energy goes into electrical energy instead of thermal energy.

Excitons, as in organic materials, provide an interesting avenue in inorganic semiconductors too. We consider one interesting nanoscale example of this. This is the multi-exciton solar cell, whose basic principal and its implication are summarized in Figure 6.41. Incident radiation generates electron-hole pairs in the higher energy states of the quantized nanocrystal, that is, quantum dots. The hot carriers relax, creating multiple excitons as shown in Figure 6.41(a). Instead of losing all the energy to the thermalizing process of phonons, the Auger emission process creates multiple charge carriers. These processes are observed in small bandgap materials, where Auger processes dominate. *PbSe* and *InAs*, with bandgaps of 0.1–0.4 *eV*, are examples of multiple exciton–generating materials where efficient multiple Auger pairs are created. If these carriers can be separated before recombining, as in organic semiconductors, then higher efficiency can be achieved, as shown by the calculated results for multiple exciton creation, as given in Figure 6.41.

See S. Tiwari, "Semiconductor physics: Principles, theory and nanoscale," Electroscience 3, Oxford University Press, ISBN 978-0-19-875986-7 (forthcoming) for a discussion of recombination and generation processes, including Auger processes.

The question of how much of the incident light energy is convertible to electrical energy is interesting: it is a function of the losses to non-electrical forms of energy, such as the heat of phonons, and the losses of the electrical form, such as to processes of electrical dissipation, as these abound in transport and at interfaces.

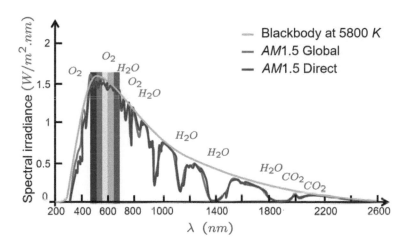

Figure 6.42: The solar intensity per unit wavelength for an airmass of 1.5 (AM1.5). Two plots, one with global averaging, and one a direct spectrum, are shown. Also shown for reference is the approximated radiation from a "blackbody sun" at a temperature of 5800 *K*. The various drops arise from absorption by specific molecular species in the atmosphere.

Sunlight has a broad spectrum of wavelengths; only some of the band of wavelengths/frequencies is collected by an optoelectronically active material, and different bandgap materials will show a spectral dependence, since only finite, usually single electron-hole pairs, are produced. The Shockley-Queisser limit describes this for single electron-hole pair generation in a constant bandgap semiconductor. The Shockley-Queisser limit for efficiency posits the highest efficiency possible if a single electron-hole pair is emitted in single bandgap material, under the assumption that the single electron-hole pair is produced by the radiation so long as $E \geq E_g$, none for $E < E_g$, that carriers thermalize and that these are collected under conditions of short circuit, that is, with no bias voltage across the junction. So, a mathematical treatment of this photoelectric conversion process needs to include the incident energy, the carrier generation, their separation and their extraction at the contact.

Blackbody radiation and solar illumination under conditions of an atmospheric 1.5 ($AM1.5$) mass——an approximation of the average mass that absorbs on Earth, which has different degrees of illumination depending on the season, latitude, longitude and time of day—is shown in Figure 6.42. The blackbody energy density shown in Figure 6.42 is an approximated fit where the surface of the sun as a blackbody is at 5800 K. The incident radiation on the earth's surface is subject to absorption in the atmosphere. This is largely due to absorption by various molecules: water vapor, carbon dioxide and oxygen being among the predominant ones, causing the spectral dips shown in Figure 6.42.

Photons at energies higher than the bandgap can create electron-hole pairs. Lucky photons at energy lower than bandgap can too, by extracting energy from the crystal—phonon assisted—as well as through donor-acceptor transitions, or exciton creation, et cetera. Because of their unlikelihood, these lucky photons can be effectively ignored, as is done in the Shockley-Queisser limit that we will discuss. So, we account for loss mechanisms of photon energy above the bandgap. This is the excess energy lost to lattice dynamics and any other coupled mechanisms that exist together with the single electron-hole pair postulated to be generated. All excess energy above bandgap is translated to dissipation. At thermal equilibrium, a detailed balance must hold. This balance at thermodynamic equilibrium is reflected in a balance between the spontaneous emission of photons by the radiative recombination of an electron and a hole, and their emission. The energy density of blackbody radiation, ignoring zero point energy, is given by

$$u(v, T) = \frac{8\pi v^3}{c^3} \frac{1}{\exp(hv/k_B T) - 1} = \frac{8\pi}{\lambda^3} \frac{1}{\exp(hc/\lambda k_B T) - 1}. \quad (6.93)$$

The thermodynamic reasoning behind the Shockley-Queisser limit of energy conversion efficiency, with its lowering of efficiency in conversion, can now be seen as follows. Let the photon gas consist of N photons, with a total energy of U_1 and an entropy of S_1. When a photon is absorbed, this photon gas now has $N - 1$ photons, an energy U_2 and entropy S_2. $U_2 < U_1$, and $S_2 < S_1$, with the loss of the photon. Free energy available for electrical work has to decrease. The change in the Helmholtz free energy is

$$\mathcal{F} = (U_1 - U_2) - T(S_1 - S_2) < (U_1 - U_2). \quad (6.94)$$

Not all energy is available for electrical conversion. This problem is further compounded by the electrochemical energetics of the junction transport itself. Maxima in current and voltage cannot occur simultaneously, so there are additional inherent losses in the device. The

The sun's core, where nuclear processes lead to the tremendous energy radiation that eventually comes to us as electromagnetic radiation, whose density peaks in the visible range, is estimated to be at nearly 10^7 K. Particles are generated and absorbed innumerable times as the energy flows from the center to the surface. Energy from the core is estimated to spend nearly 10^{11} s splitting and exchanging through particle processes before appearing as a photon at the surface on its way for an 8-minute journey to Earth.

Ozone absorption is at the short wavelength end. Ozone in the upper atmosphere is crucial to our protection on Earth. This is the part that has been depleted by man-made refrigerants and causes the ozone holes to appear over the poles. We create ozone in automobile exhaust, and this stays on Earth's surface. Ozone, being a strong oxidant—strong enough that it is used as the low-energy-damage burner of resists in semiconductor processes—is damaging in the upper atmosphere, and ozone gain is damaging in the lower. A double man-made whammy. But, then we still have these climate-effect deniers—facts not fitting the desires leads to quite amazing adult behavior. It is not limited to children.

total efficiency is the product of the probability of electron-hole pair production by a photon, the fraction of the bandgap that appears as an operational voltage, that is, the region between short circuit and open circuit and where the device operates when in use, and an impedance matching factor of this photoelectric generator.

Let the short-circuit current density be J_{sc}, and the incoming photon flux be ϕ_{inc}, with absorptance $A(E)$ being the fraction of light, at energy E, being disappeared. Per this set of assumptions,

$$A(E) = \begin{cases} 1 & \forall E \geq E_g \\ 0 & \forall E < E_g \end{cases} \tag{6.95}$$

represents the idealized absorption, causing the maximum short-circuit current density to be

$$J_{sc}^{max} = q \int_0^\infty A(E)\phi_{inc}(E)\, dE = q \int_{E_g}^\infty \phi_{inc}(E)\, dE. \tag{6.96}$$

What this states is that photons at energies higher than the band-edge energy progressively cause a decrease in efficiency, since all that energy goes into dissipation—an incoming photon creates an electron-hole pair, and excess energy is lost to an unaccounted spread of degrees of freedom of the crystal's atomic oscillations. The highest short-circuit current density occurs at the smallest bandgap, as it allows the maximum photoelectric conversion, as shown in Figure 6.43.

But, there also exists the possibility that the photons may reemit as radiation, and we must include this in our analysis, through the use of a detailed balance at thermal equilibrium. Thermal equilibrium specifies that all absorption and emission must balance in detail, for example, at all wavelengths. This translates to a balance between emissivity ϕ_{em} and absorptance (ϕ_{abs}) in *dark conditions* as

$$\phi_{abs} = A(E)\phi_{bb}(E, T), \tag{6.97}$$

where ϕ_{bb} is the photon flux incident from the blackbody. The photoelectric conversion device—the photoelectric cell—is our blackbody that emits and absorbs because it is at temperature T and is in thermal equilibrium with its environment, which too is at temperature T. The radiative recombination current (J_{rad}) can now be written under biased conditions, still in the dark but under a disturbance from thermal equilibrium by a bias voltage of V, as

$$\begin{aligned} J_{rad} &= q \int_0^\infty A(E)\phi_{bb}(E, T) \exp\left(\frac{qV}{k_B T}\right) dE \\ &= q \int_{E_g}^\infty \phi_{bb}(E, T) \exp\left(\frac{qV}{k_B T}\right) dE. \end{aligned} \tag{6.98}$$

This equation reflects the increase in the radiative recombination current arising from the increase in carrier population, an increase

In reference to "disappeared" as an obfuscating phrase, a comment is in order on the abuse of language by introducing ambiguity. Wars and really much of uncivil activity, have a way of creating words that attempt to hide the information behind them—a posterior-to-prior transformation to confuse—precisely the opposite of what information is about, as defined in our starting discussion of information in this book. "Being disappeared," a phrase that did eventually acquire informational content, caught hold during the mid-to-last decades of the 20th century, between what happened to people protesting in the deeply left-wing countries of the Soviet Union, to the deeply right-wing countries of Latin America. This euphemism, transformed to "extraordinary rendition" during the war initiated by Bush and Blair in the early 21st century and later joined by European countries, still seems accepted in a meaningless way at the time of writing of this text. *WWII* brought the "comfort women" in service of the Imperial Japanese Army. "Collateral damage" in the vernacular of today is the death of innocents: children and ordinary folks on the street. The art of reversing the meaning of a word is an art form for Wall Street: "hedge funds" increase the risk of limits on bets instead of hedging; "bailout," a word originating in bailing out water from a leaking boat, is now financial forgiving; securitization is turning anything into a risk proposition, rather than securing it; austerity is a morality play of cuts that hurt anybody but the proponent; and credit is the word used for debt. Information disappearing or reversing meaning through the communication practices of various societies is a high Orwellian tradition practiced through the ages. Information content remains a complex notion with a continuing unfolding of ambiguities. Beware of anybody who proclaims a "win-win." Thermodynamics doesn't care for it.

which is by a Boltzmann factor, due to any injection associated with the bias voltage that changed the quasi-Fermi energies of the system. Although we have disturbed the thermal equilibrium with the bias, the photoelectric conversion is still in the dark; Equation 6.98 simply states that the radiative recombination current also increases in the dark at the bias voltage V, since the carrier population has increased, even as all participating entities—electrons, holes, material—remain at a temperature of T. The emission increases with the short circuit, corresponding to all electrons and holes recombining radiatively.

Now, we illuminate this photoelectric cell. Under bias, with illumination, the total current is a superposition of the radiative combination current J_{rad} resulting from disturbing the thermal equilibrium and the maximum short-circuit current calculated earlier:

$$
\begin{aligned}
J(V) &= J_{rad} - J_{sc}^{max} \\
&= q\int_{E_g}^{\infty} \phi_{bb}(E,T)\exp\left(\frac{qV}{k_BT}\right)dE - q\int_{E_g}^{\infty}\phi_{inc}\,dE. \quad (6.99)
\end{aligned}
$$

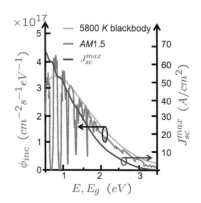

Figure 6.43: Photon flux (ϕ_{inc}) at 100 mW/cm^2—the approximate radiation intensity from the sun on the earth's surface—and the corresponding short-circuit current density for the varying bandgap energies of the photoelectric converter.

In Equation 6.99, on the right hand side, the first term is from the photoelectric cell and its environment at temperature T under the bias voltage V—J_{rad}. The second term is from the conversion of all the incident photons that generate electron-hole pairs leading to the current J_{sc}^{max}—a process that is only effective in creating one electron-hole pair from photons of energy E_g and above.

What is the total incident flux? It is due to the photoelectric cell's blackbody environment at temperature T, and the incident radiation from the sun ϕ_{sun} as a blackbody at temperature T_{sun}. So,

$$
\phi_{inc} = \phi_{bb}(E,T) + \phi_{sun}(E,T_{sun}), \quad (6.100)
$$

and one may write the total current density as

$$
J(V) = q\int_{E_g}^{\infty}\phi_{bb}(E,T)\left[\exp\left(\frac{qV}{k_BT}\right)-1\right]dE - q\int_{E_g}^{\infty}\phi_{sun}\,dE, \quad (6.101)
$$

which is valid for any bias V with the photoelectric cell and its environment at temperature T under illumination from the sun. The short-circuit current is at $V=0$, which we have derived as J_{sc}^{max}. We can also find the open-circuit voltage V_{oc} by forcing $J=0$.

$$
\begin{aligned}
V_{oc} &= \frac{k_BT}{q}\ln\left[\frac{\int_{E_g}^{\infty}\phi_{sun}(E,T)\,dE}{\int_{E_g}^{\infty}\phi_{bb}(E,T)\,dE}+1\right] \\
&= \frac{k_BT}{q}\ln\left(\frac{J_{sc}^{max}}{J_0}+1\right), \quad (6.102)
\end{aligned}
$$

where J_0 is the radiative recombination current associated with darkness and no bias. The photoelectric cell is in thermal equilibrium

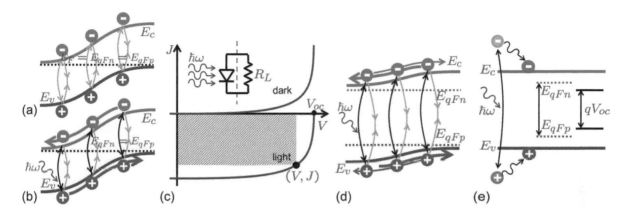

Figure 6.44: Statics and dynamics of a photoelectric cell. (a) shows dynamics in thermal equilibrium, that is, no bias and in the dark. (b) shows a short-circuit condition under illumination. (c) shows the current-voltage characteristics in dark and lighted conditions. (d) shows a more general (J, V) condition of operation in light, when photogeneration happens together with bias-modulated flow of carriers. (e) shows the open-circuit condition where net current vanishes as the different sources balance.

with its environment at temperature T, with the recombination and generation processes balancing, with each of magnitude J_0.

Figure 6.44 summarizes this discussion and its essence pictorially. In thermal equilibrium, the short-circuited photoelectric cell in dark, a p-n junction as shown in Figure 6.44(a), has a detailed balance for each process. This example shows the generation and recombination of this cell as a blackbody in thermal equilibrium with its surroundings. Figure 6.44(b) shows this short-circuited device with light. Electron-hole pairs are generated in excess if the photon has energy higher than the bandgap E_g, and this results in the current J_{sc} arising from the flow of electrons and holes. Because of the short circuit, the quasi-Fermi energies are aligned. Light energy is converted to electric energy, which flows out of the contacts. Figure 6.44(c) shows electrical characteristics. In the dark, the junction shows diode-like characteristics with current flowing into the p contact when the bias voltage, the voltage between the p and the n contacts, is positive. In the presence of light, the characteristics shift down the ordinate axis: current now flows out of the p contact towards the n contact. It is in opposite direction w.r.t. the diode forward current. This is because, at any bias point along the light curve (J, V), within the device, there are some carriers that are photogenerated, and some that are bias injected from the contacts. One can view the net current as the sum of these two components, as shown in Figure 6.44(d). The short-circuit current is being reduced by the bias injected current. In the absence of light, in the transition region of forward-biased diode, the diffusion flux of carriers exceeds the drift flux ever so slightly. The difference between the two is the the forward current. Presence of light generates carriers that are being swept by the field, and this generated current is in the opposite direction. At the open-circuit bias voltage, that is, a forced condition where no current flows, there is balance between the currents. Under these conditions, Figure 6.44(e) shows the relationship between the open-circuit voltage V_{oc} and the quasi-Fermi energies. The opening in the quasi-Fermi energy is larger than qV_{oc}. This is because, as Equation 6.102 indicates, photogenerated carriers are in excess of the thermal distribution. The quasi-Fermi level therefore must separate farther.

The Shockley-Queisser theory therefore tells us the idealized characteristics determined by thermodynamics which can be suitably applied over a range of bandgaps. Figure 6.45 sketches some of these predicted idealized characteristics.

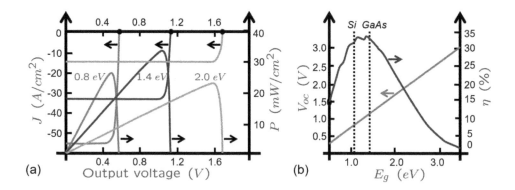

Figure 6.45: The current density and the output power per unit area for 5800 K blackbody illumination normalized to 100 mW/cm^2 as a function of photoelectric cell voltage is shown in (a). (b) plots the open-circuit voltage V_{oc} and the efficiency as a function of the bandgap energy for $AM1.5G$ illumination normalized to 100 mW/cm^2.

As shown in Figure 6.45(a), a small bandgap 0.8 eV material, for example, has a larger short-circuit current density, since photons over a broader energy range are collected. But, the small bandgap also means that the open-circuit voltage is small, since the contact injection with the forward bias is significant. Conversely, when the bandgap is large, 2.0 eV, for example, the short-circuit current density decreases since a smaller fraction of photons have energy in excess of the bandgap. But, the open-circuit voltage is significant. However, in both of these cases, the maximum power is smaller than that of a material with a bandgap of 1.4 eV—the $GaAs$ bandgap—where a suitable compromise exists between the current density and the operating voltage. Both the small bandgap and the large bandgap have a penalty in too small a voltage or too little charge carrier generation. This is reflected in the open-circuit voltage and the efficiency plotted in Figure 6.45(b). The maximum efficiency in the Shockley-Queisser limit, given by

$$\eta_{max} = -\frac{(JV)_{max}}{\int_{E_g}^{\infty} E\phi_{sun}(E)\,dE},$$

(6.103)

occurs over the *Si* to *GaAs* bandgap, that is, the range of 1.1 *eV* to about 1.5 *eV*, even as the open-circuit voltage possible continues to increase with increasing bandgap.

A schematic illustration of the underlying loss mechanisms is shown in Figure 6.46. At small bandgaps, an excessive amount of photon energy is lost to the crystal during relaxation to the band-edges. Small bandgap materials also have additional non-radiative recombination losses due to Auger processes. At large bandgaps, losses are due to the photons below the bandgap, as their energy is not captured by the photoelectric cell.

A small bandgap nanocrystal with multi-exciton processes is one example where this Shockley-Queisser limit does not apply, since multiple electrons and holes are ultimately created. Likewise, any structure where heat can be recycled and converted into electrical energy, thus recovering another fraction of the energy loss, also circumvents the Shockley-Queisser limit. Another example is a tandem cell device. These devices are photoelectric cells where multiple bandgap subcells are stacked in one assembly with tunneling contacts in between. The largest bandgap structure serves as the first absorber, passing on the lower bandgap photons to be absorbed by subsequent subcells in the scalar assembly. The challenge in such a structure is designing them so that a similar current can flow through all of them in series so that the output voltage is now the sum of the stack. Such structures, made by single-crystal growth of a combination of compound and elemental semiconductors, reach the highest efficiencies by capturing most of the photons across the energy spectrum of sun radiation and are employed when a large concentration of light can be achieved, to exploit the efficiency.

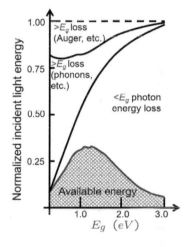

Figure 6.46: Schematic illustration of losses in a photoelectric cell across a range of bandgaps.

6.6   Lasing by quantum cascade

We tackle one nanoscale-specific example of light emission to complement the light absorption discussion: lasing in semiconductors using unipolar processes where quantum confinement of levels and transitions between them from the same band are employed— a complement of intersubband absorption employed in quantum well detectors. This requires intersubband gain. Lasing employs a single carrier type, using a superlattice structure that makes photon emission between two subband levels the rate-limiting step. Using a cascade of such multiple photon emissions via the transmission of each electron allows this structure to achieve sufficient gain for lasing in resonant geometries. Such lasers are quantum cascade lasers. Their uniqueness is in the use of unipolar transport and the emission based on it from subbands—single carrier energy levels and subbands—to make a complement of the solid-state four-level or three-level lasers of doped impurities in insulating crystals. These are semiconductor-based unipolar intersubband lasers.

Figure 6.47 show two common forms that lasers take. The first, in panel (a), is a solid-state laser employing atomic transitions such as in transition metal–doped crystals as a four-level system. Only electron transitions are involved, where pump light at higher energy is employed, causing a transition between two levels that are strongly coupled to two energy levels between which the photon emission occurs. This is unipolar. The $|1\rangle \rightarrow |4\rangle$ is the pumped transition. If the transition $|4\rangle \rightarrow |3\rangle$ is fast, that is, τ_{43}—a time constant—is small for dropping into the lasing transition level and so is the transition $|2\rangle \rightarrow |1\rangle$, that is, τ_{21} is small, then the electron can be continuously extracted for continuing photon generation and hence gain. Fast mechanisms for τ_{43} and τ_{21}, loss mechanisms to the crystal, are essential for achieving lasing. The second example here, in panel (b), shows how traditional semiconductor diode lasers operate. A radiative electron-hole recombination occurs between states of conduction and the valance band—an interband transition where the electron from the conduction band drops into an empty state in the valence band—a direct transition, where the hole is the quasiparticle of the collective excitation. So, in the bipolar semiconductor laser—interband and intersubband—it is the joint density of states that is important. For high efficiency interband quantum well lasers, again bipolar and based on intersubband transitions, the gain is related to the radiative lifetime ($\sim n$ss) and the two-dimensional joint density of states. For the solid-state laser, it is the atomic-like joint density of states with a very short lifetime ($\sim p$ss) that is important. What makes semiconductor diode lasers powerful is that they are electrically pumped and the gain-loss mechanics is such that quantum well–based structures allow very compact assemblies.

The cascade in quantum cascade lasers, achieved through the quantized energy levels in a confined heterostructure, so a unipolar intersubband transition, is the cascade of photons produced by each electron. The term quantum here refers to the centrality of quantization to the operation of the laser. Compared to solid-state lasers, however, there are many differences. Semiconductor-based lasers—

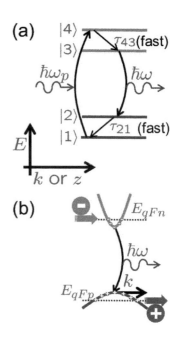

Figure 6.47: (a) shows the transitions of a four-level solid-state laser employed for light emission. (b) shows the same for a bipolar diode laser.

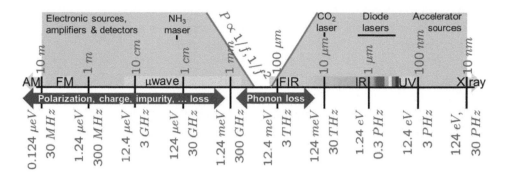

Figure 6.48: A view of the electromagnetic spectrum in common usage: the wavelengths, energies and frequencies. The figure illustrates where power sources suffice and the common loss mechanisms such as due to free electrons or phonons in semiconductors, ferroelectrics and other common materials. The THz energy of $\sim 10~meV$, which is around the natural frequency of crystals, makes the task of compact or even large sources difficult.

bipolar and unipolar—can be electrically pumped: electrons are injected and extracted, with photons being generated within this process. Solid-state lasers are pumped by light. The second important difference, and this is particularly of use, is that of electron energy levels in quantized semiconductor structures. In bipolar lasers, this is in the eV range: nitrides make many eV—blue and green—lasers possible, and antimonides, several 100s of meV—mid-infrared—lasers possible. In interband transitions, the energy is less than a few 100 meV. This makes terahertz and long and medium infrared lasers feasible. This is a frequency range for which power sources and detectors are relatively difficult to obtain. This is illustrated in Figure 6.48, across the electromagnetic spectrum that we commonly use.

At the low frequency and long wavelength end ranging up to at least a cm in wavelength or 30 GHz in frequency, electronics is quite efficient. Even ranging up to 300 GHz, useful generation, amplification and detection can be accomplished. Likewise if one reaches well further out to visible light, so eVs in energy, 400–1570 nm in wavelength, diode lasers exist. In the several μm wavelength range too, molecular lasers, for example, that of CO_2 exist. At very short wavelengths, charge particle accelerators of various types, although not usually compact, suffice. But, near a terahertz and the far infrared, the power capabilities dwindle. A frequency of 1 THz has a wavelength of 300 μm in vacuum, and its photon has an energy of 4.1 meV. For comparison, a mid-range microwave of 100 GHz is 0.3 cm wavelength light with a 0.41 meV photon. Very long infrared light of 100 μm wavelength has a 12.3 meV photon and 3 THz frequency. Lasing at wavelengths of $\sim 10~\mu m$—mid-infrared—are quite

As of the Julian calendar year 2016, only electron-based—not hole-based—quantum cascade lasers have been realized. Quantum cascade lasers, as we shall see, also suffer from numerous loss mechanisms. Between the small energies, their proximity to thermally activated loss mechanisms, and the efficiency of different necessary transitions, advancing room temperature operation and power are both non-trivial.

feasible through gas lasers such as of CO_2. But, getting below this wavelength and into the extreme infrared and into true terahertz is non-trivial. Any photon below ~ 100 meV is subject to a variety of losses through interaction within the crystal. The longitudinal optical (LO) phonon of $GaAs$, for example, has ~ 36 meV energy which is at 8.7 THz. These vibrations and mobile carrier interactions are the primary source of loss. Neither good sources nor, good amplifiers, nor selective detectors are common place in the terahertz to mid-infrared spectrum. Achieving quantization energy separation of 10s of meV is quite possible in many heterostructure systems. A bipolar laser has a long radiative lifetime—nss—and the losses depend on the temperature-dependent scattering processes. At low bandgaps, necessary as one proceeds deeper into the infrared, Auger recombination—a reverse impact ionization process where the energy of a recombination event, instead of being radiatively emitted, is transferred to another carrier—becomes significant. So, antimonide lasers, for example, or those of indium-containing arsenides have poor threshold dependence on temperature—a low characteristic temperature T_0. Quantum cascade lasers' gain depends on transition times that are in pss. They are relatively temperature immune by making the two-level transition as the dominant mechanism that is temperature independent. Temperature dependence appears through the non-radiative scattering that competes, particularly the optical phonon–based scattering processes. These become seriously consequential at the really low end of terahertz energies.

The fast processes for injection and extraction between the lasing energies of quantum cascade lasers depend on the ability to couple to these energies efficiently. Optical phonon processes—the major energy-loss mechanism in crystals—and the transition processes of the injection and extraction of electrons determines this. We will discuss both of these considerations.

Superlattices are the heterostructure assemblies, as an extension of the quantum well, that make the injection and extraction characteristics possible. Superlattices, as a coupled extension of heterostructure quantum wells, derive their interesting characteristics through the overlap of wavefunctions. When short leaky quantum wells, that is, a quasi-bound structure made using a heterostructure with a small discontinuity in the conduction or valence band or both, are assembled together, superlattice behavior arises in the longitudinal direction where the Bloch states now also have the periodicity of the superlattice constructed using multiple quantum wells. The quasi-bond states now become extended states due to small discontinuity, small widths and consequent leakiness that leads to wavefunction overlap. This makes conduction possible across the periodic discontinuities. A

For an in-depth discussion of the properties of superlattices that make injection and extraction possible, see S. Tiwari, "Semiconductor physics: Principles, theory and nanoscale," Electroscience 3, Oxford University Press, ISBN 978-0-19-875986-7 (forthcoming). We will assume this knowledge in our discussion here and provide only a brief, conceptual introduction.

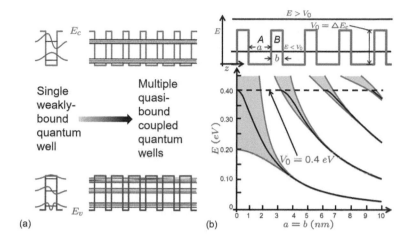

(a)

(b)

Figure 6.49: (a) shows a weakly confined quantum well formed from a heterostructure on the left with two bound states for electrons from the conduction band, and three for holes from the valence band. When this structure is assembled with periodic, narrow barriers in energy, as in the heterostructure superlattice on the right, extended states are formed because of the weak confinement and breakdown of degeneracy. Two electron states at a single energy are now twelve electron states in six energetically close energies—a miniband. This is a superlattice, where the electron, or the hole, can extend over the entire structure in the miniband. (b) shows a Kronig-Penney periodic potential with $V_0 = \Delta F_c$, a period of $a + b$ as the sum of well and barrier widths and the energy of allowed solutions for a barrier with $V_0 = 0.4\ eV$, with the free electron mass as $a = b$ is are varied. At large barriers and well widths, only discrete levels—4 of them—exist. As the size is reduced, minibands form, allowing a broadband of propagating states in energy. The single well of width a and $b \to \infty$ discrete level solution up to $E = V_0 = 0.4\ eV$ is shown, together with the superlattice solution.

schematic of this in the longitudinal transport direction (z or $\perp$ in our notation) is shown in Figure 6.49(a). Bloch states in the transverse direction are still "free electron"–like but, in the longitudinal direction, form minibands.

The example shown in Figure 6.49(a) illustrates the following physical behavior. When the discontinuities are small, the smaller confinement potential results in deeper penetration of the wavefunction of confined particles in classically disallowed regions. On the left, two electron wavefunctions and energies, each allowing $\pm 1/2$ spin, are shown. The holes have been chosen to have three allowed energies, out of deference to the larger hole mass. Both the electron and the hole wavefunction penetrate in the larger bandgap material over a size scale of the order of the quantum well width, as shown. On the right, six of these wells are brought together with a small barrier region. The degeneracy of the six separate wells is removed due to the interactions, and the new states also represent the periodicity of the structure. There are six energy levels, instead of one, quite close to each other and extending out across the periodic structure. A miniband of six energy levels capable of holding twelve carriers has been formed. The carriers can travel across this periodic structure, and, depending on the boundary conditions on either side, out. This is a superlattice with an effective new band structure consisting of these minibands—a band structure—with their own unique properties.

The Kronig-Penney model used as a toy example in introductory texts is an example of a superlattice model. Since it illustrates the importance and the relationship between energy of the barrier, width of the barrier and the quantum well, and their comparable magni-

The superlattice described here as a layered periodic structure through which electrons and holes can travel is an artificial construct. But, one could just as well view compound semiconductors as layered periodic structures. *Ga* planes and *As* planes stacked in specific forms form the *GaAs* crystal. It is a superlattice of sorts. The distinction is only in the energies that bind. And *GaAs* conducts!

tude to that of Bohr radius in the periodic arrangement, we employ it here. Figure 6.49(b) shows a spatially periodic structure with a barrier $V_0 = \Delta E_c$, where the wells are of width a for region A, and the barriers of width b for region B. The structure has a periodicity of $a + b$, so a Brillouin zone width of $2\pi/(a + b)$. We assume that both the well and the barrier are isotropic with identical mass, to simplify this calculation. The Schrödinger equation in the two regions with plane wave propagating modes in the xy plane is

$$-\frac{\hbar^2}{2m_A^*}\left(k_x^2 + k_y^2 + \frac{d^2}{dz^2}\right)\psi_A(z) = E\psi_A(z), \text{ for } z \in A \text{ and}$$

$$-\frac{\hbar^2}{2m_B^*}\left(k_x^2 + k_y^2 + \frac{d^2}{dz^2} + V_0\right)\psi_B(z) = E\psi_B(z) \text{ for } z \in B. \quad (6.104)$$

We are looking for the propagation properties in the longitudinal, that is, perpendicular z direction. The boundary condition is the continuity of the wavefunction and of $\partial\psi_A/m_A^*\partial z = \partial\psi_B/m_B^*\partial z$ at the A/B boundary, where we will take $m_A^* \approx m_B^*$ to simplify. These two boundary conditions are that of continuity of probability and of energy at the interface. Referencing the energy to the bottom of the conduction band in the smaller band material, $E = \hbar^2 k_1^2/2m_A^*$, and $E - V_0 = \hbar^2 k_2^2/2m_A^*$ for $E > V_0$, and $V_0 - E = \hbar^2\kappa^2/2m_A^*$ for $E < V_0$, when $k = i\kappa$ is imaginary. The traditional solution techniques for this lead to the following implicit equation using the boundary conditions

$$\cos kd = \cos(k_1 a)\cos(k_2 b) - \frac{k_1^2 + k_2^2}{2k_1 k_2}\sin(k_1 a)\sin(k_2 b)$$

$$\text{for } E > V_0, \text{ and}$$

$$\cos kd = \cos(k_1 a)\cosh(\kappa b) - \frac{k_1^2 - \kappa^2}{2k_1\kappa}\sin(k_1 a)\sinh(\kappa b)$$

$$\text{for } E < V_0. \quad (6.105)$$

Here, k_1 is the wavevector in region A, and k_2 and κ are the wavevector and extinction coefficients, respectively, in region B for energies that are higher or lower than the bandedge potential. What these equations imply is that energies lower than V_0 may have a transmitting solution when to appropriate potential and width conditions exist that allow sufficient coupling between the wells. It is when this happens that minibands form and the degeneracy of coupled wells is removed.

Figure 6.49(b) shows the result of this calculation for $V_0 = 0.4\ eV$, where $m^* = m_0$, and a symmetric structure where the well and barrier regions have identical width that is varied. The electron mass is assumed to be the free electron mass. The discrete levels, 4 of them in this example, exist for the isolated well of width a. When the periodic structure is formed, the minibands—a broadband of allowed transmissive energy states—appears. These appear at higher

Again, we refer to S. Tiwari, "Quantum, statistical and information mechanics: A unified introduction," Electroscience 1, Oxford University Press, ISBN 978-0-19-875985-0 (forthcoming) and S. Tiwari, "Semiconductor physics: Principles, theory and nanoscale," Electroscience 3, Oxford University Press, ISBN 978-0-19-875986-7 (forthcoming), where these techniques are exploited at length.

width in the highest quasi-bound states first. Compound semiconductor heterostructure systems such as $(Al, In)As/(Ga, In)As$, $(Al, Ga)As/GaAs$, et cetera, all have discontinuities in this 0.4 eV range. What this figure shows is that, with wells and barriers of the order of a few nms, minibands form, and transmission occurs through these miniband states. This is a resonant and elastic transport exemplifying coherent tunneling. It is no different than what one observes in periodic film gratings for light. Optical transmission and reflection bands form. The Kronig-Penney model is a toy model, quite simplified, but it is instructive. The conclusions drawn are quite useful with electrons. For holes, things are quite a bit more complicated because of anisotropy and the various idiosyncratic hole bands. The Kronig-Penney model, for example, assumes carrier interaction extending over several unit cells so that the semiconductor picture of an effective mass, a discontinuity, et cetera, are all applicable. At the smallest well and barrier width sizes, this is inappropriate. In more complex and real situations, in the presence of this periodicity, we must resort to various bandstructure calculation techniques in the presence of this periodicity. These techniques adopt a supercell approach for superlattices, and we will not dwell on this here.

In any case, it is the ability to provide these transmissive properties and the selective changing of dimensions of wells and barriers for increased confinement and transition selection that is essential to the operation of a quantum cascade laser. It makes the photon-emitting charge carrier transmission possible. $(Ga, In)As/(Al, In)As$, particularly, $(Al, Ga)As/GaAs$, antimonides, other compounds and even nitrides have all been employed for quantum cascade lasers.

To understand the operation of a quantum cascade laser, first, consider the possible approaches to the use of heterostructure quantum confinement for radiative emission as shown in Figure 6.50. Panel (a) here shows interband transition through two subbands. If a is the superlattice periodicity of this assumed isotropic idealized semiconductor assembly, the Brillouin zone has a width of $2\pi/a$ in the reciprocal space. Since carriers occupy lower energies preferentially, the radiative recombination through these subbands is between states as shown and the "inversion" for lasing occurs via extraction from the bottom subband and injection into the top. The subband is composed of states where the in-plane direction has propagating states, as shown through the lateral parabolic energy distribution. This form, however, has not led to the high gain necessary to overcome loss so that efficient lasing can be achieved. Such gains and losses are a function of the properties of the materials, of confinement, of the modes, of the electrical dissipative losses and of the radiative and non-radiative gain and loss processes that come with them. Fig-

Bandstructure calculation approaches are discussed in S. Tiwari, "Semiconductor physics: Principles, theory and nanoscale," Electroscience 3, Oxford University Press, ISBN 978-0-19-875986-7 (forthcoming) as are the techniques of supercell-based modeling.

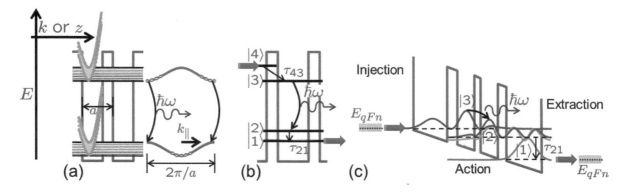

Figure 6.50: Examples of some possibilities for achieving radiative recombination in quantized structures. (a) shows the use of minibands of a superlattice for radiative recombination in a unipolar form. (b) is the quantized semiconductor analog of a 4-level solid-state laser illustrating radiative recombination where $|4\rangle \rightarrow |3\rangle$, and $|2\rangle \rightarrow |1\rangle$ are fast non-radiative transitions for injection and emission. (c) shows a realistic example of unipolar emission using efficient injection and scattering by phonon emission and extraction for providing the fast secondary processes. The three levels essential to the operation are shown, together with the probability density $|\psi|^2$.

ure 6.50(b) shows a heterostructure-based quantum active region as an intersubband analog of the 4-level system formed using minibands in a superlattice. The radiative transition $|3\rangle \rightarrow |2\rangle$ is made efficient by suitable radiative gain properties and by providing an efficient extraction of electrons from $|2\rangle$. A useful method for this is to design the structure in a way that $|2\rangle \rightarrow |1\rangle$ transition occurs efficiently by scattering such as via optical phonon emission. So, the subband separation is an optical phonon energy. Optical phonon emission can be very efficient, and this allows a rapid scattering out of electron from $|2\rangle$. But $|1\rangle$ needs to be depleted of electrons, and this requires an efficient extraction. It is in both this extraction and the injection that subbands are very useful. A major difference between panels (a) and (b) is that, in panel (a), the subbands—quantum confined therefore—have to be simultaneously amenable to population inversion, active gain and injection and extraction. This is difficult. Panel (b), however, is more realistic. The simultaneous optimization of several characteristics has now been separated. This is shown in Figure 6.50(c) under biased conditions for the radiative active region—a common form for quantum cascade lasers: efficient injection is through the design of subbands and the injection region. Depopulating of $|2\rangle$ to $|1\rangle$ requires efficient scattering, and $|1\rangle$ is emptied efficiently by another efficient coupling to the subbands. Subbands have higher density of states and thus are good reservoirs for injection and extraction.

We now assemble these preliminary thoughts into a design for a quantum cascade laser structure with potential for lasing, and into a discussion of the difficulties in achieving room temperature operation near terahertz and in efficiency.

Most quantum cascade lasers, especially as one reaches into the very low single-digit terahertz region, are low temperature lasers. It is in the farther-away infrared range that the losses and gains have let room temperature operation be possible.

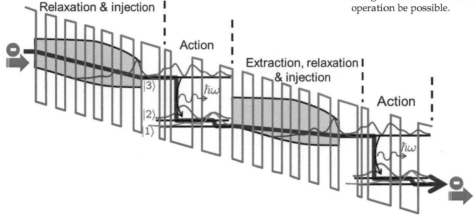

Figure 6.51: A quantum cascade laser illustrating the salient operational physical characteristics and mechanisms. The electron probability densities $|\psi|^2$ are shown for the three active levels of the action region and process. Here, longitudinal optical phonon coupling is employed to reduce τ_{21} by scattering for rapid depletion of $|2\rangle$ to $|1\rangle$. Electrons from $|1\rangle$ are extracted by employing the subband. Also shown is the path that the electron takes in this transmissive emission.

Figure 6.51 is a schematic of an arrangement of superlattices and quantum wells. A superlattice, formed with heterostructures, some of whose regions can be selectively doped, provides a miniband—the large collection of states that can be useful for injection into the upper level of a radiatively active action region. In Figure 6.51, under the self-consistent field and current in use, this miniband is aligned with a low injection resistance—high transmittance—to the upper state $|3\rangle$. The device contact for electron injection too needs to be low resistance.

The action region, with radiative transition $|3\rangle \rightarrow |2\rangle$ and fast scattering of $|2\rangle \rightarrow |1\rangle$, here is designed in such a way that $E_{21} = E_2 - E_1 \approx \hbar\omega_q$—optical phonon energy—and $E_{32} = E_3 - E_2 = \hbar\omega$—the desired lasing frequency. τ_{21} can be made fast in the absence of a phonon bottleneck. If electrons from $|1\rangle$ can be efficiently extracted, again using a coupled miniband, one may create conditions that are conducive to radiative gain. With optical emission efficient, and no bottleneck, making the $E_{32} = E_3 - E_2 = \hbar\omega$ an efficient process becomes possible with a large radiative transition matrix element. So, $\tau_{32} \ll \tau_{43}, \tau_{21}$, where τ_{32} characterizes the radiative process, and τ_{43} and τ_{21} are non-radiative time constants corresponding to getting electrons into the higher lasing energy state and out of the

The generated optical phonons too need to propagate out. If these accumulate, phonon emission may get suppressed because of the lack of states to scatter to. This is known to happen in small bandgap materials, such as $PbSe$, when they are multi-dimensionally confined, in the multi-exciton processes that we discussed for photovoltaics. This has been called phonon bottleneck.

lower lasing energy state. We will discuss these presently. During
the transit through the superlattice miniband, the extracted electron
relaxes excess energy and may now again be employed for injec-
tion to the state $|3\rangle$ of the next action region. Note that through this
cascading, state $|1\rangle$ has been connected back to state $|3\rangle$ of the next
action region. This process may be repeated to form a photon cascade
through unipolar intersubband transitions. Figure 6.51 shows a pictorial representation of the path that the electron takes through the
transmission during which the radiative emissions occur.

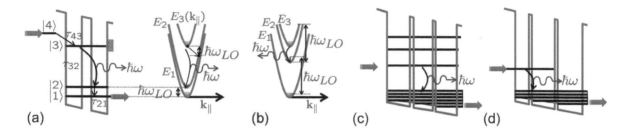

Figure 6.52: (a) shows a visualization of the active region of a quantum cascade laser where resonant phonon emission
is employed to scatter electrons from state $|2\rangle$ to state $|1\rangle$, from which they can efficiently extracted. The right side of
(a) is for $E_{32} > \hbar\omega_{LO}$—lasers of mid-infrared and shorter wavelengths. (b) shows when $E_{32} < \hbar\omega_{LO}$—lasers at terahertz frequencies. (c) shows an example where the adjacent regions are chirped superlattices leading to bound higher
energy states and a miniband of lower energy states. Extraction is possible from the lower energy states. (d) shows a
bound-to-continuum transition, again employing a design of the cladding superlattice region for efficient extraction.

To understand the basics and constraints of the quantum cascade
laser, consider the expansion of Figure 6.50(b) and (c) in the 4-level
intersubband semiconductor analog as redrawn in Figure 6.52. Fig-
ure 6.50(c) uses two wells, not one, so that the energy subbands $|3\rangle$,
$|2\rangle$ and $|1\rangle$ may be formed with controllable properties for the tran-
sitions $|3\rangle \rightarrow |2\rangle$, and $|2\rangle \rightarrow |1\rangle$, to come about. The quantum well
drawn on the left in Figure 6.52(a) has three energy subbands that
extend across the active region. This quantum well is clad with superlattices and injection and extraction interface regions and is a
repeated assembly. Within this, $|4\rangle$ is a state shrinking from a mini-
band transitioning from a state $|1\rangle$. So, $|1\rangle$ and $|4\rangle$ couple through
the minibands. The interface for the $|4\rangle \rightarrow |3\rangle$ transition must be
designed for efficient tunneling. This is equivalent to a fast non-
radiative transition time $\tau_{43} = 1/\Gamma$, where Γ is the leakage rate of
tunneling.

The periodic structure of Figure 6.50(c) has to be such that no transition occurs through the well to region beyond. So, the states beyond
must be forbidden. Also, since the energy barriers are small and
temperatures finite, injection into the propagating states at higher

energy—the thermionic and field emission—must also be kept very limited. The time constant τ_{43} of course will depend on the other processes in this sequence of events that are interconnected. For example, if $|3\rangle$ is filled with electrons because the electrons do not transition to $|2\rangle$ or elsewhere, then the time constant τ_{43} too will become long since less states are available in $|3\rangle$ to transition to. τ_{21} is made significantly smaller by resonating it with the optical phonon energy. The radiative gain here is proportional to the sheet electron density that can transition, so $g \propto \Delta n_s = J/e(\tau_3 - \tau_2)$, where τ_3 and τ_2 are the net time constants of these levels, inclusive of all the processes by which electrons flow through the level. The upper level time constant τ_3 is a function of temperature—thermionic leakage effects as well tunneling are temperature dependent through the occupation probabilities. τ_2 is constrained by an $\sim 0.4\ ps$ time constant for scattering-dominated τ_{21}. As one comes close to the phonon energy in lasing, one has to tackle the consequences of thermal backfilling from $|1\rangle$ to $|2\rangle$, and $|2\rangle$ to $|3\rangle$, since $\hbar\omega_{LO} \approx k_B T \approx E_2 - E_1 \approx E_3 - E_2$.

Figure 6.52(b) shows a contrasting example when the LO phonon energy is now significantly larger than the intersubband energy. When this happens, because $k_B T \ll \hbar\omega_q, \hbar\omega_q - \hbar\omega$, the backfilling is reduced due to the non-radiative scattering of $|2\rangle \rightarrow |3\rangle$, as that of $|1\rangle \rightarrow |2\rangle$ is reduced due to the occupational argument. The LO phonon in GaN is 92 meV, and one can conceive of structures where this picture is valid. Figure 6.52(c) and (d) show other possibilities, but these are not usually as effective as the active region subband designs. Panel (c) shows an example where a chirped superlattice, that is, one with progressively changing widths, is employed to have a miniband for lower energy and bound states for the higher energy. Panel (d) shows a bound-to-continuum transition where the active region is designed for radiative activity in one region of the quantum wells, while the cladding superlattice region allows extraction through the optimization of an additional quantum well.

This scattering involving optical phonons is strong in compound semiconductors, since it includes the electrostatic field arising in the ionization charge. The scattering rate and time constant satisfying energy and momentum conservation are

$$
S(\mathbf{k} \rightarrow \mathbf{k}') = \frac{1}{\tau_{21}} = \frac{\pi}{\hbar} \frac{e^2 \hbar \omega_{LO}}{V|\mathbf{k} - \mathbf{k}'|^2} (n_q + 1) \frac{1}{\epsilon_0} \left(\frac{1}{\epsilon_{r\infty}} - \frac{1}{\epsilon_{r0}} \right)
$$
$$
\times W(\mathbf{k}, \mathbf{k}') \delta \left[E(\mathbf{k}') - E(\mathbf{k}) + \hbar\omega_q \right] \quad (6.106)
$$

for bulk material. Here, $W(\mathbf{k}, \mathbf{k}')$ is the scattering matrix element. These are modified in heterostructure quantum wells, but we will not dwell on it, since this is sufficient to illustrate the order of magnitude.

$\hbar\omega_{LO} \approx 36\ meV$ for $GaAs$, corresponding to 8.5 THz. This means that the description with the large intersubband separation and the phonon energy shown in figures up to this point is really valid only for lasers around and above 10 THz—20 μm and below mid-infrared wavelengths.

If $\hbar\omega_{LO} \gg k_B T$, then $\tau_{32}^{-1} \approx \tau_{LO}^{-1} \exp[(\hbar\omega - \hbar\omega_{LO})/k_B T]$. However, even though GaN is favorable from this phonon energy and backfilling perspective, it also has a very large polarization. This makes bandstructure manipulation, as well as interaction at interfaces, effects that need to be tackled.

As discussed in S. Tiwari, "Semiconductor physics: Principles, theory and nanoscale," Electroscience 3, Oxford University Press, ISBN 978-0-19-875986-7 (forthcoming), this is the Fröhlich interaction. The Fröhlich interaction has a stronger effect than the deformation potential. It is also stronger than for acoustic phonons that give rise to the piezoelectric interaction.

$\tau_{21} \approx 0.4\ ps$—a fraction of a ps. In the transition $|3\rangle \rightarrow |2\rangle$, with a large energy change E_{32}, an optical phonon also involves a large momentum transfer of $\mathbf{q}$. $\mathbf{k}$ and $\mathbf{k}'$ are significantly far apart and hence the scattering rate scales as $1/q^2$. So, the optical phonon–constrained time constant is large. Designed properly, the radiative time constant prevails, even as it is significantly larger than τ_{43} and τ_{21}. Figure 6.51 shows the following optimizations: (a) a miniband design that allows the coupling of $|1\rangle$ with $|4\rangle$; (b) a high injection efficiency coupling $|4\rangle$ with $|3\rangle$, keeping τ_{43} fast and minimizing thermionic emission above the barrier; (c) radiative recombination-dominated τ_{32} relaxation that is made fast by high wavefunction overlaps; and (d) optical phonon–assisted fast τ_{21}. A structure with these characteristics, in the presence of electric field and hence electron flow, creates a cascade of the photons generated by the electrons in the action region. Radiative τ_{32} is several ps. Electrons in the active region can also thermally escape, a parasitic process τ_{para}, which is usually near but $< 10\ ps$.

The closeness of the intersubband energy and the demanding requirements for inversion gain and minimizing losses at these energies makes achieving good operating characteristics difficult, particularly as one reaches closer to the THz frequencies where thermal filling effects and competing loss mechanisms proliferate. In a $GaAs$ $36\ meV$ phonon system, the thermal effects have a three fold manifestation: (a) electrons from the conduction state $|1\rangle$ backfill state $|2\rangle$, (b) the electron injection efficiency to state $|4\rangle$ from the previous cascade step decreases and (c) the lifetime of the non-radiative recombination of thermally excited electrons decreases to compete with the radiative lifetime. At low temperatures, where THz is achievable, free electron losses still exist, and this limits the fraction of photons that do not recombine before exiting the laser.

6.7   Summary

THE INTERACTIONS OF PHOTONS IN MATTER are of significant utilitarian consequences and also have important basic foundations. Solar cells, the converters of sunlight to electricity useful in everyday life, and lasers, important to communications and elsewhere, are the utilitarian result of this interaction. The photon, as a unique bosonic particle with interaction characteristics that are very different from those of the electron, provides the ability to probe and control at the limits—this leads to a variety of the fundamental themes addressed in this chapter.

This chapter dwelt on the variety of these effects at the nanoscale basis, starting with those of fundamental consequence and ending

For low energy terahertz lasers, electron-electron scattering, given the large population existent for inversion, can also be significant. Electron-electron scattering would become important for closely spaced subbands. It is also important for miniband transport, where it enhances scattering between miniband states.

As of 2016, room temperature operation exists at continuous W-level emission around the 3–5 μm wavelength and decreases away from it. In the 1–5 THz range, maximum operational temperatures remain sub-room temperature.

with those of use. The Casimir-Polder effect is an example of one of
the many ways that vacuum fluctuations unfold in our world. This
electromagnetic effect was explored here from a few different perspectives. Parallel plates and other geometries allowed the writing of
energetics of modes, and from this we derived both attractive and repulsive effects of significant measurable magnitude at the nanoscale.
This makes it an interesting tool for ultrasensitive measurements.
We even made the connection between this and the van der Waals
interaction.

Photons also allow ultrasensitive measurement. These employ
optomechanic—electromagnetic-mechanic—interactions. Resonance
provides a means of coupling energy in and out of systems. When
we play on a swing, it is the periodic interaction with our physical
action, which is not rotationally symmetric, that couples the energy
that allows the swinging amplitude and the energy in it to rise. We
do the same through a reverse throttling physical action to remove
energy and stop. This interaction of removing energy in a nanoscale
system can be achieved through the variety of means—electrical,
magnetic, optical and mechanical—we have at our disposal. This
is particularly of use in reaching the fundamental limits of nature.
Fluctuations, or noise in the form we generally deal with it, can be
reduced through such couplings. This is different from the example
of reduction of modes in our single electron, nanoscale and ballistic
point contact discussion. The oscillations of a crystal—phonons—can
be coupled to the oscillations of electromagnetics—photons—whose
frequencies are determined by the electromagnetic cavity. This means
that the thermal modes can be removed if one is sufficiently close to
a low number of such modes by lowering the temperature and the
mechanical energy of the system. This employs momentum exchange
via the radiation pressure. There are two different aspects of this behavior. One is a dynamic back-action. A photon has an action and a
reaction with a mechanical element. A mechanical mirror and radiation pressure exhibit a dynamic back-action. This is very useful in
precision measurement of displacements. It is a condition in which
there is a large-enough density of photons. There is also a low photon number limit of this behavior—a quantum back-action, where
fluctuations due to the discreteness of the interaction event itself become important. These interactions have provided a means to make
measurements in fundamental limits: of displacement in ams and of
forces in fNs of single phonon and single photon events, and it has
allowed other measurements exploiting these sensitivities.

A particularly important use of these optomechanic interactions
has been the use of nanoscale structures, for example, dielectric periodic phonon and photon waveguides, where allowed modes can be

constrained and specific interactions enhanced. Phonons are typically at GHz, so this light and mechanical wave interaction then has the potential to provide modulation and in-wave processing at frequencies that are of interest to us. Efficient optical transmission in optical fibers is at the 1.57 μm wavelength, which is ~ 200 THz. GHzs is where much of real world signals are usually measured and manipulated. This phonon-photon direct manipulation connection makes this nanoscale optomechanic possibility simultaneously utilitarian and intellectually appealing.

We also looked at particle-electromagnetic interactions. For single molecules to 10s of nm-sized particles, this interaction helps the creation of optical traps that are used for measurement of forces, and other subtle measurements. Optical beads on a DNA strand allow one to perform biological physical studies of transduction. Simple and powerfully useful examples also include the confocal microscope, with which one may work beyond diffraction limits and perform cross-sectional measurements or a near-field measurement at an interface. Atomic clocks, with precision timing, are another example of the use that is gain dependent on nanoscale features. We also explored how the traditional large-scale apparatus, such as particle accelerators and vacuum microwave beam instruments, can be made compact through electromagnetic-particle interactions.

This charge-electromagnetic interaction also led us to the exploration of plasmonics. We looked at this through the dipole polarization interaction of a free electron as well as through an assembly of free electron charge, such as in a Drude picture. This interaction in semiconductors and metals gives rise to coupled charge-photon modes that we called plasmons. Plasmons can be static, for example, plasmons of a particle that are manifested through the enhanced response of a particle that makes them colorful and observable in longer wavelength light. Plasmons can also be dynamic, for example, in the surface plasmon polariton propagation along metal-dielectric interfaces. Both of these allow wavelengths that are smaller than those of the source photons to be active and allow measurements as well as enhanced energy exchange possible.

We also explored this energy exchange interaction in organic and inorganic structures with an aim to understanding their optoelectronic use. Organic semiconducting structures, with hole transport, acceptor transport, excitons—quasi-bound electron-hole pairs—transport and interface interactions provide a means for light emission and light absorption. This optoelectronic conversion in light emitting diodes, where emission is localized at an interface and exciton diffusion is important, and in solar cells, such as Gräzel cell, where a light-sensitive dye and high workfunction material allow separation of charge were two examples for non-traditional, non-inorganic materials. Energy exchange in molecular processes, such as in organic materials, is either by electron exchange—Dexter, which is a short-range process—or by dipole interaction—Coulombic or Förster, which are long range. Since the hole and electron transport properties are very poor, these organic structures, with their transverse transport, are by necessity thin. They also, at least for the emission structures, attempt to utilize the superior exciton diffusion characteristics of many organic semiconductors.

Excitons also appear in an unusual use in inorganic semiconductors. Multi-dimensional confinement enhances exciton formation, since it forces electrons and holes together in confined regions whose size is of the order of the Bohr radius. Small nanoscale-sized particles of small bandgap semiconductors, $PbSe$, for example, can have more than one electron-hole pair generated. A large-energy photon creating an electron and a hole far up in energy in the quantum-confined structure can generate multiple electrons and multiple holes via a phonon-assisted process. These structures have multiple excitons generated from a single photon, and more of the energy of the photon is converted into the electrical energy of the charge particles.

The other limit of this photoelectric conversion is when a single electron-hole pair is generated by a photon. The spectrum of the sun, an approximately $T = 5800\ K$ blackbody, has radiation that is both higher in energy than the bandgap of a semiconductor and lower. If only one electron-hole pair is generated, then one can determine the efficiency using thermodynamic arguments. This takes into account both the absorption by the solar cell material but also reemission as a blackbody source itself. This limit is the Shockley-Queisser limit, whose underlying foundations we explored to determine the optimum bandgap where single electron-hole generation provides the largest efficiency. This bandgap is ~ 1.00–$1.45\ eV$. As an aside, here we also discussed the possibility of working around this limit as is accomplished in tandem and multi-exciton cells.

As an example of stimulated emission, our discussion of lasing by quantum cascade brought together concepts related to intersubband emission in unipolar semiconductor lasers. Minibands formed in superlattices are useful for obtaining low impedance injection and extraction in such structures. The coupling from subbands to the minibands design is through a quantum barrier and well interface that must provide the efficient coherent injection under the conditions of bias and current flow in the structure. The same needs to be also true for the extraction. The optically active subband levels are designed in a quantum well geometry—the active region—where the wavefunction overlap is large. Fast injection to the higher energy subband, together with fast extraction from the lower energy subband, is necessary, with the radiative lifetime of the optically active transition dominating this chain. The lower optical subband level's lifetime shortening is achieved by increasing scattering out of it. Optical phonon emission to another subband level that can then be coupled to the extraction miniband is an effective method for this lifetime shortening objective. Quantum cascade lasers are exemplary in bringing together a number of quantum, photon-matter and periodic nanoscale features together for obtaining lasing at long wavelength reaching from infrared to terahertz.

6.8   Concluding remarks and bibliographic notes

MAXWELL'S UNIFICATION OF ELECTRICAL AND MAGNETIC ENERGY FORMS in his equations dates back to 1862. Planck's introduction of the quanta of radiation and Einstein's incorporation and elaboration of it as a light quantum—our photon—was an essential step in the birth of quantum mechanics. de Broglie connected the wave ideas to matter. Condensed matter theory's development followed from these early events and continues to this day, when the bringing together of light and matter in understanding limits and uses through light remains an important subject. This chapter discussed some of the key modern and relevant ideas from this. So, the history is hundreds of years old, with a long and deep developmental path, and innumerable excellent articles and books could be cited that form the basis of undergraduate and graduate teaching and research.

The Casimir-Polder effect has fascinated scientists from very early days of its first discussion by Casimir. Casimir was interested in it as an industrial problem in trying to understand why certain particles coagulate while others do not, depending on the medium they are in. Casimir, who, at an early stage of his life, spent time with Bohr at his institute in Copenhagen, discusses Bohr's thoughts on this subject. Fortunately, this subject is now much clearer, and we can look at it from both a rigorous mathematical and physically intuitive viewpoint. Serneilius's book[1] gives a comprehensive treatment. Bordag's edited compendium[2] is an excellent source for discussion of its many facets. Lamoreau's review[3] is a very readable treatment at a graduate level.

Mechanics nearly goes back to the birth of modern science in the middle of the millennium. Observational instruments of the early astronomers were optomechanic. It was the precision of optical measurements of Copernicus, Brahe and Kepler that led to the birth of Kepler's laws and eventually to the field we call classical mechanics.

Maxwell's original writing, the notations used and the incredibly long form it takes are quite dissatisfying. The accomplishment, however, stands as one of the greatest of intellectual achievements. Mathematics gets its enormous power across all humanity through its universality and its universal language. Here, notations do matter. Some have speculated that the poor calculus notations of England held it back while continental Europe prospered. When we write $\frac{\partial}{\partial y}$ operating on a function f as $\frac{\partial}{\partial y} f$, it acquires a direct and active operational meaning. The use of ∂_y as a notation for this—as I see many in Europe doing—is worthy of adoption for its simplicity and elegance without loss of the founding principles. I have reluctantly not used it, to minimize the disturbance it will cause this side of the Atlantic.

[1] B. E. Sernelius, "Surface modes in physics," Wiley-VCH, ISBN 3-527-40313-2 (2001)

[2] M. Bordag (ed.), "The Casimir effect 50 years later," World Scientific, ISBN 981-02-3820-7 (1999)

[3] S. K. Lamoreaux, "The Casimir force: Background, experiments and applications," Reports on Progress in Physics, 68, 201–236 (2005)

William Hershel, the great German-English astronomer, used the mechanical precision of his creations for the systematic observation of the heavens and his numerous discoveries. *LIGO* is a continuation of this great tradition employing mechanics for detecting gravitational waves and for the development of new techniques for observation.

The reaching into quantum limits using optomechanics, together with the use of coupled resonance phenomena for it, has been a rich subject of investigation since the early part of this century. The study of nonlinearities, as well as unusual aspects of quantum field theory, also underlies this interest. Cavities in nanoscale assemblies are essential to much of this. This is a subject that is not dealt with in our chapter, but the reader would find the book by Kavokin et al.[4] quite suitable. A number of articles taken together are suitable for an exploration of this theme. Kippenberg et al.[5] discuss back-actions. Thompson et al.[6] apply these in measurements, using mechanical membrane mirrors. Eichenfield et al.[7] comprehensively discuss the use of the periodic dielectric structure for coupling the mechanical and the electromagnetic domains.

This particle-wave interaction for miniaturization of high energy radiation sources, not normally accessible through low energy semiconductor devices, behooves a study of the vacuum high energy apparatus as well as the optical-microsystem apparatus. Wiedemann's text[8] discusses the physics of large particle accelerators, while Breuer and Hommelhoff[9] show how one may use compact lasers for doing some of this in microsystems.

The subject of particle and matter surface interaction with electromagnetic radiation became particularly interesting during the period of nanoparticle development in chemistry in 1990s onward, as did metamaterials, where the sign reversals of relative permittivity and permeability permit novel avenues for design. The review by Alù and Engheta[10] and Engheta and Ziolkowski's book[11] teach the essential concepts. Reader will also find Maier's book[12] comprehensive and detailed.

There is a large ensemble of literature on organic emitters, and energy converters and solar cells in general. It is a rapidly changing, or shall we say, practical subject where every decimal point of efficiency improvement matters and gets reported. We did not discuss chalcogenide solar cells, but these too—semiconductor-like with interesting interface-based issues—are of current interest. Much of our discussion with silicon does apply, but their solution-based processing and polycrystallinity brings the interface-based considerations to them. The paper that is perhaps most interesting from a fundamental viewpoint is the original one from Shockley and Queisser[13].

The study of quantum cascade lasers is a subject area still in its

[4] A. V. Kavokin, J. J. Baumberg, G. Malpuech and F. P. Laussy, "Microcavities," Oxford, ISBN 978-0-19-922894-2 (2007)

[5] T. J. Kippenberg and K. J. Vahala, "Cavity optomechanics: Back-action at the mesoscale," Science, 321, 1172–1176 (2008)

[6] J. D. Thompson, B. M. Zwickl, A. M. Jayich, F. Marquardt, S. M. Girvin and J. G. E. Harris, "Strong dispersive coupling of a high-finesse cavity to a micromechanical membrane," Nature, 452, 72–75 (2008)

[7] M. Eichenfield, J. Chan, R. M. Camacho, K. J. Vahala and O. Painter, "Optomechanical crystals," arXiv:0906.1236v1 (2009)

[8] H. Wiedemann, "Particle accelerator physics," Springer, ISBN-13 978-3-540-49043-2 (2007)

[9] J. Breuer and P. Hommelhoff, "Laser-based acceleration of nonrelativistic electrons at a dielectric structure," Physical Review Letters, 111, 134803-1–5 (2013)

[10] A. Alù and N. Engheta, "Achieving transparency with plasmonic and metamaterials coatings," Physical Review, E 72, 016623-1–9 (2005)

[11] N. Engheta and B. W. Ziolkowski, "Metamaterials: Physics and engineering explorations," Wiley-Interscience, ISBN-13 978-0-471-76102-0 (2006)

[12] S. A. Maier, "Plasmonics: Fundamental and applications," Springer, ISBN 0-387-33150-6 (2007)

[13] W. Shockley and H. Queisser, "Detailed balance limit of efficiency of *p-n* junction solar cells," Journal of Applied Physics, 32, 510–519 (1961)

infancy. It stresses much in accuracy of theory, as well as of practice. After all, all the processes must be understood under the fields, in the presence of the electrons, photons and dopants, and with the various modes of energy loss, including the dominant one of phonons. The review by Jirauschek and Kubis[14] is very suitable for an in-depth understanding of the interaction dynamics and the care that is needed in the exploration of this subject.

[14] C. Jirauschek and T. Kubis, "Modeling techniques for quantum cascade lasers," Applied Physics Reviews, **1**, 011307 (2014)

6.9   Exercises

1. We eliminated the conundrum of self-energy in understanding electrons in a medium, by stating that an electron only acts on its surroundings, and the surroundings act on it. The electron does not act on itself. Does a similar issue exist with determining the electromagnetic energy in a box, which is a form of cavity for the electromagnetic wave? If one quantizes the harmonic oscillator in a box, each oscillator has an energy $(1/2)\hbar\omega$ as its zero point energy. There are an infinite number of such modes in the box. The box has infinite energy. Explain in brief where there is an error in this thinking.   **[S]**

2. Compare the Casimir force between two metal plates that are a μm apart with the electrostatic force between the two plates when a bias voltage of $10\ V$ is applied between them.   **[S]**

3. The Hamiltonian for atomic hydrogen in uniform and static magnetic induction **B** that is oriented along z-axis is

$$\mathscr{H} = A\mathbf{S}_e \cdot \mathbf{S}_p + \frac{e}{m_e c}\mathbf{S}_e \cdot \mathbf{B}.$$

   The first term here is the hyperfine term, $\mathbf{S}_e$ is the electron spin and $\mathbf{S}_p$ is the proton spin. Solve this for energy levels for all four states. Show that

$$E = -A\frac{\hbar^2}{4}\left\{1 \pm 2\left[1 + \left(\frac{eB}{m_e c\hbar A}\right)^2\right]^{1/2}\right\}$$

   for the singlet with $m_s = 0$ and for the triplet with $m_s = 0$. And, that it is

$$E = A\frac{\hbar^2}{4} \pm \frac{eB\hbar}{2m_e c}$$

   for the triplet with $m_s = \pm 1$.   **[A]**

4. Why does a ground state with $n = 1$ of 2 electrons have no triplet state?   **[S]**

5. A beam of highly relativistic electrons has a bunch with 10^{10} electrons uniformly distributed in a cylindrical slug that is 1 *mm* long and 100 *nm* in radius. Determine

   - the electrical and magnetic field strength at the surface of the beam,

   - the peak electrical current of the bunch, and

   - if two of these beams at an energy of 500 *GeV* pass each 10 *μm* center to center, what is the deflection angle of each beam due to the other beam? [M]

6. Au^{+14} ions are at a kinetic energy of 72 *MeV* per nucleon moving in a ring of circumference 807.1 *m*.

   - What is the ion velocity?

   - If the accelerator was designed to get protons to 28.1 *GeV*, what is the maximum kinetic energy per nucleon for Au^{+14}?

   - The beam contains 6×10^9 ions. What is the beam current at injection and at maximum energy, assuming no losses in acceleration?

   - Why does beam current increase even as the circulating charge is a constant? [M]

   This example speaks to the specifics of the Brookhaven alternating gradient synchrotron.

7. Argue why the following relationships as effective forces and torques in terms of effective moments arising in induced electrostatic fields due to a particle are meaningful:

$$\overline{\mathbf{F}}(t) = \overline{\mathbf{p}} \cdot \nabla \overline{\mathcal{E}}(t), \text{ and}$$
$$\overline{T}(t) = \overline{\mathbf{p}} \times \overline{\mathcal{E}}(t),$$

with the variables as instantaneous functions of time. [S]

8. Show that the dielectrophoretic force in an electric field $\mathcal{E}$ on a lossless dielectric sphere of radius R and permittivity ϵ_2 in a medium of permittivity ϵ_1 is given by

$$\overline{\mathbf{F}} = 2\pi\epsilon_1 R^3 \frac{\epsilon_2 - \epsilon_1}{\epsilon_2 + \epsilon_1} \nabla \mathcal{E}^2.$$

[S]

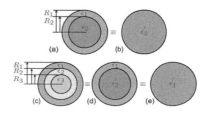

9. Following Figure 6.53, show that a layered lossless spherical shell in an uniform electric field, shown in panel (a), can be reduced to the form shown in panel (b), with an effective permittivity of

$$\epsilon_2' = \epsilon_2 \frac{(R_1/R_2)^3 + 2(\epsilon_3 - \epsilon_2)/(\epsilon_3 + 2\epsilon_2)}{(R_1/R_2)^3 - 2(\epsilon_3 - \epsilon_2)/(\epsilon_3 + 2\epsilon_2)}.$$

Figure 6.53: (a) shows a cross-section of a sphere with a shell of two different permittivity materials. (b) shows its single permittivity equivalent. (c) through (f) show the extension of this procedure to multilayered lossless spherical shell spheres.

The new effective permittivity of an equivalent homogeneous sphere of radius R_1 leads to identical fields. Extend this to argue that multilayered lossless spherical shell spheres can be extended through an iterative use of this relationship as shown in panels (c) and (d). [S]

10. Can plasmonic effects be observable in metal films as thin as a single atomic layer? Can there be a scale length limiting this phenomenon? If so, what might it be like? [S]

11. Calculate the number of photons per second in a 1 W radiation at 470 nm, 530 nm and 630 nm wavelengths. [S]

12. Let us look at blackbody radiation and how it determines the solar radiation.

   • Plot the black body-specific radiative intensity (power per unit area of an emitting surface per unit solid angle per unit frequency) as function of frequency ν for temperatures of 1500 K, 2500 K, 3500 K, 4500 K and 5500 K.

   • In the above, what is the effective color of the emission at each of the temperatures and the wavelength at the maximum specific intensity?

   • What is the power that is in the 400 nm to 700 nm range? [S]

13. An abrupt *p-n* junction solar cell has an optical generation rate of 5×10^{19} $cm^{-3} \cdot s^{-1}$ under uniform sunlight. It has an area of 100 cm^2, a depletion depth of 3 μm, a reverse saturation current density of $J_0 = 10^{-11}$ $A \cdot cm^{-2}$, and a carrier lifetime of 2×10^{-6} s. Determine the

   • optically generated current in the depletion region,

   • the total optically generated current,

   • the short-circuit current,

   • the open-circuit voltage and

   • the maximum power available if the fill factor is 0.75. [S]

14. For the solar cell described in Exercise 13, determine

   • the saturation currents at 225 K and at 375 K, compared to the value at 300 K, and

   • the maximum output power at 225 K and at 375 K, compared to the value at 300 K. [S]

15. A silicon solar cell of 100 cm^2 area is formed out of a junction at $N_A = 5 \times 10^{18}$ cm^{-3}, $N_D = 5 \times 10^{16}$ cm^{-3} at a carrier lifetime of

5×10^{-6} s. Let 7×10^{21} $cm^{-3}s^{-1}$ electron-hole pairs be generated uniformly. What is, at 300 K,

- the optically generated current and the assumptions underlying the calculation,
- the short-circuit current,
- the open-circuit voltage,
- the open-circuit voltage at 375 K and at 225 K, and
- the maximum power available if the fill factor is 0.8.   [S]

A

Information from the Shannon viewpoint

OUR STATE OF KNOWLEDGE is affected by additional evidence, which serves to change the probability distribution that we assign. It depends on what we already know. It is the subjectiveness that needs to be made objective for scientific pursuit. Establishing a measure of uncertainty to quantify the state of ignorance in a bitstream is due to Shannon. He tackled this for a message—the order and arrangement of symbols transmitted and received. There is a level of uncertainty that is removed as a result of receipt of message. The message has the ability to change the state of uncertainty in the receiver. This scale of uncertainty has certain requirements:

- Uncertainty about an outcome of an event is a function of the probability assigned to the outcome. With $\mathfrak{p}_i$ as the probability of the ith outcome out of n possible outcomes, let uncertainty H describe the choice in the selection of events or how uncertain we are of the outcome. H is a function of all $\mathfrak{p}_i$s:

$$H = H(\mathfrak{p}_1, \ldots, \mathfrak{p}_n).\qquad (A.1)$$

We are also employing the symbol $\mathfrak{P}()$ to describe a probability distribution over a continuous or discrete space, while $\mathfrak{p}$ indicates the probability of a specific continuous or discrete event occurrence.

- If all outcomes are equally likely, then the uncertainty must increase with n. More possibilities increase uncertainty. H is a monotonically increasing function of n.

- If A and B are two independent events, then the compound event of A and B ($C = A \cap B$) has an uncertainty measure

$$H(A \cap B) = H(A) + H(B).\qquad (A.2)$$

The symbols $\cap$, $\cup$, $\wedge$ and $\vee$ must be distinguished. $\cap$ and $\cup$ are intersection and unions of sets. $\wedge$ and $\vee$ are mathematical logic operations.

- The quantitative magnitude of H should be independent of how a problem is set up.

This last requirement states that intervening situations should not affect the probability of the outcome. In Figure A.1, the outcomes are

A, B, C and *D*, and *S* is the event. *E* and *F* are intervening situations. The probability of *A*, given event *S*, should not change because of intervening *E* and *F*. The probability of *A* is 1/4 independent of directly going to *A* or going through *E*. Figure A.1(b) shows *A* and *B* as the possible outcomes through intervening *E*, and *C* and *D* through the other intervening event *F*. $\mathfrak{p}(E|S) = 1/2$ with $\mathfrak{p}(A|E) = 1/2$.

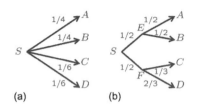

Figure A.1: (a) shows the four possible outcomes *A–D* of an event *S*. (b) shows the same four possible outcomes through two possible intervening events *E* and *F*.

Take the situation where all outcomes have equal probability $\mathfrak{p}_i$. Suppose there are *m* junction events that lead to the outcomes. Let there be ζ choices at each of these situations. There will then be ζ^m outcomes. In Figure A.1(b), *m* = 2 and ζ = 2. The second of our conditions is that $H = f(n)$—a monotonically increasing function with $\mathfrak{p}_i = 1/n$. The conditions imply that

$$f(\zeta^m) = mf(\zeta). \tag{A.3}$$

$f(\zeta)$ follows from these conditions. Equation A.3 differentiated w.r.t. *m*, since $d\zeta^m/dm = \zeta^m \ln m$, gives

$$\frac{df(\zeta^m)}{d\zeta^m}\zeta^m \ln \zeta = f(\zeta), \tag{A.4}$$

and differentiated w.r.t. ζ gives

$$\frac{df(\zeta^m)}{d\zeta^m}m\zeta^{m-1} = m\frac{df(\zeta)}{d\zeta}. \tag{A.5}$$

Eliminating $df(\zeta^m)/d\zeta^m$ between these two equations results in

$$\frac{df(\zeta)}{d\zeta} = \frac{d\zeta}{\zeta \ln \zeta}, \tag{A.6}$$

whose solution is $\ln f(\zeta) = \ln(\ln \zeta)$, and a constant, that is, $f(\zeta) = k \ln \zeta$. Since $H = f(n)$ is monotonically increasing in *n* when $\mathfrak{p}_i$s are all equal,

$$H = k \ln n, \tag{A.7}$$

and, hence, with equally likely outcomes, the uncertainty is

$$H = -k \ln \mathfrak{p}_i. \tag{A.8}$$

The additive properties are satisfied since independent outcomes with probability $\mathfrak{p}_i$ for *A* and $\mathfrak{p}_j$ for *B* lead to an uncertainty of

$$H(A, B) = -k \ln \mathfrak{p}_i \mathfrak{p}_j = -k \ln \mathfrak{p}_i - k \ln \mathfrak{p}_j = H(A) + H(B). \tag{A.9}$$

The information content of the union of *A* and *B* is the sum of individual sets iff they are independent and have nothing in common.

Now, we ask how much information is conveyed by an average symbol of a message? This depends on how much the receiver already knows. A simple example is, let there be three symbols: *x, y* and z, each of which is equally likely. So, for a message M that we know contains one of these symbols, $\mathfrak{p}(x|M) = \mathfrak{p}(y|M) = \mathfrak{p}(z|M) = 1/3$. Our uncertainty is $H(M) = k\ln 3$. Now, if one had known beforehand that the symbol x or y represented by a—or z, which we will call b to complement a—were to be sent, then $\mathfrak{p}(a|M) = 2/3$, and $\mathfrak{p}(b|M) = 1/3$. Upon receiving, say b, uncertainty has been vanquished—it is zero. But, if we get a, it could still be x or y. So, uncertainty has been reduced by $k\ln 2$. The information content of a and that of b are different:

$$
\begin{aligned}
I(a|M) &= k\ln\frac{3}{2} = -k\ln\mathfrak{p}(a|M), \text{ and} \\
I(b|M) &= k\ln 3 = -k\ln\mathfrak{p}(b|M).
\end{aligned}
\tag{A.10}
$$

The information content of a set of symbols that have not yet been transmitted and received, that is, the expected information content is

$$
I = -k\sum_i \mathfrak{p}_i \ln\mathfrak{p}_i,
\tag{A.11}
$$

where $\mathfrak{p}_i$s are the probabilities for the ith symbol. This is the uncertainty, and the average information—or the measure of uncertainty in it—then is

$$
H = -k\sum_i \mathfrak{p}_i \ln\mathfrak{p}_i.
\tag{A.12}
$$

This is Shannon's information measure. From a thermodynamic viewpoint it is a negative entropy—negentropy—since, for any specific situation, it characterizes how far one is from maximum entropy.

If events A and B are not independent, then one may quantify a mutual information by

$$
H(A:B) = H(A) + H(B) - H(A,B).
\tag{A.13}
$$

$H(A:B)$ measures the degree of interdependence of A and B.

Shannon's message theorem:

For ε as the error probability of a single bit, and a message of length M, a coding scheme exists that generates a coded message of length M_c, which allows correction of the errors, that is subject to the inequality

$$
\frac{M}{M_c} \leq 1 - \left[\varepsilon\log_2\left(\frac{1}{\varepsilon}\right) + (1-\varepsilon)\log_2\left(\frac{1}{1-\varepsilon}\right)\right].
\tag{A.14}
$$

The equality defines the most bit-efficient code.

Rigorous algebraic and geometric proofs can be found many standard texts on information theory.

PROOF: Our derivation focuses on the physical essence and on simplicity by sacrificing strict mathematical rigorousness. Let the messages—actual and coded—be long. Let k be the average number of errors in the coded message, that is, let $k = \varepsilon M_c$. The actual errors will have a statistical distribution, but the coding is designed to correct for the average expectation. These errors occur with

$$^{M_c}C_k = \frac{M_c!}{k!(M_c - k)!} \qquad \text{(A.15)}$$

possibilities in the coded message. Let there be l code bits that suffice to decode M in the presence of the k average number of errors in M_c. l code bits provide 2^l ways by which the entire M message data bits and the location of all the possibilities by which the average number of errors occur in the coded message are accounted for. Including the possibility of redundancy in bits, $l \le M_c - M$, so,

$$2^{M_c-M} \ge \frac{M_c!}{k!(M_c - k)!}. \qquad \text{(A.16)}$$

With long messages, we may use the Sterling approximation for the factorials, that is, $\ln n! \approx n \ln n - n$. Equation A.16, with $k = \varepsilon M_c$, reduces to

$$(M_c - M)\ln 2 \ge \ln M_c! - \ln k! - \ln(M_c - k)!$$
$$\therefore \ (M_c - M)\ln 2 \ge \ln M_c! - \ln(\varepsilon M_c)! - \ln[(1-\varepsilon)M_c]!$$
$$\therefore \ (M_c - M)\ln 2 \ge M_c \ln M_c - M_c - \varepsilon M_c \ln \varepsilon M_c + \varepsilon M_c$$
$$-(1-\varepsilon)M_c \ln(1-\varepsilon)M_c + (1-\varepsilon)M_c$$
$$\therefore \ (M_c - M)\ln 2 \ge -\varepsilon M_c \ln \varepsilon - (1-\varepsilon)M_c \ln(1-\varepsilon)$$
$$\therefore \ (M_c - M)\ln 2 \ge M_c\left[\varepsilon \ln\left(\frac{1}{\varepsilon}\right) + (1-\varepsilon)\ln\left(\frac{1}{1-\varepsilon}\right)\right]$$
$$\therefore \ \frac{M}{M_c} \le 1 - \left[\varepsilon \log_2\left(\frac{1}{\varepsilon}\right) + (1-\varepsilon)\log_2\left(\frac{1}{1-\varepsilon}\right)\right]. \qquad \text{(A.17)}$$

The Shannon-Hartley theorem:
The information carrying capacity of any signal channel is determined by

$$C = B \log_2\left(1 + \frac{S}{N}\right), \qquad \text{(A.18)}$$

where C is the channel capacity, B the channel bandwidth, S the signal power and N the noise power.

PROOF: Let V_N^2 and V_S^2 be the mean square noise and signal voltages. Since signal and noise have finite power, they are limited in range, and since the system is assumed to be efficient, the signal and the noise should have similar statistical properties. Their peak to root mean square voltage ratios should therefore be similar. Let

η be this ratio. An efficient information transmission system would operate with signal power S at some maximum power which would be across different messages. Signal and noise are uncorrelated. The total power is

$$P_T = S + N, \tag{A.19}$$

so

$$V_T^2 = V_S^2 + V_N^2 \tag{A.20}$$

because both signal and noise operate in the same impedance environment.

The combined signal and noise in an efficient system will be in a voltage range of $\pm \eta V_T$. Let b be the number of bits being transmitted. We divide this voltage spread into 2^b equal-sized bands of size $\Delta V = 2\eta V_T / 2^b$ over the spread. Any specific one of these ΔV-sized bands is identifiable by a b bit number. ΔV should be larger than V_N, which is the root mean square noise voltage, to avoid randomization of the actual voltage by V_N. So, the maximum number of bits of information of an instantaneous voltage is determined by

$$
\begin{aligned}
2^b &= \frac{V_T}{V_N} = \frac{\sqrt{V_N^2 + V_S^2}}{V_N} = \left(1 + \frac{S}{N}\right)^{1/2} \\
\therefore b &= \log_2\left[\left(1 + \frac{S}{N}\right)^{1/2}\right].
\end{aligned}
\tag{A.21}
$$

Over time T, one makes M of b bit measurements. The number of accumulated bits is $M \times b$. The information transmission rate $\mathscr{I}$ (b/s) is

$$\mathscr{I} = \frac{M \times b}{T} = \frac{M}{T}\log_2\left[\left(1 + \frac{S}{N}\right)^{1/2}\right]. \tag{A.22}$$

From the Nyquist rate, we know that for a channel of bandwidth B, the sampling rate of independent measurements is $2B$. This is M/T for our problem. Therefore, the maximum information transmission rate C is

$$C = 2B\log_2\left[\left(1 + \frac{S}{N}\right)^{1/2}\right] = B\log_2\left(1 + \frac{S}{N}\right). \tag{A.23}$$

This is the channel's information carrying capacity.

B

Probabilities and the Bayesian approach

ANALYSIS OF REAL WORLD SIGNALS and our ability to extract information from them is a subject that has much to do with finding patterns and of causality. This is a problem in analysis of all that the real world presents to us: audio signals, visual signals and other sensory signals, to which even a robot responds, or statistical data of all different types that economists, pollsters, bankers and others try to learn from.

This tackling of reasoning has at least three uncertainties. the first is ignorance: not knowing due to a limit of knowledge. Games of cards, for example, involve guessing what the opponent has. The second is derived from randomness and indeterminism. Even with knowing everything, one may not be able to certainly predict the outcome of a fair coin toss. The third is vagueness. What is bravery or right or wrong in war is a matter of opinion, not mathematically exact. This is a subjective notion that Shannon's information does not catch. The moment ``now″ for two observers is a range of time— not a precise time t to which a value can be ascribed. It is within this milieu that the analysis must tackle uncertainty. Pattern finding and statistical inference are major areas in engineering, science and statistics, ones where the reduction of the unknown to known is a problem with natural appeal.

The Bayesian approach is a tool for finding the causal relationships in the data where other variables—hidden and not known—lurk. Just because there appears to be a correlation or association between two variables, it does not necessarily imply that the cause of one is the other. The relationship between the two variables needs to be isolated from outside effects in order to fix the connection between the two variables. Probabilities employing randomization permit the characterization of uncertainty. Bayes's approach takes the parameter as a random variable. Bayes shows us how inverse probability may be employed to calculate the probability of antecedent events once

``The Church of the Flying Spaghetti Monster″—a tongue-in-cheek website whose followers are ``Pastafarians″— displays a graph showing the strong correlation between the decline in number of pirates, and global warming. One doesn't necessarily imply the other. Unraveling the connections is essential to determining the efficacy of pharmaceutical treatments.

an event has been observed. In a general problem, the true values of parameters are not known. So, they are random variables. Probability statements about parameters are statements of beliefs. Prior distributions are subjective. Priors of two different observers are generally different. The observers will give different relative weights to each parameter's value. This prior distribution is a measure of the plausibility assigned by an observer to each parameter value. Upon observation of a data, the observer revises her beliefs. This the posterior distribution. Bayes's theorem gives the tool that allows the observer to form a posterior distribution by using the prior distribution and the observed data.

Statistics also has a frequentist approach where the parameter is considered fixed, though not known. Only asymptotic relative frequency as a measure of probability is allowed. The sampling distribution is then employed to determine the fixed parameter, which is examined to see how it distributes over all the different repetitions of the experiment.

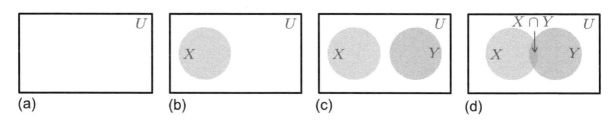

Figure B.1: (a) shows the event space U. (b) shows the subset X of events in U; this subset has a probability of $\mathfrak{P}(X)$. (c) shows the two mutually exclusive subsets X and Y in U—those with no common events—and (d) shows an X and Y that have events in common.

Here, we summarize the key axioms of probability calculus, following Kolmogorov's approach, and its use in the Bayesian approach. Let U be the space of all possible events as shown in Figure B.1. If we are uncertain about which of the possibilities are true, then the maximum probability must apply to the true event lying within U. We set maximum probability to unity.

Axiom 1:
$$\mathfrak{P}(U) = 1. \tag{B.1}$$

The probabilities are distributed over U. The event space U is pictorially shown in Figure B.1(a). Let X be a region in this space U, as seen in Figure B.1(b). Then

Axiom 2:
$$\mathfrak{P}(X) \geq 0 \ \forall X \subset U. \tag{B.2}$$

The probabilities of the combined events X and Y, if the two are mutually exclusive, are additive, as shown in Figure B.1(c), that is,

Axiom 3:
$$\text{If } X \cap Y = \varnothing, \text{ then } \mathfrak{P}(X \cup Y) = \mathfrak{P}(X) + \mathfrak{P}(Y) \ \forall X, Y \subset U. \tag{B.3}$$

When events overlap in two sets, that is, sets X and Y overlap, as seen in Figure B.1(d), then

Theorem 1:

$$\mathfrak{P}(X \cup Y) = \mathfrak{P}(X) + \mathfrak{P}(Y) - \mathfrak{P}(X \cap Y) \ \forall \ X, Y \subset U. \qquad (B.4)$$

This intuitively arises as a way to exclude double counting of $\mathfrak{P}(X \cap Y)$ in $\mathfrak{P}(X \cup Y)$ when X and Y subspaces overlap.

Conditional probability is the probability that an event X will occur given that event Y has occurred or will occur. The conditional probability for X on Y is written as $\mathfrak{P}(X|Y)$.

Definition 1: Conditional probability.

$$\mathfrak{P}(X|Y) = \frac{\mathfrak{P}(X \cap Y)}{\mathfrak{P}(Y)}. \qquad (B.5)$$

Conditional probability is only definable if the event Y has a non-zero probability. Probability conditional on Y, following Figure B.1(d), is viewable as the collapse of U to Y, and then finding what is left of X relative to what is left of Y. This is the ratio of the probability of the common events area $X \cap Y$ to that of the probability of events in the area of Y.

Two events, X and Y, are probabilistically independent, written as $X \amalg Y$, when the conditioning of one leaves the probability of the second unchanged.

Definition 2: Independence.

$$X \amalg Y \equiv \mathfrak{P}(X|Y) = \mathfrak{P}(X). \qquad (B.6)$$

Independence is symmetric, that is $X \amalg Y \equiv Y \amalg X$. Rolls of fair dice are independent: the outcome of the second event is not dependent on the first. Events in card games are examples of dependent events: the probability of drawing a diamond flush is affected by the cards that have already been drawn, since the numbers of what there is in the deck changes. The letter q is almost always followed by the letter u.

Qazi—the name of a a judge in the Caliphate and Moghal eras—is among the few exceptions. We have added qbit to this set.

Definition 3: Conditional independence.

$$X \amalg Y|Z \equiv \mathfrak{P}(X|Y, Z) = \mathfrak{P}(X|Z). \qquad (B.7)$$

When the set Z is a null set ($\oslash$), then conditional independence reduces to marginal independence. When event Z tells everything that event Y does about X, and maybe more, then knowledge of Y provides no new information. For example, a graduation transcript contains the information of a specific year's transcript, making the latter not additionally informative. Z here has screened off X from Y.

Theorem 2: Total probability.

For a set of events $\{A_i\}$ that are subset of U, with $\bigcup_i A_i = U$, and if for all distinct is and js, $A_i \hat{A}_j = \oslash$, then

$$\mathfrak{P}(U) = \sum_i \mathfrak{P}(A_i). \tag{B.8}$$

This is a statement of total probability over the distinct set of events that are possible in U. If we partition the probability over a particular event B rather than the whole event space of U, with $A_i \neq \oslash \; \forall i$, then

$$\mathfrak{P}(B) = \sum_i \mathfrak{P}(B|A_i). \tag{B.9}$$

Theorem 3: Chain rule.

Given three events A, B and C, with conditional probabilities defined,

$$\mathfrak{P}(C|A) = \mathfrak{P}(C|B)P(B|A) + \mathfrak{P}(C|\overline{B})P(\overline{B}|A). \tag{B.10}$$

The probabilistic connection between C and A has been split over the two different states that were possible for the third variable B. The chain rule is generalizable to any arity of B.

Theorem 4: Bayes's theorem.

Given three events A, B and C, with conditional probabilities defined, Bayes's theorem states that

$$\mathfrak{P}(A|BC) = \mathfrak{P}(A|C)\frac{\mathfrak{P}(B|AC)}{\mathfrak{P}(B|C)}. \tag{B.11}$$

This follows in the light of $\mathfrak{P}(A|C) + \mathfrak{P}(\overline{A}|C) = 1$ and that

$$
\begin{aligned}
\mathfrak{P}(AB|C) &= \mathfrak{P}(A|BC)\mathfrak{P}(B|C) \\
&= \mathfrak{P}(B|AC)\mathfrak{P}(A|C).
\end{aligned} \tag{B.12}
$$

Rewritten in another form that is useful for the purpose of modifying probabilities of the guess (an hypothesis) by using new information available (an evidence), this relation is

$$\mathfrak{P}(h|e) = \frac{\mathfrak{P}(e|h)\mathfrak{P}(h)}{\mathfrak{P}(e)} = \frac{\mathfrak{P}(e|h)\mathfrak{P}(h)}{\sum_x \mathfrak{P}(e|x)\mathfrak{P}(x)}. \tag{B.13}$$

The probability of a hypothesis h conditioned on evidence e is equal to its likelihood, that is, $\mathfrak{P}(e|h)$, times its probability prior to any evidences $\mathfrak{P}(h)$ and normalized by $\mathfrak{P}(e)$ so that the conditional probabilities of all the hypotheses add up to unity. Equation B.13 is a statement that follows directly from Bayes's theorem and the expansion of conditional probability. The sum in the denominator expresses the probability of evidence given all possible hypothesis and hidden variables.

Definition 4: Conditionalization.

As a result of the evidence $\mathfrak{P}(e)$, the probability $\mathfrak{P}(h|e)$ is the posterior belief in h—the new expectation of h. This belief—posterior probability—is the belief updating via probabilities conditioned upon the evidence. If joint priors do not exist over the hypothesis and evidence space, then this conditionalization is unemployable. The evidence e is all the evidence and the only evidence. It is the total evidence.

C
Algorithmic entropy and complexity

ALGORITHMIC ENTROPY OR COMPLEXITY provides the physical configurational basis in microstates for entropy, in a manner similar to
what configurational description does for other thermodynamic parameters. An example of the latter is temperature for kinetic energy.
In the Shannon information perspective, the source and the ensemble
of all possible events or sequences comprise the basis from which
the Shannon entropy is derived. The alternative view is to treat any
sequence of symbols, whether random or deterministic, as an object
independent of any other associations for exploring its content.

The complexity $K(x)$ of a sequence $s(x)$ is the number of bits of
the shortest size program that can produce the sequence. Algorithmic
complexity, or algorithmic entropy, or Kolmogorov complexity, is

$$K(s) = |s^*|, \tag{C.1}$$

where s^* is the length of the shortest sequence. A universal Turing
machine is an unambiguous approach to finding this complexity,
since it is a sequence-based machine working bit by bit. If the universal Turing machine halts, we know the length of the number of
bits of the program. The Church-Turing thesis discussed in Chapter 1 stated that all problems that have a mathematical procedure—an
algorithm—and, given sufficient storage and computing time, can be
translated to this machine.

A universal Turing machine takes a
program—a symbol string of finite
length with a delimiter symbol and
the input string—as its input from the
tape. So, all the information is together
on tape. Such a Turing machine is
capable of simulating any other Turing machine. It is a universal Turing
machine.

Algorithmic complexity therefore has a very straightforward connection to the information discussion.

In the Shannon view, for a source event $x \in X$, the information is

$$I(x) = -\log_2 \mathfrak{p}(x). \tag{C.2}$$

This is the minimum number of bits that describe the event. The
Shannon entropy $H(X)$ is the statistical average of source informa-

tion, that is,

$$H(X) = \langle I(x) \rangle = \sum_{x \in X} p(x) I(x) = - \sum_{x \in X} p(x) \log_2 p(x). \qquad (C.3)$$

Complexity $K(s)$ is a measure of the totality of information content of a specific event from a source—the sequence of bits $s(x)$ where $x \in X$—while the Shannon entropy is the average of the information content from the entirety of the event source. $K(s)$ can be associated with any symbol string $s - s(x)$ on the universal Turing machine. If it is non-universal, the complexity will increase. This simply follows from the argument that the universal Turing machine is also capable of simulating any non-universal Turing machine. With this being the case, the sequence of the program $s_v(x)$ of the non-universal machine, together with the program to simulate the non-universal machine in the universal machine, say s_u, is needed to simulate the non-universal machine within the universal machine. The complexity through this simulation will be the complexity K_v of the non-universal machine, plus another complexity associated with the specific details of simulating the non-universal machine. So, the universal machine's complexity operating directly is less than or equal to that of any non-universal machine.

Complexity has several properties that are interesting. Let $l(s)$ be the known length of a string s. In our example, the length of a program in the universal machine was $|s(x)| = l(x) + c$, where c accounted for the non-universal machine simulation. This states that complexity obeys

$$K[x|l(x)] \leq l(x) + c, \qquad (C.4)$$

with conditional complexities setting the upperbound.

Now, consider a string whose length is not known but is defined by some algorithm. This needs a program of length $\leq l(x) + c$. If one wants a machine to output a string of length $l(x) = n$ specifically, the program needs to instruct this. Equation C.4 needs to be changed. To be able to specify a string of length $l(x)$, and for the machine to output it, we need to code n into $\log_2 n$ bits and provide delimiter bits, for example, 01 or 10. This specifies

$$
\begin{aligned}
|s_o(x)| &= K[x|l(x)|] + c + 2\log_2 l(x) + 2 \\
&\approx K[x|l(x)|] + c + 2\log_2 l(x) \qquad (C.5)
\end{aligned}
$$

in the limit of long sequences. So,

$$K(x) \leq K[x|l(x)|] + c + 2\log_2 l(x) \qquad (C.6)$$

is also an upperbound. Equations C.4 and C.6 are the two upperbounds for complexity.

Some strings may be simpler to simulate than others through this simulation process.

```
READ K
DO 10 n =1, Infinity
DEFINE all s FOR l(s)=n
IF K(s) ≥ K, WRITE (6,*) s AND STOP
n=n+1
10      CONTINUE
```

Figure C.1: A program that finds at least one string with the same complexity as that of string s, that is, $K(s) \equiv K$.

Can a universal machine determine the complexity $K(s)$? Consider the argument in Figure C.1, which is based on the hypothesis that there exists a universal machine that can find $K(s)$ for any $s(x)$. This program is of length $p(x)$, and it outputs $K(x)$ $\forall x$. We write a program $r(x)$ of Figure C.1, that given $s(x)$, finds one or more strings that have the same complexity $(K(s) \equiv K)$. This program p has a length $|r| = |p| + 2\log_2 l(K) + c$. The program grows logarithmically with K and outputs a string s of complexity $K(s) \geq K > |r|$. Since, by definition the complexity $K(s)$ is the minimum description length of s, no machine program outputs a string s shorter than $K(s)$. Ergo, the program p does not exist.

By showing this contradiction, we have established that only the upperbound of Kolmogorov complexity can be defined. The Kolmogorov complexity of the sequence $s(x)$ cannot be computed. It is undecidable!

There are a few properties of $K(s)$ that are decidable. Given complexity K, one may determine how many strings s have complexity less than K, that is, $K(s) < K$, and, for any string of length n, we may define the upperbound analytically for $K[x|l(x)]$. For example, the sum of all possible binary strings of length $k < K - 1$ is

$$\sum_{k=0}^{K-1} n(k) = 2^K - 1 < 2^K, \tag{C.7}$$

and each of these strings is a machine program. Less than 2^K programs exist smaller than K. So, there are fewer than 2^K strings of complexity less than K. In general, the total length of the program $p(x)$ is

$$\begin{aligned}
|p(k,n)| &= 2\log_2 k + 2 + \log_2 \frac{n!}{k!(n-k)!} + c \\
&\equiv 2\log_2 k + \log_2 {}^nC_k + c'. \tag{C.8}
\end{aligned}$$

The combinatorial term arose from the inclusion of all strings, given n and k, which represent the ways to have k possible 1s in n positions. Applying Sterling's formula to the combinatorial term gives

$$|p(k,n)| = 2\log_2 k + nf\left(\frac{n}{k}\right) + c'', \tag{C.9}$$

where

$$f(u) = -u\log_2(u) - (1-u)\log_2(1-u). \tag{C.10}$$

The program length is the upperbound of the complexity of string s with k 1s. k is the sum of all the bits, since it is the number of 1s in the n-long string. This observation leads to the simplification

$$K(x|n) \leq 2\log_2 n + \log_2 {}^kC_n + c. \tag{C.11}$$

The Kolmogorov complexity satisfies a number of other properties similar to the definitions in Appendix B.

Joint complexity:

If two strings s_x and s_y are algorithmically independent, then the joint complexity is

$$K(s_x, s_y) = K(s_y, s_x) = K(s_x) + K(s_y), \qquad (C.12)$$

the sum of individual complexities.

Mutual complexity:

For two strings that are algorithmically dependent, the mutual complexity is

$$K(s_x : s_y) = K(s_x) + K(s_y) - K(s_x, s_y). \qquad (C.13)$$

These expressions have forms that are parallel to what we found for Shannon entropy. The difference between the Kolmogorov measure $K(s)$ for individual string $s(x)$ and the Shannon entropy $H(X)$ for an event source X is conceptual. One is for an object in hand. The other is for the possibilities that exist on an average for the source. But, there are many similarities between the two measures of information.

Let X, the source of x_i—the random events with probabilities $\mathfrak{p}(x_i)$—be a binary source, for simplicity, with $x_0 = 0$, and $x_1 = 1$, and where $\mathfrak{p}(x_2) = 1 - \mathfrak{p}(x_1)$. Take a sequence of n such events, that is,

$$x = x_i^1 x_i^2 \cdots x_i^n \text{ with } i = 1, 2. \qquad (C.14)$$

The upperbound of the conditional complexity is

$$
\begin{aligned}
K(x|n) &\leq 2\log_2 n + nf\left(\overset{n}{\underset{j=1}{\Sigma}} nx_i^j\right) + c \\
\therefore \langle K(x|n)\rangle &\leq \left\langle 2\log_2 n + nf\left(\frac{1}{n}\sum_{j=1} nx_i^j\right) + c\right\rangle \\
&\leq 2\log_2 n + nf\left(\frac{1}{n}\sum_{j=1} n\langle x_i^j\rangle\right) + c \\
&\leq 2\log_2 n + c + nf(\mathfrak{p}) \\
&\leq 2\log_2 n + c + nH(X). \qquad (C.15)
\end{aligned}
$$

Normalized to n, this is

$$\frac{\langle K(x|n)\rangle}{n} \leq \frac{2\log_2 n}{n} + \frac{c}{n} + H(X). \qquad (C.16)$$

When n is large, $\langle K(x|n)\rangle/n$ is approximately $H(X)$. The average per bit complexity of a random bit string—$\langle K(x|n)\rangle/n$—asymptotically approaches the Shannon entropy. The average complexity of a bit string is upperbounded by the Shannon entropy $nH(X)$ of the source that generates it.

D
Classical equipartition of energy

IN THE CLASSICAL PICTURE—independent particles subject to a local force—a complete description of any particle, say the ith particle, is provided by the position $\mathbf{q}_i$ and momentum $\mathbf{p}_i$ coordinates. The probability distribution is subject to the Maxwell-Boltzmann distribution function. The energy of a system of n particles is $U = U(q_1, \ldots, q_n; p_1, \ldots, p_n)$. Since the energy is additive and the particles are independent in energy, this energy term can be rewritten as

$$U = u_i(q_i, p_i) + U'(q_1, \ldots, q_{i-1}, q_{i+1} \ldots q_n; p_1, \ldots, p_{i-1}, p_{i+1}, \ldots, p_n).$$
(D.1)

The energy of the particle here arises in the kinetic form.

Equipartition of energy tell us the expectation value of this energy associated with the motional freedom.

Consider a common potential energy for all the particles as our reference. The kinetic energy of all particles is second order in the power of momentum, and the system is isotropic and homogeneous, i.e., $u_i = p_i^2/2m = u_i(p_i)$. The average measurement of this energy is

$$\langle u_i \rangle = \frac{\int_{-\infty}^{\infty} \exp[-U(q_1, \ldots, p_n)/k_B T]\, u_i \, dq_1. \cdots. dp_n}{\int_{-\infty}^{\infty} \exp[-U(q_1, \ldots, p_n)/k_B T]\, dq_1. \cdots. dp_n}$$

$$= \left\{ \int_{-\infty}^{\infty} u_i \exp\left(-\frac{u_i}{k_B T}\right) dp_i \right.$$

$$\times \int_{-\infty}^{\infty} \exp\left[-\frac{U'(\text{no } q_i, \text{no } p_i)}{k_B T}\right] \cdots dp_{i-1} dp_{i+1} \cdots \Bigg\}$$

$$\times \left\{ \int_{-\infty}^{\infty} \exp\left(-\frac{u_i}{k_B T}\right) dp_i \right.$$

$$\times \int_{-\infty}^{\infty} \exp\left[-\frac{U'(\text{no } q_i, \text{no } p_i))}{k_B T}\right] \cdots dp_{i-1} dp_{i+1} \cdots \Bigg\}^{-1}$$

$$= \frac{\int_{-\infty}^{\infty} u_i(p_i) \exp\left[-u_i(p_i)/k_B T\right] dp_i}{\int_{-\infty}^{\infty} \exp\left[-u_i(p_i)/k_B T\right] dp_i}$$

$$= -\frac{\partial}{\partial(1/k_B T)} \left\{ \ln\left[\int_{-\infty}^{\infty} \exp\left(-\frac{u_i}{k_B T}\right) dp_i \right] \right\}. \tag{D.2}$$

To evaluate this integral, let $p_i = \sqrt{k_B T} x$. Then, $dp_i = \sqrt{k_B T} dx$, and $u_i(p_i) = p_i^2/2m = k_B T x^2/2m$. One can now evaluate, by substitution, the average energy as

$$
\begin{aligned}
\langle u_i \rangle &= -\frac{\partial}{\partial(1/k_B T)} \left\{ \ln\left[\int_{-\infty}^{\infty} \exp\left(-\frac{x^2}{2m}\right) \sqrt{k_B T}\, dx \right] \right\} \\
&= -\frac{\partial}{\partial(1/k_B T)} \left[\ln(\sqrt{k_B T}) + \ln \int_{-\infty}^{\infty} \exp\left(-\frac{x^2}{2m}\right) dx \right] \\
&= \frac{1}{2} k_B T, \tag{D.3}
\end{aligned}
$$

where only the first term contributes.

When energy has a quadratic dependence on a single canonical variable, e.g., momentum here, and the particle energy is separable in the energy of the ensemble by separation of its canonical parameter because of the independence of the canonical variables from each other, and Boltzmann statistics are a good description of the ensemble, that is, a classical non-interacting particle description holds, then the average energy of the particle in the ensemble is $k_B T/2$ per motional degree of freedom. To lowest order, for air molecules or electrons in non-degenerate conditions with room to move around, these constraints hold, and the average energy of each particle—its kinetic energy—with three degrees of freedom in the momentum coordinate will be $3k_B T/2$.

E
Probability distribution functions

PROBABILITIES OF DISCRETE EVENTS, that is, variables such as coin tosses or letters of the alphabet, are quite familiar to us. In the limit of a large number of such events, that is, with the value that the variables can have, one may view probabilities through a distribution of continuous variables subject to all the constraints of probabilities that we have discussed.

A uniform distribution is an example of such a distribution. A useful random number generator will have a uniform distribution of the probabilities of generating the numbers.

Distribution functions can be exponential, for example, $\mathfrak{P}(t) = (1/\tau)\exp(-t/\tau)$ for $t \geq 0$. Examples of this type of distribution function are trapping and detrapping. A historically and safety centric important example is radioactive decay.

The Poisson distribution is a distribution where, for each interval, there is a fixed average of events with a small vanishing probability. Shot noise in an electronic device, due to discrete electron events, or in an optical device, due to discrete photon events, is a consequence of the Poisson distribution. The distribution of dopants in nanoscale devices is an additional example of this.

Another common distribution function is the Gaussian normal distribution. In the Gaussian distribution, events occur with a finite non-vanishing probability, in contrast to events in the Poisson distribution. Gaussian normal distributions are encountered w.r.t. the velocity of gas molecules, the velocity of electrons in non-degenerate conditions in semiconductors, the distribution of position in long random walks, and as a limit for the sums of independent random variables. The Poisson distribution and the Gaussian distribution are of particular interest to us since they appear so often in device problems.

That the sum of independent random variables, regardless of the underlying distribution of the independent random variables, gravitate towards the Gaussian normal distribution is a remarkable conclusion from probability theory and in nature's behavior. The Gaussian normal is the great attractor for much of what we observe. Probabilistic and statistical methods that work for the Gaussian normal distributions are also useful for other types of distributions.

E.1   *The Poisson distribution*

THE POISSON DISTRIBUTION is the law of small numbers. It describes the probability distribution of discrete events subject to a binomial probability density function and with a known average in the limit of vanishing probability.

Proposition: The probability of random discrete events $k = 0, 1, 2, \ldots$, given a number of events in a fixed interval, with a known average $\lambda > 0$, and independent of any prior, is

$$f(k, \lambda) = \frac{\lambda^k}{k!} \exp(-\lambda). \tag{E.1}$$

PROOF: For a binomial distribution $X \equiv B(n, \mathfrak{p})$, the probability density function is

$$\mathfrak{P}(X = k) = {}^nC_k \mathfrak{p}^k (1 - \mathfrak{p})^{n-k}, \quad k = 0, 1, 2, \ldots \tag{E.2}$$

where the average is $\lambda = n\mathfrak{p}$. n is the size of the sample, and $\mathfrak{p}$ is the probability of each event k. In the limit $n \to \infty$, with $\lambda = n\mathfrak{p}$ staying constant, $\mathfrak{p} = \lambda/n \to 0$. In this limit, the binomial distribution,

$$
\begin{aligned}
\lim_{n \to \infty} \mathfrak{P}(X = k) &= \lim_{n \to \infty} {}^nC_k \mathfrak{p}^k (1 - \mathfrak{p})^{n-k} \\
&= \lim_{n \to \infty} \frac{n!}{k!(n-k)!} \left(\frac{\lambda}{n}\right)^k \left(1 - \frac{\lambda}{n}\right)^{n-k} \\
&= \frac{\lambda^k}{k!} \lim_{n \to \infty} \frac{n!}{(n-k)!(n-\lambda)^k} \left(1 - \frac{\lambda}{n}\right)^n. \tag{E.3}
\end{aligned}
$$

The two terms of the large sample size limit can be evaluated separately. For the first term,

$$
\begin{aligned}
\lim_{n \to \infty} \frac{n!}{(n-k)!(n-\lambda)^k} &= \lim_{n \to \infty} \frac{n(n-1)\cdots(n-k+1)}{(n-\lambda)^k} \\
&= \lim_{n \to \infty} \frac{n^k}{n^k} = 1. \tag{E.4}
\end{aligned}
$$

For the second term, since

$$\exp(1) = \lim_{x \to \infty} \left(1 + \frac{1}{x}\right)^x, \tag{E.5}$$

we get, by substituting $x = -n/\lambda$,

$$\lim_{n \to \infty} \left(1 - \frac{\lambda}{n}\right)^n = \lim_{x \to \infty} \left(1 + \frac{1}{x}\right)^{-\lambda x} = \exp(-\lambda); \tag{E.6}$$

therefore,

$$\mathfrak{P}(X = k) = {}^nC_k \mathfrak{p}^k (1 - \mathfrak{p})^{n-k} \tag{E.7}$$

reduces to

$$f(k, \lambda) = \frac{\lambda^k}{k!} \exp(-\lambda) \qquad \text{(E.8)}$$

in the limit of small probabilities, with a constant mean for the sampling.

E.2 The Gaussian normal distribution

THE GAUSSIAN NORMAL DISTRIBUTION, in contrast to the Poisson distribution, may be viewed as the law of large numbers.

Proposition: The probability of a variable x of mean μ and variance σ is given by

$$f(x) = \frac{1}{\sqrt{2\pi}\sigma} \exp\left[-\frac{(x-\mu)^2}{2\sigma^2}\right]. \qquad \text{(E.9)}$$

PROOF: The probability of random discrete events $k = 0, 1, 2, \ldots$, given a number of events in a fixed interval and with a known probability $\mathfrak{p}$, is

$$\mathfrak{P}(X = k) = {}^nC_k \mathfrak{p}^k (1-\mathfrak{p})^{n-k} = \frac{n!}{k!(n-k)!}, \quad k = 0, 1, 2, \ldots. \qquad \text{(E.10)}$$

$\mathfrak{p}$ is finite. For large n, Stirling approximation for the factorial states is

$$n! \approx \sqrt{2\pi n}\, n^n \exp(-n). \qquad \text{(E.11)}$$

This gives

$$\mathfrak{P}(X = k) = \frac{1}{2\pi n}\left(\frac{k}{n}\right)^{-k-1/2}\left(\frac{n-k}{n}\right)^{-n+k-1/2} \mathfrak{p}^k(1-\mathfrak{p})^{n-k}. \qquad \text{(E.12)}$$

Using $x = k - n\mathfrak{p}$,

$$\left(\frac{k}{n}\right)^{-k-1/2} = \mathfrak{p}^{-k-1/2}\left(1 + \frac{x}{n\mathfrak{p}}\right)^{-k-1/2}, \text{ and}$$

$$\left(\frac{n-k}{n}\right)^{-n+k-1/2} = (1-\mathfrak{p})^{-n+k-1/2}\left(1 - \frac{x}{n(1-\mathfrak{p})}\right)^{-n+k-1/2} \qquad \text{(E.13)}$$

Substituting, this allows some simplification to

$$\begin{aligned}\mathfrak{P}(X = k) &= \frac{1}{2\pi n\mathfrak{p}(1-\mathfrak{p})} \exp\Big\{\left(-k - \frac{1}{2}\right)\ln\left(1 + \frac{x}{n\mathfrak{p}}\right) \\ &\quad + \left(-n + k - \frac{1}{2}\right)\ln\left[1 - \frac{x}{n(1-\mathfrak{p})}\right]\Big\}. \qquad \text{(E.14)}\end{aligned}$$

Now, we employ $\ln(1 + x) \approx x$ when $|x| \ll 1$, so that

$$\mathfrak{P}(X = k) = \frac{1}{2\pi n\mathfrak{p}(1-\mathfrak{p})} \exp\left[\frac{x}{n\mathfrak{p}(1-\mathfrak{p})}\left(-x + \mathfrak{p} - \frac{1}{2}\right)\right]. \qquad \text{(E.15)}$$

We may now covert back by substituting $x = k - np$:

$$\mathfrak{P}(X = k) \approx \frac{1}{\sqrt{2\pi np(1-p)}} \exp\left[-\frac{(k-np)^2}{2np(1-p)}\right]. \tag{E.16}$$

This is the Gaussian normal distribution function, where mean (μ) and variance (σ) are related to the parameters k, n and p:

$$\mu \ = \ np,$$
$$\sigma \ = \ \sqrt{np(1-p)}, \ \text{and}$$
$$f(x) \ = \ \frac{1}{\sqrt{2\pi}\sigma} \exp\left[-\frac{(x-\mu)^2}{2\sigma^2}\right]. \tag{E.17}$$

The distribution function

$$f(x) = \frac{1}{\sqrt{2\pi}} \exp\left(-\frac{x^2}{2}\right) \tag{E.18}$$

is known as the normal distribution function. It is a Gaussian normal distribution function with σ normalized to 1 and the mean—a reference—shifted to 0.

F
Fluctuations and noise

RANDOM MICROSCOPIC EVENTS are the sources of fluctuations at
the macroscopic level. Noise is the measurement of these fluctuations
in macroscopic parameters: voltage, current or power for electrical
measurements, or other parameters, such as photon count for light
sources. For electrical noise, these effects are tied through the charge
of the particle in the random events. The random phenomena—
unaccounted degrees of freedom—also show up in irreversibility and
entropy. Electrical noise is an electrical manifestation of the fluctuations. The classical particle kinetic energy fluctuation has $k_B T/2$
associated with each of the independent coordinates of motion. The
motion carried by a charged particle is the current carried by the
particle, so current fluctuates and has electrical energy fluctuation
associated with it. It exists for resistors, but it also exists for capacitors. A macroscopic capacitor, for example, under classical conditions
will have a noise energy of $k_B T/2$ related to the classical charge particle fluctuation. Different appearances and characteristics of these
fluctuations in different systems and circumstances, in frequency,
in the macroscopic electric parameters, have many different names
that have been employed historically. While the mathematics of the
description can become elaborate, in its basic mechanism, it is quite
simple. It arises from fluctuations that are random.

Our discussion here is a summary employing the machinery of
statistical mechanics. Classical thermodynamics, with its increase in
entropy as a central notion, fails. Fluctuations are the less probable
states of a system. They decrease entropy.

For any parameter of interest, we are interested in finding, statistically, when we make a measurement of the parameter, what is the
probability that the measured value departs from the most probable
value?

The answer to this question is due to Gibbs who posits the ensemble approach to evaluating most probable behavior. The Gibbs

To bring any system into equilibrium,
one must connect it to a reservoir for
particle and energy exchange so that
electrochemical and thermal equilibrium exists. If the system is classical—
and a macroscopic capacitor is—then
$k_B T/2$ of energy exists for each motional degree of freedom. It is when this
system is not macroscopic anymore,
for example, a single electron–sensitive
nanoscale dot–based capacitor, that
one will see suppression of these fluctuations. The electrostatic energy of
the single electron now dominates the
classical fluctuation.

Deterministic effects that interfere with
signals are also sometimes, unfortu-
nately, called noise. Using the word
noise, such as for cross-talk, which is
unwanted signals that couple in different electrical paths, is unwise. It can
be determined given a description of
the system and can be neutralized if
one is willing to put an effort in design. It is not random. It is cross-talk,
interference, or some other term, *but not
noise*.
In the obfuscation and confusion that
this term causes, it is similar to entropy,
its physical cousin.

The reader should follow the detailed
discussion in S. Tiwari, "Semiconductor
physics: Principles, theory and
nanoscale," Electroscience 3, Oxford
University Press, ISBN 978-0-19-
875986-7 (forthcoming) if interested
in completeness.

approach is to statistically look at the ensemble—microscopic or macroscopic—to predict the most probable behavior, using the phase-space Hamiltonian description, so of $(\mathbf{q}, \mathbf{p})$ of the particles. A characterization of this is summarized in Figure F.1. A canonical ensemble is one which approximates real experiments and can make corresponding predictions. A microcanonical ensemble is a large assembly—a heat bath—insulated from everything else so that its energy and temperature are "constant"—minuscule energy (heat) may be exchanged so that the mean energy U_0 is known within a macroscopic range of ΔU_0. A canonical ensemble exists within this microcanonical ensemble—an unfortunate nomenclature, given the hierarchy. The canonical ensemble is only in contact with the microcanonical ensemble. And this contact allows heat flow but not particle flow. The number of particles in the canonical ensemble is constant. The energy of the microcanonical ensemble is constant but that of the canonical ensemble is not. Gibbs also defines a grand canonical ensemble. The grand canonical ensemble and the microcanonical ensemble may exchange both energy (heat) and particles. So, the microcanonical ensemble has vanishing fluctuations in its energy, the canonical ensemble has fluctuations in energy, and the grand canonical has fluctuations in energy and particles. This partitioning lets us tackle an ensemble's behavior statistically as well as connect it to the real situations.

Let F_k and F_e be two physical parameters of properties of interest. If we are interested in how far these depart from the most probable value, then the moments that are of interest to us are $(\langle F_k - \langle F_k \rangle \rangle^2)$ and $(\langle F_k - \langle F_k \rangle \rangle) \times (\langle F_e - \langle F_e \rangle \rangle)$—variance and correlation. The relative fluctuation $(\langle F_k - \langle F_k \rangle \rangle / \langle F_k \rangle)$ too is of interest. If they depend only on $\mathbf{p}$—current being an example—then the distribution function suffices.

With these parameters dependent on position $\mathbf{q}$, the situation gets complicated. Consider a microcanonical ensemble with its energy constant at U within ΔU. Let S_0 be the entropy at equilibrium—the maximum entropy. This is the entropy of the most probable state. Let the system internally be dependent on some parameter ζ whose equilibrium magnitude is ζ_0. S_ζ is the entropy when the internal state of the system is characterized by ζ. Measurements made on the system will average to S_0 and ζ_0, but fluctuations will cause deviations in the individual measurements. The fluctuations are the result of the internal mechanics of the microcanonical ensemble. The Boltzmann distribution lets us write the probabilities in terms of ζ.

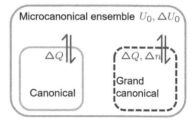

Figure F.1: The use of canonical ensembles to statistically model the most probable behavior. A microcanonical ensemble is the reservoir of free energy U_0 known to a precision of ΔU_0. A canonical ensemble may exchange heat (ΔQ), but not particles, with this reservoir. A grand canonical ensemble may exchange both heat (ΔQ) and particles (Δn) with the reservoir.

For example, energy and number density follow from this distribution function—see S. Tiwari, "Quantum, statistical and information mechanics: A unified introduction," Electroscience 1, Oxford University Press, ISBN 978-0-19-875985-0 (forthcoming)—because they are only functions of $\mathbf{p}$.

$$d\mathfrak{p} \propto \exp\left(\frac{S_\zeta - S_0}{k_B}\right) d\zeta = \exp\left(\frac{\Delta S}{k_B}\right) d\zeta, \quad \text{where} \quad \int d\mathfrak{p} = 1. \quad \text{(F.1)}$$

ΔS is negative, since S_0 is the maximum entropy. Now, we use this for a canonical ensemble with energy transport and which is a part of the microcanonical ensemble that is the reservoir at T_0. A change in the parameter ζ for the canonical ensemble happens with work dW_ζ performed on it. The entropy change of $\Delta S = \Delta S_0 + \Delta S_s$ represents the change in entropy of the reservoir (the microcanonical ensemble) and that of the canonical subsystem. The result for $d\mathfrak{p}$ is for the entirety. No particles are exchanged. The entropy change in the subsystem is therefore its free energy change, the work it performed, and any other terms for change not involving particle exchange, so the term $p_0 \Delta V_s$, which is the due to change in volume at the pressure p_0, can be introduced:

$$\Delta S_s = \frac{\Delta U_s + p_0 \Delta V_s - \Delta W_\zeta}{T_0}, \text{ and}$$

$$\Delta S_0 = \frac{\Delta U_0 + p_0 \Delta V_0}{T_0}. \tag{F.2}$$

Total energy and volume are conserved, so $\Delta U_s = -\Delta U_0$, and $\Delta V_s = -\Delta V_0$; so,

$$\Delta S_s = -\Delta S_0 - \frac{dW_\zeta}{T_0}, \tag{F.3}$$

and our relative probability equation reduces to the form

$$d\mathfrak{p} \propto \exp\left(-\frac{dW_\zeta}{k_B T_0}\right). \tag{F.4}$$

$dW_\zeta = \Pi_\zeta - \Pi_{\zeta 0}$ is the work done—al potential Π_ζ, which is the cause of the force and the field that does the work. For differences, from Taylor expansion of the potential around ζ_0, the first significant term that contributes is the second order term $(1/2)\Pi_{\zeta 0}^2 (\zeta - \zeta_0)^2$, with the expansion around the equilibrium of $\Pi(\zeta_0) = \Pi_{\zeta 0}$, which is the minimum. So,

This symbol change for potential from V to Π is restricted to only this discussion. It is meant to avoid confusion with the volume V that we have also used throughout.

$$d\mathfrak{p} = \frac{\exp\left[-\Pi_{\zeta 0}^2 (\zeta - \zeta_0)^2 / 2k_B T_0\right]}{\int \exp\left[-\Pi_{\zeta 0}^2 (\zeta - \zeta_0)^2 / 2k_B T_0\right] d\zeta}, \tag{F.5}$$

taking into account the normalization over all possibilities.

We can now extract the fluctuations and the distribution function of ζ:

$$\langle (\zeta - \zeta_0)^2 \rangle = \sigma_\zeta^2 = \frac{\int (\zeta - \zeta_0)^2 \exp\left[-\Pi_{\zeta 0}^2 (\zeta - \zeta_0)^2 / 2k_B T_0\right] d\zeta}{\int \exp\left[-\Pi_{\zeta 0}^2 (\zeta - \zeta_0)^2 / 2k_B T_0\right] d\zeta}$$

$$\approx \frac{k_B T_0}{\Pi_{\zeta 0}^2}, \tag{F.6}$$

where the integral can be expanded to ∞ in both directions since the exponential falls rapidly. So,

$$
\begin{aligned}
d\mathfrak{p} &= \frac{1}{\left[2\pi\langle(\zeta-\zeta_0)^2\rangle\right]^{1/2}} \exp\left[-\frac{(\zeta-\zeta_0)^2}{2(\zeta-\zeta_0)^2}\right] d\zeta \\
&= \frac{1}{(2\pi\sigma_\zeta^2)^{-1/2}} \exp\left[-\frac{(\zeta-\zeta_0)^2}{2\sigma_\zeta^2}\right] d\zeta.
\end{aligned}
\tag{F.7}
$$

This is a significant result. The change in probabilities as specified in the Boltzmann relationship reduces to a Gaussian normal relationship in terms of the internal state parameter ζ. The distribution function's width is the second moment. Equation F.6 gives the variance and directly corresponds to many fluctuations of interest. For example, it says in its straightforward proportionality with temperature T_0 that fluctuations increase with temperature. Large fluctuations, that is, large σ_ζ, will mean that the exponential will dominate over the inverse σ_ζ dependence of the prefactor.

But, the most significant implication is that we have found Gaussian normal distribution function to be a central probability distribution function using statistical mechanics capable of describing these higher order properties such as fluctuation. Fluctuation, as we started in the beginning, is one of the moments of the properties that are of interest to us.

The interesting corollary here is that, in measurements, we often find $\langle x \rangle$, $\langle x^2 \rangle$, et cetera, and are interested in finding the probability distribution function $f(x)$ that this corresponds to, or at least a useful approximation to it. The moment-generating function and the characteristic function are two useful tools for this. The moment generating function is

$$
M(\alpha) = \int_{-\infty}^{\infty} \exp(\alpha x) f(x)\, dx,
\tag{F.8}
$$

where α is a real number, and $f(x)$ is the probability distribution function. $M(\alpha)$ is the average of $\exp(\alpha x)$. So,

$$
\begin{aligned}
M(\alpha) &= \langle\exp(\alpha x)\rangle = 1 + \alpha\langle x\rangle + \frac{1}{2!}\alpha^2\langle x^2\rangle \\
&\quad + \cdots + \frac{1}{n!}\alpha^n\langle x^n\rangle + \cdots,
\end{aligned}
\tag{F.9}
$$

leading to

$$
\frac{dM(\alpha)}{d\alpha} = \langle x\rangle + \frac{1}{2!}2\alpha\langle x^2\rangle + \cdots + \frac{1}{n!}n\alpha^{n-1}\langle x^n\rangle + \cdots,
\tag{F.10}
$$

which, recursively, with $\alpha = 0$, gives

$$
\left.\frac{dM(\alpha)}{d\alpha}\right|_\alpha = 0 = \langle x\rangle,
$$

$$\frac{d^2 M(\alpha)}{d\alpha^2}\bigg|_\alpha = 0 = \langle x^2 \rangle,$$

$$\frac{d^3 M(\alpha)}{d\alpha^3}\bigg|_\alpha = 0 = \langle x^3 \rangle, \tag{F.11}$$

and so on. These are the moments of the distribution, and $M(\alpha)$ is the moment generating function.

The characteristic function is

$$C(\alpha) = \int_{-\infty}^{\infty} \exp(i\alpha x) f(x) \, dx. \tag{F.12}$$

It is the Fourier transform of the probability distribution function. So,

$$f(x) = \frac{1}{2\pi} \int_{-\infty}^{\infty} C(\alpha) \exp(-i\alpha x) \, d\alpha. \tag{F.13}$$

Similar to the case for the moment generating function,

$$C(\alpha) = \langle \exp(i\alpha x) \rangle, \tag{F.14}$$

with $C(0) = 1$. The MacLaurin expansion form can be summarily written as

$$C(\alpha) = 1 + \sum_{n=1}^{\infty} \frac{1}{n!} (i\alpha)^n m_n, \quad \text{where } m_n = \frac{1}{i^n} \frac{d^n C(\alpha)}{d\alpha^n}\bigg|_{\alpha=0}. \tag{F.15}$$

Another form for writing this is by expanding around $\alpha = 0$ as

$$C(\alpha) = \exp\left[\sum_{n=1}^{\infty} \frac{(i\alpha)^n}{n!} k_n\right], \tag{F.16}$$

where the cumulant k_n and the moments are related:

$$
\begin{aligned}
k_1 &= m_1, \text{ the mean of the probability distribution function (pdf)}, \\
k_2 &= m_2 - m_1^2 = \langle f^2(x) \rangle - \langle f(x) \rangle^2, \text{ the variance of the (pdf)}, \\
k_3 &= m_3 - 3m_1 m_2 + 2m_1^3, \tag{F.17}
\end{aligned}
$$

and so on. k_1 is the mean of the *pdf*, and k_2 is the variance. So, the standard deviation is

$$\sigma(x) = k_2^{1/2} = \left[\langle f^2(x) \rangle - \langle f(x) \rangle^2\right]^{1/2}. \tag{F.18}$$

If we make a measurement and know k_1 and k_2, then the first approximation for the characteristic function is

$$C(\alpha) \approx \exp\left(i\alpha k_1 - \frac{\alpha^2}{2} k_2\right). \tag{F.19}$$

The inversion of this, via the Fourier transform, is

$$
\begin{aligned}
f(x) &\approx \frac{1}{2\pi} \int_{-\infty}^{\infty} \exp(-i\alpha x) \exp\left(i\alpha k_1 - \frac{\alpha^2}{2} k_2\right) d\alpha \\
&= \frac{1}{(2\pi k_2)^{1/2}} \exp\left[-\frac{(x - k_1)^2}{2k_2}\right], \tag{F.20}
\end{aligned}
$$

a Gaussian distribution.

When two, and only two, cumulants are known, the Gaussian distribution is the most likely distribution.

We have now established the primacy of Gaussian distribution. We have also made the connection between fluctuations and the statistical mechanics–derived probability distribution function, and between measurements and, from them, finding probability distribution function approximations.

A number of distribution functions have been encountered in this text: the binomial distribution, which corresponds to events where there are only two possible outcomes; the Poisson distribution, which is arrived at from the binomial distribution in the limit of very small probabilities in large event numbers whose average is finite; and the Gaussian distribution, which has the form $x \exp(-\alpha x^2)$. The latter appears all around: in statistical mechanics and nearly anywhere where one may need to employ probabilities. This text has encountered it in many places. The binomial distribution transits to the Gaussian distribution in the limit of large numbers, such as those in statistical mechanics. But, it also is a good approximation with small numbers. This, however, needs a more careful peek than warranted in this appendix.

We now have the tools to understand noise arising from fluctuations in equilibrium and in non-equilibrium. In equilibrium, the fluctuations arise because the second moment is finite, even if the long-time average vanishes. When a canonical system is approaching equilibrium, and the departure from equilibrium is small, so that our discussion above applies, then fluctuations that lead to increase in entropy are more probable than those that decrease it, according to Equation F.1. The subsystem approaches equilibrium because fluctuations bring the subsystem along to equilibrium, on average. In our discussion in Chapter 5, we touched on the frequency decomposition of this. When the time is long, we may view the situation as a stationary random process as analyzed here. Noise also arises from different causes. Fluctuation is one. But, noise can also occur because of what is fed into the system, or because of discrete events that adhere to a mean but have low probability: electrons arriving at a collecting electrode, bit packages at a receiver, photons from a laser, and so on.

Stochastic theory and its use of both correlation in time and the fluctuation-dissipation theorem are the primary tools for analyzing noise in equilibrium.

Strictly speaking, we write the Gaussian distribution as

$$\mathfrak{P}(x) = \frac{1}{\sqrt{2\pi}\sigma_x} \exp\left[-\frac{(x - \langle x \rangle)^2}{2\sigma_x^2}\right].$$

The normal distribution—a unit normal distribution—is

$$\mathfrak{P}(x) = \frac{1}{\sqrt{2\pi}} \exp\left(-\frac{x^2}{2}\right).$$

Note that the Poisson distribution of the dopants' effects in the threshold voltage distribution of transistors in real measurements is modeled by the Gaussian distribution just fine. And the theory pointed to the small discrepancies.

A stationary random process is one in which energy over a long time interval is independent of when the time interval began. Cross-talk, atmospheric discharge, intermittent effects of poor contact, et cetera, are not noise.

F.1   Thermal noise

First consider passive elements such as resistors, inductors and capacitors. Consider the arrangement (shown in Figure F.2) of two resistances R connected through a lossless transmission line of characteristic impedance Z_0, so all matched. Under classical conditions, so ones where $\hbar\omega \ll k_B T$, one may consider the transverse electric (TE), transverse magnetic (TM) and transverse electromagnetic (TEM) modes for this system in thermal equilibrium. Consider only TEM modes—the argument with TE and TM is no different. The modes of the line are

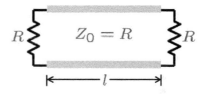

Figure F.2: Two resistors of resistance R connected through a lossless transmission line of characteristic impedance R.

$$\nu_n = n\frac{c}{2l} = n\nu_1, \quad \text{where} \quad n = 1, 2, \dots \tag{F.21}$$

where the line is of length l, c is the phase velocity, and $\nu_1 = c/2l$ is the fundamental resonance frequency. In any interval $\Delta\nu$ the number of resonances is $\Delta N = \Delta\nu/\nu_1 = (2l/c)\Delta\nu$. The electrical and magnetic fields of each mode contribute $k_B T/2$, of energy according to the equipartition energy of classical limits. So, the total energy in the frequency band $\Delta\nu$ is $(2l/c)k_B T\Delta\nu$. Since it takes l/c time to travel the length of the lossless line, the power is $P = 2k_B T\Delta\nu$. Each resistor contributes half of this; therefore,

$$\langle \Delta V_n^2 \rangle = 4k_B T R\Delta\nu. \tag{F.22}$$

This derivation is unsatisfactory in many respects. We chose certain modes, we chose resonances as the reservoirs of the $k_B T$ of energy, and there is no connection to the source of the noise, which magically appears through the classical equipartition rule—all justifiable assumptions but with contortions.

One may use a stochastic argument to arrive at the result more satisfactorily. Consider a resistor-inductor circuit as shown in Figure F.3, where $V(t)$ is the stochastic noise source characterizing the resistor. The Kirchoff current equation is

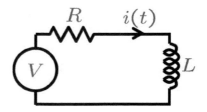

$$V(t) = L\frac{di}{dt} + iR \quad \therefore \quad \frac{di}{dt} + \frac{i}{\tau_c} = \frac{V(t)}{L}, \quad \text{where} \quad \tau_c = \frac{L}{R}. \tag{F.23}$$

Figure F.3: A noise-equivalent voltage source driving a resistor that is the source of the noise, and an inductor.

For $V(t)$ as a stochastic process, $\langle V(t) \rangle = 0$, and $\langle V^2(t) \rangle \neq 0$. Integration of Equation F.23 leads to

$$i = i_0 \exp\left(-\frac{t}{\tau_c}\right) + \exp\left(-\frac{t}{\tau_c}\right) \int_0^t \exp\left(\frac{t'}{\tau_c}\right) \frac{V(t')}{L} dt'. \tag{F.24}$$

The current-correlation function can now be written:

$$\langle i(t)i(t+\tau)\rangle = \left\langle \left[i_0 \exp\left(-\frac{t}{\tau_c}\right) + \exp\left(-\frac{t}{\tau_c}\right) \int_0^t \exp\left(\frac{t'}{\tau_c}\right) \frac{V(t')}{L} dt' \right] \right.$$

$$\times \left[i_0 \exp\left(-\frac{t+\tau}{\tau_c} \right) \right.$$
$$\left. + \exp\left(-\frac{t+\tau}{\tau_c} \right) \int_0^{t+\tau} \exp\left(\frac{t''}{\tau_c} \right) \frac{V(t'')}{L} \, dt'' \right]\Big\rangle$$
$$= i_0^2 \exp\left(-\frac{2t+\tau}{\tau_c} \right)$$
$$+ \exp\left(-\frac{2t+\tau}{\tau_c} \right) \int_0^t dt' \int_0^{t+\tau} \exp\left(\frac{t'+t''}{\tau_c} \right) \frac{\langle V(t')V(t'')\rangle}{L^2} \, dt''.$$

$$(\text{F.25})$$

$\langle V(t')V(t'')\rangle$, which we will call $K_v(s)$, is the voltage autocorrelation function of the source driving the circuit. It depends only on $s = t'' - t'$. It is symmetric, peaking at $s = 0$ and then decaying rapidly away. The double integral therefore can be simplified by using a change in coordinates and by changing the limits to $\pm\infty$:

$$\langle i(t)i(t+\tau)\rangle = i_0^2 \exp\left(-\frac{2t+\tau}{\tau_c} \right)$$
$$+ \frac{\tau_c}{2L^2}\left[1 - \exp\left(-\frac{2t}{\tau_c} \right) \right] \int_{-\infty}^{\infty} K_v(s) \, ds. \quad (\text{F.26})$$

In the limit of $t \to 0$, this is

$$\langle i(0)i(\tau)\rangle = i_0^2 \exp\left(-\frac{|\tau|}{\tau_c} \right) = i_0^2 \exp\left(-\frac{R|\tau|}{L} \right), \quad (\text{F.27})$$

which is the current autocorrelation function of the circuit; and, in the limit of $t \gg \tau$,

$$\langle i^2(t)\rangle = i_0^2 \exp\left(-\frac{2t}{\tau_c} \right) + \frac{\tau_c}{2L^2}\left[1 - \exp\left(-\frac{2t}{\tau_c} \right) \right] \int_{-\infty}^{\infty} K_v(s) \, ds$$
$$\approx \frac{\tau_c}{2L^2} \int_{-\infty}^{\infty} K_v(s) \, ds = \frac{1}{2LR} \int_{-\infty}^{\infty} K_v(s) \, ds. \quad (\text{F.28})$$

If we employ the equipartition theorem here, so $(1/2)L\langle i^2(t)\rangle = (1/2)k_B T$, this gives

$$R = \frac{1}{2k_B T} \int_{-\infty}^{\infty} K_v(s) \, ds. \quad (\text{F.29})$$

A resistance is the dissipation consequence of fluctuations. This equation represents the fluctuation-dissipation theorem's consequence in this circuit. It represents the consequence of friction for current. In the limit of $t \to \infty$,

$$K_i(\tau) = \langle i(0)i(\tau)\rangle = \frac{k_B T}{L} \exp\left(-\frac{|\tau|}{\tau_c} \right). \quad (\text{F.30})$$

We can now employ the Wiener-Khinchin theorem:

We encountered correlations in our discussion of the sensitivity limits of force measurement with probes, while discussing electromechanics. Noise, as we remarked there, is time fluctuation, and the Brownian motion that affects the sensitivity of measurement is a spatial observation. Figure 5.16 there is the geometric projection of the correlation integral of this current autocorrelation equation.

In invoking the equipartition theorem, we are restricting this analysis to systems subject to classical constraints.

See S. Tiwari, "Semiconductor physics: Principles, theory and nanoscale," Electroscience 3, Oxford University Press, ISBN 978-0-19-875986-7 (forthcoming) for a discussion of Wiener-Khinchin, fluctuation-dissipation and their connections to the Kramers-Kronig relationship.

$$\phi_i(\omega) = \frac{2k_BT}{2\pi L}\int_0^{\infty} \exp\left(-\frac{|\tau|}{\tau_c}\right)\exp(-i\omega\tau)d\tau$$

$$= \frac{k_BT}{\pi L}\frac{\tau_c}{1+\omega^2\tau_c^2} = \frac{k_BT}{\pi}\frac{R}{R^2+\omega^2L^2}. \qquad (F.31)$$

The frequency dependence of correlations—mean square fluctuations—then follows as

$$\langle i^2(\omega)\rangle = 2\phi_i(\omega)\,d\omega = \frac{4k_BT}{R}\frac{d\nu}{1+(\omega L/R)^2}, \quad \text{and}$$

$$\langle V^2(\omega)\rangle = 4k_BTR\frac{\omega L/R}{1+(\omega L/R)^2}d\nu. \qquad (F.32)$$

In the limit of $L \to 0$, $\langle i^2(\omega)\rangle = 4k_BTGd\nu$, where $G = 1/R$ is the conductance. This may be represented as a current noise source in parallel with the resistor, as shown in Figure F.4(a). One may transform this to a voltage form through Thevenin equivalence. In this case it is a voltage noise source in series with resistor of $\langle V^2(\omega)\rangle = 4k_BTRd\nu$ as shown in Figure F.4(b).

The thermal noise source has now been placed on a clear mathematical probabilistic basis. The ability to associate noise sources with simple relationships to describe correlations is quite remarkable. However, we must place limitations that arise from the assumptions made. When using equipartition energy, one is also associating a spectral power density at unlimited frequencies. Quantum mechanics tells us the nature of mode density and places limits on it. Our correlation was derived using the time constant of the circuit ($\tau_c = L/R$, in the specific case). In Chapter 5, the time constant was related to mass and viscosity and approximated the time between collision events. In the Boltzmann transport equation in Chapter 2, we had similarly employed momentum relaxation time. It is this collision time that will specify the noise characteristics arising from fluctuations for the resistor. The high frequency limit, with the quantum constraint, is

$$\langle U\rangle = \left[\frac{1}{2} + \frac{1}{\exp(h\nu/k_BT)-1}\right]h\nu, \qquad (F.33)$$

where $(1/2)h\nu$ is the zero point energy. The resistor then has the noise spectrum

$$S_N(\nu) = 4G\left[\frac{1}{2}h\nu + \frac{h\nu}{\exp(h\nu/k_BT)-1}\right]. \qquad (F.34)$$

The spectrum of noise extends to a high frequency but then collapses with an $h\nu/\exp(h\nu/k_B)$ dependence.

We have derived this thermal noise in equilibrium conditions. This thermal noise does not tell us anything about the conductor. It only tells us what the resistance or conductance is. The resistor is the

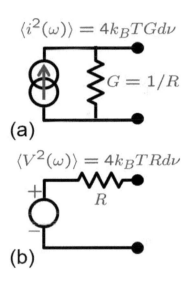

$$\langle i^2(\omega)\rangle = 4k_BTGd\nu$$

$$G = 1/R$$

(a)

$$\langle V^2(\omega)\rangle = 4k_BTRd\nu$$

$$R$$

(b)

Figure F.4: The Thevenin equivalence of current noise source (a) and voltage source (b).

The energy, mode density and other quantum-specified considerations are discussed at length in S. Tiwari, "Quantum, statistical and information mechanics: A unified introduction," Electroscience 1, Oxford University Press, ISBN 978-0-19-875985-0 (forthcoming).

source of thermal noise. An inductor or a capacitor is a noise source in equilibrium only in as much as its resistance is. Equilibrium here arises in the electrical element being in contact with a reservoir that establishes the temperatures and exchanges electrons—the particles being exchanged in this equilibrium—and which are the source of fluctuations.

F.2 $1/f$ noise

A second type of equilibrium fluctuation arises from the time-dependent fluctuation of resistance. An example of this is the $1/f$ or other, similar, inverse-frequency dependences in noise, as we encountered in the form of the random telegraph signal in Chapter 2. Since the current fluctuates proportionally to the resistance fluctuation, the power spectrum then is proportional to i^2. These processes, of which there are many, are usually thermally activated. The power spectrum and the resistance fluctuation therefore decrease with decreasing temperature. Both thermal and $1/f$ noise are examples of noise that is probed by the power spectrum—or current or voltage—arising from resistance fluctuations within. $1/f$ noise is also called flicker noise in association with its low frequency signature.

F.3   Shot noise

Shot noise is an example of non-equilibrium fluctuation. It is generated by the passing of current rather than probed by it, and it arises from the discreteness of the particle, whose charge flux is composed of discrete charge pulses in time. If a material has a current density of 1 A/cm^2 flowing through it, this corresponds to an average of $(1.6 \times 10^{-19})^{-1} \approx 6 \times 10^{18}$ electrons passing through a unit area in a second. An absolutely uniform flow will be at an arrival spacing of 1.6×10^{-19} s between the electrons. So, fluctuations from this uniformity will appear as noise. This is the shot noise. It is a fluctuation in the signal from the timing, while thermal noise is noise in the amplitude. Injection from electrodes will have features of a similar nature. Transit through p/n junction regions as the carriers are collected in the quasineutral regions will have this effect. So, it is quite ubiquitous. No contact, no current: it is this discreteness in passage in time that leads to the noise being called ``shot noise''. The noise, is in the current $\Delta i(t) \equiv i(t) - I_0$. As with thermal noise, we may find shot noise's power spectrum using current correlation, so

Time and amplitude can just as well be seen as canonical coordinates to define the signal. So, shot noise and thermal noise are complementary and have many equivalences.

$$S_N(\omega) = \int_{-\infty}^{\infty} \langle \Delta i(t + \tau)\Delta i(\tau)\rangle \exp(i\omega t)\, dt. \qquad (F.35)$$

The discreteness of the carriers crossing the boundary, together with its low probability in time, when a mean current flow is maintained, means that we may model the probability distribution in time by using the Poisson distribution function. We take a situation where electrons are being injected discretely into a region and then being collected. Current continuity means that the transit of the electron is associated with the current while the electron is flowing between the electrodes. The total current is the sum of the current pulses, which are discretely shifted in time:

$$i(t) = \sum_k \breve{i}(t - t_k), \qquad (F.36)$$

where $\breve{i}(t)$ is current pulse associated with the transfer of an electron at $t = 0$. t_k is the time when the kth electron was injected into the transit region. Let τ_w be the width of the one electron current pulse, and consider a time interval $[-T/2, T/2]$ where $T \gg \tau_w$. We now define the probability density $\mathfrak{P}^n_T(t_1, \ldots, t_n)$ for n electrons to be transmitted at the times $t_1, \ldots, t_n$. Our Poisson distribution can be written as

$$\mathfrak{P}^n_T(t_1, \ldots, t_n) = \frac{\lambda^n}{n!} \exp(-\lambda T) = \frac{(\lambda T)^n}{n!} \exp(-\lambda T) T^{-n}. \qquad (F.37)$$

λ is the average number of electrons injected per unit time. So $I_0 = e\lambda$. The average current over the ensemble at time t in the interval $[-T/2 + \tau_w, T/2 - \tau_w]$ is

$$
\begin{aligned}
\langle i(t) \rangle &= \sum_{n=0}^{\infty} \int_{-T/2}^{T/2} \left[\mathfrak{P}^n_T(t_1, \ldots, t_n) \sum_k \breve{i}(t - t_k) \right] dt_1 \cdots dt_n \\
&= \sum_{n=0}^{\infty} \frac{(\lambda T)^n}{n!} \exp(-\lambda T) \frac{n}{T} \int_{-T/2}^{T/2} \breve{i}(t - t') \, dt' \\
&= e\lambda \quad \because \sum_n n \frac{(\lambda T)^n}{n!} \exp(-\lambda T) = \lambda T, \qquad (F.38)
\end{aligned}
$$

which is the average current result expected.

The shot noise power now follows from the correlation

$$
\begin{aligned}
\langle \Delta i(t) \Delta i(0) \rangle &= \sum_{n=0}^{\infty} \int_{-T/2}^{T/2} P^n_T(t_1, \ldots, t_n) \left[\sum_{k=1}^{n} \breve{i}(t - t_k) - I_0 \right] \\
&\quad \times \left[\sum_{l=1}^{n} \breve{i}(t_l) - I_0 \right] dt_1 \cdots dt_n \\
&= \sum_{n=0}^{\infty} \frac{(\lambda T)^n}{n!} \exp(-\lambda T) \left[\frac{n}{T} \int_{-T/2}^{T/2} \breve{i}(t - t') \breve{i}(-t') \, dt' \right. \\
&\quad \left. + \frac{n(n-1)e^2}{T^2} + \frac{2ne^2\lambda}{T} + e^2\lambda^2 \right]
\end{aligned}
$$

As discussed in S. Tiwari, "Device physics: Fundamentals of electronics and optoelectronics," Electroscience 2, Oxford University Press, ISBN 978-0-19-875984-3 (forthcoming) in connection with signal delay in the base-collector region or the base-emitter region of the transistor, the total current is the sum of displacement and particle current. Current continuity is for this sum current, so the form of signal delay is a function of the transit dynamics. At constant velocity, it is half the transit time, since an electron after being half-way through the region has the field terminating at the collecting electrode. The collecting electrode feels the electron once it is past the half-way point.

$$= \lambda \int_{-T/2}^{T/2} \breve{\imath}(t - t')\breve{\imath}(-t') \, dt'$$

$$\because \sum_n n(n-1) \frac{(\lambda T)^n}{n!} \exp(-\lambda T) = (\lambda T)^2. \qquad \text{(F.39)}$$

This simple final equation form says that each current pulse is only correlated with itself. Equation F.35 and Equation F.39 can now be put together. $T \gg \tau_w$ and $|t| > \tau_w$, so the integral limits can be stretched to $\pm\infty$ to get the shot noise power spectrum as

$$S_N(\omega) = 2\lambda |\breve{I}(\omega)|^2 = 2\lambda \left| \int_{-\infty}^{\infty} \breve{\imath}(t) \exp(i\omega t) \, dt \right|^2. \qquad \text{(F.40)}$$

$\breve{I}(\omega)$ is the Fourier transform of a single current pulse. If the current pulse is very narrow, so $\tau \ll \omega^{-1}$—a δ-function—its Fourier spectrum is broad and constant, and we obtain $\breve{I}(\omega) = e$, so that

$$S_N(\omega) = 2\lambda e^2 = 2eI_0, \qquad \text{(F.41)}$$

and

$$\langle i^2(t) \rangle = 2eI_0 \Delta\nu. \qquad \text{(F.42)}$$

Shot noise has a broad spectrum—nearly white. It is only at $\omega > 1/\tau_w$ that the shot noise vanishes.

Having discussed electrical conduction behavior at the mesoscale as a scattering problem in Chapter 3, we can now relate the noise discussion here for its implications for the mesoscale and the nanoscale. The conductance is determined by a transmission matrix of electrons at the Fermi energy, with phase coherence in these conditions. The approach for understanding this behavior at the mesoscale was to bring in the essential quantum constraints and yet work with relatively large number of particles, so that a completely quantum approach would be impractical. Noise at the mesoscale and at the nanoscale will be different from that at the classical limits, just as the aggregate device behavior was. Such noise is very interesting but beyond our scope. So, we will make only a few comments here.

Just as current can be written in terms of quantum conductance and transmission matrix elements, so can noise power. In general, the conductance at zero temperature in an N-port sample is

$$G = \frac{e^2}{h} \text{Tr}(tt^\dagger) = \frac{e^2}{h} \sum_{n=1}^{N} \mathcal{T}_n. \qquad \text{(F.43)}$$

The thermal noise, determined at equilibrium, i.e, $V = 0$, has the power spectrum

$$S_N = 4k_B T \frac{e^2}{h} \text{Tr}(tt^\dagger) = 4k_B T \frac{e^2}{h} \sum_{n=1}^{N} \mathcal{T}_n, \qquad \text{(F.44)}$$

and the shot noise power spectrum can be written as

$$S_N = 2\frac{e^2}{h} \sum_{n=1}^{N} \left[2k_B T \mathscr{T}_n^2 + \mathscr{T}_n(1 - \mathscr{T}_n)e|V| \coth\left(\frac{e|V|}{2k_B T}\right) \right]. \qquad \text{(F.45)}$$

Here, $\mathscr{T}_n$ is the eigenvalue of $tt^\dagger$, which is the off-diagonal transmission term of the scattering matrix. One can see the correspondence between the Nyquist classical result and the mesoscale result for thermal noise, in terms of the conductance or resistance of the sample. Shot noise too can be seen to have proportionality to current, as it does in the classical result.

Shot noise is also very interesting in quantum point contacts. It is only between the conductance plateaus conductances which step by $2e^2/h$, that the shot noise appears. At the plateaus, it vanishes, since the channels are occupied, and modes for transport fluctuations are unavailable.

G
Dimensionality and state distribution

A PARTICLE IN A BOX is confined. Its wavefunction is a standing wave with nodes at the boundaries of the box, with an infinite potential barrier. The time-independent Schrödinger equation,

$$-\frac{\hbar^2}{2m^*}\nabla^2\psi + V\psi = E\psi, \tag{G.1}$$

where ψ is the wavefunction, V is the potential, and E is the energy, describes the quantum mechanics of the system, that is, its solution provides us with the particle wavefunction useful for determining the observables of the system, including the energy states that the particle can have. Since this is a second order differential equation, one can write a solution in the form

$$\psi(\mathbf{r}, t) = A\exp[+i(\mathbf{k}\cdot\mathbf{r} - \omega t)], \tag{G.2}$$

with the **k**s quantized appropriately to the boundary conditions. The solutions are standing waves formed from opposite and equal **k** wavevector solutions.

For each confined direction, $k = \pm n\pi/L$, where $n = 1, 2, 3, \ldots$. Figure G.1 shows this confinement in the z direction. The standing wave is formed by two counter-propagating waves of quantized opposite wavevectors. These are uniformly distributed. This relationship implies, that the larger the confined space is, the smaller the wavevector and the energy spacing of the state, since energy is $E = \hbar^2 k^2/2m^*$. At large dimensions, such as for a wafer, or a classical dimension, such as source-to-drain spacing, this L dimension is large—the states are close together, much closer than thermal energy ($k_B T$), except at near absolute zero temperatures—and one may view the distribution of the states as being continuous. The particle is confined to this large crystal box, and the electrons easily propagate in directions where the states are relatively unconfined. In the example shown in Figure G.1, this freedom exists in the x and the y directions. In

For a rigorous discussion, see S. Tiwari, "Quantum, statistical and information mechanics: A unified introduction," Electroscience 1, Oxford University Press, ISBN 978-0-19-875985-0 (forthcoming).

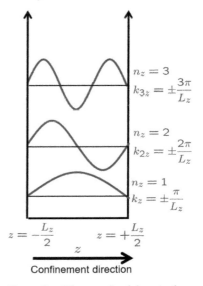

Figure G.1: When confined, here in the z direction, with infinite energy barriers, the states available for occupation are noticeably separated in energy—noticeably in the sense that the energy separation may be larger than $k_B T$. This figure shows the wavefunction formed as a standing wave composed of two counter-propagating waves of wavevector k_z in the confinement direction of z.

the confined direction, that is, in a dimension where the L is now quite small—of the de Broglie wavelength scale of the electron—the movement is restricted in the direction of this confinement. In Figure G.1, the electron's probability distribution's peak is displaced only marginally when moving from the ground to the first state—within the small, confined, well dimension, with k_x and k_y remaining constant. The electron remains confined in the z direction, but it just spreads out a little more within the well. This simple picture for the allowed states for a single effective mass m^* holds true whether L is large or small. If it is large, the states are close together, and an electron may travel around in the real and reciprocal space under the influence of energetic interactions. If it is small, it is restricted from moving in that direction. It is a standing wave.

Density of states is a measure of the number of states per unit wavevector or energy, in unit spatial coordinates. It is three dimensional ($\mathscr{G}_{3D}(E)$ or $\mathscr{G}_{3D}(k)$) if unconfined, two dimensional ($\mathscr{G}_{2D}(E)$ or $\mathscr{G}_{2D}(k)$) if confined in one dimension, and one dimensional ($\mathscr{G}_{1D}(E)$ or $\mathscr{G}_{1D}(k)$) if confined in two dimensions. It is the number of freedom of movement directions that are employed for the subscript. If all directions are confined, it cannot move, the expectation of the spatial coordinate is a constant, and it is confined with no states nearby for motion at dimensions larger than the wavelength of the particle.

When one of the dimensions is confined, z in our example, the wavevector of this confined mode is $k_z = \pm n_z \pi / L_z$. These states are π / L_z apart. For confinement at a small L, these may be quite separated, but, with a larger L, they come closer together. The state density in k-space is

$$\frac{dn_z}{dk_z} = g_s \frac{1}{2} \frac{L_z}{\pi}, \tag{G.3}$$

which represents a continuum approximation to the discrete $\Delta n_z / \Delta k_z$. Here, $g_s = 2$ is the spin degeneracy ($s = 1/2$; $m_s = \pm 1/2$), which accounts for electrons of opposite secondary spins—a different quantum number—occupying the specific n_z quantum number states. The factor 2 in the denominator indicates that both a positively and a negatively directed k state are required to make the standing wave shown in Figure G.1. By normalizing to unit dimensions, we obtain

$$\frac{dn_z / L}{dk_z} = g_s \frac{1}{2\pi}, \tag{G.4}$$

which states that the per unit length number of states per unit reciprocal wavevector varies as $g_s / 2\pi$. No effective mass enters here. This expression just speaks to the fitting of waves in a space.

This simple relationship states that, for each confinement direction, the density of states has a $1/2\pi$ factor arising from a standing wave in **k**-space. In determining the density of states for different dimensionalities of confinement, the spin degeneracy is common, and this $1/2\pi$ factor arises from each confining dimension. The density of states in ν-dimensional wavevector space is $(2\pi)^{-\nu}$. The unconfined states, or states where the separation in energy is small compared to k_BT, provide for electron movement through a viscous or free flow of electrons in response to an energy input to the system. We can also express the density of state relationship in energy. The density, in general, normalized to unit spatial extent and energy, is:

$$\mathscr{G}(E) = \frac{dn}{dE} = \frac{dn}{dk}\frac{dk}{dE}. \tag{G.5}$$

The density of states in **k**-space and E-space can be related since dE/dk is known through the Schrödinger equation's dispersion solution of $E = \hbar^2 k^2/2m^*$ in the isotropic constant mass approximation used here.

We now extend this simple description to different degrees of freedom, as shown in Figure G.2. In the wavevector coordinate space shown in Figure G.2, the $3D$ unconfined description corresponds to finding the number of states within a shell of sphere, at energy E, in the band dE for the density of states. In the $2D$ description, with one confinement direction, it is the number of states within a slice of this spherical shell where k_x, k_y or k_z is discretized. A section of the shell—a planar ring—is shown in the plane intersection, assuming that the x direction is the confinement direction. In the $1D$ description, with two confined directions, it is the number of states along the intersection line of these quantization planes. This is the line along which the one-dimensional freedom of movement exists. In Figure G.2, the state $k_x^0, -k_y, k_z^0$ is shown in the band dE at energy E, where k_x^0 and k_z^0 are quantized by the strong dimensional confinement in those orientations.

The densities of states that relate to the volume of the shell in a $3D$ distribution, to the area of a cross-section in a $2D$ distribution, and along a line in a $1D$ distribution, within this single mass description, are as follows:

No confinement, 3 dimensions of freedom:

$$\mathscr{G}_{3D}(E) = 2\frac{1}{(2\pi)^3}\frac{4\pi k^2 dk}{dE} = \frac{1}{\pi^2\hbar^3}\sqrt{2m^{*3}E}; \tag{G.6}$$

1-dimension confined, 2 dimensions of freedom:

$$\mathscr{G}_{2D}(E) = 2\frac{1}{(2\pi)^2}\frac{2\pi k_\parallel dk_\parallel}{dE} = \frac{m^*}{\pi\hbar^2}; \text{ and} \tag{G.7}$$

2-dimensions confined, 1 dimension of freedom:

$$\mathscr{G}_{1D}(E) = 2\frac{1}{2\pi}\frac{2dk_y}{dE} = \frac{1}{\pi\hbar}\sqrt{\frac{2m^*}{E}}. \tag{G.8}$$

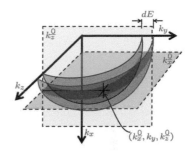

Figure G.2: The constant energy surfaces in the reciprocal space. The figure shows only a quadrant of the three-dimensional space. For three dimensions, the region of allowed **k** is in the spherical shell for a spread dE in energy at E. When confined in the x direction to $k_x = k_x^0$, the states are in a circular, two-dimensional, areal slice. When additional confinement is introduced to z, so $k_z = k_z^0$, the states allowed are along the extended line shown, with one particular state, (k_x^0, k_y, k_z^0), identified within the dE span at energy E. At this energy, there is an additional reverse momentum state at $k_x^0, -k_y, k_z^0$, which is the reflection point of the (k_x, k_z) plane running through the origin.

These relations establish the densities of states for electrons that have some freedom of movement while being circumscribed due to confinement in some of the degrees of freedom. When these states are occupied by electrons, the permitted freedom of movement leads to current when an external force is applied on the system. It is because of this density of states that electrons are available with freedom of movement when the energy separation—a kinetic energy separation—of the states is small. The channels of conduction arise in the free direction of movement. The density of states determines the channels available to give rise to current.

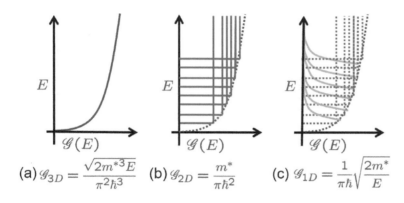

(a) $\mathscr{G}_{3D} = \dfrac{\sqrt{2m^{*3}E}}{\pi^2\hbar^3}$ (b) $\mathscr{G}_{2D} = \dfrac{m^*}{\pi\hbar^2}$ (c) $\mathscr{G}_{1D} = \dfrac{1}{\pi\hbar}\sqrt{\dfrac{2m^*}{E}}$

Figure G.3: The 3D, 2D and 1D density of states, that is, normalized to unit spatial dimension density of the states with $\pm 1/2$ secondary spin states available for conduction. (a) is for 3D, (b) for 2D and (c) for 1D. In the $\mathscr{G}_{2D}(E)$ distribution, the confinement arises from one dimension of confinement resulting in a ladder from the related discretized ks shown here as step function in (b). If the confinement dimension expands, the steps merge closer, and their distribution approaches that seen in a 3D distribution of states. The 1D density of states, $\mathscr{G}_{1D}(E)$, arises from two-dimensions of confinement and approaches 2D when one of the confinement dimensions is relaxed and 3D when both are relaxed.

Figure G.3 shows this density of states in terms of energy. In the 3D distribution, assuming isotropic constant mass, a constant spatial distance exists between all the **k** states. In $\mathscr{G}_{3D}$, since the volume of the spherical shell of width dE at energy E has this equi-spaced density, the number of states increases as the surface area per radius in the **k**-space, that is, with k, or the square root of energy. In $\mathscr{G}_{2D}$, which is a planar section of the shell, perpendicular to the direction of the confinement, the number of states increases at the same rate as the allowed ks, so the density of states in energy is a constant. Along a line, in the $\mathscr{G}_{1D}(E)$ distribution, the number of states per unit length is a constant, the k spacing varies as $\sqrt{E}$, and, therefore, in energy distribution, at higher energies, the density of states varies inversely w.wr.t. $\sqrt{E}$.

In the two-dimensional system, the freedom of movement is in two directions. One direction is confined and forms a ladder consisting of subbands of constant $\mathscr{G}_{2D}$ density. The argument associated with Figure G.1, originating in the Schrödinger equation, applies to both the confined (L small; of the order of the de Broglie wavelength) and the unconfined (L large) limits. For the small-length case, what is required is that the effective mass approximation hold, and that means that the box is at least a few unit cells long for the electron to be aware of the crystalline environment it is in, as reflected in the effective mass. The integrated number of states in these subbands over the energy range of interest provides the number of states available for conduction in the two dimensions. The ladder of multiple subbands have, as their minimum the quantized wavevectors of the confined direction. If the confinement dimension is relaxed, the subbands come closer, and in the limit, the distribution approaches a three-dimensional density of states. Confine one more dimension, and it is now a one-dimensional system. The freedom of movement now is in the one remaining unconfined direction. The density of states available for the transport in this unconfined direction varies as $\mathscr{G}_{1D}$, and integrating it, for all the subbands over the energy range of relevance, provides the total number of states available for one-dimensional conduction.

H
Schwarz-Christoffel mapping

ANALYSIS OF FUNCTIONAL RESPONSE in several geometries under specific constraints can benefit from mapping transformations that can reduce a seemingly complex form to a more palatable analytic or computational form. Schwarz-Christoffel transformation using conjugate functions is an example of this. It is particularly useful for solving the equations employed to describe semiconductor device behavior, such as the Laplace equation and the drift-diffusion equation in good approximations of realistic geometries. The reason for this is that stream lines of flux—such as current—and force lines—such as electric fields—often have a simplicity of arrangement. Lines are continuous throughout the region of interest and vortices— trapped spinning trajectories around an axis—are absent. This is a consequence of the absence of divergence and curl in the region. The sources and sinks, such as for electric field or current, are outside the region of interest for the modeling. So, even though divergences and curls, in electric and magnetic field and currents have appeared throughout device problems in this text, many of the regions of interest—in the steady state—have been subject to continuity in current. Current flow through a conducting contact region that is changing in size or has a high field change in the vicinity of a sharp corner is still a region of continuity in the steady state. These are regions where effects of interest—parasitic resistance or capacitance—arise and an analytic or simple computational determination is of value.

A lack of divergence in a region represents an absence of initiation and termination of the quantity whose divergence is absent. In a region with zero divergence of current density, there is no source and no sink of charged particles. Current flows into the region and out of the region, with the two flows balancing each other. The flow of this electric current is continuous. It may spread out or narrow down, that is, the current density may decrease or increase, but all the cur-

See T. A. Driscoll and L. N. Trefethen, "Schwarz-Christoffel Mapping," Cambridge, ISBN- 0-521-80726-3 (2002) for a rigorous exposition.

We are assuming steady state here. So, there is no displacement current.

rent entering the region also exits it. Pictorially, we represent this as lines that are continuous throughout the region. They do not branch and they do not have loose ends. Lines do not begin or end within the region. Likewise, when there is no curl, we conclude that there is an absence of vorticity. The reason is that, for a prescribed equipotential surface as the boundary of the region, there is a unique pattern of the lines representing the position of all other equipotential surfaces and stream lines. This makes it possible to find an expression for V—the potential at each point—and F—the flux vector at each point. We discuss here this approach for the two-dimensional forms, for which the approach is particularly rewarding. Sharp corners can be tackled.

To understand the reason behind this utility consider the following. Let $w = f(z)$ be an analytic function of the complex quantity z. $f(z)$, in general, is complex. For $z = x + iy$, let

$$f(z) = \phi + i\psi. \tag{H.1}$$

Each point in the z-plane ($\Re(z)$ on the abscissa, and $\Im(z)$ on the ordinate axis) is transformed into a point in the w plane, which too is a complex plane.

Now consider how w changes with z:

$$\frac{dw}{dz} = \frac{d\phi + id\psi}{dx + idy} = \frac{\frac{\partial \phi}{\partial x}dx + \frac{\partial \phi}{\partial y}dy + i\frac{\partial \psi}{\partial x}dx + i\frac{\partial \psi}{\partial y}dy}{dx + idy}$$

$$= \frac{\frac{\partial \phi}{\partial x} + i\frac{\partial \psi}{\partial x} + \left(\frac{\partial \phi}{\partial y} + i\frac{\partial \psi}{\partial y}\right)\frac{dy}{dx}}{1 + i\frac{dy}{dx}} = \frac{a + b\frac{dy}{dx}}{1 + i\frac{dy}{dx}},$$

where $\quad a = \dfrac{\partial \phi}{\partial x} + i\dfrac{\partial \psi}{\partial x}, \quad$ and $\quad b = \dfrac{\partial \phi}{\partial y} + i\dfrac{\partial \psi}{\partial y}. \tag{H.2}$

This expression for dw/dz has an important implication. If $b = ia$, then dw/dz is independent of dy/dx. This is to say that dw/dz is independent of the direction of dz in the z plane. Changes in iy with x are not reflected in changes in the function $f(z)$.

What does this $b = ia$ mean?

$$b = ia$$

$$\therefore \quad \frac{\partial \phi}{\partial y} + i\frac{\partial \psi}{\partial y} = i\left(\frac{\partial \phi}{\partial x} + i\frac{\partial \psi}{\partial x}\right)$$

$$\therefore \quad \frac{\partial \phi}{\partial y} + i\frac{\partial \psi}{\partial y} = -\frac{\partial \psi}{\partial x} + i\frac{\partial \phi}{\partial x}. \tag{H.3}$$

This last equation sets the condition for the form of $f(z)$ to be

$$\frac{\partial \phi}{\partial x} = \frac{\partial \psi}{\partial y}, \quad \text{and} \quad \frac{\partial \phi}{\partial y} = -\frac{\partial \psi}{\partial x}. \tag{H.4}$$

A pictorial representation of this is shown in Figure H.1. When this condition holds true, dw/dz is independent of dy/dx, as stated earlier. So, if dy/dx is large, say, as for a corner in the z plane, it will

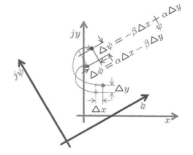

Figure H.1: A pictorial representation of the transformation of $z(= x + iy) \mapsto f(z)(= \phi + i\psi)$, with $\partial\phi/\partial x = \partial\psi/\partial y = \alpha$, and $\partial\phi/\partial y = -\partial\psi/\partial x = \beta$. This example is for $\alpha = 3$, and $\beta = 1$.

appear as gently continuous for $f(z)$ in the w plane. A sharp, highly conducting metal boundary can be transformed to a smooth, gently continuing change in the w plane. A function $f(z)$ that satisfies this criterion is called an analytic function. $\phi(z)$ and $\psi(z)$, which satisfy Equation H.4, are conjugate functions.

As an example, consider $f(z) = z^2$, so that $\phi = x^2 - y^2$, and $\psi = 2xy$—two hyperbolae. In the xy plane, these two curves are orthogonal to each other, as Equation H.4 prescribes. Figure H.2 shows a few parameterized curves. $\psi = 0$ implies that $2xy = 0$, that is, either x or y vanishes. It is the ordinate-abscissa, with its sharp corner.

For a semi-infinite, uniform, conducting sheet resistance 1 *Ohm* per square, when an electric current is fed in and extracted at a constant $\pm\phi$, the stream lines of the current will be along the constant ψ curves. The orthogonality relationship remains, even as one approaches that sharp corner in the z plane at the origin, and, away from it, the curves smoothly spread out, maintaining this orthogonality. This is an example of equipotential and streamlines in the divergence-less and curl-less condition. A short elaboration of the characteristics of this transformation follows.

Take two parallel plates with equipotential lines and streamlines like those of electrons flowing in a constant channel thickness, or water flowing in a pipe, or thermal conduction of heat along a rod, as shown in Figure H.3(a). Shrink the object to a line at one end, and expand it at the other end, and the new form of equipotential lines and streamlines will be as in Figure H.3(b). This problem is one where the points shown move in this transformation; the function $f(z)$ that describes this transformation is an analytic function, and the solutions for equipotential lines (ϕ) and streamlines (ψ) are orthogonal to each other. In an infinitesimally small volume, with orthogonality still true for each and every point, the distorted shapes appear closer and closer to the rectangular shape in Figure H.3(a). One could rotate BC π *radians* out to be in line with AD, and this relationship would still hold. Now, the form is an open one made from the straight line BA and a semicircular equipotential line—the arc BA.

The origin of the power of the conjugate function use for our examples is that a two-dimensional $ABCD$ form is transformed to a linear $ABCD$ form. The difficulties involved in analyzing continuity equations in a z-plane geometry with sharp corners or other complexities are eliminated, as the boundary conditions are now located along a straight line. This is illustrated in Figure H.4. And the orthogonality of the equipotential and streamline relationship is still maintained.

The point P in Figure H.4(a) is enclosed in the rectilinear region

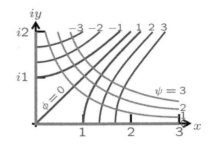

Figure H.2: Curves in the z plane, for constant values of the conjugate functions ϕ and ψ. The two are orthogonal to each other, and the first quadrant abscissa and ordinate axes form the curves for $\psi = 0$.

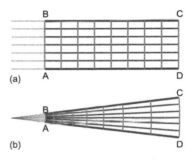

Figure H.3: (a) shows equipotential lines and streamlines in a hypothetical semi-infinite parallel plate (AD and BC) system through which a fluid, such as of electrons, flows. The lines parallel to AB or CD are equipotential, and those to AD and BC are streamlines. When the plates are brought together at one end—A and B—and the other ends of the plates are placed apart, as shown in (b), the conjugate function relationship still holds at every point.

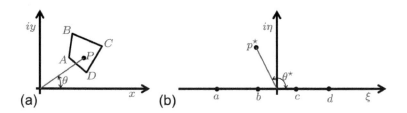

Figure H.4: (a) shows a rectilinear region $ABCD$ with a point P in it in the z plane. (b) shows a transformation where this rectilinear region is stretched along a line. The point P now is transformed to $p^{\star}$ in the κ plane.

$ABCD$. $ABCD$ is now transformed into a straight line, as shown in (b), through a magnitude and angular change. For any infinitesimally small change dz in moving from any point, say, P, to another nearer point P' in the z plane, the change of the rectilinear shape $ABCD$ to another plane with orthogonality maintained—say, the κ plane—is given by

This is similar to what we do in Fourier analysis. If a signal $f(t)$ with a Fourier transform—a complex transform–of $F(\omega)$ is fed through a system of a transform function of $Z(\omega)$, then output in time is the Fourier inverse of $F(\omega)Z(\omega)$.

$$
\begin{aligned}
\frac{dz}{d\kappa} = {} & \mathcal{K}(\kappa - a)^{-1+\alpha/\pi}(\kappa - b)^{-1+\beta/\pi} \\
& \times (\kappa - c)^{-1+\gamma/\pi}(\kappa - d)^{-1+\delta/\pi}.
\end{aligned} \tag{H.5}
$$

Here, the angles $\alpha, \beta, \ldots$, are the angles by which these corner points are to be transformed.

Let us illustrate this approach by tackling the streamlines as a flow occurs around a corner. This is the transformation from Figure H.3(a) to Figure H.2. It is an example of a $\pi/2$ angular folding rather than the π folding of a straight line; its form is

$$
\frac{dz}{d\kappa} = \mathcal{K}(\kappa - a)^{-1+\alpha/\pi} = \mathcal{K}\kappa^{-1+1/2} = \mathcal{K}\kappa^{-1/2}, \tag{H.6}
$$

reflecting that the folding was affected at $a = 0$ and the angle was $\pi/2$. Equation H.5 specifies the form that maintains the boundary condition's line-oriented ratios, so the dx and dy ratios, and tells us the form that the inside points will take. $\mathcal{K}$ and $a, b, \ldots$, et cetera, must still be found so that the analytic function and their conjugate functions are determined for the problem of interest. For Equation H.6, the solution is

$$
z = 2\mathcal{K}\kappa^{1/2}, \quad \text{or} \quad \kappa = \frac{1}{4}\frac{1}{\mathcal{K}^2}z^2, \tag{H.7}
$$

by choosing the origin to make the constant of integration vanish. So, with J as a normalized edge current,

$$
\begin{aligned}
\phi + i\psi = {} & J(\xi + i\eta) = J\kappa = \frac{J}{4\mathcal{K}^2}(x + iy)^2 \\
= {} & \frac{J}{4\mathcal{K}^2}(x^2 - y^2) + i\frac{J}{4\mathcal{K}^2}2xy. \tag{H.8}
\end{aligned}
$$

For simplicity, consider this normalized. The sheet has a resistance of
1 *Ohm* per square, and J is in A/cm. This is the solution form shown
in Figure H.2. Streamlines of constant ψ are curves of rectangular hyperbolae that asymptote to the ordinate and abscissa axes. Constant
potential lines—constant ϕ curves—are orthogonal to the streamlines and are curves of hyperbolae with the ordinate and abscissa as
the axes of symmetry. If we know the current and the voltage at the
boundary, then we know J and $\mathcal{K}$, so we have the unique solution to
this problem.

Conformal mapping in general, and Schwarz-Christoffel mapping
in particular, are effective tools for a broad range of problems. The
Laplace equation, in piecewise constant boundary conditions, as well
as in homogeneous domains with derivative conditions, appears in a
fair number of electrical, magnetic and thermal problems. Standard
problems of electrical resistance, capacitance and electric or magnetic
potential are all amenable to the mapping process. Waveguides,
magnetic motors, device contact regions, and fields at emission tips
are all examples where the technique can be robustly applied. It can
also be employed in inverse problems, through iteration.

I
Bell's inequality

EARLY IN THE DEVELOPMENT OF QUANTUM MECHANICS, among several of Schrödinger's important papers were two that specifically point to entanglement as a defining feature of quantum mechanics, and this feature can lead to counter-intuitive results. As seen in Chapter 3, the mathematical formalism of quantum mechanics has been incredibly successful. So, does it represent reality? One may articulate two diametrically opposite answers to this. The first is that this formalism tells us nothing about a quantum reality. The approach only allows us to calculate the probabilities of various realities possible. The opposite of this is the view that a unitary evolution of the quantum state completely describe the reality—an interpretation first articulated by Everett. The Everettian view states that all quantum alternatives exist in superposition. The puzzle in here is what this unitary evolution, as through Schrödinger's wave equation or Heisenberg's equation of motion, is, and what the meaning of the measurement intervention—the quantum state reduction—is when an observation takes place. The first view here is essentially the Copenhagen interpretation. Bohr viewed $|\psi\rangle$ not as a quantum reality but as information about the quantum system. The observer acquires more information through the observation and the wave-function collapse—a Bayesian thought. According to the Copenhagen interpretation, one may only accept the reality of the classical world, where the observer and his observational apparatus exist. It distinguishes between a quantum world and a classical world. The Everettian view is that $|\psi\rangle$ represents reality. No observer or measuring apparatus is necessary. We will not dwell on this—as yet—unresolved quantum foundational issue.

But, we will focus on one that relates to classical locality and to local hidden variables.

Are probabilities, such as in our classical world's daily observations' statistical expectations and those of quantum mechanics in the

Schrödinger's cat is the entanglement of a cat with a radioactive atom. von Neumann too has argued that entanglement between a measurement apparatus and the system being measured is a superposition of all possible outcomes. And this does not require any reference to probabilities. Everett's many-worlds interpretation can then be viewed as a unitary evolution where there is no collapse—no exalted observer or observing apparatus and no Copenhagen interpretation required—and no randomness. One does have to wonder where this Copenhagen thing stops and the other one begins. Or is there really this distinction?

EPR and other superposition situations, identical or distinguishable? Take the classic example due to Bohm. A pair of spin 1/2 particles in a combined spin state of 0—so, of opposite spins—are created and travel away from each other, say left L and right R, towards two separate detectors. Once the particles are far enough apart, one decides to measure the spin in a specific direction. The question is whether it is possible to get the result provided by quantum mechanics through any model in which the two particles are disconnected from each other, do not communicate with each other in this state of separation and behave as independent classical entities.

Bell's theorem—inequalities—states that it is not possible to reproduce the prediction of quantum theory in any such alternative. The joint probability of two physically separated measurements with classical expectations—or example, the values that would be expected in a situation where particles are behaving as separate entities after having been created together—is violated by quantum-mechanic expectations. The inequality is a demonstration of a quantum-mechanic effect —of entanglement—that may not be explained by any model that treats particles as unconnected and independent. Entanglement has information and Bell's inequality underlines it.

What this states is that any theory with locality and local hidden variables is inconsistent with the general statistical predictions of quantum mechanics.

Bell's inequality is a way to distinguish between the classical probabilistic statistical expectations of independent observations, that is, the expectations that we encounter in the undergraduate curriculum, and the probabilistic statistical expectations underlying quantum mechanics. Quantum entanglement is an example of one of the hidden connections embedded within the entangled state in quantum mechanics, and they are not local. Bell's inequalities are the test that invalidate any local hidden variables hypothesis. As a corollary, Bell's inequality theorem sets the test that shows that the assumption of locality in classical day-to-day experience is contradicted by quantum mechanics. The theorem's experimental conclusion is that locality is incompatible with the statistical prediction of quantum mechanics.

Bell's inequalities state constraints that probability distributions satisfy. A local—not global, that is, nonlocal—hidden variable would fail these tests for two entangled quantum systems. Bell's inequalities also help qualify and quantify entanglement.

Let a quantum experiment produce a pair of particles. Let Anarkali have particle A, and Balaji, B, the second of these particles. Anarkali and Balaji have two sets of measuring instruments—A_1 and A_2, for Anarkali, and B_1 and B_2 for Balaji. They have also agreed ahead that they will each choose one apparatus simultaneously and indepen-

There is a collection of these Bell inequalities. We are focusing on one particular one that is very pertinent to our discussions.

The work by John Clauser and Stuart Freedman and later on by others showed quantum entanglement as a reality.

dently to make a measurement. Let the outcome of a measurement on these apparatus be either $+1$ or -1. For example, the particle pair contains two spin $1/2$ particles that can be either in an up-state—a $+1 \equiv \hbar/2$ outcome—or in a down state— $-1 \equiv -\hbar/2$ outcome. Anarkali and Balaji are using the two apparatuses they each have to make this measurement, which could be a measurement of spin along different orientations. For any set of observables, hidden variable theories assign a set of eigenvalues that are a function of both the observable and the hidden variables. In the spin experiment, the observables could be the x-, y- or z-directed components of the electron's spin (normalized to $\hbar/2$). For any hidden variable λ, the measurements would be either $+1$ or -1. Different λs determine these plus and minus 1s.

Let $\mathfrak{E}(A_i B_j)$ be the expectation with Anarkali measuring using A_i and Balaji using B_j. The measurement uncovers the objective physical property: an objective value that existed before the measurement. This is what our classical day-to-day experience says. A definitive value exists—it's just that we don't know it. Since Anarkali and Balaji make the measurement simultaneously (as well as independently), the measurements have no effect on each other since information may not travel faster than light. For a hidden variable theory, the expectation over all the possibilities from the different measurement apparatus follows as

$$\begin{aligned}
&\mathfrak{E}(A_1 B_1) + \mathfrak{E}(A_1 B_2) + \mathfrak{E}(A_2 B_1) - \mathfrak{E}(A_2 B_2) \\
&- \mathfrak{E}(A_1 B_1 \mid A_1 B_2 + A_2 B_1 - A_2 B_2) \\
&= \mathfrak{E}(A_1 B_1 + A_1 B_2 + A_2 B_1 - A_2 B_2) \\
&= \mathfrak{E}\left[(A_1(B_1 + B_2) + A_2(B_1 - B_2)\right].
\end{aligned} \tag{I.1}$$

Only ± 1 outcomes are possible for each of the measurements. For simplicity, we use the same symbol for the actual measurement as for the apparatus. If $B_1 = B_2$, then $B_1 - B_2 = 0$, and $B_1 + B_2 = \pm 2$, so the sum over all measurement possibilities is ± 2. If, on the other hand, $B_1 = -B_2$, then $B_1 + B_2 = 0$, and $B_1 - B_2 = \pm 2$; so, again, the sum over all measurement possibilities is ± 2. From this, the constraint on the expectation over all probabilistic possibilities is

$$\sum_{A_1, A_2, B_1, B_2} \mathfrak{p}(A_1, A_2, B_1, B_2) \times (A_1 B_1 + A_1 B_2 + A_2 B_1 - A_2 B_2) \leq 2. \tag{I.2}$$

This result strictly has nothing in it from physics, even if we used spin as the observable. This relationship, a Bell inequality, and it will show up in different forms for different conditions—we could have had three measurement apparatuses, for example, with a different form for the inequalityfor each—has strictly come about from the mathematics of counting of these values in the different categories. In this specific case, the expectation of the measurement under our two assumptions may not exceed 2.

However, quantum-mechanically, the inequality is violated. For our spin example, the strongest choice for violation is an *EPR* pair where all directions are coplanar, with Anarkali measuring along 0 and $\pi/4$ angles and Balaji along $\pi/8$ and $3\pi/8$ angles. Using $\sigma_1 = |1\rangle\langle 0| + |0\rangle\langle 1|$, with $\sigma_2 = i|1\rangle\langle 0| - i|0\rangle\langle 1|$, and $\sigma_3 = |0\rangle\langle 0| - |1\rangle\langle 1|$—the Pauli matrices—on the Bloch sphere, Anarkali is measuring our *EPR* state (a net spin of 0 and composed of two particles) along σ_3 and $\sigma_3 + \sigma_1$, that is, the 0 and $\pi/4$ angles, while Balaji measures along $(1/4)\sigma_3 + (3/4)\sigma_1$ and $(3/4)\sigma_3 + (1/4)\sigma_1$, that is, $\pi/8$ and $3\pi/8$. The quantum-mechanical expectation now is

$$\mathfrak{E}(A_1 B_1) + \mathfrak{E}(A_1 B_2) + \mathfrak{E}(A_2 B_1) - \mathfrak{E}(A_2 B_2)$$
$$= (1/2) \times \left[3\sqrt{2} + (-1)(-\sqrt{2}) \right]$$
$$= 2\sqrt{2}, \text{ that is, } > 2. \tag{I.3}$$

This *EPR* pair has violated the Bell inequality and our classical statistical expectation. More generally, all pure entangled states violate Bell's inequalities. States that are separable do not violate Bell's inequalities. This follows quite directly from the fact that, when there are separable states for Anarkali and Balaji, one does not affect the other, and the joint state of their measurement is simply a product state. When one looks at this from a probability perspective, it doesn't change the independence. So, the expectations are a product of the independent operations. When this is the case, one may call them disentangled states.

Bell's inequality, also often referred to as Bell's theorem, is therefore a good tool for quantifying the level of entanglement in mixed states.

J

The Berry phase and its topological implications

THE ADIABATIC APPROXIMATION states that when the time scale over which a time-dependent Hamiltonian varies is much longer than the eigenstates' oscillation period, then one may introduce a simple dynamical phase factor $\exp[-(i/\hbar)\int E_n(t)\,dt]$ to the eigenstate $\psi_n(\mathbf{r},t)$. The system, if it begins in an eigenstate, will remain in the same eigenstate during an adiabatic perturbation. This view ignores a topological phase factor. The evolution is in two coordinates—t and $\mathbf{r}$—and what this means is that when the time-dependent Hamiltonian of the system returns at $t = t_f$ adiabatically to the form it had at $t = t_i$, there is nor only a phase factor θ_n arising from the time excursion—the multiples of 2π—but also a phase factor ϑ_n, due to the geometric excursion. This is the Berry or geometrical phase. It is a topological consequence. The phase of the instantaneous eigenstate may not, in general, be adjusted arbitrarily. There will be circumstances where upon return to a previous Hamiltonian form, an additional $\exp[i\vartheta_n(t)]$ phase factor will exist.

Let $\mathscr{H}(\mathbf{R})$, where $\mathbf{R}$ is a vector in variables $R_1, R_2, \ldots$, be the Hamiltonian under discussion. We will consider this problem in three-dimensional space, for convenience, even though the result is more general and valid for n-dimensional space. Our system obeys

$$\mathscr{H}(\mathbf{R},t)\,|\psi_n[\mathbf{R}(t)]\rangle = E_n[\mathbf{R}(t)]\,|\psi_n[\mathbf{R}(t)]\rangle = -\frac{\hbar}{i}\frac{\partial}{\partial t}\,|\psi_n[\mathbf{R}(t)]\rangle. \quad \text{(J.1)}$$

When the Hamiltonian returns to its previous state adiabatically, that is, $\mathbf{R}(t_f) = \mathbf{R}(t_i)$, with $t_f - t_i = T$, an evolution that does not involve any degeneracies and hence their superpositions, will be in the eigenstate $\exp(i\vartheta_n)\exp[-(i/\hbar)\int E_n(t)\,dt]|\psi_n(\mathbf{R})(t = t_f)\rangle$. We have chosen this to be the period T, so $\mathbf{R}(t = t_f) = \mathbf{R}(t = t_i)$. The time phase factor is what we had already written. The geometrical phase factor is

$$\vartheta_n = i\int_T \langle\psi_n[\mathbf{R}(t)]|\frac{d}{dt}|\psi_n[\mathbf{R}(t)]\rangle dt. \quad \text{(J.2)}$$

Shivaramakrishnan Pancharatnam, in his work in optics, discussed the implications for the parameter space of the Hamiltonian arising in the geometrical phase difference acquired over the course of a cycle in a cyclic adiabatic processes. This work, dating 1956, appeared in Proceedings of the Indian Academy of Sciences, and is largely forgotten. Michael Berry pointed out the geometric phase again in 1984 and emphasized its implications for quantum mechanics. This is sixty years into quantum mechanics' development, so it is remarkable that it took this long. It is now called Berry phase in most of the literature. A case of difficulty with pronouncing Pancharatnam—a wonderful well-meaning name translated as "five jewels"—in the West, or another example of Matthew effect at work? The implications of Berry phase are nearly everywhere—graphene layers get many of their properties from it, and vortices such as skyrmions in magnetics and elsewhere are another example, as are phase shifts in polarized light, and possibly neutrino decay. It appears in our discussion in this text through the quantum Hall effect, the quantum spin Hall effect and in topological insulators.

The time evolution can be simplified as

$$
\begin{aligned}
\frac{d}{dt}|\psi_n[\mathbf{R}(t)]\rangle &= \sum_i \frac{\partial}{\partial R_i}|\psi_n(\mathbf{R})\rangle \frac{dR_i}{dt} \\
&= \nabla_\mathbf{R}|\psi_n(\mathbf{R})\rangle \cdot \frac{d\mathbf{R}}{dt} \\
&= |\nabla_\mathbf{R}\psi_n(\mathbf{R})\rangle \cdot \frac{d\mathbf{R}}{dt},
\end{aligned}
\tag{J.3}
$$

and, therefore,

$$
\begin{aligned}
\vartheta_n &= i\int_T \langle\psi_n(\mathbf{R})|\nabla_\mathbf{R}\psi_n(\mathbf{R})\rangle \cdot \frac{d\mathbf{R}}{dt}\, dt \\
&= i\oint_C \langle\psi_n(\mathbf{R})|\nabla_\mathbf{R}\psi_n(\mathbf{R})\rangle \cdot d\mathbf{R}.
\end{aligned}
\tag{J.4}
$$

The Berry phase factor is now written as an integral over a curve in space that loops back. Recall the discussion of the Aharanov-Bohm effect, where a vector potential and gauging became necessary. One could introduce a vector potential here, in a similar vein, as

$$
\mathbf{A}_n(\mathbf{R}) = i\langle\psi_n(\mathbf{R})|\nabla_\mathbf{R}\psi_n(\mathbf{R})\rangle.
\tag{J.5}
$$

The phase of a moving particle in a closed loop in the presence of a magnetic field is describable through the spatial vector potential—this resembles and gives rise to the Aharonov-Bohm phase and relates to flux quantization. The phase is now gauge invariant, using a vector potential which itself is not gauge invariant. By analogy, we can now explicitly write fields through the cross product, so a magnetic field–like term, which is $\mathbf{B}_n^* = \nabla_\mathbf{R}\times\mathbf{A}_n(\mathbf{R})$, with

$$
\vartheta_n = \oint_C \mathbf{A}_n(\mathbf{R})\cdot d\mathbf{R} = \oint_S \mathbf{B}_n^*(\mathbf{R})\cdot d\mathbf{S}.
\tag{J.6}
$$

This use of Stoke's theorem explicitly shows that $\mathbf{B}_n^*(\mathbf{R})$, a field, is a curvature.

Now consider a curved surface in three-dimensional space, and choose a coordinate system where x and y are the two directions of principal curvature. At the origin, the surface is

$$
z = -\frac{1}{2}(\kappa_1 x^2 + \kappa_1 y^2),
\tag{J.7}
$$

where κ_1 and κ_2 are the principal curvatures, with the Gaussian curvature as

$$
\kappa = \kappa_1\kappa_2 \propto \nabla_2\cdot\nabla_1 - \nabla_1\cdot\nabla_2,
\tag{J.8}
$$

Equation 3.74 in Chapter 3. This follows from the relationship of the normal to the surface:

$$
\hat{\mathbf{n}} = \hat{\mathbf{z}} + \kappa_1 x\hat{\mathbf{x}} + \kappa_2 y\hat{\mathbf{y}},
\tag{J.9}
$$

This curvature attribute in a multidimensional space has many implications. One important one is that the vanishing of divergence, similar to that of true magnetic field, means that, when there are singularities at points of degeneracy, that is, infinities of the field, one may associate a magnetic monopole-like source to this field. These are not, however, the only sources.

with $d\hat{\mathbf{n}}/dx = \kappa_1\hat{\mathbf{x}}$; $d\hat{\mathbf{n}}/dy = \kappa_2\hat{\mathbf{y}}$; and the product

$$\frac{d\hat{\mathbf{n}}}{dx} \times \frac{d\hat{\mathbf{n}}}{dy} = \kappa_1\kappa_2\hat{\mathbf{x}} \times \hat{\mathbf{y}} = \kappa_1\kappa_2\hat{\mathbf{z}}. \tag{J.10}$$

Integration of the Gaussian curvature over a closed surface is the product of 2π and the Euler characteristic χ, which is the number of vertices plus the number of edges minus the number of faces. Polygonization of surfaces gives their Euler characteristic. Table J.1 shows some examples of Euler characteristics. So, a torus has a genus of 1—one handle. A double torus has two handles ($g = 2$), and a sphere has no handle ($g = 0$).

For a closed orientable surface, the genus g, which intuitively is the number of handles on the object, is

$$\chi = 2 - 2g. \tag{J.11}$$

Gauge invariance was the way we brought the geometric argument in here. Flux quantization appeared through the field divergence. It represents a vortex. And these vortices are ubiquitous. The quantization that appears, such as of flux, appears from a topological source that we derived through mathematical argument heretofore. We look at it now physically. Topology is the study of global properties of spaces. A coffee cup and a donut are similar in that they have the same invariant property—that of the Euler characteristic for an object with one handle. In our problem of understanding the nature of system response, given the Hamiltonian, one may view the body or many-body response through a physical space where the phase point traces a path in an internal symmetry space. The internal symmetry space is the space of phase factors that transform under the gauge. An external gauge potential $\mathbf{A}$ interacts with the particle which we interpret as making a geometrical or topological connection to the internal symmetry space.

An example of this is flux quantization in the presence of a magnetic field. The phase of the wavefunction is a geometrical coordinate in the internal symmetry space. If a magnetic field exists, it attempts to rotate the local phase of the wavefunction. In a superconductor, the Cooper pair is locked in phase. This means that either it gets unlocked, such as at high field, and the material is no more a superconductor, or the vector potential $\mathbf{A}$ causes a change in phase consistent with the surrounding superconductor. This example serves to show that some of the physical properties one observes in this system are purely geometrical, that is, topological.

Figure J.1 shows a topological picture of the internal symmetry space around a vortex. If there is no trapped flux, the closed path can be arbitrarily foreshortened and the torus vanishes. The trapped

	$\chi = V - E + F$
Interval	1
Circle	0
Disk	1
Sphere	2
Torus	0
Double torus	−2
Möbius strip	0
Klein bottle	0
Disconnected spheres	$2 + 2 = 4$

Table J.1: Euler characteristics of some common objects. V is the number of vertices; E, the number of edges; and F, the number of faces. A torus arises as a product of two circles seen through two orthogonal cross-sections. A double torus is two joined toruses.

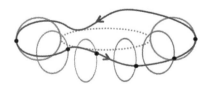

Figure J.1: A closed path in internal symmetry space around a vortex. Flux is trapped, and, topologically, the path on the surface of the torus follows an intrinsic phase relationship.

flux makes a hole, and this path cannot be arbitrarily shrunk. This is the Aharonov-Bohm effect, where the particle may not enter the region with magnetic field. Viewing this system topologically as a torus is our way of arriving at this flux quantization through the Euler characteristic reflected in the Gauss-Bonnet relationship—the relationship connecting the curvature of an object's surface with the object's topology.

The consequence of this connection between flux, topology and the phase change caused by the vector potential $\mathbf{A}$ is immediate. In the quantized conductance $1/\rho_{xy} = \sigma_{xy} = \nu e^2/h$, the consequence of the Gauss-Bonnet theorem is that

$$\nu = \frac{1}{2\pi} \int_{BZ} \boldsymbol{\nabla}_{\mathbf{k}} \times \mathbf{A}(k_x, k_y) \, d^2k, \tag{J.12}$$

where

$$\mathbf{A} = -i \langle \psi_n | \boldsymbol{\nabla}_{\mathbf{k}} | \psi_n \rangle. \tag{J.13}$$

K
Symmetry

SMALL CAPS: SYMMETRY IN SPACES AND SYMMETRY IN THE PHYSICAL LAWS
are of fundamental importance in science. So is their breaking. One
would expect exact symmetry to be the exception rather than the
rule. Yet, most physical theories and interactions depend on a symmetric physical structure. By symmetry here, we mean that there
exist congruent transformations which leave the physical form or the
law unchanged. It just permutes the component elements. Use of a
balance to measure weight is the use of symmetry. Special relativity stands on symmetry of space and time. Coulomb interactions or
gravity posits a symmetry—a $|\mathbf{r}|^2$ inverse dependence—of physical
structure that is independent of the angle! Yet, small perturbations
cause broken symmetry whose profound consequences include the
phase transitions that have been an important pursuit of this text.
The human body is largely bilaterally symmetric w.r.t. many parts—
hands, eyes, ears, and others—but has only one heart and one brain,
and they are essential for living. The observation that there are symmetry groups, that is, groups of transformations under which an
object remains invariant, underlies the idea that symmetry is a property. Its effect transcends all energy or size or other scales that we
have often referred at. The Aharanov-Bohm effect is a demonstration
of the effect of symmetry in electromagnetism, based on gauge transformation involving differentiation. A variety of these consequences
of symmetry happen all over the scalar and vector spaces that we
employ in science and engineering.

The notion of symmetry goes back to ancient times, when symmetry and the principle of conservation were considered tied to each
other—the reason why the earth was in equilibrium, so the Greeks
said. Symmetry pervades architecture and art. In the sciences, symmetry and its breaking can be seen as an essential element in the
causation that leads to observed effects.

In considering the role of symmetry or its breaking in physical

There is an old Panchtantra story of
a monkey and two cats, known to
Asian Indian children. A monkey is the
arbiter for dividing equally the bread
that the cats have found. He attempts
the balancing goal by biting off from
the heavier part and then measuring.
And this process is repeated till all is
gone. Breaking symmetry for personal
gains seems to be the hallmark of the
world we inhabit. Perhaps the Wallace-
Darwinian trait is a consequence. A
nephew of mine in London, who is
too young to remember, recalls an
expression ``Maggie Thatcher. Bread
snatcher″ as a vestige of the past in his
school playgrounds. Churchill seems
to have stood for freedom for his kind,
this being the goal of World War II,
although not for the people of color
of the imperial domain, people who
fought and sacrificed in large numbers
in what was their war too. Millions of
Indians died in Churchill's great famine
of Bengal—a man-made famine, not
unlike Stalin's in Russia at the same
time. The French walked right back to
Indochina as soon as the war finished.
What makes Alexander the Great and
Attila the Hun? The North American
holiday of Thanksgiving is a celebration
of illegal immigration. Now the tables
are turned. For better or worse, this is
the human story of symmetry breaking
and polarization.

To be accurate, general entities on a
manifold—Reimann again.

A mildly amusing fourteenth-century
story, which is probably not true, is
told about Buridan's ass. Jean Buridan
was the rector of Sorbonne and is said
to have argued that if an ass were
placed exactly between two identical
hay piles, the ass would die of hunger.
His reasoning was that the ass, being
rational, has no reason at all to prefer
one direction over the other.

laws, treating force and response in their vector forms provide a very direct way of interpreting Ørsted's findings on electromagnetic interaction. Ørsted's observations is also connected to the Aharonov-Bohm paradox in several interesting ways. Fields—representing forces— are either polar or axial. An electric field is terminated at each end by opposite charges. It is polar. A magnetic field, on the other hand, is axial, arising from the orbital flow of current. If a rod is rotating, and one looks at it from the two ends, it will appear to be rotating clockwise at one end, and anticlockwise at the other. An infinitely thin rod is a directed segment—an axial vector.

Polar vectors are symmetrical with respect to reflection in a parallel plane, and antisymmetrical with respect to reflection in a perpendicular plane, as shown in Figure K.1. Axial vectors are precisely the opposite. Ørsted's experiment showing that current in a wire interacts with the magnetic field of a compass needle is a demonstration of an interaction between polar and axial vectors. The current vector is polar, and the magnetic vector is axial. It also is an experiment in determining symmetry.

The demonstration of nuclear parity non-conservation was another experiment showing symmetry in action—similar Ørsted's, yet not. When the electron-emitting element ^{60}Co is placed in a solenoid that establishes a magnetic field, the spins of ^{60}Co nuclei become oriented normal to the plane of the loop current of he solenoid, that is, along the solenoid field. Spins are axial vectors, so if Ørsted's reasoning was applicable in this situation, a symmetric effect relative to the nuclear spin should be seen in electron decay through the γ rays it produces. A world and its mirror image would behave in the same way, with left and right, and up and down, reversed. This is the symmetry shown in the bottom part of Figure K.1(b). This is parity. But, when the experiment was finally conducted, parity violation was observed. This experiment—a rather major event in history of science—showed that parity is violated with weak nuclear forces, even if it holds true with electromagnetic forces, so Ørsted force, and strong nuclear forces.

To understand symmetry operations and, through them, what happens in crystals, we start with a simpler question: what are the symmetries of a square? One is the rotational symmetry. A right angle ($\pi/2$) rotation leaves the square unchanged. In complex number form, this is motion in an $i\times$ operation. A vertex occupies $(1,0)$, $(0,i)$, $(-1,i)$ and $(0,-i)$ in (x,iy) space. Powers of i represent rotation. And, in this case, there are four of them: $i^0 = 1$; $i^1 = i$; $i^2 = -1$; and $i^3 = -i$, with the fourth power bringing us back to the starting point. This is a 4-fold rotational symmetry, and the four elements form a group—a set of elements together with four operational laws:

Axial vectors are actually antisymmetrical tensors of rank 2. One may even argue that they are not vectors at all, but that is beyond our scope of this discussion.

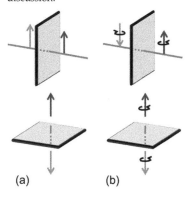

(a)   (b)

Figure K.1: (a) shows polar vectors under reflection through a parallel and perpendicular plane. They are symmetrical with the parallel plane, and antisymmetrical with the perpendicular plane. (b) shows axial vectors under reflection through a parallel and a perpendicular plane. They are antisymmetrical with the parallel plane, and symmetrical with perpendicular plane—precisely the opposite of what was seen with polar vectors.

the associative law of multiplication $(a(bc) = (ab)c)$, commutation $(ab = ba)$, identity $(1a = a1)$ and the inverse identity operation $(a^{-1}a = aa^{-1} = 1)$. The rotational symmetries of the square in the complex plane $\mathbb{C}$ are C, Ci, $-C$ and $-Ci$, where C is complex conjugation. Squares also have another type of symmetry. This is the symmetry of reflections and is orientation reversing. Call this operation $\mathscr{C}$. Its multiplication laws are

$$\mathscr{C}i = (-i)\mathscr{C}, \quad \mathscr{C}(-1) = (-1)\mathscr{C}, \quad \mathscr{C}(-i) = i\mathscr{C}, \quad \text{and} \quad \mathscr{C}\mathscr{C} - 1. \quad \text{(K.1)}$$

The multiplication laws for the entire group of symmetry operations for the square are

$$i^4 = 1, \quad \mathscr{C}^2 = 1, \quad \text{and} \quad Ci = i^3C. \quad \text{(K.2)}$$

Because of the last non-commutative relationship, this is a non-Abelian group. The number of distinct elements of a group is called the order of the group. This group has an order of 8. The group is also a finite group. Groups can be continuous too. For example, say that, in place of a square, we had a sphere. In this case, the number of rotational symmetries id infinite. The symmetry group has an infinite number of elements. In 3-dimensional space, we can rotate at any angle about any axis. If we now include reflections, they bring an additional set of symmetries to group $O(3)$. This is a continuous group—a Lie group.

This discussion was meant to illustrate the connections, from the simple to the complex, that the group approach to symmetry brings, and how it unifies observations of low energy condensed matter with those of high energy particle physics.

For the materials of interest to us, there are a number of these symmetry operations and groups that have implications for us.

Our interest is in the solid state—in the energetic transformations there and in the crystals resulting from them. A crystal arises by using the basis—the constitutive units of the solid—together with the lattice, which is the geometric construction of points. The following are the symmetry implications of interest in crystals. The unit cell is specified by the lengths of the edges that define the translation operation, and by the angles between them. The cell is a unit since each point of the unit cell is shared, but the total of all these fractionally shared points add to unity. Translation by the unit cell leaves one in a surroundings where one cannot distinguish one's locale from other positions that can be generated by the translation operation.

A plane lattice—a lattice in two dimensions—has six possible unit cells: a square $(a = b; \alpha = \pi/2)$, a rectangle $(a \neq b; \alpha = \pi/2)$, a rhombus $(a = b; \alpha \neq \pi/2)$, a pentagon $(a = b; \alpha = 3\pi/5)$, a hexagon $(a = b; \alpha = \pi/3)$ and a parallelogram $(a \neq b; \alpha \neq \pi/2)$. The pentagon,

Groups that are commutative are Abelian—named after the Norwegian mathematician Niels Henrik Abel. Any group which is represented by multiplication with complex numbers are Abelian, since complex number multiplication commutes.

This is a 3-manifold denoted by $\mathscr{R}$—a group called $SO(3)$.

however, cannot form a continuous two-dimensional structure and so is not a plane lattice building block. A plane lattice can have rotational symmetry, mirror symmetry and a center of symmetry. The possible rotational symmetries depend on the unit cell, but $360/n$, where n is the fold, must leave the unit cell indistinguishable from before. And one must employ the minimal set of these operations. A square, for example, has 4-fold rotational symmetry—2-fold works but is excluded since it is a composite of two 4-fold rotation operations. A rectangle or parallelogram, however, has $n = 2$. A 1-fold rotation exists for any irregular arrangement. Since pentagons can not reproduce the entire lattice through translation, we exclude 5-fold rotation.

But, one must remark that 5-fold symmetry is certainly possible and shows up in quasicrystals.

These symmetry operations are described geometrically through the axis, plane or point that serves as the reference for operation. Rotation axes, mirror planes, and centers of symmetries are the symmetry elements of these operations. Additionally, one may have more complex symmetry operations. An example is rotation-inversion or so-called roto-inversion, which denotes the rotation and inversion of a lattice point. Hermann-Mauguin notations describe these symmetries, based on the operation being used. Rotational symmetries are represented by the corresponding n numerical symbol, so $1, 2, 3, 4, 5$ and 6. Mirror symmetry is represented by the symbol m. That the object has a center of symmetry is represented by $\bar{1}$, which signifies that, for every point, there exists an identical point equidistant from the center, on the opposite side and in an inverted state. Table K.1 shows the symbol for the different roto-inversion symmetries.

Recall here our polar-axial vector discussion of symmetry.

A space lattice—a lattice in three dimensions—has a lot more possibilities. Lattice planes are now identified by the Miller indices hkl from the intercepts on the axes of the lattice. In space lattices, there are five rotational symmetry elements, denoted 1, 2, 3, 4 and 6; one center of symmetry element, denoted $\bar{1}$; three rotation-inversion elements, denoted $\bar{3}$, $\bar{4}$ and $\bar{6}$; and one of mirror plane element, m, which is the same as $\bar{2}$. With space lattices, 1-fold rotation followed by inversion—$\bar{1}$—is the same as a center of symmetry and 2-fold rotation followed by inversion—$\bar{2}$—is the same as a mirror plane rotation. So, they are not double counted.

n-fold rotation + inversion	Notation
$1 + \bar{1}$	$\bar{1}^*$
$2 + \bar{1}$	$\bar{2}^*$
$3 + \bar{1}$	$\bar{3}^*$
$4 + \bar{1}$	$\bar{4}^*$
$6 + \bar{1}$	$\bar{6}^*$

Table K.1: Composite Rotation and inversion operation and its Hermann-Mauguin notation.

When symmetry elements of a space lattice are also present at the macroscopic scale in the lattice of the crystal, they are referred to as macroscopic symmetry elements. When we see symmetries of various arrangements of facets of a crystal, we are observing these macroscopic symmetries. This is a reflection of the symmetry at the point at the center of the crystal. This is any point in an infinite arrangement. One need only specify at any point in the lattice. This collection of these symmetry elements at any point of the lattice is the point group of symmetry.

In general, there are 32 point groups of symmetries; and these are listed in Table K.2, together with their interpretations.

Symbol	Symmetry interpretation
1	A 1-fold rotation axis. identity symmetry
$\bar{1}$	A center of symmetry
2	A 2-fold rotation axis
$m(\bar{2})$	A single mirror plane
$2/s$	A 2-fold rotation axis + mirror plane $\perp$ to it
222	Three 2-fold rotation axes $\perp$ to each other
$2mm$	Two mirror planes $\perp$ to each other + a 2-fold rotation axis along their intersection line
$2/m\,2/m\,2/m$	Three mirror planes $\perp$ to each other + a 2-fold rotation axis along their intersection line
3	A 3-fold rotation axis
$\bar{3}$	A 3-fold roto-inversion axis
$3m$	Three mirror planes $\pi/3$ to each other, intersecting along a 3-fold rotation axis
32	A 3-fold rotation axis $\perp$ through the intersection of three 2-fold axes $\pi/3$ to each other
$\bar{3}\,2/m$	Three mirror planes $\pi/3$ to each other, intersecting along a3-fold rotary inversion, with three 2-fold axes $\perp$ to the rotary inversion and midway between the mirror planes
4	A 4-fold rotation axis
$4/m$	A 4-fold rotation axis with a mirror plane $\perp$ to it
$\bar{4}$	A 4-fold roto-inversion axis
422	A 4-fold rotation axis $\perp$ through the intersection of four 2-fold rotation axes at $\pi/4$ to each other
$4mm$	Four mirror planes $\pi/4$ to each other, with a 4-fold rotation axis along the line of intersection
$4/m\,2/m\,2/m$	Four mirror planes $\pi/4$ to each other, with a 4-fold rotation axis along each mirror plane intersection, with another mirror plane $\perp$ to the 4-fold axis intersecting other mirror planes along four 2-fold rotation axes
$\bar{4}\,2m$	Two mirror planes at $\pi/4$ to each other intersecting along a 4-fold roto-inversion, with each mirror plane containing a 2-fold rotation axis $\perp$ to the roto-inversion
6	A 6-fold rotation axis
$\bar{6}$	A 6-fold roto-inversion axis
$6/m$	A 6-fold rotation axis with a mirror plane $\perp$ to it
$6\,mm$	Six mirror planes $\pi/6$ to each other, with a 6-fold rotation axis along the line of intersection
622	A 6-fold rotation axis $\perp$ to the intersection of six 2-fold rotation axes at $\pi/6$ to each other
$\bar{6}\,2m$	Three mirror planes at $\pi/3$ to each other, intersecting along a 6-fold roto-inversion, with each mirror plane containing a 2-fold rotation axis $\perp$ to roto-inversion
$6/m\,2/m\,2/m$	Six mirror planes $\pi/6$ to each other, with a 6-fold rotation axis at their intersection; each mirror plane contains a 2-fold rotation axis $\perp$ to the 6-fold axis in a mirror plane also $\perp$ to the 6-fold axis
23	Three 2-fold axes $\perp$ to each other and $\parallel$ to the edges of the cube, with four 3-fold axes $\parallel$ to the body diagonals
$2m\,\bar{3}$	Three mirror planes $\parallel$ to the cube face, intersecting along three 2-fold axes $\parallel$ to the edges of the cube, with four 3-fold rotary inversion axes parallel to the body diagonals of the cube
$\bar{4}\,3m$	Three 4-fold rotary-inversions $\parallel$ to the edges of the cube, with four 3-fold rotation axes $\parallel$ to the body diagonals, and six mirror planes intersecting at the face diagonals
432	Three 4-fold rotation axes $\parallel$ to th edges of the cube, with four 3-fold rotation axes $\parallel$ to the body diagonals and six 2-fold rotation axes $\parallel$ to the face diagonals
$4/m\,\bar{3}\,2/m$	Three 4-fold rotation axes $\parallel$ to the edges of the cube, with four 3-fold rotary inversion axes $\parallel$ to the body diagonals, six 2-fold rotation axes $\parallel$ to the face diagonals, and nine mirror planes—three $\parallel$ to faces, and six containing face diagonals

Table K.2: The 32 point groups of a three-dimensional lattice.

L
Continuum elasticity

MECHANICAL RESPONSE IN CONTINUUM APPROXIMATION assumes that all properties are defined continuously. In determining the response of a large-enough object, where any atomic or small-scale effects are only minor irritants, this is a good approximation. The mechanical response under small forces will be linear. This is the elastic limit where one may define the lowest order constants of response. Nanoscale materials, however, can have anisotropic responses, even if the forces have not crossed over to a magnitude where nonlinearity must be considered.

Consider a material with an elemental volume of $\Delta x \Delta y \Delta z$ and centered at (x_0, y_0, z_0); at time t, it responds to an applied force $\mathbf{F}$. by undergoing a displacement (u_x, u_y, u_z), leading to a displacement response from the unstressed position of $\mathbf{u}(x - u_x, y - u_y, z - u_z; t)$. For small displacements,

$$\mathbf{u}(x - u_x, y - u_y, z - u_z; t) = \mathbf{u}(x_0, y_0, z_0; t)$$
$$- \frac{\partial \mathbf{u}}{\partial x} u_x - \frac{\partial \mathbf{u}}{\partial y} u_y - \frac{\partial \mathbf{u}}{\partial z} u_z + \cdots, \quad \text{(L.1)}$$

where we now ignore the higher order terms by limiting ourselves to the elastic limit. Force causes a time response—embodied in Newton's law as $\mathbf{F} = \rho \partial^2 \mathbf{u}/\partial t^2$, where ρ is a constant of proportionality—and the force is a gradient of the stress.

Any component of displacement may now be written, for example, the displacement in the y direction, due to the stress component T_{yz}, is

$$\rho \frac{\partial^2 u_y}{\partial t^2} = \frac{\partial T_{yz}}{\partial z}. \quad \text{(L.2)}$$

The connection between deformation and applied stress is established through strain. Strain is the perturbational relationship between deformation and displacement. A normal strain in each of the

In the text, we have employed u as the specific displacement in the orientation of interest. This is notationally clearer to see and distinguishes it from all the position and displacement coordinates that need to be considered and included.

Forces due to free charges and free currents can also be written through stress tensors, just as mechanical forces can be.

axis direction is the elongation per unit length. So,

$$e_{xx} = \frac{\partial u_x}{\partial x},$$ (L.3)

and equivalent expressions for e_{yy} and e_{zz} are the linear normal strains. A shear strain is a deformation where there is angular deflection of the edges of elemental cube volume due to shear stress. It is a distortion of the angle, with lengths being maintained, for example,

$$e_{xz} = \frac{1}{2}\left(\frac{\partial u_x}{\partial z} + \frac{\partial u_z}{\partial x}\right)$$ (L.4)

is the point relationship due to shearing stress in the xz plane. $e_{zx} = e_{xz}$ by symmetry, and this expression also reduces to that for normal strain. The three displacement and three Cartesian directions make a total of nine components of strain. Since displacement is a vector, the strain is a tensor.

The strain-stress relationship is the property of the material. Deformation is the resulting displacement of an object under stress and is a function of its geometry as well as the material's stress-strain relationship. Pure translation and pure rotation do not involve strain deformation.

A three-dimensional object in stress will elongate and contract along different directions. Figure L.1 shows xy cross-sections of a block under two different stress conditions. When a normal stress T_{yy} is applied, as shown in Figure L.1(a), the block elongates along the y direction while contracting along the x direction and the z direction, where the latter is not shown in the figure. A uniform strain e_{yy} appears along the block. When a compressive stress is applied along the x direction, as shown in Figure L.1(b), again, an elongation will appear along the y direction, with a contraction along the x direction. The distribution of strain in both these cases is shown in Figure L.1c, which shows that the effects of T_{yy} and $-T_{xx}$ are similar. They both cause the strain e_{yy}, but the proportional relationships of that strain to the two different types of stress will be different. A similar argument holds for the z direction, which is symmetric to the x direction for this discussion's conditions. While all of these stresses and the strain along the block are constant in magnitude along y, the displacement along the y direction (u_y) is not as shown in Figure L.1(d). The center of the beam doesn't displace, while the edges have maximum displacement. We write the linear behavior of normal strain with normal stress as

$$e_{yy} = \frac{\partial u_y}{\partial y} = \frac{1}{Y}\left[T_{yy} - \nu(T_{zz} + T_{xx})\right],$$ (L.5)

where Y is the Young's modulus—a modulus of elasticity—and ν is the Poisson's ratio. A similar form holds true for e_{xx} and e_{zz}. We

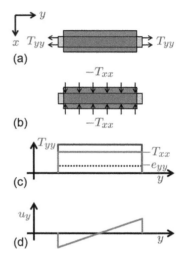

Figure L.1: (a) shows deformation under the uniform normal stress T_{yy}—elongation in the y direction, and contraction in the x direction. (b) shows an identical consequence from a normal stress of $-T_{xx}$. (c) Uniform distribution of normal strain in the y direction, resulting from the uniform T_{yy} and $-T_{xx}$. (d) Displacement in the y direction (u_y), as a function of position.

should also ask what about shear stress's effect, that is, what happens when T_{yz} or T_{xy}, et cetera, exist? Shear stress will cause shear strain. Shear strain is also proportional to shear stress, and one may write shear strain, for example, for the xy plane, as

$$e_{xy} = \frac{1}{2G} T_{xy}, \tag{L.6}$$

with similar relations for the other shear combinations. G is the shear modulus. The shear modulus can be shown to be related to Young's modulus, and Poisson's ratio because, in an isotropic material, the results must be independent of the choice of orientation of reference frame. This leads to

$$G = \frac{Y}{2(1+v)}. \tag{L.7}$$

The off-diagonal terms of stress tensors do not have a normal contribution, while the diagonal terms have both normal and shear contributions.

With all this taken together, the stresses and displacements in the linear isotropic limit may be written as

$$T_{xx} = G\left(\frac{\partial u_x}{\partial x} + \frac{\partial u_x}{\partial x}\right) + \lambda\frac{\partial u_z}{\partial z} = 2G\frac{\partial u_x}{\partial x} + \lambda\frac{\partial u_z}{\partial z},$$

$$\text{where } \lambda = \frac{vY}{(1+v)(1-2v)}, \text{ and}$$

$$T_{xy} = G\left(\frac{\partial u_x}{\partial y} + \frac{\partial u_y}{\partial x}\right). \tag{L.8}$$

The other tensor terms follow from the symmetry of these relationships.

The relationship of strain with stress, as expressed in the Hooke's law form, is a reformulation of the above derivations as

$$e_{xy} = \frac{1}{2G} T_{xy}, \text{ and}$$

$$e_{xx} = \frac{1}{2G} T_{xx} - \frac{v}{Y} T_{yy} - \frac{v}{Y} T_{zz}, \tag{L.9}$$

with the other relations following, again, by symmetry.

Written directly as force relationships, again with similar comments on symmetry, the displacements follow

$$\rho\frac{\partial^2 u_x}{\partial t^2} = \frac{\partial T_{xy}}{\partial y} + \frac{\partial T_{xz}}{\partial z} + F_x, \tag{L.10}$$

where F_x is the external force. This equation is in tensor form. We may write it in vector form as

$$\rho\frac{\partial^2 \mathbf{u}}{\partial t^2} = (2G + \lambda)\nabla(\nabla \cdot \mathbf{u}) - G\nabla \times (\nabla \times \mathbf{u}) + \mathbf{F}. \tag{L.11}$$

Because we have assumed isotropy, shear strain cannot be caused by normal stress. Isotropy implies a normal-to-normal causal relationship.

M
Lagrangian dynamics

MODELING ELECTROMECHANICAL SYSTEMS WITH MULTIPLE FORCES
is best tackled using Hamilton's principle. Deriving forces—vectors—
from scalars is naturally more general. This appendix summarizes
the physical approach of Lagrangian dynamics, with some insights
that are more generally applicable.

Variational forms of equations of dynamics, such as the vector
form of Newton's laws, follow from the principle of virtual work.
This is extensible to dynamics through d'Alembert's principle, from
which Hamilton's principle becomes easy to understand and from
which the Lagrange equations can be derived. Hamilton's principle
is, in this sense, an alternative to Newton's laws.

A moving particle with a linear momentum **p**, according to New-
ton's law, is subject to changes according to $\mathbf{F} = d\mathbf{p}/dt$. To stay
simple without losing generality here, we write the one-dimensional
form of the work increment on the particle:

$$F\,dz = \frac{dp}{dt}dz = \frac{dp}{dt}v\,dt = v\,dp, \tag{M.1}$$

where $v = dz/dt$ is the velocity of the particle. We define a kinetic
energy function $T(p)$ as the total work by F in increasing the momentum from zero. So,

$$T(p) = \int_0^p v\,dp. \tag{M.2}$$

T is a function of instantaneous momentum p, according to this and
it is associated with a velocity v given by

$$v = \frac{dT}{dp}. \tag{M.3}$$

Nowhere have we written an explicit connection between momentum
p and velocity v.

d'Alembert's principle states that
the sum of the differences between
forces acting on a system of mass
particles and the time derivatives of
the momenta of the system itself is
zero along any virtual displacement
consistent with the constraints of the
system, that is,

$$\sum_i (\mathbf{F}_i - m_i\ddot{\mathbf{r}}_i) \cdot \delta\mathbf{r}_i = 0.$$

Here i identifies the particle and the
rest of the terms have their usual
meaning.

Newtonian mechanics introduces $p = mv$ as a constitutive equation, and this, in turn, leads to

$$T(p) = \frac{p^2}{2m}. \tag{M.4}$$

For our Lagrangian derivation, we introduce a complementary kinetic state function—the kinetic coenergy function

$$T^\star(v) = \int_0^v p\,dv = pv - T(p), \tag{M.5}$$

which too is independent of any assumption of velocity-momentum relation. This is an example of a Legendre transformation, where one swapped one independent variable (p) for another (v) without causing any informational loss. A complementary coenergy function is the complement whose addition makes the sum whole—pv here. Since

$$dT^\star = p\,dv + v\,dp - \frac{dT}{dp}dp = p\,dv, \tag{M.6}$$

it follows that

$$p = \frac{dT^\star}{dv}. \tag{M.7}$$

In the Newtonian approach, using the constitutive relation $p = mv$ with the starting definition of kinetic coenergy, the first form of Equation M.5 gives

$$T^\star = \frac{1}{2}mv^2, \tag{M.8}$$

which is the common engineering form given for kinetic energy valid for Newtonian conditions. $T(p)$ and $T^\star(v)$ are identical for Newtonian conditions. But, they are not in general. For conditions of velocities close to that of speed of light c, we have $p = m_0 v/(1 - v^2/c^2)^{1/2}$.

In any reference frame, the minimum set of coordinates that allows a full geometric description of the evolution of the system is the set of its generalized coordinates. Different sets of generalized coordinates can exist, and they do not all necessarily have a straightforward physical meaning, for example, the coordinates for the amplitude of an eigenmode of an oscillating beam. The degrees of freedom of a system are the minimum number of coordinates necessary for a full geometric description. Figure M.1 gives an example of generalized coordinates for a rolling disk. There are four—the orientation of the disk is defined by two angles (θ, ϕ) and the two position coordinates (x, y) of the point of contact in the plane. Alternatively, one could have chosen Cartesian coordinates or polar coordinates (ρ, φ) to define the point of contact. These are sufficient to describe all the geometric possibilities of the disk on the plane—all the points of contact on the plane, and the orientations for the disk. The time

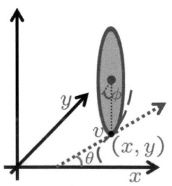

Figure M.1: A disk rolling on a plane, and a minimum set of coordinates to describe it.

derivatives are not independent. The rolling condition means that $v = r\dot\phi$, $\dot{x} = v\cos\theta$; and $\dot{y} = v\sin\theta$. These are expressible in the independent constraint equations

$$dx - r\cos\theta d\phi = 0, \text{ and}$$
$$dy - r\sin\theta d\phi = 0. \tag{M.9}$$

The paths are restricted. Motion, that is, the connection of one configuration to other, is restricted. This set of equations represents a constraint for the generalized coordinates and is expressible as

$$f(q_1, \ldots, q_n, t) = 0. \tag{M.10}$$

We now introduce a virtual change of configuration as an infinitesimally small change, for example, $\delta q_1 = \delta x$ or $\delta q_3 = \delta\theta$, et cetera, in coordinates at any instant. This is a virtual displacement that must still obey the kinematic constraints, that is, the constraint equations.

For any system with generalized coordinates q_i obeying the constraint equations in time, the allowed variations must satisfy, independently or equivalently at all instants of time,

$$\delta f = \sum_i \frac{\partial f}{\partial q_i}\delta q_i = 0. \tag{M.11}$$

Consider a one particle system in which the particle is only allowed to move on a surface defined by coordinates (x, y, z) that have a defined relationship. We have $f(x, y, z) = 0$, and the virtual displacements satisfy

$$\frac{\partial f}{\partial x}\delta x + \frac{\partial f}{\partial y}\delta y + \frac{\partial f}{\partial z}\delta z = 0, \tag{M.12}$$

which is expressible as

$$(\boldsymbol{\nabla} f)^T \cdot \delta\mathbf{r} = 0. \tag{M.13}$$

$\boldsymbol{\nabla} f$ is normal to the surface, so virtual displacements must be in the plane tangent to the surface. The particle moves on the surface under a reaction force F. In a smooth and frictionless system, this reaction force is normal to the surface:

$$\mathbf{F} \cdot \delta\mathbf{r} = \mathbf{F}^T \cdot \delta\mathbf{r} = 0. \tag{M.14}$$

Application of a force causes a constraining force, that is, $\mathbf{F}_i$, causes a constraining force $\mathbf{F}'_i$ that, under static conditions, balances and keeps the system from moving. When reversibility exists, for example, in the case where there is no friction, then the virtual work of the constraint forces on any virtual displacements is 0. The principle of virtual work is the variational expression of the static equilibrium, absent lossy forces such as friction. A generalization of this statement is that if there are N particles at position $\mathbf{r}_i$, where $i = 1, \ldots, N$, static equilibrium implies, for every force $\mathbf{F}_i$ on any particle i, that there is a constraining force $\mathbf{F}'_i$ with a resultant force $\mathbf{R}_i$ for each particle such that $\mathbf{R}_i \cdot \mathbf{r}_i = 0$ for all virtual displacements, or

$$\sum_{i=1}^{N} \mathbf{R}_i \cdot \mathbf{r}_i = 0. \tag{M.15}$$

The resultant force is due to the externally applied force $\mathbf{F}_i$ and its reaction force $\mathbf{F}'_i$, that is, $\mathbf{R}_i = \mathbf{F}_i + \mathbf{F}'_i$, or

$$\sum_{i=1}^{B} \mathbf{F}_i \cdot \mathbf{r}_i + \sum_{i=1}^{B} \mathbf{F}'_i \cdot \mathbf{r}_i = 0. \tag{M.16}$$

In a reversible, frictionless system, the virtual work performed by the constraint force, that is, reaction, vanishes, so

$$\sum_{i=1}^{B} \mathbf{F}_i \cdot \mathbf{r}_i = 0. \tag{M.17}$$

This is the principle of virtual work; it states that the virtual work of the external applied forces on virtual displacements compatible with the kinematics vanishes. So, under static and other constraints, the reaction forces are removable in equilibrium, the static equilibrium is now transformable to kinematic terms, and one can write, using generalized coordinates,

$$\sum \mathbf{Q}_i \cdot \delta \mathbf{q}_i = 0, \tag{M.18}$$

where $\mathbf{Q}_i$ is a generalized force associated with the generalized coordinate $\mathbf{q}_i$.

Consider the arrangement shown in Figure M.2, which depicts a hinged assembly with a force $\mathbf{F}$, for example, an electrostatic force, used to move the y-constrained motion of the free end. The kinematic conditions are $x = 5l \sin \theta$, and $y = 2l \cos \theta$; therefore, $\delta x = 5l \cos \theta \, \delta \theta$, and $\delta y = -2a \sin \theta \, \delta \theta$. The principle of virtual work implies that

$$F \, \delta x + G \, \delta y = 0 \quad \therefore \quad F = \frac{2}{5} \cos \theta G, \tag{M.19}$$

since the expression is true for all $\delta \theta$.

The principle of virtual work is useful under static conditions and very specific constraints. d'Alembert's principle extends it to dynamic conditions by adding inertial forces, so by adding $-m\ddot{\mathbf{r}}$ to the applied and constraining forces, thus for every particle, by including this term, one can define a resultant force of

$$\mathbf{R}_i = \mathbf{F}_i + \mathbf{F}'_i - m_i \ddot{\mathbf{r}}_i = 0, \tag{M.20}$$

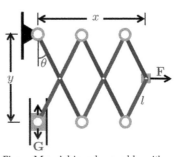

Figure M.2: A hinged assembly with a force $\mathbf{F}$ driving a constrained orthogonal movement.

which satisfies the Newtonian description. The summation of the virtual work performed by the constraint force, from all the virtual displacements over all the particles, can now be written under dynamic conditions as

$$\sum_{i=1}^{N} (\mathbf{F}_i') \cdot \mathbf{r}_i = \sum_{i=1}^{N} (\mathbf{F}_i - m_i \ddot{\mathbf{r}}_i) \cdot \mathbf{r}_i = 0. \tag{M.21}$$

This sum of the applied force and the inertia force is an effective force. The virtual work performed by the effective force on the virtual displacement vanishes under the constraints of reversibility and Newtonian dynamics. This equation is based on vectors with an inertial frame reference, that is, reference frames in a state of constant rectilinear motion with respect to each other. d'Alembert's principle formulates dynamic equilibrium with vector forces and in position coordinates of the particles. As the latter are not necessarily independent, this equation cannot be written in generalized coordinates, unlike the equation for virtual work under static conditions. While virtual work with its static conditions could be written in generalized coordinates, this cannot be. Absent explicit time in the constraints, this equation reduces to

$$\sum_{i=1}^{N} \mathbf{F}_i \cdot \mathbf{r}_i - \sum_{i=1}^{N} m_i \ddot{\mathbf{r}}_i \cdot \dot{\mathbf{r}}_i \, dt = 0. \tag{M.22}$$

Now, if a conservative force is applied, Equation M.22 can be recast as

$$\sum_{i=1}^{N} m_i \ddot{\mathbf{r}}_i \cdot \dot{\mathbf{r}}_i \, dt = \frac{d}{dt} \left(\frac{1}{2} \sum_{i=1}^{N} m_i \dot{\mathbf{r}}_i \cdot \dot{\mathbf{r}}_i \right) dt = dT^\star, \tag{M.23}$$

resulting in

$$d(T^\star + V) = 0, \tag{M.24}$$

that is, $T^* + V$ is a constant—the law of conservation of energy, with the expression strictly restricted to potential not depending explicitly on time and the time independence of kinematic constraints.

To achieve a generalized coordinate form in dynamic conditions, one employs the definite integral form of energy, a scalar function and its stationarity. This requires reformulating Equation M.21. $\delta W = \sum \mathbf{F}_i \cdot \delta \mathbf{r}_i$ is the virtual work of applied forces. The second term can be reformed as

$$\ddot{\mathbf{r}}_i \cdot \mathbf{r}_i = \frac{d}{dt} (\dot{\mathbf{r}}_i \cdot \delta \mathbf{r}_i) - \dot{\mathbf{r}}_i \cdot \delta \dot{\mathbf{r}}_i = \frac{d}{dt} (\dot{\mathbf{r}}_i \cdot \delta \mathbf{r}_i) - \frac{1}{2} \delta (\dot{\mathbf{r}}_i \cdot \delta \dot{\mathbf{r}}_i), \tag{M.25}$$

which leads to

$$\sum_{i=1}^{N} m_i \ddot{\mathbf{r}}_i \cdot \delta \dot{\mathbf{r}}_i = \sum_{i=1}^{N} m_i \frac{d}{dt} (\dot{\mathbf{r}}_i \cdot \delta \mathbf{r}_i) - \delta T^\star. \tag{M.26}$$

A conservative force is one where the work performed on a particle that moves from one place to another is the same regardless of the path the particle takes, that is, the total work depends only on the end points. The work done is then equal and opposite to the change in another energy quantity—the potential V. Forces can be expressed as gradients of potentials for conservative forces. One can expand this for a situation where V is explicitly dependent on time t. Here, one must include a partial derivative in time.

δ and time derivative operations are order independent, that is, they commute.

The scalar recasting of the d'Alembert principle is

$$\delta W + \delta T^\star = \sum_{i=1}^{N} m_i \frac{d}{dt}(\dot{\mathbf{r}}_i \cdot \delta \mathbf{r}_i), \qquad (M.27)$$

which, when integrated over a time interval that lies between t_1 and t_2 and where the system is known, that is, where the virtual displacements $\delta \mathbf{r}_i(t_1)$ and $\delta \mathbf{r}_i(t_2)$ vanish, leads to

$$\int_{t_1}^{t_2} (\delta W + \delta T^\star)\, dt = \sum_{i=1}^{N} m_i\, \dot{\mathbf{r}}_i \cdot \delta \mathbf{r}_i \big|_{t_1}^{t_2} = 0. \qquad (M.28)$$

Let some of the forces be conservative and some non-conservative, then $\delta W = -\delta V + \delta W_{nc}$, where δW_{nc} is virtual work by non-conservative forces. The variational consequence of this is a term sometimes called the variational indicator:

$$\int_{t_1}^{t_2} [\delta(T^\star - V) + \delta W_{nc}]\, dt = 0,$$

$$\text{or} \quad \int_{t_1}^{t_2} (\delta \mathscr{L} + \delta W_{nc})\, dt = 0, \qquad (M.29)$$

where $\mathscr{L} = T^\star - V$ is the Lagrangian of the system. In the evolution of the system, between time t_1 and t_2, all arbitrary variations of the path between these starting and ending times must be compatible with the kinematic constraint of $\delta \mathbf{r}_i(t_1) = \delta \mathbf{r}_i(t_2) = 0$, where $\mathbf{r}_i$ is a virtual displacement between the actual path and a perturbation of $\delta \mathbf{r}_i$ on it at any instant of time. Only at the fixed ends do we know this position, but the path that the system will follow is the one that satisfies Equation M.29. The potential here can be explicitly dependent on time since its virtual variation is taken at constant time, that is, $\delta V = \nabla V \cdot \delta \mathbf{r}$, as opposed to the total change $dV = \nabla V \cdot \delta \mathbf{r} + (\partial V / \partial t)\, dt$. Equation M.29 is Hamilton's principle, showing the constraint that the dynamical system must satisfy in its evolution in time from t_1 to t_2. It determines, through the virtual displacement implications of d'Alembert's principle, the path that the system will take. It makes no reference to any specific coordinate system, and it relies on scalar energy and work. So, the system is expressible in generalized coordinates q_j, with independent virtual displacements in coordinates of δq_j.

This equation embodying Hamilton's principle can be recast, by using the variational indicator into Lagrange's equations. Let the system have generalized coordinates q_j, where $j = 1, 2, \ldots, n$. Any point of an object may then be described. As an example, without losing generality,

$$x_i = x_i(q_1, \ldots, q_n; t) \qquad (M.30)$$

describes a specific point of the system in terms of generalized coordinates, with an explicit time dependence. This point moves with a velocity

$$\dot{x}_i = \sum_j \frac{\partial x_i}{\partial q_j} \dot{q}_j + \frac{\partial x_i}{\partial t}. \tag{M.31}$$

The velocity lets us write the kinetic coenergy as

$$
\begin{aligned}
T^\star &= \frac{1}{2} \sum_i m_i \dot{x}_i \cdot \dot{x}_i \\
&= T^\star(q_1, \ldots, q_n, \dot{q}_1, \ldots, \dot{q}_n; t) = T_2^\star + T_1^\star + T_0^\star. \tag{M.32}
\end{aligned}
$$

The different subscripted $T^\star$s are homogeneous and because they result as a product from Equation M.31, which is a sum series, they can be of order 2, 1 or 0 in the generalized velocities $\dot{q}_i$. $T_0^\star$ is independent of $\dot{q}_i$, so it is a potential. $T_1^\star$ is linearly dependent on $\dot{q}_i$; it causes gyroscopic forces. $T_2^\star$ is quadratic in $\dot{q}_i$. The potential is only a function of the generalized coordinate and time. So, the Lagrangian is

$$\mathcal{L} - T^\star - V = \mathcal{L}(q_1, \ldots, q_n, \dot{q}_1, \ldots, \dot{q}_n; t). \tag{M.33}$$

The virtual work by non-conservative forces is

$$
\begin{aligned}
\delta W_{nc} &= \sum_i F_i \cdot \delta x_i = \sum_i \sum_k F_i \frac{\partial x_i}{\partial q_k} \delta q_k \\
&= \sum_k Q_k \delta q_k, \quad \text{where} \quad Q_k = \sum_i F_i \frac{\partial x_i}{\partial q_k} \tag{M.34}
\end{aligned}
$$

is a generalized force of energy dimensions, so an energy conjugate associated with the generalized coordinate q_k. This is now applicable to Hamilton's principle of conservation of total energy. The variational integral is

$$
\begin{aligned}
\delta \mathfrak{I} &= \int_{t_1}^{t_2} \left[\delta \mathcal{L}(q_1, \ldots, q_n, \dot{q}_1, \ldots, \dot{q}_n; t) + \sum_i Q_i \delta q_i \right] dt \\
&= \int_{t_1}^{t_2} \left[\sum_i \left(\frac{\partial \mathcal{L}}{\partial q_i} \delta q_i + \frac{\partial \mathcal{L}}{\partial \dot{q}_i} \delta \dot{q}_i \right) + \sum_i Q_i \delta q_i \right] dt. \tag{M.35}
\end{aligned}
$$

Since the $\delta \dot{q}_i$ term can be recast through integration by parts as

$$\frac{\partial \mathcal{L}}{\partial \dot{q}_i} \delta \dot{q}_i = \frac{d}{dt} \left(\frac{\partial \mathcal{L}}{\partial \dot{q}_i} \delta q_i \right) - \delta q_i \frac{d}{dt} \left(\frac{\partial \mathcal{L}}{\partial \dot{q}_i} \right), \tag{M.36}$$

it follows that

$$\delta \mathfrak{I} = \sum_i \frac{\partial \mathcal{L}}{\partial \dot{q}_i} \delta q_i \Big|_{t_1}^{t_2} - \int_{t_1}^{t_2} \sum_i \left[\frac{d}{dt} \left(\frac{\partial \mathcal{L}}{\partial \dot{q}_i} \right) - \frac{\partial \mathcal{L}}{\partial q_i} - Q_i \right] \delta q_i \, dt = 0. \tag{M.37}$$

On the right hand side, the first term vanishes, since these are fixed points where the virtual displacement vanishes. Then, the second term can vanish iff

$$\frac{d}{dt}\left(\frac{\partial \mathscr{L}}{\partial \dot{q}_i}\right) - \frac{\partial \mathscr{L}}{\partial q_i} = Q_i \ \forall \ i = 1, \dots n. \tag{M.38}$$

If there are no non-conservative forces, then

$$\frac{d}{dt}\left(\frac{\partial \mathscr{L}}{\partial \dot{q}_i}\right) - \frac{\partial \mathscr{L}}{\partial q_i} = 0 \ \forall \ i = 1, \dots n. \tag{M.39}$$

These are the Lagrange equations for the system.

Lagrange equations can also be written with constraints, for example, when the n generalized coordinates are not independent, and one may wish to satisfy m constraint equations of the form

$$\sum_k a_{lk}\delta q_k = 0 \ \forall \ l = 1, \dots, m. \tag{M.40}$$

This states that the number of degrees of freedom of the system is $n - m$. One now needs to employ Lagrange multipliers because the variations δq_i can no longer be arbitrary. So, a variational approach in linear combination needs to be introduced, for example,

$$\sum_{l=1}^{m} \lambda_l \left(\sum_{k=1}^{n} a_{lk}\delta q_k\right) = \sum_{k=1}^{n} \delta q_k \left(\sum_{l=1}^{m} \lambda_l a_{lk}\right) = 0, \tag{M.41}$$

with λ_l as unknown. This can be added to our variational equation Equation M.37, as follows:

$$\int_{t_1}^{t_2} \sum_{k=1}^{n} \left[\frac{d}{dt}\left(\frac{\partial \mathscr{L}}{\partial \dot{q}_k}\right) - \frac{\partial \mathscr{L}}{\partial q_k} - Q_k - \sum_{l=1}^{m} \lambda_l a_{lk}\right] \delta q_k \, dt = 0. \tag{M.42}$$

This set of equations has $(n - m)$ of independent δq_ks that can be selected as the independent variables. For each of these, the bracketed term must vanish, leaving m terms consisting of the dependent δq_ks. We select m of the Lagrange multipliers to zero this contribution. As a result, we now have

$$\frac{d}{dt}\left(\frac{\partial \mathscr{L}}{\partial \dot{q}_k}\right) - \frac{\partial \mathscr{L}}{\partial q_k} = Q_k + \sum_{l=1}^{m} \lambda_l a_{lk} \ \forall \ k = 1, \dots n. \tag{M.43}$$

The n equations have $n + m$ unknowns of the generalized coordinate q_k and the multiplier λ_l. We also have the m constraint equation

$$\sum_k a_{lk}dq_k + a_{l0}dt = 0, \tag{M.44}$$

so a set of $n + m$ equations with $n + m$ unknowns.

N

Phase space portraiture

To visualize some of the important properties of non-linear systems, it is convenient to observe the state evolution in phase space, in which all the states of the system are represented. This appendix is a summary of this phase portraiture for some of the important features that are observable in nonlinear systems, with particular emphasis on the electromechanical ones.

A vector field defines the trajectory that a phase point takes. In this sense, it interprets a differential equation that describes a system. So, an equation $\dot{\mathbf{y}} = f(\mathbf{y})$ describes, say, the velocity of a particle. The position $\mathbf{y}$ of a particle has associated with it a velocity $\dot{\mathbf{y}}$. The equation represents a vector field. We can sketch this vector field in $(\mathbf{y}, \dot{\mathbf{y}})$ coordinates—the phase space. Points where the particle is not moving, that is, $\dot{\mathbf{y}} = 0$ is a fixed point are the fixed points of the system. These points may be stable or unstable. Stable points in phase space are ones towards which the vector field is directed from nearer points, that is, small disturbances around it. Upon disturbance, the restorative forces return the system back to the stable point. Unstable points are the opposite. The vector fields are now directed away from the point and its surroundings even at small disturbances. A simple illustration of this can be provided by a simple RC circuit driven by a battery, such as that shown in Figure N.1(a). If $Q(t)$ is the charge on the capacitor at any instant of time, $I = \dot{Q}$ is the current, and the voltage around the loop can be written as

$$-V = R\dot{Q} + \frac{Q}{C} = 0. \tag{N.1}$$

In the formal canonical conjugate form for the phase space, we can write this as

$$\dot{Q} = \frac{V}{R} - \frac{Q}{RC} = f(Q). \tag{N.2}$$

This is drawn in Figure N.1(b). The fixed point of $f(Q) = 0$ occurs at $(Q^\star, 0)$. Any perturbation around it points the flow towards it, so it is

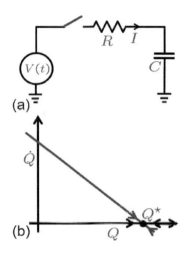

Figure N.1: (a) shows an RC circuit charging a capacitor when the switch is closed at $t = 0$. (b) shows its phase space trajectory. A is a fixed point that is also globally stable.

a stable fixed point. Since it is a point that the system approaches no matter what the initial condition, it is a globally stable fixed point.

In first order systems, fixed points dominate the dynamics. Trajectories either approach these or diverge away, since the trajectories are being forced to either increase or decrease monotonically. Geometrically, a phase point never reverses direction.

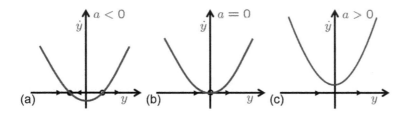

Figure N.2: The phase space trajectory is a parabola for $\dot{y} = a + y^2$. But, the three cases, $a < 0$ of (a), $a = 0$ of (b), and $a > 0$ of (c), will have quite different characteristics.

Now consider a first order system where

$$\dot{y} = a + y^2. \tag{N.3}$$

Depending on the value of a, there are three possible scenarios: $a < 0$, in which case there is one stable fixed point and one unstable fixed point, as shown in Figure N.2(a); $a = 0$, in which case the fixed point is at the origin and is half-stable, as shown in Figure N.2(b); and $a > 0$, in which case there are no fixed points, as shown in Figure N.2(c). In the second case, just a small change in a makes the fixed point vanish. This sudden change is what we call bifurcation. The behaviors of $a < 0$ and that of $a > 0$ are qualitatively very different and diverge at this saddle node—an unstable point. The two fixed points from the first case collide and annihilate each other.

This saddle-node bifurcation, also called a fold bifurcation or a turning-point bifurcation, is usually drawn as shown in Figure N.3, where the dependence on a in the characteristic behavior of this system is emphasized. It is the contour of the stable and unstable fixed points as a function of the varying parameter, which is a in our case.

Bifurcations can be of a variety of types. For example, in a solid-state laser, the rate of change of population in a rough approximation can be written in terms of the gain due to the pumping and the losses in the medium and at facets. This is expressed in an equation of the form

$$\dot{n} = Gn(N_0 - \alpha n) - kn = (GN_0 - k)n - \alpha Gn^2, \tag{N.4}$$

where n is the photon density, $G \geq 0$ is the gain coefficient, $k > 0$ is a rate constant related to the lifetime of photon, and N_0 is the population of atoms excited by the pumping, which is reduced by the photon creation at the rate of $\alpha > 0$. Figure N.4 shows the vector field,

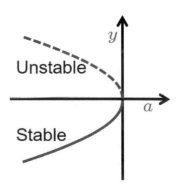

Figure N.3: A bifurcation diagram showing the stable and unstable fixed points as a function of the parameter of the system. Here, it is a in the first order system described by $\dot{y} = a + y^2$.

with the pumping strength N_0 as the parameter. In Figure N.4(a), $n^\star = 0$ is a stable point. In Figure N.4(b), there occurs a transcritical bifurcation with $N_0 = \varkappa = k/G$. In Figure N.4(c), with $N_0 > \varkappa$, the origin loses stability and the fixed point shifts to $n^\star = (GN_0 - k)/\alpha G$ positive. The laser lases at the kink point as shown in the bifurcation diagram in Figure N.4(d) at the kink point.

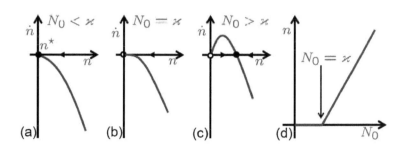

Figure N.4: Vector field of the photon population density. $\varkappa = k/G$ is a constant of the problem that determines the different fixed point behavior observed. (a), (b) and (c) show the vector fields for three different conditions, and (d) shows the bifurcation diagram.

Now let us take a simple extension of our first example to a nonlinear condition. Let

$$\dot{y} = ay - y^3. \tag{N.5}$$

This equation is invariant in inversion operation on y. This is reflected in the vector field drawn in Figure N.5. At $a < 0$, the fixed point is at the origin and is stable. At $a = 0$, the fixed point is at the origin and is stable but only weakly so. Linearization disappears at the origin. This, in turn, implies that decay is polynomial rather than exponential. For $a > 0$, one gets two stable points and one unstable point. The resulting behavior, as shown in Figure N.5(d), is called a pitchfork bifurcation, as it shows a confluence of three branches. At $a > 0$, we have two stable branches of solutions ($y^\star = \pm a^{1/2}$) to this system. So, there exist two stable fixed points of solutions for $a > 0$.

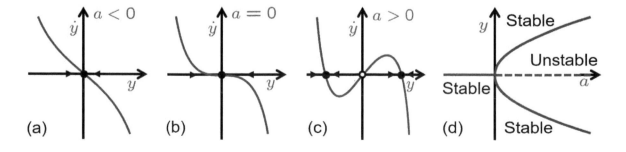

Figure N.5: The vector field for a nonlinear system described by $\dot{y} = ay - y^3$. (a), (b) and (c) show the field for three different conditions of parameter a, and (d) shows the pitchfork bifurcation of the system.

It is useful to see the correspondence between this system response and our usual way of looking at energy-configuration coordinates to find stable points—points of lowest energy. $\dot{y}$ is proportional to $-dV/dy$, so we may write

$$V(y) = -\frac{1}{2}ay^2 + \frac{1}{4}y^4. \tag{N.6}$$

The results for the three different cases are shown in Figure N.6. At $a < 0$, the minimum is a quadratic minimum. At $a = 0$, the minimum is flatter. It is quartic. At $a > 0$, the symmetry breaks, with the formation of two new minima—the stable fixed points—and a local maximum—the unstable point. Here, the cubic term of the vector field is stabilizing. As it increases, it created the bifurcation to two stable branches.

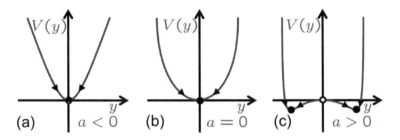

(a) $a < 0$ (b) $a = 0$ (c) $a > 0$

Figure N.6: The potentials for the three distinct cases shown in Figure N.5.

Now, consider what would happen when the sign of the cubic term of the vector field equation is reversed, that is, when $\dot{y} = ay + y^3$. The pitchfork would invert. Non-zero fixed points of $x^\star = \pm(-a)^{1/2}$ exist for $a < 0$ and are unstable—imaginary. The solution at the origin is stable for $a < 0$ and unstable for at $a > 0$—it is a locally stable point. Instability for $a > 0$ drives the system on a trajectory that explodes. So, this system is precariously stable at $y(0) = 0$, but for any other starting point, it will go to $\pm\infty$, depending on the starting position's sign.

The only way this system can be made stable is through higher order terms that compensate for the cubic term. For symmetric systems, this means the next term will be y^5:

$$\dot{y} = ay + y^3 - y^5. \tag{N.7}$$

Figure N.7 shows the bifurcation diagram for this equation. Again, the origin is a fixed point, but it is an unstable fixed point. At a_s two additional fixed points come about. In the region $a_s < a < 0$, the stable states are large amplitude, that is, large y states. $a = 0$ is locally stable in the sense that a large perturbation is needed to get it out of the state. It is not globally stable.

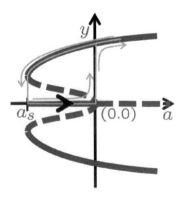

Figure N.7: The bifurcation diagram for the fifth order polynomial in Equation N.7.

So, if one were to vary the parameter a, starting with a system in fixed state $y^\star = 0$, and increase a, so following the arrow on the horizontal axis, up to $a = 0$, the state remains in $y = 0$. A small perturbation at $a = 0$ now causes it to jump to the large-amplitude branch. If it is in the large-amplitude branch with $a > 0$, then when a is lowered, the state continues to stay in the large-amplitude branch. It jumps only when a gets lower than a_s and then it jumps to the $y = 0$ branch. This is hysteresis, as is shown in the figure. At a_s, a saddle-node bifurcation point, stable and unstable fixed points are created and annihilated. In this bifurcation picture, one can see the mathematical connections of some of the hysteresis phenomena that we have explored in the text.

This type of bifurcation is connected to second order phase transitions. We saw similar forms for the order parameter with temperature. It is a supercritical pitchfork, or forward bifurcation. The earlier bifurcation that we looked at with the third power term was a subcritical, or inverted, or backward bifurcation. It is connected to first order phase transitions. For Equation N.7, we can find a condition of the parameter a where the potential has three minima that are all equal, since the potential is in the third power of y. This is a first order phase transition where the system may be found in three different possible states, such as in water freezing to ice, with gas, liquid and solid as the three different possible states.

These were all one-dimensional phase space examples. Trajectories moved monotonically or remained constant. How could one visualize oscillation? For this, one may use a form such as $\dot{\theta} = f(\theta)$, again still one dimensional, but the vector fields on a circle can return to their starting points. So, a circular orbit provides a simple way of showing oscillations in phase space. Higher dimension phase space will show richer phenomena since there is a larger coordinate space for maneuvering. Again, one can visualize the flow picture through the vector field. So, consider, a vector field for

$$\begin{aligned} \dot{y}_1 &= f_1(y_1, y_2), \text{ and} \\ \dot{y}_2 &= f_2(y_1, y_2), \end{aligned} \qquad (\text{N.8})$$

written compactly as

$$\dot{\mathbf{y}} = \mathbf{f}(\mathbf{y}), \qquad (\text{N.9})$$

where $\mathbf{y} = (y_1, y_2)$, and $\dot{\mathbf{y}} = (f_1(\mathbf{y}_1), f_2(\mathbf{y}_2))$. These, in general, can be written as arrays. What this equation tells us is how a phase point—a starting point—traces out a solution of $\mathbf{y}(t)$ by flowing along a vector field. It traces a trajectory that winds through the phase plane.

Figure N.8 shows examples of some possible phase portraits, examples of which we encounter in systems and which are particularly important to understanding electromechanical systems. *A*, *B* and *C* show fixed points, i.e, where $\mathbf{f}(\mathbf{y}^\star) = 0$. These show examples of various situations of bifurcation from which stationary trajectories evolved. *D* shows an oscillatory closed orbit, that is, one that satisfies $\mathbf{y}(t + T) = \mathbf{y}(t)$ $\forall t$ and a constant finite *T*. Flow patterns can be different. *B* and *C* are similar, but *E* and *G* are different. *H* shows a pattern with positional variation in speed. This portrait also shows stability and instability. *D*, the closed orbit, is stable, but the fixed points are unstable, since nearby trajectories are moving away.

Now we connect the phase portraiture to the system behavior. Consider a conservative system defined by our vector field relationship $\dot{\mathbf{y}} = \mathbf{f}(\mathbf{y})$. As discussed before, the term "conservative" implies that no work has been done when one returns to the original point. Energy has been conserved. In a conservative system, the trajectories are defined by a constant energy, so $U = T + V$. A conservative system cannot have any attracting fixed points because if there is an attracting point, but trajectories are composed of constant energy, then there is an inherent contradiction. Note, for example, in Figure N.8, that orbit *D*, an oscillatory orbit, has no attracting point. In stability analysis, one can identify a number of such properties. An attractor, also called a sink, is attractive. It has a negative real part in eigenvalues. A repeller, also called a source, has eigenvalues that have a positive real part. Saddle points are where one eigenvalue is positive and another negative.

Let us now consider the double-well potential, which has appeared many times in the text. We take a simple case,

$$V(y) = -\frac{1}{2}y^2 + \frac{1}{4}y^4, \tag{N.10}$$

which results in a force form of

$$F = -\frac{dV}{dy} = \ddot{y} = y - y^3, \tag{N.11}$$

where we have normalized to unit mass. When we write this equation in vector field form, with *z* representing velocity, we obtain

$$\dot{y} = z, \text{ and}$$
$$\dot{z} = y - y^3. \tag{N.12}$$

Any disturbance around a fixed point can be analyzed using linearization. Briefly, if

$$\dot{y} = f(y, z) \text{ and}$$
$$\dot{z} = g(y, z), \tag{N.13}$$

with fixed points such as $f(y^\star, z^\star) = 0$, and $g(y^\star, z^\star) = 0$, then if $u = y - y^\star$ and $v = z - z^\star$, one may expand these equations by using

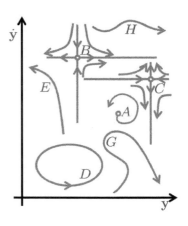

Figure N.8: Some examples of phase portraits describing different situations of vector flow.

the calculus of variations:

$$\dot{u} = u\frac{\partial f}{\partial y} + v\frac{\partial f}{\partial y} + \mathcal{O}(u^2, v^2, uv), \quad \text{and}$$

$$\dot{v} = u\frac{\partial g}{\partial y} + v\frac{\partial g}{\partial y} + \mathcal{O}(u^2, v^2, uv). \tag{N.14}$$

The evolution of (u, v) is

$$\begin{bmatrix} \dot{u} \\ \dot{v} \end{bmatrix} = \begin{bmatrix} \partial f/\partial y & \partial f/\partial z \\ \partial g/\partial y & \partial g/\partial z \end{bmatrix} \begin{bmatrix} u \\ v \end{bmatrix} + \mathcal{O}(n^2). \tag{N.15}$$

In a linearized system with small deviations, we can use the Jacobian matrix

$$[A] = \begin{bmatrix} \partial f/\partial y & \partial f/\partial z \\ \partial g/\partial y & \partial g/\partial z \end{bmatrix}_{(y^\star, z^\star)} \tag{N.16}$$

and write the evolution as

$$\begin{bmatrix} \dot{u} \\ \dot{v} \end{bmatrix} = [A] \begin{bmatrix} u \\ v \end{bmatrix}. \tag{N.17}$$

The equilibrium points for our double potential well problem are described by Equation N.12, so where $(\dot{y}, \dot{z}) = (0,0)$, then $(y^\star, z^\star) = (0,0)$ and $(\pm 1, 0)$. The Jacobian is

$$[A] = \begin{bmatrix} 0 & 1 \\ 1 - 3y^2 & 0 \end{bmatrix}. \tag{N.18}$$

$(0,0)$ is a saddle point, and $(\pm 1, 0)$ are centers. Constant energy trajectories are defined by

$$U = \frac{1}{2}z^2 - \frac{1}{2}y^2 + \frac{1}{4}y^4. \tag{N.19}$$

Figure N.9 shows representative trajectories and a schematic energy representation. Close to either of the centers that are neutrally stable, there are closed orbits, which are shown in both panels of the figures. And farther out, there are closed orbits surrounding all the three fixed points. This system's orbits are periodic. At the neutrally stable fixed points, there is no motion. At the origin, two trajectories start and end. But, this approach is in the limit of $t \to \pm\infty$.

Homoclinic orbits are trajectories that start and end at the same fixed point, such as the two trajectories through the origin in Figure N.10. Conservative systems sport these orbits. The systems are not periodic in that they complete the orbit only in the limit of time.

One should also remark here about reversibility. Any system whose dynamics is symmetric and therefore indistinguishable for a transformation of $t \mapsto -t$ is time-reversal symmetric. A reversible

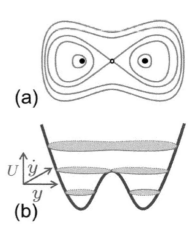

Figure N.9: Trajectories for the double-well potential of Equation N.10 in (a), and its energy landscape in (b).

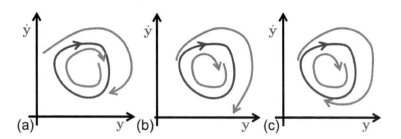

Figure N.10: Stable, unstable and half-stable limit cycles.

system is one that is invariant under $t \mapsto -t$, and $\dot{y} \mapsto -\dot{y}$, that is, time and velocity reversal. Such systems show twin saddle points connected by trajectories that are called heteroclinic trajectories. So, a homoclinic trajectory connects back to the original equilibrium point, and a heteroclinic trajectory connects two different ones. An isolated closed trajectory is a limit cycle. Isolated implies that neighboring trajectories are open. A trajectory that is surrounded by others, each of which approach it, is a stable or attracting limit cycle. Others may be unstable or half-stable. These limit cycles, shown in Figure N.10, are inherent to nonlinear systems. Linear systems have closed orbits, but the orbits are not isolated, as is shown by the examples in Figure N.10.

O
Oscillators

OSCILLATIONS AND OSCILLATORS CONSTITUTE A VERY BASIC MECHANISM of energy storage and exchange. We have encountered them in many forms, and it is these interactions that make much of what is interesting in devices possible.

The harmonic oscillator is a canonical form for quantum mechanics whether we think of a photon, or the vibrating atoms of a crystal (a phonon), or of the electromagnetic interactions in solids. Plasmons, polarons and polaritons are oscillation forms. Even the introduction of classical mechanics start with an oscillating pendulum.

In electromechanics, we encountered the Duffing oscillator—forced oscillations in nonlinear conditions in a double-well system. Chaos was an important manifestation in many forms originating in nonlinearity. Oscillations in circuits employing nonlinearity, such as a van der Pol oscillator, originate in triode nonlinearity. The adiabatic discussion included a rotary Möbius clock that used shunt-connected inverters. An odd number of inverters in a loop form a ring oscillator. If negative dynamic resistance exists under certain conditions, one can force oscillations.

These examples of oscillators range from a simple harmonic oscillator of the simplest of potential wells and representing a stable system—the harmonic potential with oscillator strengths or resonance and quality factors representing efficient coupling over a finite energy or frequency range—to their use in signal manipulation, transmission and extraction, employing a variety of mechanisms. Oscillators can be viewed through the eyes of quantum mechanics as well as through those of classical mechanics. Quantum oscillators will be discussed in Appendix P.

Here, we summarize the characteristics of a few classic oscillators that are instructive and from which one can draw lessons for other oscillators that one encounters.

It is instructive to visualize oscillations from their underlying mechanisms.

A relaxation oscillation results from the system returning to an equilibrium state even as energy continues to be supplied. This happens because there exists a nonlinear energy leakage path. A capacitor charging and discharging repeatedly, such as through a transfer or loss mechanism that gets initiated, is a relaxation oscillation. An LC network can oscillate. This is so because, when the system is disturbed from their thermal equilibrium by the supply of energy, oscillation is a response that is an attempt at relaxation. A seesaw with a bucket for filling water at one end and a counterweight on the other will undergo relaxation oscillation as water is filled continuously into the bucket in a steady stream; upon gaining sufficient weight, the end with the bucket moves down, dumps the water and then immediately returns to its filling state. This system is nonlinear.

A parametric oscillation results from a periodic varying of a parameter or parameters that are accompanied by energy input. A varactor oscillates under bias because the capacitance varies with the voltage across it. The impact avalanche and transit time ($IMPATT$) diode and other negative resistance diodes, such as the Gunn diode or the limited space charge accumulation diode, are parameter driven. Parametric oscillations happen in mechanical systems because of parameter excitation. A child swings by pumping a swing periodically—parametrically.

A harmonic oscillation is the natural response of a system that achieves its stability through restoring forces and has natural periods of oscillations. A pendulum, for example, is a harmonic oscillator. A harmonic oscillator's fundamental importance to the description of natural world through quantum mechanics can scarcely be overemphasized.

In addition, we may have examples of oscillations with both relaxation and parametric features.

In electronics, the most common method for achieving oscillation in circuits is via the use of positive feedback together with a gain, as shown in Figure O.1. Since $V_0 = A(V_i + \beta V_o)$,

$$V_o = \frac{A}{1 - \beta A} V_i, \tag{O.1}$$

which diverges at $\Re[\beta A] = 1$, and $\Im[\beta A] = 0$. This is the Barkhausen criterion for oscillations; the second condition specifically allows for phases of $0, 2\pi, 4\pi, \ldots$. βA is the open loop gain of feedback when the summing is disconnected. In general, phase contributions come from both the amplifier and the feedback element, as well as from the delays in the paths. A real oscillator with $\beta A > 1$ will start and sustain oscillations triggered by noise, and one would need

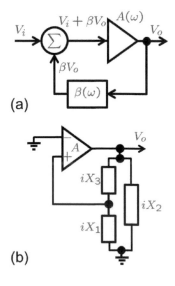

Figure O.1: (a) shows an amplifier of gain $A(\omega)$ with a positive feedback of $\beta(\omega)$. (b) shows an LC-based feedback network represented here through general impedances Zs in an oscillator.

mechanisms for automatic gain control to stabilize gain under the different conditions that the system sees.

Phase stability too is important. Oscillators may employ phase shift introduced by phase shifting filters, for example, RC networks. A Wien bridge achieves the conditions of gain magnitude and phase through differential technique. An LC oscillator, such as that shown in Figure O.1(b), can be shown to have

$$\beta(\omega_0) = \frac{X_1}{X_1 + X_3} = -\frac{X_1}{X_2}, \text{ and } A_{open} = -A\left(-\frac{X_1}{X_2}\right), \quad (O.2)$$

since real β comes about at $X_1 + X_2 + X_3 = 0$. To oscillate requires X_1 and X_2 to be of the same sign, and X_3 to be of the opposite sign, given the positive gain and additive feedback. So, here an opposite combination in the dividing arm is needed to satisfy these conditions.

A Colpitts oscillator uses two capacitors and one inductor. A Hartley oscillator uses two inductors and one capacitor. Crystal oscillators used to be quite common in tuners. They employed piezoelectricity-based mechanical resonance with its high Q, $\sim 10^4$, together with an LCR circuit to exploit its sharp mechanical resonance. In each of these types of oscillators, shifts in phase will cause frequency instability. So, phase fluctuations introduce noise into frequency. At low frequency, the $1/f$ noise is particularly irritating because of this frequency consequence. While these are all old examples, electronic oscillators still employ these basic principles, even up to microwave frequencies.

O.1 Relaxation oscillators

RELAXATION OSCILLATORS ARE ENCOUNTERED ACROSS THE SCIENCES including in biology besides the traditional mechanical and electrical domain. Relaxation oscillators operate with bistable states that arise from a nonlinearity. A relaxation process causes transitions between the stable states and the oscillation period arises from the time constant of the relaxation process.

The van der Pol oscillator that we referred to in Chapter 5 is an example of a relaxation oscillator. The van der Pol oscillator in its original form employed the nonlinearity of a triode. A classic electric example is a discharge tube and a parallel capacitor together connected in series to a direct voltage source through a resistor. The tube glows when voltage across it is sufficient. When the tube turns on, the resistance drops, causing a short circuiting of the parallel capacitor, so the tube turns off. Now, the tube has a high resistance, and the capacitor charges to a voltage where the tube can ignite again.

The blinking of fluorescent light happens for a number of reasons. One of these is the nonlinearities at the end of life when emission has slight rectification during the alternating excursion of the applied bias from the ballast.

Figure O.2 shows a practical example of the use of such nonlinearity: using a gain element configured as a Schmitt trigger and combined with an RC discharge. The gain element has a high input impedance, so the RC time constant charges and discharges the capacitor from the output. If the trigger is set to $-V/2$ and $+V/2$, then V_c will either charge or discharge in this range. If it is discharging, when it reaches $V_c = -V/2$, the output V_0 flips, and the capacitor starts charging. When the voltage rises high enough, the Schmitt triggers flips again, and the capacitor discharges. The Schmitt trigger here worked as the nonlinear element, and the result is a square-wave oscillation.

The important consideration to note in this relaxation oscillation example or that of the seesaw mentioned earlier is that there is nothing periodic about the energy supply process. It is the presence of the triggering by the Schmitt trigger, or the angular rotation of the bucket that causes it to discharge, that imposes a limit on the extent to which the energy gain may proceed and yet, at the same time, provides for the removal of energy. It is this combination of bistability coupled to nonlinearity that makes the periodicity possible. The special property that results from this process is that systems undergoing relaxation oscillations have the ability to utilize energy from a nonperiodic source to maintain the periodicity.

A harmonic oscillator is not capable of extracting this unidirectional energy supply. So, if it is lossy, it suffers extinction due to dissipation. A pendulum will eventually stop swinging. And one cannot apply a constant force to provide energy. This will only cause the center of oscillation to displace and keep displacing. Over a cycle, the force does not provide a balancing energy. That only comes by partitioning out energy at source and synchronizing it in proper phase relationship to the receiving entity. A damped harmonic oscillator will need a suitable relaxation oscillator to continue operating.

Other properties too distinguish harmonic and relaxation oscillators. When a periodic force is applied, the harmonic oscillator response at the force's input is one of small amplitude. Periodic force works well only when it resonates at the harmonic oscillator's natural frequencies. And the response nonlinearly decays away from these frequencies. In relaxation oscillation, the response has a much broader bandwidth. It is limited by the interval in time that determines the two bistable upper and lower values of the potential driving the system. Because it is broad, it is also possible to lock relaxation oscillation, as happens in mechanical clocks.

A Schmitt trigger is a high amplification of the input, following saturation in voltage. So, depending on the polarity of the floating input, the output either goes to a high voltage or a low voltage.

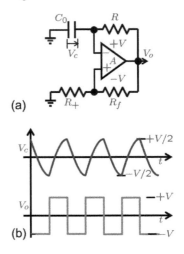

(a)

(b)

Figure O.2: (a) a high gain amplifier set as a Schmitt trigger, with the RC feedback elements shown. (b) shows voltage across the capacitor and at the output as a function of time.

The mechanical clock is an example of a damped harmonic oscillator coupled to a relaxation oscillator. The supply of energy from the spring or weight is regulated by an impulse mechanism. There exists an anchor lever that has bistability through its stops and that transfers energy when the harmonic oscillator is moving at maximum speed, with the driving force differing in phase from the displacement by a quarter cycle. The anchor lever and the harmonic oscillator work together with the relaxation-harmonic coupling. In this case neither can work on its own.

O.2   Parametric oscillators

PARAMETRIC OSCILLATORS—oscillators employing varying of parameters in time—too are common across science and engineering since parameter variation is often convenient to implement in design. In optics, this approach has been particularly powerful in making pump and nonlinear fiber-based implementations standard techniques. In electronics, varactors—bias voltage–dependent capacitors—allow one to make a varactor parametric oscillator, where the parameter is varied periodically to induce oscillations. These devices have been used across the breadth of the microwave spectrum in tank circuits, in distributed circuits and together with differential negative resistance elements. Often, since a reactance (in this case, the varactor's) is varied, thermal noise is low, and this makes the use of varactor-based parametric oscillation very appealing. Another reason for varactor's appeal is that their variable capacitance is easy to achieve in semiconductor technology. Varying capacitance with bias and with large capacitance changes is possible through doping changes that are available in technologies. Yet another reason is simplicity. Figure O.3 shows two varactors—back to back, as shown by the diode-capacitance combination symbol—in parallel with an LC tank circuit. A bias voltage V provides static energy, and an alternating source acts as the pump. With a suitable combination of the inductor, the capacitor, and the varactor's rapidly changing bias-dependent capacitance, this circuit will oscillate at its natural frequency. Optical parametric oscillators employ a similar two-source approach for oscillation, exploiting third order susceptibility (χ^3). This approach of parametric oscillation can also be employed for amplification.

Amplification through varactors—with their modulated reactance—works as follows. A changing bias voltage changes capacitance, that is, the amount of energy in the capacitor. This is electrical work performed by a pump force which becomes available for the alternating signal. Mathematically, it can be seen through the following argument. The charge on the capacitor at time t is $q(t) = C(t)v(t)$. The energy of the capacitor $U_{cap} = q^2/2C$. The power is

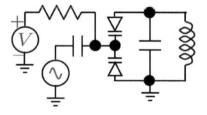

Figure O.3: A varactor diodes—based parametric oscillator.

$$
\begin{aligned}
P_{cap} &= \frac{dU_{cap}}{dt} = \frac{\partial U_{cap}}{\partial q}\bigg|_C \frac{dq}{dt} + \frac{\partial U_{cap}}{\partial C}\bigg|_q \frac{dC}{dt} \\
&= \frac{q}{C}i + \frac{\partial U_{cap}}{\partial C}\bigg|_q \frac{dC}{dz}\frac{dz}{dt} \\
&= vi + \frac{\partial U_{cap}}{\partial z}\frac{dz}{dt},
\end{aligned}
\tag{O.3}
$$

where the boundary z is in the thickness direction that forms this capacitance. The boundary change must be accounted for. The total power integrated over time consists of work to store the energy in the medium and the work associated with the movement of the boundaries. The total power can be exchanged between these two, and this is the basis of parametric operation.

A nonlinear capacitance with two signals—an input signal $v_i(t)$ at frequency v_i, and a pump signal $v_p(t)$ at frequency v_p—superimposed on it will generate harmonics: the primary, that is, the first term, will be $v_0 = v_p \pm v_i$. Expanding the function, we obtain

$$q(v) = \alpha_1 v + \alpha_2 v^2 + \cdots, \qquad (O.4)$$

which is a Taylor series form, and by considering the first two terms, we obtain

$$i(t) = \frac{dq(t)}{dt} = \alpha_1 \frac{dv(t)}{dt} + 2\alpha_2 v(t) \frac{dv(t)}{dt} = [C_0 + C_v(t)] \frac{dv(t)}{dt}, \quad (O.5)$$

where C_0 is the linear dependence of the capacitance, and $C_v(t)$ is the voltage-dependent second order term. The nonlinear capacitor behaves as a time-varying linear capacitance.

The optical analog of this is to split a pump photon at v_p into two in a nonlinear medium, creating two frequency photons (v_s for signal and v_i for idler). Phase matching is achieved using birefringence which also compensates for dispersion and specifies the wavelength combination for phase matching. Quasi-phase matching, where domain inversion is employed in the crystal, can also be employed to match combinations of signal and idler photons. The result is that high power can be generated at longer wavelengths—in regions where such power may not easily be available.

P
Quantum oscillators

FIELD QUANTIZATION is essential for understanding quantum optical interactions, and ladder operators, also called creation and annihilation operators, provide a very compact and convenient mechanism to tackle the elementary excitation changes that are pervasive with such interactions. Here, we summarize the preliminary elements of this approach, with the goal of describing the coupling between mechanical oscillations and the field oscillations of interest to us.

The potential $V = m\omega^2 x^2/2$ describes a classical simple harmonic oscillator. Using just one spatial dimension x, the position $\hat{x}$ and momentum operators $\hat{p}$ satisfy the commutation relation $[\hat{x}, \hat{p}] = i\hbar$, and the Hamiltonian of the simple harmonic oscillator is

$$\mathscr{H} = \frac{\hat{p}^2}{2m} + \frac{m\omega^2 \hat{x}^2}{2}. \tag{P.1}$$

To write energy in terms of quanta, we use a normalization that allows the positional and momentum operators to be suitably mapped to energy forms. These normalizations for position and momentum are

$$x_0 = \sqrt{\frac{\hbar}{2m\omega}} \quad \therefore \hat{x}_0 = \frac{\hat{x}}{2x_0}, \quad \text{and}$$

$$p_0 = \sqrt{\frac{m\omega\hbar}{2}} \quad \therefore \hat{p}_0 = \frac{\hat{p}}{2p_0}, \tag{P.2}$$

which reduces the Hamiltonian to

$$\mathscr{H} = \hbar\omega(\hat{x}_0^2 + \hat{p}_0^2), \tag{P.3}$$

with the position and momentum energy contributions on identical energy-normalized numerical footing and a form that is more symmetric than that of Equation P.1. We now define two non-Hermitian operators,

$$\hat{a} = \hat{x}_0 + i\hat{p}_0 = \frac{1}{2}\sqrt{\frac{2m\omega}{\hbar}}\hat{x} + i\frac{1}{2}\sqrt{\frac{2}{m\omega\hbar}}\hat{p}, \quad \text{and its conjugate}$$

$$\hat{a}^\dagger = \hat{x}_0 - i\hat{p}_0 = \frac{1}{2}\sqrt{\frac{2m\omega}{\hbar}}\hat{x} - i\frac{1}{2}\sqrt{\frac{2}{m\omega\hbar}}\hat{p}, \qquad (\text{P.4})$$

whose operation results in the normalized amplitude and its complex conjugate for the classical oscillator. These ladder operators have a number of interesting properties. The following relations hold:

$$\begin{aligned}
[\hat{a}, \hat{a}^\dagger] &= \mathbb{I}, \\
\hat{x}_0 = \frac{1}{2}\sqrt{\frac{2m\omega}{\hbar}}\hat{x} &= \frac{1}{2}(\hat{a} + \hat{a}^\dagger), \\
\hat{p}_0 = \frac{1}{2}\sqrt{\frac{2}{m\omega\hbar}}\hat{p} &= \frac{i}{2}(\hat{a}^\dagger - \hat{a}), \text{ and} \\
\mathcal{H} &= \hbar\omega(\hat{a}^\dagger\hat{a} + \frac{1}{2}).
\end{aligned} \qquad (\text{P.5})$$

Looking at the Hamiltonian in Equation P.5, one sees $\hat{a}^\dagger\hat{a} = \hat{n}$—a number operator—and a fractional term of $\hbar\omega/2$—the vacuum fluctuation energy. $\hat{a}$ is the annihilation operator, and the $\hat{a}^\dagger$ is the creation operator. If the fluctuation energy, in an analysis of the problem, appears as a constant shift and does not have a direct implication, one could just work with a modified Hamiltonian, of shifted origin, of

$$\mathcal{H}_s = \hbar\omega\hat{a}^\dagger\hat{a} \propto \hat{n}. \qquad (\text{P.6})$$

The shifted Hamiltonian $\mathcal{H}_s$ or the number and ladder operators settle for us nearly all of the spectrum-related properties of the quantum oscillator and is thus useful in coupled-energy problems. They have a number of interesting properties:

$$\begin{aligned}
\hat{n}|n\rangle &= n|n\rangle, \\
[\hat{a}, \hat{n}] &= \hat{a}, \\
[\hat{a}^\dagger, \hat{n}] &= -\hat{a}^\dagger, \\
\hat{a}|n\rangle &= \sqrt{n}|n-1\rangle, \\
\hat{a}^\dagger|n\rangle &= \sqrt{n+1}|n+1\rangle, \text{ and} \\
|n\rangle &= \frac{\hat{a}^{\dagger n}}{\sqrt{n!}}|0\rangle.
\end{aligned} \qquad (\text{P.7})$$

$|n\rangle$ is a Fock state that forms the ladder in n, as described by the first of the operations in Equation P.7. These constitute an orthonormal basis set in the Hilbert space. The shifted Hamiltonian eigenenergy is $n\hbar\omega$—it is the energy associated with n of $\hbar\omega$ quanta, for example, the quanta of phonons. The number operator and the ladder operators do not commute, as stated by the second and third relationship. The annihilation operator removes an elementary excitation, and the creation ladder operator adds an elementary excitation. And the final relationship states that Fock states may be generated by repeated use of the creation operator from the ground state $|0\rangle$.

Measurable properties of particles—mass, charge, magnetic moment—are "dressed" by the vacuum fluctuations that are omnipresent and omniscient. They cannot be switched off. However, they may in most situations cancel out in the balance of the equation modeling the physical phenomenon. Renormalization is needed otherwise, as it was in settling the ultraviolet catastrophe.

In the Schrödinger view, Fock states are stationary states—probability amplitudes evolve with time-dependent, eigenenergy-specified phases. In the Heisenberg view, oscillator states are stationary, and the operators evolve as

$$\frac{d\hat{a}}{dt} = i\hbar[\hat{a}, \mathscr{H}] = -i\omega[\hat{a}, \hat{n}] = -i\omega\hat{a} \tag{P.8}$$

so that

$$\hat{a}(t) = \hat{a}\exp(-i\omega t). \tag{P.9}$$

This description applies to the field mode of a cavity such as a Fabry-Pérot cavity, where the quanta are that of the photon, and the frequency $\omega = \omega_{cav}$, the cavity mode's angular frequency; so, in our notation, $\mathscr{H}_s |n\rangle = n\hbar\omega_{cav}$. The electric field is expressible as

$$\mathcal{E}(\mathbf{r}) = i\mathcal{E}_0[u(\mathbf{r})\epsilon\hat{a} - u^*(\mathbf{r})\epsilon^*\hat{a}^\dagger], \tag{P.10}$$

with the following descriptions: $u(\mathbf{r})$ is a function describing the normalized amplitude spatial dependence of the field mode, ϵ is the unit polarization vector, $\mathcal{E}_0$ is a normalization factor, and the form of the equation is consistent with that of the Helmholtz equation. Since field is related to vector potential as $\mathcal{E} = -\nabla\psi - \partial\mathbf{A}/\partial t$, including the operator's time evolution,

$$\hat{\mathbf{A}}(\mathbf{r}, t) = \frac{\mathcal{E}_0}{\omega_{cav}}[u(\mathbf{r})\epsilon\hat{a}\exp(-i\omega_{cav}t) + u^*(\mathbf{r})\epsilon^*\hat{a}^\dagger\exp(i\omega_{cav}t)] \tag{P.11}$$

in the Heisenberg representation, and

$$\hat{\mathbf{A}}(\mathbf{r}) = \frac{\mathcal{E}_0}{\omega_{cav}}[u(\mathbf{r})\epsilon\hat{a} + u^*(\mathbf{r})\epsilon^*\hat{a}^\dagger] \tag{P.12}$$

in the Schrödinger representation. The Fock state $|0\rangle$, the ground state, is the vacuum state. The expectation values of the field and the vector potential vanish in this state. But, field fluctuations do not. The energy is the square of the field, so terms resulting from $\hat{a}^\dagger\hat{a} \neq \hat{a}\hat{a}^\dagger$ are non-zero. The field is zero, but a finite energy density also exists showing up as fluctuations, uncertainty and exchange between kinetic and potential energies in this uncertainty. The normalization can be thought of in terms of mode volume. Since the energy of state $|n\rangle$ is $\hbar\omega_{cav}(n + 1/2)$,

$$\langle n| \int \epsilon_0|\mathcal{E}|^2 d^3\mathbf{r}|n\rangle = \langle n|\epsilon_0\mathcal{E}_0^2\mathcal{V}(2n + 1)|n\rangle$$
$$= \hbar\omega_{cav}\left(n + \frac{1}{2}\right), \tag{P.13}$$

with $\mathcal{V} = \int |\mathbf{u}(\mathbf{r})|^2 d^3\mathbf{r}$ being the mode volume. Our field normalization is then related to the energy and mode volume through

$$\mathcal{E}_0 = \sqrt{\frac{\hbar\omega_{cav}}{2\epsilon_0\mathcal{V}}}, \tag{P.14}$$

which depends only on the frequency and cavity geometry. This captures the essence of the quantization of field in the finite-sized cavity. If mode is propagating and is a single mode, then one may imagine it as a mode of a single-mode circulating cavity, such as a ring laser, of a size larger than the size of the system of interest. Here, the vacuum field amplitude is very small. Spontaneous emission conditions can now be taken to be an expansion over a continuum of modes in a volume that goes to infinity.

Glossary

A SYMBOL GENERATED BY using a tilde sign on a symbol, for example, $\tilde{a}$ from a, is used to signify, explicitly, the complex time-varying quantity. The real part of this has a sinusoidal time variation. The phasor, or the amplitude of this time-varying component, is denoted by using the hat sign on the symbol, for example, $\hat{a}$ for a. This notation is used in the context of small-signal variation. An exception to this nomenclature is the use of the hat symbol to denote a unit normal vector, for example, $\hat{n}$ to denote the unit normal vector perpendicular to a surface. A hat is also employed, together with calligraphic, mathematics-specific or normal font, to denote a quantum-mechanical operator. A lowercase subscript to an uppercase letter denotes a quantity which may have both a static and a time-varying component. An uppercase subscript or an overline represents quasistatic quantities. Any other exceptions have been pointed out in context. This list defines the most frequently used symbols. Système international d'unités—SI—units are employed in the text. Any exceptions are either pointed out in context or follow from dimensionality.

Symbol	Symbol definition	Unit
↑	Spin/polarization up	—
↓	Spin/polarization down	—
$\Delta U_{xlr}(r)$	Energy of cluster formation	J
$\Delta U_{lc,\Omega}(r)$	Gibbs free energy difference	J
$1/4\pi\epsilon_0$	Prefactor of SI units	$8.99 \times 10^9 \ V \cdot m/A.s$
$\alpha_\uparrow$	Spin-up fraction of electrons	—
α_i	Fine structure constant $(e^2/4\pi\epsilon_0\hbar c)$	$1/137.04$
a_B	Bohr radius $(4\pi\hbar^2/m_0e^2)$	$0.529 \times 10^{-10} \ m$
$\mathbf{A}$	Vector potential	$V.s \cdot m^{-1}$
$\mathscr{A}^*$	Richardson constant	A/m^2K^2
A	Absorptance	—
$\mathbf{B}$	Magnetic induction $(\mu_r\mu_0\mathbf{H}, \ \mu\mathbf{H})$	$V.s/m^2$, that is, T
c	Speed of light in free space $(1/(\mu_0\epsilon_0)^{1/2})$	$2.998 \times 10^8 \ m/s$

χ	Magnetic susceptibility	—
C_{dc}	Drain coupling capacitance	F
C_{dg}	Drain-to-gate capacitance	F
C_{ds}	Drain-to-source capacitance	F
C_{gs}	Gate-to-source capacitance	F
Δ	Exchange energy	J, eV
Δ_s	Splitting energy	J, eV
$\mathbf{D}$	Displacement $(\epsilon\epsilon_0\mathcal{E})$	$A.s/m^2$
$\mathcal{D}$	Diffusion coefficient	m^2/s
ϵ	Permittivity of material	$A.s/V \cdot m$
ϵ_0	Permittivity of free space $(1/\mu_0 c^2)$	$8.854 \times 10^{-12}\ A.s/V \cdot m$
ϵ_r	Dielectric (relative) constant	
e	Absolute electron charge	$1.602 \times 10^{-19}\ A.s\ (\equiv Coulomb)$
e/m_0	Electron charge to mass (ω/B)	$1.759 \times 10^{11}\ rad/s \cdot T$
$\mathcal{E}, \mathbf{\mathcal{E}}$	Electric field	V/m
E_c	Conduction bandedge	eV
E_v	Valence bandedge	eV
E_g	Bandgap	eV
E_F	Fermi energy	eV
E_{qF}	Quasi-Fermi energy	eV
$E_{qF\uparrow}$	Spin-up quasi-Fermi energy	eV
$E_{qF\downarrow}$	Spin-down quasi-Fermi energy	eV
E_R	Rydberg energy $(m_0 e^4/2\hbar^2(4\pi\epsilon_0)^2)$	
f	Frequency	s^{-1}
F	Force	N
F_{es}	Electrostatic force	N
$\mathcal{F}$	Fermi-Dirac distribution function	—
$\mathcal{F}$	Helmholtz free energy	J, eV
$\mathscr{F}_\nu$	Fermi integral of order ν	—
$\mathcal{F}$	Slowly varying force	N
$\mathfrak{F}$	Rapidly varying force	N
φ_B	Barrier height	eV
ϕ_{abs}	Absorptance	$m^{-2} \cdot s^{-1} \cdot eV^{-1}$
ϕ_B	Barrier height in units of V	V
ϕ_{bb}	Blackbody photon flux	$m^{-2} \cdot s^{-1} \cdot eV^{-1}$
ϕ_{em}	Emissivity	$m^{-2} \cdot s^{-1} \cdot eV^{-1}$
ϕ_{inc}	Incident photon flux	$m^{-2} \cdot s^{-1} \cdot eV^{-1}$
ϕ_{sun}	Photon flux from sun	$m^{-2} \cdot s^{-1} \cdot eV^{-1}$
$\mathbf{F}_C, F_C$	Casimir force	$N, kg \cdot m/s^2$
g_{bj}	Band degeneracy	—
g_c	Contact conductance	$S \cdot m$
g_m	Transconductance	A/V
g_{mi}	Intrinsic transconductance	A/V
g_{mx}	Extrinsic transconductance	A/V

g_q	Quantum conductance ($2e^2/h$)	$\sim 80 \ \mu S$
g_s	Spin degeneracy	2
g_s	Sheet conductance of semiconductor	S
$\mathcal{G}$	Gibbs free energy	J, eV
$\mathcal{G}_{3D}(E)$	Three-dimensional density of states	$m^{-3}eV^{-1}$
$\mathcal{G}_{2D}(E)$	Two-dimensional density of states	$m^{-2}eV^{-1}$
$\mathcal{G}_{1D}(E)$	One-dimensional density of states	$m^{-1}eV^{-1}$
$\mathcal{G}_\uparrow(E)$	Up spin density of states	$m^{-3}eV^{-1}$
$\mathcal{G}_\downarrow(E)$	Down spin density of states	$m^{-3}eV^{-1}$
Γ	Electron leakage rate from dot	$1/s$
GST	$Ge_x Sb_{1-x} Te$	
h	Planck's constant	$6.626 \times 10^{-34} \ kg \cdot m^2/s$ or $4.136 \times 10^{-15} \ eV \cdot s$
$\hbar$	Reduced Planck's constant	$1.055 \times 10^{-34} \ kg \cdot m^2/s$ or $6.582 \times 10^{-16} \ eV \cdot s$
H	Average information content $\equiv$ uncertainty	b
H	Boltzmann H-factor	—
$\mathbf{H}$	Magnetic field ($\mathbf{m}/V$)	A/m
$\mathbf{H}_b$	Exchange magnetic field ($\mathbf{m}/V$)	A/m
$\mathcal{H}$	Enthalpy	J, eV
$\mathscr{H}$	Hamiltonian operator	—
$\mathfrak{H}$	Hadamard operation	—
i	Current (small-signal)	A
i_{di}	Internal drain current (small-signal)	A
i_{dx}	Extrinsic drain current (small-signal)	A
I	Information content	b
$I, \mathbf{I}$	Current	A
I	Moment of inertia	m^4
$J, \mathbf{J}$	Current density	A/m^2
J_{rad}	Radiative recombination current density	A/m^2
J_{sc}	Short-circuit current density	A/m^2
J_{sc}^{max}	Maximum short-circuit current density	A/m^2
J_{te}	Thermionic current density	A/m^2
κ	Anisotropy constant	$J/m^3, eV/m^3$
$\mathbf{k}$	Wave vector	m^{-1}
k_B	Boltzmann constant	$1.38 \times 10^{-23} \ J/K$
$\mathbf{k}_s$	Spring constant	N/m
$\mathbf{k}_F$	Wave vector at Fermi energy	cm^{-1}
$\mathcal{K}(s)$	Correlation function	N^2
$\mathfrak{K}$	Knudsen number	—
χ	Susceptibility	—
ℓ_{eff}	Length scale of travel at Bethe condition	m
ℓ	Magnetic length scale	m
$\lambda^\uparrow$	Mean free path for $\uparrow$ electrons	m
$\lambda^\downarrow$	Mean free path for $\downarrow$ electrons	m
λ_{deB}	de Broglie wavelength	m

$\lambda_{\mathbf{k}}$	Mean free path	m
λ_w	Energy relaxation length	m
$\mathcal{L}$	Diffusion length	m
$\mathscr{L}$	Lagrangian function	J
L_c	Contact length	m
$\mathbf{m}$	Magnetic moment	$V \cdot s \cdot m$
$\mathbf{M}$	Magnetization $(\mathbf{m}/V)$	A/m
$\mathfrak{M}$	Moment	$N \cdot m$
m_0	Free electron mass	$9.1 \times 10^{-31}\ kg$
$m_0 c^2$	Electron rest energy	$0.819 \times 10^{-13}\ V \cdot A/s = 0.5111\ MeV$
m^*	Effective mass	kg
μ	Mobility	$m^2/V.s$
μ	Permeability	$V.s/A \cdot m$
μ_0	Permeability of free space $(1/\epsilon_0 c^2)$	$4\pi \times 10^{-7}\ V.s/A \cdot m$
μ_r	Relative permeability of material	
μ_B	Bohr magneton $(e\hbar\mu_0/2m_0)$	$1.165 \times 10^{-29}\ V \cdot m \cdot s$
ν	Frequency	$1/s$
ν	Dimensionality	—
N	Noise energy	J
$N_\uparrow$	Up spin carrier density	m^{-3}
$N_\uparrow$	Down spin carrier density	m^{-3}
n^{xs}	Excess spin electron density	$1/m^3$
N_A	Avogadro's number	6.02214×10^{23}
NA	Numerical aperture	—
$\mathcal{N}_c$	Effective density of states	m^{-3}
$p, \mathbf{p}$	Momentum	$kg \cdot m/s$
$\mathbf{p}$	Electric dipole moment	$A.s \cdot m$
$\mathfrak{p}$	Probability	—
$\mathfrak{P}$	Probability distribution	—
PL	Plate line	—
q	Fundamental charge $(-e)$	$-1.602 \times 10^{-19}\ A.s$
Q	Heat energy	J
$\mathcal{Q}$	Heat energy flux	$J/cm \cdot s$
$\mathscr{Q}$	Heat energy flux density	$J/m^2 \cdot s$
Q	Quality factor	—
r_c	Critical radius	m
r_e	Classical electron radius $(e^2/4\pi\epsilon_0 m_0 c^2)$	$2.818 \times 10^{-15}\ m$
r_H	Hall factor	—
ρ	Resistivity	$\Omega.cm$
ρ	Volume charge density	C/m^3
ρ_s	Sheet resistance	Ω
ρ_{chs}	Channel sheet resistance	Ω
ρ_{jns}	Source sheet resistance	Ω
ϱ	Reflectivity	—

ϱ_{MR}	Magnetoresistance ratio	—
ϱ_J	Spin current polarization ratio	—
R_d	Drain resistance	Ω
R_{ds}	Drain-to-source resistance	Ω
R_g	Gate resistance	Ω
R_H	Hall constant	m^3/C
R_i	Intrinsic resistance	Ω
R_s	Source resistance	Ω
R_{ss}	Spreading resistance	Ω
$\mathfrak{R}$	Reynold's number	—
σ	Variance	—
σ	Conductivity	S/cm
σ_e	Thomson cross-section $(8\pi r_e^2/3)$	$0.665 \times 10^{-28}\ m^2$
s	Spin quantum number	$\pm 1/2$
s_k	Collection of microstates	—
S	Entropy	J/K
S	Signal energy	J
$\mathscr{S}$	Statistical ensemble	—
SNR	Signal-to-noise ratio	W/W
$SNR|_{dB}$	Signal-to-noise ratio dB	dB
τ	Scattering time	s
τ	Electron lifetime on dot	s
τ_c	Capture time constant	s
τ_d	Signal transit delay	s
τ_e	Emission time constant	s
$\tau_{\mathbf{k}}$	Momentum relaxation time	s
τ_{sf}	Spin flip relaxation time	s
τ_w	Energy relaxation time	s
τ^*	Correlation time	s
T	Temperature	K
T	Specific time	s
T	Torque	$N \cdot m$
T	Kinetic energy	J
$\mathscr{T}$	Transmission coefficient	—
T_{ix}	Injected spin exchange torque	$N \cdot m$
T_c	Critical temperature	K
T_C	Curie temperature	K
T_g	Glass temperature	K
T_K	Kondo temperature	K
T_m	Melting temperature	K
T_N	Néel temperature	K
U	Internal energy	J, eV
U_{es}	Electrostatic energy	J, eV
u_i	Internal energy of ith particle	J, eV

$u, \mathbf{u}$	Displacement	m
v	Velocity	m/s
v_{sat}	Saturation velocity	m/s
v_F	Fermi velocity	m/s
v_g	Group velocity	m/s
v_{gs}	Gate-to-source small-signal voltage	V
v_{gsx}	External gate-to-source small-signal voltage	V
v_{gsi}	Internal gate-to-source small-signal voltage	V
v_S	Effective source velocity	m/s
v_θ	Thermal velocity	m/s
V_{sa}	Spin accumulation voltage	V
V_{DD}	Supply voltage	V
V_D	Static drain voltage	V
V_S	Static source voltage	V
V_{sa}	Spin accumulation voltage	V
V_G	Static gate voltage	V
W	Kinetic energy density	J/m^3

OTHER UNITS, IN POPULAR USAGE BECAUSE OF HISTORY, as well as because of the insight they give, can be understood through the following relationships provided here for reference.

Unit	Conversion
Oersted (Oe)	$= 10^3/4\pi \ A/m = 79.59 \ A/m$
Tesla (T)	$= N/A \cdot m = V.s/m^2 = kg/s^2 A = 10^4 \ Gauss$
Ohm (Ω)	$= V/A$
Coulomb (C)	$= A.s$
Newton (N)	$= V \cdot A.s/m = kg \cdot m/s^2$
Kilogram (kg)	$= V \cdot A.s^3/m^2$
Farad (F)	$= A.s/V$
Henry (H)	$= kg \cdot m^2/s^2 A^2$
Joule (J)	$= N \cdot m = V \cdot A.s = 10^7 \ erg$
Watt (W)	$= V \cdot A = J/s$
eV	$= 1.602 \times 10^{-19} \ V \cdot A.s$
eV/k_B	$= 1.1605 \times 10^4 \ K$
eV/h	$= 2.418 \times 10^{14} \ Hz$
eV/hc	$= 8066 \ cm^{-1} = 8066 \ Kayser$
$h\nu \ (eV)$	$= 1249.852/\lambda$ in nm
μ_B/μ_0	$= 0.578 \times 10^{-4} \ eV/T$
barn (b)	$= 1 \times 10^{-28} \ m^2$
deg $(°)$	$= \pi/180 \ rad = 17.45 \ mrad$
arcmin	$= 1/60° = 290.0 \ \mu rad$

ACRONYMS have been employed sparingly in the text, but a few do
slip in for compactness. These are listed here.

Acronym	Full form
ASCII	American standard code for information exchange
ATP	Adenosine triphosphate
ADP	Adenosine diphosphate
BAR	Bulk acoustic resonator
BAW	Bulk acoustic wave
BCC	Body-centered cubic
BD	Blue ray disk
BL	Bit line
BLAS	Basic linear algebra subprograms
cdROM	Compact disk read-only memory
CERN	Conseil européenne pour la recherche nucléaire
CMOS	Complementary metal oxide semiconductor
DIBL	Drain-induced barrier lowering
DNA	Deoxyribonucleic acid
DOS	Density of states
dRAM	Dynamic random access memory
fBAR	Film bulk acoustic resonator
FCC	Face-centered cubic
feRAM	Ferroelectric random access memory
FL	Field line
FSM	Finite state automaton
GIDL	Gate-induced drain lowering
HCP	Hexagonal close-packed
HOMO	Highest occupied molecular orbital
IMPATT	Impact avalanche and transit time
LNA	Low noise amplifier
LIGO	Laser interferometer gravitational wave observatory
LUMO	Lowest unoccupied molecular orbital
KDP	Potassium dihydrogen phosphate
mRAM	Magnetic random access memory
nMOS	n-Type metal oxide semiconductor
NA	Numerical aperture
NASA	National Aeronautics and Space Administration
NIST	National Institute of Standards and Technology
NSA	National Security Agency
pMOS	p-Type metal oxide semiconductor
pcRAM	Phase change random access memory
pdf	Probability distribution function

PET	Positron emission tomography
PL	Plate line
RNA	Ribonucleic acid
ROM	Read-only memory
$rRAM$	Resistive random access memory
SAT	Satisfiability
SAW	Surface acoustic wave
SBT	Strontium bismuth tantalate
SI	Système international d'unités
$sRAM$	Static random access memory
SS	Subthreshold swing
$stRAM$	Spin torque random access memory
TE	Transverse electric
TEM	Transverse electromagnetic
TM	Transverse magnetic
WL	Word line

SCALES as a parameter that approximately guides the validity of a formalism have been employed throughout the book. Particularly crucial in this have been those associated with length scales of an important phenomenon. This table is a representative list of those emphasized.

Scale parameter	Meaning
a_B	Bohr radius
a_B^*	Effective Bohr radius
ℓ_{eff}	Length scale of travel at a barrier in Bethe condition
ℓ	Magnetic length scale
λ	Wavelength
λ	Off state scaling length
λ	Coulombic screening length
λ_{deB}	de Broglie wavelength
λ_D	Debye screening length
λ_ϕ	Phase coherence length
$\lambda_\mathbf{k}$	Momentum relaxation length
λ_{mfp}	Mean free path $\equiv$ momentum relaxation length
λ_{scatt}	Scattering length
λ_{scr}	Screening length
λ_{TF}	Thomas-Fermi screening length
$\mathcal{L}$	Diffusion length
$\tau_\mathbf{k}$	Momentum relaxation time
τ_w	Energy relaxation time

VECTOR IDENTITY RELATIONSHIPS used in the text are provided here for reference.

Identities:

$$
\begin{aligned}
\mathbf{A} \times \mathbf{B} \cdot \mathbf{C} &= \mathbf{A} \cdot \mathbf{B} \times \mathbf{C}, \\
\mathbf{A} \times (\mathbf{B} \times \mathbf{C}) &= \mathbf{B}(\mathbf{A} \cdot \mathbf{C}) - \mathbf{C}(\mathbf{A} \cdot \mathbf{B}), \\
\nabla(f + g) &= \nabla f + \nabla g, \\
\nabla \cdot (\mathbf{A} + \mathbf{B}) &= \nabla \cdot \mathbf{A} + \nabla \cdot \mathbf{B}, \\
\nabla \times (\mathbf{A} + \mathbf{B}) &= \nabla \times \mathbf{A} + \nabla \times \mathbf{B}, \\
\nabla(fg) &= f\nabla g + g\nabla f, \\
\nabla \cdot (f\mathbf{A}) &= \mathbf{A} \cdot \nabla f + f\nabla \cdot \mathbf{A}, \\
\nabla \cdot (\mathbf{A} \times \mathbf{B}) &= \mathbf{B} \cdot \nabla \times \mathbf{A} - \mathbf{A} \cdot \nabla \times \mathbf{B}, \\
\nabla \cdot \nabla f &= \nabla^2 f, \\
\nabla \cdot \nabla \times \mathbf{A} &= 0, \\
\nabla \times \nabla f &= 0, \\
\nabla \times (\nabla \times \mathbf{A}) &= \nabla(\nabla \cdot \mathbf{A}) - \nabla^2\mathbf{A}, \\
(\nabla \times \mathbf{A}) \times \mathbf{A} &= (\mathbf{A} \cdot \nabla)\mathbf{A} - \frac{1}{2}\nabla(\mathbf{A} \cdot \mathbf{A}), \\
\nabla(\mathbf{A} \cdot \mathbf{B}) &= (\mathbf{A} \cdot \nabla)\mathbf{B} + (\mathbf{B} \cdot \nabla)\mathbf{A} \\
&\quad + \mathbf{A} \times (\nabla \times \mathbf{B}) + \mathbf{B} \times (\nabla \times \mathbf{A}), \\
\nabla \times (f\mathbf{A}) &= \nabla f \times \mathbf{A} + f\nabla \times \mathbf{A}, \\
\nabla \times (\mathbf{A} \times \mathbf{B}) &= \mathbf{A}(\nabla \cdot \mathbf{B}) - \mathbf{B}(\nabla \cdot \mathbf{A}) \\
&\quad + (\mathbf{B} \cdot \nabla)\mathbf{A} - (\mathbf{A} \cdot \nabla)\mathbf{B}.
\end{aligned}
\tag{Q.1}
$$

MAJOR EQUATIONAL RELATIONSHIPS IN SI UNITS employed in the text are provided here for reference.

Maxwell's equations:

$$
\begin{aligned}
\nabla \cdot \mathbf{D} &= \rho, \\
\nabla \cdot \mathbf{B} &= 0, \\
\nabla \times \mathcal{E} &= -\frac{\partial \mathbf{B}}{\partial t}, \text{ and} \\
\nabla \times \mathbf{H} &= \mathbf{J} + \frac{\partial \mathbf{D}}{\partial t}
\end{aligned}
\tag{Q.2}
$$

Constitutive relationships of Maxwell equations:

$$
\mathbf{D} = \epsilon_0\mathcal{E} + \mathbf{P} = \epsilon_0\mathcal{E} + \chi\epsilon_0\mathcal{E} = \epsilon_r\epsilon_0\mathcal{E} = \epsilon\mathcal{E}, \text{ and}
$$
$$
\mathbf{B} = \mu_0(\mathbf{H} + \mathbf{M}) = \mu_0\mathbf{H} + \chi\mu_0\mathbf{H} = \mu_r\mu_0\mathbf{H} = \mu\mathbf{H}.
\tag{Q.3}
$$

Gauss's theorem:

$$
\begin{aligned}
\lim_{\Omega \to 0} \frac{1}{\Omega} \int_S \mathbf{B} \cdot \hat{\mathbf{n}}\, d^2r &= \nabla \cdot \mathbf{B}, \text{ or} \\
\int_S \mathbf{B} \cdot \hat{\mathbf{n}}\, d^2r &= \int_\Omega \nabla \cdot \mathbf{B}\, d\Omega.
\end{aligned}
\tag{Q.4}
$$

Stokes' theorem:

$$\lim_{S \to 0} \frac{1}{S} \oint_r \mathbf{H} \cdot d\mathbf{r} = \hat{\mathbf{n}} \cdot (\nabla \times \mathbf{H}), \text{ or}$$

$$\oint_r \mathbf{H} \cdot d\mathbf{r} = \int_S \hat{\mathbf{n}} \cdot (\nabla \times \mathbf{H}) \, d^2 r. \quad (Q.5)$$

Semi-classical conservation equations for electronics:

$$\nabla \cdot \mathbf{D} = \rho,$$

$$-\frac{1}{q} \nabla \cdot \mathbf{J}_n = \mathcal{U} = \mathcal{G} - \mathcal{R}, \text{ and}$$

$$\frac{1}{q} \nabla \cdot \mathbf{J}_p = \mathcal{U} = \mathcal{G} - \mathcal{R}. \quad (Q.6)$$

Semi-classical constitutive equations for electronics:

$$\mathbf{D} = \epsilon_r \epsilon_0 \mathcal{E} = -\epsilon_r \epsilon_0 \nabla V,$$

$$\rho = q \left(p - n + N_D^+ - N_A^- \right),$$

$$\mathbf{J}_n = nq\mu_n \mathcal{E} + q\mathcal{D}_n \nabla n,$$

$$\mathbf{J}_p = pq\mu_p \mathcal{E} - q\mathcal{D}_n \nabla p,$$

$$\mathcal{G} = \mathcal{G}(n, p, \ldots), \text{ and}$$

$$\mathcal{R} = \mathcal{R}(n, p, \ldots). \quad (Q.7)$$

Index

0th law of thermodynamics, 3
1st law of thermodynamics, 3, 22
$1/f$ noise, 88, 598
1T1C, 269
1T2C, 272
2nd law of thermodynamics, 3, 15, 29
2N-2P inverter, 38
2T2C, 272
3rd law of thermodynamics, 3
$ASCII$, 176
$As-Te-Br$, 395
$As-Te-I$, 395
BD, 381, 383
$BLAS$, 171
CD, 380, 383
C_{60}, 529
Co, 348
$CoCrPt$, 358
CoO, 293
$CoPt$ segregated, 358
Cs, 504
$DIBL$, 70
DVD, 381, 383
EPR, 172, 618
Fe, 348
FeO, 293
$GIDL$, 70
GMR head
   Current-in-plane, 345
   Perpendicular-to-plane, 345
$GaAs$, 94, 119
$Ga_{1-x}In_xAs$, 94, 119, 123, 126
$Ga_{1-x}Mn_xAs$, 360, 361
Gd, 338
$Ge_xSb_{1-x}Te$, 379
$HOMO$ level, 529
$HgTe$, 200
$InAs$, 94, 119, 123, 126
$LIGO$, 492

$LUMO$ level, 529
$LaCu_xMn_{1-x}O_3$, 305
$LaTiO_3$, 295
$LaVO_3$, 295
MnO, 293
NA, 383
$NAND$, 10, 13, 32, 35, 157
$NIST$, 177
NOR, 13, 32, 35, 157
NOT, 206
   Completely random, 165
   Deterministic, 165
   Nondeterministic, 165
   Quantum, 167
$NOTCOPY$, 159
NSA, 177
Nb_2O_3, 532
Ni, 348
NiO, 293, 295, 400
$PZT(PbZr_{0.52}Ti_{0.48}O_3)$, 260
$Pb_xZr_{1-x}TiO_3$, 305
ROM, 380
RSA, 166
Rb, 504
S-like curve, 300
SAT, 242
SBT ($SrBi_2Ta_2O_9$), 261
$SHIFT$, 168
SS, 75
$SWAP$, 159, 168
SnO_2, 532
$SrRu_xTi_{1-x}O_3$, 305
$THROUGH$, 168
TiO_2, 531
VO_2, 295, 296
   Density of states, 297
   Monoclinic, 297
   Tetragonal, 297
V_2O_3, 295

XOR, 159, 176
XOR function ($\oplus$), 159
$YTiO_3$, 295
ZnO, 532
$ZnTe$, 361
$Zn_{1-x}Cr_xTe$, 360
$cNOT$, 159, 168, 205
$ccNOT$, 159, 168
$dRAM$, 269
$feRAM$, 268
$mRAM$, 337, 373
   Astroid, 375
   Robustness, 375
   Switching energy, 377
$rRAM$, 392
   "reset", 393
   "set", 393
   Apolar, 395
   Electrode interaction, 393
   Interface region, 391
   Joule heating, 392
   Polar, 395
   Thermochemical, 392
$rROM$, 380
$rwROM$, 380
$stRAM$, 337, 373, 377
   Critical current I_c, 378
   Write time, 378
$NOT^{1/2}$, 166
Ørsted's law, 307
Ørsted, H. C., 307
Ørsted, H. C., 624
"Magnetic" charge, 309
$1/f$ noise, 598

Abel, N., 625
Abelian group, 625
Abgrall, P., 475
Absolute zero temperature, 3

Absorptance, 539
Absorption
    Intersubband, 543
Accelerator
    Cyclotron, 509
    Electrooptical, 510
    Linear, 509
    Synchrotron, 511
Accommodation coefficient α, 471
Acheson, D., 475
Acoustic waves, 455
Action, 241
Adiabatic approximation, 619
Adiabatic charging, 34
Adiabatic circuits, 35
Adiabatic continuity, 283
Adiabatic process, 26, 30
Adlung, S., viii
Aharonov-Bohm effect, 180, 622
Aharonov-Bohm phase, 195
Akbar, 175
Akdŏgan, E. K., 475
Alù, A., 559
Albert, R., 56
Algorithm
    Grover, 171
    Shor, 170, 177
Algorithmic complexity, 577
Algorithmic entropy, 22, 24, 577
Algorthmic randomness, 23
Allee effect, 478
Allowed transitions, 317
Altmann, S. L., 405
Amorphous–crystalline transition,
    243, 391
Amperère's law, 306
Analytic functions, 611
Anarkali, 175
Anderson localization, 220, 287
Anderson transition, 286
Anderson, P., 242, 402
Angular momentum
    Orbital, 312
    Spin, 311
Anharmonicity of elasticity, 448
Anisotropic exchange interaction, 326
Anisotropic magnetoresistance, 337
Anisotropy
    Crystalline, 352
    Dipolar shape, 378
    Magnetocrystalline, 342

Mangetocrystalline, 349
    Shape, 349, 352
    Uniaxial, 342
    Unidirectional, 343
Anisotropy coefficient, 358
Anisotropy constant κ, 352
Annihilation operator, 655, 656
Anomalous Zeeman effect, 327
Antdamping torque ($\mathbf{T}_{nd}$), 370
Anti-Stokes shift, 494
Antiferrimagnetism, 326
Antiferromagnetic order, 288
Antiferromagnetism, 326, 342
    Superexchange, 331
Antipodal state, 167
Anyon, 204
Apolar switching, 396
Archimedes' principle, 480
Arrow of time, 3
Ashcroft, J., 258
Atland, A., 406
Atomic clock
    Fountain, 505
Atomic clocks, 504
Attractor, 468, 646
Automaton
    Cellular, 8, 49
    Finite state, 6
    Non-pushdown, 8
    Pushdown, 7

Back injection, 129
Back propagation, 199
Back-action
    Dynamic, 492
    Quantum, 493
Bader, S., viii
Ballistic limit, 65
Ballistic transistor
    Contact, 139
    Device barrier length scale λ_B, 131
    Dissipation in, 117
    Effective length, 131
    Effective length ℓ_{eff}, 129
    Linear condition, 115
    Saturation condition, 114
    Thermalization, 126
    Virtual cathode, 131
Ballistic transport as biased walk, 98
Band
    Majority spin, 336

Majority spin band, 332
Minority spin, 336
Minority spin band, 332
Band conductor, 301
Band degeneracy (g_{bj}), 116
Band insulator, 285
Bandgap
    Phonon, 497
    Photon, 498
    Pseudo, 499
Barábasi, A.-L., 56
Barkhausen criterion, 650
Barnett, S. M., 57
Bayes's theorem, 18, 574
Bayesian approach, 571
Beam
    Forced and damped, 436
    Free vibration, 416
    Mechanical response, 412
    Moment, 413
    Moment of inertia ($\mathfrak{M}$), 413
    Strain, 413
    Stress, 413
Beating phenomenon, 468
Becker-Döring model, 388
Belief propagation, 242
Bell basis, 165
Bell inequalities, 616
Bell inequality, 617
Bell, J. S., 165, 174, 233
Bennett, C., viii, 16, 55, 160
Bernoulli effect, 435
Bernstein, D. H., 474
Berry phase, 195, 619
Berry, M., 619
Berz, J., 147
Bethe condition, 129
Bethe lattice, 324
Bethe, H., 104, 318
Bhat, N., viii
Bifurcation, 460, 642
    Forward, 645
    Global, 460
    Hopf, 460, 468
    Local, 460
    Pitchfork, 643
    Saddle-node point, 645
    Supercritical pitchfork, 645
    Transcritical, 643
    Turning point, 642
Billiard ball

Adiabatic, 160
Billiard ball computing, 32
Binomial distribution, 586
Binomial distribution function, 82
Biot-Savart law, 306, 374
Bipolar transistor, 63
Black, C., vii
Blackbody radiation, 538
Bloch sphere, 162, 618
Bloch state, 546
Bohm, D., 174, 616
Bohr magneton (μ_B), 312
Bohr radius
    (a_B), 81
    Effective (a_B^*), 81, 282
Bohr, N., 485, 615
Boltzmann entropy, 13, 23
Boltzmann transport equation, 80, 91,
    98
    0th moment, 99
    1st moment, 99
    2nd moment, 99
Boltzmann's H-factor, 13
Boltzmann's constant, 13
Boltzmann, L., 3, 9
Bordag, M., 558
Born, M., 281, 405
Bouwmeester, D., 233
Brahe, T., 474, 558
Braid group, 204
Braiding, 204
Breuer, J., 559
Bright state, 507
Brillouin, L., 16
British East India Company, 175
Broadening
    Gaussian, 187
Broken symmetry, 238, 239, 241, 623
    Inversion, 244
    Time reversal, 244
    Translation, 240
    Translational, 379
Brown relaxation, 354
Brownian fluctuations, 442
Brownian motion, 105, 441
    Damping, 448
    Dissipation, 448
Bubble memory, 356
Buchanan, D., vii
Bulk acoustic wave, 455
Butterfly curve, 40

Canonical ensemble, 590
Capacitance
    Quantum, 97, 119
Carbon
    Graphene, 94, 123
    Nanotube, 123
Carnot cycle, 27
Carrier capture, 87
Carrier continuity equation, 100
Carrier emission, 87
Casimir effect, 482
Casimir, H., 485
Casimir-Polder effect, 484
Causal force, 447
Causality, 571
Causseé force, 490
Cbit gate
    NOT, 168
    THROUGH, 168
Cellular automaton, 49
Chalcogenides, 379, 392
Chaos, 457, 461, 462
Characteristic function, 84, 592, 593
Charge control, 63
Charge density wave, 285
Charge gap, 209, 293
Charge injection, 63
Charging of capacitor, 29
Chemical potential, 103
Chern number, 197
Chikasum, S., 406
Chiral edge states, 196
Chiral molecule, 239
Chirality, 241
Church, A., 5
Church-Turing thesis, 5
Cipher
    RSA, 177
    Caesarean, 176
    Modulo 2 addition, 176
    Substitution, 176
    Vernam, 176
Circuits
    Adiabatic, 35
    Clocks, 39
    Differential logic, 38
    Dissipation, 35
    Fluctuations, 43
    Inverter, 40
    Quasi-adiabatic, 37, 38
    Sequential logic, 39

Clamped nanotube, 469
Classical bit (cbit), 161
Classical mechancis
    Hamiltionian, 7
Classical mechanics
    Lagrangian, 7
Clauser, J., 616
Clausius, R., 3
Clocks, 39
Closed orbit, 465
Cobden, D. H., 147
Coherence effect, 287
Coherent backscattering, 287
Coherent tunneling, 549
Coleman, P., 406
Collective excitation response, 237
Collins, R., viii
Colossal magnetoresistance, 337, 339
Colpitts oscillator, 651
Complexity, 577
    Joint, 580
    Kolmogorov, 580
    Mutual, 580
Composite fermion, 195
Computing
    Adiabatic, 160
    Billiard ball, 32
    Reversible, 32
Conductance parameter (e^2/h), 65
Configuration space, 7
Confocal microscopy, 503
Conjugate functions, 609, 611
Conservation, 25
    Energy, 25
    Momentum, 25
    Of angular momentum, 241
    Of energy, 241
    Of momentum, 241
Conservation of energy, 3
Conservative force, 25
Constant mobility, 64
Contact
    Conductance (g_c), 137
    Resistance (ρ_c), 137
    To ballistic transistor, 139
Contact conductance (g_c), 137
Contact length (L_c), 138
Contact resistance
    Ballistic limit, 141
    Diffusive limit, 142
    Long contact limit, 142

Short contact limit, 142
Contact resistance (ρ_c), 137
Continuity equation, 99
Continuum
    Approximation, 629
    Elasticity, 629
Continuum approximation, 155
Continuum mechanics, 411
Continuum theory, 413
Conway, J., 8
Coordinates
    Generalized, 634
Copenhagen interpretation, 1, 615
Copernicus, N., 558
Correlated electron effect, 292
    Caution, 288
Correlation
    Electron-electron, 288
    Spin, 306
    Strong, 286, 288, 320
Correlation function, 446
    As measure of dispersion, 446
Correlation time (τ^*), 444
Coulomb blockade, 156, 208, 258
    Electron localization, 226
    Mean field caution, 226
    Phase space, 225
    Resonance point, 226
    Triple point, 225
Coulomb blockade oscillation, 212, 217
Coulomb diamond, 215
Coulomb staircase, 210
Coulombic screening length, 81
Coupled analysis, 426
Courant, R., 405
Creation operator, 655, 656
Critical temperature (T_c), 245
Cross-talk, 589
Cryptography, 25
    Quantum, 177
Crystalline–amorphous transformation, 379
Curie temperature (T_c), 247, 268, 342
Curie's law, 258, 326
Curie, P., 247
Curie-Weiss law, 258
Current
    Diffusion, 109
    Radiative recombination, 539
    Short circuit, 542

Thermionic, 110
Current continuity, 67
Cycle
    Carnot, 27
    Information, 27
Cyclotron accelerator, 509
Cyclotron coherence, 185
Cyclotron frequency (ω_c), 183, 184

d'Alembert's principle, 633
Damping parameter, 448
Damping relaxation process, 368
Damping torque ($\mathbf{T}_d$), 314, 370
Dark state, 507
Data, 3
de Broglie paradox, 174
de Broglie wavelength, 101, 102, 442, 604
de Broglie, L., 174
De Morgan's laws, 50
De Morgan, A., 50
Debye screening length (λ_D), 81
Debye screening length (λ_D), 289
Decoherence, 17
Deflection
    Electrostatic, 511
    Magnetic, 511
Deformation, 629
Degrees of freedom, 634
Delay machine, 6
Demagnetization, 310
Demokritov, S. O., 407
Denker, J. S., 56
Dennard, R., vii
Density matrix, 17, 53
Density of states
    1D, 93, 605
    2D, 93, 605
    3D, 93, 605
    Effective, 110
Depolarization field, 276
Depolarizing magnetic field, 309
Desurvire, E., 55
Deterministic view, 7
Deutsch-Jozsa algorithm, 170
Devoret, M. H., 234
Dexter process, 532
Diamagnetic material, 308
Dickinson, A. G., 56
Dielectric electrophoresis, 519
Dielectric function, 513, 519

Dielectric material, 246, 308
Diffusion coefficient ($\mathcal{D}$), 97
Diffusion current, 109
Diffusion length
    Spin flip, 365
Diffusion length ($\mathcal{L}$), 110
DiMaria, D., vii
Dimensionality, 93
Dipole
    Electric, 246
    Magnetic, 246
Dipole rules of transition, 316
Dirac point, 237
Direct recombination, 527
Direct transitions, 527
Disallowed transitions, 317
Discrete mechanics, 411
Disk drive, 337
Dissipation
    In switching, 43
Dissipation in circuits, 35
Dissipationless computing
    Fredkin gate, 160
    Toffoli gate, 160
Distribution
    Poisson, 88
Distribution function
    Binomial, 82, 586, 594
    Exponential, 585
    Fermi-Dirac, 92
    Gaussian, 82, 594
    Gaussian normal, 585, 587
    Maxwell-Boltzmann, 92
    Moments, 98, 593
    Normal, 588, 594
    Poisson, 82, 88, 585
    Uniform, 585
    Wigner-Dyson, 219
Dittrich, T., 234
Domain wall, 343
    Bloch, 350
    Energy, 353
    Motion, 351
    Néel, 350
Domains
    In-plane polarization, 350
    Out-of-plane polarization, 350
Dopant
    Shallow hydrogenic, 81
Dopant effects, 80
Dot, 207

Exchange energy, 219
Double dot
  Charge gap, 224
  Electron localization, 226
  Phase space, 225
  Resonance point, 226
  State diagram, 224
  Triple point, 225
Double exchange, 346
Drawing inference, 21
Drift-diffusion equation, 80, 99, 100
Drude model, 91, 183, 281
Duffing oscillator, 468
Dutch East India Company, 175
Dutta, P., 147
Dwell time, 89
Dynamic back-action, 492, 555
Dynamic random access memory, 44
Dyson, F., 491

Easley, D., 57
East, J., viii
East, M., viii
Eastman, L., ix
Easy axis, 348
Effect
  Allee, 478
  Coherence, 287
  Fano, 221
  Gibbs-Thomson, 398
  Kondo, 220
  Spin torque, 367, 369
  Zeeman, 323, 331
Effective Bohr radius
  Silicon, 289
Effective density of states, 110
Effective field, 260
Effective length ℓ_{eff}, 129
Effective mass theorem, 260
  As a mean field result, 260
Effective velocity, 125
Ehrenfest, P., 246
Eichenfeld, M., 559
Eigenmode analysis, 416
Eigenstate
  Energy width, 214
Einstein relationship, 97
Einstein, A., 172, 281, 405
Einstein-Podolsky-Rosen pair (EPR),
  172
Elastic response, 412

Electric dipole, 246
Electric field
  Polar, 306
Electric polarization, 208, 243
Electrochemical potential, 103
Electron correlation, 281
Electron-electron correlation, 288
Electron-electron coupling, 242
Electron-electron interaction, 189
Electron-phonon coupling, 242
Electrostatic energy
  Infinity cancellation, 229
  Metallic conductor and discrete
    charge, 228
Emergent property, 186
Emissivity, 539
Energy
  Enthalpy, 26, 245
  Free internal, 26
  Gibbs, 245
  Helmholtz, 26, 245
  Of cluster formation ($\Delta U_{xlr}(r)$), 387
  Recovery, 37
Energy conversion, 526
Energy cost
  Kinetic, 282
  Potential, 282
Engheta, N., viii, 559
Engine
  Information, 27
  Thermodynamic, 27
Ensemble
  Canonical, 590
  Grand canonical, 590
  Microcanonical, 590
Entangled state, 162, 164
Entanglement, 52, 173
  Schrödinger's cat, 615
Entanglement entropy, 52
Enthalpy, 26
Entropy, 1, 3, 249
  Algorithmic, 22, 24
  Boltzmann, 2, 13, 23
  Entanglement, 52
  Increase in closed system, 3
  Information, 3
  Macroscopic, 24
  Microscopic, 24
  Pattern, 23
  Physical, 4
  Shannon, 2, 23

  Statistical, 24
  von Neumann, 52
Equation
  Boltzmann transport, 80, 91, 98
  Carrier continuity, 100
  Continuity, 99
  Drift-diffusion, 80, 99, 100
  Energy transport, 99
  Euler-Lagrange, 474
  Fokker-Planck, 91, 354
  Hydrodynamic, 80
  Kolmogorov forward, 354
  Komlogorov differential, 354
  Lagrange, 633, 640
  Landau-Lifshitz, 315
  Landau-Lifshitz-Gilbert, 315, 367
  Laplace, 75
  Liouville, 98
  Momentum balance, 100
  Navier-Stokes, 99, 436, 470
  Poisson, 73
  Quantum Boltzmann, 91
  Quantum Liouville, 80
  Schrödinger's, 603
Equipartition of energy, 45, 105, 583
Erdös, P., 18
Errors in computing, 40
Esaki, L., vii
Euler characteristic, 621
Euler, L., 11, 49
Euler-Bernoulli equation, 454
Everett, H., 615
Exchange
  Bias field, 341
  Heisenberg interaction isotropy, 348
Exchange bias, 341
  Pinning, 341
Exchange coupling
  Co-Pd, 359
  Co-Pt, 359
  Interlayer, 340
Exchange energy, 219
Exchange energy (Δ), 336
Exchange field, 342
Exchange integral, 219, 319
Exchange interaction, 313, 316, 320
  Anisotropic, 326
Exchange splitting, 333
Exchange torque
  Injection, 368
Exciton, 529

Excitons, 536
Exclusion zone, 441
Exponential classical operation, 169
Extensive state function
    Entropy, 28

Förster mechanism, 483
Förster process, 532
Fabry-Pérot cavity, 493
Fang, F., viii
Fang-Howard function, 151
Fano effect, 221
Faraday's law, 308
Farrell, E., viii
Fatou set, 463
Feigenbaum δ, 463
Fermi gas, 155, 282, 289
Fermi integral ($\mathscr{F}_v$), 98
Fermi liquid, 155, 283
Fermi surface, 254, 285, 293, 333, 338,
    339
    Co, 333
    Cu, 333
Fermi velocity, 92, 93, 120, 125
Fermi wavevector, 93
Fermi-Dirac function, 105
Fermionic systems, 283
Ferrite core memory, 356
Ferroelectric
    Depolarization field, 276
    Fatigue, 267
    Imprint, 267
    Read disturb, 272
Ferroelectric material, 246
    Nonlinearities, 260
Ferroelectricity, 242, 260
    Critical thickness, 256
Ferromagnetism, 326
Feynman, R., 174, 233, 258
Fick's law, 365
Field
    Axial, 624
    Polar, 624
Field control, 63
Field quantization, 655
Film bulk resonant acoustic resonator,
    456
Finch, W., 175
Fine structure
    Cs, 505
Fine structure constant, 490

Fischetti, M., vii
Fitzgerald, F. S., 174
Flatband voltage (V_{FB}), 118
Flicker noise ($1/f$), 598
Fluctuation, 227
Fluctuation dissipation theorem, 329
Fluctuation-dissipation, 596
Fluctuation-dissipation theorem, 259
Fluctuations, 40, 257, 589
    In circuits, 43
    Mean thermal energy, 45
Fluid flow, 435
Fluid viscosity η, 471
Fluidic interface, 470
Fluidic mesoscale, 470
Flux quantization, 621
Flux quantum, 65
Flux quantum h/e, 182
Fock state, 656, 657
Forbidden transitions, 528
Force
    Casimir-Polder, 484
    Caussé, 490
    Conservative, 25
    Gyroscropic, 639
    Lorentz, 511
    Minimum measurable, 451
    Rapidly varying, 444
    Slowly varying, 444
Force on magnetic dipole, 314
Forward bifurcation, 645
Fowler, A., viii
Fröhlich condition, 519
Fröhlich interaction, 553
Fractional charge
    Quasiparticle, 191
Fractional quantum Hall effect, 156
Frank, D., vii, 146
Fredkin gate, 158, 169
    AND, 159
    $NOTCOPY$, 159
    OR, 159
Free energy
    Enthalpy, 26, 245
    Gibbs, 26, 245
    Helmholtz, 22, 26, 245, 285
Freedman, S., 616
Friction, 240, 422
Function
    Characteristic, 84, 592, 593
    Conjugate, 609

Dielectric, 513
    Fang-Howard, 150
    Lagrangian, 241
    Langevin, 348
    Moment generating, 592
    One-way, 177
Functional
    Ginzburg-Landau, 256
Functionally complete
    $NAND$, 159
    $ccNOT$, 159
Functions
    Analytic, 611
    Conjugate, 611
Funneling effect
    Real space, 108, 135
    Reciprocal space, 108

g factor, 312
Gabor, D., 16
Gammaitoni, L., 475
Gardner, M., 8, 18
Gate
    $NAND$, 10, 13, 32, 35
    NOR, 13, 32, 35
    NOT, 206
    $THROUGH$, 168
    $cNOT$, 205
    $NOT^{1/2}$, 166
    Fredkin, 158, 169
    Hadamard, 170
    Inverter, 35
    Reversible, 158
    Toffoli, 159
    Unit wire, 158
Gauge transformation, 191
Gauss's law, 308
Gauss, J. C. F., 17, 405
Gauss-Bonnet theorem, 196, 622
Gaussian broadening, 187
Gaussian distribution function, 82
Gaussian normal distribution, 587
Gebhard, F., 406
Generalized coordinates, 634
Geometric phase, 195, 619
Germanium, 123
Ghosh, A., viii
Giant magnetoresistance, 337, 339,
    345, 366
Gibbs energy, 26
Gibbs free energy, 26

Gibbs, J. W., 3, 9, 589
Gibbs-Thompson potential, 398
Gibbs-Thomson effect, 398
Ginzburg, V., 485
Ginzburg-Landau functional, 256
Gitterman, M., 406
Glass temperature, 385
Go, 8
Goldberg, L., ix
Golden rule, 26, 535
Goldstone theorem, 242
Gosper, R. W., 8
Gräzel cell, 531
Grabert, H., 234
Gradual channel approximation, 66, 72
Grand canonical ensemble, 590
Grandy, W. T., 56
Graph theory, 49
Graphene, 132, 237
Guha, S., vii
Gunn diode, 650
Gutzwiller, M. C., 475
Gyromagnetic ratio γ, 312, 314, 340
Gyroscopic force, 639

Hadamard gate, 168, 170
Hall effect, 183
    Classical, 183, 338
    Electron-electron interaction, 189
    Fractional quantum, 189, 283
    Integral quantum, 186, 197, 283
    Localized electron, 192
    Role of inhomogeneity, 186
    Spin, 196, 197
    Spin degeneracy, 189
    Spin in $3D$, 197
    Zeeman doublet, 189
Hall factor (r_H), 184
Hall mobility, 184
Halperin, B. A., 234
Halpern, V., 406
Halprin, M., 18
Halting problem, 5
Hamiltion, W. R., 474
Hamilton's principle, 415, 633
Hamiltonian, 15
Han, M., 475
Handle, 197
Hard axis, 348
Harmonic oscillator, 187

Harrison, W., viii
Hartley oscillator, 651
Hartley, R. V. L., 41
Hasan, M. Z., 234
Hauser, W., 474
Head
    Read-write, 310, 337
Heat capacity (C_V), 28
Heat energy flow, 99
Heat flux, 180
Heinzel, T., 234
Heisenberg Hamiltonian, 322
Heisenberg model, 323
Heisenberg, W., 281, 317
Heitler-London calculation, 320
Helm, G., 13
Helmholtz free energy, 26
Helmhotz free energy, 22
Hermann-Mauguin notation, 626
Hermitian operator, 4, 16
Hershel, W., 559
Heteroclinic orbit, 466
Heteroclinic trajectory, 648
Heterogeneous-nucleated growth, 390
Hey, A. J. G., 56
Hilbert, D., 405
Hilborn, R. C., 475
Hillebrands, B., 407
Homeomorphism, 196
Hommelhoff, P., 559
Homoclinic orbit, 466, 647
Homogeneous nucleation kinetics, 386, 387
Hooke's law, 631
Hopf bifurcation, 460, 468
Horn, P. M., 147
Howe, R., viii
Hu, E., viii
Hubbard band, 291, 293
Hubbard mechanism, 379
Hubbard model, 290, 323
Hubert, A., 406
Hund's rules, 287
Hydrodynamic equation, 80
Hysteresis, 468, 645
    Memory of path, 252

Ihn, T., 234
Ilic, B., viii
Image charge, 227
Imry, Y., 234

Inaccuracies
    Time constant, 100
Incompressible fluid, 189, 285
Incomputable functions, 5
Indirect recombination, 527
Indirect transitions, 527
Inelastic scattering, 66
Information, 1–4
    Algorithmic, 23
    And quantum, 52
    Averaged content (H), 11
    Capacity, 10, 569
    Diffusive flow in network, 48
    Dynamic, 1
    Human interpretation, 2
    Ignorance, 565
    In superposition, 4
    Incomplete, 21
    Is physical, 4, 9
    Lack of, 11
    Measure (I), 9
    Measure of uncertainty (H), 565
    Mutual, 567
    Nature of representation, 160
    Negentropy, 567
    Observation, 4
    Reversibility, 4
    Search via network, 48
    Self-, 11
    Uncertainty reduction, 9
    Wavefunction, 4
Information cycle, 27
Information engine, 27
Insulator
    Antiferromagnetic, 294
    Band, 285, 301
    Charge gap, 293
    Charge transfer, 294
    Mott-Heisenberg, 294
    Mott-Hubbard, 331
Insulator-semiconductor interface, 127
Intensive state function
    Pressure, 28
    Temperature, 28
Interaction
    Fröhlich, 553
    Spin-orbit, 287
Interface
    Contact-semiconductor, 141, 146
    Insulator-semiconductor, 72, 73, 80,

81, 83, 87, 118, 127
  Metal-semiconductor, 104
  Semiconductor-semiconductor, 104
  Source-channel, 128, 131, 136
Interfacial spin scattering, 369
Internal energy, 26
Internal free energy, 26
Intersubband absorption, 543
Invariance with reference frame
  Galilean, 238
  Lorentzian, 238
  Newtonian, 238
Invariant of system, 25
Inverter, 35
Invisibility cloak, 519
Ionic movement, 391
Irreversibility
  Degrees of freedom formulation,
    227
Irreversible logic, 13
Irreversible process, 25, 26
Ising model, 52, 323, 328
Ising, E., 329
Isothermal process, 26
Iyer, S., vii

Jackson, J. D., 229
Jackson, T., vii
Jaynes , E. T., 406
Jaynes, E. T., 56
Jellium model, 81, 282
Jirauschek, C., viii, 560
Jones, T. B., 475
Julia set, 463

Königsburg, 49
Kane, C., 234
Kash, J., viii
Kavokin, A. V., 559
Kepler, J., 474, 558
Khomskii, D. I., 406
Khosla, R., ix
Kinetics
  Boundary growth, 386
  By crystalline clustering, 389
  Diffusion-dominated, 389
  Heterogenous-nucleated, 390
  Homogeneous nucleation, 386, 387
  Nucleation and growth, 385
  Nucleation-dominated, 386
Kippenberg, T. J., 559

Kirchner, P., viii
Klein, F., 405
Kleinberg, J., 57
Knowledge, 1, 2
Knudsen number ($\mathfrak{K}$), 471
Kolmogorov complexity, 577
Kolmogorov forward equation, 354
Kolmogorov probability approach,
    572
Kolmogorov, A. N., 56, 233
Komogorov differential equation, 354
Kondo effect, 220
Kondo temperature (T_K), 220
Konishi, H., viii
Kramer, A., 56
Kramers-Kronig principal value, 487
Kramers-Kronig relations, 259
Kubis, T., 560
Kumar, A., vii

Ladder operators, 655
Lagrange equation, 633
Lagrange multipliers, 19
Lagrange, G. L., 474
Lagrangian, 638
Lamb shift, 312
Lamoreaux, S. K., 558
Landau level, 185
Landau, Lev, 185, 245
Landau-Lifshitz equation, 315
Landau-Lifshitz-Gilbert equation, 315,
    367
Landauer, R., vii, 4, 16, 55
Langevin function, 348
Laplace equation, 75
Laplace, P.-S., 75
Larmor precession frequency, 314,
    367, 374
Larmor relationship, 511
Laser
  Active gain, 550
  Bipolar interband, 544
  Bipolar intersubband, 544
  Bipolar quantum well, 544
  Population inversion, 550
  Quantum cascade, 544
  Radiative gain, 553
  Unipolar intersubband, 544
Laughlin wavefunction, 190
Laux, S., vii, 147
Law

Hooke's, 631
Leakage current, 63, 87
Least action
  Path off, 339
Legendre polynomials, 517
Legendre transformation, 20, 22, 634
Leibniz, G. W., 238
Length scale
  Device barrier, 131
  Magnetic, 187
  Screening (λ_{scr}), 282
Length scale of travel, 129
Lennard Jones potential, 470
Lenz's law, 308
Lenz, W., 329
Li, M., 56
Lifshitz point, 257
Lifshitz, E., 485
Light
  Absorption, 526
  Emission, 526
Light emission diode, 531
Light-sensitive dye, 531
Likharev, K. K., 147
Limit cycle, 465, 468, 647
Limited space charge accumulation,
    650
Linear accelerator, 509
Linear velocity-field, 66
Liouville equation, 98
Local hidden variable hypothesis, 616
Localization
  Anderson, 220
Localized plasmon, 516
Logic
  Invertible, 158
  Irreversible, 13
  Reversible, 158
London, F., 483
Long-range interactions, 82
Lorentz force, 183, 306, 511
Lorentz spectrum, 90
Lorentz, E., 461
Lu, W., viii
Lugli, P., viii
Luttinger liquid, 285
Luttinger's theorem, 285

Mézard, M., 57
Mach, E., 13
MacLaurin series, 593

Macroscopic state, 1
Maekawa, S., 407
Magnetic anisotropy, 325
Magnetic coercivity, 343, 353
Magnetic dipole, 246
Magnetic domain, 339, 349
Magnetic domain boundary, 315
Magnetic domains, 349
Magnetic field
    Axial, 306
    Chiral, 306
Magnetic fluctuations
    Brown, 46
Magnetic flux density (**B**), 307
Magnetic induction (**B**), 307
Magnetic length scale, 187
Magnetic media
    $CoCrPt$, 358
    Assistance layer, 359
    Granular texture, 358
    Patterned, 359
    Segregated, 357
    Segregated $CoPt$, 358
    Textured, 357
    Underlayer of $Ru(0002)$, 359
Magnetic memory, 373
Magnetic moment
    Intrinsic, 306
Magnetic moment (**m**), 311
Magnetic moment pinning, 343
Magnetic permeability (μ_0), 307
Magnetic polarization (**M**), 243, 306,
    307
Magnetic recording media, 310
Magnetic storage, 337
    Polarization, 348
Magnetic susceptibility (χ), 308, 325,
    329
Magnetic yoke, 310
Magnetism, 243
    Chiral dependence, 326
    Spiral change, 257
    Stability, 348
Magnetite perovskites, 345
Magnetites, 339
Magnetization
    Easy axis, 348
    Hard axis, 348
    Interparticle energetics, 354
Magnetocrystalline anisotropy, 342,
    349

Magnetoresistance, 183, 337, 338
    Anisotropic, 337, 340
    Colossal, 337, 339, 345
    Giant, 337, 339, 345, 366
    Tunneling, 345, 347
Magnetoresistance ratio (ϱ), 338
Magnetostatic energy, 349, 353
Magnetostriction, 349
Magnon
    Spin wave, 330
Magyari-Köpe, B., 407
Maier, S. A., 559
Majlis, N., 406
Majority spin, 332
Majority spin band, 336
Mandelbrot set, 463
Mangetoresistance
    Anisotropic, 339
Manifold, 197
Markov process, 354
Mathieson's rule, 363
Maxwell relations, 27
Maxwell's demon, 3, 15, 160
Maxwell, J. C., 15
McGroddy, J., ix
McLuhan, M., 4
Mean, 593
Mean field, 226
    Effective mass theorem, 260
Mean field approximation, 327
Mean field caution
    Nonlinearity, 227
Mean free path, 92, 129
Melnikov method, 465
Melting temperature, 385
Memories
    $sRAM$, 157
Memory
    1T1C, 269
    1T2C, 272
    2T2C, 272
    $FLASH$, 44
    $dRAM$, 44, 269
    $feRAM$, 44, 268
    $mRAM$, 337, 373
    $rRAM$, 379, 392
    $sRAM$, 41
    $stRAM$, 337, 373, 377
    Broken translational symmetry, 379
    Chain ferroelectric, 280
    Electrochemical, 44

Electrolytic, 397
Electron correlation, 391
Error, 44
Ferroelectric, 260
Ionic movement, 395
Mott-Hubbard, 394
Optical, 379
Peierls, 394
Phase change, 379, 391
Random walk, 44
Resistive, 379, 391
State transition rate, 43
Stay time, 44
Mendez, E., vii
Mermin, N. D., 161, 233, 258
Mermin-Wagner theorem, 325, 328
Mesoscale, 155, 180
Metal–insulator transition, 286, 291
Metals, 281
    Transition, 288
Metamaterials, 520
Metastability and rate of cooling, 385
Method of
    Reduction, 5
    Lagrange multipliers, 19
    Lagrangian, 415
    Melnikov, 465
Microcanonical ensemble, 590
Microscopic state, 1
Miller feedback, 134
Miniature accelerators, 508
Miniband, 548
Minkowski, H., 405
Minority spin, 332
Minority spin band, 336
Misewich, J., vii
Mobility, 66, 92
    Constant, 64
    Hall, 184
Mode filtering, 101, 104, 109
Mode squeezing, 496
Model
    Drude, 91, 281
    Hubbard, 290
    Sommerfeld, 281
    Stoner-Wohlfarth-Slater, 335
    Two-current, 340, 360
Modeling
    Monte Carlo, 127
Modes of optomechanic beam, 500
Moment

Orbital magnetic, 312
Spin magnetic, 312
Moment generating function, 592
Moments of distribution, 98, 593
Momentum balance equation, 100
Momentum relaxation time, 91, 183
Mondale, W., ix
Monoatomic layer materials, 132
Monolayer dichalcogenides
MoS_2, 132
WSe_2, 132
Montanari, A., 57
Monte Carlo approach, 99
Monty Hall problem, 18
Mooney, P., vii
Moore, J., 234
Morrison, P., 238
Mott-Anderson transition, 288
Mott-Heisenberg transition, 288, 293
Mott-Hubbard interface, 301
Filter, 303
Mott-Hubbard transition, 288, 290
Multiferroics, 257
Mutual information, 567

Néel relaxation, 354
Néel temperature, 288, 294, 342
Néel, L., 294
Néel-Brown relationship, 46
Nair, R., vii
Nanoscale dot, 207
Narayanamurti, V., ix
Nathan, M., viii
National Institute of Standards and
    Technology (NIST), 166
National Security Agency (NSA), 166
Natori, K., 147
Navier-Stokes equation, 436
Nayak, C., 234
Negative damping, 371
Negentropy, 11, 14, 567
Nernst potential, 397
Network, 48
    Connectors, 49
    Edges, 49
    High-degree nodes, 49
    Input port, 128
    Long-range connection, 49
    Low-degree nodes, 49
    Nodes, 49
    Short-range connection, 49

Neuenschwander, D. E., 405
Newtonian mechanics, 634
Nishi, Y., viii
Nishimori, H., 406
No-cloning theorem, 178
Node
    Saddle, 642
Noether's theorem, 241
Noether, A., 241, 405
Noise, 442, 589
    $1/f$, 88, 598
    Capacitor, 589
    Flicker $(1/f)$, 598
    Random telegraph, 87
    Shot, 40, 86, 88, 492, 598
    Thermal, 492, 595
Non-Abelian group, 625
Non-Abelian quasiparticle, 203
Non-conservation, 25
Non-serial machine, 8
Nonlinear system, 641
Nonlinearity, 456
Nonlocal belief, 51
Nonlocality, 173
Normal distribution function, 588
Nuclear magnetic moment, 313
Nucleation-dominated kinetics, 386
Nyquist sampling rate, 569
Nyquist, H., 41

Observation, 2
Off state, 63
    Parabolic dependence, 73
    Scaling length, 75, 76
    Series solution, 76
Off-equilibrium, 67
Offset charge, 209
Ohm's law, 180, 184
    Breakdown, 186
On state, 63
Onsager universality, 325
Onsager, L., 325, 329
Open-circuit voltage, 532
Operators
    Annihilation, 655, 656
    Creation, 655, 656
    Ladder, 655
Optical memory, 380
Optical phonon scattering, 64
Optical recording
    Contrast, recordability and stability,

384
Optical storage, 379
Optical storage media, 379
Optical trap, 503
Optical tweezer, 502
Optomechanic
    Heating and cooling, 493
Optomechanics, 492
Orbit
    Heteroclinic, 466
    Homoclinic, 466
    Periodic, 467
Orbital angular momentum, 312
Orbital magnetic moment, 312
Order parameter, 51, 244, 245
Organic semiconductor, 526
    Acceptor transport layer, 530
    Donor transport layer, 530
    Electron transport, 530
    Hole transport, 530
Organic semiconductors, 529
Oscillator, 468
    IMPATT, 650
    LC, 651
    Barkhausen criterion, 650
    Colpitts, 651
    Crystal, 651
    Duffing, 468
    Harmonic, 650
    Hartley, 651
    Nonlinearity, 650
    Parametric, 650, 653
    Relaxation, 650
    van der Pol, 468, 649, 651
    Varactor, 650, 653
Oscillator strength, 527–529
Oscillators
    Energy exchange with, 649
    Energy storage in, 649
    Parametric, 653
    Quantum, 655
    Relaxation, 651
Ostwald, W., 13
Oxidation-reduction, 391

Pancharatnam, S., 619
Paraelectric material, 308
Paramagnetic material, 246, 308
Paramagnetic metallic state, 294
Parametric oscillator, 653
Parametric oscillators, 653

Parasistic resistance, 133
Parity, 241
Parity violation, 241
Particle in box, 603
Partition function
    Ising, 323
Path of least action, 339
Patterns, 571
Pauli matrix
    X, 167, 179
    Y, 179
    Z, 179
Pauli, W., 172
Pearl, J., 50, 56
Peierls instability, 241, 286, 297
Peierls transition, 286, 379
Peierls, R., 241, 242, 402, 405
Pelesko, J. A., 474
Percolation model
    Long range, 51
Percolation of mobile charge, 83
Peres, A., 233
Period cycle, 468
Period doubling, 462
Periodic orbit, 467
Periodic oscillations, 182
Permalloy Py, 307
Perovskites, 294
Perpendicular recording head, 358
Peshkin, M., 234
Phase change memory, 380
Phase coherence length, 81
Phase contrast microscopy, 259
Phase diagram
    Equilibrium, 384
Phase space, 7, 15, 641
    Incompressibility, 1
    Incompressible, 98
Phase space portraiture, 465
Phase transition, 227, 242, 243
    2nd order, 328, 645
    Amorphous–crystalline, 246
    Ferroelectric, 249
    First order, 245
    Hysteresis, 251
    Ice–water, 246
    Metal–insulator transition line, 254
    Mott-Hubbard, 208
    Overlaying structure, 257
    Quantum, 202, 283
    Second order, 245, 246, 249

Spinodal hysteresis, 252
Spinodal point, 251
Supercooling, 251
Superheating, 251
Thermodynamic, 237, 283
Tricritical point, 252
Triple point, 252
Without symmetry change, 253, 283
Phonon bandgap, 497
Phonon bottleneck, 551
Phononic crystal, 497
Photon bandgap, 498
Photon-phonon coupling, 501
Photovoltaics, 526
Piezoelectricity, 244, 255
    Stress and strain, 255
Pilkey, W. D., 474
Pinned magnetic moment, 343
Pitchfork bifurcation, 643
Planck, M., 92
Plasma damping, 513
Plasmon, 285, 481, 512
    Localized, 516
    Surface, 521
    Surface plasmon polariton, 522
    Two-dimensional, 521
Plasmon frequency, 513
Plasmon wave, 285
Podolsky, B., 172
Poincaré, H., 460, 461
Point
    Fixed, 641
    Globally stable fixed, 642
    Stable, 641
    Unstable, 641
Poisson distribution, 88
Poisson distribution function, 82, 88
Poisson equation, 73
Poisson's ratio, 411, 630
Polar switching, 396
Polariton, 481
    Surface plasmon, 522
Polarizability α, 518
Polarization
    Electric, 42, 208, 243
    Magnetic, 42, 243
    Remnant, 262
Polder, D., 485
Polynomial quantum operation, 169
Porod, W., viii
Positron emission tomography, 509

Possible configurations Ω, 13
Power gain, 63
Power spectral density, 90
Power spectrum, 444, 449
    Shot noise, 600
Poynting vector, 487
Pratap, R., viii
Precession
    Damping relaxation, 368
    Sampling orientation, 368
Preumont, A., 475
Price, P., vii
Prime number factorization, 177
Principle
    Archimedes', 480
Principle of
    d'Alembert, 415, 633
    Detailed balance, 107, 538
    Equipartition of energy, 440
    Hamilton, 415, 633
    Least action, 241, 256
    Maximum entropy, 17
    Virtual work, 415, 429, 633
Probability, 9
    Bayes's theorem, 574
    Chain rule, 574
    Conditional, 573
    Conditional independence, 573
    Conditionalization, 575
    Independence, 573
    Kolmogorov's approach, 572
    Total probability, 574
Process
    Adiabatic, 26, 30
    Carrier capture, 87
    Carrier emission, 87
    Dexter, 532
    Förster, 532
    Irreversible, 25, 26
    Isothermal, 26
    Quasi-reversible, 26
    Quasistatic, 26
    Redox, 531
    Reversible, 25, 26
    Stochastic, 35
Pseudo bandgap, 499
Pull-in voltage (V_π), 431
Purcell, E., 472, 483
Putterman, S., viii

Qbit, 52

Qbit gate
 SHIFT, 168
 Hadamard, 168
Quadrupole
 Compound lens, 510
 Magnet, 510
Quality factor (Q), 424
Quality factor (Q), 36
Quantized conductance, 102
Quantized Hall effect, 65
Quantum
 Coherence, 156
 Communication, 157
 Computation, 157
 Description, 155
 Entanglement, 156
 Superposition, 156, 157
Quantum NOT, 167
Quantum back-action, 493
Quantum bit
 Superposition, 161
Quantum bit (qbit), 160, 161
Quantum capacitance, 97, 119
Quantum cascade laser, 543, 544
Quantum conductance, 186
Quantum Liouville equation, 80
Quantum logic
 SWAP, 159
 cNOT, 159
 Measuring function, 159
Quantum mechanics
 Linear theory, 157
 Reversibility, 157
 Reversible, 174
Quantum number
 Azimuthal (m_l), 312
 Orbital (l), 312
 Principal (n), 312
 Secondary spin (m_s), 311
 Spin (s), 311
Quantum oscillators, 655
Quantum point contact, 601
Quantum register, 163
Quantum-mechanical theorems
 No-cloning, 17, 178
 No-deletion, 17
 No-measurement, 17
Quantum-mechanical viewpoint
 Causality, 17
 Correlations, 163
 Field, 180

Heisenberg representation, 16
Irreversible process, 17
Linear theory, 16
Potential, 180
Reversible process, 16
Schrödinger representation, 16
State reduction, 17
Wavefunction collapse, 17
Quasi-adiabatic circuits, 37
Quasi-reversible process, 26
Quasiantiparticle, 284
Quasiparticle
 Fractional charge, 191
 Non-Abelian, 203
 Response, 283
 Scattering, 283
Quasistatic process, 26
Quaternions, 405
Queisser, H., 559
Querlioz, D., viii

Rössler, U., 407
Röthig, C., 234
Rabe, K., 406
Rabi oscillation, 490, 506
Radiative recombination current, 539
Raghavan, S., viii
Ramsey theorem, 49
Random potential, 287
Random telegraph
 Signal, 88
Random telegraph noise, 87
Random walk
 First moment ($\langle x \rangle$), 98
 Second moment ($\langle x^2 \rangle$), 98
Raoux, R., 407
Rashba effect, 361
Rayleigh wave, 455
Read-write head, 310
Reagan, R., ix
Recombination
 Auger, 100
 Direct, 527
 Hall-Shockley-Read, 100
 Indirect, 527
Recording head
 Perpendicular, 358
Recursive function theory, 5
Redox process, 531
Reference frame
 Field analysis, 191

Inertial, 637
Laboratory, 191
Reflection coefficient, 127, 128
Relative permeability (μ_r), 307
Relativistic electron, 237
Relaxation
 Brown, 354
 Néel, 354
Relaxation oscillators, 651
Relaxation time
 $\tau_\uparrow$ and $\tau_\downarrow$, 362
 Carrier lifetime (τ_n, τ_p), 100
 Energy (τ_w), 100
 Inter-carrier (τ_n, τ_p), 100
 Momentum ($\tau_\mathbf{k}$), 100, 183
Relaxation time approximation, 98
Remnant
 Polarization, 262
Renormalization, 242
Repeller, 646
Reservoir, 1
Resistance
 Of quantized channels, 139
Resistive memory, 391
 Apolar, 396
 Electochemical redox, 398
 Electroforming, 399
 Filament formation, 397
 Ion transport, 397
 Phase change of TiO_2, 399
 Polar, 396
 Redox reaction, 397
 Reset operation, 395
 Set operation, 395
 Thermochemical redox, 398
 Valence change, 398
Resonance dispersion, 490
Retarded back-action, 496
Reversible computing, 32
Reversible gate, 32, 158
Reversible logic, 157
Reversible process, 25, 26
Reversible system, 648
Reversibly complete, 169
Reynold's number ($\Re$), 471
Richardson constant ($\mathscr{A}^*$), 106, 109
Riemann, B., 405
River Pregel, 49
Rogers, D., viii
Rosen, N., 172
Russer, J., viii

Russer, P., viii
Rutz, R., viii

Saddle point, 465
Saddle-node bifurcation point, 645
Safari, A., 475
Salpeter, E., 318
Sano, N., 146
Satisfiability (SAT), 5, 242
Saturated velocity, 66, 67, 126
Scalar potential, 181
Scaling
  Constant field, 70
  Constant potential, 70
Scattering, 64, 91
  Acoustic phonon, 127
  Bragg, 526
  Cascade, 128
  Coulombic, 127
  Due to $3d$, 340
  Field direction dependence, 340
  Finite, 68
  Inelastic, 66, 69
  Interfacial spin, 369
  Length $\langle \lambda_{scatt} \rangle$, 64, 68
  Of $4s$ electrons, 340
  Optical phonon, 64, 127, 553
  Rate, 64
  Spin, 313
  Spin flip, 333, 339, 360
  Spin preserving, 340
  Spin selective, 369
  Time, 68
Scattering length
  $\lambda_\uparrow$ and $\lambda_\downarrow$, 361, 362
Scattering matrix, 104, 127
Scattering time
  Momentum, 91
Schäfer, R., 406
Schmitt trigger, 652
Schottky barier height, 301
Schrödinger's cat, 615
Schrödinger, E., 615
Schwarz-Christoffel transformation, 135, 609
Schwinger, J., 485
Screening length
  Coulombic, 81
  Debye (λ_D), 81
  Debye (λ_D), 289
  Metal, 282

Thomas-Fermi (λ_{TF}), 265, 289, 290, 299
Self-energy, 229
Self-information, 11
Self-similarity, 463
Senturia, S. D., 474
Sequential logic circuit, 39
Serial machine, 7
Sernelius, B. E., 558
Shakespeare, W., 10
Shallow hydrogenic dopant, 81
Shannon entropy, 23, 578
Shannon's message theorem, 12, 567
Shannon, C., 9, 41
Shannon, C.., 55
Shannon-Hartley theorem, 12, 568
Shanon, C, 565
Shape anisotropy, 349
Shear mode wave, 455
Shear modulus, 631
Sheet conductance (g_s), 137
Sheet resistance (ρ_s), 137
Shivashankar, S., viii
Shockley, W., 559
Shockley-Queisser limit, 537
Short-circuit current, 532, 542
Short-range interactions, 82
Shot noise, 40, 86, 88, 492, 598
Shubnikov-de Haas oscillations, 186
Siegmann, H. C., 407
Signal, 3
Signal-to-noise ratio (SNR), 12
Silicon, 123
Simons, B. D., 406
Single electron, 207
Single electron effect
  $0D$ system, 207
  Asymmetry, 208
  Bimodal behavior, 218
  Bistability point, 212, 217
  Charge gap, 209, 224
  Confinement and dwelling, 215
  Coulomb blockade, 208
  Coulomb blockade oscillation, 212, 217
  Coulomb diamond, 215
  Coulomb staircase, 208, 210
  Crossover point, 212
  Double dot, 221
  Electron leakage rate, 214, 219
  Electron lifetime on dot, 214

Energy width, 219
Exchange asymmetry, 218
Exchange energy, 219
Extended region of conduction, 216
Lead coupling, 219
Non-ohmic constraint, 207
Offset charge, 209
Phase diagram, 215
Semi-classical, 207
Threshold voltage, 210
Single electron interaction, 292
Singlet state, 317, 318
Sink, 646
Skilling, J., 56
Skyrmion, 203, 619
Slip coefficient (b), 471
Snell's law, 109
Snowden, E., 166
Solar cell
  Grazel, 531
  Maximum efficiency, 543
  Multi-exciton, 536
  Tandem, 543
Solar illumination, 538
Solomon, P., vii, 147
Solomon, Paul, 147
Solomonoff, R., 56
Solvable problem, 8
  Symbolic manipulation form, 8
Sommerfeld model, 281
Sommerfeld, A., 281
Source, 646
Source electron wave flux, 128
Specific heat
  Constant pressure, 249
Spin
  Crystal interaction, 367
  Diffusion, 364
  Flipping, 313
  Majority, 332
  Minority, 332
  Precession time, 314
  Scattering, 313
Spin accumulation potential, 363
Spin angular momentum, 311
Spin correlation, 306
Spin flip diffusion length, 365
Spin flip scattering, 333, 339, 360
Spin Hall effect, 196
Spin magnetic moment, 312
Spin polarized current, 315

Spin torque effect, 315, 367, 369, 371
Spin torque memory, 373
Spin valve, 345, 347
Spin wave, 242, 258, 285, 330, 368
Spin-crystal interaction, 367
Spin-orbit coupling, 197, 332, 339, 352
Spin-orbit interaction, 287, 349
Split-level charge recovery logic
    (SCRL), 37
Spontaneous emission losses, 507
Spontaneous polarization
    Electric, 242
    Magnetic, 243, 306
Spontaneous processes, 526
Stöhr, J., 407
Stable point, 465
State
    Chiral edge, 196
    Entangled, 164
    Excited, 319
    Extended, 187
    Macroscopic, 1
    Many-body bound, 190
    Microscopic, 1
    Singlet, 317, 318
    Triplet, 317, 318
State function, 161
State machine, 5
State space, 15
Stathis, J., vii
Static random access memory
    (sRAM), 41
Statistical simulation, 127
Sterling approximation, 568
Sterling relation, 21
Stern, F., vii
Stern-Gerlach experiment, 314
Stimulated processes, 526
Stochastic process, 35
Stokes shift, 494
Stokes's theorem, 306
Stokes-Einstein relationship, 389
Stoner gap (Δ_s), 336
Stoner model, 335
Stoner-Wohlfarth-Slater model, 335
Stopa, M., viii
Strain, 629
Stress, 629
Stress-strain relationship, 630
Strogatz, S. H., 475
Strong correlation, 286, 288, 292, 320

Subthreshold swing (SS), 70, 75, 87
Supercritical pitchfork bifurcation,
    645
Superexchange, 331, 346
Superexchange interaction, 316
Superlattice, 546
    Miniband, 548
Superlensing, 261
Superparamagnetic limit, 375
Superparamagnetism, 352, 354
Superresolution, 261
Surface acoustic wave, 455
Surface plasmon, 521
Surface plasmon polariton, 522
    TM, 523
    Charge density wave, 523
Susceptibility, 259
    Magnetic, 325, 329
Switching
    Apolar, 396
    Polar, 396
Symmetry, 237, 623
    32 point groups, 627
    Broken, 238, 239, 241
    Center of, 626
    Mirror, 241, 626
    Rotation, 241, 626
    Rotation-inversion, 626
    Space translation, 241
    Time translation, 241
Symmetry group, 623
Synchronous logic, 39
System
    Fixed point, 641
    Reversible, 648
Szilard, L., 3, 15, 16

Telefacsimile, 178
Teleportation, 178
Temperature
    Critical, 245
    Curie (T_c), 247, 268
    Glass (T_g), 385
    Melting (T_m), 385
    Néel, 288, 294, 342
Teraro, M., 407
Ternary phase diagram
    Ge-Sb-Te, 386
Terry, E., 175
Theorem
    Fluctuation-dissipation, 259

Gauss-Bonnet, 196
Goldstone, 242, 330
Luttinger, 285
Noether's first, 241
Ramsey, 49
Stokes's, 306
Wiener-Khinchin, 596
Theory
    Shockley-Queisser, 538
Thermal current, 180
Thermal equilibrium, 3
Thermal escape, 554
Thermal noise, 492, 595
Thermal resistance, 180
Thermal velocity, 92, 125, 126
Thermionic current, 106, 110
Thermionic emission, 104, 554
Thermocouple, 3
Thermodynamic engine, 27
Thermodynamics
    0th law, 3
    1st law, 3, 22
    2nd law, 3, 29
    3rd law, 3
    Cost of decoherence, 17
Thermoelasticity, 453
Thiaville, A., 407
Thomas-Fermi screening length (λ_{TF}),
    289, 290, 299
Thompson, J. D., 559
Thought, 2
Thouless, D. J., 234
Threshold, 210
    Variance, 86
Tight-binding approximation, 291
Timoshenko, S. P., 474
Tiwari, K., viii
Tiwari, M., viii
Toffoli gate, 159
    ccNOT, 159
Tokamak, 511
Tonomura, A., 234
Topological insulator, 156, 196
Topological phase relationship, 203
Topological protection, 199
Topology, 195, 196
Torque
    Antidamping ($\mathbf{T}_{nd}$), 370
    Damping ($\mathbf{T}_d$), 370
Torque in magnetic field, 313
Transcritical bifurcation, 643

Transformation
  Legendre, 634
Transistor
  DIBL, 70
  GIDL, 70
  SS, 75
  Ballistic, 65, 69
  Ballistic, thermalization in, 69
  Bulk, 78
  Constant potential, 70
  Cylindrical polar form, 71
  Dennard scaling, 70
  Dissipation, 117
  Double gate, 72
  Field scaling, 70
  Figures of merit, 134
  Leakage current, 87
  Long-channel, 65
  Long-range interactions, 82
  Macroscopic view, 82
  Microscopic view, 82
  Near-ballistic, 65, 68, 127
  Negative feedback, 133
  Off state, 69
  On state, 69, 111
  Pinch-off, 67
  Rectangular Cartesian form, 71
  Short-channel, 65, 67
  Short range interactions, 82
  Source resistance, 133
  Source spreading resistance, 136
  Subthreshold swing, 87
  Subthreshold swing *SS*, 70
  Threshold variance, 86
  Threshold voltage, 80
  Unity current gain frequency (f_T), 134
  Unity maximum power gain frequency (f_{max}), 134
  Velocity saturation, 67
Transistors, 63
  Dopant effects, 80
Transition
  Anderson, 286
  Mott, 286
  Mott-Anderson, 288
  Mott-Heisenberg, 288, 293
  Mott-Hubbard, 288, 290
  Peierls, 286
Transition kinetics, 384
Transition metal oxides, 294, 335, 380

Of *Ti* through *Cu*, 380
Transition metals, 288, 335
Transitions
  Direct, 527
  Indirect, 527
Transmission
  Energy aperture, 104, 108, 126
Transmission coefficient, 128
Transmission coefficient $\mathscr{T}$, 103
Transmission line, 137
Transport
  Ballistic, 64, 91, 104, 111, 141, 147
  Diffusive, 141
  Drift, 142
  Drift diffusion, 100
  Drift-diffusion, 67, 91, 99, 109
Traveling salesman problem, 5
Triplet state, 317, 318
Tsang, J., viii
Tufte, E., viii
Tunneling
  Elastic, 213
  Inelastic, 213
  Resonant, 213
  Sequential, 213
  Superconducting, 65
Tunneling magnetoresistance, 345, 347
Turing machine, 5, 7, 23
  Non-universal, 578
  Universal, 577
Turing, Alan, 5
Turning-point bifurcation, 642
Two current model, 360
Two-current model, 340
Type III heterostructure, 200

Ulam, S., 457
Undercooled liquid, 389
Undulator, 511
Uniaxial anisotropy, 342
Unit wire, 158
Universal Turing machine, 23

Vacuum fluctuation energy, 656
Valence change, 391
van der Pol, B., 468
van der Waals interaction, 482
van der Walls, J., 482
van der Wiel, W. G., 234
van Hove singularity, 333

Varactor oscillator, 650, 653
Variance, 593
Variational indicator, 638
Varshney, U., ix
Vector field, 641
Vector potential, 180, 307
Vedral, V., 57
Velocity
  Effective, 120, 125
  Fermi, 92, 96, 120, 125
  Kinetic energy–defined, 94
  Optical phonon–limited, 94
  Saturated, 126
  Thermal, 69, 92, 125, 126
Velocity overshoot, 68
Velocity-field relationship
  Linear, 66
  Saturated, 66
Venn diagram, 14
Vibrational precession
  Néel, 46
Vibrations
  Circular plate, 419
  Forced and damped beam, 436
Vinyl record, 381
Virial of Clausius, 45
Vitányi, P., 56
Volmer-Weber model, 388
von Neumann entropy, 52
von Neumann, J., 17, 615
Vortex, 195, 203

Wang, W., vii
Waser, R., 407
Wave
  Charge density, 285
  Spin, 285, 330
Wave vector
  Fermi, 96
Wavefunction
  Laughlin, 190
Weak correlation, 290
Weakly interacting fermions, 283
Weiss field, 368
Welser, J., vii
Wetting angle (θ), 390
Weyl, H., 405
Wheeler, J. A., 4, 56, 174, 233
White, R. M., 406
Wiedemann, H., 559
Wien bridge, 651

Wiggler, 511

Wigner crystal, 284, 288

Wigner, E., 241, 405

Wigner-Dyson distribution function, 219

Williams, C. P., 233

Wolfram, S., 8

Wong, P., viii

Woodall, J., viii

Workfunction, 302

World line, 204

Wuttig, M., 407

Yoke, 310

Young's modulus (Y), 411, 413, 630

Younis, M. I., 475

Yu, E., viii

Yu, P., 475

Zeeman effect, 323, 331

Zeeman splitting, 327

Zener damping, 449

Zener effect, 453

Zero bandgap, 132

Zero point energy, 483

Ziolkowski, B. W., 559

Zurek, W. H., 56, 233